# MATHS IN FOCUS

## MATHEMATICS STANDARD (PATHWAY 2)

**5TH EDITION**

**Klaas Bootsma**
**Sarah Hamper**
**Margaret Willard**
**Robert Yen**
**Series editor: Robert Yen**

**Maths in Focus 11 Mathematics Standard (Pathway 2)**
**5th Edition**
**Klaas Bootsma**
**Sarah Hamper**
**Margaret Willard**
**Robert Yen**
**ISBN 9780170497886**

Product manager: Robert Yen
Content developer: Helen Kaminski
Content manager: Renee Tome
Associate Content project manager: Kubra Ameen
Project designer: Nikita Bansal
Text designer: Cengage Creative Studio
Cover designer: Leigh Ashforth (Watershed Art & Design) and Cengage Creative Studio
Editor: Tanya Smith, Nikki M Group Pty Ltd
Proofreader: Jane Fitzpatrick, Nikki M Group Pty Ltd
Permissions/Photo researcher: Liz McShane/Lumina Datamatics
Cover: Nikita Bansal
Typeset by: KGL

For product information and technology assistance,
in Australia call **1300 790 853**;
in New Zealand call **0800 449 725**

For permission to use material from this text or product, please email
**aust.permissions@cengage.com**

**National Library of Australia Cataloguing-in-Publication Data**
A catalogue record for this book is available from the National Library of Australia

**Cengage Learning Australia**
Level 7, 80 Dorcas Street
South Melbourne, Victoria Australia 3205

For learning solutions, visit **cengage.com.au**

Printed in Malaysia by Papercraft.
1 2 3 4 5 6 7 30 29 28 27 26

# PREFACE

*Maths in Focus 11 Mathematics Standard (Pathway 2)* is the new name for the successful *New Century Maths 11* book. This 5th edition has been written for the new Mathematics Standard syllabus (2024).

This book retains the clear and abundant explanations and syllabus coverage of previous editions, which allow students to achieve self-paced learning and develop a strong understanding of the theory behind each syllabus topic. Exercise questions are now colour-coded according to level of difficulty. The theory follows a logical order, although some topics may be learned in any order. Syllabus coverage is shown by the table of contents and the syllabus reference grid on the next pages.

The syllabus describes one Year 11 Mathematics Standard course that splits into 2 courses at Year 12:

- **Mathematics Standard 1**: for students heading towards the workforce or further training after school, providing practical mathematical skills for life.
- **Mathematics Standard 2**: for students heading towards the HSC exam, ATAR and university studies.

We strongly believe that students studying Standard 1 and Standard 2 have specific learning needs, so we have published 2 levels of texts for Year 11 as well, both covering the same syllabus content, but at different depths and emphases.

This *Pathway 2* book covers the Year 11 course for students who intend to study Mathematics Standard 2 in Year 12. Mathematics Standard 2 is designed for students who will work or study in fields that require a mathematical or statistical background, and we have written a practical text that captures the spirit of the course, providing meaningful examples of mathematics being used in society and industry.

The *Nelson MindTap* online learning platform contains an eBook, worksheets, videos, topic tests and worked solutions. We have designed this book to be user-friendly and uncomplicated so that you can pick it up and use it straight away. We wish all teachers and students using *Maths in Focus* every success in embracing this revised senior mathematics course.

## About the authors

**Klaas Bootsma** was head teacher of Mathematics at Ambarvale High School and has taught at Lurnea and Grantham high schools. He was a senior HSC marker and has worked on the HSC Advice Line. Klaas has been the lead author of *New Century Maths 9–10* for over 30 years.

**Sarah Hamper** is head teacher of Mathematics at Cheltenham Girls' High School and has taught at Abbotsleigh, Meriden and Tara Anglican schools. Her expertise is in using modelling, problem solving and ICT in mathematics learning, specialising in gifted and talented students and girls' education. Sarah co-wrote *New Century Maths 9–10.*

**Margaret Willard** has extensive experience writing units of work designed for distance education and was Manager at TAFE's distance education unit. She has served on the executive of MANSW (the Mathematical Association of NSW), managed its post-secondary programs and has presented at both MANSW and TAFE conferences.

**Robert Yen** taught at Hurlstone Agricultural High School. He co-edited *Reflections*, the MANSW journal, and wrote General Mathematics HSC study guides for *The Sydney Morning Herald*. Robert has been writing for *New Century Maths 7–12* for over 30 years and now works for Nelson Cengage as the mathematics publisher and series editor.

## Contributing authors

**Megan Boltze** and **Kuldip Khehra** wrote and edited many of the *NelsonMindTap* worksheets.

**John Drake**, **Katie Jackson**, **Joanne Magner**, **Brad Pathuis** and **Scott Smith** created the video tutorials.

**Trisha Goss** wrote the topic tests.

**Roger Walter** wrote the testbank questions.

**Andrew Siu, Jayanthi Viswanathan, Anna Ma, Bela Adash** and **Tracey Macbeth-Dunn** wrote the worked solutions to all exercise sets.

# CONTENTS

# SYLLABUS REFERENCE GRID

| | Area of study and focus area | *Maths in Focus 11 Mathematics Standard* (Pathway 2) chapter |
|---|---|---|
| **ALGEBRA** | | |
| MST-11-01 | **Formulas and equations**<br>Algebraic techniques | 2 Formulas and equations<br>3 Driving safely |
| MST-11-02 | **Linear relationships**<br>Linear modelling<br>Direct variation | 9 Linear functions |
| **FINANCIAL MATHEMATICS** | | |
| MST-11-03 | **Earning money**<br>Ways of earning<br>Taxation | 4 Earning money and taxation |
| MST-11-04 | **Managing money**<br>Purchasing goods<br>Budgeting | 6 Managing money<br>7 Owning a car |
| **MEASUREMENT** | | |
| MST-11-05 | **Applications of measurement**<br>Practicalities of measurement<br>Perimeter, area and volume | 5 Measurement |
| MST-11-06 | **Time and location**<br>Positions on the Earth's surface<br>Time and time differences | 5 Measurement<br>11 World locations and times |
| **NETWORKS** | | |
| MST-11-07 | **Networks, paths and trees**<br>Network concepts<br>Shortest paths and spanning trees | 8 Networks |
| **STATISTICS** | | |
| MST-11-08 | **Data analysis**<br>Statistical investigation process<br>Population and sample<br>Data classification<br>Display and interpret grouped and ungrouped data<br>Measures of centre and spread<br>Quartiles and interquartile range<br>Five-number summary and box plots<br>Clusters and outliers | 1 Collecting and presenting data<br>3 Driving safely<br>10 Analysing data |

# ABOUT THIS BOOK

## At the beginning of each chapter

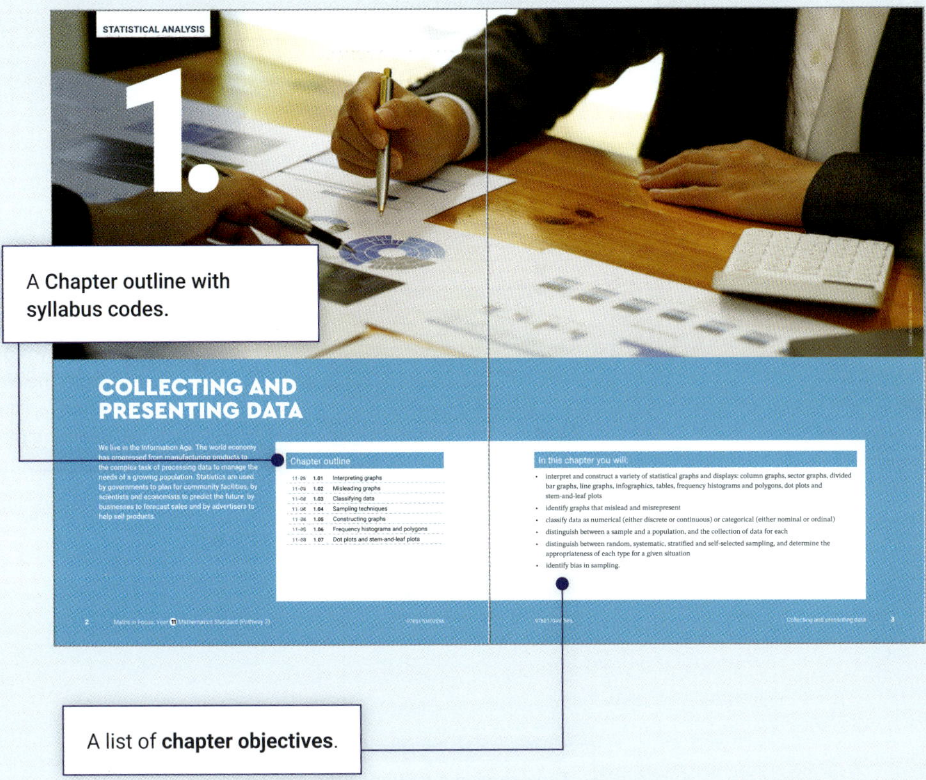


A Chapter outline with syllabus codes.

A list of **chapter objectives**.

A listing of *Nelson MindTap* chapter resources.

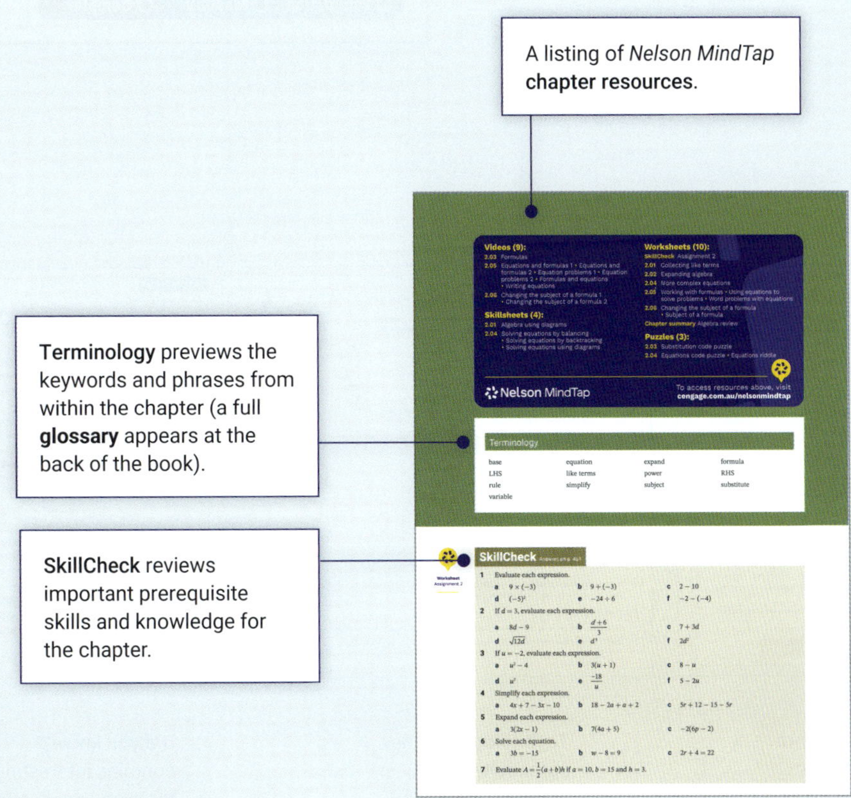


**Terminology** previews the keywords and phrases from within the chapter (a full **glossary** appears at the back of the book).

SkillCheck reviews important prerequisite skills and knowledge for the chapter.

## In each chapter

Glossary terms are printed in **red**.

Graded exercises are linked to worked examples and include HSC exam-style problems and realistic applications.

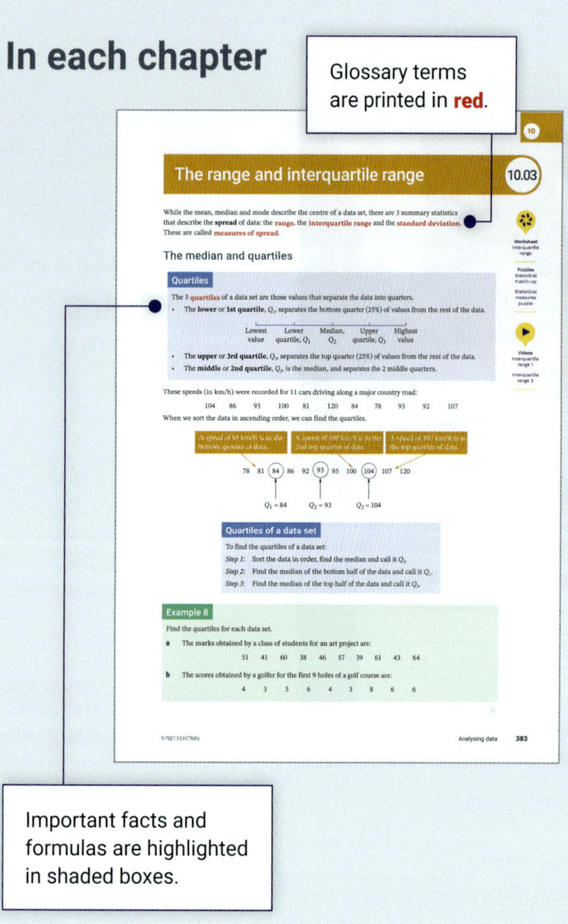


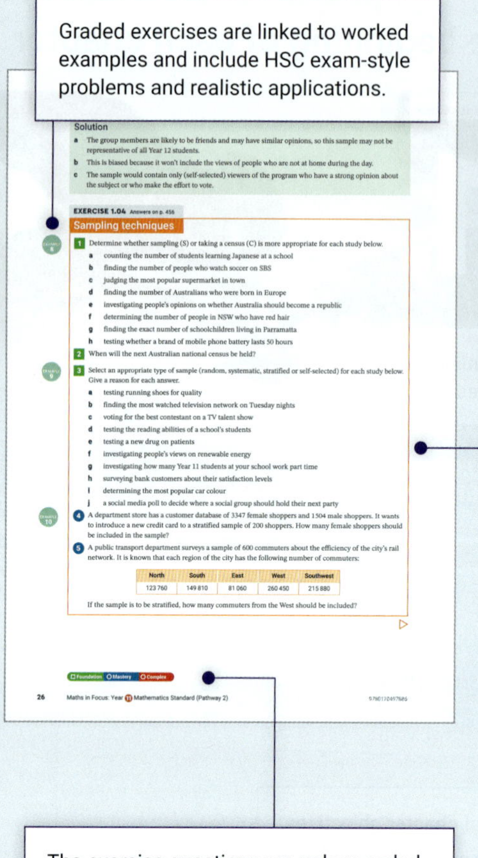


Worked solutions to all questions are downloadable from the *Nelson MindTap* online learning platform.

Important facts and formulas are highlighted in shaded boxes.

The exercise questions are colour-coded for level of difficulty:

**Foundation** | **Mastery** | **Complex**

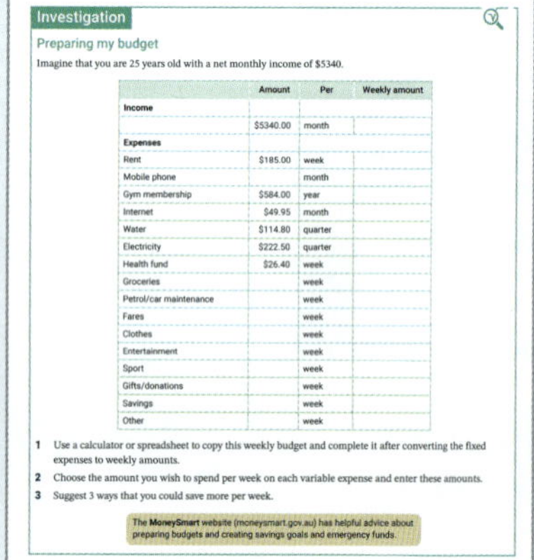


**Investigations** explore the syllabus in more detail, providing ideas for modelling activities and assessment tasks.

**Technology** includes spreadsheets, graphing software and the internet.

**Did you know?** contains interesting facts and applications of the mathematics learned in the chapter.

## At the end of each chapter

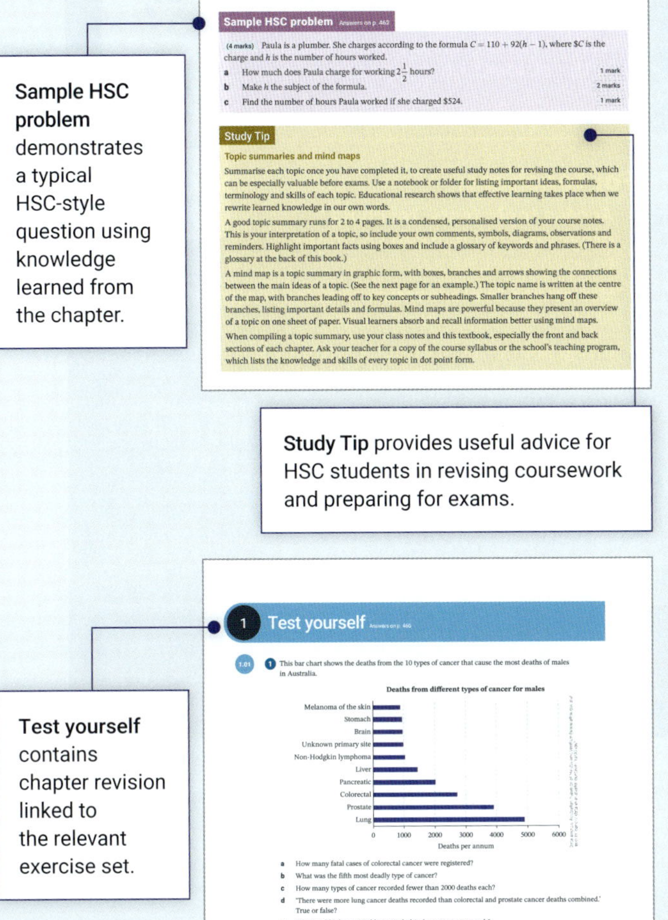

**Sample HSC problem** demonstrates a typical HSC-style question using knowledge learned from the chapter.

**Study Tip** provides useful advice for HSC students in revising coursework and preparing for exams.

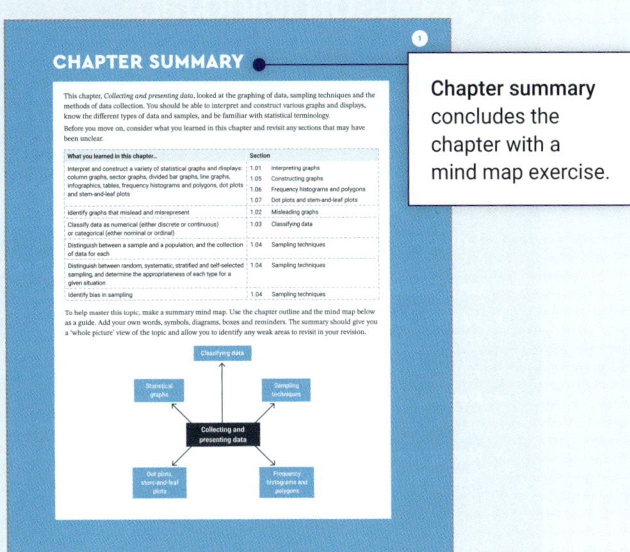

**Chapter summary** concludes the chapter with a mind map exercise.

**Test yourself** contains chapter revision linked to the relevant exercise set.

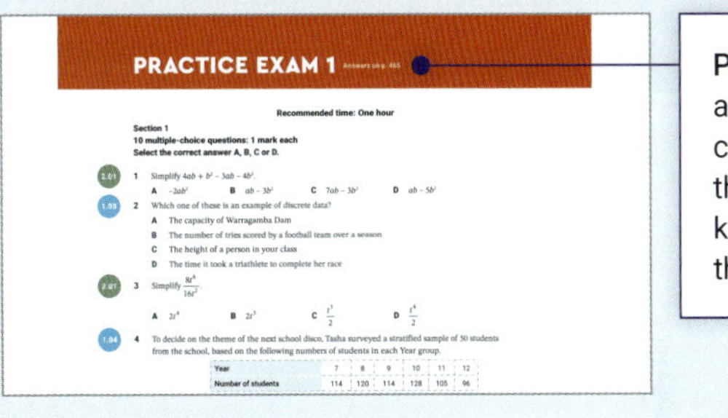

**Practice exams** after every 3 chapters revise the skills and knowledge of those chapters.

## At the end of the book

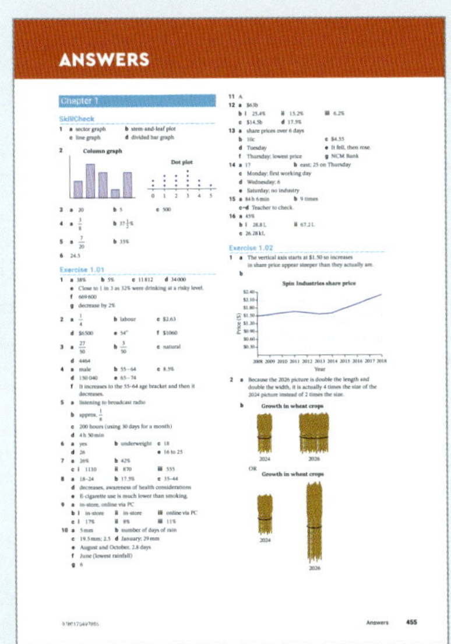

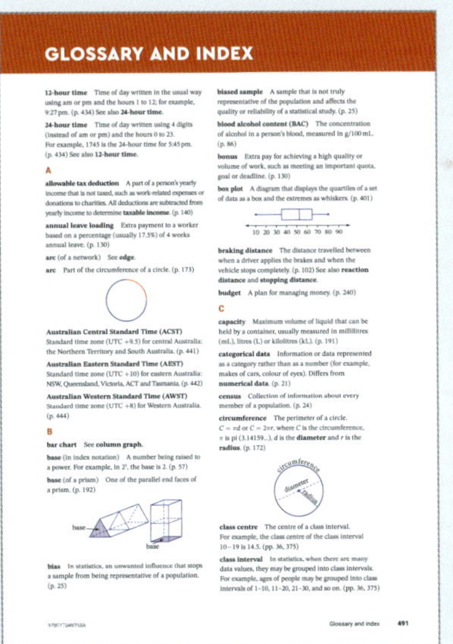

**Answers** (worked solutions provided on *Nelson MindTap* for teachers to allocate to students)

**Glossary and index** includes a comprehensive dictionary of course terminology.

ABOUT THIS BOOK

## Nelson MindTap

**Nelson MindTap** is an online learning space that provides students and teachers with access to additional learning and teaching resources. Margin links in the student book signpost multimedia student resources found on Nelson MindTap*.

Videos

Skillsheets

Worksheets

### For students:

- **Engage** with the **online eBook** by adding notes, highlights, and bookmarks, and using the **Search** and **Read Aloud** (in Australian voice) functions.
- **Watch videos** featuring expert teacher advice to unpack worked examples and deepen your understanding.
- **Revise** using **skillsheets** and **worksheets** to strengthen your skills and build your confidence.
- **Navigate** your own learning path, accessing the content and support as you need it.

### For teachers:

- **Access topic tests, teaching programs** and **worked solutions** to each exercise set.
- **Use course customisation** to tailor content to groups of students or the whole class.
- **Integrate** content directly within your school's LMS for ease of access.
- Help build your students' exam readiness with **Cognero Assess** – a test bank containing hundreds of questions and answers to create, assign or export formative and summative tests.

* Complimentary access to these resources is only available to schools that use this book as part of a class set, book hire or booklist. Not available for single purchases. Contact your Cengage Learning Consultant for information about access and conditions.

### Security & privacy:

Nelson MindTap joined the Safer Technologies 4 Schools (ST4S) Product Badge Program in 2024. The annual ST4S assessment is part of our commitment to supporting the online security and safety of students and schools. Learn more at st4s.edu.au

# STUDY SKILLS

There is no right or wrong way to learn. Different styles of learning suit different people. There is also no magical number of hours a week that you should study, because this will be different for every student. But just listening in class and taking notes is not enough, especially when you are learning material that is totally new.

If a skill is not practised within the first 24 hours, up to 50% can be forgotten.

If it is not practised within 72 hours, up to 90% can be forgotten!

It is really important that, whatever your study routine, new work must be studied soon after it is presented to you.

With a constant flow of new work to learn and retain, this is a challenge.

The good news is that you don't have to study for hours on end!

## In class

|  Leave distractions and issues outside the classroom. |  Listen to what you're being taught. |  Take notes. |  Highlight main points. |  Ask questions. |
|---|---|---|---|---|

## At home

To best remember, revise

*Today*

*Tomorrow*

*In a week*

*In a month.*

|  Set a realistic timetable. |  Revise when you are fresh and have energy. |  Study in small chunks instead of marathons. |
|---|---|---|
|  Create an environment and routine that helps you focus. |  Revise previous topics as well as new work. |  Balance study with other activities and commitments. |

## Learning skills

Try different learning styles:

*Seeing*

*Hearing*

*Doing.*

| | |
|---|---|
| <br>Summarise on cue cards.<br>Draw mind maps or pictures.<br>Use colours to highlight main points. | <br>Read notes out loud.<br>Discuss work with a friend. |
| <br>Create songs or rhymes to<br>remember facts and formulas.<br>Revise by teaching someone else. | <br>Once you have mastered a topic,<br>do practice exams under exam conditions. |

## During exams

| | |
|---|---|
| <br>During reading time, skim the whole paper<br>to know how many questions there are.<br>Read each question carefully.<br>Underline or highlight key words. | <br>Divide time so you can answer each question.<br>Allow time at the end to check answers<br>or complete harder questions.<br>Write legibly.<br>Don't spend too much time on one question:<br>move on if you are genuinely stuck.<br>Show all working out and include<br>diagrams and formulas.<br>Cross out mistakes with a single line<br>so that they can still be read. |

## Finally…

Study involves knowing what you don't know, and putting in a lot of time focussing on these areas. This is a positive way to learn. Rather than just saying, 'I can't do this', say 'I can't do this *yet*', and use your teachers, friends, textbooks and other ways of finding out.

Some students hardly find time to study while others give up their outside lives to devote their time to study.

The ideal situation is to balance study with other aspects of your life, including going out with friends, working, and keeping up with sport and other activities that you enjoy.

*Good luck with your studies!*

## New Century Maths / Maths in Focus 7–12 series

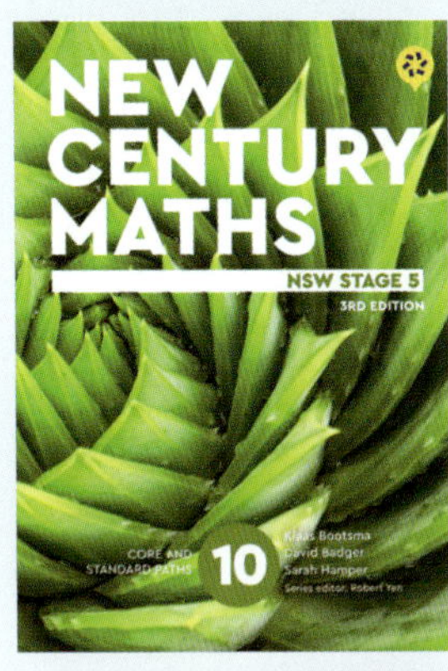

# MATHEMATICAL VERBS

## A glossary of 'doing words' commonly found in mathematics problems

**analyse**: study in detail the parts of a situation

**apply**: use knowledge or a procedure in a given situation

**calculate**: find a numerical value

**classify**: identify and sort into a category

**comment**: express an observation or opinion about a result

**complete**: fill in detail to make a statement, diagram or table correct or finished

**compare**: show how two or more things are similar or different

**construct**: draw an accurate diagram

**convert**: change from one form to another, for example, from a fraction to a decimal, or from kilograms to grams

**decrease**: make smaller

**describe**: state the features of a situation

**estimate**: make an educated guess for a number, measurement or solution, to find roughly or approximately

**evaluate**, **calculate**: find the value of an expression, for example, $3 \cdot 8^2$ or $4x + 1$ when $x = 5$

**expand**: remove brackets in an algebraic expression, for example, expanding $3(2y + 1)$ gives $6y + 3$

**explain**: describe why or how

**give reasons**: show the rules or thinking used when solving a problem.

**graph**: display on a number line, number plane or statistical graph

**hence find/prove**: find an answer or prove a result using previous answers or information supplied

**identify**: state the type, name or distinguishing feature of an item or situation

**increase**: make larger

**interpret**: find meaning in a mathematical result

**justify**: give reasons or evidence to support your argument or conclusion.

**measure**: determine the size of something; for example, use a ruler to find the length of a pen

**prove**: see **show that**

**recall**: remember and state

**show that**, **prove**: (in questions where the answer is given) use calculation, procedure or reasoning to prove that a result is true

**simplify**: give a result in its most basic, shortest, neatest form, for example, simplifying a ratio or algebraic expression

**sketch**: draw a rough diagram that shows the general shape or ideas, less accurate than **construct**

**solve**: find the value(s) of an unknown pronumeral in an equation or inequality

**state**: see **write**

**substitute**: replace a variable by a number and evaluate

**verify**: check that a solution or result is correct, usually by substituting back into the equation or referring back to the problem

**write/state**: give the answer, formula or result without showing any working or explanation (This usually means that the answer can be found mentally, or in one step.)

# SYMBOLS AND ABBREVIATIONS

| Symbol | Meaning | Symbol | Meaning |
|---|---|---|---|
| $=$ | is equal to | $\sqrt{\ }$ | square root, radical sign |
| $\neq$ | is not equal to | $\sqrt[3]{\ }$ | cube root |
| $\approx$ | is approximately equal to | $P(E)$ | the probability of event $E$ occurring |
| $<$ | is less than | $P(\tilde{E})$ | the probability of event $E$ not occurring |
| $>$ | is greater than | LHS | left-hand side |
| $\leq$ | is less than or equal to | RHS | right-hand side |
| $\geq$ | is greater than or equal to | % | percentage |
| ( ) | parentheses, round brackets | cos | cosine ratio |
| [ ] | (square) brackets | sin | sine ratio |
| { } | braces | tan | tangent ratio |
| $\pm$ | plus or minus | $\bar{x}$ | the mean |
| $\pi$ | pi = 3.141... | $\Sigma$ | the sum of |
| $0.\dot{1}5\dot{2}$ | the recurring decimal 0.152152 ... | $Q_1$ | first quartile or lower quartile |
| $\circ$ | degree | $Q_2$ | median (second quartile) |
| $\angle$ | angle | $Q_3$ | third quartile or upper quartile |
| $\triangle$ | triangle | IQR | interquartile range |
| $\|\|$ | is parallel to | $\alpha$ | alpha |
| $\perp$ | is perpendicular to | $\theta$ | theta |
| $\therefore$ | therefore | $\mu$ | micro-, mu |
| $x^2$ | $x$ squared, $x \times x$ | | |
| $x^3$ | $x$ cubed, $x \times x \times x$ | | |

# 1. COLLECTING AND PRESENTING DATA

We live in the Information Age. The world economy has progressed from manufacturing products to the complex task of processing data to manage the needs of a growing population. Statistics are used by governments to plan for community facilities, by scientists and economists to predict the future, by businesses to forecast sales and by advertisers to help sell products.

## Chapter outline

## In this chapter you will:

- interpret and construct a variety of statistical graphs and displays: column graphs, sector graphs, divided bar graphs, line graphs, infographics, tables, frequency histograms and polygons, dot plots and stem-and-leaf plots
- identify graphs that mislead and misrepresent
- classify data as numerical (either discrete or continuous) or categorical (either nominal or ordinal)
- distinguish between a sample and a population, and the collection of data for each
- distinguish between random, systematic, stratified and self-selected sampling, and determine the appropriateness of each type for a given situation
- identify bias in sampling.

**Videos (4):**

**1.03** Categorical and numerical data • Types of data

**1.06** Frequency histograms • Frequency histograms and polygons

**Worksheets (9):**

**SkillCheck** Assignment 1

**1.01** Every picture tells a story

**1.04** Student survey form • Census questions

**1.05** Australian statistics

**1.06** Frequency tables • Frequency distribution tables • Histograms

**1.07** Stem-and-leaf plots

**Puzzle (1):**

**1.03** Statistical data match-up

**Spreadsheet (1):**

**1.05** Statistical graphs

Nelson MindTap

To access resources above, visit **cengage.com.au/nelsonmindtap**

## Terminology

| | | | |
|---|---|---|---|
| bias | biased sample | categorical data | census |
| class centre | class interval | cluster | continuous data |
| discrete data | frequency histogram | frequency polygon | infographic |
| nominal data | numerical data | ordinal data | population |
| random sample | sample size | sector graph | self-selected sample |
| stem-and-leaf plot | stratified sample | survey | systematic sample |

Worksheet
Assignment 1

## SkillCheck Answers on p. 455

**1** Match each statistical display to its correct name from this list.

column graph (bar chart) divided bar graph dot plot
line graph sector graph stem-and-leaf plot

**a**

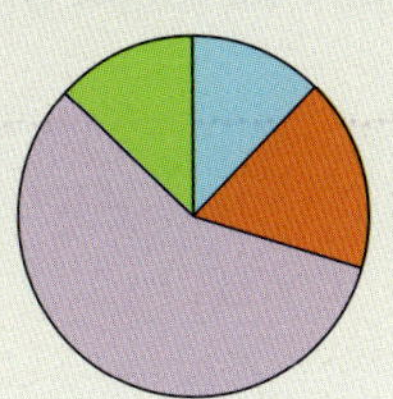

**b**

| | |
|---|---|
| 3 | 0 4 4 |
| 4 | 3 6 6 6 8 |
| 5 | 2 4 5 7 |
| 6 | 1 3 3 |

**c**

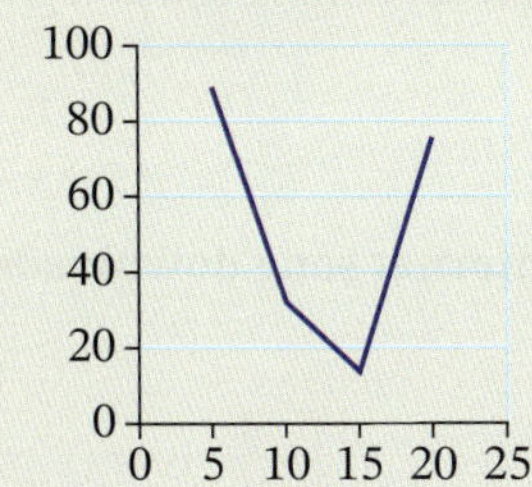


**d**

**2** Sketch an example of each of the 2 statistical displays from the list in Question **1** that were not illustrated in the question.

**3** Write the size of one interval on each vertical scale.

**a**

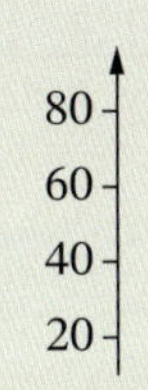


**b**

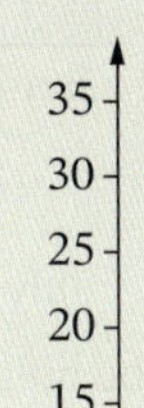


**c**

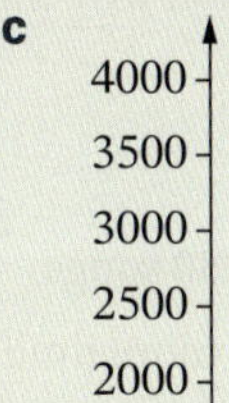


**4** **a** What fraction of this rectangle is shaded?

**b** What percentage of this rectangle is shaded?

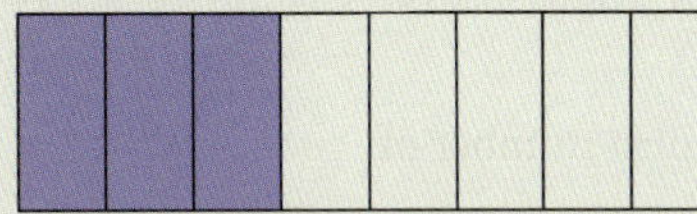

**5** **a** What fraction of this circle is shaded?

**b** What percentage of this circle is shaded?

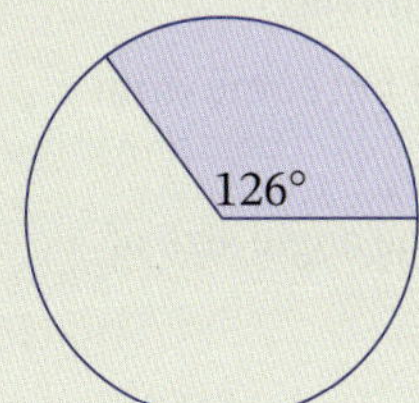


**6** The ages of the patients at a hospital were grouped into the class intervals 0–9, 10–19, 20–29 and so on. What is the class centre of the interval 20–29?

# 1.01 Interpreting graphs

Worksheet
Every picture tells a story

Graphs are used to display statistical information in an attractive and meaningful way. Four simple types of graph are:

- **column graph**, also known as a **bar chart**
- **line graph**
- **sector graph**, also known as a **pie chart**
- **divided bar graph**.

Another popular way to present information and **data** is to use an **infographic**, combining text and graphics.

## Example 1

This column graph shows the number of songs downloaded from the internet by a group of teenagers each month over 7 months.

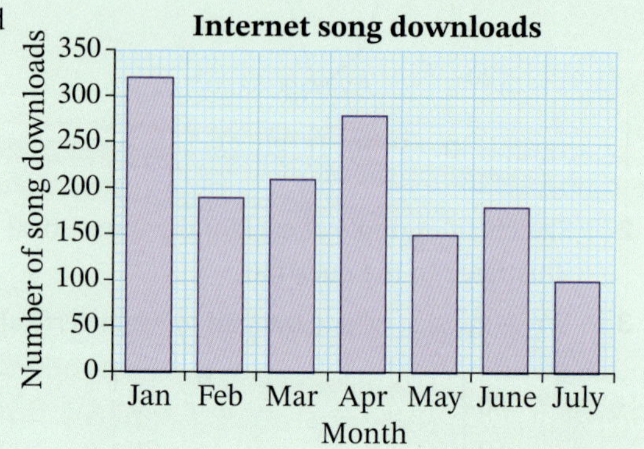


**a** What is the size of one interval on the vertical axis?

**b** How many songs were downloaded in April?

**c** In which month were 100 songs downloaded?

**d** In which month were the highest number of songs downloaded? Suggest a reason why this may be so.

### Solution

**a** 10 songs **b** 280 songs **c** July

**d** January; Teenagers may have had more time or money during the holidays to download songs.

## Example 2

This sector graph shows the countries of origin of overseas students studying in Australia in 2024.

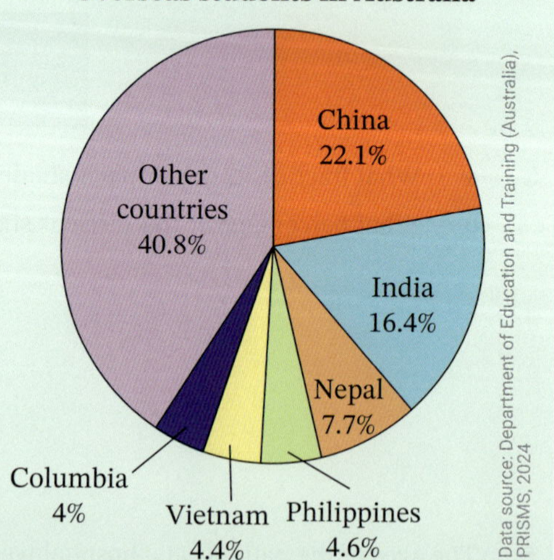


**a** Which country had the second greatest number of students studying in Australia?

**b** If 853 000 overseas students studied in Australia, approximately how many came from:

**i** China? **ii** Columbia?

**c** What percentage of overseas students did not come from China or India?

**d** Calculate, correct to the nearest degree, the angle size of the 'Other countries' sector.

### Solution

**a** India

**b** **i** $22.1\% \times 853\,000 = 188\,513$ students came from China.

**ii** $4.0\% \times 853\,000 = 34\,120$ students came from Columbia.

**c** $100\% - 22.1\% - 16.4\% = 61.5\%$ or $7.7\% + 4.6\% + 4.4\% + 4.0\% + 40.8\% = 61.5\%$

**d** $40.8\% \times 360° = 146.88° \approx 147°$.

## Example 3

This infographic compares some prices and salaries between Sydney and London.

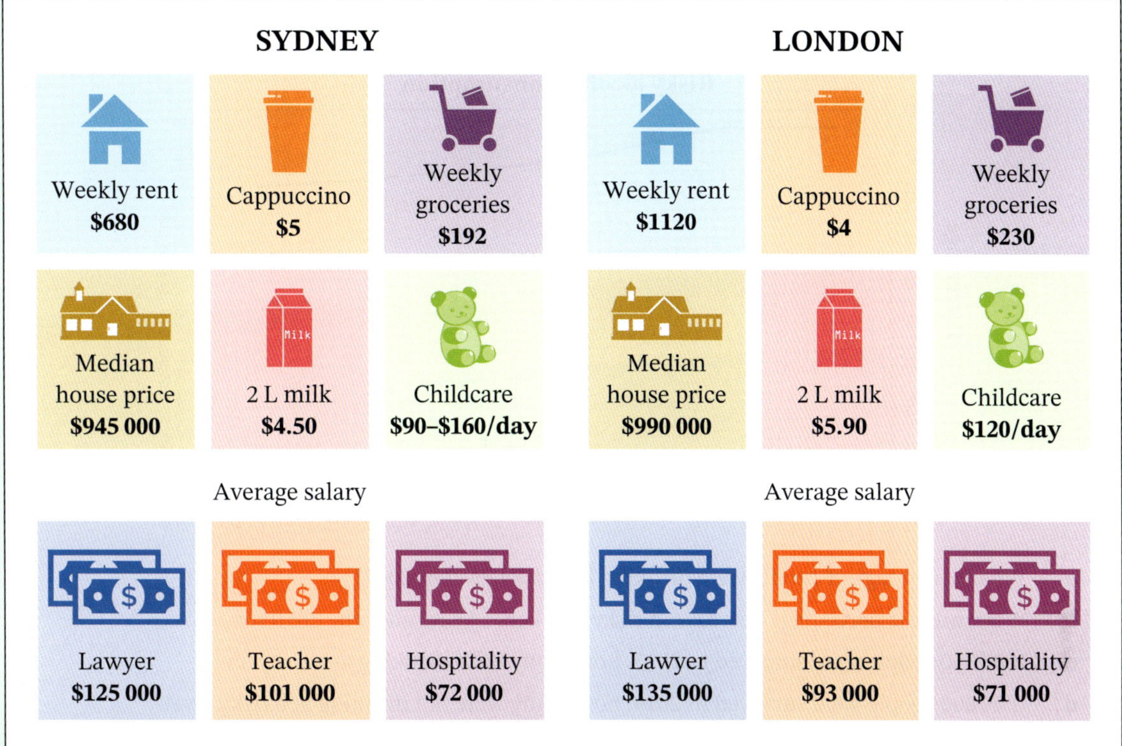


Note: Amounts are given in Australian dollars.

**a** Which items are more expensive in London than in Sydney?

**b** How much extra rent per year would you pay on average in London than in Sydney?

**c** Ben works in hospitality and Anita is a teacher living in Sydney. They have 2 children and are considering moving to London for 2 years. How much less are they likely to earn per year?

**d** Ben and Anita rent their home and pay $130 per child for childcare for 48 weeks per year. Determine how much more expensive it will be for them to live in London for 2 years.

### Solution

**a** rent, groceries, house and milk

**b** Extra rent per year = ($1120 − $680) × 52
= $22 880

**c** Difference in salary = ($72 000 + $101 000) − ($71 300 + $93 000)
= $8700 per year

**d** Living expenses in Sydney for 2 years:

Expenses = rent + groceries + childcare for 2 children for 2 years
= ($680 + $192) × 52 × 2 + $130 × 2 × 48 × 2
= $115 648

Living expenses in London for 2 years:

Expenses = ($1120 + $230) × 52 × 2 + $120 × 2 × 48 × 2
= $163 440

Difference = $163 440 − $115 648
= $47 792

It will be approximately $47 792 more expensive to live in London for 2 years, assuming costs and salaries stay the same.

**EXERCISE 1.01** Answers on p. 455

## Interpreting graphs

**1** This line graph shows the percentage of Australians aged 14 and over drinking alcohol at a 'risky' level, from 2007 to 2022–2023.

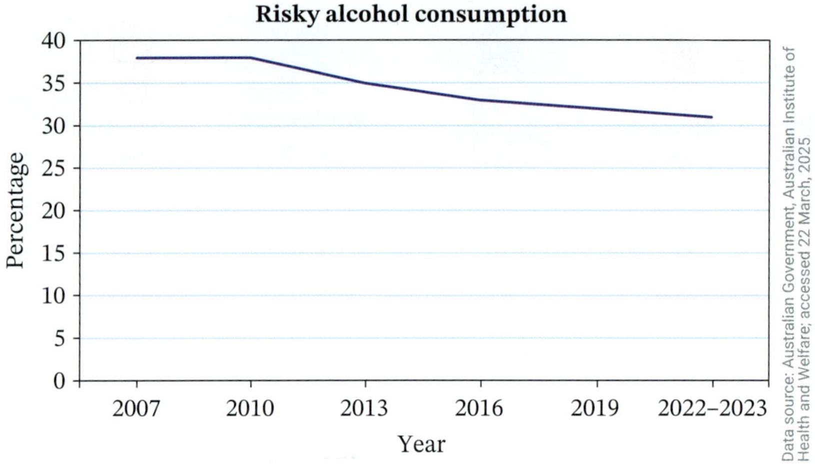


**a** What percentage of people were drinking at a risky level in 2007?

**b** What was the percentage decrease in risky drinkers 6 years after the 2010 survey?

**c** If 33 750 people were surveyed in 2013, how many were drinking at a risky level?

**d** If 50 000 people were surveyed in 2019, how many were not drinking at a risky level?

**e** Comment on the statement: 'In 2022–2023, about 1 in 3 people aged 14 and over consumed alcohol in ways that put their health at risk.' (Australian Institute of Health and Welfare, 2024)

**f** If the number of people in Australia aged 14 and over in 2023 was 21.6 million, how many were drinking at a risky level?

**g** Predict what might happen in the results of the next survey.

**2** This sector graph shows the budget of Frank's Pizza, including expenses and the profit margin.

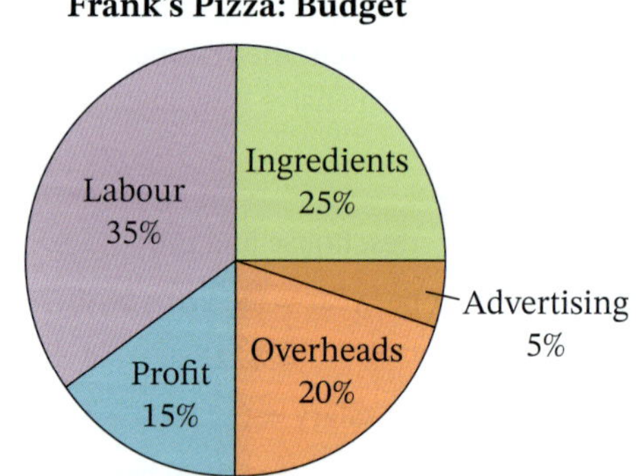


**a** What fraction of his budget does Frank spend on ingredients?

**b** What is the largest expense?

**c** For each family-size pizza Frank sells for $17.50, how much of this is profit?

**d** If $325 per week is spent on advertising, what is his total weekly budget?

**e** What is the angle size of the 'Profit' sector on the graph?

**f** Overheads are business expenses, such as rent.
Calculate how much Frank spends on overheads if his weekly budget is $5300.

□ Foundation ○ Mastery ○ Complex

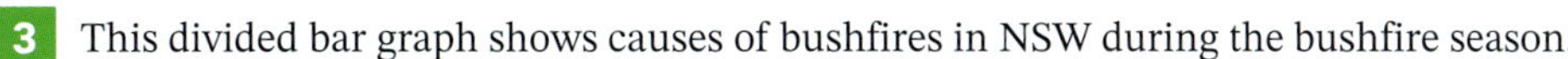

**3** This divided bar graph shows causes of bushfires in NSW during the bushfire season.

**Causes of bushfires in NSW**

| Natural 18% | Deliberate (inc. cigarettes) 54% | Accidental 6% | Debris burning (inc. campfires) 9% | Undetermined 13% |
|---|---|---|---|---|

Data source: NSW EPA, NSW State of the Environment Report, Causes of fire 2019–2020; https://www.soe.epa.nsw.gov.au/all-themes/land/fire#causes-of-fire-pressures, accessed 22 March, 2025

**a** What fraction of bushfires were deliberately lit?

**b** What percentage of bushfires were accidental?

**c** Which cause was linked to 18% of bushfires?

**d** If there were 24 800 bushfires during summer, how many were started by natural causes?

**4** This graph shows the percentage of people in each age group and gender who are smokers.

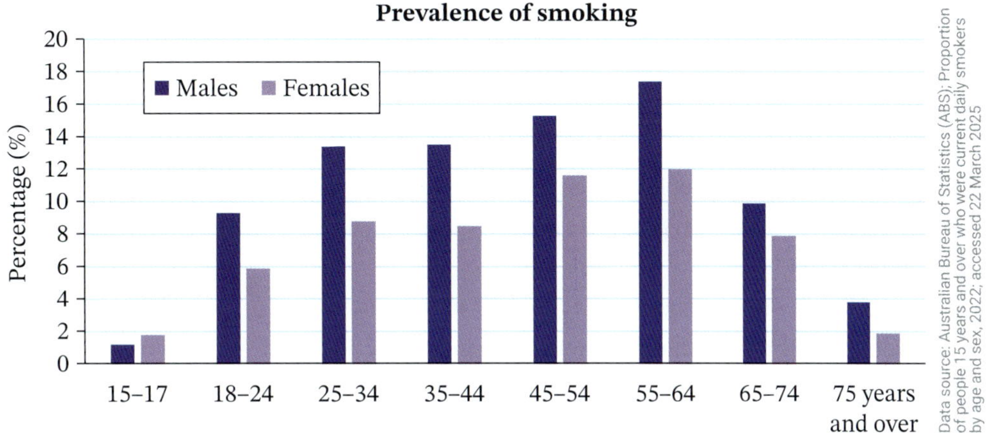


Data source: Australian Bureau of Statistics (ABS); Proportion of people 15 years and over who were current daily smokers by age and sex, 2022; accessed 22 March 2025

**a** Which gender has the higher percentage of smokers?

**b** Which age group contains the highest percentage of male smokers?

**c** What percentage of females aged 35 to 44 are smokers?

**d** If 1 705 000 females were surveyed in the 25–34 age group, how many were smokers?

**e** In which age group were 10% of men smokers?

**f** What generally happens to the percentage of smokers as age increases?

☐ Foundation ○ Mastery ○ Complex

5 This infographic shows the average amount of time per day Australian internet users spend with different media and devices.

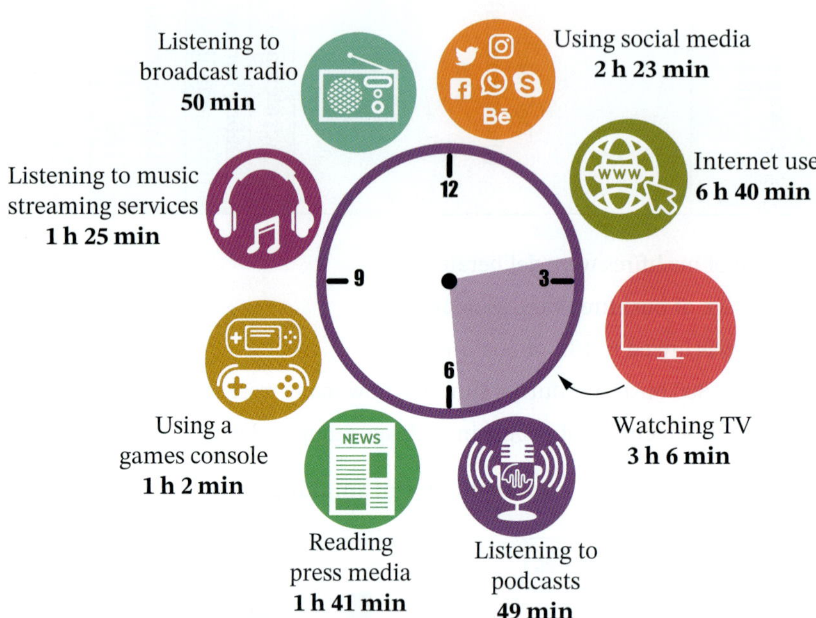

**a** Which media usage do people use for the least amount of time?

**b** What fraction of the day do people on average spend watching TV? Express your answer to 2 decimal places.

**c** How much time does average internet use take up in a month?

**d** According to this infographic, how much time in one day does Seth spend on media, if he uses social media, listens to music streaming services and plays on a game console?

6 The graphs below illustrate one approach to measuring healthy weight ranges, according to sex, age and body mass index (BMI). A person's BMI is calculated by dividing weight (in kilograms) by height (in metres) squared.

**BMI healthy weight range**

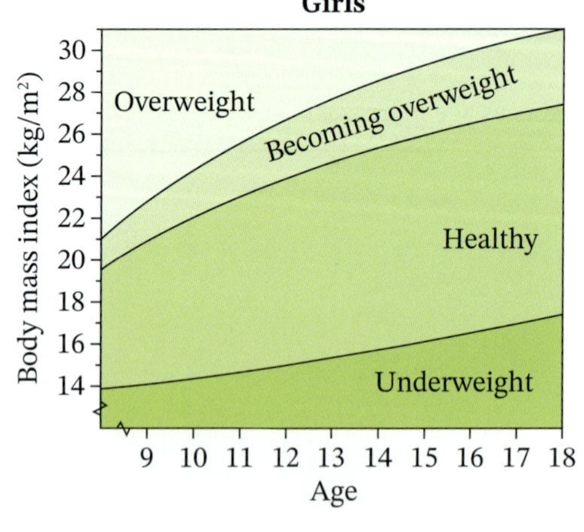


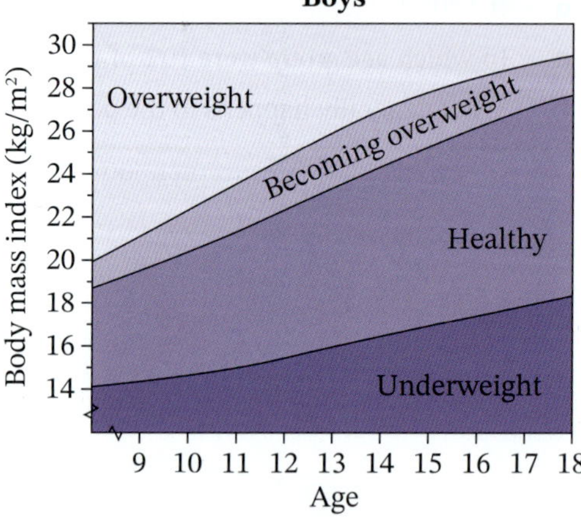

**a** Is a 16-year-old girl with a BMI of 20 considered to be in the healthy weight range?

**b** Describe the health of a 13-year-old boy with a BMI of 14.

**c** What is the minimum BMI for a 17-year-old boy to be considered in the healthy weight range?

**d** Jake is aged 15, weighs 78 kg and is 173 cm tall. Calculate his BMI.

**e** Write the lowest and highest BMI for a 14-year-old girl to be considered in the healthy weight range.

**f** Calculate your BMI and use the graph to find your state of health.

> The BMI is not a good indicator of health for athletes, children, pregnant women and others. See page 67 for more details on the limitations of this model.

**7** A large number of office workers were surveyed about their work and work needs. The results are presented in this infographic.

**JOB SATISFACTION**

I am happy with my current job

**JOB NEEDS**

Compared to the job I currently have, I would rather have

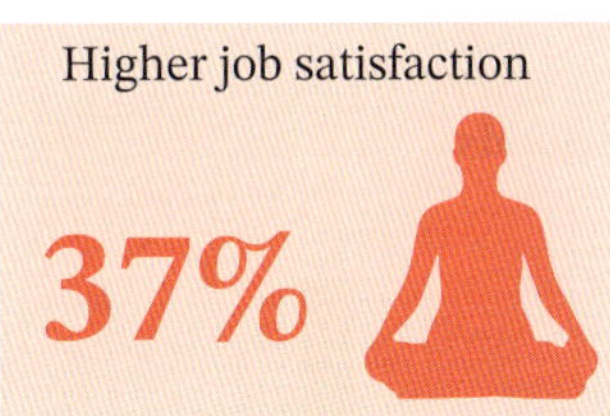


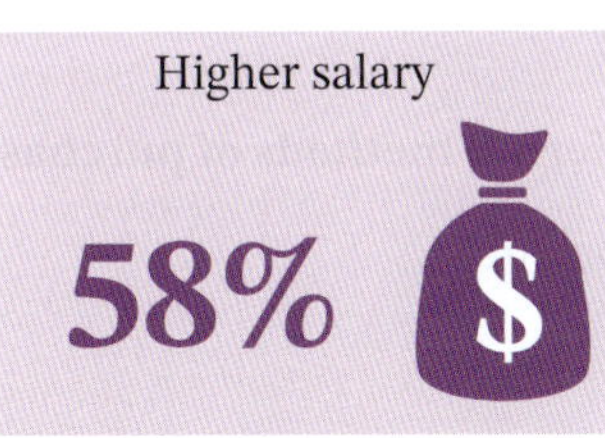

**a** What percentage of office workers are dissatisfied with their jobs?

**b** What percentage are satisfied with their salary?

**c** If 1500 office workers were surveyed, how many:

- **i** were satisfied with their job?
- **ii** would like a higher salary?
- **iii** would like higher job satisfaction?

□ Foundation ○ Mastery ○ Complex

**8** This graph shows the percentage of people in each age group who have used e-cigarette or vaping devices at least once.

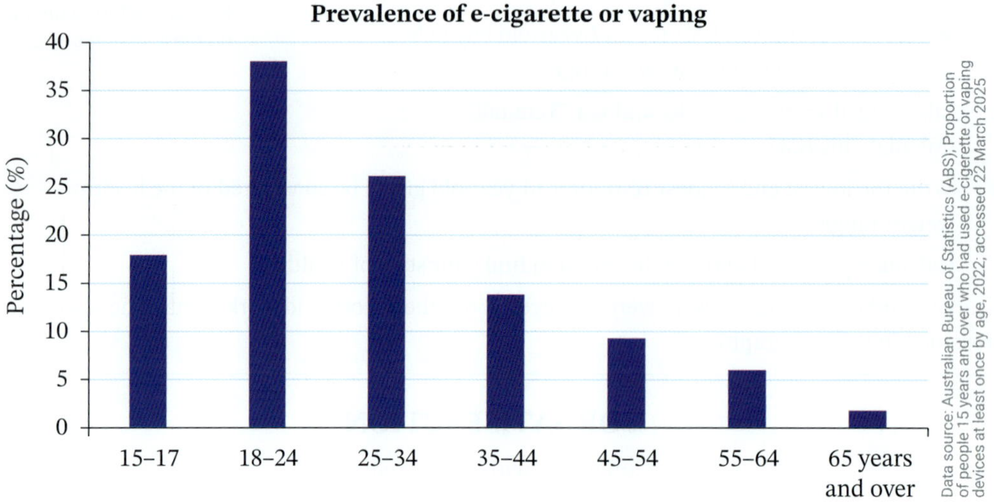


**a** Which age group has the highest percentage of users of vaping devices?

**b** What percentage of people aged 15–17 have used a vaping device?

**c** In which age group have 14% used an e-cigarette or vaping device?

**d** What happens to the percentage of e-cigarette or vaping users as age increases? Give possible reasons for this trend.

**e** How does the use of e-cigarette or vaping devices compare with that of people smoking at age 55–64 (see Question **4**)?

**9** This **clustered column graph** shows the different ways people shop for sports/outdoor equipment, groceries and home entertainment.

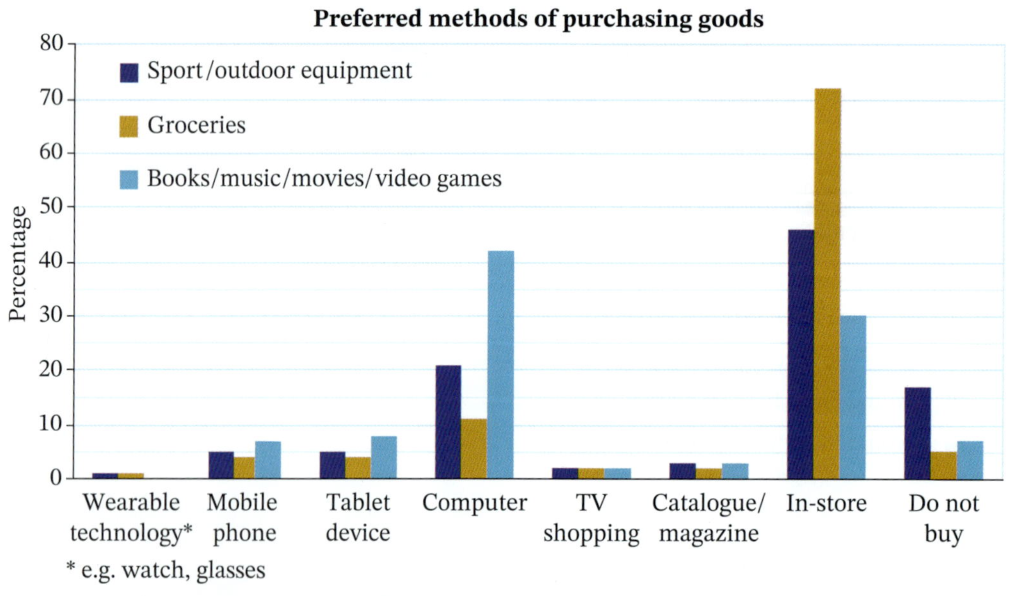


**a** Name the 2 most popular methods of purchasing goods.

**b** What is the most popular method for purchasing:

**i** groceries?

**ii** sports/outdoor equipment?

**iii** home entertainment?

**c** What percentage of people:

**i** do not buy sports/outdoor equipment at all?

**ii** buy home entertainment items using their tablets?

**iii** buy groceries using their computers?

**10** This graph shows the average rainfall and number of rainy days per month for Broken Hill.

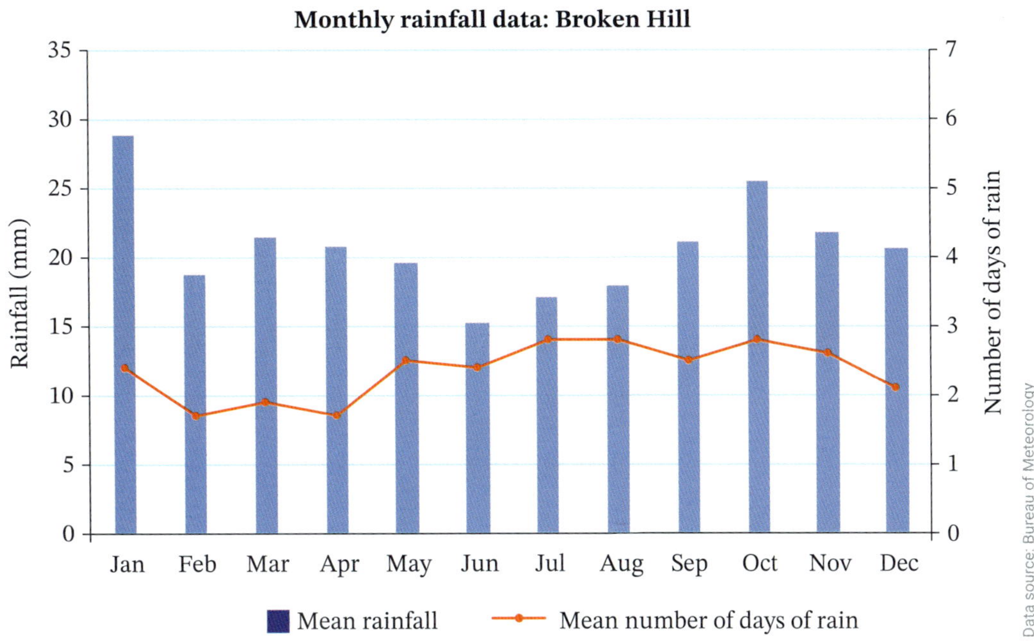


**a** What is the size of one unit on the 'Rainfall' axis?

**b** What is graphed on the line graph?

**c** What is the average rainfall and the number of days with rain for May?

**d** Which month has the highest rainfall? How many millimetres fall on average in that month?

**e** Which month has the most days of rain? How many days of rain does it have?

**f** What is the driest month?

**g** How many days in summer have rain?

**11** A group of people was surveyed about their favourite Australian state or territory to visit. The results are shown on the sector graph.

If 2475 tourists chose Western Australia, which of the following is the total number of people who participated in the survey? Select **A**, **B**, **C** or **D**.

**A** 27 500 **B** 8800

**C** 5775 **D** 25 025

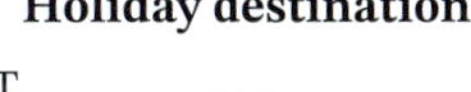


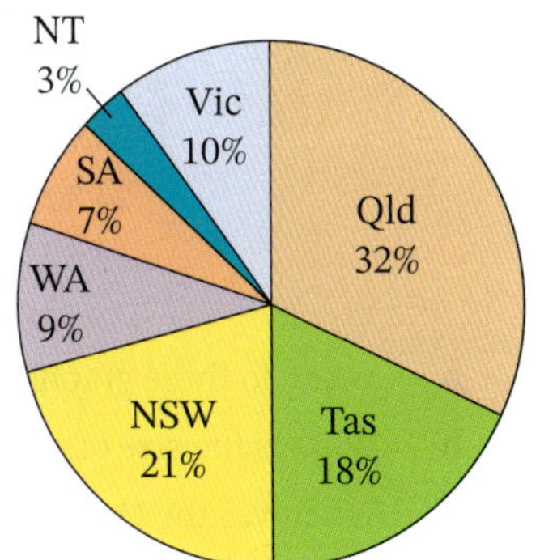

12 This infographic shows the amount that Australians spend on online shopping in a year, in billions of dollars.

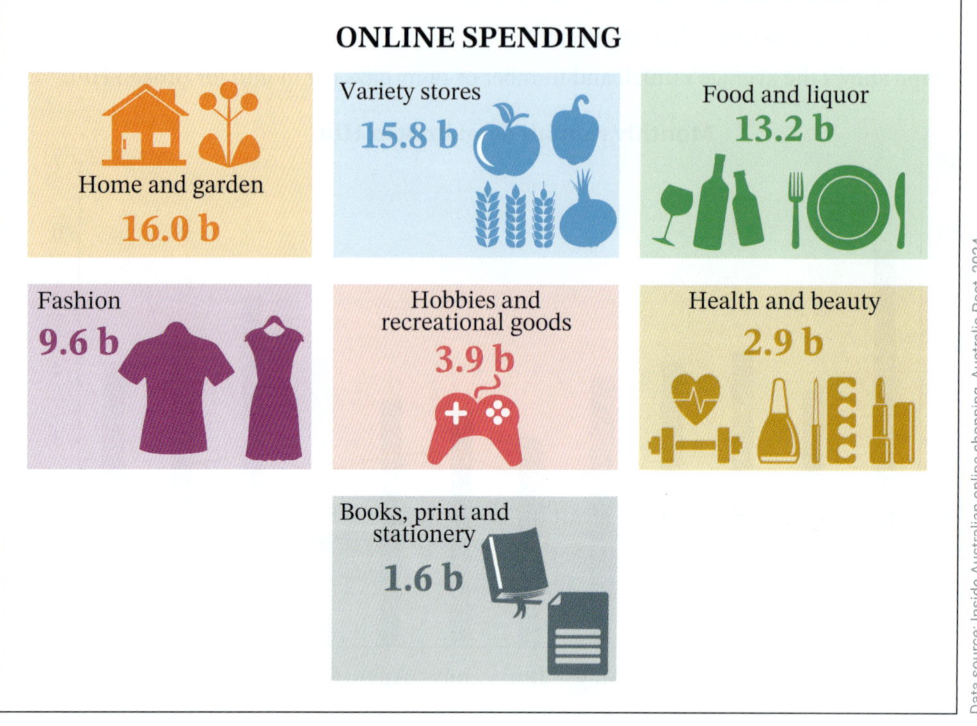


Data source: Inside Australian online shopping, Australia Post, 2024.

**a** What was the total amount spent online?

**b** What percentage of online spending was on:

**i** home and garden? **ii** fashion? **iii** hobbies and recreational goods?

**c** The online spending for variety stores represented a 9.1% increase on the previous year. What was the amount for variety store online spending the previous year?

**d** If total spending on retail goods was \$361 b, what percentage of this is online spending?

13

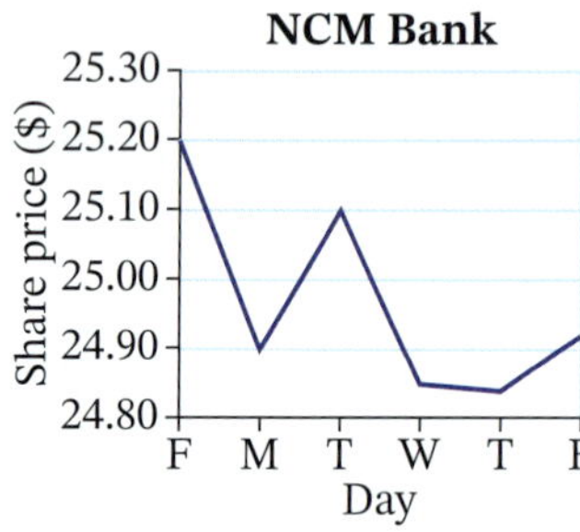


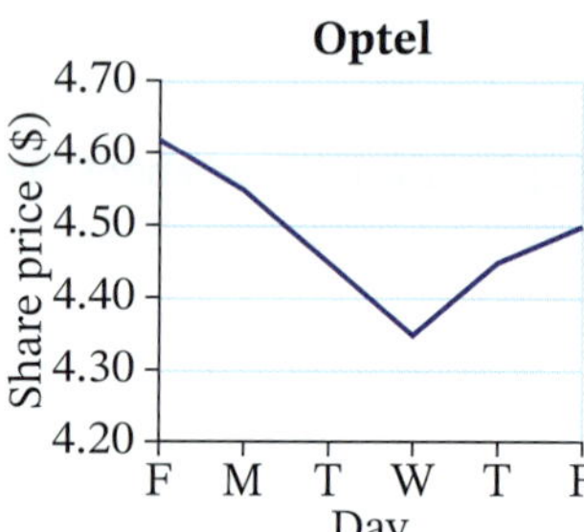


**a** What do these graphs illustrate?

**b** What is the size of one interval on the vertical axis of the NCM Bank graph?

**c** What was the Optel share price on Monday?

**d** On which day was the NCM Bank share price \$25.10?

**e** Describe what happened to the Optel share price over the week.

**f** What would have been the worst day to sell NCM Bank shares? Why?

**g** Which company's shares changed more suddenly in price over the week?

☐ Foundation ○ Mastery ○ Complex

**14** These line graphs show the pollution index in 3 Sydney regions over a week.

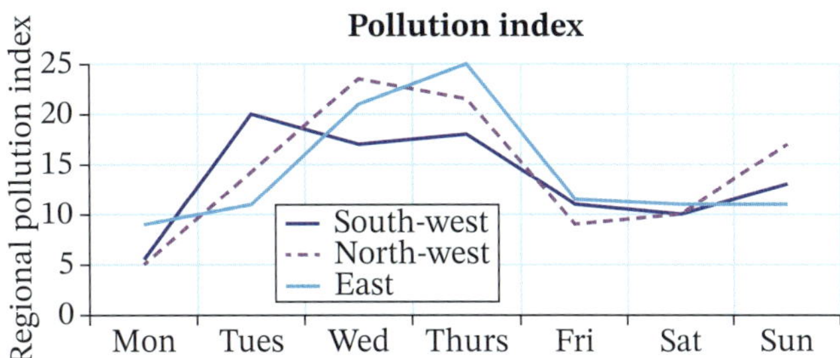


- **a** What was the south-west's pollution index on Wednesday?
- **b** Which region had the highest index for the week? What was this reading and on which day did it occur?
- **c** Which day showed the lowest index for all 3 regions? Why do you think it did?
- **d** Which day showed the greatest difference in readings between the south-west and north-west? What was this difference?
- **e** On which day were the readings about the same for all 3 regions? Why do you think it was?

**15** The average monthly times Australians spend on social media platforms are shown in this infographic.

- **a** Calculate the monthly amount of time Kristy uses social media if she uses TikTok, YouTube and Facebook for the average times shown.
- **b** Approximately how many times more often is TikTok used than Messenger (what multiple)?
- **c** Write down the amount of time per day you spend on each platform and calculate your monthly usage.
- **d** Compare your results with the information given in the infographic.

**16** This infographic shows household water use for a family. The amount of water used is 480 L per day.

**a** What percentage of water is used for showers and toilets?

**b** How much water per day is used for:

**i** cleaning? **ii** tap use?

**c** How much water (in kL) is used for doing the laundry in 1 year?

## Investigation

### Infographics

**1** Here is an infographic from the Australian Bureau of Statistics.

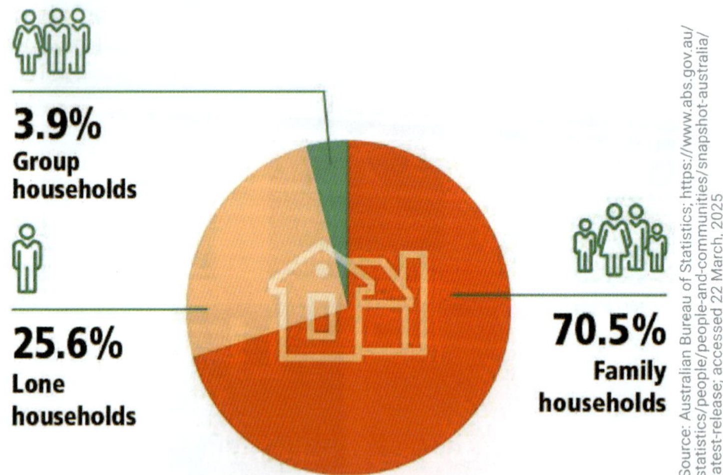


Source: Australian Bureau of Statistics; https://www.abs.gov.au/statistics/people/people-and-communities/snapshot-australia/latest-release; accessed 22 March, 2025

**a** Explain how data and information is displayed in this infographic. What strategies are used to present the statistics in an interesting way?

**b** What percentage of homes are occupied by only one person?

**c** What is the meaning of 'Group households'?

**d** What general conclusion can you make from reading this infographic?

**2** **a** Select an infographic from Exercise 1.01 and explain how the data and information is shown in the infographic. What strategies are used to present the statistics in an interesting way?

**b** How does the infographic compare to the infographic shown in this investigation?

# Misleading graphs 1.02

Statistical graphs are sometimes used to misrepresent information, and to falsely support a particular opinion or claim. For example, an advertisement may use a **misleading graph** to promote a product, or a company may use one to exaggerate its achievements and profits.

## Example 4

Three misleading graphs are shown below, used by the marketing manager of a computer store to highlight the increase in the sales of notebook computers over 5 weeks.

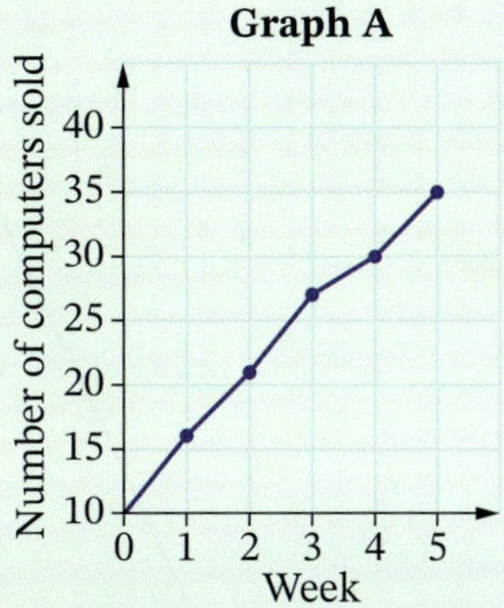


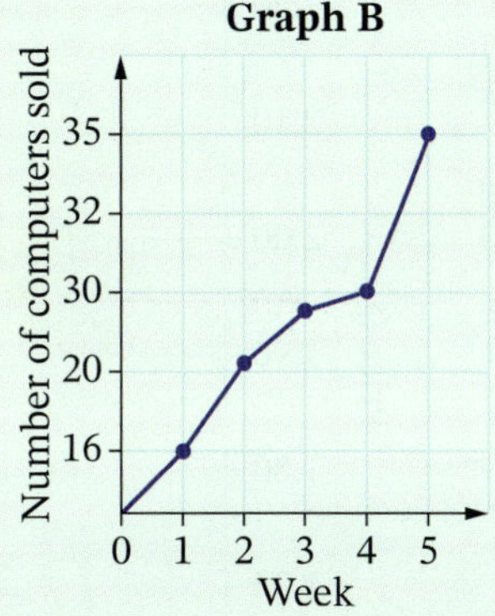


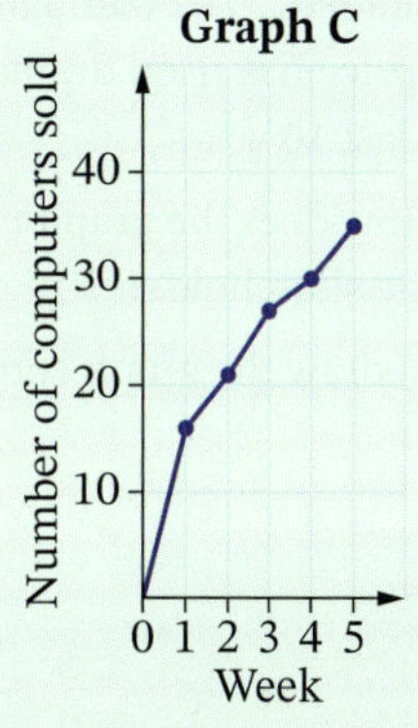


The sales figures are shown in this table:

| Week | 1 | 2 | 3 | 4 | 5 |
|---|---|---|---|---|---|
| Computers sold | 16 | 21 | 27 | 30 | 35 |

**a** Explain what is incorrect and misleading about each graph.

**b** Draw the correct graph to represent the sales data.

### Solution

**a** In Graph A, the vertical axis begins at 10 instead of 0, so the section of the graph between 10 and 40 has been exaggerated or 'stretched up' to give the impression that sales have increased greatly.

In Graph B, the scale on the vertical axis is not even, which gives the false impression that the increases in sales are greater than they actually are.

In Graph C, the horizontal axis has been compressed or 'shrunk' to make it look like the sales have increased more quickly.

**b** The correct graph has an even scale on both axes with the vertical axis beginning at zero. It correctly shows the sales pattern.

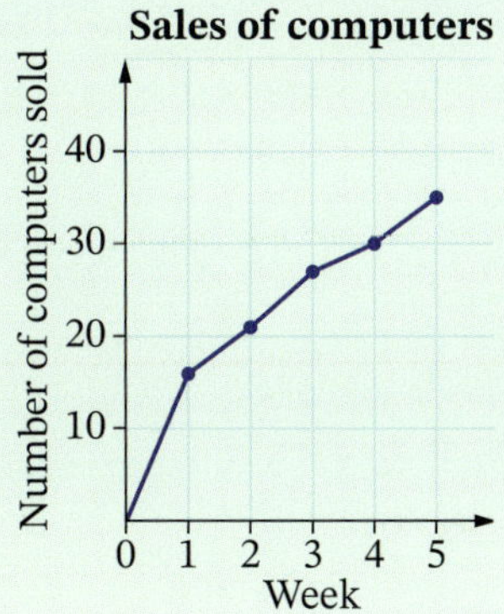

## Misleading graphs

A misleading graph can give a wrong impression by:

- not having a scale
- having an uneven scale or showing only part of the scale
- not showing the correct position of zero on the scale.

**EXERCISE 1.02** Answers on p. 455

## Misleading graphs

EXAMPLE 4

**1** Spin Industries used this graph to make the following claim to shareholders: 'The company's share price has risen dramatically in the last 10 years'.

**a** How does the graph mislead the shareholders?

**b** Redraw the graph correctly.

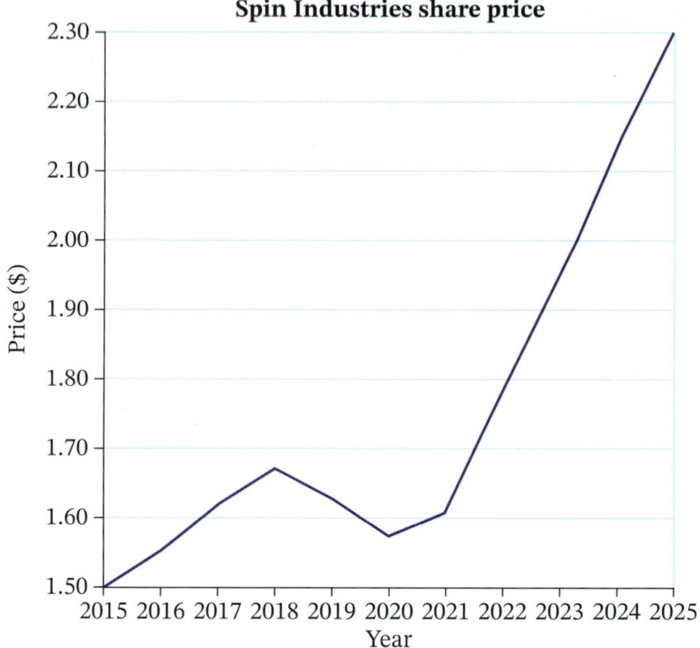


**2** Using pictures in graphs can be misleading if they are badly scaled to exaggerate differences. This graph should show that the wheat crop has doubled in 2 years, but the 2026 wheat picture is double the length and double the width of the 2024 wheat picture.

**a** How is the graph misleading?

**b** Redraw the graph for 2026 so it is correct.

**Growth in wheat crops**

2024

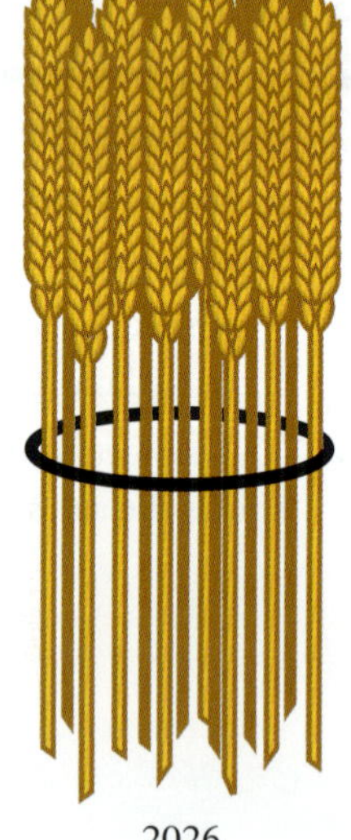

2026

Foundation Mastery Complex

3 This sales graph was presented to a meeting of shareholders by the ex-director of a company.

a What does the graph describe?

b What message is the ex-director trying to convey to the shareholders?

c What information is missing from the graph?

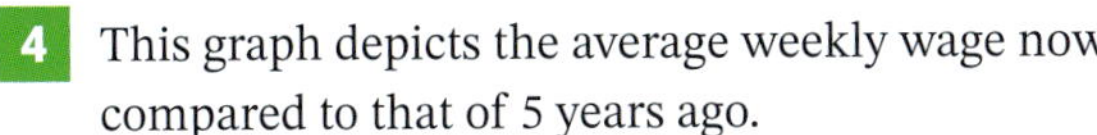

4 This graph depicts the average weekly wage now compared to that of 5 years ago.

a How is the graph misleading to the reader?

b Redraw the graph correctly.

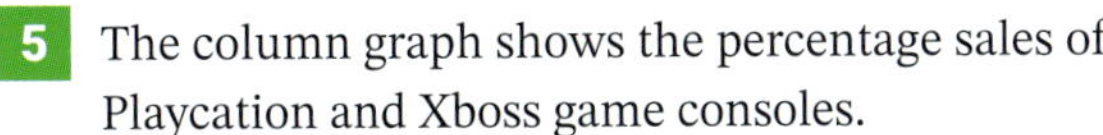

5 The column graph shows the percentage sales of Playcation and Xboss game consoles.

a How are Playcation sales exaggerated on this graph?

b How many times greater do Playcation sales appear to be than the sales of Xboss?

c Redraw the graph correctly.

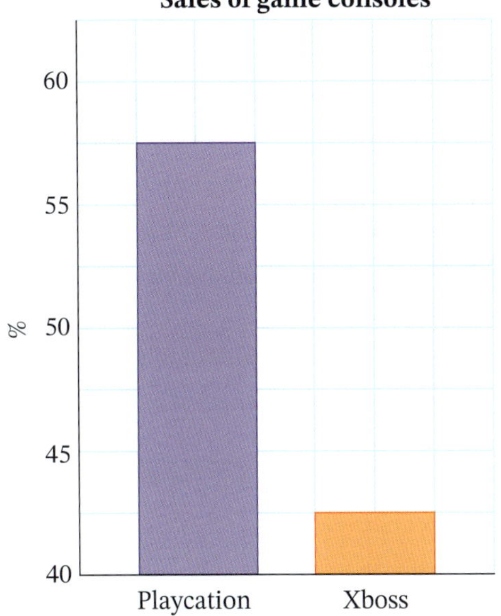


6 Oil prices have increased so much that $100 buys only half as much oil as it did 10 years ago. A newspaper article used the following graph to illustrate this.

a How is this graph misleading?

b Redraw the graph so that it represents the situation accurately.

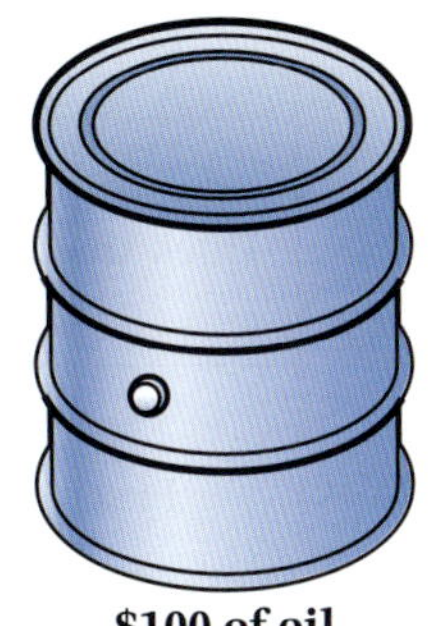

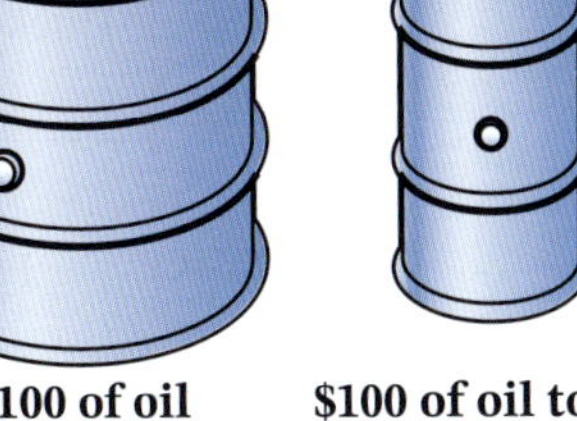

7 Jay drew this bar chart to illustrate how he spends a typical weekday.

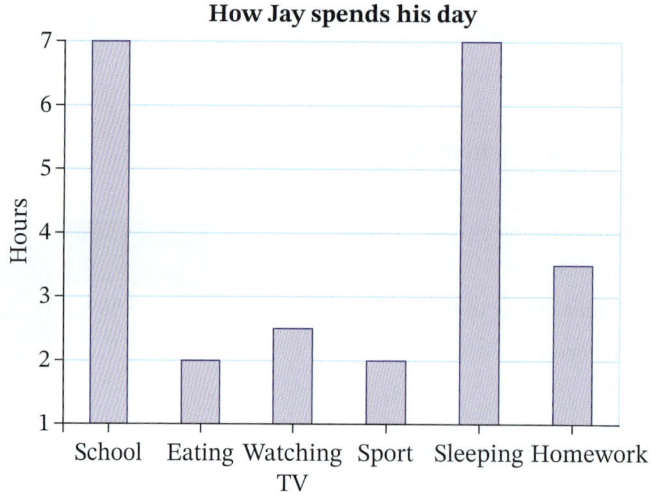


a How many hours does Jay spend sleeping?

b Explain what is wrong with this graph.

c What misleading impression does this graph give?

d Redraw this graph to show accurately how Joe spends a typical weekday.

8 This sector graph shows the annual sales of different makes of car.

a Explain how this graph is misleading.

b Use a spreadsheet to redraw the graph so that the sizes of the slices can be clearly compared.

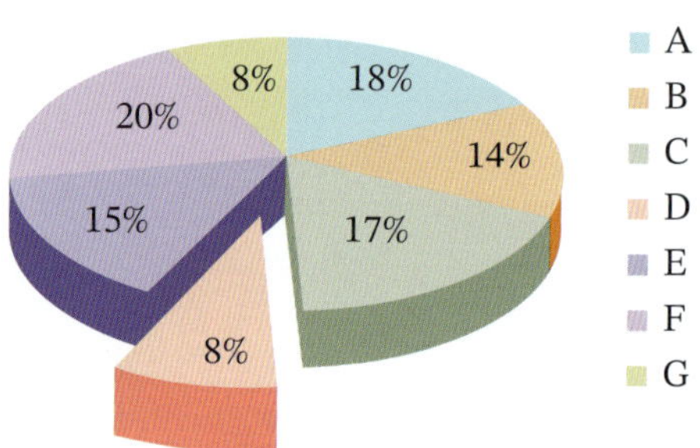


9 This picture graph shows the number of sales of different items at a fast-food store.

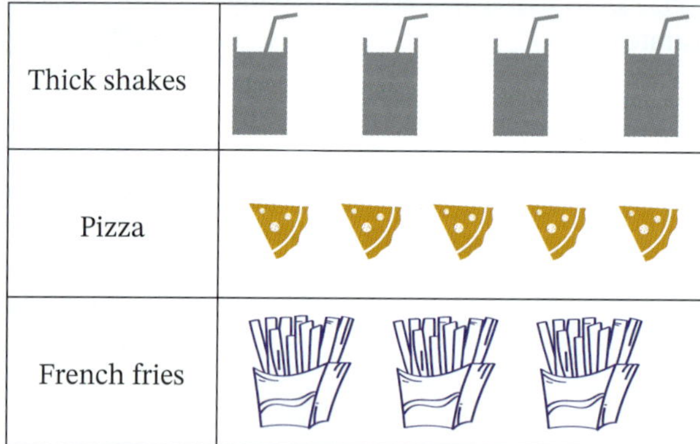


a Explain how this graph is misleading.

b What should be added to the graph to give more information about the number of sales?

# Classifying data

1.03

There are 2 main types of statistical **data**:

- **categorical data** are represented by categories, usually in words or symbols
- **numerical data** are represented by quantities or numbers.

**Videos**
Categorical and numerical data

Types of data

**Puzzle**
Statistical data match-up

## Example 5

Classify each type of data as categorical or numerical.

**a** number of tries scored by a football team  **b** colour of hair

**c** month of birth  **d** amount of petrol in cars.

### Solution

**a** numerical  For example, 0 or 3

**b** categorical  For example, blonde or brown

**c** categorical  For example, May or September

**d** numerical  For example, 32.5 L or 40 L

**Categorical data** can be further divided into **nominal data** and **ordinal data**, while **numerical data** can be further divided into **discrete data** and **continuous data**.

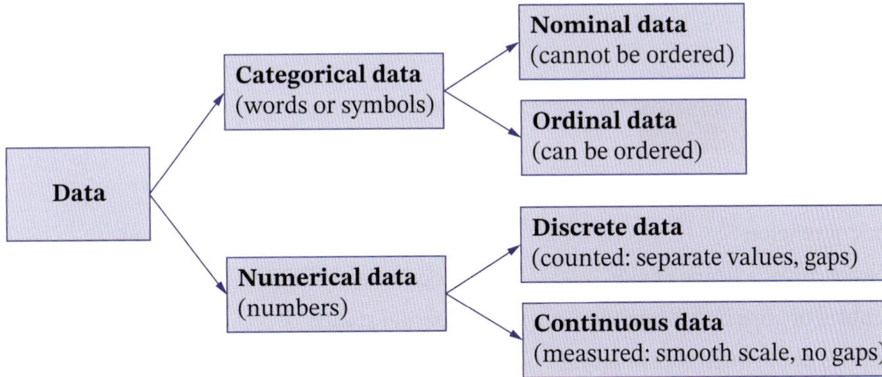


Colour of hair is an example of **nominal data** because the categories cannot be ordered; for example, blonde does not come before or after brown.

Month of birth is an example of **ordinal data** because the categories can be ordered; for example, May comes before September in the calendar year.

Number of tries scored in a rugby game is an example of **discrete data** because the numbers are obtained by *counting*. Discrete values are separate numbers (in this case, whole numbers) with gaps between the possible values. You cannot have $2\frac{1}{2}$ tries or 1.8 tries scored.

Amount of petrol is an example of **continuous data** because the values are obtained by *measuring*. Continuous values lie on a smooth scale with no gaps between the possible values. You can have 43.1 L, 43.15 L or 43.02 L of petrol.

## Example 6

Classify each type of categorical data as nominal or ordinal.

**a** method of travel to work  **b** report grade

**c** home suburb  **d** fan setting of air conditioner

### Solution

**a** nominal

**b** ordinal  For example, outstanding, satisfactory

**c** nominal

**d** ordinal  For example, low, medium, high

## Example 7

Classify each type of numerical data as discrete or continuous.

**a** amount of electricity used in homes  **b** numbers of contacts on a phone

**c** shoe size  **d** weekly profits of a business

### Solution

**a** continuous  smooth scale of values

**b** discrete  separate values (for example, 32 and 58)

**c** discrete  separate values (for example, 7 and $8\frac{1}{2}$)

**d** continuous  smooth scale of values

**EXERCISE 1.03** Answers on p. 456

## Classifying data

**1** Classify each type of data as categorical (C) or numerical (N).

**a** types of religion  **b** numbers of computers in homes

**c** types of driver's licence held  **d** populations of cities

**e** numbers of apps bought online  **f** marital status

**g** methods of payment for shopping  **h** movie classifications

**i** patients' blood pressures  **j** swimmers' finishing places in a race

**k** amounts of time spent on the internet per week

**l** towns or suburbs where people live

**2** What is the difference between categorical data and numerical data? Give an example of each type.

**3** Classify each type of categorical data as nominal (N) or ordinal (O).

**a** brand of car

**b** level of manager

**c** level of satisfaction with the government

**d** marital status

**e** favourite football team

**f** colour of Olympic medal

**g** size of soft drink sold at a fast food store

**h** a person's gender

Foundation Mastery Complex

EXAMPLE 7

**4** Classify each type of numerical data as discrete (D) or continuous (C).

- **a** amount of rainfall each day
- **b** mass of an athlete
- **c** length of a mobile phone call
- **d** running time for a 100 m race
- **e** star rating of a hotel
- **f** height of the world's tallest building
- **g** number of seats on a bus
- **h** loudness of a lawn mower

**5** Give one example of:

- **a** nominal data
- **b** ordinal data
- **c** discrete data
- **d** continuous data.

**6** Give the correct name for:

- **a** data that are usually described by numbers
- **b** data that are usually described by words
- **c** data that are usually measured on a smooth scale
- **d** categorical data that cannot be sorted in order.

**7** Which statement about categorical data is true? Select **A**, **B**, **C** or **D**.

- **A** They are sometimes represented by numbers that indicate size or an amount.
- **B** They are sometimes represented by symbols.
- **C** They are represented by numbers that take on separate values.
- **D** They are represented on a smooth scale.

**8** Each of these qualities can be both numerical and categorical. Explain how.

- **a** shirt size
- **b** exam results
- **c** fan speed

**9** The maximum temperatures for different parts of Sydney are shown.

| Location | Maximum temperature (°C) |
|---|---|
| Sydney CBD | 26 |
| Penrith | 30 |
| Liverpool | 28 |
| Richmond | 30 |
| Parramatta | 27 |
| Campbelltown | 28 |
| Bondi | 23 |

Is the data numerical or categorical?

# 1.04 Sampling techniques

## Sampling versus census

In **statistics**, a **population** means *all* of the items being studied, whether they are people, fish, cars or buildings. A **census** is a survey of a population in which every item is included.

In Australia, a national census takes place every 5 years, in a year ending with a 1 or a 6, such as 2026 and 2031.

If the whole population is too large or too difficult to survey, a **sample** of items is selected from the population. The results are then used to draw conclusions about the whole population. For example, a market researcher investigating the TV viewing habits of Australians will survey a representative sample of Australians.

| Sampling | Taking a census |
|---|---|
| Surveys a representative group of items from a population | Surveys all items in a population |
| Provides an estimate or approximate information about the population | Provides exact information about the population |
| Simple and not expensive | Complex and expensive |

### Example 8

Determine whether sampling or taking a census is more appropriate for each study below. Give a reason for your choice.

**a** Finding the most popular radio station in Newcastle

**b** Conducting a survey on whether Australians want a new national flag

**c** Finding the exact number of young women living on the South Coast, with the view to building a new maternity hospital

**d** Selecting politicians to form a new state government

### Solution

| | | |
|---|---|---|
| **a** | sampling | It would be impractical to survey everyone in the population. |
| **b** | sampling | It would be impractical to survey everyone in the population. |
| **c** | census | An exact number is required. |
| **d** | census | This decision is important and requires everyone eligible to vote. |

## Types of samples

If a sample is to be truly representative of a population, then each member of the population must have an equal chance of being chosen. Also, the **sample size** is important. The larger the sample, the more accurate the results of the study will be. There are 4 types of sample, described below.

In a **random sample**, each item is chosen completely at random from the population (for example, selecting names out of a box, or setting a computer to create random numbers to choose people or items).

In a **systematic sample**, the first item is chosen at random and all other items are chosen at regular intervals. For example, every 20th mobile phone is taken from a factory for testing, or the driver of every 8th car on the road is selected for random breath testing. 'Systematic' means a 'system' is used to select items.

Dwight Smith/Shutterstock.com

In a **stratified sample**, the population is divided into strata (layers) according to some characteristic (for example, gender or age group) and a random sample is taken from each layer using representative proportions or percentages. For example, if 28% of all shoppers in Australia are male, then a stratified sample of shoppers must be 28% male. 'Stratified' means 'in layers', like stratified rock (as shown in the photo).

In a **self-selected sample**, people choose to participate in a survey, so the sample is not randomly chosen nor necessarily representative of the population. For example, people who vote in an online poll about changing the national anthem could form an unrepresentative or biased sample if only people with strong opinions either way participated. People may be biased towards a particular view that has been promoted by a newspaper, radio station, news channel or social media website.

### Example 9

Which type of sample would you use to investigate each issue?

**a** The opinion of Year 11 students on a new senior school uniform

**b** The quality of Buzz rechargeable batteries

**c** The amount of time spent on the internet by Australian primary schoolchildren

**d** The support for a new government tax from the listeners of a radio talkback show

**e** The popularity of the Prime Minister of Australia

#### Solution

**a** A random sample will provide a fair representation of Year 11.

**b** A systematic sample will ensure that batteries produced at different times will be tested.

**c** A stratified sample will ensure that children of all types and areas are included.

**d** A self-selected sample is required as listeners have to call in to express their views.

**e** A stratified sample will ensure that people of all types and areas are included.

### Example 10

At Newcent High School, there are 126 senior students and 572 junior students. The student council wants to survey 80 students about the school's environmental program. How many senior students should be selected for a stratified sample?

#### Solution

Total number of students = 126 + 572 = 698

Number of senior students = 126

Sample of students = 80

Number of senior students in sample is $\frac{126}{698} \times 80 = 14.44...$

$\approx 14$

14 senior students should be selected for the sample.

## Bias in sampling

A **bias** is an unwanted influence that favours a particular section of the population unfairly. A **biased sample** is not truly representative of the population and negatively affects the quality or reliability of a statistical study.

### Example 11

Explain the bias in each sample.

**a** A group of Year 12 students sitting together in the playground was surveyed on their preferences for a school formal venue.

**b** A random selection of homes was contacted by telephone at midday for a survey on political opinions.

**c** A current affairs TV program asked viewers to participate in an online poll on gun laws.

### Solution

**a** The group members are likely to be friends and may have similar opinions, so this sample may not be representative of all Year 12 students.

**b** This is biased because it won't include the views of people who are not at home during the day.

**c** The sample would contain only (self-selected) viewers of the program who have a strong opinion about the subject or who make the effort to vote.

**EXERCISE 1.04** Answers on p. 456

## Sampling techniques

**1** Determine whether sampling (S) or taking a census (C) is more appropriate for each study below.

- **a** counting the number of students learning Japanese at a school
- **b** finding the number of people who watch soccer on SBS
- **c** judging the most popular supermarket in town
- **d** finding the number of Australians who were born in Europe
- **e** investigating people's opinions on whether Australia should become a republic
- **f** determining the number of people in NSW who have red hair
- **g** finding the exact number of schoolchildren living in Parramatta
- **h** testing whether a brand of mobile phone battery lasts 50 hours

**2** When will the next Australian national census be held?

**3** Select an appropriate type of sample (random, systematic, stratified or self-selected) for each study below. Give a reason for each answer.

- **a** testing running shoes for quality
- **b** finding the most watched television network on Tuesday nights
- **c** voting for the best contestant on a TV talent show
- **d** testing the reading abilities of a school's students
- **e** testing a new drug on patients
- **f** investigating people's views on renewable energy
- **g** investigating how many Year 11 students at your school work part time
- **h** surveying bank customers about their satisfaction levels
- **i** determining the most popular car colour
- **j** a social media poll to decide where a social group should hold their next party

**4** A department store has a customer database of 3347 female shoppers and 1504 male shoppers. It wants to introduce a new credit card to a stratified sample of 200 shoppers. How many female shoppers should be included in the sample?

**5** A public transport department surveys a sample of 600 commuters about the efficiency of the city's rail network. It is known that each region of the city has the following number of commuters:

| North | South | East | West | Southwest |
|---|---|---|---|---|
| 123 760 | 149 810 | 81 060 | 260 450 | 215 880 |

If the sample is to be stratified, how many commuters from the West should be included?

**6** A school's gardening program is run by 25 students, 5 teachers and 8 parents. A stratified sample of 8 people is selected to attend an environmental conference. Which of the following is the number of teachers who should be in the sample? Select **A**, **B**, **C** or **D**.

**A** 0 **B** 1 **C** 2 **D** 5

**7** Which one of the following is an example of a systematic sample? Select **A**, **B**, **C** or **D**.

**A** The first 40 customers in a supermarket are chosen.

**B** Every 10th customer to walk into a supermarket was chosen.

**C** 20 males and 20 females are chosen as they enter the shopping mall.

**D** 40 people at a ticketed concert are chosen by a random number generator.

**8** Explain why each sample is biased. EXAMPLE 11

**a** A sample of visitors to girlfriend.com.au was surveyed to investigate teenagers' views on road safety.

**b** Diners at a restaurant can give feedback about their meal by filling in a survey and posting it to the restaurant.

**c** For a survey about canteen food, the school council interviewed the first 20 students to arrive at school this morning.

**d** At a factory, a sample of executives was surveyed about safety in the workplace.

**e** A sample of audience members at a film premiere was surveyed about their views on the film.

**f** Triple J radio listeners were asked to vote for the best song of the year.

**g** After a plane flight, passengers were emailed a customer experience survey to complete.

**9** Determine which samples in Question **8** are self-selected.

**10** A sample of shoppers is to be chosen for a survey about customer service at a new shopping centre.

**a** How would you choose the shoppers to ensure that you have an unbiased sample?

**b** List 2 ways of choosing shoppers that would result in a biased sample.

## Did you know?

### How TV ratings are measured

Each week, research company OzTAM investigates the TV viewing habits of Australians, based on a sample of over 5000 homes. OzTAM, which stands for Australian Television Audience Measurement, investigates which free-to-air channels and programs are being watched by a sample of Australian households stratified according to the ages and number of people living in each home. Participants in the survey have their television sets linked to a black box called a 'peoplemeter' that records the details of all programs being watched each day (including pay television). The results are sent overnight by modem to OzTAM.

During one week in October 2023, the Australian television networks had the following percentage shares of viewers in Sydney, Melbourne, Brisbane, Adelaide and Perth, from 6:00 am to 12:00 midnight.

| Television channel | Percentage share |
|---|---|
| Nine | 21.6% |
| Seven | 19.0% |
| ABC | 12.1% |
| 10 | 10.0% |
| 7two | 4.2% |
| 7mate | 3.7% |
| SBS | 3.3% |
| 9 Gem | 3.0% |
| 10 Bold | 2.6% |
| 10 Peach | 2.5% |
| ABC Family | 2.4% |
| ABC News | 1.9% |
| SBS Viceland | 1.9% |

Data source: www.comparetv.com.au/australia-tv-ratings

**What types of people would be interested in the television viewing habits of Australians and why?**

Complex

## Investigation

### Estimating populations

How can you estimate how many students at your school:

**a** have red hair? **b** have brown eyes? **c** are left-handed?

**d** are vegetarian? **e** walk to school? **f** have a dog as a pet?

A survey is a common method for such estimations. Surveys are often conducted using questionnaires, which ask for details about a person and their lifestyle, beliefs, attitudes and behaviours. The table in the next investigation describes the features of a good questionnaire, giving examples.

Estimate the answers to these questions by surveying a sample of 50 students from your school and making appropriate calculations. How could you improve on your estimates?

**Worksheets**
Student survey form

Census questions

## Investigation

### Designing a questionnaire

The table below describes the features of a good questionnaire, giving examples.

| Feature | Bad example | Good example |
|---|---|---|
| 1 **Simple language**<br>All instructions and questions should be easily understood. | 'Describe the frequency with which you consume takeaway food on a weekly basis.' | 'How many times per week do you eat takeaway food?' |
| 2 **Unambiguous questions**<br>Questions must not have more than one meaning or interpretation. | 'What type of car do you own?'<br>Does the question mean:<br>• the make of car (for example, Toyota)?<br>• the style of car (for example, hatchback)?<br>• the size of car (for example, small)? | 'What make of car do you own? Tick one box from the following list: ...' (This is followed by a list of different makes of cars.) |
| 3 **Respect for privacy**<br>A survey should be anonymous and not require personal details. Questions that are too personal should be optional. | 'What is your email address?'<br>People are concerned about receiving junk email if they give out their email address. | 'In case we need to contact you further regarding this survey, please supply your email address. This is optional and your details will not be forwarded to other parties or used for promotional purposes.' |
| 4 **Freedom from bias**<br>Questions should not suggest or influence a person's response. | 'What do you think of changing the Australian flag, given that it is part of our history and our soldiers have died for it in the past?'<br>Too emotional, not objective, influencing people to vote against a new flag. | 'Do you think the current Australian flag should be changed?' |
| 5 **Consideration of a number of choices**<br>A question may have many different answers, so either make the question very specific or provide a list of answers from which to choose. | 'How did you enjoy your meal?' (from a restaurant survey).<br>Too open ended. There is a wide range of possible answers that would be difficult to organise and analyse. | 'How satisfied were you with your meal? Choose one of the following:<br>Very satisfied ☐<br>Satisfied ☐<br>Not satisfied ☐<br>Very unsatisfied ☐<br>No opinion ☐' |

1 Explain what is wrong with each survey question and write a better question for each.

  a Are politicians paid too much?

  b What is your favourite food?

  c How often do you smoke?

  d Which mode of transport do you employ to commute to your place of work?

  e Are you happy at school?

  f How satisfied are you with the level of customer service at the New Century Bank?

2 In groups of 2 to 4 students, design a questionnaire of 10–12 questions about one of the following topics:

  - TV viewing habits of the general population
  - spending habits of teenagers
  - students' use of smartphones and/or tablets
  - family and household details
  - favourite sports to watch and play.

  Decide on the type of facts you wish to collect, then determine suitable wording for each question and the format of each answer. Make sure that your questions reflect the features of an effective questionnaire.

# Constructing graphs 1.05

**Worksheet** Australian statistics

**Spreadsheet** Statistical graphs

When displaying a set of data, remember to:

- select the most appropriate graph or display
- choose a format that catches the reader's attention
- show enough detail, including a title, labels, key or scale
- avoid giving a misleading impression.

| | Useful for | Strengths | Weaknesses |
|---|---|---|---|
| **Column graph** | • Categorical or numerical data<br>• Showing differences in amounts | • Shows exact values<br>• Can graph a large amount of data | • Takes a long time to draw |
| **Line graph** | • Continuous numerical data<br>• Showing changes over time | • Shows exact values<br>• Can graph a large amount of data<br>• Shows differences and patterns clearly<br>• Can be used to estimate values between points | • Can only be used to graph continuous data |
| **Sector graph and divided bar graph** | • Categorical data<br>• Comparing parts of a whole | • Shows parts of a whole | • Doesn't show exact values<br>• Doesn't show small proportions or a large number of categories well |

## Example 12

The 2021 Census results for the religions of Australian people are shown in the table.

| Religion | Frequency |
|---|---|
| Buddhism | 615 823 |
| Christianity | 11 148 814 |
| Hinduism | 684 002 |
| Islam | 813 392 |
| Judaism | 99 956 |
| Other religions | 325 421 |
| No religion | 9 886 957 |
| Not stated or inadequately described | 2 848 428 |
| Total | 26 422 793 |

Data source: Australian Bureau of Statistics (ABS), 2021 Census

Represent this data on:

**a** a sector graph

**b** a divided bar graph.

### Solution

**a** First calculate the angle of each sector, rounded to the nearest degree.

Buddhism: $\frac{615\ 823}{26\ 422\ 793} \times 360° \approx 8°$

Write each religion as a fraction and multiply by 360°.

Christianity: $\frac{11\ 148\ 814}{26\ 422\ 793} \times 360° \approx 152°$

Hinduism: $\frac{684\ 002}{26\ 422\ 793} \times 360° \approx 9°$

Islam: $\frac{813\ 392}{26\ 422\ 793} \times 360° \approx 11°$

*Note*: Due to rounding, the sector angles may not add exactly to 360°.

Judaism: $\frac{99\ 956}{26\ 422\ 793} \times 360° \approx 1°$

Other religions: $\frac{325\ 421}{26\ 422\ 793} \times 360° \approx 4°$

No religion: $\frac{9\ 886\ 957}{26\ 422\ 793} \times 360° \approx 135°$

Not stated: $\frac{2\ 848\ 428}{26\ 422\ 793} \times 360° \approx 39°$

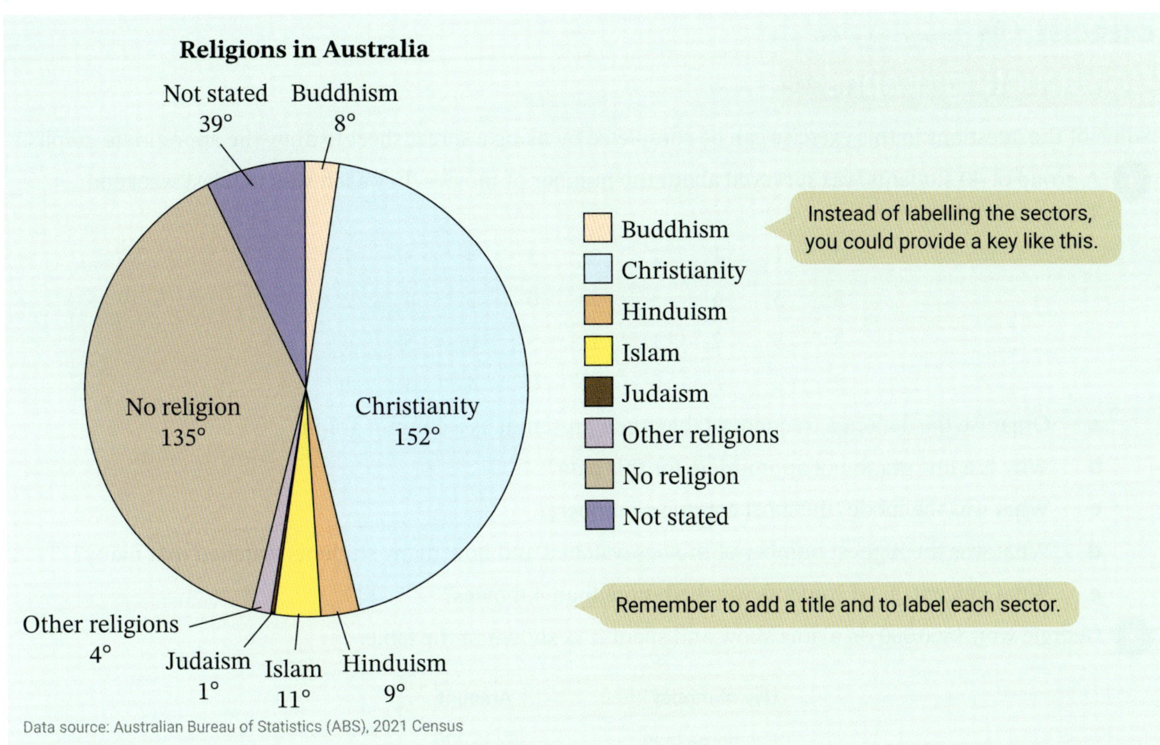


**b** Draw a long rectangle (for example, 12 cm long), then calculate the length of the section representing each religion, rounded to 2 decimal places.

Buddhism: $\frac{615\,823}{26\,422\,793} \times 12\text{ cm} = 0.28\text{ cm}$

Write each religion as a fraction and multiply by 12 cm.

Christianity: $\frac{11\,148\,814}{26\,422\,793} \times 12\text{ cm} = 5.06\text{ cm}$

Hinduism: $\frac{684\,002}{26\,422\,793} \times 12\text{ cm} = 0.31\text{ cm}$

Islam: $\frac{813\,392}{26\,422\,793} \times 12\text{ cm} = 0.37\text{ cm}$

Judaism: $\frac{99\,956}{26\,422\,793} \times 12\text{ cm} = 0.05\text{ cm}$

"Judaism" will be too small to show as a section.

Other religions: $\frac{325\,421}{26\,422\,793} \times 12\text{ cm} = 0.15\text{ cm}$

*Note*: Due to rounding, the sections may not add exactly to 12 cm.

No religion: $\frac{9\,886\,957}{26\,422\,793} \times 12\text{ cm} = 4.49\text{ cm}$

Not stated: $\frac{2\,848\,428}{26\,422\,793} \times 12\text{ cm} = 1.29\text{ cm}$

Remember to add a title and to label each section.

Or, instead of labelling the sections, you could provide a key.

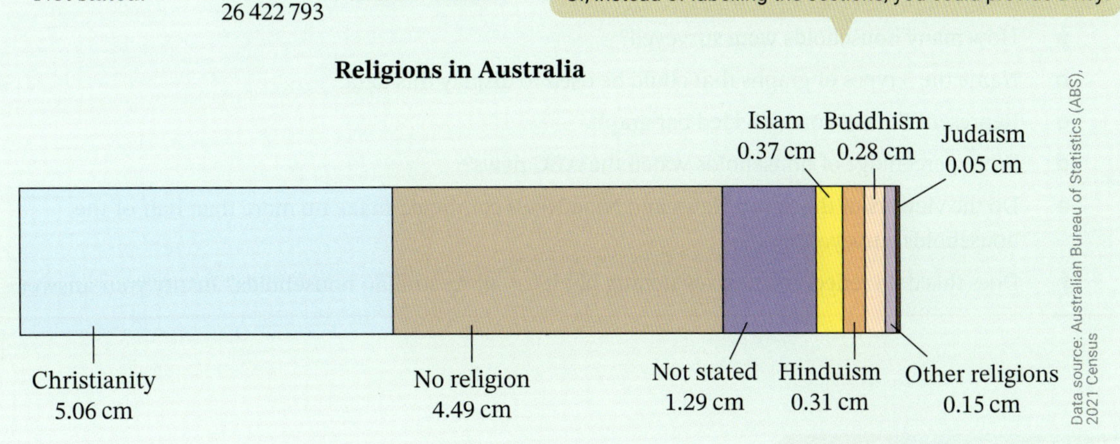

**EXERCISE 1.05** Answers on p. 457

## Constructing graphs

Some of the questions in this exercise can be completed by using a spreadsheet to draw the appropriate graphs.

**1** A group of 40 students was surveyed about the number of movies they each watched last weekend. The results were:

| | | | | | | | | | |
|---|---|---|---|---|---|---|---|---|---|
| 2 | 1 | 1 | 2 | 3 | 3 | 4 | 3 | 1 | 4 |
| 3 | 3 | 6 | 3 | 2 | 0 | 3 | 4 | 2 | 2 |
| 3 | 3 | 2 | 1 | 1 | 3 | 0 | 5 | 4 | 4 |
| 2 | 2 | 2 | 5 | 4 | 0 | 4 | 3 | 1 | 3 |

**a** Organise the data in a frequency table, then present it as a column graph.

**b** Why is a line graph not appropriate for this data?

**c** What was the mode (the most common number)?

**d** What was the highest number of movies watched and how many students watched that many?

**e** What percentage of students watched more than 4 movies?

EXAMPLE 12

**2** Georgie won $800 000 on a quiz show and spent it as shown in the table.

| Use of money | Amount |
|---|---|
| Pay home loan | $235 000 |
| New car | $ 96 000 |
| Home renovations | $144 000 |
| Travel | $125 000 |
| Savings | $200 000 |

**a** Why is a sector graph or divided bar graph suitable for this information?

**b** Draw a sector graph for this information.

**c** What did Georgie do with one-quarter of her winnings?

**d** What fraction of Georgie's winnings was spent on home renovations?

**e** What percentage of Georgie's winnings was spent on a new car?

**3** A sample of Sydney households was surveyed about which TV network they watched for the evening news. The results are shown in the table.

| Network | Number of households |
|---|---|
| ABC | 3460 |
| SBS | 855 |
| Seven | 4140 |
| Nine | 4215 |
| Ten | 2300 |

**a** How many households were surveyed?

**b** Name the 3 types of graphs that could be used to display this data.

**c** Represent the data on a divided bar graph.

**d** What percentage of households watch the ABC news?

**e** Do the viewers of the Seven News and Nine News combined make up more than half of the households surveyed?

**f** Does this data reflect the news-watching habits of all Australian households? Justify your answer.

 Foundation  Mastery ○ Complex

**4** The teachers at Summer Bay High School were surveyed on the number of cups of coffee they drank yesterday.

| Cups of coffee | 0 | 1 | 2 | 3 | 4 | 5 | 6 |
|---|---|---|---|---|---|---|---|
| Frequency | 7 | 5 | 11 | 21 | 15 | 9 | 4 |

**a** How many teachers are there at the school?

**b** What type of graph is best used to illustrate this data? Give a reason for your answer.

**c** What percentage of the teachers did not drink coffee that day? Answer correct to one decimal place.

**d** If the recommended healthy coffee intake is no more than 3 cups of coffee per day, what fraction of teachers had a healthy intake of coffee?

**5** Serena wants to graph the results of a survey about people's favourite ice-cream flavour. Which of the following graphs can she use? Select **A**, **B**, **C** or **D**.

**A** column graph

**B** sector graph

**C** divided bar graph

**D** all of the above

**6** According to the 2021 Census, there are 167 Aboriginal and Torres Strait Islander languages used at home by 76 978 Aboriginal and Torres Strait Islander peoples. The most widely used language groups are shown in the table.

| Language groups | Percentage |
|---|---|
| Arnhem Land and Daly River Region Languages | 14.5% |
| Western Desert Languages | 10.9% |
| Torres Strait Island Languages | 12.0% |
| Yolngu Matha | 8.5% |
| Arandic | 7.4% |

Data source: Australian Bureau of Statistics (ABS); 2021 Census

**a** What is the percentage of languages not covered by the language groups in the table?

**b** Display this data in a sector graph and in a divided bar graph. Use 'Other' for the language groups not shown in the table.

**7** This table shows the percentage of Australians who use some popular social media platforms.

| Platform | Percentage |
|---|---|
| Facebook | 78.2 |
| Messenger | 69.9 |
| Instagram | 62.4 |
| WhatsApp | 44.8 |
| TikTok | 40.0 |
| Snapchat | 33.0 |
| X | 29.6 |
| LinkedIn | 28.9 |

Data source: Meltwater; https://www.meltwater.com/en/blog/social-media-statistics-australia; accessed 22 March 2025

**a** Explain why a sector graph and divided bar graph cannot be used to display this data.

**b** Use an appropriate graph to display the data.

**8** This table shows the percentage of Australians in each generation according to the 2021 Census.

| Generation | Ages | Percentage |
|---|---|---|
| Gen Alpha | 0–9 years | 12.0% |
| Gen Z | 10–24 years | 18.2% |
| Millennials | 25–39 years | 21.5% |
| Gen X | 40–54 years | 19.3% |
| Baby Boomers | 55–74 years | 21.5% |
| Interwar | 75 years and over | 7.5% |

Data source: Australian Bureau of Statistics (ABS); 2021 Census

**a** Which graphs can be used to illustrate this data? Give reasons for your answer.

**b** Represent the data on a column graph.

**c** If Australia's population is 25 760 867, find the number of Australians who are:

**i** millennials **ii** interwar.

**9** This table lists the top 8 countries of the world by population.

**a** What is the percentage for 'Others'?

**b** Draw a divided bar graph to illustrate this data.

**c** India surpassed China in 2023 as being the world's most populous nation.
It has a population of 1 450 935 791.
Use the table to find the population of:

**i** China **ii** Indonesia **iii** the world.

| Country | Percentage of world population |
|---|---|
| India | 17.78 |
| China | 17.39 |
| USA | 4.23 |
| Indonesia | 3.47 |
| Pakistan | 3.08 |
| Nigeria | 2.85 |
| Brazil | 2.6 |
| Bangladesh | 2.13 |
| Others | |

Data source: worldometer; Countries in the world by population (2025)

**10** In a 2-week period, Australian Customs recorded the number of immigrants arriving at Sydney airport, as shown in the table.

**a** Draw a divided bar graph to represent this data.

**b** 'More than half of the immigrants came from the UK or New Zealand.' True or false?

**c** What fraction of immigrants came from Asia?

**d** What percentage (to the nearest whole number) of immigrants came from the Middle East?

**e** Which region represents 10.75% of the immigrants?

| Country of origin | Number of immigrants |
|---|---|
| United Kingdom | 20 |
| New Zealand | 35 |
| Middle East | 7 |
| India | 10 |
| Africa | 3 |
| Asia | 12 |
| Other regions | 6 |

☐ Foundation ○ Mastery ⬡ Complex

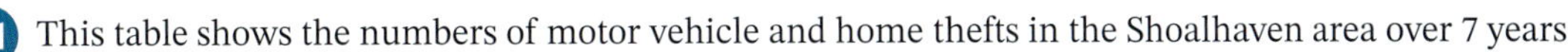

**11** This table shows the numbers of motor vehicle and home thefts in the Shoalhaven area over 7 years.

| Year | 2018 | 2019 | 2020 | 2021 | 2022 | 2023 | 2024 |
|---|---|---|---|---|---|---|---|
| Motor vehicle theft | 155 | 171 | 135 | 147 | 129 | 185 | 221 |
| Home theft | 360 | 355 | 295 | 264 | 245 | 255 | 302 |

**a** Draw line graphs showing both types of theft on the same axes. Include labels or a key.

**b** In which year was home theft highest?

**c** In which year was motor vehicle theft lowest?

**d** Which type of theft is generally more common?

**e** Write a sentence to describe what happened to the number of home thefts over the 7 years.

**f** What do you think happened to motor vehicle theft in 2025–2026? Justify your answer.

## Technology

### Drawing sector graphs

We can use **Insert** and **Chart** in a spreadsheet to draw different types of graphs.
A sample of 800 commuters was surveyed about the way they travelled to work.

1 Enter the data from the survey into a spreadsheet as shown.

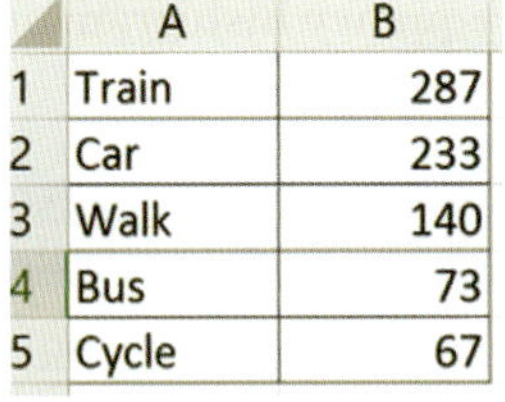

| | A | B |
|---|---|---|
| 1 | Train | 287 |
| 2 | Car | 233 |
| 3 | Walk | 140 |
| 4 | Bus | 73 |
| 5 | Cycle | 67 |



2 Select (highlight) all the cells A1:B5 and click **Insert** and **Pie** or **Doughnut Chart**.

3 Click **Quick Layout** from the **Design** menu and choose **Layout 6**, which shows a **Legend** and **Chart Title**. Percentages and sector names are also now shown on the sector graph.

4 Click on **Chart Title** and replace with the title given below.

**Method of travelling to work**

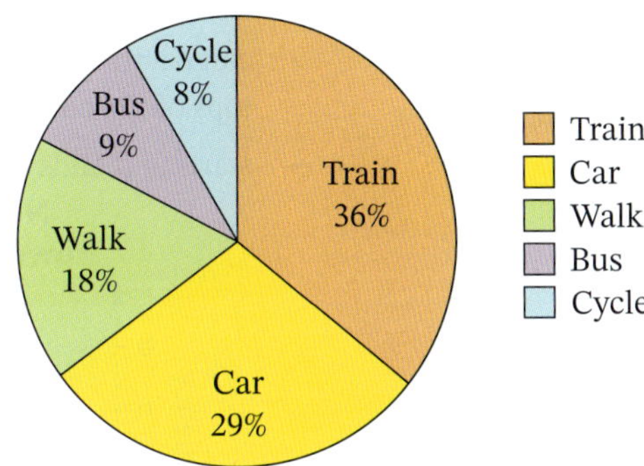


5 Format any features of your graph, including colour and design, by clicking on it and using **Format** or **Design** from the Toolbar. You may also resize the graph.

### Further investigations

- Repeat the steps above for other data, creating different sector graphs and layout options.
- Graph the same data as a column graph by clicking **Design** and **Change Chart Type**.

Foundation Mastery Complex

## Did you know?

### Bias in drafting US soldiers

Bias can affect life-or-death situations, as some young Americans found out during the Vietnam War. In 1970, men aged 19 to 26 were randomly drafted into the armed forces according to their birthdates (this also occurred in Australia). The 366 dates of the year were written on plastic capsules and put into a large bowl. Capsules were then picked out and the men who had the selected birthdates were drafted. However, it was later discovered that birthdates in November and December had a slightly higher chance of being drawn because the capsules were placed in the bowl in order and not mixed properly, and the ones chosen were often from the top of the pile.

**Suggest 2 ways the bias could have been eliminated during this selection process.**

# 1.06 Frequency histograms and polygons

**Videos**
Frequency histograms
Frequency histograms and polygons

**Worksheets**
Frequency tables
Frequency distribution tables
Histograms

A **frequency histogram** or **frequency polygon** graphs numerical data, including data grouped into **class intervals**.

## Class intervals and centres

- In statistics, when there are many values in a **data set**, they may be grouped into class intervals. For example, ages of people may be grouped into class intervals of 1–10, 11–20, 21–30 and so on.
- A **class centre** is the centre of a class interval. For example, the class centre of the class interval 10–19 is 14.5.

**Frequency histogram**

- no gaps between columns
- has a half-column gap at the start.

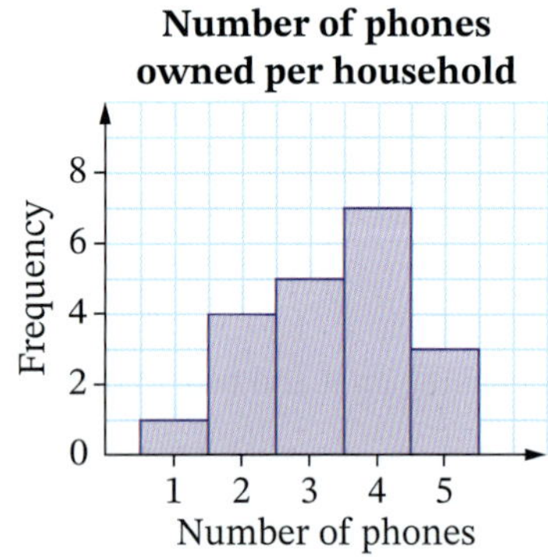


**Frequency polygon**

- can be drawn by joining the midpoints of the tops of the columns of a histogram
- starts and ends on the horizontal axis.

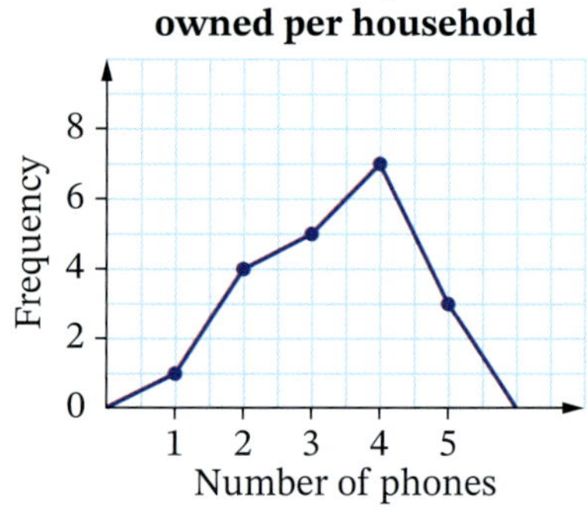


| | Useful for | Strengths | Weaknesses |
|---|---|---|---|
| Frequency histogram and polygon | • numerical data<br>• data grouped into class intervals<br>• showing the shape of a set of data; how the values are spread out | • are easy to read<br>• show differences and patterns clearly<br>• can graph a large amount of data<br>• show how the data are spread out | • takes a long time to draw |

## Example 13

The ages of the patients at a medical centre in one afternoon were recorded:

| | | | | | | | | | | | | | |
|---|---|---|---|---|---|---|---|---|---|---|---|---|---|
| 37 | 12 | 19 | 65 | 44 | 30 | 1 | 8 | 14 | 36 | 25 | 57 | 21 | 16 |
| 4 | 30 | 40 | 49 | 38 | 69 | 33 | 55 | 16 | 4 | 8 | 16 | 6 | 28 |
| 29 | 54 | 43 | 48 | 33 | 18 | 9 | 59 | 21 | 70 | 22 | 33 | 2 | 60 |

**a** Copy and complete this frequency distribution table for the data.

| Class interval | Class centre | Tally | Frequency |
|---|---|---|---|
| 0–9 | | | |
| 10–19 | | | |
| 20–29 | | | |
| 30–39 | | | |
| 40–49 | | | |
| 50–59 | | | |
| 60–69 | | | |
| 70–79 | | | |
| | | **Total** | |

**b** Construct a frequency histogram for this data.

**c** How many patients went to the medical centre?

**d** How many of these patients were aged 50 or over?

### Solution

**a**

These are discrete class intervals.

| Class interval | Class centre | Tally | Frequency |
|---|---|---|---|
| 0–9 | 4.5 | ~~IIII~~ III | 8 |
| 10–19 | 14.5 | ~~IIII~~ II | 7 |
| 20–29 | 24.5 | ~~IIII~~ I | 6 |
| 30–39 | 34.5 | ~~IIII~~ III | 8 |
| 40–49 | 44.5 | ~~IIII~~ | 5 |
| 50–59 | 54.5 | IIII | 4 |
| 60–69 | 64.5 | III | 3 |
| 70–79 | 74.5 | I | 1 |
| | | **Total** | **42** |

**b** When graphing data that are grouped into discrete class intervals, either the class intervals or class centres are shown on the horizontal axis.

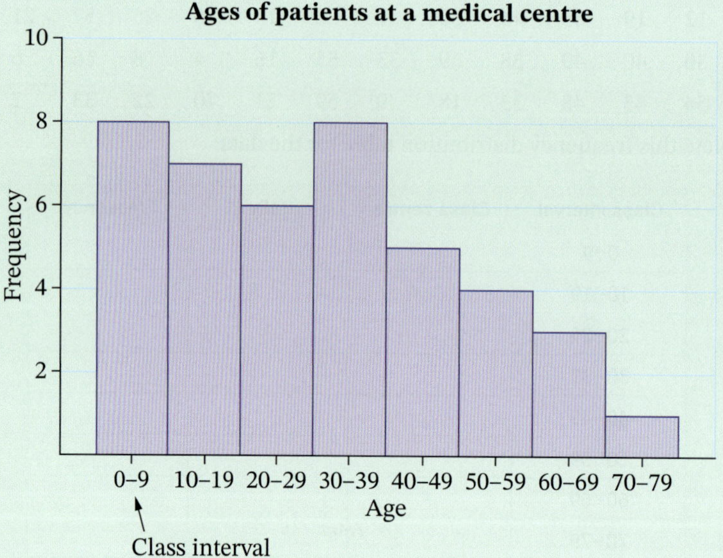


OR

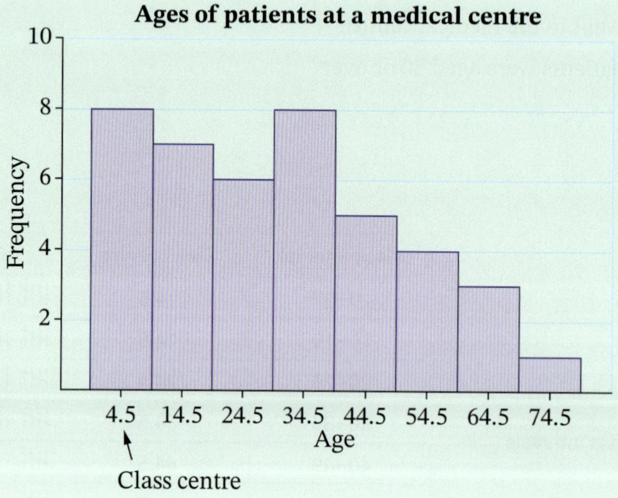


**c** 42 patients went to the medical centre. total frequency

**d** Number of patients aged 50 or over = $4 + 3 + 1 = 8$

## Example 14

The monthly call costs for a group of 120 mobile phone users were recorded and grouped into class intervals as shown.

These are continuous class intervals.

| Call cost ($) | Class centre | Number of users |
|---|---|---|
| 0–<20 | 10 | 6 |
| 20–<40 | 30 | 8 |
| 40–<60 | 50 | 13 |
| 60–<80 | 70 | 17 |
| 80–<100 | 90 | 23 |
| 100–<120 | 110 | 20 |
| 120–<140 | 130 | 16 |
| 140–<160 | 150 | 10 |
| 160–<180 | 170 | 4 |
| 180–<200 | 190 | 3 |
| | **Total** | **120** |

**a** Construct a frequency histogram and polygon for this data.

**b** Which call costs were the most common?

**c** Which call costs were used by exactly 10 people?

**d** What percentage of phone users had call costs below $60?

### Solution

**a** When graphing data grouped into continuous class intervals, either the class boundaries or class centres are shown on the horizontal axis, like this:

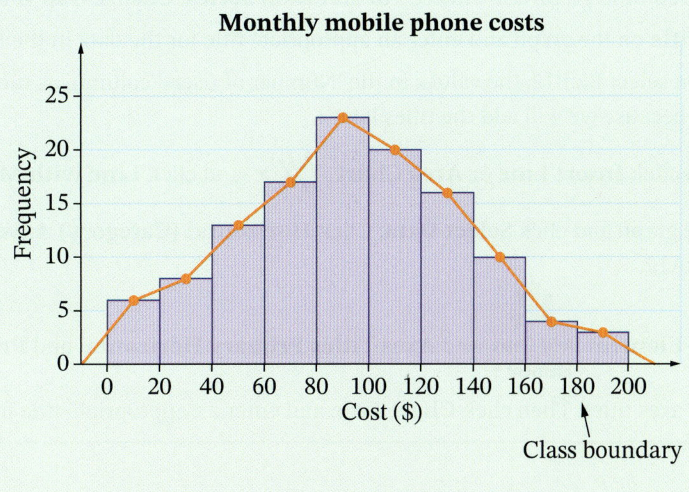


or like this:

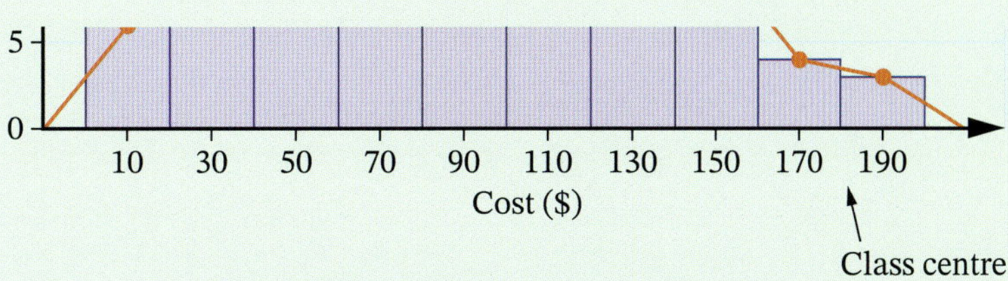


**b** $80–<$100 had the highest frequency (23)

**c** $140–<$160 had a frequency of 10

**d** Number of users with call costs below $60 = 6 + 8 + 13 = 27

Percentage of users with call costs below $60 $= \frac{27}{120} \times 100\% = 22.5\%$

## Technology

### Drawing frequency histograms and polygons

We can use **Insert** and **Chart** in a spreadsheet to draw a frequency histogram and polygon.

**1** Enter the data about the cost of mobile phone calls from Example 14 into a spreadsheet but add a last row (210, 0) as shown.

This allows you to draw a frequency polygon that touches the *x*-axis.

**2** Select B2:B12, the values in the 'Number of users' column, as the '*y* values'. (Do not select B1 because we will add the titles later.)

| | A | B |
|---|---|---|
| 1 | Call cost ($) | Number of users |
| 2 | 10 | 6 |
| 3 | 30 | 8 |
| 4 | 50 | 13 |
| 5 | 70 | 17 |
| 6 | 90 | 23 |
| 7 | 110 | 20 |
| 8 | 130 | 16 |
| 9 | 150 | 10 |
| 10 | 170 | 4 |
| 11 | 190 | 3 |
| 12 | 210 | 0 |



**3** Click **Insert** then click **Insert Column** or **Bar Chart** and **Clustered Column**.

**4** Right-click on the graph and click **Select Data**. Click **Horizontal (Category) Axis Labels** then **Edit** and highlight A2:A12.

**5** Click **Add Chart Element** Add Chart Element and **Axes Titles Primary Horizontal and Primary Vertical** and enter appropriate axes titles.

**6** Right-click on the column graph and choose **Format Data Series**. Change **Gap Width** to 0%.

**7** Click on **Chart Title** on the graph and enter an appropriate title for the data in the frequency histogram.

**8** To make a polygon select B2:B12, the values in the 'Number of users' column, as the '*y* values'. (Do not select B1 because we will add the titles later.)

**9** Click **Insert** then click **Insert Line** or **Area Chart** and click **Line with Markers**.

**10** Right-click on the graph and click **Select Data**. Click **Horizontal (Category) Axis Labels** then **Edit** and highlight A2:A12.

**11** Click **Add Chart Element** Add Chart Element and **Axes Titles Primary Horizontal** and **Primary Vertical** and enter appropriate axes titles. Then click **Chart Title** and enter an appropriate title for the polygon.

**EXERCISE 1.06** Answers on p. 458

## Frequency histograms and polygons

**1** Forty people were surveyed about the number of times they posted photos online in one day. The results are shown below.

| | | | | | | | | | |
|---|---|---|---|---|---|---|---|---|---|
| 2 | 1 | 5 | 1 | 3 | 4 | 3 | 5 | 5 | 4 |
| 3 | 4 | 5 | 2 | 1 | 0 | 3 | 4 | 1 | 1 |
| 2 | 3 | 2 | 0 | 1 | 5 | 4 | 5 | 3 | 3 |
| 1 | 1 | 1 | 4 | 4 | 5 | 2 | 3 | 1 | 5 |

**a** Organise this data into a frequency distribution table.

**b** Construct a frequency histogram for the data.

**2** The masses (in grams) of a sample of 48 eggs were measured and recorded.

| | | | | | | | | | | | |
|---|---|---|---|---|---|---|---|---|---|---|---|
| 57 | 58 | 61 | 59 | 62 | 59 | 59 | 56 | 60 | 64 | 58 | 58 |
| 56 | 59 | 64 | 57 | 60 | 62 | 58 | 60 | 64 | 57 | 61 | 58 |
| 59 | 57 | 64 | 58 | 59 | 57 | 58 | 64 | 60 | 58 | 60 | 57 |
| 61 | 64 | 58 | 60 | 61 | 62 | 62 | 58 | 60 | 61 | 57 | 58 |

**a** Construct a frequency distribution table for this data.

**b** Draw a frequency polygon for the data.

**c** If these eggs are labelled as being 60 grams, would you say that the label is misleading? Give reasons for your answer.

**3** The table shows the number of road fatalities according to age group on a section of the Pacific Highway over 6 months.

| Age group | Road fatalities |
|---|---|
| 0–9 | 6 |
| 10–19 | 14 |
| 20–29 | 28 |
| 30–39 | 11 |
| 40–49 | 8 |
| 50–59 | 3 |

**a** How many people died on this section of the highway during the period?

**b** Draw a frequency histogram and polygon for this data.

**c** What observations can you make about road fatalities from the graphs?

**4** The number of words in each sentence of a magazine article was counted, with the results shown below.

| | | | | | | | | | | | |
|---|---|---|---|---|---|---|---|---|---|---|---|
| 27 | 22 | 15 | 8 | 14 | 7 | 9 | 25 | 15 | 17 | 5 | 24 |
| 9 | 11 | 22 | 8 | 5 | 15 | 25 | 18 | 10 | 21 | 24 | 13 |
| 9 | 14 | 18 | 11 | 9 | 23 | 15 | 19 | 10 | 8 | 14 | 17 |

**a** Is this data discrete or continuous?

**b** Using class intervals of 1–5, 6–10 etc., complete a frequency table for the data.

**c** Construct a frequency polygon for the data.

**d** What was the modal class (the class with the highest frequency)?

Foundation | Mastery | Complex

**5** Here are the ages of the employees at Burger Heaven.

| | | | | | | | | | |
|---|---|---|---|---|---|---|---|---|---|
| 18 | 19 | 18 | 17 | 20 | 20 | 24 | 15 | 24 | 19 |
| 15 | 40 | 21 | 17 | 20 | 22 | 23 | 21 | 24 | 23 |
| 34 | 19 | 45 | 20 | 15 | 21 | 24 | 27 | 19 | 33 |
| 34 | 24 | 16 | 18 | 30 | 21 | 26 | 31 | 16 | 25 |
| 49 | 21 | 21 | 35 | 16 | 22 | 15 | 25 | 44 | 23 |

**a** Organise the data into a frequency table using classes 15–19, 20–24 etc.

**b** How many people work at Burger Heaven?

**c** Draw a frequency histogram and polygon for the data.

**d** Comment on the shape of the data.

**6** A group of students was surveyed on the number of mobile phone calls made in a week. The results are shown in this histogram.

**a** How many students made 22 calls?

**b** How many students were surveyed?

**c** What percentage of students made fewer than 20 calls? Answer correct to one decimal place.

**d** Can you tell how many students made 32 calls? Why?

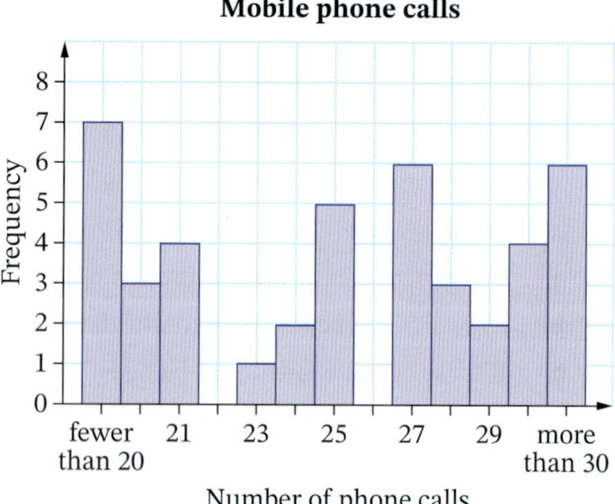

**7** The weight of tomatoes in kilograms picked from each plant after adopting a "no dig" vegetable garden is shown below.

| | | | | | | | | | |
|---|---|---|---|---|---|---|---|---|---|
| 1.8 | 2.5 | 4.3 | 6.5 | 2.7 | 4.6 | 11.0 | 10.8 | 0.3 | 8.2 |
| 2.1 | 3.8 | 4.4 | 5.8 | 1.6 | 5.9 | 7.6 | 9.3 | 4.8 | 3.4 |
| 12.5 | 4.6 | 2.5 | 6.9 | 7.5 | 3.5 | 4.8 | 12.2 | 4.3 | 3.7 |
| 0.0 | 0.9 | 2.6 | 7.8 | 4.9 | 7.4 | 9.8 | 10.4 | 2.6 | 8.2 |

**a** Is this data discrete or continuous?

**b** Complete a frequency table for this data using class intervals 0–<2, 2–<4 etc.

**c** Draw a frequency histogram for this data.

**d** What is the modal class?

**e** Use your histogram to comment on the effectiveness of the "no dig" garden.

**8** The heights (in centimetres) of new applicants to a modelling agency are:

| | | | | | | | | |
|---|---|---|---|---|---|---|---|---|
| 168 | 182 | 187 | 185 | 178 | 176 | 174 | 163 | 164 |
| 165 | 189 | 185 | 165 | 162 | 173 | 164 | 173 | 178 |
| 172 | 186 | 189 | 175 | 176 | 180 | 170 | 168 | 165 |

**a** Using class intervals of 160–<165, 165–<170 etc., complete a frequency table for this data.

**b** Draw a frequency polygon for the data.

**c** What is the modal class interval?

**d** What is the lowest class interval?

**e** Which class interval is least common?

☐ Foundation ○ Mastery ○ Complex

# Dot plots and stem-and-leaf plots



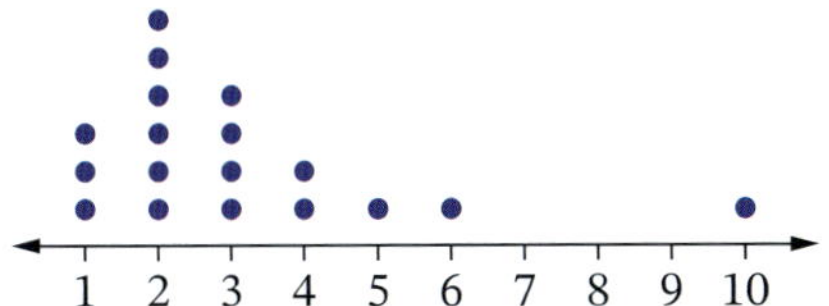


| Stem | Leaf |
|---|---|
| 7 | 9 5 8 1 6 9 |
| 8 | 4 2 5 0 0 8 0 3 2 4 2 8 |
| 9 | 0 8 4 2 2 0 0 8 0 5 6 |
| 10 | 5 4 |

**Worksheet**
Stem-and-leaf plots

**Dot plot**

- a simple type of column graph
- uses a dot, cross or symbol to display each data value
- also called a line plot.

**Stem-and-leaf plot**

- actual data values are listed, usually in order
- the stem shows the first digit(s)
- the leaf shows the last digit(s)
- also called a stem plot.

| | Useful for | Strengths | Weaknesses |
|---|---|---|---|
| Dot plot and stem-and-leaf plot | • numerical data<br>• small sets of data<br>• showing gaps and **clusters**<br>• showing **outliers** (values that are very different from the rest) | • easy to read and draw<br>• shows differences and patterns clearly<br>• shows the spread of data<br>• stem-and-leaf plot shows actual data values | • hard to display large sets of data |

## Example 15

This dot plot shows the temperatures of a sample of hospital patients.

**Patients' temperatures**

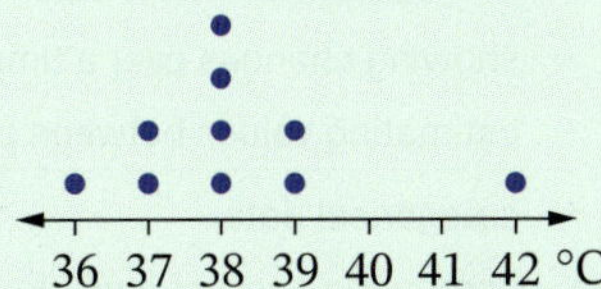


**a** How many patients are in the sample?

**b** How many patients have a normal body temperature of 37°C?

**c** What is the mode (the most common value)?

**d** Where are the temperatures clustered?

**e** Find the outlier and describe the patient who has this temperature.

### Solution

**a** 10 — 10 dots in total

**b** 2 patients — there are 2 dots at 37°C

**c** 38°C — the value with the most dots

**d** From 37°C to 39°C

**e** The outlier temperature is 42°C and a patient with this temperature would be very hot and ill.

## Example 16

This stem-and-leaf plot shows the exam marks of a class of students.

| Stem | Leaf |
|---|---|
| 4 | 3 7 |
| 5 | 0 2 2 4 8 |
| 6 | 1 3 3 3 6 6 9 |
| 7 | 5 7 8 |
| 8 | 4 6 |
| 9 | 2 |

The stem shows the tens digit, and the leaf shows the units digit: 6 | 1 means 61.

**a** What is the highest mark?

**b** How many students are in the class?

**c** What is the most common mark?

**d** What percentage of students scored over 60?

### Solution

**a** 92

**b** 20 — the number of values in the Leaf column

**c** 63 — this data value occurs the most, 3 times

**d** Number of students who scored over 60 = 13

Percentage of students who scored over 60 $= \frac{13}{20} \times 100\% = 65\%$

| Display | Useful for |
|---|---|
| Column graph | • categorical or numerical data<br>• showing differences in amounts |
| Line graph | • continuous numerical data<br>• showing changes over a time period<br>• estimating values between points |
| Sector graph and divided bar graph | • categorical data<br>• comparing parts of a whole<br>• a small number of categories |
| Frequency histogram and polygon | • numerical data<br>• data grouped into class intervals<br>• showing the shape of a set of data, how the values are spread out |
| Dot plot and stem-and-leaf plot | • small sets of numerical data<br>• showing clusters, gaps and outliers<br>• showing the shape of a set of data, how the values are spread out |

**EXERCISE 1.07** Answers on p. 459

## Dot plots and stem-and-leaf plots

EXAMPLE 15

**1** A class was surveyed to find out how many hours each student spent on homework each week. The results are shown below.

| 7 | 6 | 8 | 9 | 5 | 10 | 6 | 9 | 9 | 0 | 9 | 8 |
|---|---|---|---|---|---|---|---|---|---|---|---|
| 18 | 7 | 5 | 3 | 4 | 9 | 6 | 7 | 8 | 10 | 7 | 8 |

- **a** Draw a dot plot for this data.
- **b** How many students are in the class?
- **c** What are the outliers?
- **d** What are the clusters or gaps?

**2** This dot plot shows the shoe sizes of a sample of Year 11 students.

**Shoe size**

6 7 8 9 10 11 12

- **a** How many students are there in the sample?

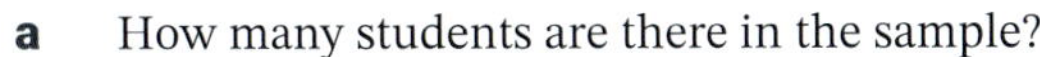

- **b** What is the mode?
- **c** Find the outlier and describe the student that has this outlier shoe size.
- **d** What fraction of students have a shoe size of 8 or 9?

**3** The Bennelong Bears scored the following numbers of goals per match in their first season.

| 4 | 0 | 2 | 3 | 7 | 1 | 2 | 4 | 5 | 0 | 3 | 5 |
|---|---|---|---|---|---|---|---|---|---|---|---|

- **a** Display the data as a dot plot.
- **b** What was the lowest score?
- **c** Comment on any clusters or outliers.

**4** A survey was conducted to determine the number of people in each car in a sample of cars on a road. The results are shown below.

| 6 | 3 | 1 | 2 | 3 | 1 | 1 | 3 | 4 | 2 |
|---|---|---|---|---|---|---|---|---|---|
| 5 | 3 | 1 | 1 | 2 | 3 | 4 | 5 | 6 | 4 |

- **a** Display the data as a dot plot.
- **b** How many cars were surveyed?
- **c** What was the highest value?
- **d** Comment on any clusters or outliers.

**5** This dot plot shows the scores (out of 10) of students competing in a spelling bee.

**Words correct**

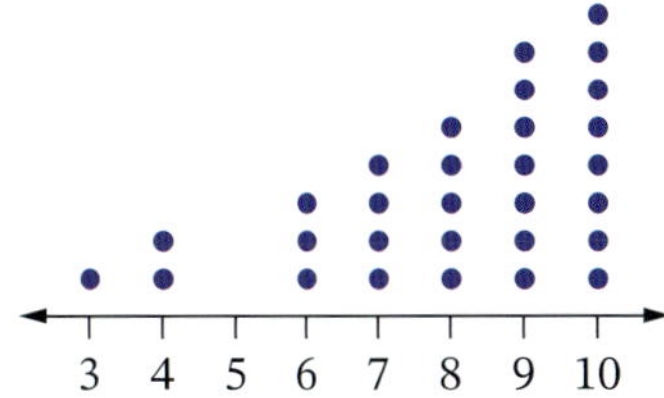


- **a** How many students competed in the spelling bee?

- **b** What percentage of students scored more than 8?
- **c** Where were the scores clustered?
- **d** Comment on the general performance of the contestants in the spelling bee.

Foundation Mastery Complex

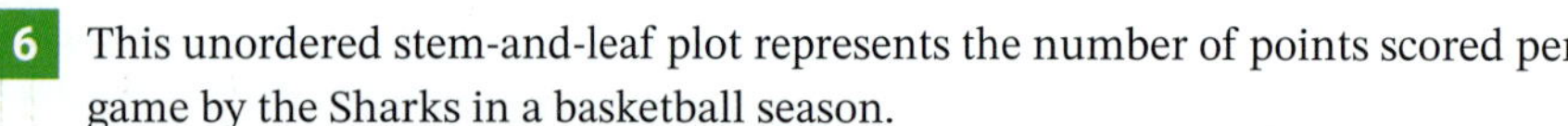

EXAMPLE 16

**6** This unordered stem-and-leaf plot represents the number of points scored per game by the Sharks in a basketball season.

| Stem | Leaf |
|---|---|
| 4 | 5 3 9 |
| 5 | 7 2 0 8 |
| 6 | 4 7 8 5 1 2 |
| 7 | 2 9 3 0 |
| 8 | 9 4 2 |

**a** Redraw the data as an ordered stem-and-leaf plot.

**b** How many games were played in the season?

**c** What was the Sharks' highest score for a game?

**d** For what percentage of games did the Sharks score below 56 points?

**7** Jeremy sells sausage sandwiches outside the hardware store on Sundays. The number of sandwiches sold each day over 30 Sundays was recorded as follows.

| | | | | | | | | | |
|---|---|---|---|---|---|---|---|---|---|
| 66 | 64 | 28 | 93 | 47 | 110 | 53 | 68 | 117 | 43 |
| 72 | 68 | 84 | 103 | 59 | 82 | 78 | 61 | 104 | 79 |
| 51 | 63 | 112 | 81 | 79 | 94 | 42 | 57 | 83 | 100 |

**a** Draw a stem-and-leaf plot to represent this data.

**b** On what percentage of days did Jeremy sell more than 50 sausage sandwiches?

**c** What is the outlier? Give one possible reason for this outlier.

**8** This plot shows the times (in seconds) in a slalom ski race.

| Stem | Leaf |
|---|---|
| 9 | 1 5 7 9 |
| 10 | 2 4 5 6 6 8 |
| 11 | 0 2 2 3 4 4 5 |
| 12 | 1 2 3 3 3 7 9 |
| 13 | 2 3 4 5 7 7 |
| 14 | 3 6 9 |
| 15 | 0 1 2 |

**a** How many skiers participated in the race?

**b** What was the winning time?

**c** If those skiers with times under 110 seconds were of Olympic standard, what percentage of skiers were of this standard?

**9** A nurse at Greenacres Hospital took the following pulse rates (in heartbeats per minute) of 40 patients.

| | | | | | | | | | |
|---|---|---|---|---|---|---|---|---|---|
| 71 | 81 | 63 | 55 | 93 | 52 | 69 | 78 | 84 | 65 |
| 72 | 80 | 68 | 74 | 85 | 79 | 90 | 84 | 76 | 68 |
| 58 | 64 | 60 | 97 | 83 | 69 | 74 | 56 | 64 | 89 |
| 94 | 81 | 63 | 60 | 76 | 72 | 110 | 83 | 90 | 64 |

**a** Draw a stem-and-leaf plot for this data.

**b** If the average pulse is from 65 to 75 beats per minute inclusive, how many patients had an average pulse?

**c** Find the outlier and describe the patient that has this pulse.

**d** A very fit person has a low resting pulse (say, less than 60). How many patients were in this category?

**10** This plot shows the prices (in \$1000s) of homes listed by a Mudgee real estate agent.

| Stem | Leaf |
|---|---|
| 2 | 04 10 17 18 26 35 37 76 77 |
| 3 | 10 26 42 54 66 76 85 85 85 |
| 4 | 02 53 62 65 73 75 76 85 |
| 5 | 03 10 24 60 68 92 |
| 6 | 14 29 32 34 35 35 54 |
| 7 | 18 52 63 85 |
| 8 | 12 25 46 72 93 |
| 9 | 38 75 |

**a** What is the price of the most expensive home?

**b** How many homes are listed?

**c** Which price is shared by 3 different homes?

☐ Foundation ○ Mastery ⬡ Complex

**11** At the school athletics carnival, runners recorded the following times (in seconds) in the 100-metre sprint.

| | | | | | | | |
|---|---|---|---|---|---|---|---|
| 12.1 | 13.6 | 11.8 | 18.1 | 12.0 | 15.6 | 13.9 | 17.2 |
| 14.5 | 18.7 | 15.7 | 14.6 | 16.3 | 11.6 | 17.7 | 15.6 |
| 14.6 | 15.4 | 12.4 | 16.5 | 17.4 | 14.6 | 16.8 | 14.3 |

**a** Represent this data on a stem-and-leaf plot using stems of 11, 12 etc.

**b** How many runners were there?

**c** What was:

**i** the best time recorded?

**ii** the slowest time recorded?

**d** What percentage of students took more than 15 seconds to run the distance? Answer correct to one decimal place.

**e** To qualify for the regional carnival, a runner's time must be less than 13.5 seconds. What fraction of students qualified for the regional carnival?

## Sample HSC problem

Answers on p. 460

6 marks A sample of 3000 rugby league fans was surveyed on their main concerns about the national competition. The results of the survey are shown by the sector graph below.

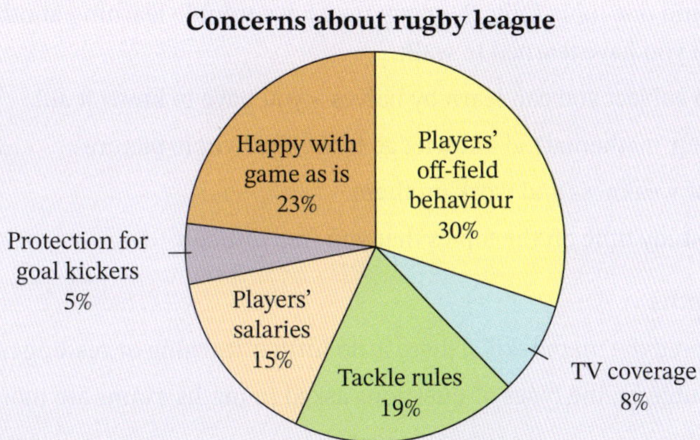


**a** Which one of the following statements is correct? Select **A**, **B**, **C** or **D**. 1 mark

**A** The data shown in the sector graph is categorical.

**B** More than one-third of fans were concerned about players' off-field behaviour.

**C** The data shown in the sector graph could also be displayed on a line graph.

**D** The least common concern of fans was TV coverage.

**b** Describe a method of selecting a stratified sample of rugby league fans. 2 marks

**c** Calculate, correct to the nearest degree, the angle size of the sector for 'Happy with the game as is'. Show your working. 1 mark

**d** Which concern was shared by 450 fans? 1 mark

**e** Write one disadvantage of using a sector graph to illustrate the data. 1 mark

**Study Tip**

## 4 practical steps in studying maths

*A journey of a thousand miles begins with a single step.*
Attributed to Lao Tzu (c. 570–490 BCE), ancient Chinese philosopher

### Step 1: Practise your maths

- Do your homework.
- Learning maths is about mastering a collection of skills.
- You become successful at maths by doing it more, through regular practice and training.
- Aim to achieve a high level of understanding.

### Step 2: Rewrite your maths

- Homework and study are not the same thing.
- Study is your private revision work for strengthening your understanding of a subject.
- Take ownership of your maths.
- Rewrite the theory and examples in your own words.
- Summarise each topic to see the 'whole picture' and know it all.

### Step 3: Attack your maths

- All maths knowledge is interconnected.
- If you don't understand one topic fully, then you may have trouble learning another topic: You cannot run until you have learned to walk.
- Mathematics is not a subject you can learn by halves – you have to know it all!
- Fill in any gaps in your mathematical knowledge to see the 'whole picture'.
- Identify your areas of weakness and work on them.
- Spend most of your study time on the topics that you find difficult.

### Step 4: Check your maths

- Once you have mastered the maths skills, there is no further learning or reading needed.
- Compared to other subjects, the types of questions asked in maths exams are more conventional and predictable.
- Test your understanding with revision exercises, practice papers and past exams.
- Develop your exam technique and problem-solving skills.
- Go back to Steps 1 to 3 to improve your study.

# CHAPTER SUMMARY

This chapter, *Collecting and presenting data*, looked at the graphing of data, sampling techniques and the methods of data collection. You should be able to interpret and construct various graphs and displays, know the different types of data and samples, and be familiar with statistical terminology.

Before you move on, consider what you learned in this chapter and revisit any sections that may have been unclear.

| What you learned in this chapter... | Section | |
|---|---|---|
| Interpret and construct a variety of statistical graphs and displays: column graphs, sector graphs, divided bar graphs, line graphs, infographics, tables, frequency histograms and polygons, dot plots and stem-and-leaf plots | 1.01 | Interpreting graphs |
| | 1.05 | Constructing graphs |
| | 1.06 | Frequency histograms and polygons |
| | 1.07 | Dot plots and stem-and-leaf plots |
| Identify graphs that mislead and misrepresent | 1.02 | Misleading graphs |
| Classify data as numerical (either discrete or continuous) or categorical (either nominal or ordinal) | 1.03 | Classifying data |
| Distinguish between a sample and a population, and the collection of data for each | 1.04 | Sampling techniques |
| Distinguish between random, systematic, stratified and self-selected sampling, and determine the appropriateness of each type for a given situation | 1.04 | Sampling techniques |
| Identify bias in sampling | 1.04 | Sampling techniques |

To help master this topic, make a summary mind map. Use the chapter outline and the mind map below as a guide. Add your own words, symbols, diagrams, boxes and reminders. The summary should give you a 'whole picture' view of the topic and allow you to identify any weak areas to revisit in your revision.

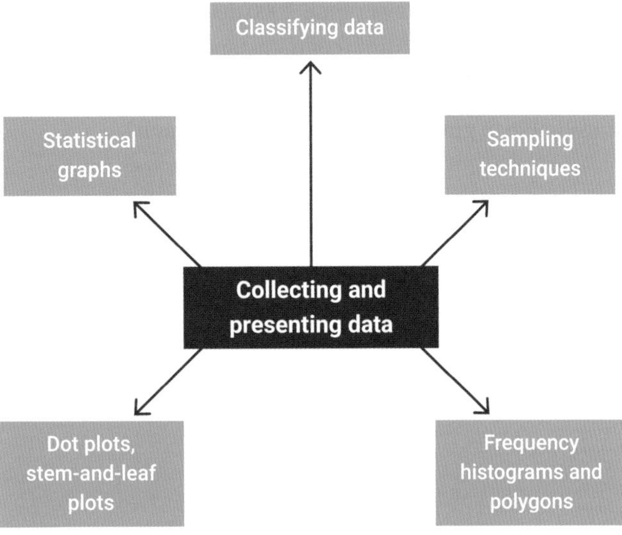

# 1 Test yourself

Answers on p. 460

1.01

**1** This bar chart shows the deaths from the 10 types of cancer that cause the most deaths of males in Australia.

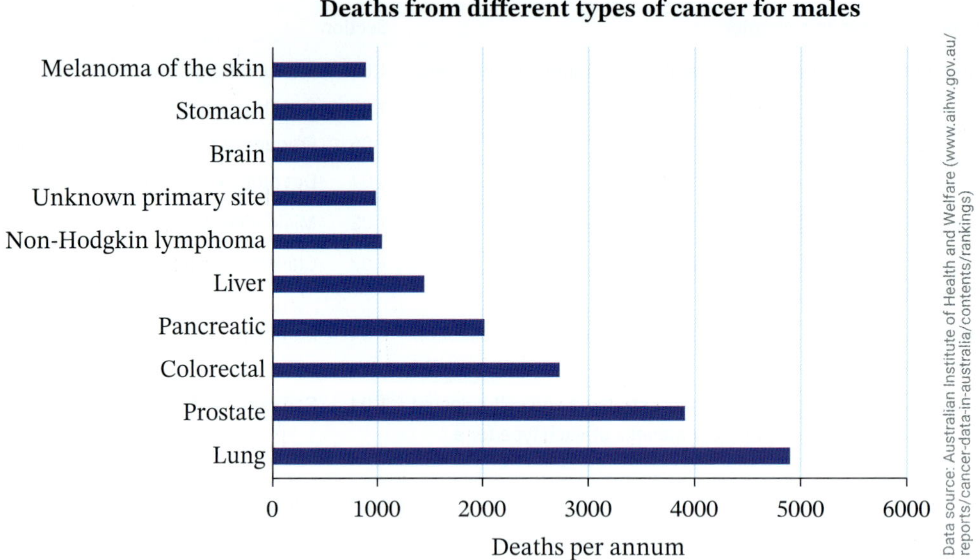


**a** How many fatal cases of colorectal cancer were registered?

**b** What was the fifth most deadly type of cancer?

**c** How many types of cancer recorded fewer than 2000 deaths each?

**d** 'There were more lung cancer deaths recorded than colorectal and prostate cancer deaths combined.' True or false?

**e** Why would it be unsuitable to graph this data on a sector graph?

□ Foundation ○ Mastery ⬡ Complex

**2** This infographic is about Australia. 1.01

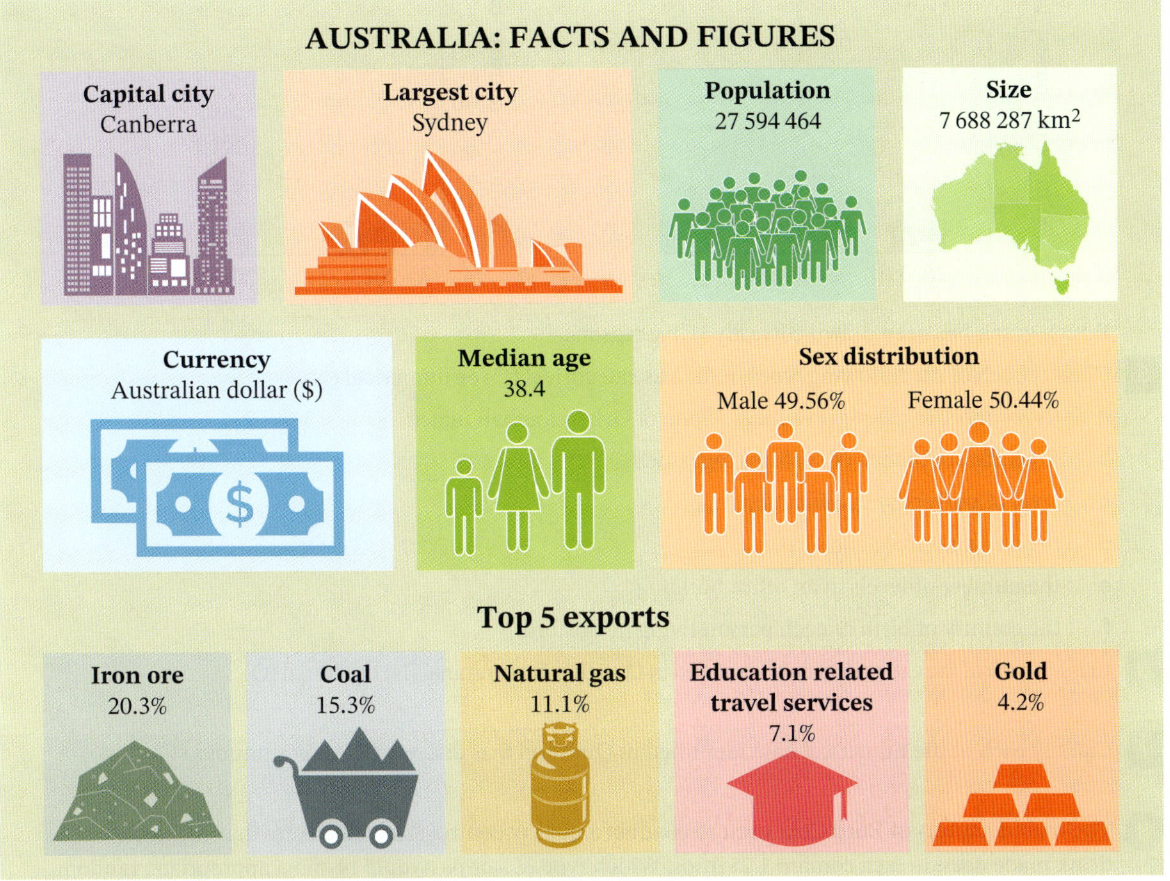


**a** Calculate the female population of Australia.

**b** Find an estimate for world land area if Australia's land area is about 5% of the world land area.

**c** The top 5 exports in the infographic shows the percentage of the total value of exports. Australia's export earnings were $689 209 million. Find the export value of:

**i** coal **ii** natural gas **iii** gold

**3** A sample of consumers was surveyed about the brand of shower gel they used. This graph of the survey results was used by Green Water in their advertising campaign. 1.02

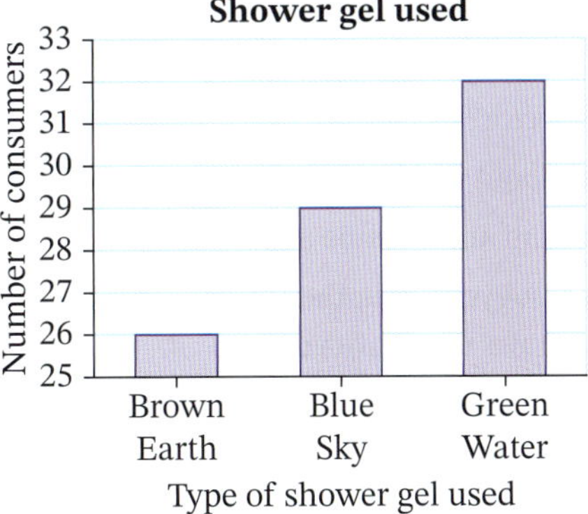


**a** Describe one way in which this graph is misleading and draw what a correct graph should look like.

**b** Name another type of graph that could be used to display this data.

Foundation Mastery Complex

1.02 **4** Graphs have been drawn to represent the number of houses constructed in 2023 and 2024.

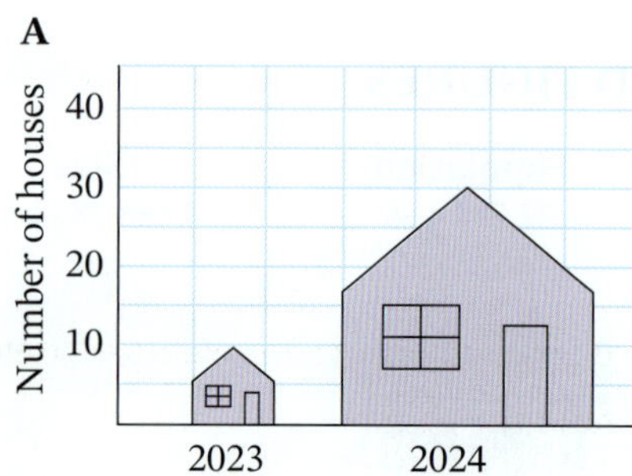


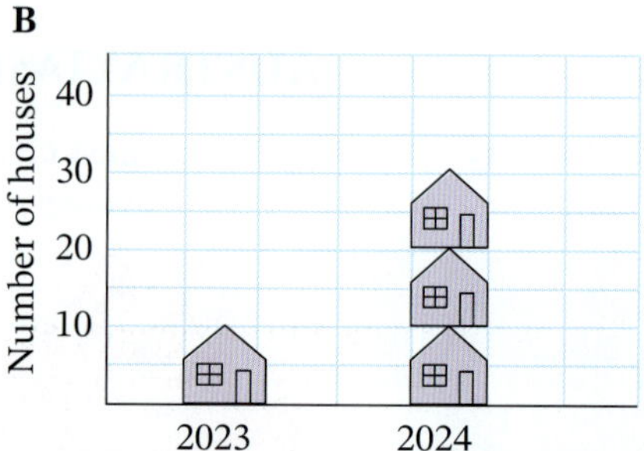


Which graph has been drawn correctly? Give reasons.

1.03 **5** Classify each of the following types of data as categorical (C) or numerical (N).

**a** the number of people watching a State of Origin football match

**b** the different brands of surfing magazines

**c** level of university qualification

**d** the temperatures at an airport

**e** the number of levels in an office building

**f** the country of birth of each person living in Gosford.

1.03 **6** Classify each of the categorical data identified in Question **5** as nominal (N) or ordinal (O).

1.03 **7** Classify each of the numerical data identified in Question **5** as discrete (D) or continuous (C).

1.04 **8** The manufacturers of Burp Cola want to conduct a quality control study to ensure that each bottle of drink made does, in fact, contain 1.25 litres. Which type of sample would be most appropriate: random, systematic, stratified or self-selected? Give a reason for your answer.

1.04 **9** **a** To investigate the types of music to play at the school dance, Tasha surveyed a sample of 50 students visiting the canteen. Why might this sample be biased?

**b** Tasha decided to use a stratified sample instead, based on the following numbers of school students in each Year group:

| Year level | 7 | 8 | 9 | 10 | 11 | 12 |
|---|---|---|---|---|---|---|
| Number of students | 114 | 120 | 114 | 128 | 105 | 96 |

If she surveys a sample of 50 students, how many Year 10 students should be in the sample?

1.04 **10** Explain what is wrong with each survey question below and write a better question for each.

**a** How often do you visit the doctor?

**b** What did you like or dislike about the film?

**c** Is it time that Australia grew up and became a republic?

1.05 **11** A sample of senior citizens was surveyed about the number of countries outside Australia they have visited. The results are shown in the table.

| Number of countries | Number of people |
|---|---|
| 0 | 3 |
| 1–3 | 8 |
| 4–10 | 32 |
| 11–20 | 28 |
| 21–30 | 17 |
| 31+ | 5 |

**a** How many people were in the sample?

**b** Graph this data on a sector graph.

**c** What is the angle of the sector representing those who have not travelled outside Australia?

**d** What percentage of the sample have visited 4 to 10 countries? Answer correct to one decimal place.

**e** 'More than half of the sample have visited 11 or more countries.' True or false?

**f** Why would it be inappropriate to graph this data on a line graph?

☐ Foundation ○ Mastery ○ Complex

**12** The masses (in kilograms) of 40 skydivers were recorded. The results are shown below. 1.06

| | | | | | | | | | |
|---|---|---|---|---|---|---|---|---|---|
| 58 | 63 | 77 | 82 | 53 | 69 | 65 | 80 | 96 | 105 |
| 79 | 63 | 52 | 90 | 104 | 85 | 65 | 87 | 68 | 105 |
| 65 | 87 | 109 | 84 | 62 | 75 | 102 | 78 | 93 | 84 |
| 68 | 105 | 74 | 59 | 68 | 74 | 88 | 66 | 70 | 62 |

**a** Are people's masses discrete data or continuous data?

**b** Organise this data into a frequency distribution table with class intervals 50–<60, 60–<70 etc.

**c** Draw a frequency histogram and polygon to represent the data.

**d** Which class interval had the highest frequency?

**e** What fraction of skydivers were in the 80–<90 kg class?

**13** The ages (in years) of a sample of children at a Wiggles concert are listed below. 1.07

| | | | | | | | | | | | |
|---|---|---|---|---|---|---|---|---|---|---|---|
| 6 | 0 | 4 | 1 | 5 | 6 | 4 | 2 | 8 | 6 | 4 | 4 |
| 3 | 6 | 5 | 5 | 2 | 5 | 3 | 1 | 5 | 6 | 3 | 3 |

**a** Draw a dot plot for this data.

**b** How many children were in the sample?

**c** What fraction of children were over 5 years old?

**d** Comment on any clusters.

**14** This stem-and-leaf plot shows the ages of visitors entering the Royal Easter Show in a 5-minute period. 1.07

| Stem | Leaf |
|---|---|
| 0 | 3 8 9 |
| 1 | 0 2 2 2 5 7 6 9 |
| 2 | 0 2 3 4 6 7 |
| 3 | 1 3 3 4 9 |
| 4 | 3 4 7 8 |
| 5 | 5 5 8 |

**a** What was the age of the youngest visitor?

**b** How many visitors entered the show during the 5-minute period?

**c** What was the most common age of the visitors?

**d** What percentage (correct to the nearest whole number) of visitors were over 30 years old?

☐ Foundation ○ Mastery ○ Complex

# 2. FORMULAS AND EQUATIONS

German scientist Albert Einstein was only 26 years old when he proposed a new theory of physics for small particles of matter (such as atoms) moving at very high speeds. In 1905, he proposed that matter (mass) could be converted to large amounts of energy, describing this relationship with the formula $E = mc^2$, where $E$ stands for energy, $m$ stands for mass and $c$ stands for the speed of light. Einstein's Theory of Special Relativity revolutionised conventional laws of physics and led to the development of nuclear energy.

## Chapter outline



*REVISION

Who is Danny/Adobe Stock Photos

## In this chapter you will:

- expand and simplify algebraic expressions
- substitute values into algebraic expressions and formulas, including Fried's, Young's and Clark's formulas for medicine dosage for children
- solve linear equations, including after substituting into a formula
- form an equation from a worded problem and solve the equation to solve the problem
- change the subject of a formula.

## Videos (9):

**2.03** Formulas

**2.05** Equations and formulas 1 • Equations and formulas 2 • Equation problems 1 • Equation problems 2 • Formulas and equations • Writing equations

**2.06** Changing the subject of a formula 1 • Changing the subject of a formula 2

## Skillsheets (4):

**2.01** Algebra using diagrams

**2.04** Solving equations by balancing • Solving equations by backtracking • Solving equations using diagrams

## Worksheets (10):

**SkillCheck** Assignment 2

**2.01** Collecting like terms

**2.02** Expanding algebra

**2.04** More complex equations

**2.05** Working with formulas • Using equations to solve problems • Word problems with equations

**2.06** Changing the subject of a formula • Subject of a formula

**Chapter summary** Algebra review

## Puzzles (3):

**2.03** Substitution code puzzle

**2.04** Equations code puzzle • Equations riddle

Nelson MindTap

To access resources above, visit **cengage.com.au/nelsonmindtap**

## Terminology

| | | | |
|---|---|---|---|
| base | equation | expand | formula |
| LHS | like terms | power | RHS |
| rule | simplify | subject | substitute |
| variable | | | |

**Worksheet** Assignment 2

## SkillCheck Answers on p. 461

**1** Evaluate each expression.

**a** $9 \times (-3)$ **b** $9 + (-3)$ **c** $2 - 10$

**d** $(-5)^2$ **e** $-24 \div 6$ **f** $-2 - (-4)$

**2** If $d = 3$, evaluate each expression.

**a** $8d - 9$ **b** $\frac{d+6}{3}$ **c** $7 + 3d$

**d** $\sqrt{12d}$ **e** $d^5$ **f** $2d^2$

**3** If $u = -2$, evaluate each expression.

**a** $u^2 - 4$ **b** $3(u + 1)$ **c** $8 - u$

**d** $u^7$ **e** $\frac{-18}{u}$ **f** $5 - 2u$

**4** Simplify each expression.

**a** $4x + 7 - 3x - 10$ **b** $18 - 2a + a + 2$ **c** $5r + 12 - 15 - 5r$

**5** Expand each expression.

**a** $3(2x - 1)$ **b** $7(4a + 5)$ **c** $-2(6p - 2)$

**6** Solve each equation.

**a** $3b = -15$ **b** $w - 8 = 9$ **c** $2r + 4 = 22$

**7** Evaluate $A = \frac{1}{2}(a + b)h$ if $a = 10$, $b = 15$ and $h = 3$.

# Simplifying algebraic expressions

2.01

An **algebraic expression** is made up of **terms** involving **variables** and numerals. For example, $3x^2 - x + 10$ has 3 terms: $3x^2$, $-x$ and 10.

**Skillsheet** Algebra using diagrams

**Worksheet** Collecting like terms

## Working with like terms

- **Like terms** have exactly the same variables.
- Only like terms can be added or subtracted.

### Example 1

Simplify each expression.

**a** $2a^2 - a + 5 + 8a$ **b** $4kr - 6pr - pr + 10kr$ **c** $2xy + 4 - y + 4yx$

#### Solution

**a** $2a^2 - a + 5 + 8a = 2a^2 - a + 8a + 5$ — $-a$ and $8a$ are like terms

$= 2a^2 + 7a + 5$ — $2a^2$ and $-a$ are not like terms

**b** $4kr - 6pr - pr + 10kr = 4kr + 10kr - 6pr - pr$ — grouping pairs of like terms

$= 14kr - 7pr$

**c** $2xy + 4 - y + 4yx = 2xy + 4yx - y + 4$ — $2xy$ and $4yx$ are like terms

$= 6xy - y + 4$

## Multiplying terms

- When multiplying terms, multiply the numbers and variables separately.
- When multiplying terms with **powers** that have the same **base**, add the powers.

### Example 2

Simplify each expression.

**a** $-3bc \times 8ab$ **b** $4p^2q \times \frac{3p^3}{2}$ **c** $(-5r)^2$

#### Solution

**a** $-3bc \times 8ab = (-3 \times 8) \times (a \times b \times b \times c)$

$= -24ab^2c$

To **simplify**, first multiply the numbers, then multiply the variables in alphabetical order.

**b** $4p^2q \times \frac{3p^3}{2} = \frac{4 \times 3}{2} \times p^2 \times p^3 \times q$

$= 6p^5q$

$p^2 \times p^3 = p^5$
Add the powers.

**c** $(-5r)^2 = (-5r) \times (-5r)$ or $(-5)^2 \times r^2$

$= (-5) \times (-5) \times r \times r$ or $25 \times r^2$

$= 25r^2$

## Dividing terms

- When dividing terms, divide the numbers and variables separately.
- When dividing terms with powers that have the same base, subtract the powers.

## Example 3

Simplify each expression.

**a** $\dfrac{8m^3n}{2m^2}$ **b** $-15w \div 3w^2$ **c** $\dfrac{6ak^3}{20ak}$

### Solution

**a** $\dfrac{8m^3n}{2m^2} = \dfrac{8}{2} \times \dfrac{m^3n}{m^2}$

$= 4mn$

$\dfrac{m^3}{m^2} = m^1 = m$
Subtract the powers.

**b** $-15w \div 3w^2 = \dfrac{-15w}{3w^2}$

$= \dfrac{-15}{3} \times \dfrac{w}{w^2}$

$\dfrac{w}{w^2} = \dfrac{\cancel{w}^{1}}{w\cancel{w}} = \dfrac{1}{w}$

$= -5 \times \dfrac{1}{w}$

$= \dfrac{-5}{w}$

**c** $\dfrac{6ak^3}{20ak} = \dfrac{6}{20} \times \dfrac{ak^3}{ak}$

$= \dfrac{3}{10} \times k^2$

$= \dfrac{3k^2}{10}$

## Example 4

Simplify each expression.

**a** $\dfrac{4a}{5} \times \dfrac{5h}{12a}$ **b** $\dfrac{3m}{10n} \div \dfrac{6m}{25n}$

### Solution

**a** $\dfrac{4a}{5} \times \dfrac{5h}{12a} = \dfrac{\overset{1}{\cancel{4}}\,\cancel{a}}{\cancel{5}} \times \dfrac{\cancel{5}h}{\underset{3}{\cancel{12}}\,\cancel{a}}$

Each fraction cannot be simplified on its own, so simplify the numerator of each fraction with the **denominator** of the other fraction.

$\dfrac{4}{12} = \dfrac{1}{3}, \dfrac{a}{a} = 1, \dfrac{5}{5} = 1$

$= \dfrac{1 \times h}{3}$

$= \dfrac{h}{3}$

**b** $\dfrac{3m}{10n} \div \dfrac{6m}{25n} = \dfrac{3m}{10n} \times \dfrac{25n}{6m}$

When dividing by a fraction, multiply by its reciprocal.

$= \dfrac{\overset{1}{\cancel{3}}\,\cancel{m}}{\underset{2}{\cancel{10}}\,\cancel{n}} \times \dfrac{\overset{5}{\cancel{25}}\,\cancel{n}}{\underset{2}{\cancel{6}}\,\cancel{m}}$

$= \dfrac{1 \times 5}{2 \times 2}$

$= \dfrac{5}{4}$

**EXERCISE 2.01** Answers on p. 461

## Simplifying algebraic expressions

EXAMPLE 1

**1** Simplify each expression.

**a** $2x^2 + 3x + x^2 + 2x$ **b** $2ab + a + 5ab + b$ **c** $10 - 4m^2 + 6m^2 - 5$

**d** $3r + 4ar - 2ar - 7r$ **e** $y^2 - y + 3y - 5$ **f** $2a^2 - 4 - a^2 + 6$

**g** $k^2 - k - 3k - 1$ **h** $3de - 6ed + 2d$ **i** $p^2 - p - p^2 - 5p$

**j** $-5 + 7u - 10 - 3u$ **k** $t^2 + at + t^2 + 2at$ **l** $15x - 5 + 3x + 5$

**2** Simplify each expression.

**a** $4ak \times 4am$ **b** $-3d \times 8cd$ **c** $\frac{1}{4}t^2w \times 10tw$

**d** $(5x^3)^2$ **e** $9n^2 \times \frac{2n^2}{3}$ **f** $7k^3p \times 2k^2p$

**g** $-6m \times (-6n)$ **h** $5e \times \frac{3e^2}{10}$ **i** $(2xy^2)^2$

**j** $10ab \times (-2ab)$ **k** $-4n^3 \times \frac{1}{4}n^3$ **l** $(-3k)^2$

**3** Simplify each expression.

**a** $\frac{4m}{2m}$ **b** $\frac{14x^2y}{2xy}$ **c** $-15e^2 \div 3e^2$

**d** $10u^2t^3 \div 5ut^2$ **e** $\frac{3x}{x^3}$ **f** $\frac{27pr}{-9pq}$

**g** $\frac{12g^2h}{20gh}$ **h** $a^2 \div a^3$ **i** $\frac{15m^4n}{9m^2n^2}$

**j** $\frac{-4de}{20e}$ **k** $u \div 4u$ **l** $6wx^2 \div 24w^2$

**4** Which of the following is the simplified expression for $\frac{2xy}{5} \times \frac{-3}{10y}$? Select **A**, **B**, **C** or **D**.

**A** $-\frac{3x}{25}$ **B** $-\frac{3x}{25y^2}$ **C** $-\frac{6x}{50}$ **D** $-\frac{3}{25y}$

**5** Simplify each expression.

**a** $\frac{9y}{4} \times 5y$ **b** $\frac{4a}{5} \times \frac{2h}{7}$ **c** $\frac{x}{3} \div \frac{x}{6}$

**d** $\frac{20r}{6} \div 4r$ **e** $\frac{4nw}{b} \times \frac{bn}{2w}$ **f** $\frac{3m}{6m} \times \frac{10n^2}{4n}$

**g** $\frac{4a}{5} \div \frac{2h}{7}$ **h** $\frac{2e}{5} \div \frac{3e}{8}$ **i** $\frac{2de}{7p} \times \frac{14p}{4e}$

**6** Simplify each expression.

**a** $2k + 8 - k^2 - 4k$ **b** $ay^2 \times \frac{1}{2}y^2$ **c** $-5t \div (-20t)$

**d** $\frac{-32f^3g^2}{4f^2g}$ **e** $-3c^2 + 7c - 7c^2 + 6c$ **f** $4z + z + 8 - 4z^2$

**g** $-3a \times (-2a^2)$ **h** $\frac{24cd}{8de}$ **i** $2 - 4m^2 - 6m^2 - 2$

**j** $\frac{20n^3x}{10nx^2}$ **k** $4x^2 - 2xy + 3x + 4xy$ **l** $6a^4m^2 \div (-2a^2m^2)$

☐ Foundation ○ Mastery ⬡ Complex

## Did you know?

### Al Gebra is Arabic

In 825 CE, the Persian mathematician **al-Khwarizmi** wrote a book called *Hisab al-jabr w'al-muqabala* (or *The Science of Equations*). The Arabic word *al-jabr* meant the process of adding the same amount to both sides of an equation but, when it was translated into Latin, it was changed to 'algebra' and became the name of a whole branch of mathematics, not just equation-solving.

**From al-Khwarizmi's name, we get the word *algorithm*. What does "algorithm" mean?**

## Investigation

### Number patterns

| 10 | 11 | 12 |
|---|---|---|
| 17 | 18 | 19 |
| 24 | 25 | 26 |

1. Choose a month from any calendar and draw a box around any block of 9 numbers. See the boxed numbers on the right for an example.
2. Add together the numbers in any line of 3 numbers going through the centre of the box (row, column or diagonal). For example, $10 + 18 + 26$.
3. Add another line of 3 numbers that goes through the centre of the box.
4. There are 2 more lines that go through the centre of the box. Find their sums as well.
5. What do you notice? Why does this pattern work? *Hint:* let the number in the top left-hand corner be $x$.

# 2.02 Expanding algebraic expressions

**Worksheet**
Expanding algebra

Expanding an expression means removing grouping symbols (brackets) by multiplying each term inside the grouping symbols by the term on the outside of the grouping symbols.

### Expanding algebraic expressions

$$a(b + c) = a \times (b + c) = ab + ac$$

### Example 5

Expand each expression.

**a** $5(2t - 3)$ **b** $2k(7 + 3k)$ **c** $-3(d^2 - 3d + 8)$

### Solution

**a** $5(2t - 3) = 5 \times 2t - 5 \times 3$
$= 10t - 15$

Multiply each term inside the grouping symbols by the 5 (outside the grouping symbols).

**b** $2k(7 + 3k) = 2k \times 7 + 2k \times 3k$
$= 14k + 6k^2$

**c** $-3(d^2 - 3d + 8) = -3 \times d^2 - (-3) \times 3d + (-3) \times 8$
$= -3d^2 + 9d - 24$

## Example 6

Expand and simplify each expression.

**a** $x(x-1)+4(x+1)$ **b** $2m(m-3)-(m-6)$ **c** $5u(2-y)+u(2y-9)$

### Solution

**a** $x(x-1)+4(x+1)=x^2-x+4x+4$ simplify by collecting the like terms $-x$ and $4x$

$=x^2+3x+4$

**b** $2m(m-3)-(m-6)=2m^2-6m-m+6$ $-(m-6)$ means $-1(m-6)$

$=2m^2-7m+6$

**c** $5u(2-y)+u(2y-9)=10u-5uy+2uy-9u$

$=u-3uy$

**EXERCISE 2.02** Answers on p. 461

## Expanding algebraic expressions

**1** Expand each expression.

**a** $3(a+2)$ **b** $5(3-2b)$ **c** $-2(2a+1)$
**d** $-6(b-2)$ **e** $3x(x-2)$ **f** $3p(p-a)$
**g** $-4(2k+4)$ **h** $2t(3-4t)$ **i** $-d(d-5)$
**j** $k(7-5k)$ **k** $-9b(b-1)$ **l** $2y(7x+4y)$

**2** Expand $-3m(2m-9)$. Select **A**, **B**, **C** or **D**.

**A** $-6m^2-27m$ **B** $-m^2+27m$ **C** $-5m^2-27m$ **D** $-6m^2+27m$

**3** Expand $4k-5(2k-3)$. Select **A**, **B**, **C** or **D**.

**A** $6k-15$ **B** $-6k+15$ **C** $14k+15$ **D** $14k-8$

**4** Expand each expression.

**a** $-6n(4-n)$ **b** $5x(rx+2r)$ **c** $-(2a^2-4)$
**d** $5b(a^2+3b-7)$ **e** $-(x^2-4x+10)$ **f** $3h(h-7e-4eh)$
**g** $y(2y+3-y^2)$ **h** $de(d^2-2+e^2)$ **i** $-3v(-3av+v-2a)$

**5** Expand and simplify each expression.

**a** $5(x+4)-2(x+3)$ **b** $3(d-4)-2(d+5)$ **c** $6(r+10)-4(r-5)$
**d** $8(f+2)-(f+7)$ **e** $3(2x-4)-5(3x+4)$ **f** $6x(x+4)-3x(x-1)$
**g** $3b(b+5)-b(b-8)$ **h** $4w(w-7)-w(w+1)$ **i** $6(k+p)+3(k+2p)$
**j** $2(a-b)+2(b+a)$ **k** $x(2v+4)-x(v+1)$ **l** $-3(t+w)-2(2t-w)$
**m** $e(3e+5)-(2e-e^2)$ **n** $-2(a+3)+4(a-3)$ **o** $p(p-q)-q(q-p)$

**6** Expand and simplify $4(2-4y)-3(2+5y)$. Select **A**, **B**, **C** or **D**.

**A** $14-31y$ **B** $2-7y$ **C** $2-31y$ **D** $14-y$

**7** **a** Evaluate $7-5$ and $5-7$. How are the 2 answers related?

**b** Evaluate $4-10$ and $10-4$.

**c** Is $a-b$ always the same as $-(b-a)$? Can you prove it algebraically?

□ Foundation ○ Mastery ○ Complex

## Investigation

### Mental multiplication by expanding

Expanding is useful for multiplying numbers mentally without using a calculator, especially if one of the numbers is close to 10, 100 or 1000.

**1** Study the following examples.

**a** $35 \times 11 = 35 \times (10 + 1)$ — think of 11 as 10 + 1
$= 35 \times 10 + 35 \times 1$ — expand
$= 350 + 35$ — simplify
$= 385$

**b** $43 \times 102 = 43 \times (100 + 2)$ — think of 102 as 100 + 2
$= 43 \times 100 + 43 \times 2$ — expand
$= 4300 + 86$ — simplify
$= 4386$

**c** $16 \times 8 = 16 \times (10 - 2)$ — think of 8 as 10 − 2
$= 16 \times 10 + 16 \times (-2)$ — expand
$= 160 - 32$ — simplify
$= 128$

**2** Use expanding to multiply each pair of numbers mentally.

| | | | | | |
|---|---|---|---|---|---|
| **a** | $25 \times 12$ | **b** | $18 \times 9$ | **c** | $6 \times 105$ |
| **d** | $87 \times 11$ | **e** | $50 \times 99$ | **f** | $45 \times 8$ |

# 2.03 Formulas

**Video** Formulas

**Puzzle** Substitution code puzzle

A **formula** is an algebraic rule that describes a mathematical relationship between variables. For example, the volume of a cylinder has the formula $V = \pi r^2 h$, where $r$ is the radius of the cylinder's base and $h$ is its perpendicular height.

Because the formula describes $V$, with $V$ on the left-hand side of the = sign, we say that $V$ is the **subject** of the formula.

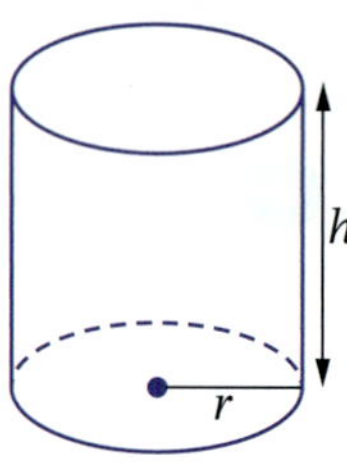


## Example 7

If a principal, $\$P$, is invested at an interest rate of $r$ **per annum** (where $r$ is written as a decimal) then, after $n$ years, it will grow to $\$A$, where $A$ is given by the compound interest formula $A = P(1 + r)^n$. Use the formula to calculate (to the nearest cent) the final amount of an investment of \$4000 after 6 years at 11% p.a.

### Solution

$P = 4000, r = 11\% = 0.11, n = 6$

$A = P(1 + r)^n$
$= 4000(1 + 0.11)^6$
$= 7481.658\,209 \ldots$
$\approx 7481.66$

The final amount is \$7481.66.

## Example 8

The correct **dosage** of a medicine for children depends on the child's age or weight. There are different formulas for calculating this, each named after the person who created it.

Fried's formula (for children aged 1 to 2):

$$\text{child dosage} = \frac{\text{age in months}}{150} \times \text{adult dosage}$$

Young's formula (for children aged 1 to 12):

$$\text{child dosage} = \frac{\text{age in years}}{\text{age in years} + 12} \times \text{adult dosage}$$

Clark's formula (for children):

$$\text{child dosage} = \frac{\text{weight in kg}}{70} \times \text{adult dosage}$$

Fried's formula assumes that the adult dose is for a person aged 12.5 years, while Clark's formula assumes that it is for a person who weighs 70 kg.

Ryan is 18 months old. The adult dosage of a drug prescribed for him is 20 g. Calculate Ryan's dosage using:

**a** Fried's formula

**b** Young's formula

### Solution

**a** Ryan's dosage using Fried's formula $= \frac{18}{150} \times 20\text{ g}$

$= 2.4\text{ g}$

Age = 18 months

**b** Ryan's dosage using Young's formula $= \frac{1.5}{1.5 + 12} \times 20\text{ g}$

$= 2.2222\ldots\text{ g}$

$\approx 2.2\text{ g}$

Age = 18 months = 1.5 years

## Example 9

Karryn is 4 years old and weighs 17.8 kg. She needs to take medicine with a recommended adult dose of 1200 mg each day. Use Clark's formula to calculate Karryn's daily dosage.

### Solution

Karryn's dosage using Clark's formula $= \frac{17.8}{70} \times 1200\text{ mg}$

$= 305.1428\text{ mg}$

$\approx 305\text{ mg}$

## Example 10

From a height of $h$ metres above sea level, an observer can see a distance of $d$ km to the horizon, where $d = 8\sqrt{\frac{h}{5}}$. What distance, correct to the nearest kilometre, can be seen from the highest point of Sydney Harbour Bridge, 134 m above sea level?

### Solution

$h = 134$

$$\begin{aligned} d &= 8\sqrt{\frac{h}{5}} \\ &= 8\sqrt{\frac{134}{5}} \\ &= 41.41497\ldots \\ &\approx 41\text{ km} \end{aligned}$$

A distance of 41 km can be seen from the top of Sydney Harbour Bridge.

**EXERCISE 2.03** Answers on p. 462

## Formulas

**1** If $a = -3$, $b = 10$ and $c = 6$, evaluate each expression.

**a** $b^2 - a^2$ **b** $\frac{b+c}{ac}$ **c** $\sqrt{c+3b}$ **d** $4(3a+9)$

**e** $8c + 4a$ **f** $\frac{2b}{5}$ **g** $b(b-4)$ **h** $\sqrt{\frac{9b}{c+4}}$

**i** $c^2 + c$ **j** $\frac{2ac}{3}$ **k** $\sqrt{a^2+7c-2}$ **l** $(a-b)^2$

EXAMPLE 7

**2** If $w = -3.625$, what is the value of $w^2 + 2w$, rounded to 3 decimal places. Select **A**, **B**, **C** or **D**.

**A** $-20.391$ **B** 5.880 **C** 5.891 **D** $-20.390$

**3** If $y = \frac{x}{9}$, find the value of $y$, correct to 4 significant figures, when $x = 14$. Select **A**, **B**, **C** or **D**.

**A** 1.555 **B** 1.556 **C** 1.5555 **D** 1.5556

**4** Calculate the volume, correct to 2 decimal places, of a cylinder with a base radius of 4.07 cm and perpendicular height of 11.58 cm, using the formula $V = \pi r^2 h$.

**5** The temperature $T$ (in °C) of the water in a kettle $t$ minutes after it is switched on is given by the formula $T = 18t + 28$. Find the temperature of the water:

**a** 4 minutes after it is turned on

**b** $1\frac{1}{2}$ minutes after it is turned on

**c** when the kettle is first turned on.

**6** On a given day, the formula for converting Australian dollars (\$A) to US dollars (\$US) is US = 0.722A. Convert the following \$A amounts to \$US, correct to the nearest cent.

**a** \$20.00 **b** \$89.50 **c** \$4800

Foundation Mastery Complex

**7** If an object is moving with speed $u$ m/s and then undergoes acceleration $a$ m/s$^2$, its speed $v$ m/s after $t$ seconds is calculated using the formula $v = u + at$.

**a** Which variable is the subject of the formula?

**b** Calculate the speed of a car after 5 seconds if its speed starts at 6 m/s and it is accelerating at 2 m/s$^2$.

**8** The number of matches, $m$, needed to make this pattern of triangles is $m = 2t + 1$, where $t$ is the number of triangles in the pattern.
How many matches are required to make:

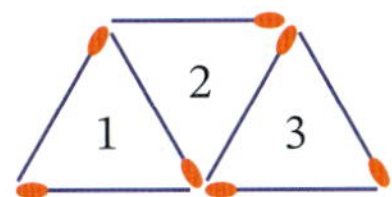


**a** 8 triangles?

**b** 40 triangles?

**c** 150 triangles?

**9** The height, $h$ metres, of a ball above the ground after $t$ seconds is modelled by the equation $h = 20t - 4.9t^2$. Find the height of the ball after 4 seconds.

**10** The volume of a sphere with radius $r$ is $V = \frac{4}{3}\pi r^3$. Calculate, correct to one decimal place, the volume of a sphere with radius 14.5 cm.

**11** The time, $T$ seconds, it takes a swing to go back and forth once is $T = 2\pi\sqrt{\frac{l}{g}}$, where $l$ is the length of the swing (in metres) and g is the gravitational acceleration (in m/s$^2$). Find $T$, correct to 2 decimal places, if $l = 2.35$ and $g = 10$.

Stephen Rees/Shutterstock.com

**12** Use the compound interest formula $A = P(1 + r)^n$ to calculate the amount to which a principal of \$5600 will grow if invested at 9.4% p.a. for 5 years.

**13** The maximum distance, $d$ m, a ball travels if thrown with speed $v$ m/s can be calculated using the formula $d = \frac{v^2}{g}$. Find the maximum distance, correct to one decimal place, if a ball is thrown at a speed of 11.5 m/s and g = 9.8.

**14** The body mass index (BMI) of an adult can be calculated using the formula $B = \frac{m}{h^2}$, where $m$ is the mass in kilograms and $h$ is the height in metres. Zoe is 1.7 m tall and weighs 60 kg.

**a** What is the subject of the formula given above?

**b** Calculate Zoe's BMI, correct to one decimal place.

**c** If a BMI between 21 and 25 is an indication of good health, then how might Zoe change her weight to improve her health?

Foundation | Mastery | Complex

**15** The formula for converting Fahrenheit temperatures (°F) to Celsius (°C) is $C = \frac{5}{9}(F - 32)$.

Convert each of the following temperatures to Celsius, correct to the nearest degree.

**a** New York 77°F **b** Los Angeles 100°F

**c** Rio de Janeiro 59°F **d** Normal body temperature 98.6°F

**16** The formula for converting a speed of $k$ km/h to metres per second (m/s) is $M = \frac{5k}{18}$.

A speed of 80 km/h is closest to which of the following speeds? Select **A**, **B**, **C** or **D**.

**A** 22 m/s **B** 32 m/s **C** 35 m/s **D** 47 m/s

EXAMPLES 8, 9

**17** Mia is 2 years old and weighs 11 kg. A nurse calculates her medicine dosage based on the adult dosage of 900 mg per day. Calculate Mia's daily dosage to the nearest milligram using:

**a** Fried's formula **b** Young's formula **c** Clark's formula.

**18** Ashleigh is 16 years old and weighs 50 kg. She takes a drug with a recommended adult dose of 1680 mg each day. Using Clark's formula, what would be Ashleigh's daily dosage? Select **A**, **B**, **C** or **D**.

**A** 1200 mg **B** 1800 mg **C** 300 mg **D** 336 mg

**19** Cowling's rule for calculating a child's dosage is

$$\text{child dosage} = \frac{\text{age at next birthday}}{24} \times \text{adult dosage}$$

If the adult daily dosage of an antibiotic is 100 mg, use Cowling's rule to find the daily dosage, correct to one decimal place, for:

**a** Brittany, aged 2 **b** Patrick, aged 6 **c** Mercedes, aged $10\frac{1}{2}$.

**20** Elena is $1\frac{1}{2}$ years old. If the adult dosage of a drug is 1000 mg per day, find, correct to the nearest milligram, Elena's appropriate daily dosage using:

**a** Young's rule **b** Fried's rule.

EXAMPLE 10

**21** The distance, $d$ km, an observer can see to the horizon from a height, $h$ m, above sea level can be calculated using the formula $d = 8\sqrt{\frac{h}{5}}$.

What distance (correct to the nearest kilometre) can a person see from the observation deck of Sydney Tower Eye at 268 m?

**22** Liam earns a weekly wage of \$450 plus commission of 11% of the value of international phonecards he sells in excess of \$900. His weekly pay is given by the formula $P = 450 + 0.11(V - 900)$, where $V$ is the total value of the cards sold. How much will Liam earn for selling \$2310 worth of cards in a week?

**23** The braking distance (in metres) of a bicycle travelling at a speed of $V$ m/s is given by $d = \frac{V(V+1)}{2}$.
Calculate the braking distance of a bicycle travelling at a speed of 6 m/s.

**24** The surface area of a cylinder of radius $r$ and height $h$ is calculated using the formula $S = 2\pi r(r + h)$.
Calculate the surface area, correct to 2 decimal places, of a cylinder with radius 3 cm and height 8 cm.

**25** The speed, $V$ m/s, required for a spacecraft to escape Earth's gravitational pull during take-off is given by $V = \sqrt{2gr}$, where g is 9.8 m/s$^2$ and $r$ is the radius of Earth (6 378 000 m). Which of the following is closest to the escape speed of a spacecraft leaving Earth's atmosphere? Select **A**, **B**, **C** or **D**.

**A** 8840 m/s **B** 11 180 m/s **C** 12 500 m/s **D** 35 000 m/s

Foundation Mastery Complex

## Did you know?

### The body mass index

The **body mass index (BMI)** is used by the World Health Organization (WHO) as an international measure of weight-related health for adults. The formula for BMI is $B = \frac{m}{h^2}$, where $m$ is a person's mass in kilograms and $h$ is the person's height in metres. This is a convenient measure that represents the weight-related health of an adult by a single value. The table describes a range of BMI scores.

| BMI | Weight-related health status |
|---|---|
| Under 18.5 | Underweight |
| 18.5 to 24.9 | Normal |
| 25.0 to 29.9 | Overweight |
| 30 and above | Obese |

According to this model, about 38% of Australian adults are overweight and 28% of Australian adults are obese, having a greater risk of heart disease, stroke, diabetes, high blood pressure and high cholesterol.

Because the BMI is an algebraic *model* that has been developed using data from particular subsets of people, the WHO acknowledges that it has limitations and inaccuracies. It does not take into account a person's frame size, muscle mass, bone density or distribution of body fat. It does not apply well to First Nations, Pacific Island and Asian people. Within the remaining groups, it should also not be applied to bodybuilders, athletes, children under 19, pregnant women or the frail and sedentary elderly.

**Why is the BMI less accurate for measuring the weight-related health of some groups of people**?

**Would the groups of people listed above (bodybuilders, athletes, children, pregnant women and the elderly) score unusually high or unusually low on the BMI scale**?

## Investigation

### Getting the right formula

Listed below are 15 commonly used formulas. As a group or individual activity, select 8 formulas and, for each one:

**a** describe what the formula is used for

**b** write the subject of the formula, and what it stands for

**c** describe what the other variables in the formula stand for.

**1** $V = \frac{1}{3}Ah$ **2** $A = 180(n - 2)$ **3** $c^2 = a^2 + b^2$

**4** $A = \pi r^2$ **5** $S = 2\pi r^2 + 2\pi rh$ **6** $I = Prn$

**7** $m = \frac{y_2 - y_1}{x_2 - x_1}$ **8** $C = 2\pi r = \pi d$ **9** $A = \frac{1}{2}xy$

**10** $S = \frac{d}{t}$ **11** $A = \frac{1}{2}(a + b)h$ **12** $A = P(1 + r)^n$

**13** $A = s^2$ **14** $V = \pi r^2 h$ **15** $V = \frac{4}{3}\pi r^3$

# 2.04 Solving equations

**Skillsheets**
Solving equations by balancing
Solving equations by backtracking
Solving equations using diagrams

**Worksheet**
More complex equations

**Puzzles**
Equations code puzzle
Equations riddle

An **equation** contains an algebraic expression and an equals (=) sign. For example, $3x - 4$ is an **expression,** while $3x - 4 = -13$ is an **equation**. An equation is **solved** when the value of the variable (for example, $x$) is found that makes the equation true.

## Solving equations

- Keep the equation balanced by doing the same operation on both sides.
- Aim to have the variable (for example, $x$) on one side of the equation and a number on the other side. For example, $x = 4$.

## Example 11

Solve each equation.

**a** $3x - 4 = -13$ **b** $\frac{b}{2} + 7 = 1$ **c** $\frac{h-10}{4} = 3$

### Solution

**a**

$3x - 4 = -13$

$3x - 4 + 4 = -13 + 4$ adding 4 to both sides

$3x = -9$

Remember: We are aiming to get $x$ on its own.

$\frac{3x}{3} = \frac{-9}{3}$ dividing both sides by 3

$x = -3$

Check by substituting $x = -3$ back into the original equation:

$\text{LHS} = 3(-3) - 4$

LHS means 'left-hand side'.

$= -13$

$= \text{RHS}$

RHS means 'right-hand side'.

**b**

$\frac{b}{2} + 7 = 1$

$\frac{b}{2} + 7 - 7 = 1 - 7$ subtracting 7 from both sides

$\frac{b}{2} = -6$

$\frac{b}{2} \times 2 = -6 \times 2$ multiplying both sides by 2

$b = -12$

Check by substituting $b = -12$ back into the original equation:

$\text{LHS} = \frac{-12}{2} + 7$

$= 1$

$= \text{RHS}$

**c**

$\frac{h-10}{4} = 3$

$\frac{h-10}{4} \times 4 = 3 \times 4$ multiplying both sides by 4

$h - 10 = 12$

$h - 10 + 10 = 12 + 10$ adding 10 to both sides

$h = 22$

## Example 12

Solve each equation.

These equations need more steps to solve.

**a** $5y + 5 = 2y + 17$

**b** $\frac{2m-7}{3} = 6$

**c** $4(1 - 2t) = 16$

### Solution

Remember: We are aiming to get $y$ on its own.

**a**

$$5y + 5 = 2y + 17$$

$$5y + 5 - 2y = 2y + 17 - 2y$$ subtracting $2y$ from both sides to get all the $y$ terms on the LHS

$$3y + 5 = 17$$

$$3y + 5 - 5 = 17 - 5$$ subtracting 5 from both sides to get all the numbers on the RHS

$$3y = 12$$

$$\frac{3y}{3} = \frac{12}{3}$$ dividing both sides by 3

$$y = 4$$

Checking:

$$\text{LHS} = 5(4) + 5 = 25$$

$$\text{RHS} = 2(4) + 17 = 25 = \text{LHS}$$

**b**

$$\frac{2m-7}{3} = 6$$

$$\frac{2m-7}{3} \times 3 = 6 \times 3$$ multiplying both sides by 3

$$2m - 7 = 18$$

$$2m = 25$$ adding 7 to both sides

$$m = \frac{25}{2}$$ dividing both sides by 2

$$= 12\frac{1}{2}$$

Checking:

$$\text{LHS} = \frac{2\left(12\frac{1}{2}\right) - 7}{3} = \frac{18}{3} = 6 = \text{RHS}$$

**c**

$$4(1 - 2t) = 16$$

$$4 - 8t = 16$$ expanding LHS first

$$-8t = 12$$ subtracting 4 from both sides

$$t = \frac{12}{-8}$$ dividing both sides by (−8)

$$t = -1\frac{1}{2}$$

**EXERCISE 2.04** Answers on p. 462

## Solving equations

**1** Solve each equation.

**a** $3d + 2 = 20$ **b** $2p - 3 = 2$ **c** $4u + 6 = 20$

**d** $5a + 3 = -12$ **e** $12b + 8 = 4$ **f** $3 - 2a = -6$

**g** $3m = m - 10$ **h** $\frac{3h}{4} = 9$ **i** $\frac{r-1}{6} = 2$

**j** $-\frac{2x}{5} = 8$ **k** $\frac{y+2}{-3} = 1$ **l** $11 - 4n = 15$

**m** $6 = 11 - y$ **n** $\frac{4c}{10} = 3$ **o** $\frac{z}{3} - 11 = 9$

**2** What is the solution to $2t - 4 = 10 + t$? Select **A**, **B**, **C** or **D**.

**A** $t = 3$ **B** $t = 6$ **C** $t = 4\frac{2}{3}$ **D** $t = 14$

EXAMPLE 12

**3** Solve each equation.

**a** $5k - 13 = 3k + 9$ **b** $8e = 2(e - 6)$ **c** $\frac{2f+7}{2} = 10$

**d** $3(x - 2) = 45$ **e** $\frac{w}{5} - 8 = 6$ **f** $4(2d - 9) = -12$

**g** $\frac{4n+7}{9} = 2$ **h** $7u + 7 = 2u - 10$ **i** $3p + 4 = 4p$

**j** $\frac{5z+8}{6} = -3$ **k** $\frac{8-2b}{2} = 7$ **l** $6(2 - 3q) = -24$

**4** In which line was an error made in solving this equation? Select **A**, **B** or **C**.

$$\begin{aligned} \text{Line 1: } \quad \frac{c-4}{8} + 2 &= 6 \\ \text{Line 2: } \quad \frac{c-4}{8} &= 8 \\ \text{Line 3: } \quad c - 4 &= 64 \\ \text{Line 4: } \quad c &= 68 \end{aligned}$$

**A** Line 2

**B** Line 3

**C** Line 4

**5** Solve $\frac{2k+3}{5} + k = 7$. Select **A**, **B**, **C** or **D**.

**A** $k = 4\frac{4}{7}$ **B** $k = 10\frac{2}{3}$ **C** $k = \frac{4}{5}$ **D** $k = \frac{3}{4}$

☐ Foundation ◯ Mastery ◯ Complex

# Formulas and equations 2.05

Sometimes, after we **substitute** values into a formula, the result is an equation that must be solved.

**Videos**
Equations and formulas 1
Equations and formulas 2
Equation problems 1
Equation problems 2

**Worksheets**
Working with formulas
Using equations to solve problems
Word problems with equations

## Example 13

The surface area of a rectangular prism of length $l$, width $w$ and height $h$ is $S = 2lw + 2lh + 2wh$.

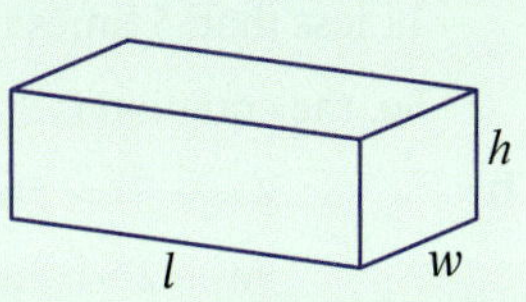


If a rectangular prism with surface area 132 cm² has length 7 cm and width 3 cm, find its height.

### Solution

$S = 132, l = 7, w = 3$

$S = 2lw + 2lh + 2wh$

$132 = (2 \times 7 \times 3) + (2 \times 7 \times h) + (2 \times 3 \times h)$ — substituting into formula

$132 = 42 + 14h + 6h$

$132 = 42 + 20h$ — collecting like terms

$90 = 20h$ — subtracting 42 from both sides

$h = \frac{90}{20}$ — dividing both sides by 20

$= 4\frac{1}{2}$

The height of the rectangular prism is $4\frac{1}{2}$ cm.

## Example 14

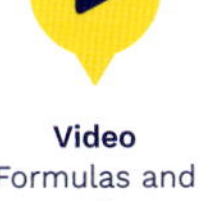

**Video**
Formulas and equations

The formula for converting Fahrenheit temperature (°F) to Celsius (°C) is $C = \frac{5}{9}(F - 32)$.

Convert 100°C to °F.

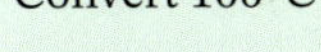

### Solution

When $C = 100$:

$100 = \frac{5}{9}(F - 32)$

$\frac{5}{9}(F - 32) = 100$ — swapping sides so $F$ is on LHS

$5(F - 32) = 900$ — multiplying both sides by 9

$5F - 160 = 900$ — expanding the LHS

$5F = 1060$ — adding 160 to both sides

$F = \frac{1060}{5}$ — dividing both sides by 5

$= 212$

$\therefore 100°\text{C} = 212°\text{F}$

Video
Writing equations

## Example 15

When the sum of a number and 7 is tripled, the result is the same as 11 less than 5 times the number.

**a** If $w$ is the number, write an equation that can be used to find $w$.

**b** Solve the equation.

### Solution

**a** The sum of $w$ and 7 when tripled is written as $3(w + 7)$.

11 less than 5 times the number is $5w - 11$.

So, the equation is $3(w + 7) = 5w - 11$.

**b**

$$3(w + 7) = 5w - 11$$

$$3w + 21 = 5w - 11 \quad \text{expanding the LHS}$$

$$3w + 21 - 5w = 5w - 11 - 5w \quad \text{subtracting } 5w \text{ from both sides}$$

$$-2w + 21 = -11$$

$$-2w + 21 - 21 = -11 - 21 \quad \text{subtracting 21 from both sides}$$

$$-2w = -32$$

$$\frac{-2w}{-2} = \frac{-32}{-2} \quad \text{dividing both sides by } -2$$

$$w = 16$$

**EXERCISE 2.05** Answers on p. 462

## Formulas and equations

EXAMPLE 13

**1** The number of matchsticks, $m$, needed to make a pattern of $s$ squares is $m = 3s + 1$.

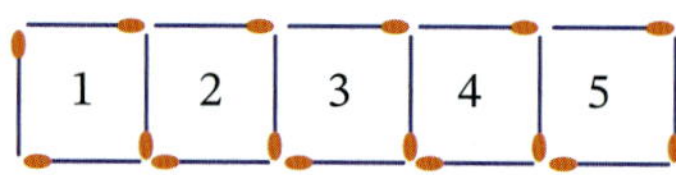


**a** How many matches are needed to make:

**i** 4 squares? **ii** 10 squares?

**b** How many squares can be made from:

**i** 22 matches? **ii** 55 matches?

**2** Find the base length, $b$, of a triangle with area 90 cm$^2$ and perpendicular height 15 cm, given the area formula $A = \frac{1}{2}bh$.

**3** The average speed of a moving object in metres per second can be calculated using the formula $s = \frac{d}{t}$, where $d$ is the distance travelled in metres and $t$ is the time taken in seconds.

**a** Find the distance travelled by a car in 20 seconds if its speed is 15 m/s.

**b** Find the time taken (to the nearest second) for a cyclist to travel 250 m if his speed is 4.2 m/s.

EXAMPLE 14

**4** The formula for the area of a trapezium is $A = \frac{h}{2}(a + b)$, where $a$ and $b$ are the lengths of the parallel sides, and $h$ is the distance between them. What is the length of one parallel side of a trapezium if the other parallel side is 7 m, the distance between them is 5 m and the area is 22.5 m$^2$?

**5** The mean, $M$, of 3 numbers, $x$, $y$ and $z$, is calculated using the formula $M = \frac{x+y+z}{3}$.

If 3 numbers have a mean of 17, and 2 of the numbers are 10 and 20, find the third number.

**6** The circumference of a circle with radius $r$ can be calculated using the formula $C = 2\pi r$. If a circle has a circumference of 50.27 cm, find its radius, correct to the nearest centimetre.

☐ Foundation ○ Mastery ⬡ Complex

7 If a principal, $P, is invested at an interest rate of $r$ per annum (where $r$ is written as a decimal) then, after $n$ years, it will grow to $A, where $A$ is given by the compound interest formula $A = P(1 + r)^n$. What principal needs to be invested at 11% p.a. for it to grow to $6000 in 3 years? Express your answer to the nearest cent.

8 A kettle is switched on to boil and the temperature, $T$°C, of the water after $t$ minutes is $T = 18t + 28$. After how many minutes is the temperature:

**a** 64°C? **b** 92.8°C?

9 The formula for converting miles ($M$) to kilometres ($K$) is $K = 1.61M$. Convert each of the following distances to miles, correct to 2 decimal places.

**a** 5 km **b** 1.5 km

10 The number of chairs, $c$, that can be seated around $t$ square tables joined together is $c = 2t + 2$.

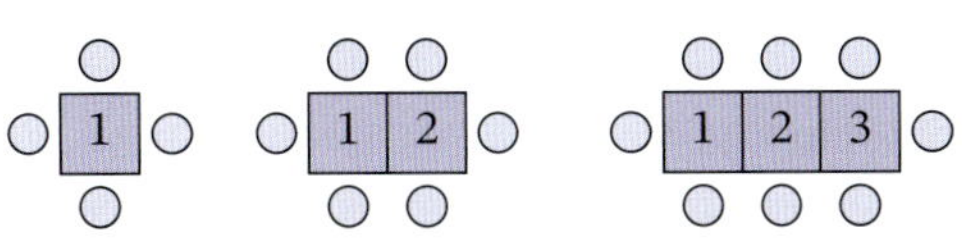


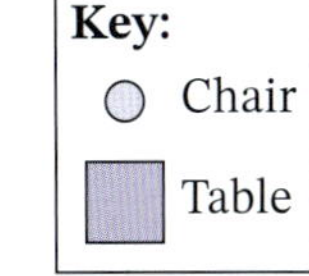


**a** How many chairs can be seated around:

**i** 5 tables? **ii** 12 tables?

**b** How many square tables are required to seat:

**i** 24 chairs? **ii** 40 chairs?

11 Rhianna earns a weekly wage of $540 plus commission of 12% on the value of cosmetics she sells in excess of $2000. Her total pay, $P$, is given by the formula $P = 540 + 0.12(V - 2000)$, where $V$ is the value of the cosmetics sold that week. What was the value of the cosmetics Rhianna sold if her total pay was $852?

12 The surface area of a rectangular prism of length $l$, width $w$ and height $h$ is calculated using the formula $S = 2lw + 2lh + 2wh$. If a rectangular prism has length 4 cm, width 2 cm and surface area 58 cm$^2$, what is its height? Select **A**, **B**, **C** or **D**.

**A** 3.5 cm **B** $4\frac{1}{6}$ cm **C** 5.5 cm **D** 7.25 cm

13 The formula for converting $k$ kilometres to $M$ miles is $M = \frac{5k}{8}$. Convert 12 miles to kilometres.

14 The number of matchsticks, $m$, needed to make this pattern of triangles is $m = 2t + 1$, where $t$ is the number of triangles in the pattern.

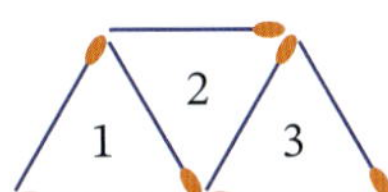


How many triangles can be made with:

**a** 37 matches? **b** 55 matches?

15 For babysitting, Kirstie charges $18 for the first hour and $14 for each hour after that. This charge, $C, can be expressed using the formula $C = 18 + 14(h - 1)$, where $h$ is the number of hours worked. If Kirstie earned $46 for babysitting 1 night, for how many hours did she work?

16 Mohammed works in a high-rise office building on the 32nd floor. At lunchtime, he travels down in the lift at a speed of 2 floors per second. This is described by the formula $F = 32 - 2t$, where $F$ is the floor being passed by the lift and $t$ is the time in seconds.

**a** Which floor is Mohammed passing after:

**i** 8 seconds? **ii** 14 seconds?

**b** After how many seconds will Mohammed pass the:

**i** 18th floor? **ii** 8th floor?

**17** According to one theory, the recommended nightly hours of sleep for a child can be calculated using the formula $S = 8 + \frac{18-a}{2}$, where $a$ is the age of the child in years. What is the age of a child who requires $14\frac{1}{2}$ hours of sleep each night? Select **A**, **B**, **C** or **D**.

**A** 2.5 years **B** 3 years **C** 3.5 years **D** 5 years

**18** If an object has an initial speed $u$ m/s and acceleration $a$ m/s$^2$, after $t$ seconds its speed (in m/s) is given by the formula $v = u + at$. Find the acceleration of an object if its initial speed is 8 m/s and its final speed after 12 seconds is 44 m/s.

**19** The charge, $\$C$, for hiring a hall for an event is given by $C = 95N + p$, where $N$ is the number of people at the event and $p$ is the fixed cost. Find the fixed cost when the charge for hiring the hall for 125 people is \$13 500.

EXAMPLE 15

**20** When 3 is added to a number $N$ and then multiplied by 5, the answer is the same as subtracting the number from 33.

**a** Write an equation that can be solved to find $N$.

**b** Solve the equation for $N$.

**21** Doubling the sum of a number $x$ and 7 will give the same result as subtracting 4 times the number from 29.

**a** Write an equation that can be solved to find $x$.

**b** Solve the equation for $x$.

**22** When 4 more than a number is halved, the answer is 5 more than 3 times the number. If $p$ is the number, which equation can be used to find $p$? Select **A**, **B**, **C** or **D**.

**A** $\frac{p}{2} + 4 = 3p + 5$ **B** $p + 2 = 3(p + 5)$

**C** $\frac{p+4}{2} = 3(p + 5)$ **D** $\frac{p+4}{2} = 3p + 5$

**23** When the number $k$ is subtracted from 15 and the result is multiplied by 4, the answer is the same as 3 more than twice the number.

**a** Write an equation to find $k$.

**b** Solve the equation.

**24** The length of a rectangle is 3 times the width, $w$. The perimeter of the rectangle is 84 cm more than 5 times the width.

**a** Write an equation that can be used to find the width, $w$, of the rectangle.

**b** Solve the equation to find the value of $w$.

**c** Find the area of the rectangle.

☐ Foundation ○ Mastery ○ Complex

# Changing the subject of a formula 2.06

In the formula $v = u + at$, $v$ is called the **subject** of the formula because it is on the LHS. To change the subject of a formula to another variable, use the same rules as for solving an equation. The answer is not a number but an algebraic equation (another formula).

**Videos**
Changing the subject of a formula 1
Changing the subject of a formula 2

**Worksheets**
Changing the subject of a formula
Subject of a formula

## Example 16

If a moving object has initial speed $u$ m/s and acceleration $a$ m/s², its final speed $v$ m/s after time $t$ seconds is given by the formula $v = u + at$. Change the subject of this formula to:

**a** $u$ **b** $t$

### Solution

**a** $v = u + at$

$u + at = v$ swapping sides so new subject $u$ appears on LHS

$u = v - at$ subtracting $at$ from both sides

**b** $v = u + at$

$u + at = v$ swapping sides so new subject $t$ appears on LHS

$at = v - u$ subtracting $u$ from both sides

$t = \dfrac{v - u}{a}$ dividing both sides by $a$

## Example 17

The volume of a cone is given by the formula $V = \dfrac{1}{3}\pi r^2 h$, where $r$ is the radius of the circular base and $h$ is the height.

**a** Make $h$ the subject of the formula.

**b** Make $r$ the subject of the formula.

### Solution

**a** $V = \dfrac{1}{3}\pi r^2 h$

$\dfrac{1}{3}\pi r^2 h = V$ swapping sides so new subject $h$ appears on LHS

$\pi r^2 h = 3V$ multiplying both sides by 3 to remove fraction

$h = \dfrac{3V}{\pi r^2}$ dividing both sides by $\pi r^2$

**b** $V = \dfrac{1}{3}\pi r^2 h$

$\dfrac{1}{3}\pi r^2 h = V$ swapping sides so $r$ appears on the LHS

$\pi r^2 h = 3V$ multiplying both sides by 3

$r^2 = \dfrac{3V}{\pi h}$ dividing both sides by $\pi h$

$r = \pm\sqrt{\dfrac{3V}{\pi h}}$ taking the square root of both sides

$= \sqrt{\dfrac{3V}{\pi h}}$ positive square root because radius, $r$, is positive

**EXERCISE 2.06** Answers on p. 462

## Changing the subject of a formula

EXAMPLE 16

**1** Make $x$ the subject of each formula.

**a** $y = 2x + 4$ **b** $T = 3x - 7$ **c** $d = \frac{x+1}{3}$

**d** $p = 18 - x$ **e** $k = 4x + r$ **f** $C = n + \frac{x}{2}$

**g** $S = 10bx$ **h** $V = \frac{x-5}{6}$ **i** $z = 12 - ax$

**2** If an object travels a distance, $d$, in a period of time, $t$, its average speed, $S$, is given by the formula $S = \frac{d}{t}$. Change the subject of the formula to:

**a** $d$ **b** $t$.

**3** Make $y$ the subject of the formula $A = \frac{4y+m}{3}$. Select **A**, **B**, **C** or **D**.

**A** $y = \frac{A-m}{4}$ **B** $y = \frac{3A}{4} - 3m$ **C** $y = \frac{3A}{4} - m$ **D** $y = \frac{3A-m}{4}$

**4** Change the subject of the formula $g = \frac{ax}{d} - 5$ to $x$. Select **A**, **B**, **C** or **D**.

**A** $x = \frac{g+5d}{a}$ **B** $x = \frac{gd+5d}{a}$ **C** $x = \frac{gd}{a} + 5$ **D** $x = dg + \frac{5d}{a}$

**5** The angle sum of a shape with $n$ sides is $A°$, where $A = 180(n - 2)$.

**a** Use the formula to find the angle sum of a shape with 12 sides.

**b** Make $n$ the subject of the formula.

**c** If the angle sum of a polygon is 1260°, how many sides does it have?

**6** The volume of a pyramid has the formula $V = \frac{1}{3}Ah$, where $A$ is the area of the base and $h$ is the perpendicular height. Which of the following is the correct formula for $A$? Select **A**, **B**, **C** or **D**.

**A** $A = \frac{1}{3}Vh$ **B** $A = \frac{3}{Vh}$ **C** $A = 3Vh$ **D** $A = \frac{3V}{h}$

**7** The area of a trapezium has the formula $A = \frac{h}{2}(a+b)$. Change the subject to:

**a** $h$ **b** $b$

EXAMPLE 17

**8** Change the subject of each formula to the variable shown in brackets.

**a** $A = \pi r^2$ [$r$] **b** $y = mx + c$ [$m$] **c** $v^2 = u^2 + 2as$ [$s$]

**d** $z = \frac{x-m}{s}$ [$x$] **e** $E = mc^2$ [$c$] **f** $K = \frac{1}{2}mv^2$ [$v$]

**g** $A = P(1 + r)^n$ [$P$] **h** $x = \sqrt{6y+3}$ [$y$] **i** $A = \frac{1}{2}bh$ [$h$]

**j** $T = 2\pi\sqrt{\frac{l}{g}}$ [$l$] **k** $V = IR - E$ [$R$] **l** $e = ir + \frac{Q}{C}$ [$Q$]

**m** $s = ut + \frac{1}{2}at^2$ [$a$] **n** $K = \frac{m+d}{w}$ [$d$] **o** $h = 4.9t^2 + 15$ [$t$]

Foundation | Mastery | Complex

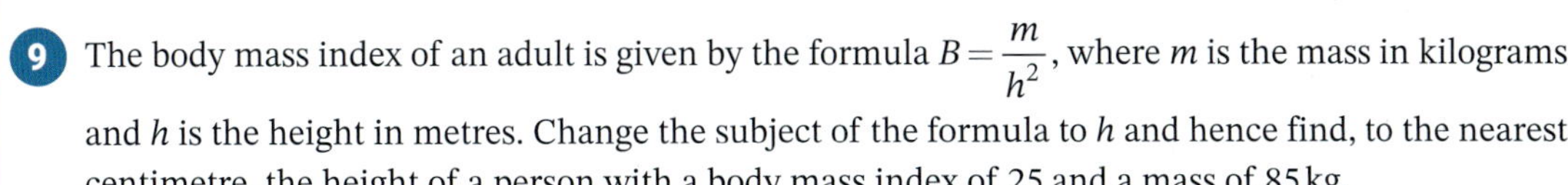

**9** The body mass index of an adult is given by the formula $B = \frac{m}{h^2}$, where $m$ is the mass in kilograms and $h$ is the height in metres. Change the subject of the formula to $h$ and hence find, to the nearest centimetre, the height of a person with a body mass index of 25 and a mass of 85 kg.

**10** The volume of a cone is given by $V = \frac{1}{3}\pi r^2 h$, where $r$ is the radius of the base and $h$ is the height. Show that this formula can be rewritten as $h = \frac{3V}{\pi r^2}$ and hence find, to the nearest centimetre, the height of a cone with volume 3539.53 cm$^3$ and a base radius of 13 cm.

**11** Make $F$ the subject of the formula $C = \frac{5}{9}(F - 32)$ and hence find $F$ when:

**a** $C = 40$ **b** $C = 100$

**12** Make $P$ the subject of the formula $M = \frac{10}{\pi hP}$, then find the value of $P$, correct to 2 significant figures, if $M = 9.8$ and $h = 0.27$.

**13** The volume of a cylinder is calculated using the formula $V = \pi r^2 h$, where $r$ is the radius and $h$ is the height of the cylinder.

**a** Change the subject of the formula to $r$.

**b** Find the radius (to the nearest millimetre) of a cylindrical tank with volume 118 m$^3$ and height 4.4 m.

**14** Make $m$ the subject of the formula $y = 3m^2 + d$.

**15** The height ($h$ metres) of a stone dropped from a cliff at time $t$ seconds is given by the formula $h = 100 - 4.9t^2$.

**a** Make $t$ the subject of the formula.

**b** At what time, correct to one decimal place, will the stone be at a height of:

**i** 80 m? **ii** 10 m?

**16** The shape of cables hanging between 2 towers 120 m apart on a suspension bridge is given by the equation $h = 15 + 0.008d^2$, where $h$ is the height in metres of the cables above ground level and $d$ metres is the horizontal distance from the centre of the bridge.

**a** Make $d$ the subject of the formula.

**b** Find the distance from the centre, to the nearest metre, when the height is:

**i** 35 m **ii** 18 m **iii** 60 m

□ Foundation ○ Mastery ○ Complex

## Sample HSC problem Answers on p. 462

(4 marks) Paula is a plumber. She charges according to the formula $C = 110 + 92(h - 1)$, where $\$C$ is the charge and $h$ is the number of hours worked.

**a** How much does Paula charge for working $2\frac{1}{2}$ hours? 1 mark

**b** Make $h$ the subject of the formula. 2 marks

**c** Find the number of hours Paula worked if she charged \$524. 1 mark

## Study Tip

### Topic summaries and mind maps

Summarise each topic once you have completed it, to create useful study notes for revising the course, which can be especially valuable before exams. Use a notebook or folder for listing important ideas, formulas, terminology and skills of each topic. Educational research shows that effective learning takes place when we rewrite learned knowledge in our own words.

A good topic summary runs for 2 to 4 pages. It is a condensed, personalised version of your course notes. This is your interpretation of a topic, so include your own comments, symbols, diagrams, observations and reminders. Highlight important facts using boxes and include a glossary of keywords and phrases. (There is a glossary at the back of this book.)

A mind map is a topic summary in graphic form, with boxes, branches and arrows showing the connections between the main ideas of a topic. (See the next page for an example.) The topic name is written at the centre of the map, with branches leading off to key concepts or subheadings. Smaller branches hang off these branches, listing important details and formulas. Mind maps are powerful because they present an overview of a topic on one sheet of paper. Visual learners absorb and recall information better using mind maps.

When compiling a topic summary, use your class notes and this textbook, especially the front and back sections of each chapter. Ask your teacher for a copy of the course syllabus or the school's teaching program, which lists the knowledge and skills of every topic in dot point form.

# CHAPTER SUMMARY

This chapter, *Formulas and equations*, revised basic algebra skills and then extended from them. Make sure you master the algebraic techniques required to expand and simplify algebraic expressions, solve equations and work with formulas.

Before you move on, consider what you learned in this chapter and revisit any sections that may have been unclear.

| What you learned in this chapter... | Section | |
|---|---|---|
| Expand and simplify algebraic expressions | 2.01<br>2.02 | Simplifying algebraic expressions<br>Expanding algebraic expressions |
| Substitute values into algebraic expressions and formulas, including Fried's, Young's and Clark's formulas for medicine dosage for children | 2.03 | Formulas |
| Solve linear equations, including after substituting into a formula | 2.04<br>2.05 | Solving equations<br>Formulas and equations |
| Form an equation from a worded problem and solve the equation to solve the problem | 2.05 | Formulas and equations |
| Change the subject of a formula | 2.06 | Changing the subject of a formula |

To help master these techniques, make a summary of this topic. Use the chapter outline and the mind map below as a guide. Add your own words, symbols, diagrams, boxes and reminders. The summary should give you a 'whole picture' view of the topic and allow you to identify any weak areas to revisit in your revision.

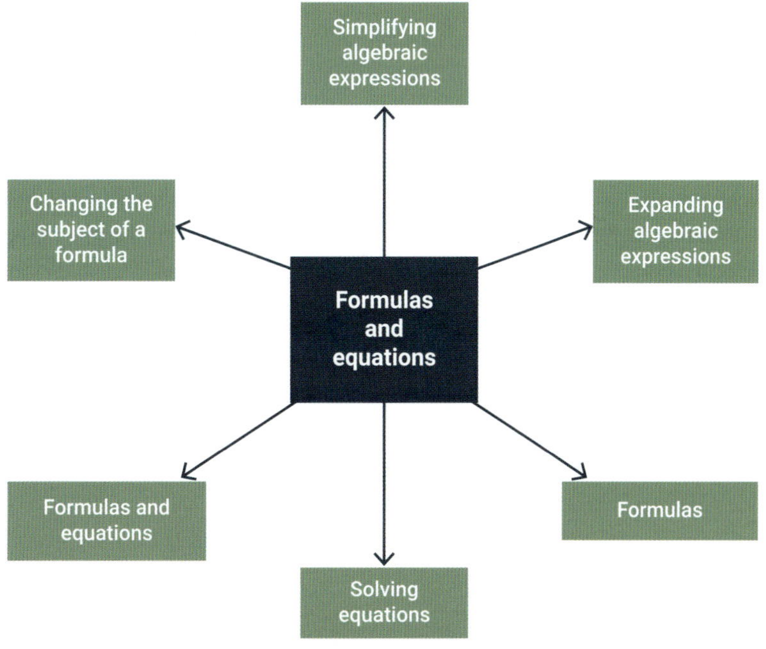

# 2 Test yourself Answers on p. 462

2.01

**1** Simplify each expression.

**a** $5ut + 2t^2 - t^2 + 4ut$ **b** $3k^2 \times 5k$ **c** $-9d \times \frac{2d}{3}$

**d** $8p^2 \times 4p^3$ **e** $-1 + 4h + 8 - 10h$ **f** $\frac{16r^3}{2r}$

**g** $(-3d^2)^3$ **h** $\frac{9n^2}{15n^2}$ **i** $-3v^2w^2 \div 21vw$

**j** $10x^2 + 7x - 2x^2 + x$ **k** $\frac{24bc}{8b^2}$ **l** $\frac{-a}{a^3}$

**m** $\frac{48r}{6} \div 4r$ **n** $\frac{4y}{3} \times \frac{5v}{10}$ **o** $\frac{10x}{18p} \times \frac{3p}{20}$

**p** $\frac{3y}{2a} \div \frac{9dy}{10d}$

2.02

**2** Expand each expression.

**a** $5(2x - 4)$ **b** $-3(a + 7)$ **c** $4(12t - y)$

**d** $-9(r^2 + 2w)$ **e** $8mn(m - n)$ **f** $-2d(4d - d^2)$

**3** Expand and simplify each expression.

**a** $3(4x + 1) + 2(x - 2)$ **b** $2n(n - 1) + (n - 1)$

**c** $6(2 - d) - 4(d - 3)$ **d** $p(p + 4) - p(p + 8)$

**e** $3(4u + 5) - (u + 7)$ **f** $h(5h - 1) + 3h(h + 9)$

2.03

**4** If $p = 4$, $q = -5$ and $r = 20$, then evaluate each expression.

**a** $3p^2 + 4r$ **b** $\frac{7r}{q}$ **c** $pqr$ **d** $\sqrt{p(r-q)}$

2.03

**5** The surface area of a cone is given by the formula $S = \pi r(r + s)$, where $r$ is the radius of the base and $s$ is the slant height of the cone. Find, correct to 2 decimal places, the surface area of a cone with base radius 5 cm and slant height 8 cm.

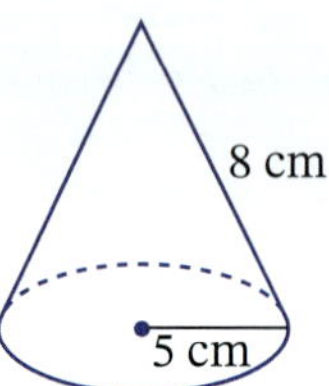


2.03

**6** Brett earns a weekly wage of \$480 plus a commission of 10% of the value of the insurance plans he sells in excess of \$1200. His total weekly pay, \$$P$, is given by the formula

$$P = 480 + \frac{V - 1200}{10}$$

where $V$ is the value of the insurance plans sold.

Calculate Brett's pay for a week in which he sold \$3400 worth of insurance plans.

☐ Foundation ○ Mastery ○ Complex

**7** Solve each equation. 2.04

**a** $5p - 4 = 21$ **b** $-2a + 6 = 8$ **c** $\dfrac{b-3}{2} = -6$

**d** $23 - 8r = 19$ **e** $\dfrac{4n}{5} = 11$ **f** $\dfrac{r}{3} + 7 = 1$

**g** $\dfrac{20 - 4n}{4} = 7$ **h** $3t + 13 = t - 12$ **i** $5(2g - 4) = -30$

**8** If an object is travelling with initial speed $u$ m/s, accelerating at a rate of $a$ m/s$^2$, and covers a distance $s$ m, then its final speed $v$ m/s follows the rule $v^2 = u^2 + 2as$. Calculate the distance travelled by a car whose speed increases from 11 m/s to 28 m/s with an acceleration of 3 m/s$^2$.

**9** According to one theory, the formula that links the surrounding air temperature, $T$ °C, to the number of chirps per minute, $C$, made by a cricket at night during summer is

$$T = \frac{C}{8} + 3.$$

How many chirps per minute are made by the cricket when the temperature is 13°C?

**10** When the number $d$ is decreased by 5 and the result is divided by 4, the answer is 7 more than twice the number.

**a** Write an equation to find $d$.

**b** Solve the equation.

**11** The average blood pressure, $P$, of a person aged $y$ years, measured in millimetres of mercury (mmHg), is given by the formula $P = 110 + \dfrac{y}{2}$. Make $y$ the subject of this formula and use it to find the age of a person whose blood pressure is 124 mmHg.

**12** Make $y$ the subject of each formula.

**a** $a - gy = h$ **b** $\dfrac{1 + 5y}{m} = 3$ **c** $4w = 3 + ky^2$ **d** $V = \dfrac{1}{2}hy^2 + 12$

**13** A ball is dropped from a window and its height $h$ metres after $t$ seconds is given by the formula $h = 50 - 4.9t^2$. 2.06

**a** Change the subject of the formula to $t$.

**b** Find the time when the ball is at a height of 15 metres.

Foundation Mastery Complex

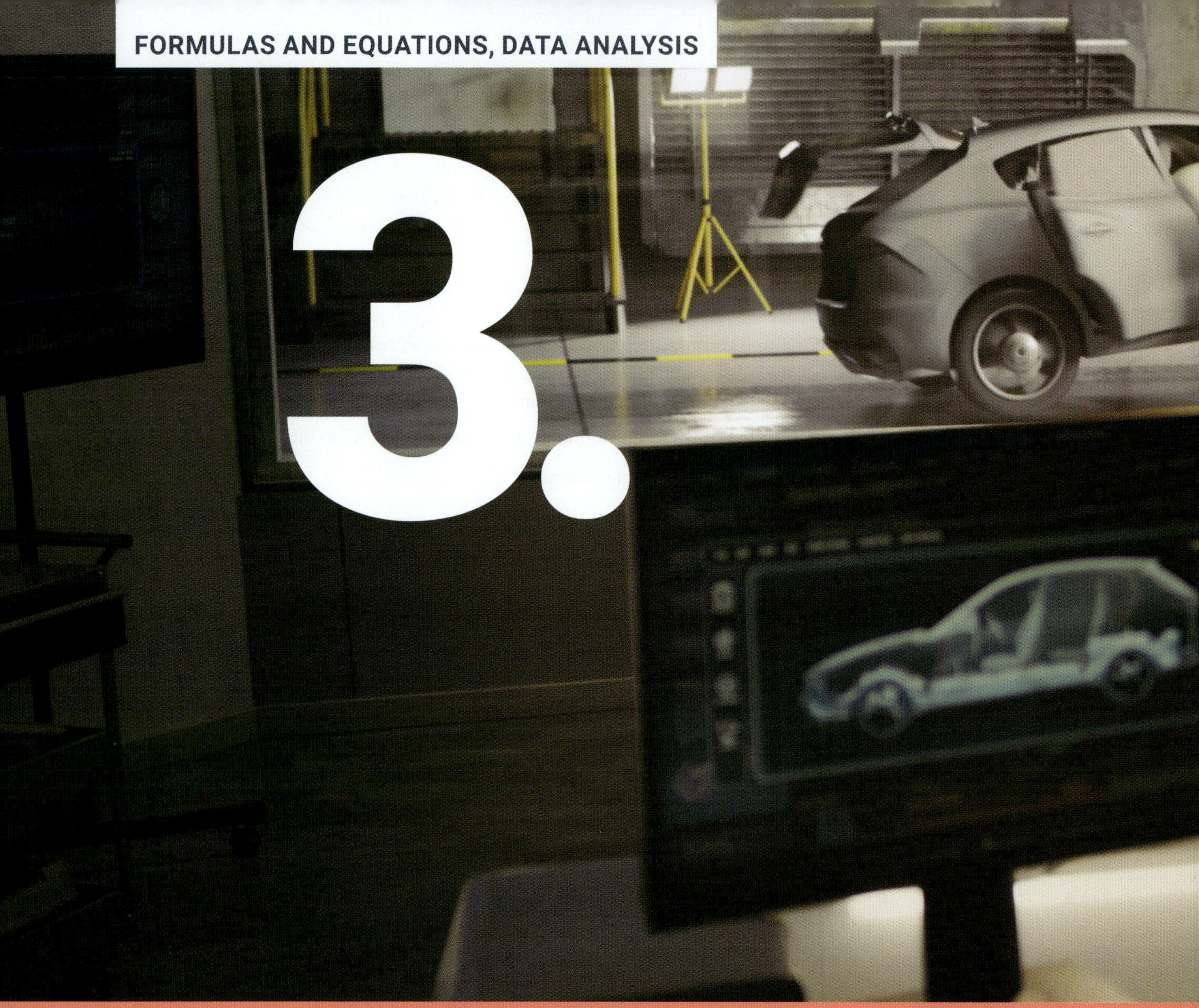

# 3. DRIVING SAFELY

ANCAP, the Australian New Car Assessment Program, crash-tests cars in 5 ways:

- front-on test, where the car hits a barrier at 64 km/h
- side impact test, where the car is hit on the driver's side by another car travelling at 50 km/h
- pedestrian test to assess head and leg injuries to pedestrians when the car is moving at 40 km/h
- pole test, where the car moves sideways at 29 km/h and hits a pole lined up with driver's head
- whiplash test, where the stationary car is rear-ended by another car travelling at 32 km/h.

The 2024 Hyundai Santa Fe Hybrid received a 5-star safety rating, scoring 84% for adult occupation protection, 86% for child occupation protection, 77% for vulnerable road user protection and 80% for safety assist.

## Chapter outline

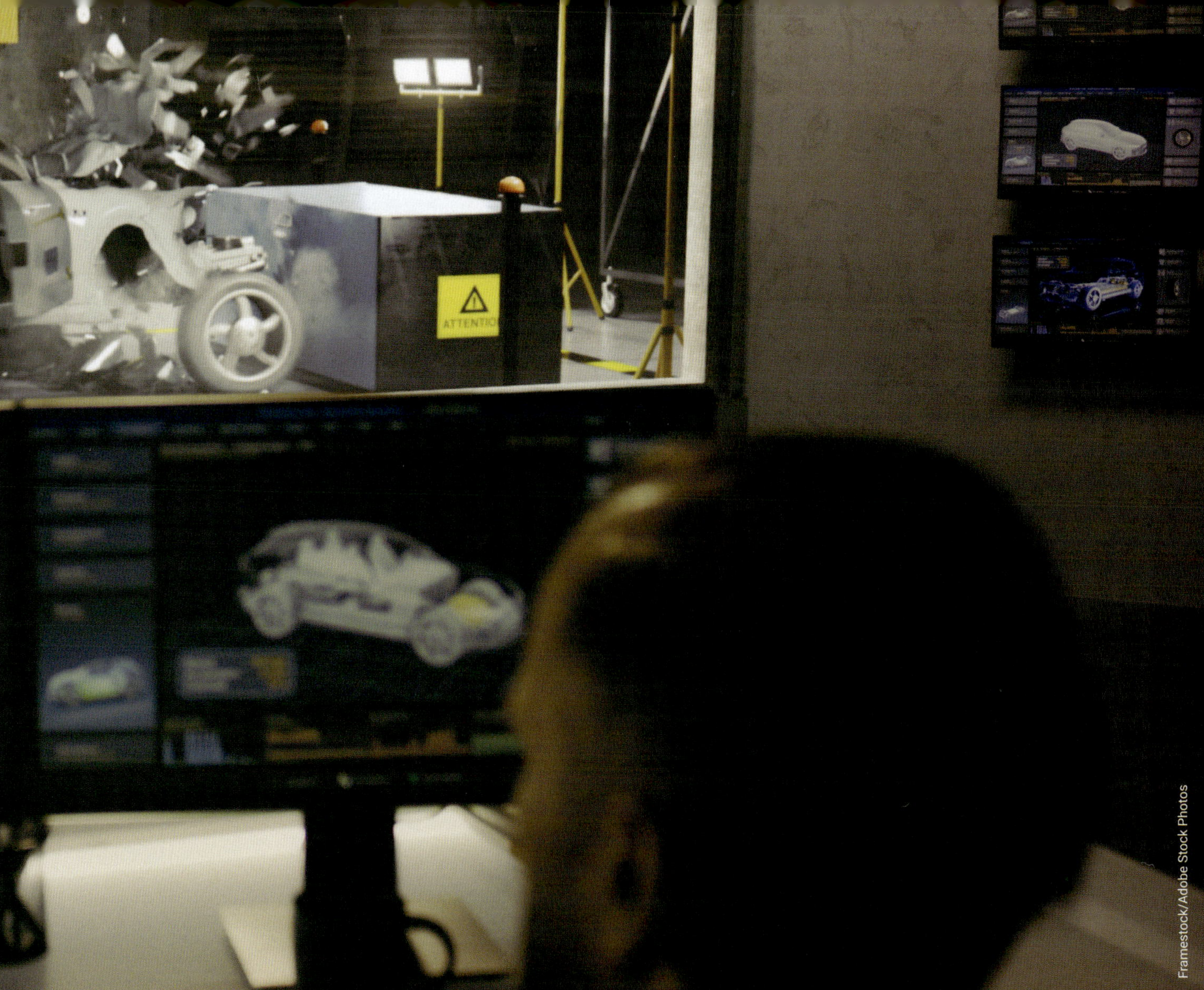

Framestock/Adobe Stock Photos

## In this chapter you will:

- use formulas and online calculators to calculate blood alcohol content (BAC) for males and females, based on number of drinks consumed, number of hours drinking and the person's mass
- interpret BAC and solve problems involving reducing BAC
- analyse data and graphs involving BAC and road accidents
- construct and interpret graphs that illustrate the level of blood alcohol over time
- solve problems involving speed, distance, speed and time
- use formulas to calculate stopping distance.

**Videos (2):**

**3.01** Blood alcohol content (BAC)

**3.03** Speed, distance and time

**Worksheets (8):**

**SkillCheck** Assignment 3

**3.01** BAC formula practice • Graphing BAC

**3.02** Investigating drivers in road fatalities • Investigating young drivers

**3.03** Speed formula practice • Speed stories • Distance, speed and time

Nelson MindTap

To access resources above, visit **cengage.com.au/nelsonmindtap**

## Terminology

| | | | |
|---|---|---|---|
| blood alcohol content (BAC) | braking distance | fatality | reaction distance |
| reaction time | standard drink | stopping distance | |

## SkillCheck Answers on p.463

Worksheet
Assignment 3

**1** Evaluate each expression, correct to 2 decimal places.

**a** $\dfrac{0.03}{0.015}$ **b** $\dfrac{0.04}{0.023}$ **c** $\dfrac{0.1}{0.034}$

**d** $\dfrac{26-15}{5.5\times 45}$ **e** $\dfrac{10\times 7-5\times 7.5}{6.8\times 64}$ **f** $\dfrac{2\times 10-18}{5.5\times 83}$

**2** **a** Copy and complete this table for the linear function $y = 4 - 0.3x$.

| $x$ | 0 | 1 | 2 | 3 | 4 | 5 |
|---|---|---|---|---|---|---|
| $y$ | | | | | | |

**b** If this function was graphed, what would be the gradient of the line?

**3** For each set of data, find the:

**i** mean **ii** median **iii** mode

**iv** range **v** interquartile range.

Answer correct to one decimal place where necessary.

**a** 5 7 8 9 10 4 5 2 4 1

**b** 10 12 14 9 0 2 8 12 1

**4** Evaluate the subject of each formula.

**a** $v = u + at$ when $u = 2.3$, $a = 8.7$, $t = 3$

**b** $s = kv(kv + b)$ when $k = 4.1$, $v = 42.3$, $b = 7.2$

**c** $S = \dfrac{D}{T}$ when $D = 128$, $T = 3.4$

**5** Convert:

**a** 25 km to m **b** 2463 m to km

**c** 3 h 45 min to hours **d** 5.2 h to hours and minutes

**e** $7.\dot{3}$ h to hours and minutes **f** 20 km/h to m/s

**g** 110 km/h to m/s **h** 3.5 m/s to km/h.

**6** A plane moves at an average speed of 840 km/h for 1 day. How many kilometres does it travel?

**7** A car travels 245 km in 3 h 45 min. What is its average speed in kilometres per hour?

**8** Find, correct to the nearest cent:

**a** 14% of \$35.80 **b** 9.5% of \$26 580 **c** 12.5% of \$298.60.

# 3.01 Blood alcohol content (BAC)

**Worksheets**
BAC formula practice

Graphing BAC

Your **blood alcohol content (BAC)** is the concentration of alcohol in your blood. A person with a BAC of 0.05 has 0.05 grams (or 50 mg) of alcohol per 100 mL of blood. In NSW, it is illegal to drive with a BAC of 0.05 or above and if you have a learner or provisional licence, you must drive with a zero BAC.

When a person consumes alcohol, skills that safe driving requires become impaired and a driver will have difficulty concentrating on multiple tasks at a time. For example, a driver may be able to monitor their **speed** but be unable to stay within their lane.

Alcohol affects a driver's:

- **judgement** – ability to reason and respond appropriately
- **concentration** – ability to focus
- **comprehension** – ability to understand a situation quickly
- **reaction time** – ability to respond to a situation quickly.

Ralf Kleemann/Shutterstock.com

## Calculating BAC

A **standard drink**, for example a middy of beer or a small glass of wine, contains 10 grams of alcohol.

istock.com/270770

istock.com/Yarn

A person's BAC depends on 4 factors:

- the sex of the person
- the number of alcoholic drinks consumed
- how quickly the drinks are consumed
- the mass of the person.

**Calculating blood alcohol content**

$$BAC_{male} = \frac{10N - 7.5H}{6.8M} \text{ and } BAC_{female} = \frac{10N - 7.5H}{5.5M}$$

where $N$ is the number of standard drinks consumed
$H$ is the number of hours drinking
$M$ is the mass in kg

Blood accounts for about 7% of a person's mass so if we assume 1 kg ≈ 1 L then an 80 kg **male driver** will have 5.6 L of blood, whereas a 60 kg **female driver** has only 4.2 L of blood. The male has a greater mass to 'soak up' the alcohol and will not get drunk as quickly.

**Example 1**

Jackson and Britney were at a party for 5 hours. Jackson, who weighs 75 kg, had 6 standard drinks and Britney, who weighs 58 kg, had 5 standard drinks.

**a** Who had the lower blood alcohol content (BAC)?

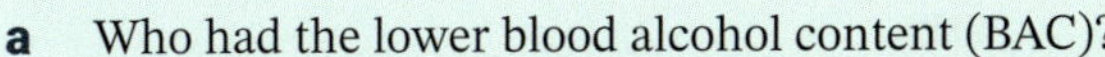

**b** If 7% of Jackson's mass is blood, what is his blood volume, correct to one decimal place?

**Solution**

**a** For Jackson: $N = 6, H = 5, M = 75$

$$BAC_{male} = \frac{10N - 7.5H}{6.8M} = \frac{10 \times 6 - 7.5 \times 5}{6.8 \times 75} = 0.0441\ldots \approx 0.044$$

For Britney: $N = 5, H = 5, M = 58$

$$BAC_{female} = \frac{10N - 7.5H}{5.5M} = \frac{10 \times 5 - 7.5 \times 5}{5.5 \times 58} = 0.0391\ldots \approx 0.039$$

Britney had the lower BAC.

**b** Jackson's blood volume = 7% × 75 = 5.25 L

## Limitations of the formulas

The BAC formulas only give an **approximate** BAC because many other factors affect BAC, such as:

- your health
- the time taken for your liver to process alcohol
- the size and type of drink
- how fast you drink (the faster you drink, the higher your BAC)
- alcohol is absorbed more slowly into the bloodstream if food is in your stomach
- the type of mixer used (water and juice slow absorption while carbonated mixers speed it up)
- the temperature of the drink (warm alcohol is absorbed quicker).

## Getting back to a BAC of zero

A healthy liver can only break down about 1 standard drink (10 g alcohol) per hour. Not even black coffee or cold showers will speed up the rate at which your body gets rid of alcohol. Once a person stops consuming alcohol and is at a certain BAC level, a normal body can only reduce the BAC by about 0.015 per hour.

### Returning to zero BAC

For a body that reduces BAC by 0.015 per hour, the number of hours before zero BAC is reached is:

$$\text{number of hours} = \frac{\text{BAC}}{0.015}$$

### Example 2

Jackson and Britney from Example 1 are both on provisional licences (P plates). They stopped drinking at 11 pm and waited until one of them had a BAC of zero and could legally drive home. Who drove home and at what time? Assume BAC is reduced by 0.015 per hour.

#### Solution

For Jackson: BAC = 0.044

$$\begin{aligned}\text{Number of hours} &= \frac{\text{BAC}}{0.015}\\ &= \frac{0.044}{0.015}\\ &= 2.9333\ldots\text{ h}\\ &= 2\text{ h }56\text{ min}\end{aligned}$$

For Britney: BAC = 0.039

$$\begin{aligned}\text{Number of hours} &= \frac{\text{BAC}}{0.015}\\ &= \frac{0.039}{0.015}\\ &= 2.6\text{ h}\\ &= 2\text{ h }36\text{ min}\end{aligned}$$

Press SHIFT ° ' " to change decimal hours to hours and minutes.

Britney drove home 2 h 36 min after 11 pm, that is, at 1:36 am.

## Graphing BAC

### Example 3

After drinking 5 standard drinks, Ben's BAC was 0.09. His BAC decreases over time according to the linear function $B = 0.09 - 0.014H$.

**a** Use the formula to complete this table of values, correct to 3 decimal places.

| Time after drinking, *H* hours | 0.0 | 0.5 | 1.0 | 1.5 | 2.0 | 2.5 | 3.0 | 3.5 | 4.0 | 4.5 | 5.0 |
|---|---|---|---|---|---|---|---|---|---|---|---|
| Blood alcohol content, *B* | 0.09 | | 0.076 | | | | | 0.041 | | | |

**b** Graph the linear function on a number plane.

**c** What is the gradient of the line and what does it represent?

**d** Use your graph to find:

**i** Ben's BAC after 3 h 15 min

**ii** when his BAC reaches 0.05.

**e** Use the formula to find when Ben's BAC reaches zero.

## Solution

**a**

| Time after drinking, *H* hours | 0 | 0.5 | 1 | 1.5 | 2 | 2.5 | 3 | 3.5 | 4 | 4.5 | 5 |
|---|---|---|---|---|---|---|---|---|---|---|---|
| Blood alcohol content, *B* | 0.090 | 0.083 | 0.076 | 0.069 | 0.062 | 0.055 | 0.048 | 0.041 | 0.034 | 0.027 | 0.020 |

**b**

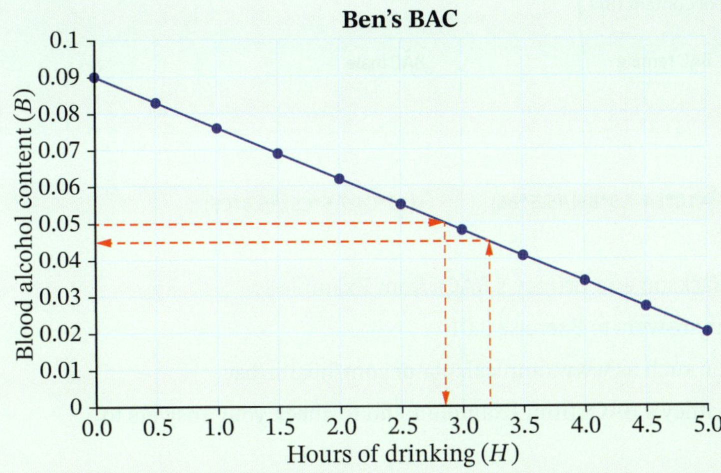


**c** Choosing 2 points on the line or from the table, (0, 0.09) and (5, 0.02):

Gradient $= \dfrac{0.02 - 0.09}{5 - 0} = -0.014.$

The gradient is the rate at which Ben's BAC is decreasing per hour.

**d** **i** BAC after 3.25 h ≈ 0.045    from graph

**ii** BAC reaches 0.05 after about 2.9 h (or 2 h 50 min)

**e** When $B = 0$:

$$0 = 0.09 - 0.014H$$
$$0.014H = 0.09$$
$$H = \frac{0.09}{0.014}$$
$$= 6.428\ldots$$
$$\approx 6\text{ h } 26\text{ min}$$

Ben's BAC reaches zero after 6 h 26 min.

## Did you know?

### Princess Diana's tragedy

In 1997, Princess Diana, the mother of Princes William and Harry, died in a car accident in a road tunnel in Paris, France. Hotel security manager Henri Paul was the driver of the Mercedes Benz that crashed into a pillar, killing himself, the princess and her partner Dodi Fayed. He had a BAC of 0.187, was driving at 105 km/h, and none of them were wearing seat belts. His BAC was over 3 times the legal limit of 0.05, equivalent to drinking 9 shots of whisky together quickly.

**If Henri Paul's BAC had dropped by 0.017 per hour, how long would it have taken his BAC to return to zero?**

## Technology

### BAC by spreadsheet and online calculator

You can use a spreadsheet or online calculator to calculate BAC.

**1** Enter the following labels into a blank spreadsheet, including the formulas in row 8.

| | A | B | C |
|---|---|---|---|
| 1 | Calculating Blood Alcohol Content (BAC) | | |
| 2 | | | |
| 3 | | BAC female | BAC male |
| 4 | *N* standard drinks | | |
| 5 | *H* hours drinking | | |
| 6 | *M* mass in kg | | |
| 7 | | | |
| 8 | BAC | =(10*B4-7.5*B5)/(5.5*B6) | =(10*C4-7.5*C5)/(6.8*C6) |
| 9 | | | |



**2** Use the spreadsheet to calculate Jackson and Britney's BACs from Example 1.

**3** Use the spreadsheet to check your answers to Exercise 3.01.

**4** Search online for a BAC calculator, such as www.omnicalculator.com/health/bac.

Use it to calculate Jackson and Britney's BACs from Example 1 and to check your answers to Exercise 3.01.

**5** If you are using the omnicalculator, find out:

**a** how long until a person is sober (BAC of zero) if their BAC is 0.06

**b** what is likely to happen to someone who has a BAC greater than 0.3

**c** 2 ways of achieving a BAC of zero.

**EXERCISE 3.01** Answers on p. 463

## Blood alcohol content (BAC)

**1** What is the main factor affecting your blood alcohol content? Select **A**, **B**, **C** or **D**.

**A** age

**B** number of drinks consumed

**C** time of day

**D** heart function

**2** Christine weighs 61 kg, Carla weighs 52 kg and Connie weighs 55 kg. If each drink 8 standard drinks over 6 hours, who has the highest BAC? Select **A**, **B**, **C** or **D**.

**A** Christine

**B** Carla

**C** Connie

**D** BAC is the same for all as they are all female

**3** Sonia and Maris drank wine at an 80th birthday party over 4 hours. Sonia, who weighs 94 kg, had 4 standard drinks and Maris, who weighed 87 kg, had 3 standard drinks.

**a** Who had the higher blood alcohol content, correct to 2 decimal places?

**b** If 7% of Sonia's weight was blood, how much blood did she have?

□ Foundation ○ Mastery ○ Complex

4 Who has the lowest blood alcohol content (BAC), correct to 2 decimal places? Select **A**, **B**, **C** or **D**.

**A** a 72 kg male who drank 8 drinks over 4 hours

**B** a 65 kg female who drank 9 drinks over 5 hours

**C** an 82 kg female who drank 7 drinks over 6 hours

**D** a 93 kg male who drank 10 drinks over 7 hours

5 This table shows the BAC for various body masses and number of drinks.

| | Number of drinks | | | | | | | | | |
|---|---|---|---|---|---|---|---|---|---|---|
| **Body mass (kg)** | **1** | **2** | **3** | **4** | **5** | **6** | **7** | **8** | **9** | **10** |
| **50** | 0.03 | 0.07 | 0.10 | 0.14 | 0.17 | 0.20 | 0.24 | 0.27 | 0.31 | 0.34 |
| **55** | 0.03 | 0.06 | 0.09 | 0.13 | 0.16 | 0.19 | 0.22 | 0.25 | 0.28 | 0.31 |
| **73** | 0.02 | 0.05 | 0.07 | 0.09 | 0.12 | 0.14 | 0.16 | 0.19 | 0.21 | 0.23 |
| **82** | 0.02 | 0.04 | 0.06 | 0.08 | 0.10 | 0.13 | 0.15 | 0.17 | 0.19 | 0.21 |
| **95** | 0.02 | 0.04 | 0.05 | 0.07 | 0.09 | 0.11 | 0.13 | 0.14 | 0.16 | 0.18 |
| **100** | 0.02 | 0.03 | 0.05 | 0.07 | 0.09 | 0.10 | 0.12 | 0.14 | 0.15 | 0.17 |

**a** How many drinks does a 73 kg person need to drink to reach the legal limit of 0.05?

**b** How many drinks does a 100 kg person need to drink to reach the legal limit of 0.05?

**c** What is the mass of a person who has a BAC of 0.16 after 7 drinks?

**d** What is the BAC of a 55 kg person after 8 drinks?

**e** Draw a line graph showing the BAC of a 50 kg person for 1 to 10 drinks.

6 A heavy drinker consumes a large quantity of alcohol most days of the week, whereas a moderate drinker consumes less alcohol less frequently. The 2 graphs show the difference in the decline of the BAC for a heavy drinker and a moderate drinker over time.

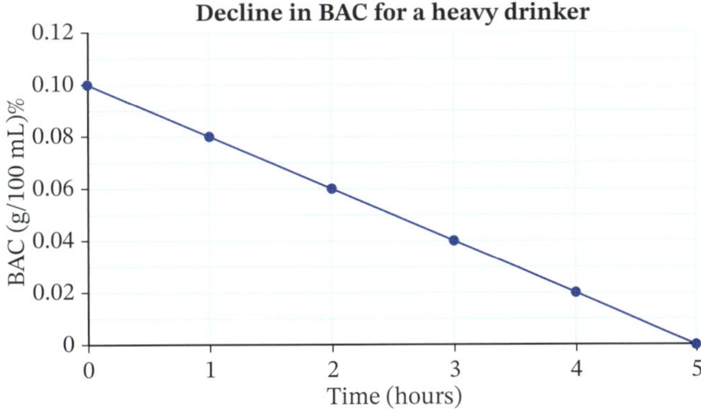


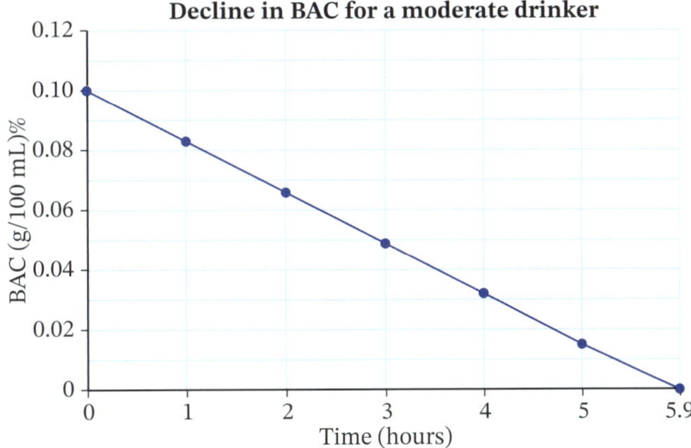


Foundation Mastery Complex

a Which drinker's BAC returns to zero quicker?

b What is the rate of decline of BAC for a:

i heavy drinker? ii moderate drinker?

c What is the BAC of a heavy drinker after 1 h 30 min?

d What is the BAC of a moderate drinker after 4 h 15 min?

e What is the difference in the BAC for a heavy drinker compared to a moderate drinker after 3 hours?

f Write 1–2 sentences describing the difference in the rate of BAC decrease for each type of drinker and possible reasons for the difference.

7 Roman is 19, weighs 72 kg and holds a provisional licence with a zero alcohol limit. He started drinking at 6 pm and had 10 schooners of full strength beer (15 standard drinks) over 6 hours. He stopped drinking at midnight.

a What was his BAC, to 2 decimal places, at midnight?

b How long did it take before his BAC was back to zero if his BAC reduced by 0.02 per hour?

c At what time could he legally drive his car?

EXAMPLE 2

8 Dasha is 17, weighs 61 kg and holds a provisional licence with a zero alcohol limit. She started drinking at 10 pm and had 6 mixer (9 standard) drinks over 4 hours.

a What was her BAC, to 2 decimal places, at 2 am when she stopped drinking?

b How long did it take before her BAC was back to zero if her BAC reduced by 0.015 per hour?

c At what time could she legally drive her car?

9 Hunter had a BAC of 0.06 when he stopped drinking and it took 5 hours for his BAC to return to zero. At what percentage rate per hour was his BAC reducing?

EXAMPLE 3

10 Chloe has a BAC of 0.08 and burns off alcohol at the rate of 0.016 per hour.

a Copy and complete this table of values showing the decline in her BAC over 5 hours.

| Time (hours) | BAC |
|---|---|
| 0.0 | 0.08 |
| 0.5 | 0.072 |
| 1.0 | |
| 1.5 | |
| 2.0 | |
| 2.5 | |
| 3.0 | |
| 3.5 | |
| 4.0 | |
| 4.5 | |
| 5.0 | |

b Use the table of values to draw a line graph representing this information.

c If Chloe stopped drinking at midnight, then estimate to the nearest 15 minutes, when Chloe's BAC is equal to:

i 0.05 ii 0.

d What is the gradient of the line and what does it represent?

☐ Foundation ○ Mastery ⬡ Complex

## Investigation

### BAC feelings and effects

| BAC | Person's feelings | Physical and mental effects |
|---|---|---|
| 0.03–0.05 | • relaxed<br>• talkative | • worse concentration and coordination<br>• less able to make a sound judgement |
| 0.05–0.1 | • blunted feelings<br>• lack of self-control | • bad reflexes<br>• unable to perceive depth<br>• decreased reasoning skills |
| 0.1–0.2 | • need to be over-expressive or act in a loud and disruptive manner | • decrease in reaction time<br>• loss of physical control<br>• slurred speech |
| 0.2–0.3 | • can't remember simple things<br>• in a daze | • loss of understanding<br>• loss of sensations<br>• possible loss of consciousness |
| Over 0.3 | • depressed<br>• think they can do anything | • can't read signs<br>• can't respond to traffic signals<br>• impaired breathing<br>• loss of bladder control<br>• unconsciousness or death |

For example, a person in a nightclub with a BAC of 0.15 might talk loudly, disrupt other patrons and express themselves in a way that is not typical of their personality. Their speech may be slurred and they may not be able to walk or dance properly.

**1** How does a person feel with a BAC of 0.24?

**2** State 3 effects of drinking to a BAC of 0.06.

**3** What is the legal BAC for:

**a** a learner driver? **b** a P-plate driver?

**4** Investigate the penalties for driving with a BAC of:

**a** 0.05 **b** 0.1 **c** 0.4

**5** Does a BAC below the legal limit mean that it is safe to drive? Explain.

## Did you know?

### BAC limits for drivers

The BAC limit in NSW (2026) for most drivers is 0.05, but lower if you are a less experienced driver or a driver of a heavy vehicle:

0.00 for all learner (L) and provisional (P) drivers

0.02 for drivers of buses, taxis, trucks over 13.9 tonnes or carrying dangerous goods

0.05 for all other drivers

Compared to a driver with a zero BAC, the risk of an accident with a BAC of 0.05 is double, 0.08 is 7 times and 0.15 is 20 times.

**How many drinks can a provisional driver have if they plan to drive? Justify your answer.**

## 3.02 Road accident statistics

**Worksheets**
Investigating drivers in road fatalities

Investigating young drivers

Alcohol, excessive speed, driver fatigue and distraction are the main causes of road accidents.

The risk of an accident resulting in death or serious injury is doubled for every 5 km/h you travel over 60 km/h. For example, a car travelling at 65 km/h is twice as likely to crash and a car travelling at 70 km/h is 4 times as likely to crash.

In a **fatal** accident, one or more persons die but there may also be others who suffer serious injuries or permanent disability.

Fatal means 'resulting in death'.

istock.com/vasiliki

### Example 4

This table shows road accidents by type and region in NSW in 2023.

| Region | Fatal crash | Serious injury | Moderate/minor injury | Total |
|---|---|---|---|---|
| City | 94 | 2496 | 5246 | 7836 |
| Country | 209 | 1822 | 3091 | 5122 |
| **Total** | **303** | **4318** | **8337** | **12 958** |

Data source: transport.nsw.gov.au

**a** What percentage (correct to one decimal place) of fatal crashes occurred in the city?

**b** What percentage of road accidents occurred in the country?

**c** What percentage of road accidents involved a serious injury or **fatality**?

**d** If 8% of fatal crashes involved alcohol, how many was this?

### Solution

**a** Percentage of fatal crashes in the city $= \frac{94}{303} \times 100\%$

$= 31.023...\%$

$\approx 31.0\%$

**b** Percentage of road accidents in the country $= \frac{5122}{12\,958} \times 100\%$

$= 39.5277...\%$

$\approx 39.5\%$

**c** Percentage of accidents involving serious injury or fatality $= \frac{303 + 4318}{12\,958} \times 100\%$

$= 35.661...\%$

$\approx 35.7\%$

**d** Number of fatal crashes involving alcohol $= 8\% \times 303$

$= 24.24$

$\approx 24$

## Example 5

This table gives the number of road fatalities in Australia over 10 years from 2014 to 2023.

| Year | 2014 | 2015 | 2016 | 2017 | 2018 | 2019 | 2020 | 2021 | 2022 | 2023 |
|---|---|---|---|---|---|---|---|---|---|---|
| Road fatalities | 1150 | 1209 | 1293 | 1225 | 1135 | 1194 | 1095 | 1123 | 1194 | 1266 |

Data source: bitre.gov.au (2024)

**a** For the 10-year period, find the mean and range of fatalities per year.

**b** Draw a column graph representing the data.

**c** Write a sentence discussing the trend in fatalities from 2014 to 2023. Suggest reasons for these trends.

### Solution

**a** $\text{mean} = \frac{\text{sum of fatalities}}{\text{number of years}}$

$= \frac{11\,884}{10}$

$= 1188.4$

$\text{range} = \text{highest value} - \text{lowest value}$

$= 1293 - 1095$

$= 198$

**b**

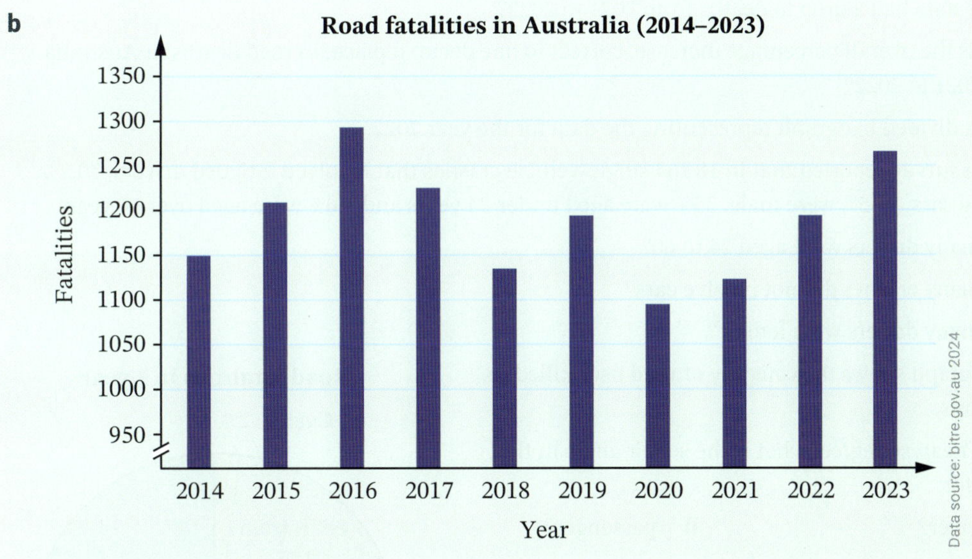


**c** The number of road fatalities rose and fell and rose again over the 10 years. It was lowest in 2020, probably due to the pandemic, which lead to there being fewer vehicles on the road.

**EXERCISE 3.02** Answers on p. 463

## Road accident statistics

**1** In one year, 440 driver deaths on Australian roads were related to alcohol. Of these driver deaths, 224 had a BAC of over 0.05 and 273 were aged under 34.

- **a** What fraction of drivers who died had a BAC over 0.05?
- **b** What percentage (to the nearest whole number) of the drivers were aged under 34?
- **c** What percentage of the drivers had a BAC of 0.05 or less?

EXAMPLE 4

**2** This table shows the number of Australian road user deaths by state and territory over 2 years.

| Year | NSW | Vic | Qld | SA | WA | Tas | NT | ACT |
|---|---|---|---|---|---|---|---|---|
| 2021 | 275 | 231 | 277 | 99 | 166 | 35 | 35 | 11 |
| 2022 | 293 | 242 | 297 | 71 | 175 | 51 | 47 | 18 |

Data source: bitre.gov.au 2024

- **a** How many deaths were there in Australia in:
  - **i** 2021?
  - **ii** 2022?
- **b** Out of all fatalities over the 2 years, what percentage, correct to one decimal place, occurred in NSW?
- **c** Which state or territory has the fewest deaths? Why do you think this was the case?
- **d** Which state had a drop in deaths from 2021 to 2022?
- **e** What is the overall percentage increase, correct to one decimal place, in road deaths in Australia from 2021 to 2022?
- **f** Draw a divided bar graph representing the data for the year 2022.

**3** An accident survey reported that in 18 764 single-vehicle crashes that involved fatigued drivers, 70.2% were driving cars, 75.5% were male, 35% were aged under 24 years and 9.6% were aged over 60 years.

- **a** How many drivers were aged 24 to 60?
- **b** How many crashes did not involve cars?
- **c** How many drivers were female?

**4** This sector graph shows the category of road user killed in road accidents.

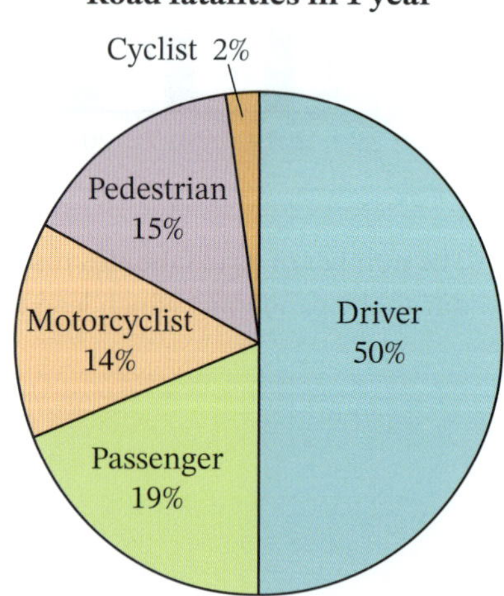


- **a** To the nearest degree, what is the sector angle in the graph for:
  - **i** driver?
  - **ii** passenger?
  - **iii** cyclist?
- **b** If there were 443 road fatalities last year, how many of the people killed were:
  - **i** passengers?
  - **ii** motorcyclists?
- **c** If there were 68 pedestrians killed this year, how many drivers were killed?

☐ Foundation ○ Mastery ○ Complex

5 This table shows the ages of drivers who died in road accidents in Australia over 4 years.

| | Year | | | |
|---|---|---|---|---|
| **Age of driver** | **2019** | **2020** | **2021** | **2022** |
| **Under 17** | 2 | 2 | 7 | 3 |
| **17–25** | 105 | 109 | 113 | 116 |
| **26–39** | 134 | 151 | 127 | 125 |
| **40–64** | 198 | 148 | 167 | 190 |
| **65–74** | 46 | 47 | 49 | 52 |
| **Over 75** | 84 | 77 | 65 | 68 |

Data source: Road Trauma Australia, 2024 statistical summary, BITRE, © Commonwealth of Australia

**a** Find the number of drivers aged 25 or less who were killed over the 4 years.

**b** Calculate, correct to one decimal place, the percentage of drivers killed in 2022 who were 17–39 years old.

**c** What percentage, correct to one decimal place, of drivers killed in 2021 were aged 65 or over?

**d** Which adult age group had the least fatalities over the 4 years? Give a possible reason.

**e** Draw a column graph showing the data for 2022.

6 This table shows the number of Australian road injuries involving different categories of road users over 5 years.

| Year | Drivers | Passengers | Pedestrians | Motorcyclists | Pedal cyclists | All road users |
|---|---|---|---|---|---|---|
| 2018 | 519 | 205 | 178 | 191 | 35 | 1135 |
| 2019 | 569 | 205 | 158 | 212 | 39 | 1186 |
| 2020 | 534 | 192 | 138 | 187 | 41 | 1097 |
| 2021 | 530 | 182 | 133 | 237 | 41 | 1129 |
| 2022 | 555 | 182 | 164 | 244 | 35 | 1194 |

Data source: Road Trauma Australia, 2024 statistical summary, BITRE, © Commonwealth of Australia

**a** Which year and category had the highest number of road injuries?

**b** How many road injuries were there in Australia over the 5 years?

**c** What is the mean annual number of injuries involving motorcyclists over the 5 years?

**d** What is the range of passenger injuries over the 5 years?

**e** Represent the data for pedestrians on a line graph.

7 This table shows the number of Australian single-vehicle and multi-vehicle crashes over 6 years.

| Year | Single-vehicle crashes | Multiple-vehicle crashes |
|---|---|---|
| 2017 | 659 | 564 |
| 2018 | 658 | 477 |
| 2019 | 701 | 485 |
| 2020 | 603 | 494 |
| 2021 | 641 | 488 |
| 2022 | 726 | 468 |

Data source: Road Trauma Australia, 2024 statistical summary, BITRE, © Commonwealth of Australia

**a** Calculate for single-vehicle crashes the mean and range of crashes over the 6 years.

**b** Why do you think there are more single-vehicle crashes than multiple-vehicle crashes?

**c** Looking at the data for multi-vehicle crashes, describe any general trends over the 6 years and possible reasons for these.

☐ Foundation ○ Mastery ○ Complex

**8** This clustered column graph compares male and female road fatalities over 5 years.

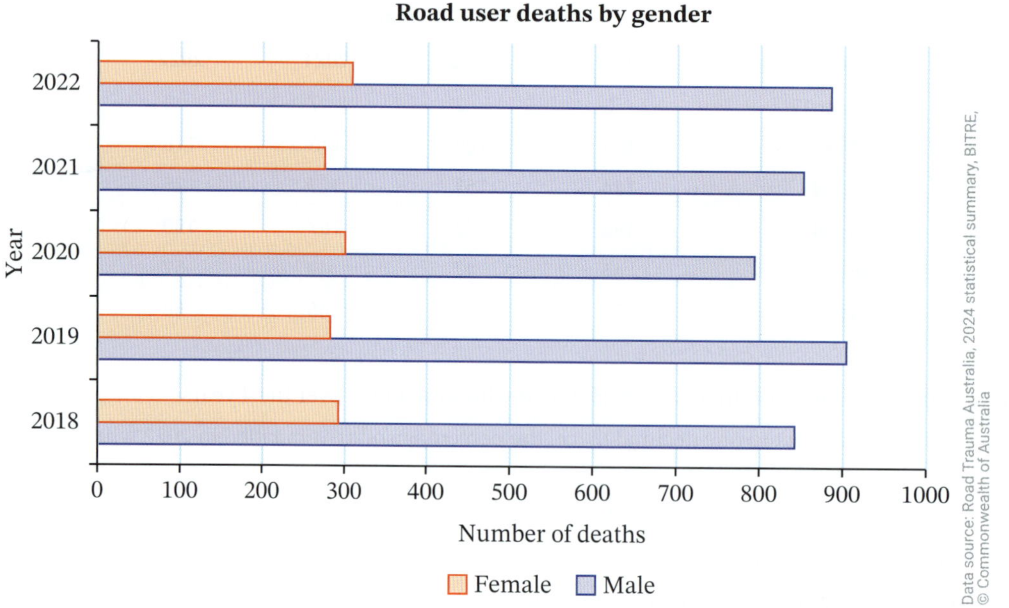


**a** How many male deaths, to the nearest 10, occurred in 2019?

**b** How many female deaths, to the nearest 10, occurred in 2021?

**c** How many road user deaths, to the nearest 10, occurred in 2022?

**d** Describe any patterns in the data over the years and any differences between numbers of deaths of males and females, stating possible reasons for these.

**9** This table shows the times of day NSW road casualties occurred one year.

| Time of day | Road casualties |
|---|---|
| 12 midnight–1:59 am | 329 |
| 2 am–3:59 am | 233 |
| 4 am– 5:59 am | 380 |
| 6 am–7:59 am | 1093 |
| 8 am–9:59 am | 1477 |
| 10 am–11:59 am | 1473 |
| 12 midday–1:59 pm | 1533 |
| 2 pm–3:59 pm | 1910 |
| 4 pm–5:59 pm | 1862 |
| 6 pm–7:59 pm | 1244 |
| 8 pm–9:59 pm | 859 |
| 10 pm–11:59 pm | 595 |

Data source: Centre for Road Safety, Transport for NSW, State of NSW (2023)

**a** Which time interval had the highest number of casualties? Give a reason why this may be so.

**b** Which time interval had the lowest number of casualties? Give a reason why this may be so.

**c** True or false: The number of road casualties after midday is more than double the number of casualties before midday.

**d** **i** Describe the pattern in the number of road casualties over the 24-hour period.

**ii** Do you think the pattern would still be the same today? Give reasons.

**10** This table shows the age and sex of speeding drivers involved in road accidents in NSW one year.

| Age | Male drivers | Female drivers |
|---|---|---|
| 17–20 | 313 | 91 |
| 21–25 | 265 | 64 |
| 26–29 | 159 | 57 |
| 30–39 | 343 | 73 |
| 40–49 | 225 | 68 |
| All ages | 1753 | 493 |

Data source: Centre for Road Safety, Transport for NSW, State of NSW (2023)

- **a** How many speeding drivers were aged 50 years or older?
- **b** How many speeding female drivers were aged under 30?
- **c** What percentage of speeding drivers aged 17–20 were male? Answer correct to 3 significant figures.
- **d** What percentage (correct to 2 decimal places) of female speeding drivers were aged 17–20?
- **e** What patterns do you notice between the numbers of male and female speeding drivers? Why do you think this is so?
- **f** What patterns do you notice in the number of speeding drivers as age increases? Why do you think this is so?

## Did you know?

### Driver fatigue

**Fatigue** is mental or physical tiredness due to lack of sleep. The only way to overcome fatigue is to sleep. Most fatigue-related road fatalities occur between midday and 6 pm, not at night. Fatigue is a factor in about 20% of head-on collisions, where a vehicle crosses to the wrong side of the road due to a driver's slower reaction time or 'microsleeping'.

Around 20–30% of road fatalities involve driver fatigue. Drivers in accidents are 4 times more likely to have been affected by fatigue than by drugs or alcohol.

Kathie Nichols/Shutterstock.com

# 3.03 Speed, distance and time

**Worksheets**
Speed formula practice

Speed stories

Distance, speed and time

Speed limits are imposed to ensure drivers travel at a safe speed for the road and surrounding environment. Some common speed limits in NSW are:

| | |
|---|---|
| School zone | 40 km/h |
| Residential area | 50 km/h |
| City street | 60 km/h |
| Highway | 100 km/h |
| Motorway | 110 km/h |

**Average speed**

$$\text{average speed} = \frac{\text{distance travelled}}{\text{time taken}}$$

$$S = \frac{D}{T}$$

This formula can also be written as $D = ST$ or $T = \frac{D}{S}$.

**Video**
Speed, distance and time

**Example 6**

Dilshan left home at 6 am to travel from Sydney to Brisbane, a distance of 1027 km. He stopped for meal breaks and fuel and arrived in Brisbane at 8:30 pm.

**a** What was his average speed for the trip, correct to the nearest km/h?

**b** How far, to nearest kilometre, could he travel at this speed in 50 minutes?

**c** How long, in hours and minutes, would it take him to drive 864 km at this speed?

**Solution**

**a** $D = 1027$ km; $T = 14.5$ h (6 am to 8:30 pm)

$$S = \frac{D}{T}$$
$$= \frac{1027}{14.5}$$
$$= 70.8275 \ldots$$
$$\approx 71 \text{ km/h}$$

Average speed was 71 km/h.

**b** $S = 71$ km/h; $D = ?$; $T = 50 \text{ min} = \frac{50}{60} \text{ h} = \frac{5}{6} \text{ h}$

$$D = S \times T$$
$$= 71 \times \frac{5}{6}$$
$$= 59.166 \ldots$$
$$\approx 59 \text{ km}$$

He could travel 59 km.

**c** $T = \frac{D}{S}$

$$= \frac{864}{71}$$
$$= 12.169 \ldots$$
$$= 12 \text{ h } 10 \text{ min } 8.45 \text{ s}$$
$$\approx 12 \text{ h } 10 \text{ min}$$

(Press [°′″] to convert decimal hours to hours, minutes and seconds.)

It would take him 12 h 10 min.

**EXERCISE 3.03** Answers on p. 464

## Speed, distance and time

EXAMPLE 6

**1** Eva drove 93 km from Wollongong to Nowra in 1 hour 30 min. What was her average speed in km/h? Select **A**, **B**, **C** or **D**.

**A** 46.5  **B** 31  **C** 62  **D** 60

**2** Stefan travelled 160 km at an average speed of 76 km/h. How long was his journey in hours and minutes? Select **A**, **B**, **C** or **D**.

**A** 2 h 10 min  **B** 2 h 6 min  **C** 2.2 h  **D** 2.1 h

**3** Larni drove 742 km from home to her farm in 9 h 30 min.

- **a** What was her average speed, correct to the nearest km/h?
- **b** How long, in hours and minutes, would it take her to drive 1026 km at this rate?
- **c** How far, to the nearest kilometre, could she travel at this rate in 3 h 10 min?

**4** Cathy drove at an average speed of 57 km/h. At this rate, how long, in hours and minutes, will it take her to drive:

**a** 507 km?  **b** 160 km?  **c** 440 km?

**5** Tan cycled 24 km to Scott's place at an average speed of 16 km/h. He cycled back at 12 km/h. Find:

- **a** the total time for Tan's complete trip
- **b** Tan's average speed, to the nearest km/h, for the trip.

**6** Justine's car journey took 9 h 45 min at an average speed of 9.5 m/s. What is her:

- **a** average speed in km/h?
- **b** distance travelled, correct to the nearest metre?

**7** How far, in metres to one decimal place, will a vehicle travel in 0.1 seconds when travelling at:

**a** 40 km/h?  **b** 60 km/h?  **c** 80 km/h?  **d** 100 km/h?

**8** Courtney drove a racecar for 402 m at 530 km/h.

- **a** How long, to the nearest second, did it take?
- **b** How far (correct to the nearest metre) could she travel in 10 seconds at this speed?

**9** Luis rode his motorbike 460 m in 6.011 seconds.

- **a** What was his speed in m/s, correct to one decimal place?
- **b** How long, to the nearest 0.1 minute, would it take to go 58 km at this rate?

**10**
- **a** Casey's average speed was calculated at 208.33 km/h for a 60.72 km motorbike race. How long, to the nearest 0.1 minute, did he take?
- **b** If he took the same time to complete a 72.4 km race, what was his average speed in km/h for this race, correct to the nearest km/h?

**11** Shellie and Roxy set off for the beach 11.5 km away. Shellie rides her motor scooter at 32 km/h and Roxy cycles at 11 km/h. How many minutes head start should Shellie give Roxy so they arrive at the beach at the same time?

**12** A car is travelling at 75 km/h. How far, to the nearest 0.1 m, will it travel in the 5 seconds it takes you to cross the road?

**13** Police investigating a car crash determined that the car covered 20.7 m in 1.2 s. At what speed was the car travelling in km/h?

# 3.04 Stopping distance

## See, think, brake

Imagine you are driving and you see a child run out onto the road in front of you. You hit the brakes to avoid an accident. What is the time and distance from when you sense the danger to when you come to a full stop?

### Stopping distance

- **Reaction time** is the time between sensing a situation and applying the brakes.
- **Reaction distance** is the distance travelled during your reaction time.
- **Braking distance** is the distance travelled from when you apply the brakes until you stop.
- **Stopping distance** = reaction distance + braking distance

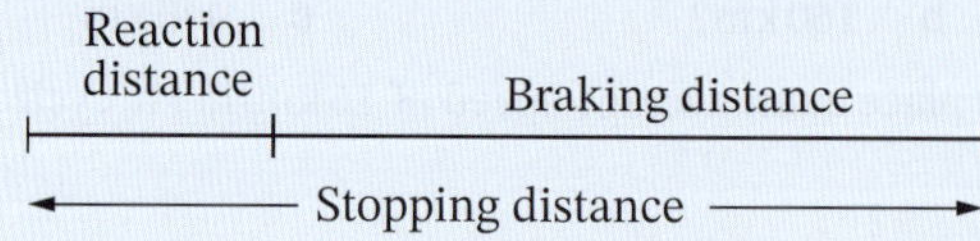


## Braking distance

In dry conditions, a car travelling at 60 km/h can stop 20 m before a car travelling at 80 km/h.

Drivers tend to underestimate stopping distance because the time taken to stop depends on many factors:

- your speed
- your reaction time
- the condition of your brakes and tyres
- the rate at which you decelerate
- the road surface and slope
- the wind speed and weather
- the weight of your vehicle
- driver's vision, including whether it is night or day.

Amy Johansson/Shutterstock.com

### Braking distance formula

The approximate braking distance ($d$ m) can be found using the formula:

$$d = kv^2$$

where $v$ km/h is the speed of the car and $k$ is a constant.

## Example 7

Ben has a reaction time of 1.5 s. He was driving at 60 km/h when he saw a tree had fallen across the road. He applied his brakes and stopped 17.4 m later.

**a** What was his reaction distance, to the nearest metre?

**b** What was his stopping distance?

**c** Use the braking distance formula $d = kv^2$ and the given values to find:

**i** the value of $k$ (correct to 2 significant figures)

**ii** the braking distance, to the nearest metre, when travelling at 84 km/h

**iii** the stopping distance, to the nearest metre, when travelling at 84 km/h

**iv** the speed, to the nearest km/h, of a vehicle with a braking distance of 100 m.

## Solution

**a** Reaction distance is the distance the car travels in 1.5 s at 60 km/h.

Convert 60 km/h to m/s first.

$60\text{ km/h} = 60\,000\text{ m/h}$ — converting km to m

$= \frac{60\,000}{3600}\text{ m/s}$ — converting h to s

$= 16.666\ldots\text{ m/s}$

$d = 16.666\ldots \times 1.5$

$= 25\text{ m}$

Reaction distance is 25 m.

**b** Stopping distance = reaction distance + braking distance

$= 25 + 17.4$

$= 42.4\text{ m}$

**c** **i** Given $d = 17.4$ m and $v = 60$ km/h,

$d = kv^2$

$17.4 = k \times 60^2$

$= 3600k$

$k = \frac{17.4}{3600}$

$= 0.004\,8333\ldots$

$k \approx 0.0048$

**ii** Given $v = 84$ km/h and $k = 0.0048$,

$d = kv^2$

$= 0.0048 \times 84^2$

$= 33.8688$

$d \approx 34$

Braking distance is 34 m.

**iii** Reaction distance $= \frac{84\,000}{3600} \times 1.5$

$= 35\text{ m}$

Stopping distance = reaction distance + braking distance

$= 35 + 34$

$= 69\text{ m}$

Stopping distance is 69 m.

**iv** Given $d = 100$ m and $k = 0.0048$,

$$d = kv^2$$
$$100 = 0.0048v^2$$
$$v^2 = \frac{100}{0.0048}$$
$$= 20833.333\ldots$$
$$v = \sqrt{20\,833.333\ldots}$$
$$= 144.3375\ldots$$
$$v \approx 144$$

Speed is 144 km/h.

## Speed and road conditions

The braking distance changes with the condition of the road. It will take longer to stop on a wet road and even longer on an icy road.

### Example 8

This table shows the different stopping distances at 4 different speeds and 2 different road conditions for a driver whose reaction time is 2 s.

| Stopping distances for various speeds and road conditions | | | | | | |
|---|---|---|---|---|---|---|
| Speed (km/h) | Road | Reaction time (s) | Reaction distance (m) | Braking distance (m) | Stopping distance (m) | The equivalent of |
| 40 | dry | 2 | 22.2 | 7.9 | 30.1 | |
| 40 | wet | 2 | 22.2 | 12.6 | 34.8 | |
| 50 | dry | 2 | 27.8 | 12.3 | 40.1 | 8 car lengths |
| 50 | wet | 2 | 27.8 | 19.7 | 47.5 | 4 bus lengths |
| 60 | dry | 2 | 33.3 | 17.7 | 51.0 | Olympic swimming pool |
| 60 | wet | 2 | 33.3 | 28.3 | 61.6 | 3 cricket pitches |
| 100 | dry | 2 | 55.6 | 49.2 | 104.8 | length of a football field |
| 100 | wet | 2 | 55.6 | 78.7 | 134.3 | width of 10 houses |

**a** What do you notice about the reaction distances?

**b** Discuss the difference in braking distance for a car travelling at:

  **i** 40 km/h on a wet road and 50 km/h on a dry road

  **ii** 50 km/h on a wet road and 100 km/h on a wet road.

**c** Discuss the difference in stopping distance for a car travelling at:

  **i** 40 km/h on a dry road and 40 km/h on a wet road

  **ii** 60 km/h on a dry road and 100 km/h on a dry road.

**d** Why should you reduce speed when travelling on a wet road?

**e** Draw a clustered column graph to represent the data and say why this is a good representation of the data.

## Solution

**a** The reaction distances increase with speed but do not depend on the road condition.

**b** **i** There is a 10 km/h difference in speed but the braking distances are about the same.

**ii** The speed is doubled but the braking distance is roughly 4 times greater.

**c** **i** At 40 km/h there is a 5 m difference in the stopping distance between a dry and wet road and this difference is significant.

**ii** There is a 40 km/h difference in speed but the stopping distance is doubled: 51.0 m and 104.8 m.

**d** You take longer to stop on a wet road so you should reduce your speed to avoid any incidents.

**e** The height of the columns shows clearly that as the speed increases, the difference between the stopping distances on dry and wet roads is greater.

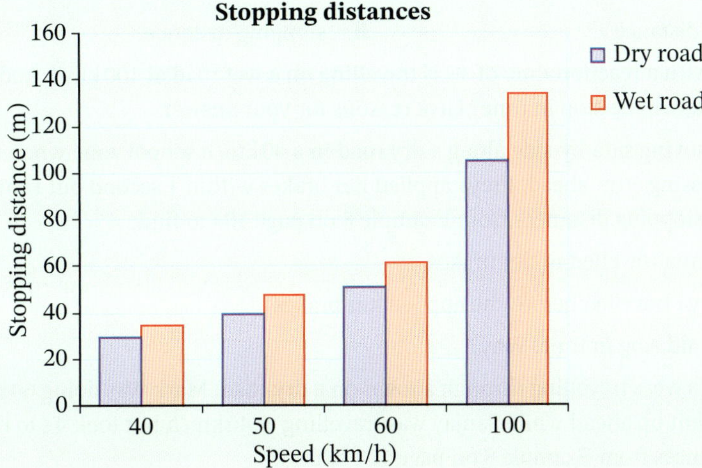


**EXERCISE 3.04** Answers on p. 464

### Stopping distance

**1** Davuth saw a car stopped 80 m in front of him. He just stopped in time after travelling 51 m while braking. What was his reaction distance? Select **A**, **B**, **C** or **D**.

**A** 29 m  **B** 80 m  **C** 51 m  **D** 30 m

**2** A van moving at 110 km/h travelled a distance of 118.4 m on a wet road before its driver applied the brakes and finally stopped another 79.3 m later. What was the van's stopping distance?

**3** A motorcyclist with a quick reaction time stopped in 15.6 m after braking for a distance of 9.8 m. How far did he travel during the time it took to react to the danger?

4 This table shows the reaction, braking and stopping distances for a driver on a wet road with a reaction time of 1 second.

| Speed (km/h) | Reaction time (s) | Reaction distance (m) | Braking distance (m) | Stopping distance (m) |
|---|---|---|---|---|
| 40 | 1 | 11.1 | B | 23.7 |
| 50 | 1 | A | 19.7 | 33.6 |
| 60 | 1 | 16.7 | C | 45.0 |
| 100 | 1 | 27.8 | 78.7 | D |

**a** Find the values of A, B, C and D.

**b** When the speed is doubled from 50 km/h to 100 km/h, find to the nearest whole number the percentage increase in:

**i** braking distance **ii** stopping distance.

**c** If a driver with a reaction time of 1 s is travelling on a wet road at 100 km/h and sees an obstacle 110 m ahead, will he stop in time? Give reasons for your answer.

EXAMPLE 8

5 Two cars were moving side-by-side along a dry road in a 40 km/h school zone when their drivers each saw children crossing 50 m ahead. Freya applied her brakes within 1 second but Hanna took 2 seconds. Use the table of stopping distances from Example 8 on page 104 to find:

**a** how far Hanna travelled under brakes

**b** how far Freya travelled before she applied her brakes

**c** if either would stop in time. Why?

6 Two truck drivers were travelling through a town on a dry road. Mark was doing 60 km/h and took 2 s to react to an accident up ahead while Sanjay was travelling at 40 km/h but took 4 s to react. Use the table of stopping distances from Example 8 on page 104 to find:

**a** how far Sanjay travelled before the brakes were applied

**b** how far Mark travelled under brakes

**c** which driver had the greater stopping distance.

7 Fatima has a reaction time of 1.6 s. When travelling at 80 km/h, she applied her brakes and travelled 62 m before stopping. Use the braking distance formula $d = kv^2$ to find:

**a** the constant $k$, correct to 4 decimal places

**b** the braking distance, to the nearest 0.1 m, when travelling at 105 km/h

**c** the speed (to the nearest km/h) of Fatima if she travels 75 m under brakes before stopping

**d** the stopping distance (to the nearest 0.1 m) from a speed of:

**i** 80 km/h **ii** 105 km/h.

8 A car travelling at 115 km/h covered 89.6 m under brakes. Use the formula $d = kv^2$ to find:

**a** the constant $k$, correct to 3 significant figures

**b** the braking distance, correct to one decimal place, if the initial speed is 95 km/h

**c** the stopping distance for a driver with a reaction time of 2 s who is travelling at 115 km/h.

9 Use the formula $d = 0.02754v^2$, where $v$ km/h is the speed of a car and $d$ m the distance the car travels under brakes, to find (correct to 3 significant figures) the braking distance of a vehicle travelling at:

**a** 35 km/h **b** 61 km/h **c** 112 km/h **d** 93 km/h.

10 A car cruising at 80 km/h travelled 41.8 m on an icy road during the time it took for the driver to see a stop sign and start braking. It then travelled 57.2 m under brakes before it stopped.

**a** What was the driver's reaction time in seconds, correct to 2 decimal places?

**b** What was the stopping distance?

**11** Use the formula $d = 0.00435v^2$, where $d$ m is the distance travelled under brakes and $v$ km/h is the speed as brakes are applied, to find the speed of a racecar, correct to the nearest km/h, with a braking distance of:

**a** 142 m **b** 91.2 m **c** 101.4 m **d** 68.9 m

**12** A car travelled 84.3 m under brakes when initially travelling at 106.3 km/h. Use the braking formula $d = kv^2$ to find (correct to 3 significant figures):

**a** the constant $k$

**b** the braking distance for a car applying brakes from a speed of 110 km/h

**c** the stopping distance of a car travelling at 110 km/h if the driver's reaction time is 2 s.

**13** The formula $d = \frac{1}{2}x^2 + 5x$, where $x$ km/h is the speed of a vehicle and $d$ m the distance travelled, is only used for very fast cars. How many metres can a racing car travel at a speed of:

**a** 100 km/h? **b** 250 km/h? **c** 300 km/h?

**14** The graph shows the stopping distances on wet and dry roads with various reaction times.

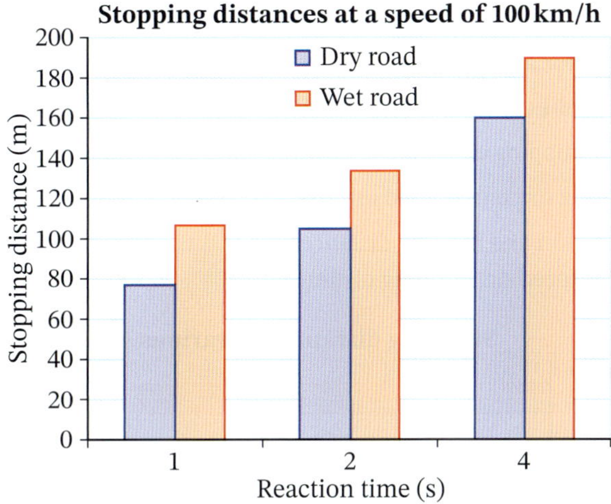


**a** Hudson driving on a dry road has a reaction time of 4 s. How far will he travel before he stops?

**b** How many metres does it take for a car to stop when travelling at 100 km/h on a wet road if the driver has a reaction time of:

**i** 1 s? **ii** 2 s? **iii** 4 s?

**c** Use the graph to copy and complete this table.

| Road condition | Reaction time | Stopping distance |
|---|---|---|
| | | |
| | | |
| | | |
| | | |
| | | |
| | | |

**d** Compare the benefits of showing data in a table with data represented on a graph.

☐ Foundation ○ Mastery ⬡ Complex

## Investigation

### Stopping distances

**1** **a** Use this table to describe the relationship between road conditions, speed and stopping distance, for a driver with a reaction time of 1.5 s. All values in the table are rounded to one decimal place.

| Road condition | Speed (km/h) | Reaction distance (m) | Braking distance (m) | Stopping distance (m) |
|---|---|---|---|---|
| Dry | 60 | 25.0 | 17.4 | 42.4 |
| | 100 | 41.7 | 48.2 | 89.9 |
| Wet | 60 | 25.0 | 27.8 | 52.8 |
| | 100 | 41.7 | 77.2 | 118.8 |
| Icy | 60 | 25.0 | 138.9 | 163.9 |
| | 100 | 41.7 | 358.8 | 427.5 |

This table of values, to one decimal place, was found using an online calculator.

**b** Find an online calculator and use it to determine the stopping distance for:

**i** a car travelling at 60 km/h on an icy road if the driver has a reaction time of 1.6 s

**ii** the difference in stopping distances at 80 km/h for reaction times of 0.9 s and 1.4 s

**iii** a car travelling on a dry or wet road at 110 km/h, if reaction time is 1.5 s.

**2** This graph shows the reaction and braking distances for cars travelling at various speeds. Use the graph to investigate the effect of speed on stopping distances.

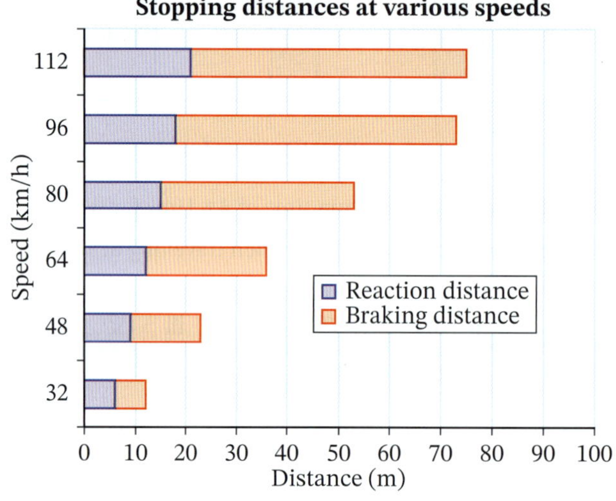


Give one trend that you have discovered by looking at stopping distances with road conditions and speed.

## Did you know?

### Will you stop in time?

These graphs were published by the Australian Federal Police to display stopping distances for different speeds and road conditions. If you are driving at 60 km/h in a 60 km/h zone and a child runs onto the road 45 metres ahead and you brake hard, will you stop in time?

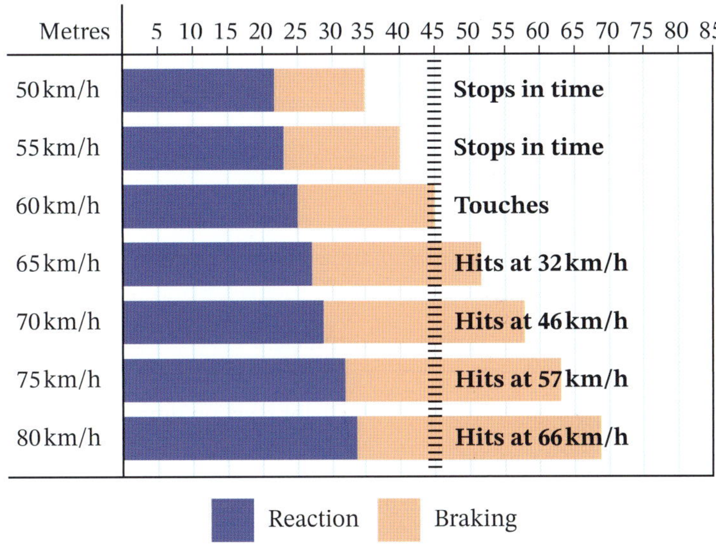


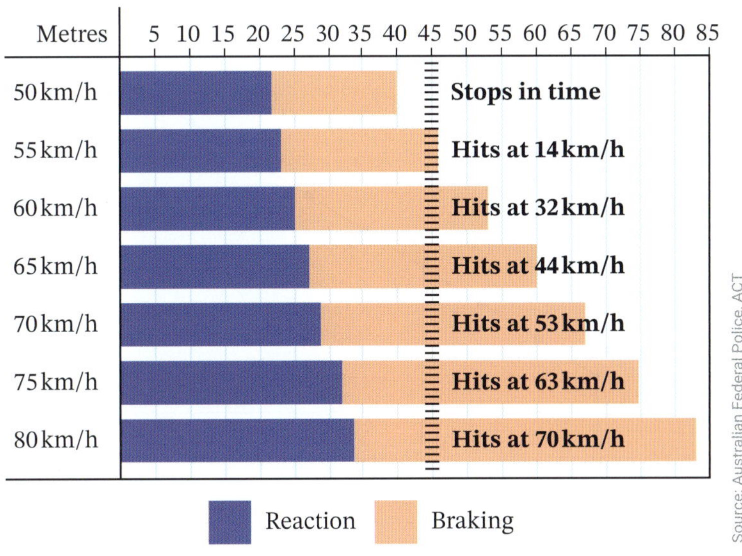


Source: Australian Federal Police, ACT

## Investigation

### Reaction time tests

1 Use an internet search engine to find a **Reaction time test** where the times you take to click the button when the light comes on are recorded and the average reaction time calculated. Work in groups of 4 or 5 and decide who has the best average reaction time.

2 Work in pairs to measure your 'reaction distance'.

- Your partner holds the top of a 30 cm ruler, at the 30 cm end.
- You sit on a chair with your thumb and forefinger around the bottom of the ruler, at the 0 cm end.
- Your partner drops the ruler at a random time.
- You catch the ruler and record the reaction distance.
- Repeat 5 times and find your average reaction distance.
- Swap places with your partner and repeat.

Which student in the class has the fastest reaction time (shortest reaction distance)?

What is the class average?

3 Name 3 activities where a fast reaction time is an advantage.

## Sample HSC problem Answers on p. 464

(4 marks) This graph shows the blood alcohol content (BAC) of a 55 kg woman according to the number of drinks she has had.

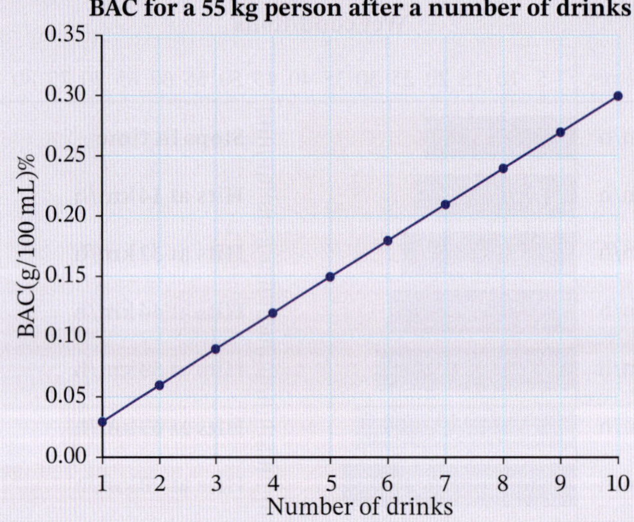


**a** Bianca is 55 kg. Estimate her BAC after 5 drinks. 1 mark

**b** How many drinks will cause her to reach a BAC beyond 0.1? 1 mark

**c** When BAC reaches 0.3, a person usually becomes unconscious. After how many drinks will this occur? 1 mark

**d** If Bianca consumes 8 drinks, how many hours will it take her to return to a BAC of zero? 1 mark

$$\left(h = \frac{\text{BAC}}{0.015}\right)$$

## Study Tip

### When and where to study

Are you an early bird or a night owl? Do you work best in the morning, in the afternoon or at night? Identify your **peak performance period** and study during this time each day, so that it becomes a habit. There will be times when you can't study due to other commitments, but it's important to develop a **regular routine** so that your family and friends know when you should not be interrupted.

Students often ask, 'How many hours should I study each night?', but it is the **quality** of study that is more important than the **quantity**. It is a question of the effort and commitment required, not the number of hours. The amount of time you should study is the amount of time necessary for you to fulfil all your study tasks and demands. Some students like to do the same amount each day, some do more or less on weekends, or have one day that is study-free. Find out what works best for you and stick to it.

Students often study by themselves in their rooms or at the library. Some prefer company and use the dining table, because they like the noise and space. Others like to sit outside in the fresh air. You can choose different places for different types of homework.

A good study place has:

- plenty of space to spread out work, such as on a big desk
- minimal noise, few distractions and interruptions
- good lighting and ventilation, and is neither too hot nor too cold
- comfortable and supportive seating.

Study is a serious business. You will be concentrating for a while, so choose a place where you won't be easily distracted. Use a good desk lamp and open the window to prevent sore eyes and drowsiness.

# CHAPTER SUMMARY

This chapter, *Driving safely*, looked at the mathematics behind safe driving, covering the areas of measurement (blood alcohol, speed and stopping distance), algebra (formulas) and data analysis (road accident statistics).

Before you move on, consider what you learned in this chapter and revisit any sections that may have been unclear.

| What you learned in this chapter... | Section |
|---|---|
| Use formulas and online calculators to calculate blood alcohol content (BAC) for males and females, based on number of drinks consumed, number of hours drinking and the person's mass | 3.01 Blood alcohol content (BAC) |
| Interpret BAC and solve problems involving reducing BAC | 3.01 Blood alcohol content (BAC) |
| Analyse data and graphs involving BAC and road accidents | 3.01 Blood alcohol content (BAC)<br>3.02 Road accident statistics |
| Construct and interpret graphs that illustrate the level of blood alcohol over time | 3.01 Blood alcohol content (BAC) |
| Solve problems involving speed, distance and time | 3.03 Speed, distance and time |
| Use formulas to calculate stopping distance | 3.04 Stopping distance |

To help master this topic, make a summary mind map. Use the chapter outline and the mind map below as a guide. Add your own words, symbols, diagrams, boxes and reminders. The summary should give you a 'whole picture' view of the topic and allow you to identify any weak areas to revisit in your revision.

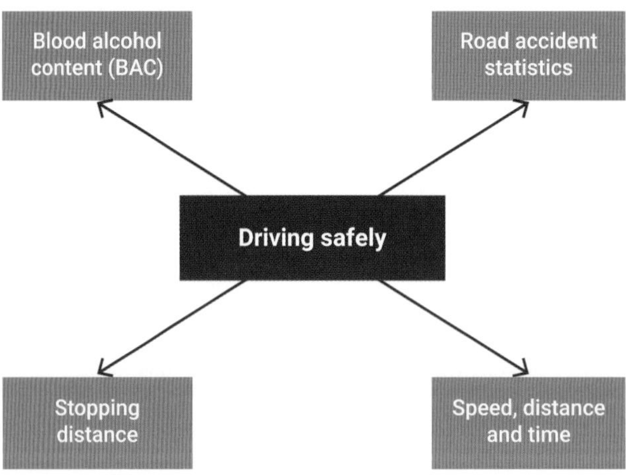

# Test yourself 3

Answers on p. 464

**1** Jennifer weighs 56 kg and drank 3 glasses of wine at a party over 3 hours. She thought she was safe to drive but when pulled over and tested, she was over the legal limit of 0.05. She later discovered that each wine glass contained $1\frac{1}{2}$ standard drinks. 3.01

Use the formulas $BAC_{female} = \frac{10N - 7.5H}{5.5M}$ and number of hours $= \frac{BAC}{0.015}$ to answer the following questions.

- **a** What was her BAC, correct to 3 decimal places, when she was tested?
- **b** How long, in hours and minutes, did it take before her BAC was back to zero?
- **c** What would her BAC be, correct to 3 decimal places, after 3 standard drinks in 3 hours?

**2** This table shows the BAC for women of different weights and number of drinks consumed in an hour. 3.01

| | Body weight (kg) | | | | | | |
|---|---|---|---|---|---|---|---|
| Drinks per hour | 45 | 55 | 64 | 73 | 82 | 91 | 100 |
| 1 | 0.05 | 0.04 | 0.03 | 0.03 | 0.03 | 0.02 | 0.02 |
| 2 | 0.09 | 0.08 | 0.07 | 0.06 | 0.05 | 0.05 | 0.04 |
| 3 | 0.14 | 0.11 | 0.10 | 0.09 | 0.08 | 0.07 | 0.06 |

Sheridan is 64 kg. At a party, she consumed 2 drinks in the first hour and 3 drinks in the second hour.

- **a** Calculate Sheridan's BAC:
  - **i** at the start of the party
  - **ii** after the first hour
  - **iii** after the second hour.
- **b** If Sheridan had no more drinks, how many hours would it be until her BAC returns to zero? Assume that her body reduces her BAC at the rate of 0.015 per hour.

**3** **a** Why is a heavier person less affected by alcohol? 3.01

- **b** Why are females more affected by alcohol?

**4** This table shows the number of people killed in road crashes in a 5-year period. 3.02

| Road fatalities in Australia for 2017–2021 | | | | |
|---|---|---|---|---|
| | Drivers | | Passengers | |
| Year | Male | Female | Male | Female |
| 2017 | 429 | 137 | 105 | 130 |
| 2018 | 420 | 101 | 103 | 101 |
| 2019 | 418 | 152 | 100 | 105 |
| 2020 | 405 | 129 | 98 | 91 |
| 2021 | 399 | 129 | 95 | 86 |

Data source: Road Trauma Australia, 2024 statistical summary, BITRE, © Commonwealth of Australia

- **a** What is the mean number of male drivers killed per year?
- **b** What is the range of female passengers killed over the 5 years?
- **c** In 2020, what percentage, correct to one decimal place, of drivers killed were male?
- **d** In 2018, what percentage, correct to one decimal place, of females killed were drivers?
- **e** What do you notice when you compare the numbers of drivers killed with the number of passengers killed each year? Give a reason why this may be so.
- **f** Is the number of drivers killed each year generally increasing or decreasing? Give a reason why this may be so.

3.02 **5** This table shows the number of Australian fatal crashes over 6 years categorised by number of vehicles and pedestrians involved.

| Year | Single-vehicle crashes | Multiple-vehicle crashes | Pedestrian crashes | Total crashes |
|---|---|---|---|---|
| 2017 | 659 | 564 | 101 | 1324 |
| 2018 | 658 | 477 | 129 | 1264 |
| 2019 | 701 | 485 | 129 | 1315 |
| 2020 | 603 | 494 | 137 | 1234 |
| 2021 | 641 | 488 | 152 | 1281 |
| 2022 | 726 | 468 | 149 | 1343 |

Data source: Road Trauma Australia, 2024 statistical summary, BITRE, © Commonwealth of Australia

**a** In 2020, what percentage, correct to one decimal place, of crashes involved pedestrians?

**b** True or false: There are always more single-vehicle crashes than multiple-vehicle crashes.

**c** In which year were there the highest number of:

**i** multiple-vehicle crashes?

**ii** crashes overall?

**d** Calculate the percentage increase (to one decimal place) in single-vehicle crashes between 2021 and 2022.

**e** Draw a sector graph to display the data for 2022.

3.03 **6** Scott took $2\frac{3}{4}$ h to drive from Canberra to Bodalla, 200 km away.

**a** What was his average speed, correct to the nearest kilometre per hour?

**b** How long, in hours and minutes, would it take him to drive 430 km to Newcastle at this speed?

**c** How far, to the nearest kilometre, could he travel at this speed in 5 hours?

3.04 **7** A truck driver was travelling in the Northern Territory at 150 km/h on a road with no speed limits. He began braking 3 seconds after he saw a sign to a truck stop. If he travelled 810 m under brakes before coming to a stop, how far did he travel after seeing the sign?

3.04 **8** Georgy is travelling on a dry road at 65 km/h and sees a cow in the middle of the road about 70 m ahead. He takes 1.2 s to apply the brakes, then travels 24 m under brakes before coming to a stop. Did he hit the cow? Give reasons for your answer.

3.04 **9** The following measurements were taken in a school zone.

| Stopping distances on a wet road in a school zone | | | |
|---|---|---|---|
| Reaction time (s) | Reaction distance (m) | Braking distance (m) | Stopping distance (m) |
| 1 | 11.1 | 12.6 | A |
| 2 | 22.2 | 12.6 | B |
| 4 | 44.4 | 12.6 | C |

The speed limit in a school zone during school hours is 40 km/h.

**a** What do you notice about the braking distances?

**b** What do you notice about the reaction distances?

**c** Find the values of A, B and C.

**d** If an 85-year-old driver in a school zone has a slow reaction time of 4 s, would he stop in time if he sees a child crossing the road 50 m ahead?

**e** Suggest 2 strategies for the driver in part **d** that would help him reduce his risk of an accident.

☐ Foundation ○ Mastery ○ Complex

**10** This graph gives the braking distance for a car travelling on a dry or wet road at various speeds. 3.04

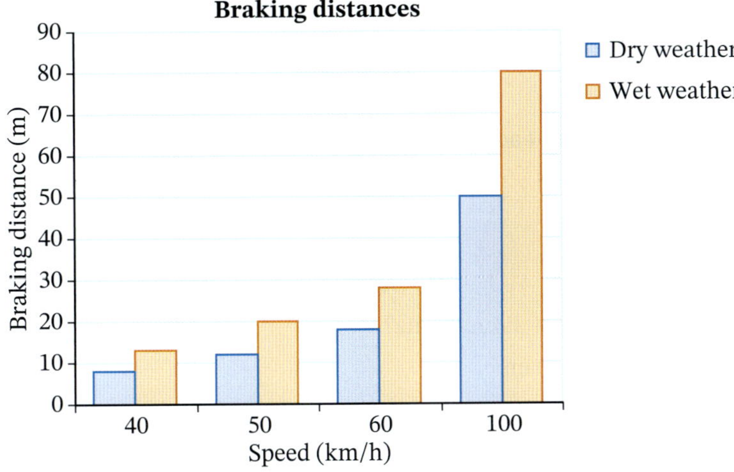


**a** What is the difference in braking distance on a dry road for 2 cars travelling at 50 km/h and 100 km/h respectively?

**b** How much further will a car travel under brakes on a wet road at 60 km/h than on a dry road at the same speed?

**c** If a car travels 50 m under brakes on a dry road, what is its approximate speed?

**d** What is the approximate speed of a car on a dry road that travels 12 m after the brakes are applied?

**e** Write a sentence noting any similarities or differences between braking distances on wet and dry roads, giving reasons for your answer.

☐ Foundation ○ Mastery ○ Complex

# PRACTICE EXAM 1

Answers on p. 465

**Recommended time: One hour**

**Section 1**
**10 multiple-choice questions: 1 mark each**
**Select the correct answer A, B, C or D.**

2.01 **1** Simplify $4ab + b^2 - 3ab - 4b^2$.

**A** $-2ab^3$ **B** $ab - 3b^2$ **C** $7ab - 3b^2$ **D** $ab - 5b^2$

**2** Which one of these is an example of discrete data?

**A** The capacity of Warragamba Dam

**B** The number of tries scored by a football team over a season

**C** The height of a person in your class

**D** The time it took a triathlete to complete her race

**3** Simplify $\frac{8t^6}{16t^2}$.

**A** $2t^4$ **B** $2t^3$ **C** $\frac{t^3}{2}$ **D** $\frac{t^4}{2}$

**4** To decide on the theme of the next school disco, Tasha surveyed a stratified sample of 50 students from the school, based on the following numbers of students in each Year group.

| Year | 7 | 8 | 9 | 10 | 11 | 12 |
|---|---|---|---|---|---|---|
| Number of students | 114 | 120 | 114 | 128 | 105 | 96 |

How many Year 12 students were in Tasha's sample?

**A** 4 **B** 7 **C** 8 **D** 14

3.01 **5** Sarah had a blood alcohol content (BAC) reading of 0.062 at 10:30 pm, at the end of a party. Determine when her BAC will be back to zero by using the formula:

$$\text{number of hours} = \frac{\text{BAC}}{0.015}$$

**A** 2:38 am **B** 2:43 am **C** 4:08 am **D** 4:13 am

1.07 **6** The dot plot shows the total number of minutes students studied to prepare for a mathematics assessment task.

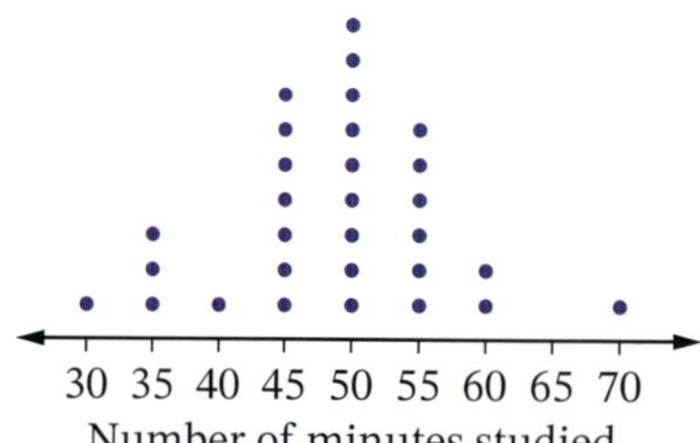


Between which 2 values is the data clustered?

**A** 35 and 55 **B** 40 and 60 **C** 45 and 55 **D** 45 and 60

**7** Joe left Sydney at 8 am to travel a distance of 1000 km to Melbourne. He arrived in Melbourne at 9 pm. What was his average speed for the journey, correct to the nearest kilometre per hour? 3.03

**A** 77 km/h **B** 83 km/h **C** 91 km/h **D** 111 km/h

**8** This frequency histogram shows the results of a survey about the number of children in each family. 1.06

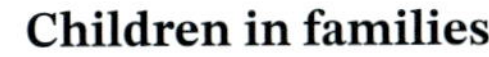


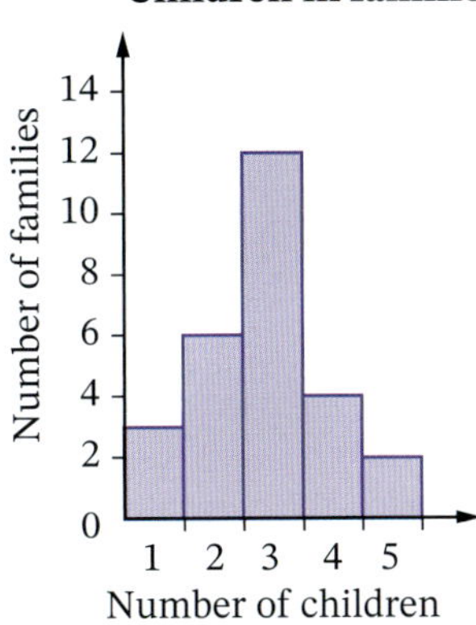


How many families were surveyed?

**A** 12 **B** 15 **C** 27 **D** 77

**9** Use the formula $m = d + 3at^2$ to find the positive value of $t$ when $m = 241$, $d = 16$ and $a = 3$. 2.05

**A** $t = 241$ **B** $t = 225$ **C** $t = 20$ **D** $t = 5$

**10** Solve $5 - 2x = -6$. 2.04

**A** $x = -5\frac{1}{2}$ **B** $x = 5\frac{1}{2}$ **C** $x = -\frac{1}{2}$ **D** $x = \frac{1}{2}$

**Section 2**
**3 questions: 10 marks each**

**Question 11** 10 marks

**a** Solve $9x - 11 = 4x + 7$. 2 marks 2.04

**b** Jayden was travelling at 70 km/h when he saw a car stopped ahead of him. He took 1.2 s to react and apply his brakes, then stopped 14 m later. 3.04

**i** What was his reaction distance, to the nearest metre? 2 marks

**ii** Use the braking distance formula $d = kv^2$ to find the value of $k$, correct to 2 significant figures. 1 mark

**iii** Find Jayden's braking distance, correct to one decimal place, if he was travelling at 90 km/h. 1 mark

*Question 11 continues on the next page.*

1.07 **c** The waiting times, in minutes, for patients at the Westvale Health Centre are shown in the stem-and-leaf plot.

| Stem | Leaf |
|---|---|
| 0 | 2 3 4 5 |
| 1 | 0 1 3 6 □ 8 |
| 2 | 1 1 3 4 4 |
| 3 | 1 2 |
| 4 | 1 |
| 5 | 1 |

**i** One entry is missing and is shown as □.
What waiting time could be represented by this entry? 1 mark

**ii** How many patients were surveyed? 1 mark

**iii** Calculate, correct to one decimal place, the percentage of patients who waited for longer than half an hour. 1 mark

**iv** What was the shortest waiting time? 1 mark

## Question 12 10 marks

3.01 **a** Tegan weighs 56 kg and drank 5 standard alcoholic drinks over 4 hours.

**i** Calculate, correct to 3 decimal places, her blood alcohol content using the formula:

$$\text{BAC}_{\text{female}} = \frac{10N - 7.5H}{5.5M}$$

where $N$ is the number of standard drinks consumed
$H$ is the number of hours drinking
$M$ is the mass in kg. 1 mark

**ii** If Tegan drank the same number of standard drinks over a longer time, how would her blood alcohol content have been affected? 1 mark

2.02 **b** Expand and simplify $3(2p + 5) - 4(p + 3)$. 1 mark

2.03 **c** Zayn needs to give his son some medicine. He uses Clark's formula:

$$D = \frac{kA}{70}$$

where $D$ is the child dosage,
$k$ is the mass of the child in kilograms and
$A$ is the adult dosage.

Zayn's son is 6 years old and weighs 22 kg. The adult dosage is 10 mL, taken every morning and every night. How many days will a 300 mL bottle of medicine last for Zayn's child? 2 marks

1.03, 1.06 **d** The masses, in kilograms, of 20 Year 11 students were recorded:

| | | | | | | | | | |
|---|---|---|---|---|---|---|---|---|---|
| 58 | 43 | 62 | 52 | 49 | 48 | 52 | 45 | 68 | 72 |
| 54 | 41 | 65 | 41 | 48 | 51 | 60 | 66 | 57 | 69 |

**i** Are people's masses discrete data or continuous data? 1 mark

**ii** Organise the data into a frequency table with classes 40–<45, 45–<50, and so on. 1 mark

**iii** Draw a frequency histogram to represent the data. 2 marks

**iv** What percentage of students were 60 kg or more? 1 mark

**Question 13** 10 marks

**a** The sum of a number, $y$, and 5 is doubled and the result is the same as subtracting 5 times the number from 44. 2.05

**i** Write an equation that can be solved to find $y$. 2 marks

**ii** Solve the equation for $y$. 1 mark

**b** The volume of a cylinder is given by the formula $V = \pi r^2 h$, where $r$ is the radius of the circular base and $h$ is the height. 2.06

**i** Make $r$ the subject of the formula. 1 mark

**ii** Find the radius of a cylinder that has a volume of $1200\,\text{cm}^3$ and a height of $15\,\text{cm}$. Express your answer correct to 3 significant figures. 2 marks

**c** The graph below shows the road fatalities in NSW for one year, sorted by age group.

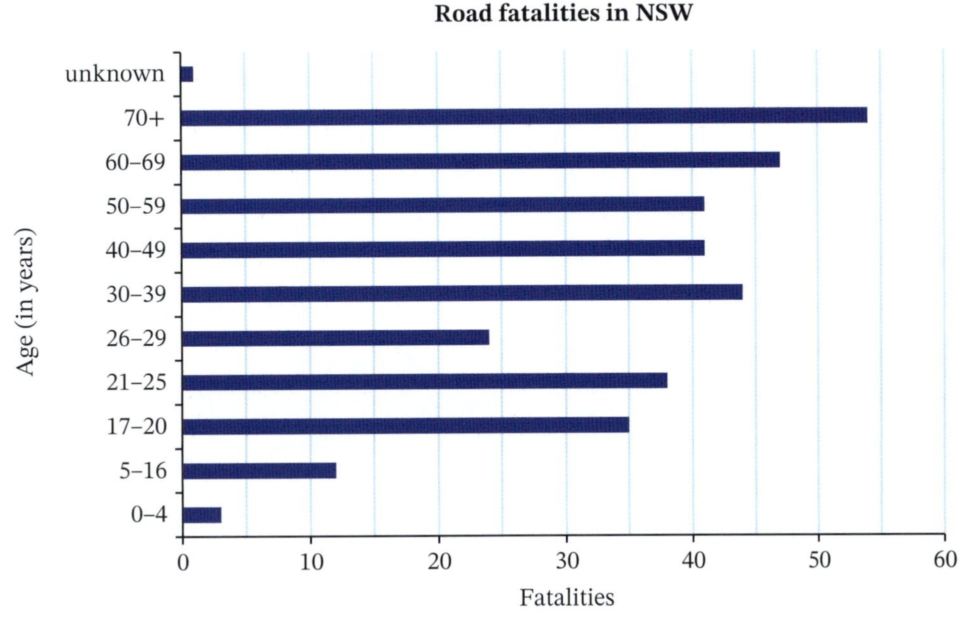


**i** What was the total number of fatalities for NSW that year? 1 mark

**ii** What percentage of total fatalities were from the 21–29 age group? 1 mark

**iii** Give a possible reason for fatalities increasing from age 60 onwards. 1 mark

**iv** Find the number of fatalities for people aged 17–25 and compare it to the other age groups. 1 mark

**END OF EXAMINATION**

# EARNING MONEY AND TAXATION

Wages and salaries are the most common types of income in Australia and people may work full-time, part-time or be casual workers. Full-time employees work 38 hours per week, while part-time employees work fewer than 38 hours per week. In 2024–2025, the median salary in Australia was $65 000.

The Australian taxation system is a progressive tax system, which means the more you earn, the higher the rate of tax you pay.

## Chapter outline

Kym McLeod/Adobe Stock Photos

## In this chapter you will:

- calculate wages, salaries and overtime pay and convert between weekly, fortnightly, monthly and yearly earnings
- calculate earnings based on commission, piecework and royalties
- calculate bonuses and annual leave loading
- calculate government allowances and pensions
- calculate net pay following deductions from gross pay
- calculate allowable tax deductions, taxable income and income tax (including Medicare levy)
- calculate Pay As You Go (PAYG) tax and tax payable or refund owing after a tax return is completed at the end of the financial year.

**Videos (8):**
4.01 Annual salaries • Overtime 1 • Overtime 2
4.02 Commission • Commission involving sliding scales
4.03 Annual leave loading
4.06 Income tax
4.07 Tax refund

**Skillsheets (2):**
SkillCheck, 4.02 Percentage calculations • Mental percentages

**Worksheets (6):**
SkillCheck Assignment 4
4.01 Wages and salaries • S-Mart payroll
4.04 Earning money
4.05 Pay day
4.06 Income tax tables

**Puzzles (2):**
4.02 Commission and piecework
Test yourself Earning and taxation crossword

**Spreadsheets (3):**
4.01 Wages
4.03 Holiday pay
4.06 Income tax

Nelson MindTap

To access resources above, visit **cengage.com.au/nelsonmindtap**

## Terminology

| | | | |
|---|---|---|---|
| allowable tax deduction | annual leave loading | bonus | commission |
| double time | expense | gross income | gross pay |
| income | income tax | Medicare levy | net pay |
| overtime | Pay As You Go (PAYG) tax | penalty rate | piecework |
| royalties | salary | sole trader | superannuation |
| tax refund | taxable income | time-and-a-half | wage |

## SkillCheck Answers on p. 465

**Worksheet** Assignment 4

**Skillsheets** Percentage calculations

Mental percentages

**1** Find the cost of:

**a** 5 calculators @ \$46.75 each

**b** 1.5 kg carrots @ \$3.10 per kg

**c** 1400 protractors @ \$1.98 each.

**2** Calculate, correct to the nearest cent:

**a** 34% of \$42 568

**b** \$153 000 ÷ 52

**c** 1.5% of \$85 629

**3** To the nearest whole number, what percentage is \$2360 of \$36 580?

**4** Evaluate:

**a** $36 \times 22.5 + 4 \times 22.5 + 3 \times 22.5 \times 2$

**b** $58\,000 + 0.45 \times (250\,000 - 180\,000)$

**5** How many hours and minutes are there between:

**a** 8:00 am and 12:00 pm?

**b** 9:00 am and 2:30 pm?

**c** 7:30 am and 5:15 pm?

# Wages, salaries and overtime

4.01

## Wages and overtime

A person's **wage** is calculated on the number of hours they worked in a given period, with **overtime** pay being paid at a higher rate for working beyond normal hours, called a **penalty rate**. Types of overtime are shown in the table on the right. People who earn a wage include waiters, mechanics and receptionists.

| Type of overtime | Meaning |
|---|---|
| Time-and-a-half | 1.5 × normal rate |
| Double time | 2 × normal rate |
| Double time-and-a-half | 2.5 × normal rate |
| Triple time | 3 × normal rate |

These are called **penalty** rates because employers pay extra for having their staff work longer than standard hours.

**Video**
Annual salaries

**Worksheets**
Wages and salaries

S-Mart payroll

**Spreadsheet**
Wages

### Example 1

This week, Christy worked at a childcare centre for 37 hours at \$34.70 per hour plus 6 hours overtime at time-and-a-half. Calculate her earnings for the week.

#### Solution

Normal pay $37 \times \$34.70 = \$1283.90$

Overtime pay $6 \times 1.5 \times \$34.70 = \$312.30$

Total earnings $\$1283.90 + \$312.30 = \$1596.20$

### Example 2

Ian works as a bartender at an RSL club. His timesheet for the 3 days he worked in April is shown below.

| Day | Start time | Finish time |
|---|---|---|
| Tuesday | 3:00 pm | 10:00 pm |
| Wednesday (Anzac Day holiday) | 9:00 am | 2:00 pm |
| Thursday | 12:00 noon | 7:00 pm |

The rate for bartenders is \$33.17 per hour between 8:30 am and 6:00 pm, time-and-a-half between 6:00 pm and 11:30 pm, and double time-and-a-half on public holidays. Calculate Ian's pay for the 3 days he worked.

#### Solution

| Day | Normal hours | Time-and-a-half | Double time-and-a-half |
|---|---|---|---|
| Tuesday | 3 | 4 | 0 |
| Anzac Day | 0 | 0 | 5 |
| Thursday | 6 | 1 | 0 |
| **Total hours** | **9** | **5** | **5** |

Normal pay $9 \times \$33.17 = \$298.53$

Time-and-a-half pay $5 \times 1.5 \times \$33.17 = \$248.775$

Double time-and-a-half pay $5 \times 2.5 \times \$33.17 = \$414.625$

Total pay $\$298.53 + \$248.775 + \$414.625 = \$961.93$

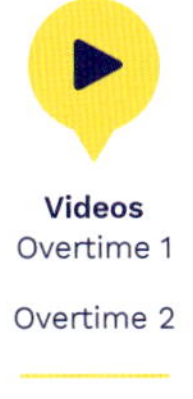

**Videos**
Overtime 1

Overtime 2

## Salaries

A **salary** is a fixed amount of annual pay (yearly pay), paid weekly, **fortnightly** or monthly. People who earn a salary include teachers, website designers and police officers.

### Time conversions

1 year = 12 months

1 fortnight = 2 weeks

1 year = 52 weeks or 26 fortnights

Strictly speaking, 1 year = 52 weeks + 1–2 days, so when converting a salary to a weekly amount, dividing by 52 will give the average pay per week rather than the actual amount you might receive if paid weekly.

Note: 1 month $\neq$ 4 weeks. To convert between weeks and months, convert to years first.

### Example 3

Siobhan earns a salary of $98 025 as an IT consultant. To the nearest cent, how much is she paid

**a** monthly?

**b** fortnightly?

#### Solution

**a** monthly pay = \$98 025 ÷ 12
= \$8168.75

**b** fortnightly pay = \$98025 ÷ 26
= \$3770.1923 ...
≈ \$3770.19

Answers to money calculations should be rounded to the nearest cent, not 5 cents. Only cash payments are rounded to the nearest 5 cents.

**EXERCISE 4.01** Answers on p. 465

### Wages, salaries and overtime

1. Ken is a gardener who works from 6:30 am to 5:00 pm, Monday to Friday, and earns $34.15 per hour. Calculate his weekly pay, correct to the nearest cent.

EXAMPLE 1

2. Sandra is paid $42.08 per hour as a computer operator. She works 9 hours per day, Monday to Friday, and 5 hours on Saturday, for which she is paid time-and-a-half. Calculate her weekly wage.

3. Rabiba is paid $31.75 per hour and is paid a penalty rate of time-and-a-half for working beyond 8 hours per day. What are Rabiba's earnings if she works 11 hours a day for 4 days? Select **A**, **B**, **C** or **D**.

   **A** \$396.88 **B** \$1158.88 **C** \$1587.50 **D** \$2095.50

4. For an annual salary of $120 000, what is the weekly amount of pay? Select **A**, **B**, **C** or **D**.

   **A** \$2307.69 **B** \$2299.73 **C** \$2307.70 **D** \$2299.75

5. If a labourer earns a weekly wage of $1219.70, how much does he earn in a year? Select **A**, **B**, **C** or **D**.

   **A** \$63 424.00 **B** \$62 433.50 **C** \$63 424.40 **D** \$63 422.40

6. Jacinta, an internet website designer, works a 38-hour week and earns $3115.35 every fortnight. Calculate her earnings:

   **a** per year **b** per week **c** per hour.

7. Liam is a trainee nurse who works 36 hours at an ordinary rate of $27.44 and 10 hours at time-and-a-half. Calculate his total pay.

8. A childcare worker earns $1179.20 per week. How much will she earn in:

   **a** a year? **b** a month?

9. On a long weekend, a waitress is paid double time on Sunday and double time-and-a-half on Monday. How much does she earn for working 9 hours on each of the days if her normal wage rate is $33.54 per hour?

☐ Foundation ◯ Mastery ◯ Complex

**10** William is a casual shift worker, working from midnight to 5 am, 5 days a week. His hourly rate of pay is $30.35. He is paid 155% of his hourly rate for shiftwork, which also includes his casual loading. Calculate how much he earns in a week.

EXAMPLE 2

**11** A chef earns $41.20 per hour, but is paid time-and-a-half for working after 8:00 pm on weekdays and any time on weekends. The timesheet below shows the times he worked last week. Copy and complete the table, then calculate the chef's total pay for the week.

| Day | Start time | Finish time | Normal hours | Overtime hours |
|---|---|---|---|---|
| Monday | 12:00 noon | 6:00 pm | | |
| Tuesday | 12:00 noon | 6:00 pm | | |
| Wednesday | 12:00 noon | 7:00 pm | | |
| Thursday | 12:00 noon | 10:00 pm | | |
| Friday | 12:00 noon | 12:00 midnight | | |
| Saturday | 4:00 pm | 12:00 midnight | | |
| Sunday | 4:00 pm | 10:00 pm | | |
| | | **Total hours** | | |

**12** Dom is a personal assistant who earns $39.65 per hour for a 35-hour week, time-and-a-half for the first 4 hours of overtime, and double time thereafter. Calculate his total earnings for a week in which he worked from 8:30 am to 5:30 pm from Tuesday to Saturday.

EXAMPLE 3

**13** A Member of Parliament earns a salary of $233 650. How much is this if paid:

**a** fortnightly? **b** monthly? **c** weekly?

**14** A law clerk earns a salary of $83 215. How much does she earn:

**a** a month? **b** a fortnight? **c** a weekday?

**15** Rachna is a computer systems engineer who earns a salary of $135 000. If she works a 42-hour week, calculate how much she earns:

**a** per hour **b** per fortnight **c** per month.

**16** The table below shows the hours worked by a hotel porter. If he earns $28.40 per hour from 0600 to 1900 (in 24-hour time) and time-and-a-half otherwise, calculate his weekly pay.

| Day | Start time | Finish time |
|---|---|---|
| Tuesday | 0630 | 1400 |
| Wednesday | 0800 | 1530 |
| Friday | 1500 | 2330 |
| Sunday | 0500 | 1330 |

**17** A newspaper printer worked for 33 hours at $26.72 per hour during the week before Good Friday, then for 6 hours at double time-and-a-half on Good Friday. Calculate the printer's total pay.

**18** A survey field-hand earns $37.18 per hour, working from 7:00 am to 4:30 pm every weekday. Calculate how much he is paid:

**a** a day **b** a week **c** a month.

Foundation Mastery Complex

**19** Ben must work on a public holiday. The penalty rate is double time-and-a-half for a minimum of 4 hours, even if 4 hours are not worked. Calculate Ben's pay for working 3 hours if he is paid $27.85 per hour. Select **A**, **B**, **C** or **D**.

**A** $222.80 **B** $278.50 **C** $208.88 **D** $208.90

**20** Which person earns the most? Select **A**, **B**, **C** or **D**.

**A** Ali: $28.70 per hour for a 35-hour week

**B** Boun: $984 per week

**C** Carlos: $4384 per month

**D** Dimitri: $51 600 salary

**21** Simon is paid $275.40 each day for working at a bakery from 4:30 am to 1:00 pm. If he works 6 days per week, how much does he earn:

**a** per hour? **b** per week? **c** per year?

**22** Each day, a storeman earns $30.72 per hour for 9 hours and time-and-a-half for any extra hours. Last week, he worked the following number of hours each day: 9, 8, 10, 14 and 12.5. Calculate his total pay for the week.

**23** Naved is a casual employee at the local supermarket and his hourly rate of pay is $18.50. Naved also receives casual loading, which is an extra payment made to casual employees, at 20% of his hourly rate.

**a** Calculate the amount of Naved's casual loading.

**b** How much will Naved earn for working 5 hours on a Sunday, which is paid at double time?

**24** Alex worked 32 hours at the normal rate and 4 hours at time-and-a-half, earning a total of $1094.40. Find her normal rate of pay per hour.

**25** Aryan is paid at an hourly rate of $42 for a 36-hour week and overtime hours are paid at time-and-a-half. Calculate the number of overtime hours Aryan worked in a week in which he earned $1827.

## Technology

### Payroll calculations

**1** Enter the weekly pay sheet into a spreadsheet as shown.

| | A | B | C | D | E | F | G | H |
|---|---|---|---|---|---|---|---|---|
| 1 | | | GLOBAL ENGINEERING | | | | | |
| 2 | Weekly pay sheet | | | | | | | |
| 3 | | | | Hours worked | | | | |
| 4 | Employee | Pay rate/hr | | Normal time | Time-and-a-half | Double time | Double time-and-a-half | Weekly pay |
| 5 | J. Dengate | $18.80 | | 38 | 2 | 0 | 0 | |
| 6 | B. Goodman | $25.75 | | 38 | 4 | 1 | 0 | |
| 7 | D. Singh | $32.40 | | 40 | 6 | 3.5 | 0 | |
| 8 | K. Lui | $56.80 | | 38 | 6 | 4 | 2 | |
| 9 | P. Elm | $43.75 | | 35 | 4 | 1 | 1 | |
| 10 | | | | | | | | |



**2** Enter a formula into cell H5 to calculate the weekly pay for J. Dengate.

**3** Fill down from H5 to H9 to find the weekly pay of each employee.

**4** Enter a formula in H10 to find the total weekly wage bill.

**5** Use this spreadsheet to check your answers for Exercise 4.01.

☐ Foundation ○ Mastery ⬡ Complex

# Commission, piecework and royalties 4.02

Some workers are not paid by the amount of time they work, but by the number of items they make or process.

- **Commission** is **income** for selling items, calculated as a percentage of the value of items sold. A real estate agent earns commission on each house sold.
- **Piecework** is income for making or processing items, calculated as a fixed rate per item. Fruit pickers and garment makers earn money this way.
- **Royalties** are income based on the number of copies sold. Musicians and authors earn royalties from sales of their music and books.

**Videos**
Commission

Commission involving sliding scales

**Puzzle**
Commission and piecework

## Example 4

Josh is paid 73c for each shirt folded and boxed.
Last week he processed 1440 shirts.
How much did he earn?

Volodymyr Krasyuk/Shutterstock.com

### Solution

Earnings = 1440 × \$0.73 = \$1051.20

## Example 5

Ally earns a retainer of \$415 per week plus commission of 8.5% on the value of Wonder mops sold.

A **retainer** is a fixed amount earned that does not depend on the number of items sold.

**a** To the nearest cent, how much does Ally earn for selling \$1595 worth of mops in one week?

**b** If Ally earned \$700 in a week, what was the value of mops sold that week, to the nearest dollar?

**Skillsheets**
Percentage calculations

Mental percentages

### Solution

**a**
$$\begin{aligned} \text{earnings} &= \$415 + 8.5\% \times \$1595 \\ &= \$550.575 \\ &\approx \$550.58 \end{aligned}$$

**b** Let $V$ be the value of mops sold.

$$\text{earnings} = \text{retainer} + 8.5\% \times V$$

$$700 = 415 + 0.085V$$

$$285 = 0.085V$$

$$V = \frac{285}{0.0085}$$

$$= 3352.9411\ldots$$

$$V \approx 3353$$

The value of the mops sold is \$3353.

Income can also be earned by a person working as a **sole trader**. A sole trader is a person who is the only owner of a small business, not a company. Examples of sole traders are electricians, mechanics, personal trainers and photographers. All income earned through the business is regarded as personal income.

## Example 6

Mia, a photographer, works as a sole trader on weekends. She charges $270 for a 1-hour shoot, $455 for 2 hours and $795 for 4 hours. On any weekend, Mia will do three 1-hour shoots, two 2-hour shoots and one 4-hour shoot.

**a** How much will Mia earn on a weekend?

**b** If Mia works 35 weekends per year, what is her annual income as a sole trader?

### Solution

**a** $\text{earnings} = 3 \times \$270 + 2 \times \$455 + \$795$

$= \$2515$

**b** $\text{annual income} = 35 \times \$2515$

$= \$88\,025$

**EXERCISE 4.02** Answers on p. 465

## Commission, piecework and royalties

EXAMPLE 4

**1** Brett is paid $10.45 for each lug of cherries he picks. If he picked 38 lugs of cherries, which of the following is the amount he earned, correct to the nearest 5c? Select **A**, **B**, **C** or **D**.

**A** $397.10 **B** $390.10 **C** $397.00 **D** $398.00

**2** Danny is paid 13.5c for each envelope he folds and addresses. If he can process 14 envelopes in 5 minutes:

**a** how many envelopes can he process in 1 hour?

**b** how much does he earn in $2\frac{1}{2}$ hours?

**3** Leila is paid $4.65 for each garment she irons. If a basket contains 18 items and she irons 3 basketsful, which of the following is the amount she earns? Select **A**, **B**, **C** or **D**.

**A** $83.70 **B** $13.95 **C** $253.80 **D** $251.10

**4** A stockbroker receives a commission of 2% of the selling price of shares.

**a** If he sold 350 shares at $4.30 each, calculate his commission.

**b** What was the value of the shares he sold if he earned $135 commission?

EXAMPLE 5

**5** A car salesman's commission is 4% for the first $30 000 worth of sales and 2.5% for the rest. If he is also paid a retainer of $420, calculate his total earnings in a week in which he sold $350 000 worth of cars.

**6** A real estate agent charges the following commission for selling properties, based on this sliding scale of property prices:

- 2.5% for the first $900 000
- 2% for the next $400 000
- 1.75% for the remaining amount.

What commission will the real estate agent earn for selling:

**a** an apartment for $642 400?

**b** a villa for $980 000?

**c** a house for $1.65 million?

**7** A songwriter earns a royalty of 47c every time her song is played on the radio. If, in one week, her song was played 434 times on radio stations, calculate her royalty payment for that week.

▷

□ Foundation ○ Mastery ○ Complex

8 An inventor is paid 1.8% of the total sales of a can-opener that uses his original design. How much does the inventor earn if 8400 can-openers are sold at $17.95 each?

9 Taylor, a singer-songwriter, earns 8% of the payment for each song downloaded and 12.5% of the sales of her latest album. In 3 months, 1 650 000 of her songs were downloaded at $2.59 each and 836 254 copies of her album were sold at $16.99 each. Calculate:

- **a** the value of Taylor's royalties from song downloads
- **b** the total value of Taylor's album sales
- **c** Taylor's total royalties from album sales and song downloads.

10 Truong earns 11 cents for every advertising brochure he delivers to letterboxes.

- **a** How much does Truong earn in a day in which he delivers 1329 brochures?
- **b** If Truong earned $275 one day, how many brochures did he deliver?

11 Nick earns $16.50 for every car he vacuums. He vacuums an average of 3.6 cars per hour. How much will he earn in a $6\frac{1}{2}$ hour shift?

12 Libby earns 18% commission for selling cosmetics door to door. What value of cosmetics did she sell this week if she earned $648?

13 An author earns 13% in royalties on the sale of her novels. If 34 700 copies of her novels sold for $32.95 each, calculate the author's royalty payment.

14 Pranathi earns $32.40 for every dress she sews. If she can make 7.5 dresses per day, calculate her earnings for a 5-day week.

15 Leon is a plumber and works as a sole trader. He charges a call-out fee of $110 and then $130 per hour. Calculate his earnings for a week in which he has 3 call-outs and works for 25 hours.

16 Aurora is a hairdresser and works 4 days per week from home. In that time, she does 19 cuts at $39 each, 5 cut-and-colours at $103 each and 4 seniors cuts at $32 each. What is Aurora's income for the week?

17 Jules is paid $4.62 for every toy he assembles. In a week, he receives an extra $1.20 for each toy he has assembled after the first 150. Calculate Jules' total earnings if he assembles 216 toys in a week.

18 Debbie earns 15% commission on all the storage containers she sells. If she earned $251.70 this week, what was the total value of the containers she sold?

19 An agent earned, as his commission, $2024 of the actor's fee in a TV commercial. What percentage commission does this represent if the actor's fee was $23 000?

20 At Christmas time, Shane earns $5.65 for every item he gift wraps.

- **a** Calculate Shane's earnings for wrapping 114 gifts in one day.
- **b** How many gifts did Shane wrap on the next day if his earnings were $265.55?

21 Salma earns $8.40 each time she taste-tests a new brand of food. How many tastings must she perform to earn over $500?

22 Hamish is a mechanic and has his own business servicing cars and doing mechanical repairs. In one week, he does 3 minor services at $260 each, 2 logbook services at $650 each and also spends 15.25 hours doing mechanical repairs charging $127 per hour. What is Hamish's weekly income?

23 Miriam is a self-employed personal trainer. She charges $55 per session, and she has a weekly client base between 20–30 people.

- **a** What is the least amount that Miriam can earn in a week?
- **b** If Miriam runs sessions for 46 weeks per year, what is the maximum amount she could earn in a year?

**24** Travis is paid the following commission rates for selling mobile phone plans:

- 4.8% for the first \$3000
- 6% for the next \$3000
- 7.5% for the remaining amount.

Calculate Travis's commission for selling plans amounting to \$8758 in value.

# 4.03 Bonuses and annual leave loading

Video
Annual leave loading

Spreadsheet
Holiday pay

Bonuses and annual leave loading are examples of extra pay, calculated as a percentage of normal pay.

A **bonus** is paid to employees who produce work of a high standard or volume, or for meeting an important quota, goal or deadline. The owners of some businesses give their employees a Christmas bonus to reward them for a productive year's work.

**Annual leave loading** (also called **holiday loading**) is extra pay given during annual leave. It is usually paid at a rate of 17.5% of 4 weeks' normal pay.

## Example 7

Hayden earns a yearly salary of \$76 598. Calculate his holiday loading, which is 17.5% of the normal pay for 4 weeks, correct to the nearest cent.

### Solution

4 weeks' pay = \$76 598 ÷ 52 × 4
= \$5892.153 ...

> Do not round your answer. Leave it on the calculator display.

holiday loading = 17.5% × \$5892.153 ...
= \$1031.126 ...
≈ \$1031.13

**EXERCISE 4.03** Answers on p. 466

## Bonuses and annual leave loading

**1** A police officer earns a salary of \$97 206. Calculate her annual leave loading if it is 17.5% of 4 weeks normal pay.

**2** At the end of the year, the owners of a small business gave their employees a Christmas bonus of 60% of a normal week's pay. Calculate Lisa's total pay for that week if her usual pay is \$1020.48.

**3** A miner earns a wage of \$38.05 per hour. If the miner works from 8:30 am to 4:00 pm each day from Monday to Friday, calculate:

**a** his total weekly income **b** his annual leave loading.

**4** A carpenter stationed at the South Pole earns a salary of \$79 801 plus a weekly allowance of \$1172.58 for working under extreme and isolated conditions. Calculate her annual leave loading.

Complex

5 For meeting a target one week ahead of schedule, the owners of a firm gave their staff a bonus equal to 15% of the normal pay for 1 month. What is Amanda's bonus if her salary is $53 045?

6 The owners of a biscuit factory gave their casual staff a bonus of $472 at Christmas, while the permanent staff received a holiday loading of 17.5% of the pay for 4 weeks. If the normal weekly pay for the permanent staff was $1152, what was the difference between their holiday loading and the bonus given to the casual staff?

7 A telephone interviewer employed by a market research company earns $27.28 per hour for a 35-hour week. Calculate her holiday loading if it is 17.5% of the normal pay for 4 weeks.

8 One year, Kirsten received a holiday loading of $999.60. What was her normal weekly wage?

9 Anneliese normally works a 40-hour week. She goes on holidays and receives 4 weeks' wages plus 17.5% holiday loading. If she earns $29.55 per hour, calculate (correct to the nearest cent) Anneliese's:

**a** holiday loading

**b** total holiday pay.

Holiday pay = 4 weeks' normal pay + holiday loading

Polina Petrenko/Shutterstock.com

10 Tameer's annual leave loading of 17.5% of his normal pay for 4 weeks was $1079.68. If Tameer works a 40-hour week, what is his hourly rate of pay? Select **A**, **B**, **C** or **D**.

**A** $38.50 **B** $38.56 **C** $38.60 **D** $39.40

11 Mia's holiday pay is made up of 4 weeks' pay plus annual leave loading. What is Mia's salary if her holiday pay was $7456.73? Select **A**, **B**, **C** or **D**.

**A** $72 568 **B** $78 600 **C** $82 500 **D** $84 500

12 Brad's pay rate is $29.40 per hour. If his holiday loading is $782.04, find the number of hours per week that Brad works.

13 Each day, an electrician is paid $406.50.

**a** If the electrician works 6 days per week, calculate his total weekly earnings.

**b** Calculate the electrician's annual leave loading if it is 17.5% of the normal pay for 4 weeks.

Foundation Mastery Complex

# 4.04 Government allowances and pensions

Worksheet
Earning money

**Centrelink** is an Australian government agency that supports people in need and assists them to become self-sufficient. Various allowances and benefits are available for the aged, the unemployed, students, parents and people with disabilities.

## Age Pension

The **Age Pension** is a fortnightly payment to seniors, aged 67 or over, who are unable to support themselves in their retirement. To be eligible, a person needs to have low income and assets. The payment and income test tables are shown below.

| Marital status | Full payment |
|---|---|
| Single | $1144.40 |
| A couple | $862.60 each |

All figures were correct at time of printing but the rates for all allowances and pensions may have changed.

| Marital status | To get full payment, earnings per fortnight* can be up to: |
|---|---|
| Single | $212 plus $24.60 per dependent child |
| A couple (combined income) | $372 plus $24.60 per dependent child |

*Income over these amounts reduces the rate of pension payable by 50 cents in the dollar (single), and 50 cents in the dollar each (for couples).

Source: How much you can get, Age Pension, Services Australia.

### Example 8

Use the tables to calculate the fortnightly Age Pension payment for each person or couple.

**a** Liza, aged 72, receives $400 per year dividend from her shares.

**b** Dulcie (aged 68) and Bert (aged 69), are legal guardians of their granddaughter and have a combined income of $280 per week.

**c** Stanley, aged 63, has no income and lives alone.

### Solution

**a** Fortnightly income $= \$400 \div 52 \times 2$

$= \$15.384\ldots$

Liza is single and earns less than $212 per fortnight, so she receives the full payment of $1144.40.

**b** Combined fortnightly income $= \$280 \times 2$

$= \$560$

To receive the full payment, the couple must earn less than:

$\$372 + \$24.60 = \$396.60$ — $24.60 for the granddaughter

The payment is reduced by 50c for each dollar earned over $396.60.

Combined pension $= (2 \times \$862.60) - 0.5 \times (\$560 - \$396.60)$

$= \$1643.50$

Dulcie and Bert receive a combined pension of $1643.50 per fortnight.

**c** Stanley is not eligible for the Age Pension because he is under 67 years old.

## Youth and study allowances

There are several allowances available for young people or students.

| Allowance | Conditions for eligibility |
|---|---|
| Youth Allowance | Aged 16–21 years, looking for full-time work or doing approved activities<br>Aged 18–24 years, studying full-time<br>Aged 16–24 years, undertaking a full-time apprenticeship<br>Aged 16–17 years, finished Year 12 or equivalent, need to live away from home to study or are independent |
| Austudy | Aged over 25 years and studying full-time or undertaking an apprenticeship |
| ABSTUDY | Aboriginal or Torres Strait Islander students or apprentices |

## Youth Allowance (for people aged 16 to 24)

| Conditions for eligibility | Maximum fortnightly payment |
|---|---|
| Single, no children, younger than 18, and live at your parent's home | $410.30 |
| Single, no children, younger than 18, and living away from your parent's home to study, train or look for work | $663.30 |
| Single, no children, 18 years or older and live at your parent's home | $472.50 |
| Single, no children, 18 years or older and need to live away from your parent's home | $663.30 |
| Single, with children | $836.60 |
| A couple, with no children | $663.30 |
| A couple, with children | $718.10 |

Source: How much you can get, Austudy, Services Australia.

## Austudy (for people aged 25 or more)

| Conditions for eligibility | Maximum fortnightly payment |
|---|---|
| Single, no children | $663.30 |
| Single, with children | $836.60 |
| A couple, no children | $663.30 |
| A couple, with children | $718.10 |

Source: How much you can get, Austudy, Services Australia.

### Example 9

Calculate the fortnightly Youth Allowance or Austudy payment for each person.

**a** Pablo, aged 17, a first-year apprentice plumber living with his parents

**b** Cassie, aged 27, studying her HSC full-time at TAFE while raising her son alone

**c** Zina, aged 32, studying part-time and living with her husband and 2 children

**d** Greg, aged 16, has left school, is homeless, and is looking for full-time employment

### Solution

Remember to check the eligibility criteria as well.

**a** Pablo is under 18 and is living at home, so he receives a Youth Allowance of $410.30.

**b** Cassie is over 25 and a single parent, so she receives an Austudy payment of $836.60.

**c** Zina is not eligible for Austudy, because she is not studying full time.

**d** Greg is under 18 and is not living at home, so he receives a Youth Allowance of $663.30.

## ABSTUDY

| Conditions for eligibility | Allowance per fortnight |
|---|---|
| **Single, no children** | |
| Under 18, at home | \$410.30 |
| 18–21 years, at home | \$472.50 |
| Under 22, not at home | \$663.30 |
| 22 or older, at home | \$778.00 |
| 22–54, not at home | \$778.00 |
| 55 or older, not at home | \$833.20 |
| **Partnered, no dependent children** | |
| Younger than 22 years | \$663.30 |
| 22 years or older | \$712.30 |
| **Single with dependent child** | |
| Younger than 22 years | \$836.60 |
| 22 years or older | \$833.20 |
| **Partnered with dependent child** | |
| Younger than 22 years | \$718.10 |
| 22 years or older | \$712.30 |

Source: Maximum rates for independent students and Australian Apprentices, Services Australia.

### Example 10

Calculate the fortnightly ABSTUDY payment for each student.

**a** Ned, aged 20, who is living with his partner and their 2-year-old son

**b** Annie, aged 14, who is living at home with her parents

**c** Jackie, aged 27, who is living with her husband and has no children

#### Solution

**a** Ned is younger than 22 and is partnered, with a dependent child, so he receives \$718.10.

**b** Annie is under 18 and is living at home so she receives \$410.30.

**c** Jackie is over 22 and is partnered, with no children, so she receives \$712.30.

**EXERCISE 4.04** Answers on p. 466

### Government allowances and pensions

**Use the tables on the previous pages and the table above to answer the questions in this exercise. Assume all income and assets requirements are met.**

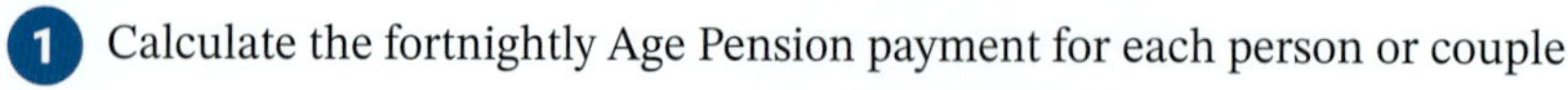

EXAMPLE 8

**1** Calculate the fortnightly Age Pension payment for each person or couple.

**a** Georgia, aged 89, receives \$1250 per year from her investments.

**b** Rodney, aged 71, and Bernadette, aged 75, care for their disabled daughter Katie full-time and have a combined income of \$470 per week.

**c** James, aged 62 years, has no savings, no job and lives alone.

**2** Why do you think that age pensioners receive less if they are partnered rather than single?

Foundation Mastery Complex

**3** The JobSeeker Payment is paid to unemployed people aged 22 years or over.
The maximum fortnightly rates are shown in the table.

| Status | Dependants | JobSeeker Payment |
|---|---|---|
| Single | No children | $778.00 |
| Single | Children | $833.20 |
| Partnered | | $712.30 |

Source: How much you can get, JobSeeker Payment, Services Australia.

**a** Hani is unemployed, and is single with no children. How much Jobseeker Payment would she receive if it was paid:

**i** fortnightly? **ii** yearly? **iii** monthly?

**b** Jay is married, unemployed and has 2 children. How much does he receive over 6 months?

**c** Does a single unemployed person receive more, less or the same as one with a partner? Do you think this is fair? Give 2 or 3 reasons for your answer.

**4** Carrie is 18 years old and is studying full-time at a senior high school. She lives with her parents and has a 2-year-old daughter. Which of the following is her fortnightly Youth Allowance? Select **A**, **B**, **C** or **D**.

**A** $395.30 **B** $639.00 **C** $836.60 **D** $691.80

**5** The Parenting Payment is paid to parents of children younger than 6 (younger than 14 for single parents). The maximum fortnightly rates are shown in the table.

| Status | Parenting Payment |
|---|---|
| Partnered | $712.30 |
| Partnered but separated due to illness, respite or prison | $833.20 |
| Single | $1007.50 |

Source: How much you can get, Parenting Payment, Services Australia.

**a** The Taylor family has 3 children: Daniel 15, Julianne 12 and Liam 8. How much does the Taylor family receive as a Parenting Payment per fortnight?

**b** The Lovejoys have 10 children under 18 at home and 2 children living away from home. Mr Lovejoy is currently in prison. What Parenting Payment does Mrs Lovejoy receive per fortnight?

**c** Jayda is separated from her partner and has a boy aged 9 years and a girl aged 7 years. What is her fortnightly Parenting Payment?

**d** Kell's wife is in hospital for the next 12 months. He is looking after their twin 1-year-old boys. How much Parenting Payment will he receive for:

**i** 1 year? **ii** 1 week?

EXAMPLE 10

**6** Shaun is a 19-year-old student living at home with his parents.
Which of the following is his fortnightly ABSTUDY allowance? Select **A**, **B**, **C** or **D**.

**A** $410.30 **B** $472.50 **C** $663.30 **D** $836.60

**7** Use the ABSTUDY table on page 134 to determine the fortnightly payment for each student.

**a** Susan is 17 years old, and is living with her partner

**b** Maxim is 28 years old, and is living at home with his wife and 2 children under 10 years

**c** Zelda is 24 years old, and is living away from her parents' home with her boyfriend

**d** Jasmine is 15 years old, and is living at home with her grandparents

☐ Foundation ○ Mastery ⬡ Complex

## Investigation

### Visit Centrelink

1. Investigate the different allowances by visiting a Services Australia/Centrelink branch or the **Services Australia** government website.
2. Research one type of allowance, looking at the eligibility conditions, residential qualifications, and income and assets tests.
3. Write a report and present this to the class.

# 4.05 Gross pay and net pay

Worksheet
Pay day

**Gross pay** is the amount a person earns, while **net pay** is the amount remaining after income tax and other **deductions** have been taken from gross pay.

**Net pay**

net pay = gross pay − tax − other deductions

Other deductions include:

- superannuation contributions (retirement fund)
- private health fund instalments
- union fees
- direct debits for bills
- savings.

### Example 11

Diane earns \$1029.60 per week. Deductions of \$303.73 for tax, \$42.17 for superannuation, \$26.34 for a health fund contribution and \$17.50 for gym fees are taken from her gross wage.

Calculate her:

**a** net pay

**b** income tax as a percentage of her gross pay, correct to one decimal place.

### Solution

**a** 
$$\begin{aligned}\text{net pay} &= \$1029.60 - \$303.73 - \$42.17 - \$26.34 - \$17.50 \\ &= \$639.86\end{aligned}$$

Diane's net pay is \$639.86.

**b** 
$$\begin{aligned}\text{percentage tax} &= \frac{\text{tax}}{\text{gross pay}} \times 100\% \\ &= \frac{\$303.73}{\$1029.60} \times 100\% \\ &= 29.4998\ldots \\ &\approx 29.5\%\end{aligned}$$

Diane's income tax is 29.5% of her gross pay.

## Example 12

Aidan earns a salary of \$67 510, pays 34% tax and has \$28.70 per fortnight deducted for union fees. Correct to the nearest cent, calculate:

**a** how much tax Aidan pays per fortnight

**b** Aidan's gross fortnightly pay

**c** Aidan's net fortnightly pay.

### Solution

**a** yearly tax = 34% × \$67 510
= \$22 953.40

fortnightly tax = \$22 953.40 ÷ 26
= \$882.8230 ...
≈ \$882.82

Aidan pays \$882.82 tax per fortnight.

**b** gross fortnightly pay = \$67 510 ÷ 26
= \$2596.5384 ...
= \$2596.54

Aidan's gross fortnightly pay is \$2596.54.

**c** net fortnightly pay = \$2596.54 – \$882.82 – \$28.70
= \$1685.02

Aidan's net fortnightly pay is \$1685.02.

**EXERCISE 4.05** Answers on p. 466

## Gross pay and net pay

**1** Alice earns \$1427.25 per week and has the following deductions from her gross wage: tax \$244.04, superannuation \$64.83, union fees \$18.20 and health fund \$26.15.Which of the following is Alice's net pay? Select **A**, **B**, **C** or **D**.

**A** \$1427.25 **B** \$1183.21 **C** \$1074.03 **D** \$1318.07

**2** Melanie earns \$1245.30 per week and has deductions of \$315.90 for tax, \$55.17 for superannuation, \$24.78 for a health fund contribution and \$17.50 for union fees. Calculate her:

**a** net pay

**b** income tax as a percentage of her gross pay, correct to one decimal place.

**3** Zhi earns a salary of \$65 674. Each week, he pays 28% tax, \$30.76 to his superannuation fund and \$86.11 in other deductions. Calculate Zhi's:

**a** gross weekly pay **b** tax **c** net weekly pay.

**4** Fatima earns a gross weekly wage of \$1035.25, from which is deducted \$34.11 for superannuation and \$7.90 for union fees.

**a** If Fatima's net pay is \$692.06, how much tax does she pay?

**b** What percentage of Fatima's gross wage is paid in tax (correct to 2 decimal places)?

**c** What percentage of Fatima's gross wage is paid in superannuation (correct to 2 decimal places)?

Foundation | Mastery | Complex

**5** Jeff pays 27.5% of his gross pay in tax and 8% in superannuation. Each week, he also has deducted $7.10 for union fees and $14.84 for life insurance premiums. If Jeff's gross weekly pay is $1238.20, calculate his:

**a** weekly tax

**b** weekly superannuation payment

**c** total weekly deductions

**d** net weekly pay.

**6** Lindy pays 22% of her gross pay in tax and 10% in superannuation. She also contributes $4.70 towards a staff social fund every week.

**a** If Lindy pays $235.52 tax each week, find her gross weekly pay.

**b** Calculate Lindy's weekly superannuation payment.

**c** Find Lindy's net weekly pay.

**7** **a** Fady earns a salary of $54 251. Calculate his fortnightly pay.

**b** Each fortnight, Fady's deductions from his pay are $307.26 tax, $145.60 for a car loan repayment and $27.14 to a health fund. What is Fady's net pay?

**c** What percentage of Fady's gross pay is paid in tax, correct to one decimal place?

**8** Melinda earns $29.38 per hour for a 39-hour week and pays 25% tax. Each week, she has $17.40 deducted from her pay for accident insurance. Calculate her:

**a** gross weekly pay **b** tax **c** net weekly pay.

**9** Matthew earns a salary of $108 275. He is paid monthly but, each payday, $4330 is deducted in tax, $1080.50 is deducted for superannuation and $107.54 is deducted for medical insurance.

**a** Calculate Matthew's monthly gross and net incomes.

**b** What percentage of Matthew's gross pay is paid in tax?

**c** What percentage of Matthew's gross pay is his net pay?

**10** Copy and complete the pay statement below, using a calculator or spreadsheet.

**Springfield Super Power**

Employee: H. Simpson | Employee number: 5521716

Pay period: 5.11.26–11.11.26 | Hourly rate: $35.48

| | Hours | Pay | Deductions | |
|---|---|---|---|---|
| Normal | 35 | ☐ | Tax (25%) | ☐ |
| Time-and-a-half | 4 | ☐ | Superannuation | $60.65 |
| Double time | 1 | ☐ | Union | $18.15 |
| **Gross wage** | | ☐ | Health fund | $24.80 |
| **Total deductions** | | ☐ | | |
| **Net wage** | | ☐ | | |

**11** Copy and complete the employee payslip below.

**State Education Board**

Employee: E. Krabappel | Serial number: 8805285

Pay period: 14.3.27–27.3.27 | Salary: $87 026

| Other deductions | | Pay for this fortnightly period | |
|---|---|---|---|
| Superannuation | $71.60 | Gross pay | ☐ |
| Union | $16.22 | Tax (31%) | ☐ |
| Loan repayment | $43.55 | Other deductions | ☐ |
| Car insurance | $20.84 | Net pay | ☐ |
| **Total deductions** | ☐ | | |

☐ Foundation ○ Mastery ○ Complex

## Technology

### Payslip spreadsheet

Lachlan is a forklift driver. He is paid an hourly rate of $20.91, for a normal 8-hour day working from 9 am until 6 pm on weekdays.

He is paid time-and-a-half for working on Saturdays and double time for any nights working from 6 pm until midnight. He is also paid an extra $8.06 per hour for each hour or part of an hour worked after 6 pm.

**a** Copy Lachlan's incomplete payslip into a spreadsheet, as shown below.

| | A | B | C | D | E | F | G | H | I |
|---|---|---|---|---|---|---|---|---|---|
| 1 | Payslip Monday 27th March - Saturday 1st April | | | | | | | | |
| 2 | | | | | | | | | |
| 3 | Employee | Lachlan O'Connell | | | Normal Hourly Pay Rate | 20.91 | | Extra pay/hr | 8.06 |
| 4 | | | | | | | | | |
| 5 | Date | 27-Mar | 28-Mar | 29-Mar | 30-Mar | 31-Mar | 01-Apr | TOTAL | |
| 6 | Hours | 9 am – 5 pm | 10:30 am – 6 pm | 6:30 pm – 10 pm | 11 am – 6:30 pm | 9:45 am – 5:30 pm | 9:30 am – 2 pm | | |
| 7 | Normal hours | | | | | | | | |
| 8 | Time-and-half hours | | | | | | | | |
| 9 | Double Time hours | | | | | | | | |
| 10 | Extra Pay for nights | | | | | | | | |
| 11 | | | | | | | TOTAL PAY | | |



Complete rows 7, 8 and 9 for the total number of hours for each pay rate.

In cells H7, H8 and H9, enter formulas for the totals. In cell D10 enter =D9*I3. In cell E10 enter =E9*I3 In cell H10 enter a formula for the total. In cell H11, enter a formula for the total amount Lachlan was paid for this week.

**b** The following week, Lachlan worked these hours:

| | | | |
|---|---|---|---|
| Mon 4 Apr | 9:30 am – 5:00 pm | Thur 7 Apr | 12:00 pm – 7:00 pm |
| Tues 5 Apr | 6:00 pm – 12:00 am | Sat 9 Apr | 10:00 am – 4:00 pm |
| Wed 6 Apr | 6:30 pm – 12:00 am | | |

Modify your spreadsheet to calculate Lachlan's total pay for this week.

# 4.06 Income tax and Medicare levy

**Video**
Income tax

**Worksheet**
Income tax tables

**Spreadsheet**
Income tax

## Taxable income and allowable tax deductions

Anyone who earns an income pays **income tax** to the government. The government uses this money to fund public services such as schools, hospitals, roads, trains, the defence services and the police force. Not all of a person's **gross income** is taxed, however. If we use some of our income for work-related expenses or to donate money to charities, these amounts are called **allowable tax deductions** and are not taxed. The remaining part of our income that is taxed is called **taxable income**.

**Taxable income**

taxable income = gross income − allowable tax deductions

Here are some examples of allowable tax deductions:

- work-related expenses, such as tools of trade, uniforms, car-related expenses, trade journals, subscriptions to professional organisations and travel to conferences
- donations to charities
- depreciation of home office equipment that is used to earn income
- bank fees on investments for which **interest** is earned.

## Income tax

Income tax is calculated on our taxable income (rounded down to the nearest dollar) for a financial year (1 July of one year to 30 June of the next year) and operates on a sliding scale so the higher our taxable income, the higher the rate of tax we pay.

| Income tax rates for Australian residents | |
|---|---|
| **Taxable income** | **Tax on this income** |
| \$0 – \$18 200 | Nil |
| \$18 201 – \$45 000 | 16c for each \$1 over \$18 200 |
| \$45 001 – \$135 000 | \$4288 plus 30c for each \$1 over \$45 000 |
| \$135 001 – \$190 000 | \$31 288 plus 37c for each \$1 over \$135 000 |
| \$190 001 and over | \$51 638 plus 45c for each \$1 over \$190 000 |

Source: Tax rates – Australian resident, Australian Taxation Office, 2025

### Example 13

Cyndi earned \$56 375 and claimed \$326.20 in work-related expenses and \$125 in donations to charities. Find the amount of income tax Cyndi should pay, using the table above.

#### Solution

Cyndi's taxable income = \$56 375 − \$326.20 − \$125
= \$55 923.80
≈ \$55 923 (rounded down to the nearest dollar)

Cyndi's income is in the 3rd tax bracket (\$45 001 – \$135 000) for which the tax is \$4288 plus 30c for each \$1 over \$45 000.

Cyndi's tax payable = \$4288 + 0.30 × (\$55 923 − \$45 000)
= \$7564.90

Cyndi should pay \$7564.90 in income tax.

## Medicare levy

The standard **Medicare levy** for Australian residents is 2% of taxable income, although some low-income earners pay a reduced levy or no levy.

### Example 14

Calculate the Medicare levy for Suyen, who has a taxable income of $62 379.

#### Solution

Medicare levy = 2% × $62 379
= $1247.58

**EXERCISE 4.06** Answers on p. 466

### Income tax and Medicare levy

**1** Find the amount of income tax payable for each person.

- **a** Kamel earned $31 425 and claimed $285 in work-related expenses
- **b** Susie earned $151 412 and claimed $1036 in travel expenses and $643 depreciation of her home office equipment
- **c** Toula received $14 615 for the year.

**2** Use the table on the previous to calculate the income tax payable on a taxable income of $88 758. Select **A**, **B**, **C** or **D**.

**A** $4288 **B** $30 913 **C** $17 415.40 **D** $20 475.50

**3** Jenni earns a salary of $91 262 and claims a tax deduction of $1810.15 for work-related expenses.

- **a** Calculate Jenni's taxable income.
- **b** Calculate Jenni's amount of income tax payable.
- **c** Calculate the amount of Jenni's Medicare levy.

**4** Catriona earns a salary of $139 215 and claims these deductions:

- travel and entertainment expenses $650.25
- depreciation of home computer $314.80
- membership of a professional association $120.50

- **a** Calculate Catriona's taxable income.
- **b** Calculate Catriona's income tax payable.
- **c** What is the amount of Catriona's Medicare levy?

**5** Luke earns $3769.24 per fortnight as a teacher. Last financial year, he also earned $286.10 bank interest, $2050.96 share dividends and $892.51 book royalties. He is able to make tax deductions of $241.60 for union fees, $345.80 for car expenses, $175.80 for professional journals and $843.50 for expenses related to the writing of his book. For the financial year, calculate Luke's:

**a** taxable income **b** income tax payable **c** Medicare levy.

**6** Ronel is an electrician who works as a sole trader. His yearly income, through his business, is $185 000. His work-related expenses include $6500 for insurance, $19 200 for his vehicle and travel, $1000 for licenses and union fees, $900 for home office expenses, $850 for clothing and protective gear, and $1750 for tools and equipment. Calculate Ronel's:

> For sole traders, all business expenses are allowable tax deductions.

**a** taxable income **b** income tax payable **c** Medicare levy.

7 Lachlan has an annual salary of $138 500 and also earns $2760 from interest on investments. His allowable tax deductions are $1560 for travel, $350 for home office supplies and $286 for union fees. Calculate the total tax Lachlan must pay, including the Medicare levy.

8 Eddie is a sole trader. He cleans windows and charges $55.80 per hour. He works 8 hours Monday to Friday and 5 hours on a Saturday, for 48 weeks in a year. His weekly expenses amount to $450 per week and replacement of equipment costs $4800 per year. Calculate Eddie's:

**a** taxable income **b** total tax payable.

9 The table shows the tax rates for Australian citizens who live in another country.

| Income tax rates for Australian foreign residents | |
|---|---|
| Taxable income | Tax on this income |
| $0–$135 000 | 30c for each $1 |
| $135 001–$190 000 | $40 500 + 37c for each $1 over $135 000 |
| $190 001 and over | $60 850 + 45c for each $1 over $190 000 |

Source: Tax rates – Foriegn resident, Australian Taxation Office, 2025.

Find the income tax payable by each Australian citizen.

**a** Claude, who lives in France and has a taxable income of $37 850

**b** Marie, who has a taxable income of $142 670 and lives in Canada

**c** Benny, who lives in Sweden and has a taxable income of $203 500.

10 Tuala has a taxable income of $225 000 and lives in a country where her tax payable is $56 595 plus $M$ cents for each $1 over $195 000. Tuala paid 31.02% of her taxable income in tax. Find the value of $M$.

## Technology

### Income tax calculator

Create and complete the spreadsheet below for calculating the income tax payable when you enter the taxable income in the correct tax bracket. Two formulas have already been given.

| | A | B | C | D |
|---|---|---|---|---|
| 1 | Tax bracket | | | |
| 2 | From | To | Taxable income | Income tax |
| 3 | $0 | $18 200 | | $0 |
| 4 | $18 201 | $45 000 | | =(C4-B3)*0.16 |
| 5 | $45 001 | $135 000 | | =4288+(C5-B4)*0.30 |
| 6 | $135 001 | $190 000 | | |
| 7 | $190 001 | and over | | |



Use your spreadsheet to check your answers to Exercise 4.06.

Create a similar spreadsheet to calculate the Medicare levy as well.

### Online income tax calculator

The **Australian Taxation Office** website has online calculators for income tax. Visit www.ato.gov.au and search 'Simple tax calculator' to find the income tax calculator for individuals.

1 Select the current financial year.
2 Enter the taxable income $83 000 as '83000' (no spaces).
3 Select 'Resident for full year' and click 'Next'.
4 The estimated tax payable will be shown on a new screen.
5 Repeat for at least 2 more taxable incomes.

☐ Foundation ○ Mastery ⬡ Complex

## Did you know?

### Income tax in other countries

Denmark has the highest income tax rates, ranging from 38% to 65%, while the United Arab Emirates (UAE) has no income tax at all. The **Worldwide Tax** website gives the current tax rates around the world. Here are the tax rates for 7 countries.

| Country | Rate |
|---|---|
| Brazil | 7.5% to 27.5% |
| China | 3% to 45% |
| New Zealand | 10.5% to 39.0% |
| Saudi Arabia | No income tax |
| Spain | 19% to 47% |
| Sweden | 0% to 54% |
| USA | 10% to 37% |

**Find the current tax rates in 5 other countries.**

# PAYG and tax returns

4.07

To avoid having to pay income tax as a huge lump sum at the end of a financial year, tax is deducted every payday from a person's gross pay and is called **Pay As You Go (PAYG) tax**. The employer estimates how much tax to take out of each pay and, at the end of the financial year, the actual income tax payable is calculated.

Every income earner must submit a **tax return** to the Australian Tax Office (ATO) at the end of the financial year. A PAYG Payment Summary, showing total gross earnings and PAYG tax paid, is issued by the employer and used by the employee to complete the tax return, including any allowable tax deductions. The ATO then calculates the income tax and Medicare levy payable. Any earner who has paid more PAYG tax than necessary will receive a **tax refund**. However, anyone who has not paid enough PAYG tax will have a **tax debt** and will need to pay the difference owing.

### Example 15

Mehek earns a gross pay of \$2025.38 per fortnight. Her deductions are PAYG tax and \$65.70 for superannuation and \$15.25 for union fees. Find Mehek's:

**a** PAYG tax using this table

| Fortnightly earnings (\$) | PAYG tax withheld (\$) |
|---|---|
| 2010.00 | 290.00 |
| 2016.00 | 292.00 |
| 2022.00 | 294.00 |
| 2030.00 | 296.00 |
| 2036.00 | 298.00 |
| 2042.00 | 300.00 |

**b** net pay per fortnight

**c** total deductions as a percentage of her gross income (correct to one decimal place)

**d** PAYG tax paid in one year

## Solution

**a** In the table, \$2025.38 falls between \$2022 and \$2030 range, so use the value for \$2022.

fortnightly PAYG tax = \$294

**b** net pay = \$2025.38 – (\$294 + \$65.70 + \$15.25)

= \$1650.43

**c** total deductions = \$294 + \$65.70 + \$15.25

= \$374.95

$$\text{deductions percentage} = \frac{\$374.95}{\$2025.38} \times 100\%$$
$$= 18.5125\ldots\%$$
$$= 18.5\%$$

**d** PAYG tax paid in one year = \$294 × 26

= \$7644

Video
Tax refund

## Example 16

Mia earns a salary of \$115 412 and claims tax deductions of \$1742.30. Last year she paid a total of \$27 768 PAYG tax. Calculate her:

**a** taxable income

**b** income tax payable

**c** Medicare levy

**d** tax refund or tax debt.

## Solution

**a** taxable income = \$115 412 – \$1742.30

= \$113 669.70

≈ \$113 669 — rounding down to the nearest dollar

Mia's taxable income is \$113 669.

**b** tax payable = \$4288 + 0.30 × (\$113 669 – \$45 000) — from the table on page 140

= \$24 888.70

Mia's income tax is \$24 888.70.

**c** Medicare levy = 2% × \$113 669

= \$2273.38

Mia's Medicare levy is \$2273.38.

**d** total taxes = \$24 888.70 + \$2273.38 = \$27 162.08

amount of PAYG tax paid = \$27 768

tax refund = \$27 768 – \$27 162.08 — refund because PAYG tax paid > total taxes

= \$605.92

Mia's tax refund will be \$605.92

**EXERCISE 4.07** Answers on p. 466

## PAYG and tax returns

**1** Jasmine paid $148 PAYG tax out of her weekly wage of $1014.65.

**a** How much PAYG tax did Jasmine pay for the year?

**b** How much did Jasmine earn for the year, after tax?

**2** Hope earned a salary of $73 489 per year. She received her weekly pay, after PAYG tax of $276 and $62.71 health insurance were taken out.

**a** How much PAYG tax did Hope pay in a year?

**b** How much was Hope's weekly pay?

**3** Find the PAYG tax payable for a gross pay of: EXAMPLE 15

**a** $6550.17 monthly, using this table

| Monthly earnings ($) | PAYG tax withheld ($) |
|---|---|
| 6530.33 | 1326.00 |
| 6539.00 | 1330.00 |
| 6552.00 | 1335.00 |
| 6565.00 | 1339.00 |
| 6582.33 | 1343.00 |
| 6595.33 | 1348.00 |

**b** $2041.70 fortnightly, using the table from Example 15 on page 143.

**Use the table of income tax rates on page 140 to answer the remaining questions.**

**4** Jacinta is on a salary of $78 389 and is paid monthly. Her monthly deductions are PAYG tax, $72.30 for superannuation, $93.40 for health insurance, and $63.80 for union membership. Find Jacinta's:

**a** monthly income

**b** monthly PAYG tax

**c** monthly net pay

**d** total deductions as a percentage of her gross income, correct to one decimal place

**e** PAYG tax paid in one year.

**5** Tyrone earns a salary of $52 469 and each fortnight has deductions of $74 for superannuation and $58.70 for health insurance.

**a** Calculate Tyrone's gross fortnightly income.

**b** How much PAYG tax does he pay per fortnight?

**c** What is Tyrone's net fortnightly income?

**d** How much PAYG tax did he pay in one year?

**6** Gabby is an Australian resident who earned a yearly salary of $204 560. She claimed tax deductions of $10 534 and paid a total of $62 376 PAYG tax. Calculate her: EXAMPLE 16

**a** taxable income  **b** income tax payable

**c** Medicare levy  **d** tax debt or tax refund.

**7** Anand had a taxable income of $278 639 and paid $97 203 in PAYG tax over the financial year. Find his:

**a** income tax payable  **b** Medicare levy  **c** tax debt or tax refund.

**8** Liam earned a wage of $1089.60 per week and paid PAYG tax of $172. He was able to claim $38.25 per week in allowable tax deductions and $150 per year in donations. Find his:

**a** taxable income  **b** income tax payable

**c** Medicare levy  **d** tax debt or tax refund.

**9** Daeho paid PAYG tax of \$6032 and had a taxable income of \$44 589. Which of the following is Daeho's tax refund? Select **A**, **B**, **C** or **D**.

**A** \$1809.76 **B** \$1102.24 **C** \$3257.12 **D** \$4222.24

**10** Kim had a taxable income of \$381 459 and paid \$131 545 in PAYG tax. Which of the following is Kim's tax debt? Select **A**, **B**, **C** or **D**.

**A** \$6249.55 **B** \$13 878.73 **C** \$9642.17 **D** \$8281.60

**11** Use the tax table for Australian foreign residents on page 142 to find the tax refund or tax debt for each of the following Australian citizens.

**a** Cecily, who has a taxable income of \$93 457 and who paid \$734 per fortnight PAYG tax

**b** Alice, who paid \$8521 PAYG tax every quarter and earned a taxable income of \$148 534

**c** Simon, who paid \$43 460 tax per month and earned a taxable income of \$1.2 million.

**12** Alex earns a salary of \$79 080 p.a. Each month, he has deductions of \$140.50 for superannuation and \$48 for union fees. He also has allowable deductions of \$864 for work-related expenses and \$154 in donations to charities.

**a** Calculate Alex's monthly gross income.

**b** Use the table in Question **3** to find the PAYG tax that is taken out of his gross pay.

**c** What is Alex's monthly net income?

**d** Calculate Alex's taxable income.

**e** What is the total tax Alex must pay, including a 2% Medicare levy?

**f** Does Alex have a tax debt or a tax refund? Give reasons.

## Technology

### Online PAYG tax calculator

Visit the **Australian Taxation Office** website at www.ato.gov.au and search 'Tax withheld for individuals calculator' to find the PAYG calculator for individuals.

Use the calculator to find the PAYG tax payable for a gross pay of:

**a** \$1370 weekly

**b** \$3985 fortnightly

**c** \$11 108 monthly

## Investigation

### Tax returns

You may use a spreadsheet to complete this task.

**1** Choose the profile of a person with a particular occupation and salary or wage. Decide on what PAYG tax that person pays each week, fortnight or quarter.

**2** Research the allowable tax deductions for this occupation. These may include travel expenses, special clothing, tools of trade, books and journals, attendance at conferences and seminars, home office expenses and self-education expenses.

**3** Add other income you wish, such as bank interest or money from a part-time job. If you do this, there may be other allowable tax deductions, such as bank fees or travel between jobs, which are claimable.

**4** Use the form from a Tax Pack to complete a tax return for the person.

**5** Calculate the person's Medicare levy.

**6** Calculate the person's tax payable and any tax refund or tax debt.

☐ Foundation ○ Mastery ○ Complex

## Sample HSC problem Answers on p. 466

(7 marks) Nathan paid PAYG tax of $237 per week from his weekly wage of $1295.20.

**a** How much PAYG tax did Nathan pay in the year? 1 mark

**b** What was Nathan's gross income for the year? 1 mark

**c** If he had $986.50 in allowable tax deductions, what was Nathan's taxable income? 1 mark

**d** Use the income tax table on page 140 to calculate Nathan's income tax and Medicare levy. 2 marks

**e** Will Nathan receive a tax refund or will he need to pay a tax debt? Calculate the amount. 2 marks

## Study Tip

### Showing your working

Why is it important to show your working when answering a mathematics question?

- To demonstrate that you know your mathematics and the steps required to solve the problem
- So that in an exam, you can receive 'part marks' for correct working, even if you get the wrong answer
- To show that you really do know how to answer the question and aren't simply guessing or bluffing
- To remind yourself how you solved the problem, especially when studying
- So that you can describe how you achieved your answer (explaining something to someone often reinforces your learning and understanding).

# CHAPTER SUMMARY

This chapter, *Earning money and taxation*, looks at the different ways people earn or receive incomes and pay income tax. Before you move on, consider what you learned in this chapter and revisit any sections that may have been unclear.

| What you learned in this chapter... | Section |
|---|---|
| Calculate wages, salaries and overtime pay and convert between weekly, fortnightly, monthly and yearly earnings | 4.01 Wages, salaries and overtime |
| Calculate earnings based on commission, piecework and royalties | 4.02 Commission, piecework and royalties |
| Calculate bonuses and annual leave loading | 4.03 Bonuses and annual leave loading |
| Calculate government allowances and pensions | 4.04 Government allowances and pensions |
| Calculate net pay following deductions from gross pay | 4.05 Gross pay and net pay |
| Calculate allowable tax deductions, taxable income and income tax (including Medicare levy) | 4.06 Income tax and Medicare levy |
| Calculate Pay As You Go (PAYG) tax and tax payable or refund owing after a tax return is completed at the end of the financial year | 4.07 PAYG and tax returns |

To help master this topic, make a summary mind map. Use the chapter outline and the mind map below as a guide. Add your own words, symbols, diagrams, boxes and reminders. The summary should give you a 'whole picture' view of the topic and allow you to identify any weak areas to revisit in your revision.

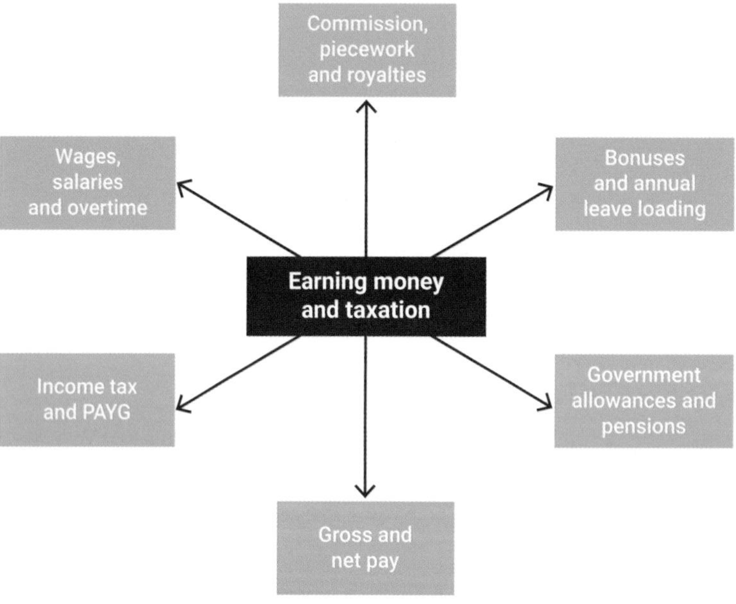

Puzzle
ning and
axation
ossword

# Test yourself

Answers on p. 466

4

**1** Convert each wage to a yearly amount, correct to the nearest cent. 4.01

**a** weekly wage of $831.15 **b** fortnightly wage of $1948

**c** monthly wage of $5786.

**2** Calculate to the nearest cent the pay amount for an annual salary of: 4.01

**a** $138 611 paid weekly **b** $98 503 paid fortnightly

**c** $263 764 paid monthly.

**3** Marley works 3 days a week as a part-time carer. She is paid $33.40 per hour up to 5 pm and time-and-a-half after that. Use the roster to calculate how much she earned this week. 4.01

| Day | Start time | Finish time |
|---|---|---|
| Monday | 1000 | 1900 |
| Wednesday | 0900 | 1300 |
| Friday | 1600 | 2200 |

**4** Riya worked 8 hours on Boxing Day and was paid double time-and-a half, earning $572.60. Calculate Riya's hourly rate of pay. 4.01

**5** Huon picked 28 buckets of grapes and was paid $7.45 per bucket. 4.02

**a** How much did Huon earn?

**b** How many buckets would Huon need to pick to earn at least $400?

**6** Maggie made cupcakes and sold them to a shop at $4.80 per dozen, or $38 for 100. How much more per cupcake did she earn by selling them by the dozen than in lots of 100? 4.02

**7** Goran earned a 10% royalty on each of his books sold. If his book sells for $45.95, how much will he earn if 2537 books are sold? 4.02

**8** Byron works in a mine and earns an hourly rate of $44.65. He is paid double time for working weekend shifts. 4.03

Calculate Byron's total income in a week he worked the following shifts:

Monday: 12 noon – 5 pm
Wednesday: 2 pm – 7 pm
Friday: 9 am – 12 noon
2 pm – 6 pm
Saturday: 10 am – 4 pm

**9** The Great Gals electrical store pays a holiday loading of 17.5% of 4 weeks' pay to each employee. How much holiday loading is paid to Nirmal, the manager, who earns a salary of $86 430? 4.03

**10** Mariette's holiday pay included her annual leave loading of 17.5% for 4 weeks and her normal pay for 3 weeks. What is Mariette's fortnightly income if her annual leave loading is $1247.89 and her holiday pay was $6595.97? Select **A**, **B**, **C** or **D**. 4.03

**A** $5348.08 **B** $1337.02 **C** $4010.06 **D** $3565.39

**11** Use the tables from Section 4.04 on page 132 to calculate the fortnightly Age Pension payment of Roger (aged 67) and Tatiana (aged 68) who are legal guardians of their 2 school-aged grandsons and have a combined income of $280 per week doing administrative work in an office. 4.04

☐ Foundation ○ Mastery ⬡ Complex

4.05 **12** Jack earns a salary of $94 762, pays $411 PAYG tax per week and has $28.70 per week deducted for union fees. Correct to the nearest cent, calculate:

**a** how much tax Jack pays per year

**b** how much tax Jack pays per fortnight

**c** Jack's gross fortnightly pay

**d** Jack's net pay per fortnight.

4.06 **13** Melissa earned a yearly salary of $142 636. She paid $1424.30 PAYG tax per fortnight and claimed tax deductions of $13 102.20 per year. Calculate her:

**a** taxable income (rounded down to the nearest dollar)

**b** income tax payable (using the table on page 140)

**c** Medicare levy

**d** tax debt or tax refund.

4.07 **14** Joshua earns a salary of $53 000.

**a** Use the table in Example 15 on page 143 to find the PAYG tax if Joshua is paid fortnightly.

**b** Joshua has allowable deductions of $1260 for work-related expenses. Use the table on page 140 to calculate the total tax Joshua has to pay, including the Medicare levy.

**c** Does Joshua have a tax debt or tax refund? Give reasons.

4.07 **15** If an employee earns more than $7305 per fortnight, the PAYG tax is $2134 plus 47 cents for each $1 of earnings more than $7305. Calculate the PAYG tax for a fortnightly income of $8450.

□ Foundation ○ Mastery ○ Complex

# MEASUREMENT

One gigalitre is a billion litres.

Sydney Harbour contains approximately 500 gigalitres of water. The entire Sydney urban area uses the equivalent of 4 Sydney Harbours each year, while the whole of Australia uses the equivalent of 50 Sydney Harbours per year.

## Chapter outline

## In this chapter you will:

- convert between metric units of length, area, volume, capacity, mass and time
- express numbers in scientific notation and round to significant figures
- solve problems involving Pythagoras' theorem
- calculate perimeters and areas of shapes, including circular and composite shapes
- calculate perimeters, areas and volumes of irregular shapes using dissection and the trapezoidal rule
- calculate surface areas, volumes and capacities of prisms, cylinders, spheres and composite solids
- calculate volumes of pyramids and cones.

**Videos (15):**

**5.02** Significant figures
**5.04** Pythagoras' theorem problems
**5.05, 5.06** Sectors • Perimeter and area of a sector
**5.06** Areas of trapeziums, circles and sectors
**5.07** The trapezoidal rule
**5.08** Volume and capacity units • Volumes of prisms and cylinders • Surface area of a prism
**5.09** Volumes of prisms and cylinders • Capacity of a cylinder • Surface area of a cylinder
**5.10** Volume of a pyramid • Volume of a cone
**5.11** Volume of a sphere
**5.12** Volumes of composite shapes

**Worksheets (21):**

**Skillcheck** Assignment 5
**5.02** Significant figures
**5.04** Applications of Pythagoras' theorem • Pythagoras' problems • Pythagorean two-step problems
**5.05** A page of circular shapes
**5.06** Areas of composite shapes • A page of circular shapes • Composite areas • Applications of area
**5.07** Offset surveys • Trapezoidal rule
**5.08** Estimating area and volume • Nets of solids
**5.09** Measurement in the home • Sweet areas and volumes • Surface areas of solids
**5.11** Measurement formulas chart • Volumes of solids
**5.12** A page of solid shapes • Volumes of water

**Puzzles (3):**

**5.03** Scientific notation puzzle
**5.11** Formula matching game • Surface area riddle

Nelson MindTap

To access resources above, visit **cengage.com.au/nelsonmindtap**

## Terminology

| | | | |
|---|---|---|---|
| arc | capacity | circumference | cone |
| hemisphere | irregular | offset survey | prism |
| pyramid | quadrant | scientific notation | sector |
| significant figures | sphere | trapezoidal rule | volume |

## SkillCheck Answers on p. 466

Worksheet
Assignment 5

**1** Evaluate without using a calculator:

**a** $20.83 \times 1000$ **b** $970.2 \div 10$ **c** $6.59 \times 10\,000$

**d** $72.5 \div 100$ **e** $10.4 \div 1000$ **f** $0.0735 \times 10$

**2** Evaluate:

**a** $10^7$ **b** $10^4$ **c** $10^{-2}$

**3** Find the area of each shape. Express the answer to part **c** to 2 decimal places.

**a**

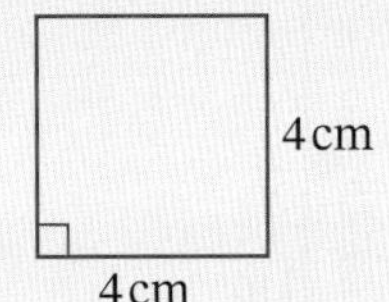


**b**

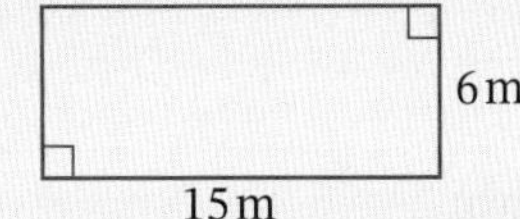


**c**

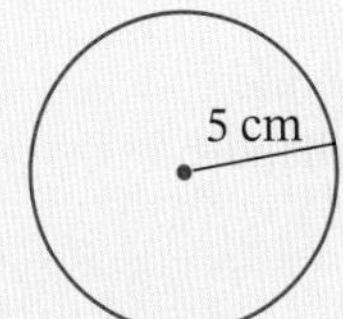


**d**

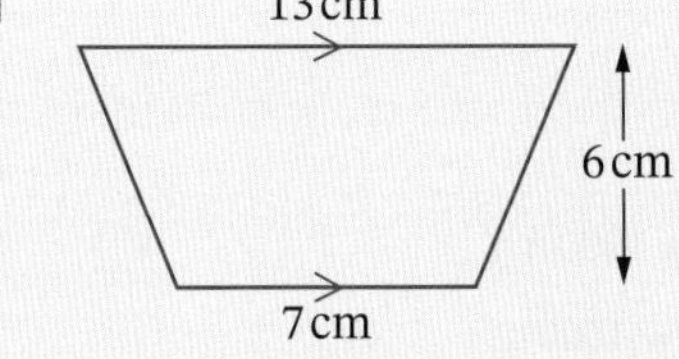


**4** **a** Find the area of this triangle.

**b** Find the value of $h$.

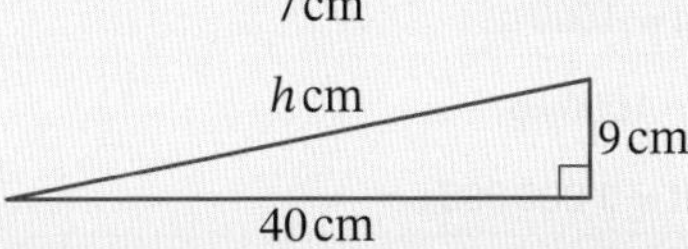


**5** **a** If bananas cost $5.10 per kilogram, how much would we pay for 250 grams?

**b** If potatoes cost $4.15 for 5 kg, how much would 3 kg of potatoes cost?

**6** Convert 210 cm to:

**a** metres **b** millimetres **c** kilometres.

**7** Convert 8.1 L to:

**a** kilolitres **b** millilitres.

# 5.01 Metric units

This table shows some common metric units for length, mass, time and capacity.

| Unit | Relationships |
|---|---|
| **Length** | |
| micrometre (μm) | |
| millimetre (mm) | |
| centimetre (cm) | 1 cm = 10 mm |
| metre (m) | 1 m = 100 cm = 1000 mm = 1 000 000 μm |
| kilometre (km) | 1 km = 1000 m |
| **Mass** | |
| milligram (mg) | |
| gram (g) | 1 g = 1000 mg |
| kilogram (kg) | 1 kg = 1000 g |
| tonne (t) | 1 t = 1000 kg |
| megatonne (Mt) | 1 Mt = 1 000 000 t |
| **Time** | |
| second (s) | |
| minute (min) | 1 min = 60 s |
| hour (h) | 1 h = 60 min = 3600 s |
| day (day) | 1 day = 24 h |
| **Capacity** | |
| millilitre (mL) | |
| litre (L) | 1 L = 1000 mL |
| kilolitre (kL) | 1 kL = 1000 L |
| megalitre (ML) | 1 ML = 1 000 000 L = 1000 kL |

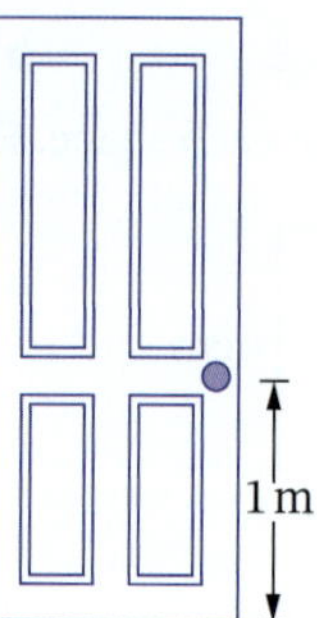


A **metre** is about the height of a door handle.

A **kilogram** is the mass of about 5 oranges.

A **litre** is the capacity of a tall carton of milk.

## Metric prefixes

- micro- means 'one-millionth' or $\frac{1}{1000000}$
- milli- means 'one-thousandth' or $\frac{1}{1000}$
- centi- means 'one-hundredth' or $\frac{1}{100}$
- kilo- means 'one thousand' or 1000
- mega- means 'one million' or 1 000 000

## Example 1

Convert:

**a** 23.5 km to metres **b** 450 L to kilolitres **c** 84 000 mg to kilograms

### Solution

**a** $23.5\text{ km} = 23.5 \times 1000\text{ m}$

$= 23\,500\text{ m}$

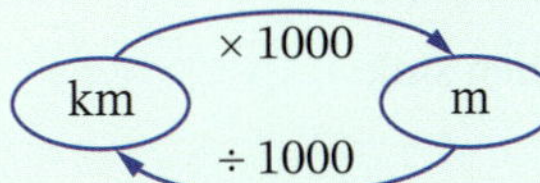


**b** $450\text{ L} = 450 \div 1000\text{ kL}$

$= 0.45\text{ kL}$

× 1000
kL
L
÷ 1000

**c** $84\,000\text{ mg} = 84\,000 \div 1000 \div 1000\text{ kg}$

$= 0.084\text{ kg}$

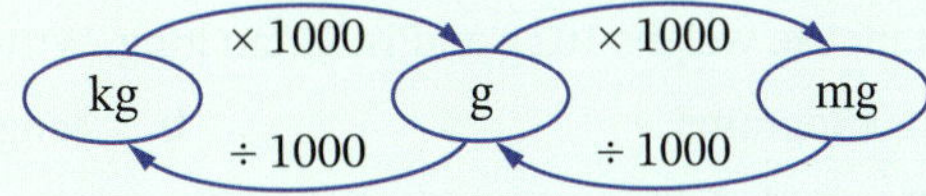


### Converting units

- To convert a larger unit to a smaller one, multiply to get more of the smaller units.
- To convert a smaller unit to a larger one, divide to get fewer larger units.

**EXERCISE 5.01** Answers on p. 467

## Metric units

EXAMPLE 1

**1** Convert:

**a** 6.3 cm to mm **b** 4.36 m to cm **c** 7200 mm to m

**d** 285 g to kg **e** 6.9 mm to μm **f** 58 000 mL to L

**g** 5.32 kg to g **h** 3400 kg to t **i** 4720 L to kL

**j** 6000 mg to kg **k** 7.5 h to s **l** 9.4 km to cm

**2** Which of the following is the number of grams in 100 kg? Select **A**, **B**, **C** or **D**.

**A** 100 000 **B** 1000 **C** 1 000 000 **D** 10 000

**3** Oliver's height is 169 cm. Convert this to:

**a** millimetres **b** metres.

**4** Mai weighs 57.5 kg. What is her mass in grams?

**5** Rhys swam in a 1500 m race. How many kilometres did he swim?

**6** Erin left home at 7:41 am and arrived at work at 9:18 am What was her travelling time in:

**a** minutes? **b** seconds? **c** hours (correct to one decimal place)?

**7** Yusuf's water consumption for July was 59 kL. How many litres of water did he use?

☐ Foundation ○ Mastery ○ Complex

8 What metric unit would you use to measure each of the following?

a your mass
b your height
c the length of your bedroom
d the distance from Perth to Sydney
e the capacity of your kitchen sink
f the mass of an ant
g the width of a sheet of paper
h the amount of water in an Olympic-sized pool
i the mass of a train
j the thickness of a strand of hair

9 Taylor's pet mouse weighs 102 g and is 5.2 cm long. Write its:

a mass in kilograms  b length in metres.

10 The Great Pyramid of Cheops (or Khufu) at Giza in Egypt contains 2 000 000 blocks, each weighing 2.45 tonnes. What is the total mass of the pyramid in megatonnes?

11 Josh competed in a sprint triathlon. He took 42 minutes for the 20 km bike ride, 18 minutes for the 750 m swim and 38 minutes for the 5.5 km run.

a What was the total distance that Josh covered, in kilometres?
b What was Josh's total time, in hours and minutes?

12 Mia's swimming pool holds 8560 L of water. How many kilolitres is this?

13 Which is the most likely measurement for each of the following? Select **A**, **B**, **C** or **D** each time.

a The height of a 10-storey building
**A** 400 m **B** 40 m **C** 4 m **D** 4000 m
b The capacity of a backyard swimming pool
**A** 30 L **B** 300 L **C** 3000 L **D** 30 000 L
c The mass of an orange
**A** 3 mg **B** 3 g **C** 300 g **D** 3000 g
d The distance from Sydney to Brisbane
**A** 100 km **B** 500 km **C** 1000 km **D** 5000 km
e The diameter of a small plate
**A** 12 cm **B** 12 mm **C** 120 m **D** 1.2 cm
f The capacity of a soft drink can
**A** 380 mL **B** 3800 mL **C** 3.8 mL **D** 38 mL

14 On a 2.5 km walk, John covers 80 cm with each step, while his daughter Anna covers 55 cm with each of her steps. How many more steps does Anna take than her father during the walk?

15 How many seconds are there in one day?

☐ Foundation ○ Mastery ○ Complex

## Did you know?

### Metric prefixes

| Prefix | Abbreviation | Meaning | Example |
|---|---|---|---|
| atto | a | $10^{-18}$ | |
| femto | f | $10^{-15}$ | |
| pico | p | $10^{-12}$ | |
| nano | n | $10^{-9}$ | nanosecond |
| micro | μ | $10^{-6} = \frac{1}{1000000}$ | microgram |
| milli | m | $10^{-3} = \frac{1}{1000}$ | millibar |
| centi | c | $10^{-2} = \frac{1}{100}$ | centilitre |
| deci | d | $10^{-1} = \frac{1}{10}$ | decibel |
| deca | da | $10^1 = 10$ | decametre |
| hecto | h | $10^2 = 100$ | hectopascal |
| kilo | k | $10^3 = 1000$ | kilojoule |
| mega | M | $10^6 = 1000\,000$ | megahertz |
| giga | G | $10^9$ | gigawatt |
| tera | T | $10^{12}$ | terabyte |
| peta | P | $10^{15}$ | |
| exa | E | $10^{18}$ | |

**Choose 4 units from the Example column in this table and find out what they are used to measure.**

# Significant figures 5.02

Numbers can be rounded to decimal places or **significant figures** to show their level of accuracy.

**Video** Significant figures

**Worksheet** Significant figures

## Example 2

Round:

**a** 3812.4, correct to 2 significant figures

**b** 1005, correct to 3 significant figures

**c** 0.003 572, correct to 2 significant figures.

### Solution

**a** <u>38</u>12.4

The first 2 significant figures are **<u>38</u>** and the next digit is **1** (which is less than 5), so we round down to **8** and use 2 0s as placeholders.

$3812.4 \approx 3800$ (correct to 2 significant figures)

**b** **1005**

The first 3 significant figures are **100** and the next digit is **5**, so we round up to **10** and use a 0 as a placeholder.

1005 ≈ 1010 (correct to 3 significant figures)

**c** 0.003572

For decimals, ignore any zeros at the front of the number. The first 2 significant figures are **35** and the next digit is **7** (which is greater than 5), so we round the **5** up to **6**.

0.003 572 ≈ 0.0036 (correct to 2 significant figures)

## Significant figures

- The first significant figure in a number is the first non-zero digit.
- When rounding to significant figures, start counting from the first digit that is not 0.
- If it is a large number, you may need to insert 0s at the end as placeholders.
- Zeros at the end of a whole number or at the beginning of a decimal are not significant.

## Significant figures in measurement

As we can only record a measurement correct to the precision of the measuring device, this means that we must use the appropriate number of significant figures.

### Example 3

Sections of 2 tape measures are shown. Tape A has a precision of 1 cm, tape B has a precision of 0.1 cm.

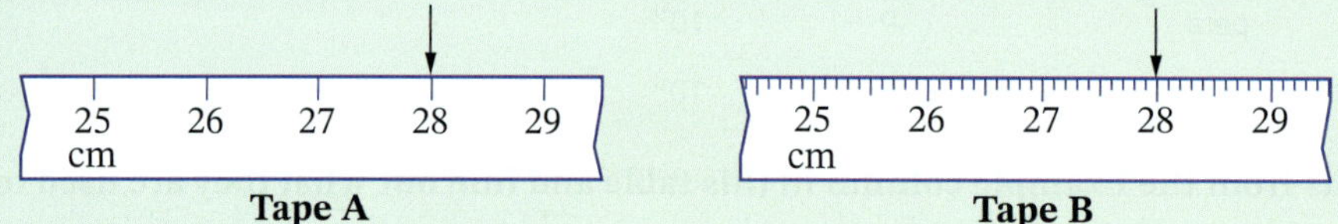


**a** Write the measurement shown on each tape and state how many significant figures each measurement has.

**b** Which of the 2 measurements is more accurate?

**c** Vincent recorded a measurement of 57.4 cm using tape A. What should he have written instead? Why?

**d** Rose recorded a measurement of 19 cm using tape B. What should she have written instead? Why?

### Solution

**a** Tape A: 28 cm, 2 significant figures

Tape B: 28.0 cm, 3 significant figures

**b** 28.0 cm is more accurate than 28 cm because it has more significant figures and is correct to the nearest 0.1 cm.

**c** 57 cm; we can only record to the nearest centimetre, using 2 significant figures.

**d** 19.0 cm; we should record correct to 0.1 cm, using 3 significant figures.

**EXERCISE 5.02** Answers on p. 467

## Significant figures

EXAMPLE 2

**1** What is 305.725 rounded to 2 significant figures? Select **A**, **B**, **C** or **D**.

**A** 306 **B** 310 **C** 300 **D** 305.73

**2** Write each number correct to 2 significant figures.

**a** 3812.4 **b** 2068 **c** 0.006 102

**d** 245 698 **e** 14 569 235 **f** 0.000 467

**3** Write each number correct to 3 significant figures.

**a** 129.805 **b** 4982 **c** 0.010 609

**d** 1 359 048 **e** 25 430 082 **f** 0.000 680 32

**4** Write each number correct to one significant figure.

**a** 2914.23 **b** 3.2548 **c** 11 950

**d** 0.005 134 **e** 20.46 **f** 0.6517

**5** Measurements made using a rain gauge marked in millimetres should be recorded correct to 2 significant figures for rainfall from 10 mm to 99 mm, and to one significant figure for rainfall less than 10 mm. State whether each rainfall measurement has been recorded correctly.

**a** 9 mm **b** 7.23 mm

**c** 24 mm **d** 6.5 mm

**e** 31 mm **f** 10 mm

**6** A micrometre (μm) or micron is one-thousandth of a millimetre and is used for very small measurements. If each of the following measurements should be recorded correct to 2 significant figures, which measurements have been recorded incorrectly?

**a** 23 μm **b** 104 μm

**c** 19.6 μm **d** 832 μm

**e** 28.6 μm **f** 3 μm

iStock.com/manfredxy

**7** Evaluate each expression, correct to 2 significant figures:

**a** $0.2 \div 0.3$ **b** $11 \div 1990$ **c** $16 \div 12$

**d** $\sqrt{0.0075}$ **e** $9\,300\,000 \times 0.085$ **f** $2.7^2$

**g** $\sqrt{560}$ **h** $\sqrt{5.6}$ **i** $3.4 \times 9.9$

**8** The population of New South Wales in 2025 was 8 659 259. Round this number to 3 significant figures.

**9** Express each measurement correct to 3 significant figures.

**a** mass of an aeroplane, 351 540 kg

**b** average depth of the Pacific Ocean, 4188 m

**c** speed of a greyhound, 67.14 km/h

**d** capacity of a tablespoon, 14.79 mL

**e** distance from Earth to the Sun, 149 573 881 km

 Foundation  Mastery  Complex

# 5.03 Scientific notation

**Puzzle**
Scientific notation puzzle

**Scientific notation** (or **standard form notation**) is a shorthand way of writing very large or very small numbers, using powers of 10.

### Scientific notation

Scientific notation has the form:

$$m \times 10^n$$

where $m$ is a number between 1 and 10
$n$ is an integer.

## Example 4

Write each number in scientific notation.

**a** 23 500 000 **b** 0.004 762

### Solution

**a** Write a decimal between 1 and 10 using the significant figures in the number: 2.35

Count how many places to move the decimal point to make the original number: 7

2 . 3 5 0 0 0 0 0

Or count the number of places after the first significant figure, 2

So, $23\,500\,000 = 2.35 \times 10^7$.

**b** Write a decimal between 1 and 10 using the significant figures in the number: 4.762

Count how many places to move the decimal point to make the original number: −3

0 0 0 4 . 7 6 2

Or count the number of 0s before the first significant figure, 4

So, $0.004\,762 = 4.762 \times 10^{-3}$.

## Example 5

Write 354 780 in scientific notation, correct to 2 significant figures.

### Solution

$354\,780 = 3.5478 \times 10^5$ writing in scientific notation

$\approx 3.5 \times 10^5$ rounding to 2 significant figures

We could also round to 2 significant figures first, then write in scientific notation.

## Example 6

Write each number in normal decimal form.

**a** $6.92 \times 10^4$ **b** $3.716 \times 10^{-5}$

### Solution

**a** $6.92 \times 10^4 = 69\,200$ moving the decimal point 4 places to the right (or make 4 places after the 6)

**b** $3.716 \times 10^{-5} = 0.000\,037\,16$ moving the decimal point 5 places to the left (or insert 5 zeros before the 3 and place the decimal point after the first zero)

## Scientific notation on a calculator

To enter a number in scientific notation on a calculator, use the ×10^x or EXP key.

## Example 7

Evaluate:

**a** $(6.8 \times 10^5) \times (7.4 \times 10^{-2})$

**b** $(3.2 \times 10^{-4}) \div (9.1 \times 10^6)$, correct to 2 significant figures.

### Solution

**a** $(6.8 \times 10^5) \times (7.4 \times 10^{-2}) = 50\,320$

**Calculator steps:** 6.8 ×10^x 5 × 7.4 ×10^x (−) 2 =

**b** $(3.2 \times 10^{-4}) \div (9.1 \times 10^6) = 3.516\,483 \ldots \times 10^{-11}$

$\approx 3.5 \times 10^{-11}$ rounding to 2 significant figures

**Calculator steps:** 3.2 ×10^x (−) 4 ÷ 9.1 ×10^x 6 =

## Scientific notation and measurements

We use prefixes to represent very large and very small numbers and units of measurement (also see Did you know? Metric prefixes, page 159).

| Prefix | Abbreviation | Meaning | Example |
|---|---|---|---|
| nano | n | $10^{-9} = \frac{1}{1\,000\,000\,000}$ (one-billionth) | nanometre |
| micro | μ | $10^{-6} = \frac{1}{1\,000\,000}$ (one-millionth) | microsecond |
| milli | m | $10^{-3} = \frac{1}{1000}$ (one-thousandth) | millilitre |
| centi | c | $10^{-2} = \frac{1}{100}$ (one-hundredth) | centigram |
| kilo | k | $10^3 = 1000$ | kilowatt |
| mega | M | $10^6 = 1\,000\,000$ (million) | megalitre |
| giga | G | $10^9 = 1\,000\,000\,000$ (billion) | gigahertz |
| tera | T | $10^{12} = 1\,000\,000\,000\,000$ (trillion) | terabyte |

## Example 8

Convert each unit, writing your answer in scientific notation.

**a** 470 m to nm **b** 5 km to mm **c** 96 μs to s **d** 1.8 GL to mL

### Solution

**a** $470\text{ m} = (470 \times 10^9)\text{ nm}$

$= 4.7 \times 10^{11}\text{ nm}$

A nanometre is $\frac{1}{10^9}$ of a metre, so 1 m = $10^9$ nm

**b** $5\text{ km} = (5 \times 1000)\text{ m}$ converting to metres first

$= (5 \times 1000 \times 1000)\text{ mm}$ converting to mm next

$= 5\,000\,000\text{ mm}$

$= 5 \times 10^6\text{ mm}$

**c** $96\,\mu\text{s} = (96 \div 10^6)\text{ s}$ or $(96 \times 10^{-6})\text{ s}$

$= 0.000\,096\text{ s}$

$= 9.6 \times 10^{-5}\text{ s}$

**d** $1.8\text{ GL} = (1.8 \times 10^9)\text{ L}$ converting to litres first

$= (1.8 \times 10^9 \times 1000)\text{ mL}$ converting to mL next

$= 1.8 \times 10^{12}\text{ mL}$

**EXERCISE 5.03** Answers on p. 467

## Scientific notation

**1** Write each number in scientific notation.

**a** 42 130 000 **b** 0.0181 **c** 3400
**d** 20 000 **e** 0.0035 **f** 0.0002
**g** 0.33 **h** 0.004 **i** 230
**j** 0.000 0723 **k** 610 000 000 **l** 0.000 000 08

EXAMPLE 5

**2** Write each number in scientific notation, correct to 2 significant figures.

**a** 53 467 892 **b** 146 089 **c** 2453
**d** 0.000 4573 **e** 0.002 652 **f** 0.102 05
**g** 0.000 033 38 **h** 0.444 44 **i** 6.5207

**3** What is 356 587 239 in scientific notation, correct to 3 significant figures? Select **A**, **B**, **C** or **D**.

**A** 357 000 000 **B** $3.57 \times 10^8$ **C** $3.6 \times 10^8$ **D** $3.56 \times 10^8$

EXAMPLE 6

**4** Write each number in normal decimal form.

**a** $7.4 \times 10^5$ **b** $3.12 \times 10^{-1}$ **c** $1.85 \times 10^3$
**d** $6.6 \times 10^{-4}$ **e** $2.54 \times 10^{-3}$ **f** $4.751 \times 10^8$
**g** $9.8 \times 10^{-2}$ **h** $3 \times 10^2$ **i** $5.497 \times 10^{-2}$
**j** $1.216 \times 10^4$ **k** $8.02 \times 10^{-1}$ **l** $6.309 \times 10^3$

**5** Write each measurement in scientific notation.

**a** The age of the universe is about 13 700 000 000 years.

**b** One micrometre (μm) is $\frac{1}{1\,000\,000}$ m.

Foundation Mastery Complex

**6** Write each measurement in normal decimal notation.

a The diameter of an atom is $3 \times 10^{-8}$ mm.

b One light year, the distance light travels in one year, is equal to $9.461 \times 10^{9}$ m.

c The diameter of the flu virus is $2 \times 10^{-6}$ mm.

d The distance of Earth from the Sun is $1.526 \times 10^{8}$ km.

**7** The world population in 2024 was 8 188 760 983. Write this number in scientific notation, correct to 3 significant figures.

**8** Evaluate each expression using scientific notation.

a $(6.7 \times 10^{4}) \times (3.2 \times 10^{2})$

b $(1.92 \times 10^{10}) \div (6 \times 10^{4})$

c $(3.75 \times 10^{4}) - (2.5 \times 10^{3})$

d $(4.8 \times 10^{-3})^{2}$

e $(8.1 \times 10^{4})^{3}$

f $\dfrac{9.4 \times 10^{-2}}{2.5 \times 10^{-3}}$

g $(4.2 \times 10^{5}) \times (3 \times 10^{-2})$

h $\sqrt{5.29 \times 10^{10}}$

**9** Evaluate each expression in scientific notation, correct to 2 significant figures.

a $(5.1 \times 10^{3}) + (2.5 \times 10^{4})$

b $(7.8 \times 10^{-5}) - (3.7 \times 10^{-3})$

c $(9.6 \times 10^{8}) \times (4.3 \times 10^{3})$

d $(3.2 \times 10^{-3})^{3}$

e $\dfrac{3.2 \times 10^{2}}{5.4 \times 10^{-3}}$

f $\sqrt[3]{3.8 \times 10^{-5}}$

g $\sqrt{8.5 \times 10^{5}}$

h $\dfrac{6 \times 10^{3}}{5.1 \times 10^{7}}$

**10** Convert 3.4 metres to nanometres.

**11** Convert 78 GL to mL.

**12** How many nanoseconds are in 375 μs? Write your answer in scientific notation.

**13** Convert $4 \times 10^{9}$ nm to Tm.

**14** Evaluate each expression, expressing the answer in normal decimal form.

a $96 \div 20\,000$

b $480\,000 \times 91\,000$

c $\dfrac{0.235}{25}$

d $13\,700^{2}$

e $(4.4 \times 10^{4}) \times (3.1 \times 10^{2})$

f $(1.9 \times 10^{7}) + (6.3 \times 10^{4})$

g $\sqrt{2.704 \times 10^{-9}}$

h $\dfrac{7.7 \times 10^{9}}{2.2 \times 10^{4}}$

**15** What is the largest number you can enter into your calculator in scientific notation? What is the smallest number?

**16** A construction company produces steel cables in lengths of 2.5 m. The manufacturer needs to know the length of steel cables produced in micrometres for precision cutting. Convert a length of 2.5 m to μm, writing your answer in scientific notation.

**17** A biotechnology company is studying the concentration of a specific enzyme in a sample. The concentration is measured as 0.0045 Mg/L. Convert this measurement to μg/L.

**18** A data centre stores large amounts of digital information. The total data storage capacity is 4 555 000 kilobytes (KB). Convert the storage capacity from kilobytes (KB) to terabytes (TB). Write your answer in scientific notation, correct to 2 significant figures.

Foundation Mastery Complex

# 5.04 Pythagoras' theorem problems

**Video**
Pythagoras' theorem problems

**Worksheets**
Applications of Pythagoras' theorem

Pythagoras' problems

Pythagorean two-step problems

## Pythagoras theorem

**Pythagoras' theorem** states that in a right-angled triangle, the square of the hypotenuse is equal to the sum of the squares of the other 2 sides.

As a formula: $c^2 = a^2 + b^2$

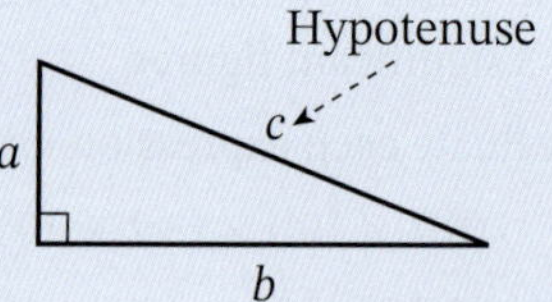


The *hypotenuse* is the longest side of a right-angled triangle.

## Example 9

A boat sailed 30 nautical miles due south, then 25 nautical miles due east. How far, correct to the nearest nautical mile, is it from its starting point?

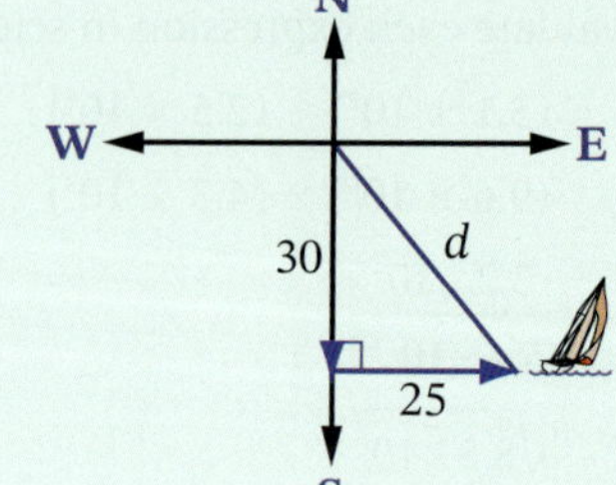


### Solution

Let the boat be $d$ nautical miles from its starting point.

$$
\begin{aligned}
d^2 &= 30^2 + 25^2 \\
&= 1525 \\
d &= \sqrt{1525} \\
&= 39.0512... \\
&\approx 39
\end{aligned}
$$

The boat is 39 nautical miles from its starting point.

From the diagram, an answer of 39 nautical miles seems reasonable.

## Example 10

A surveyor measured different distances (in metres) on the irregular field *ABCDE*, as shown on the diagram. How many metres of fencing, correct to the nearest 0.1 m, are needed to fence the field?

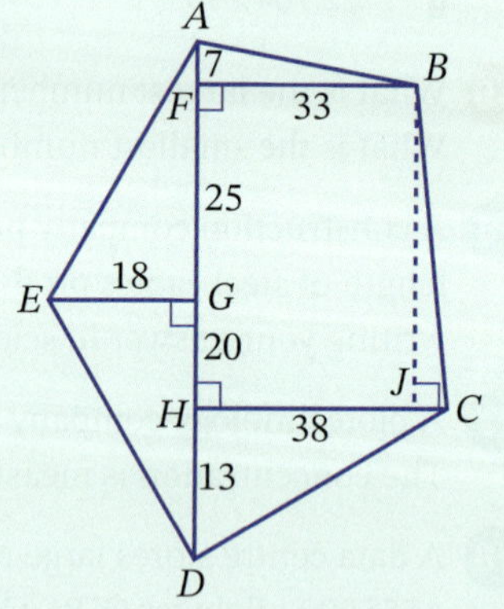

## Solution

The amount of fencing is the perimeter of the field.

In $\triangle AGE$: $AE^2 = 18^2 + 32^2$     $AG = 25 + 7 = 32$

$= 1348$

$AE = \sqrt{1348}$

$= 36.7151 \ldots$

Do not round until the end of this solution.

In $\triangle DGE$: $ED^2 = 18^2 + 33^2$

$= 1413$

$ED = \sqrt{1413}$

$= 37.5898 \ldots$

In $\triangle HDC$: $DC^2 = 13^2 + 38^2$

$= 1613$

$DC = \sqrt{1613}$

$= 40.1621 \ldots$

In $\triangle BJC$: $BC^2 = 45^2 + 5^2$

$= 2050$

$BC = \sqrt{2050}$

$= 45.2769 \ldots$

In $\triangle AFB$: $AB^2 = 7^2 + 33^2$

$= 1138$

$AB = \sqrt{1138}$

$= 33.7342 \ldots$

perimeter $= AE + ED + DC + BC + AB$

$= 193.4783 \ldots$

$\approx 193.5$

The amount of fencing needed is 193.5 m.

**EXERCISE 5.04** Answers on p. 467

## Pythagoras' theorem problems

**1** Find the value of each variable, correct to 2 decimal places.

**a**

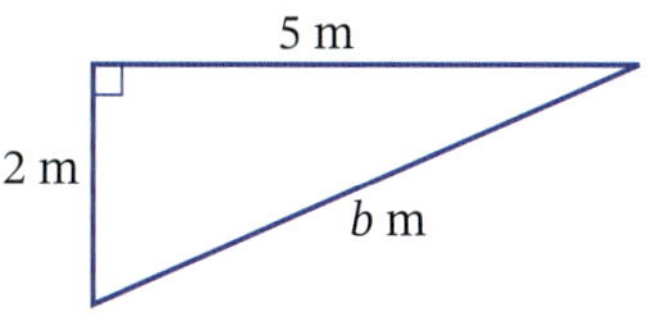


**b**

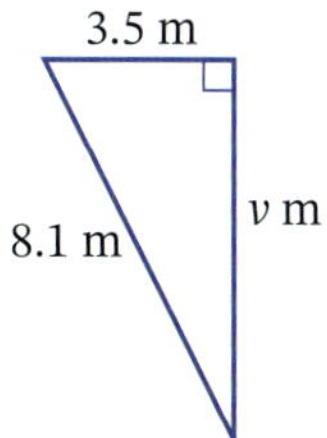


**c**

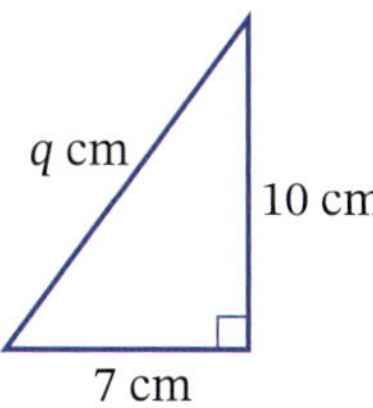


**d**

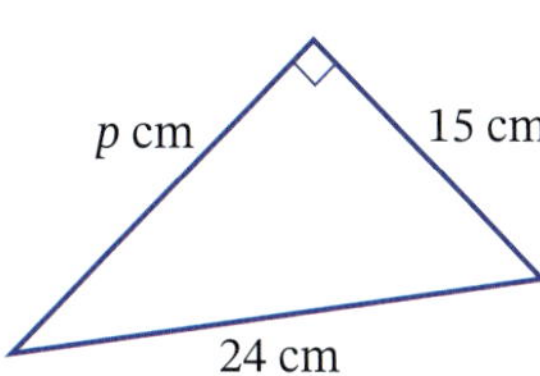


**e**

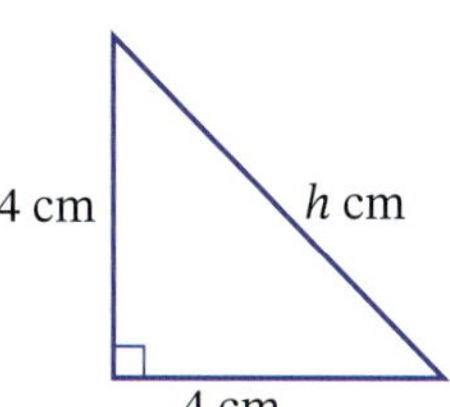


**f**

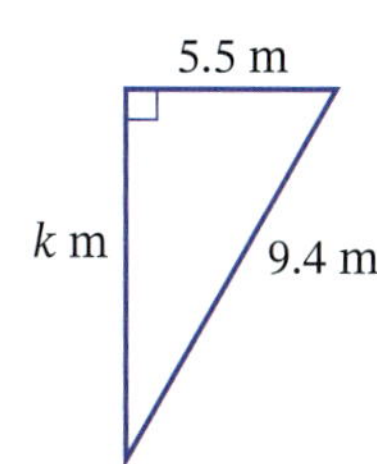

**2** Find the value of each variable, correct to one decimal place.

**a**

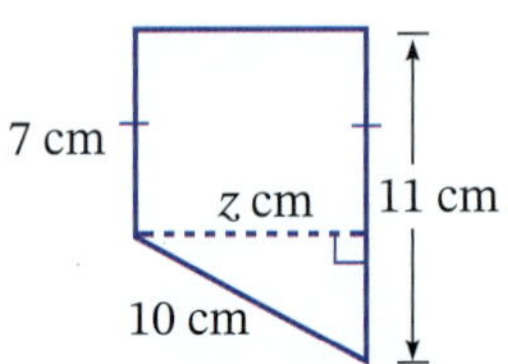


**b**

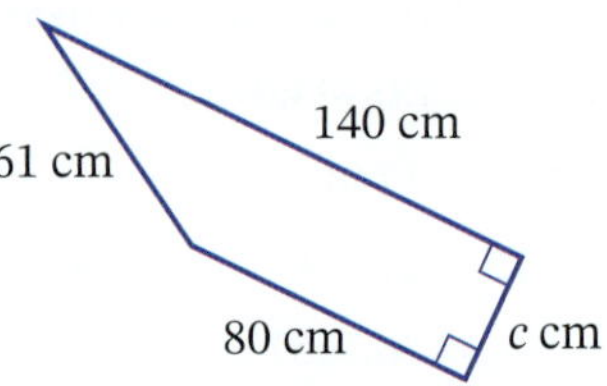


**c**

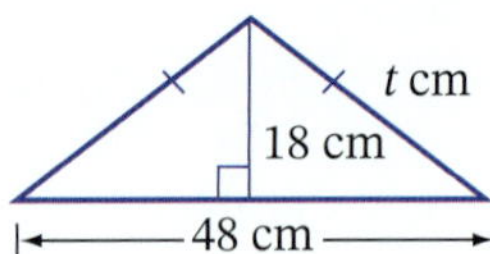


**d**

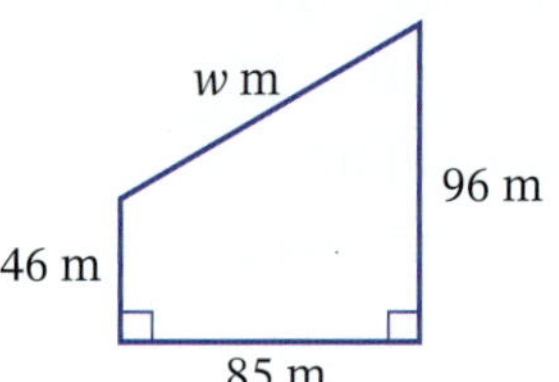


**e**

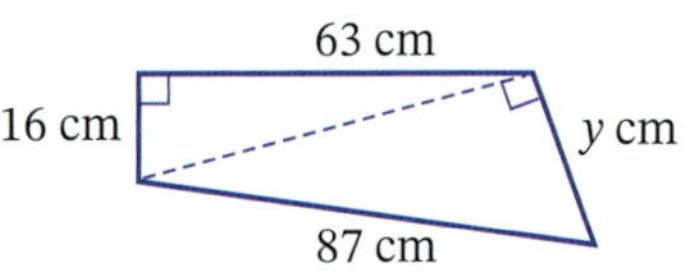


**f**

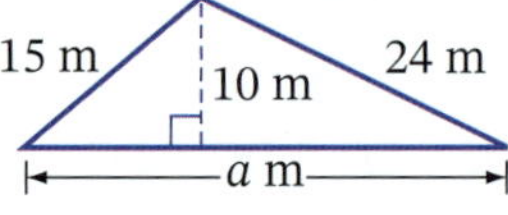


EXAMPLE 9

**3** A gate has dimensions 3.8 m by 1.8 m. What is the length of its diagonal brace (to the nearest centimetre)?

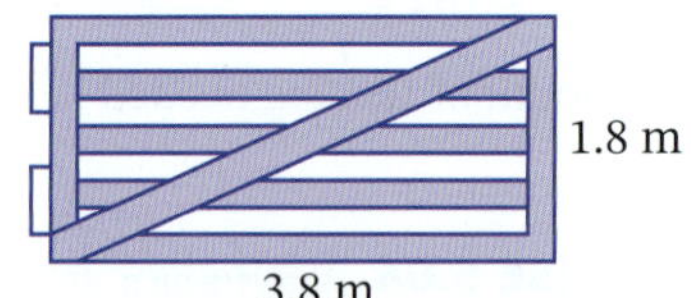


**4** Vimala calculated the distance across the lake by taking the measurements shown on the diagram.

What was the distance across the lake, to the nearest metre?

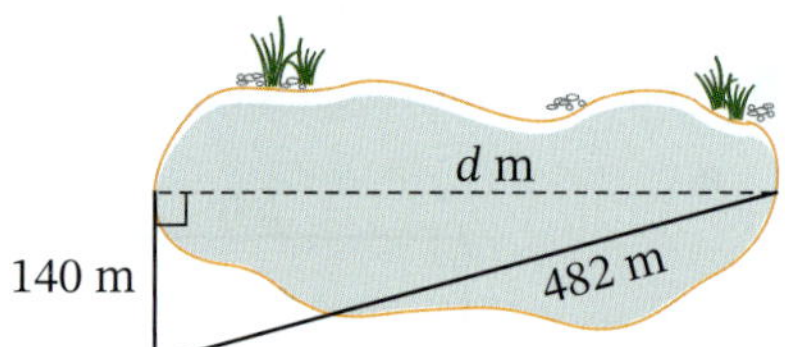


**5** Ashton is a builder. He places a 3.6 m long ladder against a wall. To be safe to use, the base of the ladder must be 0.9 m away from the wall. If the ladder is placed on flat ground, how far up the wall does it reach? Select **A**, **B**, **C** or **D**.

**A** 10.71 m **B** 3.90 m

**C** 3.49 m **D** 3.10 m

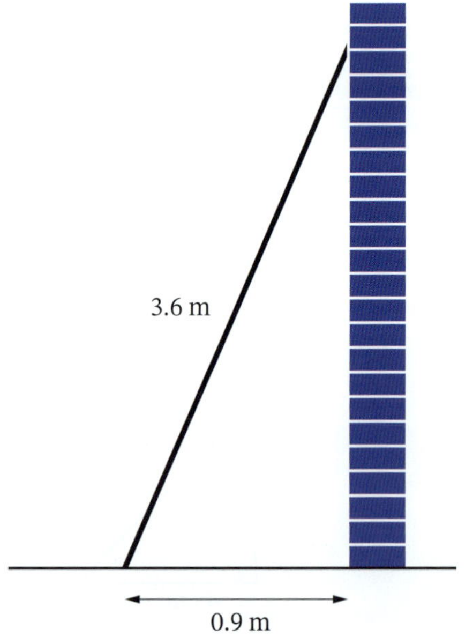

6 A yacht leaves Newcastle and sails 160 nautical miles due east. It turns and sails due north until it is exactly 200 nautical miles from Newcastle. How far north did it sail?

7 A chairlift takes skiers up to the top of a hill 270 m high and 1050 m away. How long is the chairlift's cable, correct to the nearest 10 m?

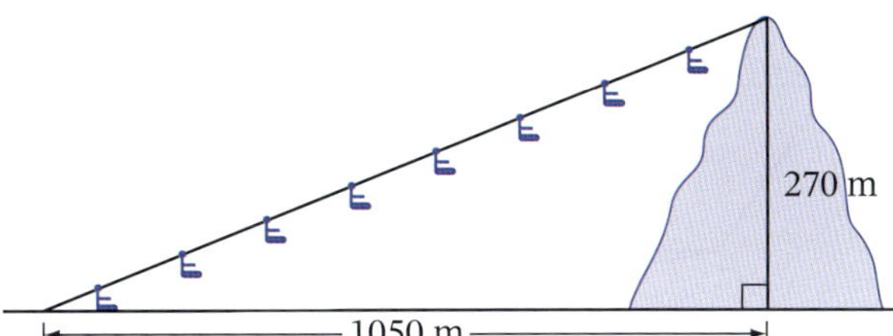


8 A playground slide is made up of 2 right-angled triangles. Find, correct to the nearest centimetre:

**a** $h$ metres, the height of the slide.

**b** $l$ metres, the length of the slide.

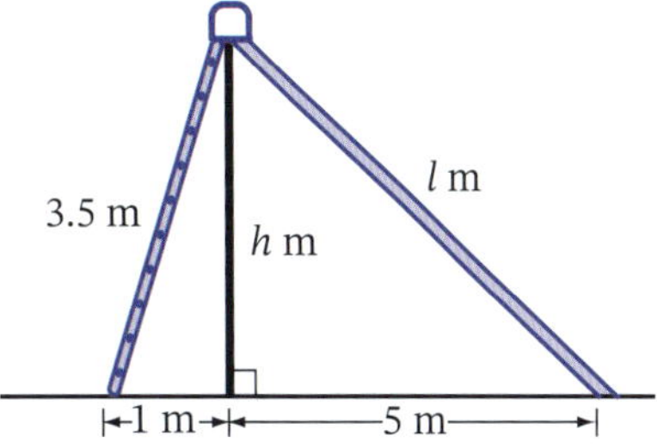


9 This diagram shows a boy flying a kite. How high is the kite above the ground, correct to one decimal place?

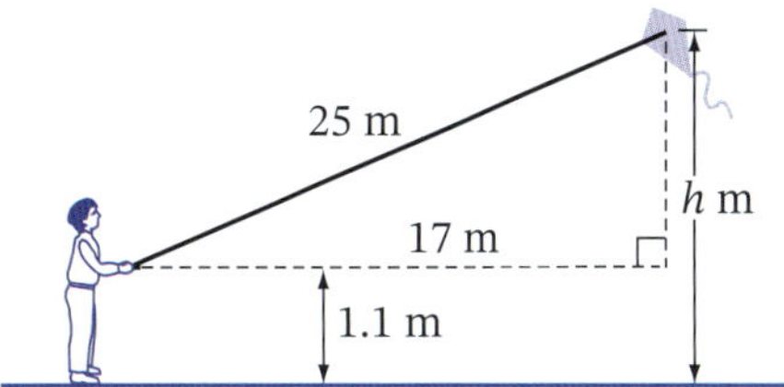


10 Jackie wants to use an old tennis ball container as a pencil case. If the container has a diameter of 7.5 cm and a height of 20 cm, what is the length of the longest pencil that will fit inside the container (to the nearest millimetre)?

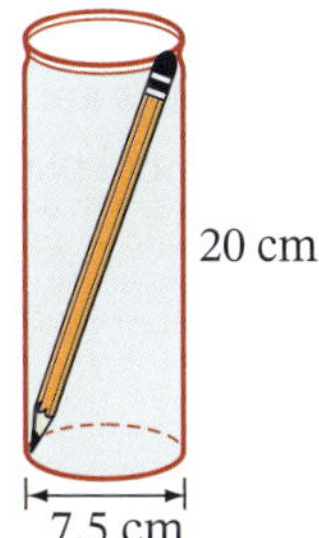


11 Find the length of the pitch line, $p$, of the roof of this house (to the nearest centimetre).

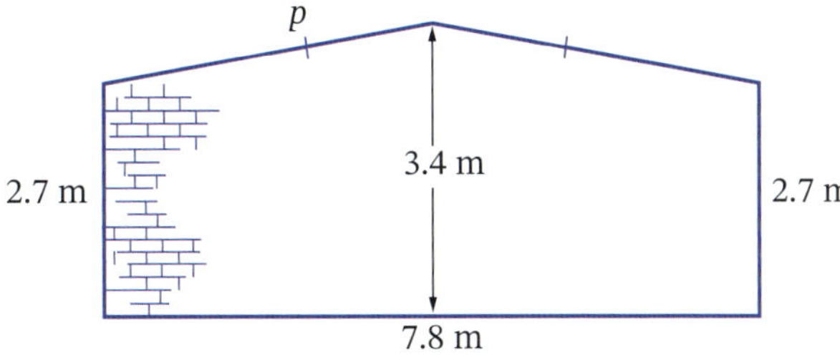

Foundation Mastery Complex

EXAMPLE 10

**12** Calculate the perimeter of the field *ABCDE* in this diagram, correct to 2 decimal places. All measurements are in metres.

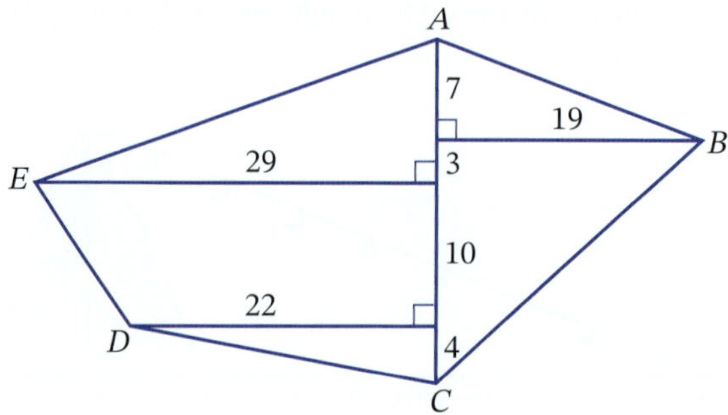


**13** Calculate the perimeter of each figure, to the nearest 0.1 m.

**a**

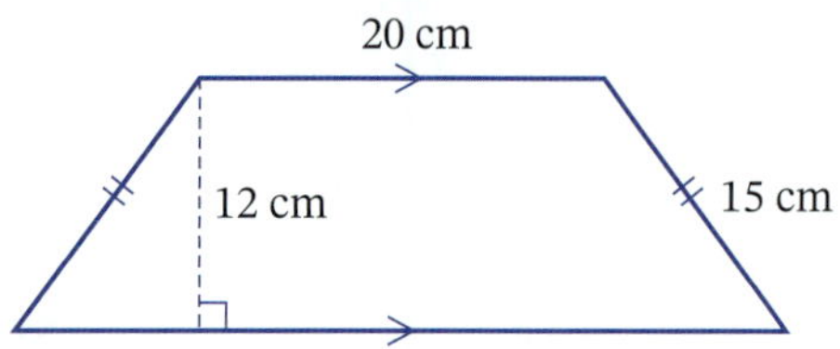


**b**

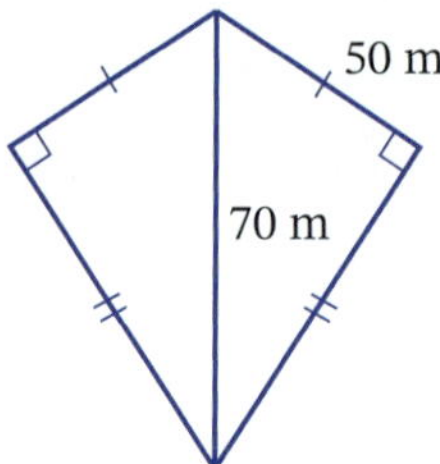


**14** If a tent pole is 2 m high and the rope is 2.4 m long, how far from the base of the pole should the rope be pegged (correct to one decimal place)?

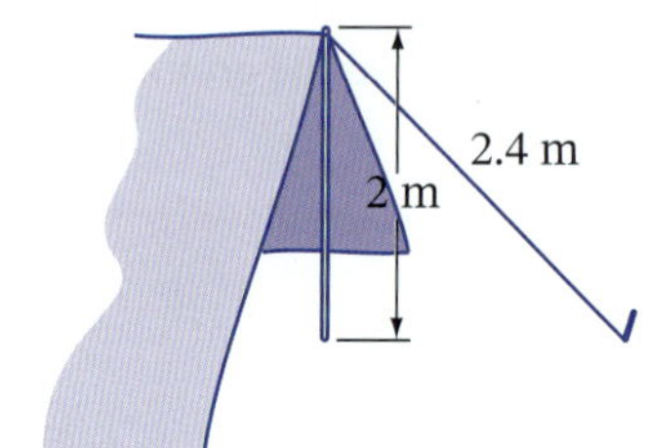


**15** When a stepladder is placed upright, its top is 2.5 m above the ground and its legs are 0.9 m apart. How long is the stepladder, correct to 2 decimal places?

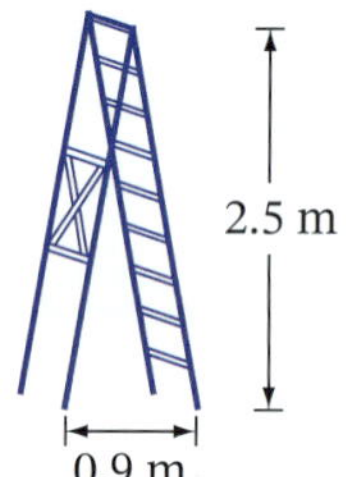


**16** Mack walks 30 m east, then turns and walks 60 m south and then 70 m east, as shown in the diagram.

Calculate the direct distance from Mack's starting point, *K*, to his final destination, *L*. Write your answer correct to 2 decimal places.

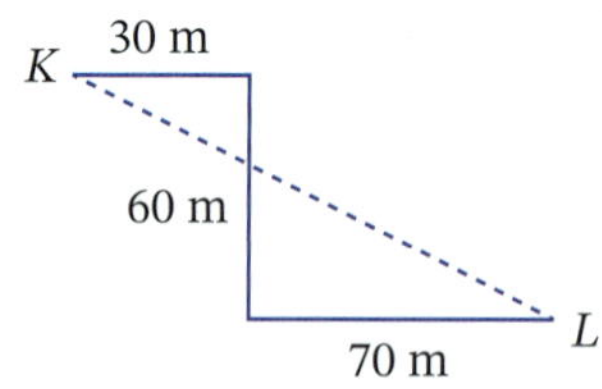

Foundation Mastery Complex

**17** Kaisleigh's ship leaves the harbour and sails 30 km on a bearing of 025° and then sails 50 km on a bearing of 115°.

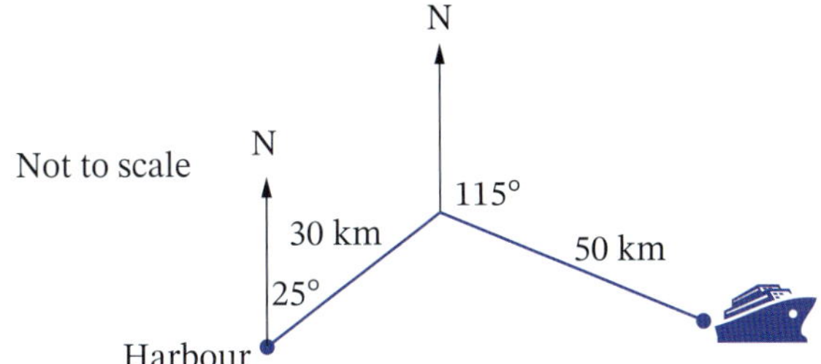


Calculate the distance (correct to 4 significant figures) Kaisleigh's ship is from the harbour.

**18** What is the longest length of pencil that can fit into a box with dimensions 20 cm × 5 cm × 7 cm? Calculate your answer correct to 2 decimal places.

**19** Keiran pedalled his pushbike along a road for 4.6 km in the direction S65°E. He then changed course to N25°E and pedalled a further 3.7 km.

**a** Copy the diagram into your book and complete it to show all of the given information.

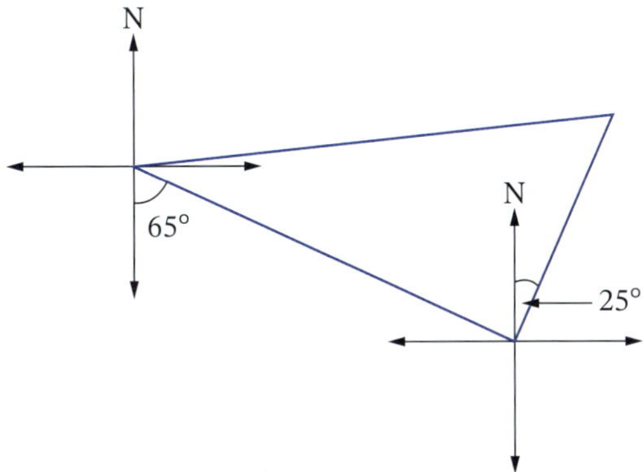


**b** Hence find the direct distance between Keiran's starting and finishing points, correct to one decimal place.

**20** A hollow sphere with a radius of 1.7 m is filled with water to a depth of $x$ m. The upper surface of the water has a diameter of 2.7 m, as shown in the diagram.

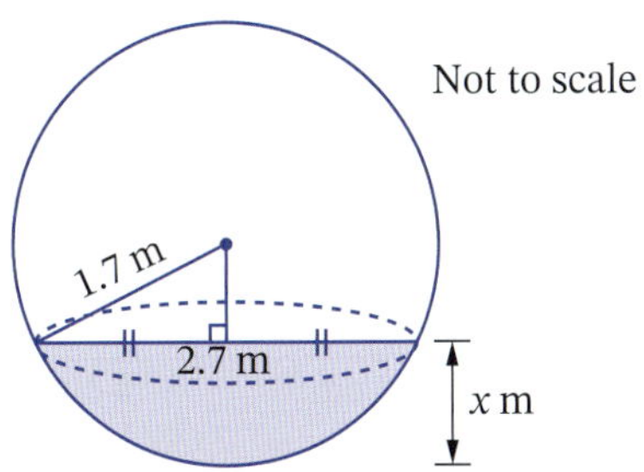


Vertical cross-section of sphere

Find the value of $x$, correct to 2 decimal places.

Foundation Mastery Complex

# 5.05 Perimeters of circular and composite shapes

Worksheet
A page of circular shapes

## Circumference

- The **perimeter** of a shape is the distance around the shape. It is the sum of the lengths of the sides of the shape.
- The perimeter of a **circle** is called its **circumference**. The **radius** is the distance from the centre of a circle to the circumference. The radius is half of the **diameter**.

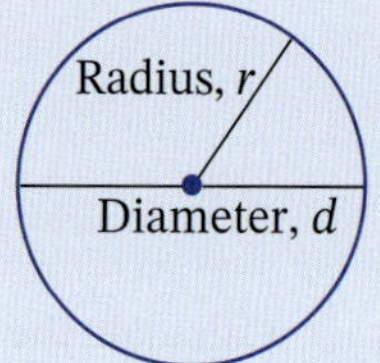


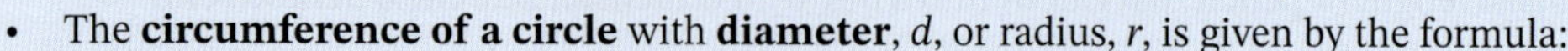

- The **circumference of a circle** with **diameter**, $d$, or radius, $r$, is given by the formula:

$$C = \pi d \text{ or } C = 2\pi r$$

## Example 11

Find the perimeter of each shape.

**a**

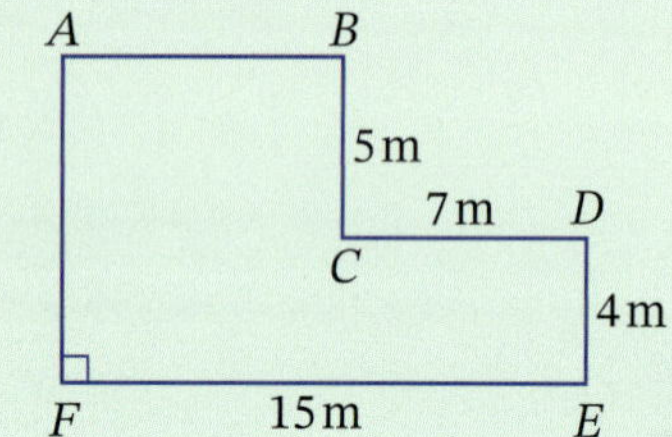


**b**

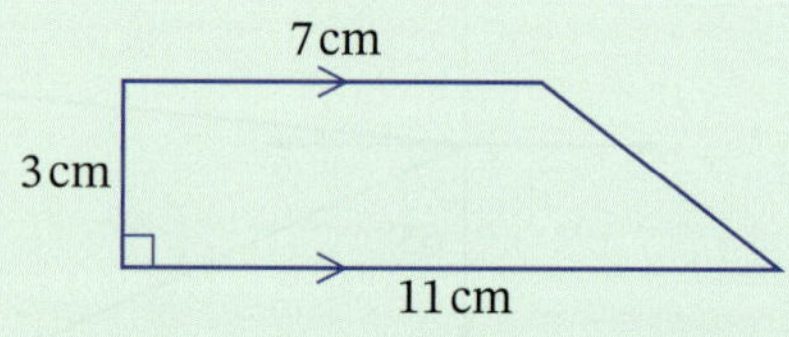


### Solution

**a** Find the lengths of the unknown sides first.

$AB = 15 - 7 = 8$

$AF = 5 + 4 = 9$

perimeter $= 8 + 5 + 7 + 4 + 15 + 9 = 48$ m

It is possible to make this shape into one big rectangle by moving the 5 m side to the right and the 7 m side upwards. This rectangle has a length of 15 m and a width of 9 m, so its perimeter (15 + 15 + 9 + 9 = 48 m) should be the same as the perimeter of the composite shape.

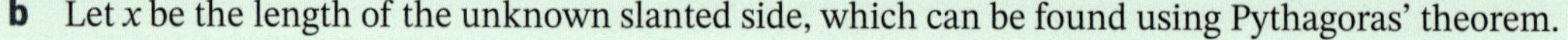

**b** Let $x$ be the length of the unknown slanted side, which can be found using Pythagoras' theorem.

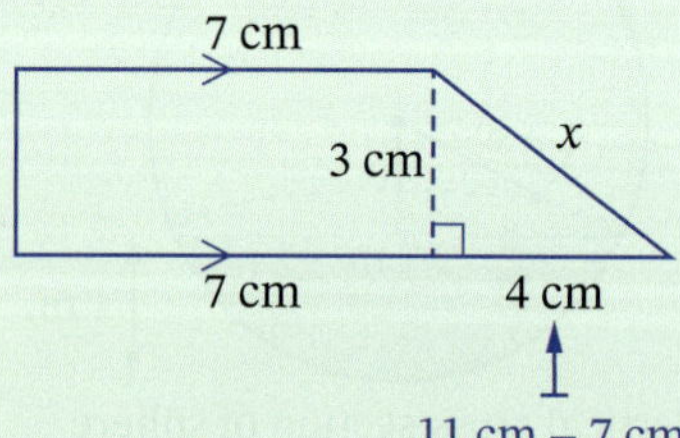


$x^2 = 3^2 + 4^2 = 25$

$x = \sqrt{25} = 5$ cm

Perimeter $= 7 + 5 + 11 + 3 = 26$ cm

## Example 12

Find, correct to one decimal place, the perimeter of each shape.

**a**

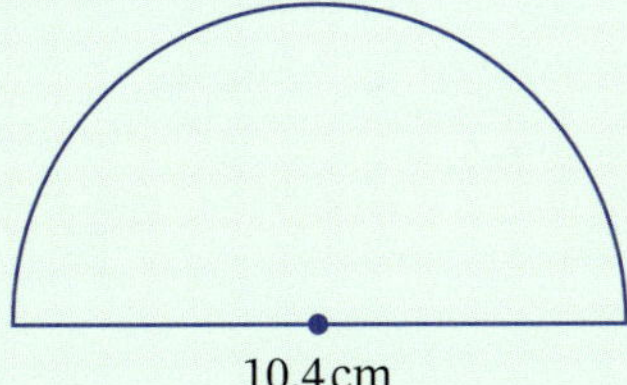


**b**

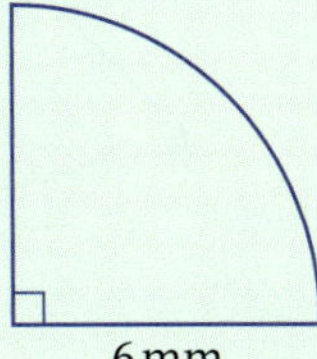


### Solution

**a** Perimeter of semicircle $= \left(\frac{1}{2} \times \text{circumference}\right) + 10.4$

$= \left(\frac{1}{2} \times \pi \times 10.4\right) + 10.4$ $\quad C = \pi d$

$= 26.73628 \ldots$

$\approx 26.7 \text{ cm}$

From the diagram, a perimeter of 26.7 cm looks reasonable.

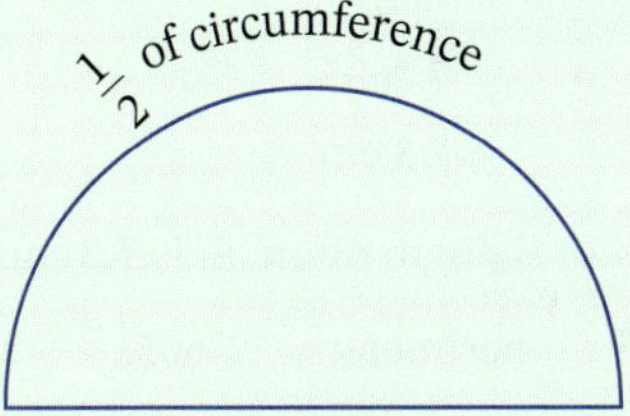


**b** Perimeter of quadrant $= \left(\frac{1}{4} \times \text{circumference}\right) + 6 + 6$

$= \left(\frac{1}{4} \times 2 \times \pi \times 6\right) + 12$ $\quad C = 2\pi r$

$= 21.424777 \ldots$

$\approx 21.4 \text{ mm}$

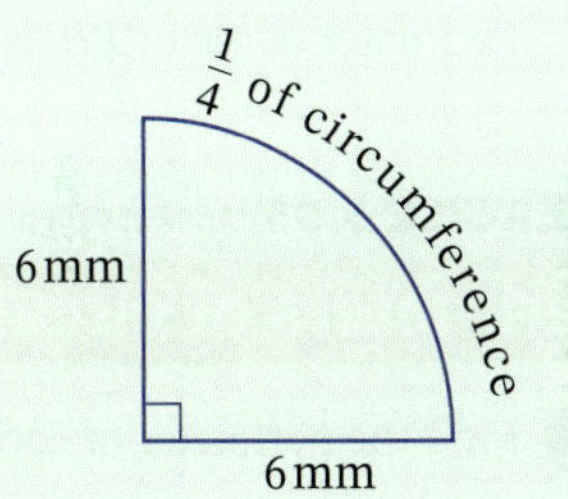


## Perimeter of a sector

A **sector** is a fraction of a circle 'cut' along 2 radii, like a pizza slice.

'radii' is the plural of 'radius': 1 radius, 2 radii

An **arc** is a part of the circumference of a circle. Its length ($l$) is a fraction of the circumference ($2\pi r$) of the circle. There are 360° in a circle.

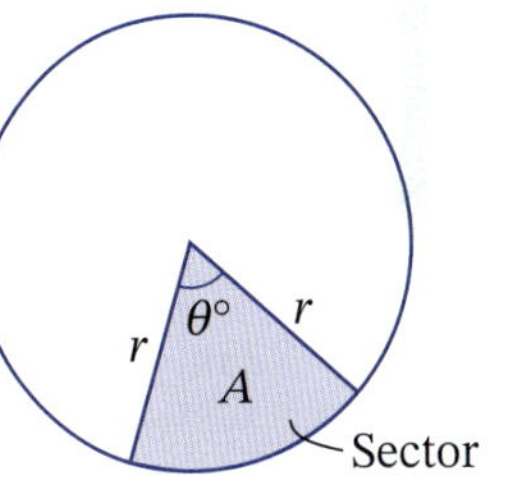


### Length of an arc

$$l = \frac{\theta}{360} \times 2\pi r$$

where $l$ = arc length

$\theta°$ = the size of the central angle

$r$ = radius

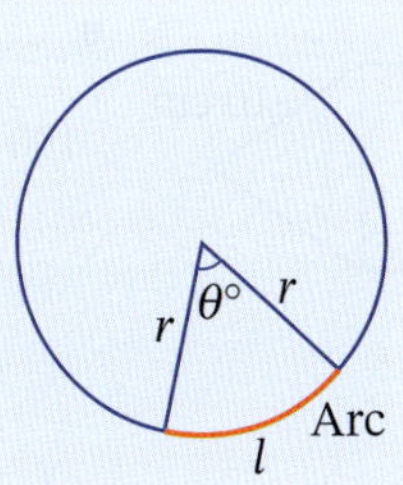


### Perimeter of a sector

$$P = \frac{\theta}{360} \times 2\pi r + 2r \quad \text{(arc length + radius + radius)}$$

Note: If $\theta = 90°$, the sector is called a **quadrant** (quarter of a circle) and if $\theta = 180°$, the sector is a **semicircle** (half of a circle).

Videos
Sectors

Perimeter and area of a sector

## Example 13

Find, correct to 2 significant figures:

**a** the length of arc *EI*

**b** the perimeter of sector *PIE*.

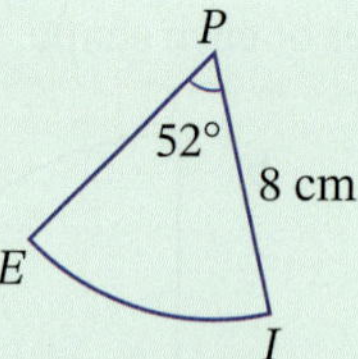


### Solution

**a**
$$l = \frac{\theta}{360} \times 2\pi r$$
$$= \frac{52}{360} \times 2 \times \pi \times 8$$
$$= 7.2605\ldots$$
$$\approx 7.3 \text{ cm}$$

Length of arc *EI* is 7.3 cm.

**b** perimeter $= 7.2605\ldots + 8 + 8$    arc length + radius + radius

$= 23.2605\ldots$

$\approx 23$ cm

Perimeter of sector *PIE* is 23 cm.

**EXERCISE 5.05** Answers on p. 467

## Perimeters of circular and composite shapes

**1** Find the perimeter of each shape.

**a**

2 m
8 m
3 m
2 m

**b**

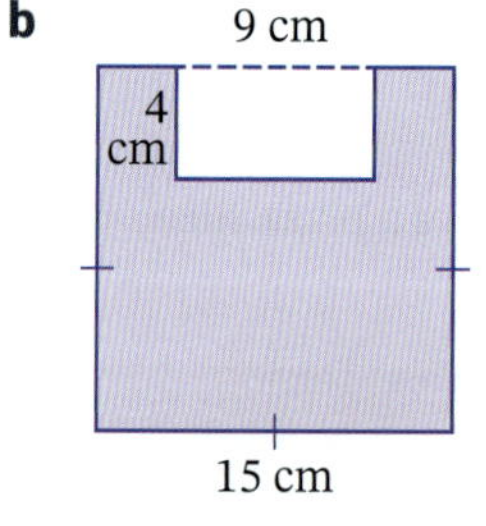


**c**

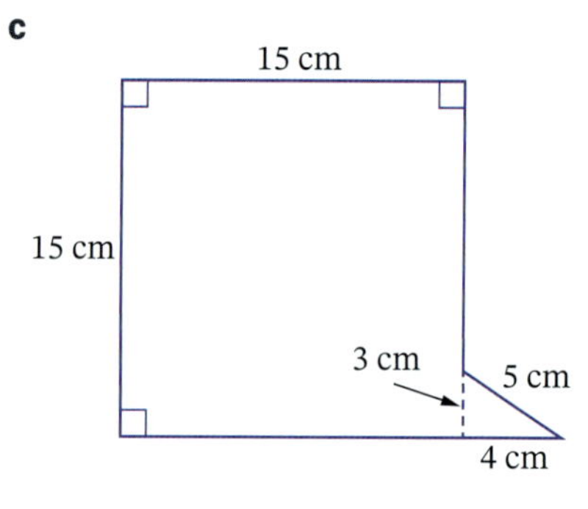


**d**

12 cm
10 cm

**e**

300 cm
100 cm
80 cm
80 cm
200 cm
100 cm

**f**

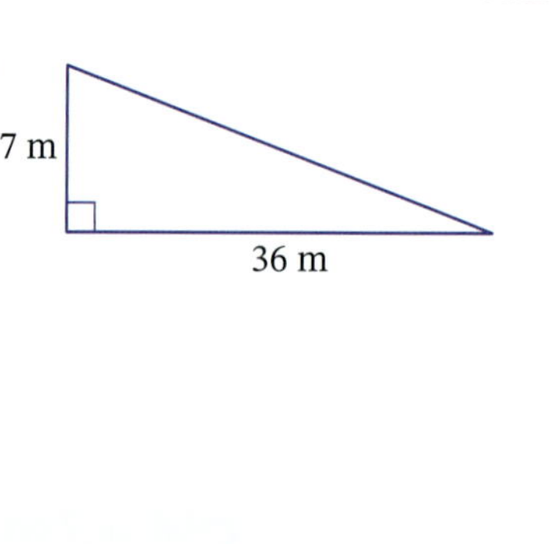


Foundation Mastery Complex

**g** 12 cm, 5 cm

**h** 8 mm, 13 mm, 36 mm

**i** 12 cm, 6 cm, 8 cm, 3 cm, 11 cm, 30 cm

**j** 75 mm, 85 mm, 25 mm

**k** 10 cm, 2 cm, 4 cm, 8 cm, 2 cm

**l** 7 cm, 6 cm, 6 cm, 13 cm, 15 cm

**2** Find, correct to one decimal place, the perimeter of each shape.

EXAMPLE 12

**a** 8.5 m

**b** 4 m

**c** 6 cm

**d** 4 cm, 6 cm

**e** 15 mm

**f** 15 cm, 8 cm, 9 cm

**g** 10 cm

**h** 26 mm, 22 mm

**i** 6 m

**j** 5 cm, 5 cm, 5 cm, 5 cm

**k** 16 m, 40°

**l** 10 cm, 20 cm

Foundation Mastery Complex

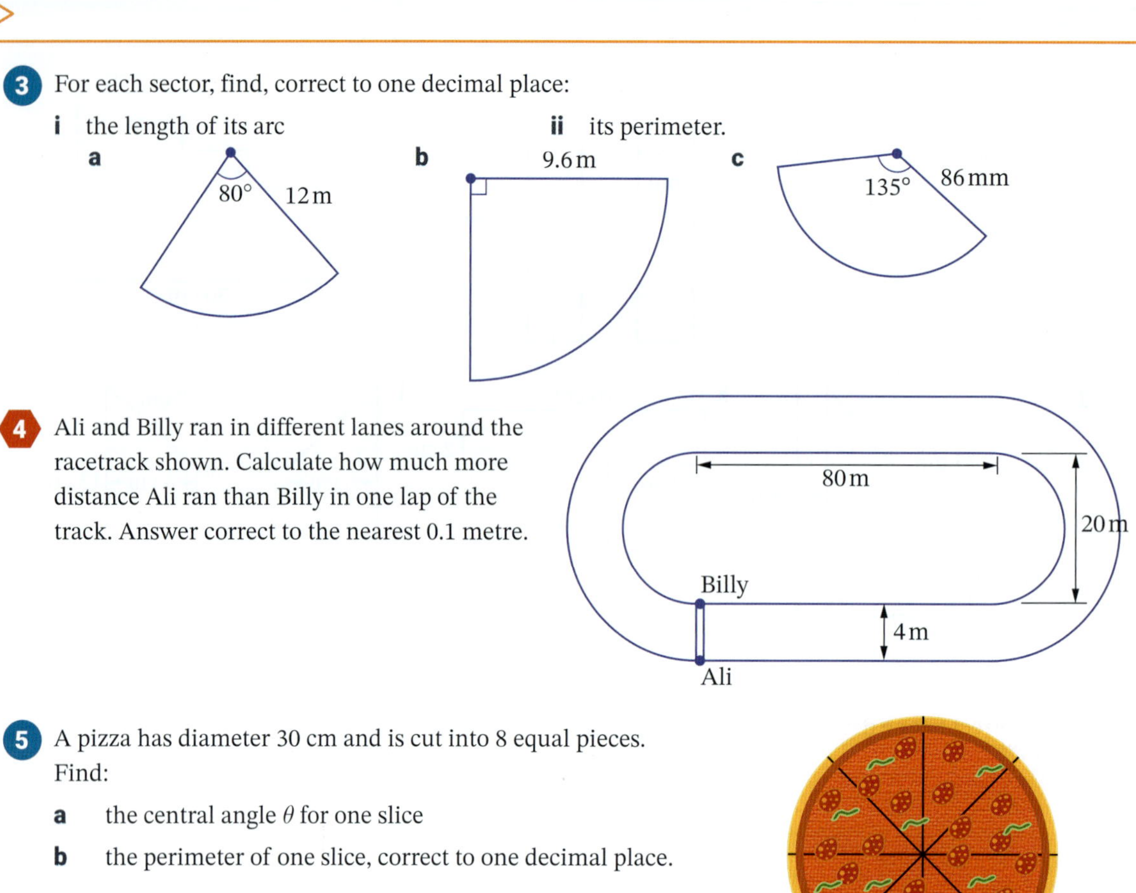

EXAMPLE 13

**3** For each sector, find, correct to one decimal place:

**i** the length of its arc **ii** its perimeter.



**4** Ali and Billy ran in different lanes around the racetrack shown. Calculate how much more distance Ali ran than Billy in one lap of the track. Answer correct to the nearest 0.1 metre.



**5** A pizza has diameter 30 cm and is cut into 8 equal pieces. Find:

**a** the central angle $\theta$ for one slice

**b** the perimeter of one slice, correct to one decimal place.

# 5.06 Area

The **area** of a plane (flat) shape is the amount of surface it occupies, measured in **square units**.

## Metric units for area

| Unit | Relationship | Size |
|---|---|---|
| square millimetre (mm²) | $1 \text{ mm}^2 = 1 \text{ mm} \times 1 \text{ mm}$ | The area of this tiny 1 mm square |
| square centimetre (cm²) | $1 \text{ cm}^2 = 1 \text{ cm} \times 1 \text{ cm}$<br>$= 10 \text{ mm} \times 10 \text{ mm}$<br>$= 100 \text{ mm}^2$ | The area of this 1 cm square |
| square metre (m²) | $1 \text{ m}^2 = 1 \text{ m} \times 1 \text{ m}$<br>$= 100 \text{ cm} \times 100 \text{ cm}$<br>$= 10\,000 \text{ cm}^2$ | The area of a 1 m square (about the size of the floor of a large shower recess) |
| hectare (ha) | $1 \text{ ha} = 100 \text{ m} \times 100 \text{ m}$<br>$= 10\,000 \text{ m}^2$ | The area of a 100 m square (about the size of 2 football fields) |
| square kilometre (km²) | $1 \text{ km}^2 = 1 \text{ km} \times 1 \text{ km}$<br>$= 1000 \text{ m} \times 1000 \text{ m}$<br>$= 1\,000\,000 \text{ m}^2$ | The area of a 1 km square (about the size of a large theme park) |

Foundation Mastery Complex

When converting units of area, we need to multiply or divide by two lots of units. For example, if 1 m = 100 cm, then $1\text{ m}^2 = (100 \times 100)\text{ cm}^2 = 10\,000\text{ cm}^2$.

## Example 14

Convert:

**a** $55\text{ m}^2$ to $\text{cm}^2$ **b** $2350\text{ m}^2$ to ha **c** $0.8\text{ km}^2$ to ha.

### Solution

**a** $55\text{ m}^2 = (55 \times 100 \times 100)\text{ cm}^2$ because $1\text{ m}^2 = 100\text{ cm} \times 100\text{ cm}$

$= 550\,000\text{ cm}^2$

**b** $2350\text{ m}^2 = (2350 \div 10\,000)\text{ ha}$ because $1\text{ ha} = 10\,000\text{ m}^2$

$= 0.235\text{ ha}$

**c** $0.8\text{ km}^2 = (0.8 \times 1000 \times 1000)\text{ m}^2$ because $1\text{ km}^2 = 1000\text{ m} \times 1000\text{ m}$

$= 800\,000\text{ m}^2$ converting to $\text{m}^2$ first

$= (800\,000 \div 10\,000)\text{ ha}$ because $1\text{ ha} = 10\,000\text{ m}^2$

$= 80\text{ ha}$

## Area formulas

| Shape | Area formula | Example |
|---|---|---|
| Rectangle 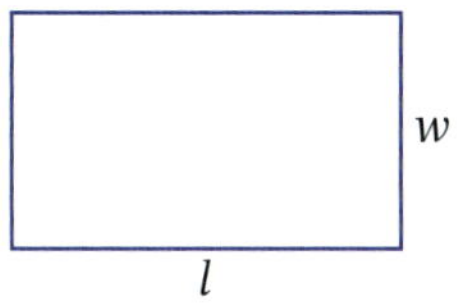  | $A = \text{length} \times \text{width}$<br>$A = lw$ | 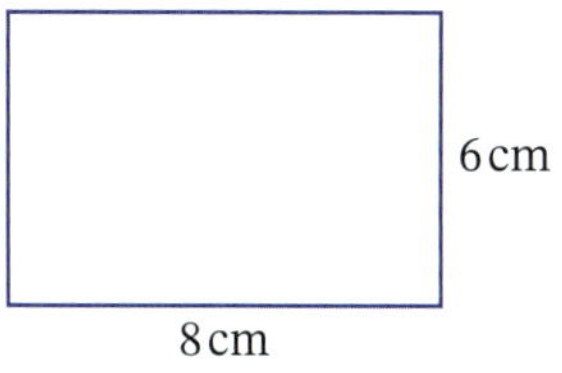  $A = 8 \times 6 = 48\text{ cm}^2$ |
| Square 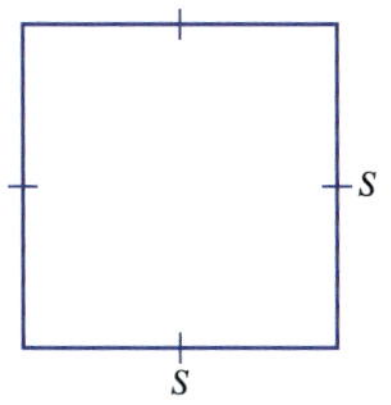  | $A = \text{side} \times \text{side}$<br>$= (\text{side})^2$<br>$A = s^2$ | 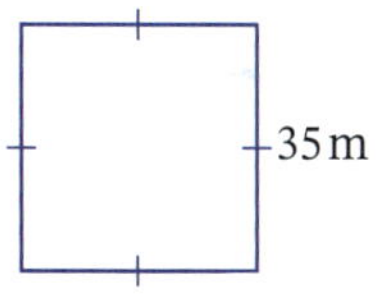  $A = 35^2 = 1225\text{ m}^2$ |
| Triangle 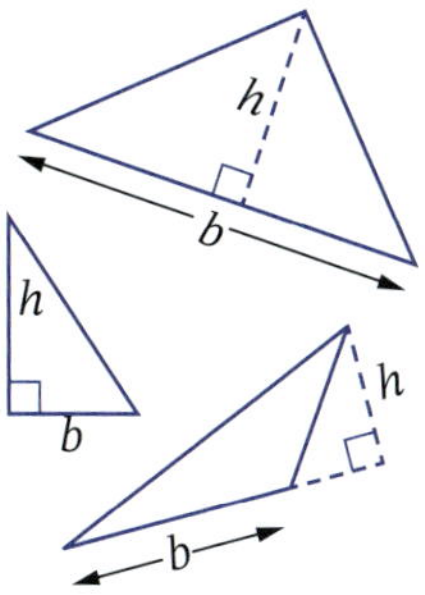  | $A = \frac{1}{2} \times \text{base} \times \text{perpendicular height}$<br>$A = \frac{1}{2}bh$ | 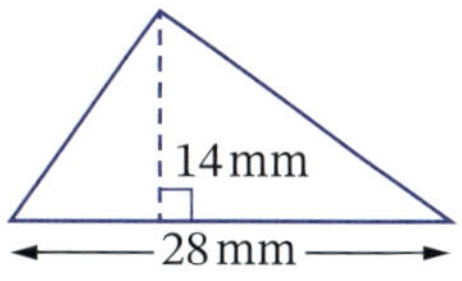  $A = \frac{1}{2} \times 28 \times 14 = 196\text{ mm}^2$ |

**Worksheets**
Areas of composite shapes
A page of circular shapes
Composite areas
Applications of area

| Shape | Area formula | Example |
|---|---|---|
| Parallelogram | $A$ = base × perpendicular height<br>$A = bh$ | $A = 12 \times 5 = 60\text{ m}^2$ |
| Trapezium | $A = \frac{1}{2} \times$ sum of parallel sides × perpendicular distance between sides<br>$A = \frac{1}{2}(a+b)h$<br>$= \frac{h}{2}(a+b)$ | $A = \frac{75}{2} \times (94 + 150)$<br>$= 9150\text{ mm}^2$ |
| Rhombus or kite | $A = \frac{1}{2} \times$ product of diagonals<br>$A = \frac{1}{2}xy$ | $A = \frac{1}{2} \times 20 \times 18 = 180\text{ cm}^2$ |
| Circle | $A = \pi \times (\text{radius})^2$<br>$A = \pi r^2$ | $A = \pi \times 8.5^2 = 227\text{ cm}^2$ |

## Area of a sector

Video: Areas of trapeziums, circles and sectors

A sector's area ($A$) is a fraction of the area ($\pi r^2$) of the circle.

### Area of a sector

$$A = \frac{\theta}{360} \times \pi r^2$$

where $A$ = area

$\theta°$ = is the size of the central angle.

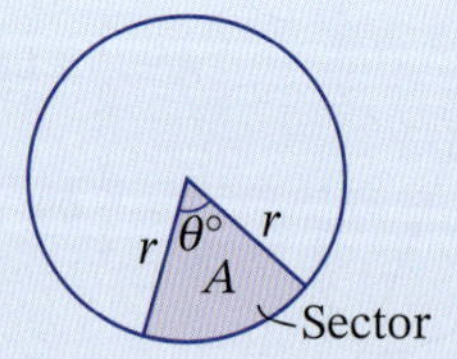

## Example 15

Find, correct to 2 significant figures, the area of sector *PIE*.

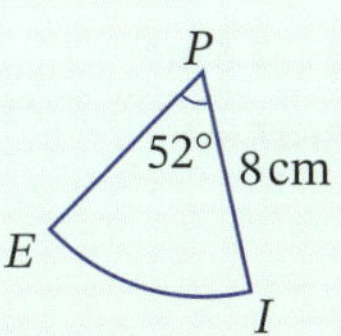


### Solution

$$A = \frac{\theta}{360} \times \pi r^2$$

$$= \frac{52}{360} \times \pi \times 8^2$$

$$= 29.0422\ldots$$

$$\approx 29 \text{ cm}^2$$

Area of sector *PIE* is 29 cm².

## Example 16

A local park has the shape of a rectangle with a semicircle on each end as shown.

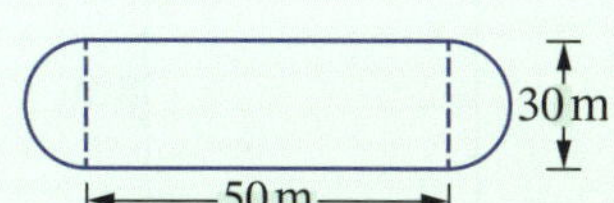


**a** What is the area of this park, correct to the nearest square metre?

**b** What would it cost to cover this park with new grass if the grass costs \$10.50 per square metre?

### Solution

**a** Area of park = rectangle + circle
$= 50 \times 30 + \pi \times 15^2$
$= 2206.858\ldots$
$\approx 2207 \text{ m}^2$

2 semicircles = 1 circle
radius of circle $= \frac{1}{2} \times 30 = 15$ m

**b** Cost of grass $= 2207 \times \$10.50$
$= \$23\,173.50$

**EXERCISE 5.06** Answers on p. 467

## Area

**1** How many square centimetres in 300 m²? Select **A**, **B**, **C** or **D**.

**A** 3 **B** 3000 **C** 3 000 000 **D** 300

**2** How many hectares in 5 km²? Select **A**, **B**, **C** or **D**.

**A** 500 **B** 50 **C** 50 000 **D** 0.5

**3** Convert:

**a** 5 m² to cm² **b** 2500 mm² to cm²

**c** 72 000 m² to ha **d** 6800 cm² to m²

**e** 3.09 km² to m² **f** 3.6 km² to ha

**g** 4.73 m² to mm² **h** 540 ha to km²

**4** The area of NSW is 801 600 km². How many hectares is this in scientific notation?

**5** Builders measure lengths in millimetres, and carpet layers quote per square metre. How many square metres of carpet are needed in a room that measures 2300 mm by 1830 mm? Answer this question using the following 2 methods:

**a** calculate the carpet area in mm², and then convert to m².

**b** convert the lengths to metres first, and then calculate the carpet area in m².

6 Find the area of this park. Select **A**, **B**, **C** or **D**.

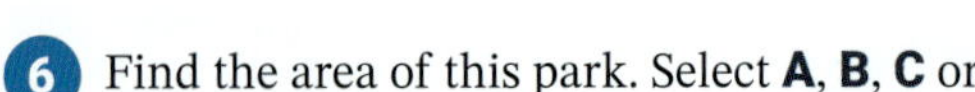

**A** $3127.5\text{ m}^2$ **B** $3944\text{ m}^2$

**C** $3645\text{ m}^2$ **D** $1972\text{ m}^2$

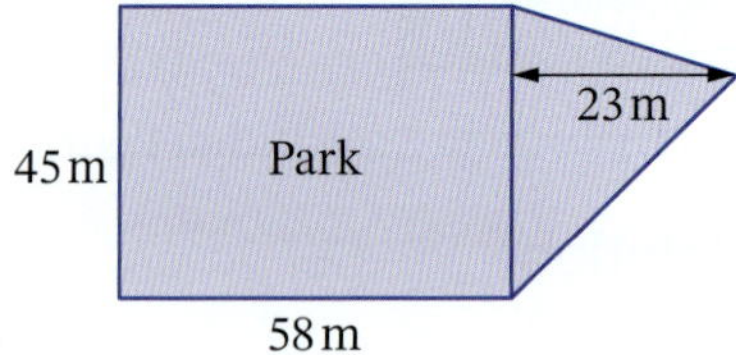

7 Find the area of this sector, correct to one decimal place. Select **A**, **B**, **C** or **D**.

**A** $3.09\text{ cm}^2$ **B** $9.7\text{ cm}^2$

**C** $39.1\text{ cm}^2$ **D** $9.8\text{ cm}^2$

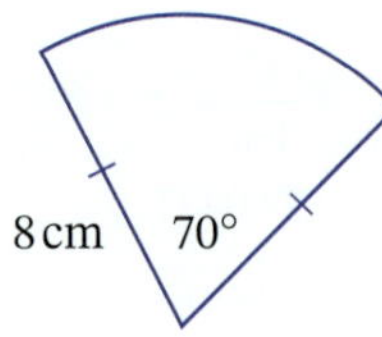


8 Find the area of each shape, correct to one decimal place where needed.
(Note: Some questions require the use of Pythagoras' theorem first.)

**a**

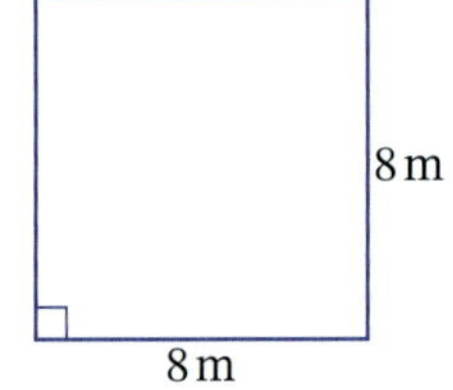


**b**

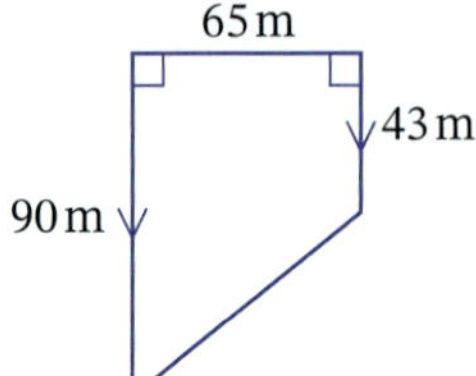


**c**

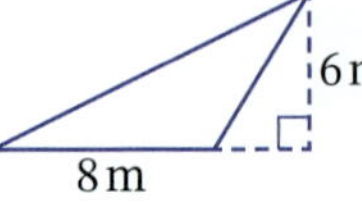


**d**

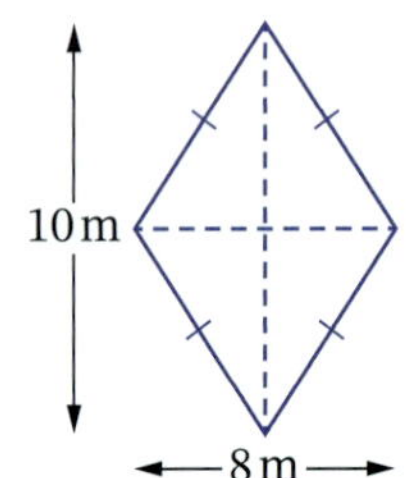


**e**

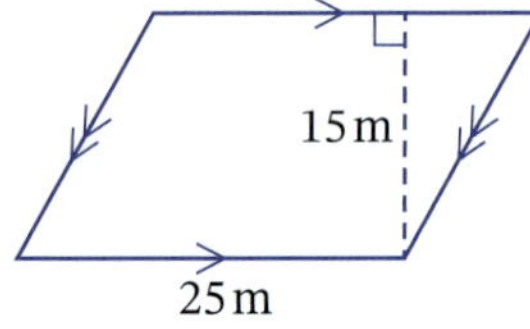


**f**

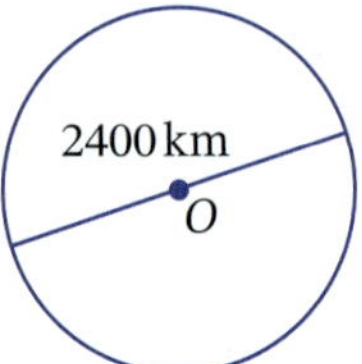


**g**

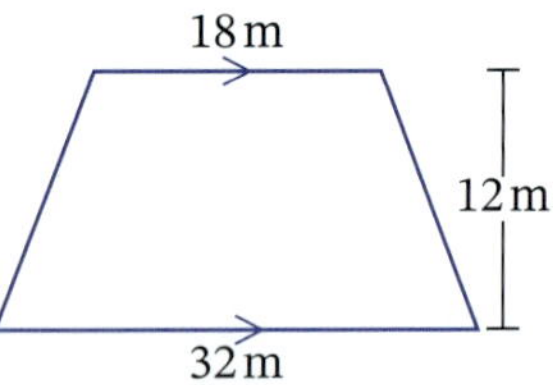


**h**

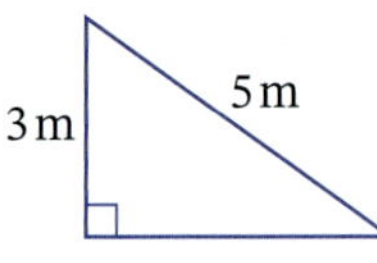


**i**

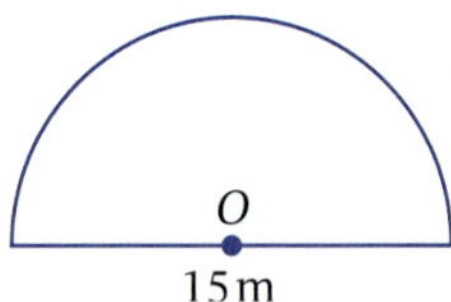


**j** Find the area of the shaded annulus.

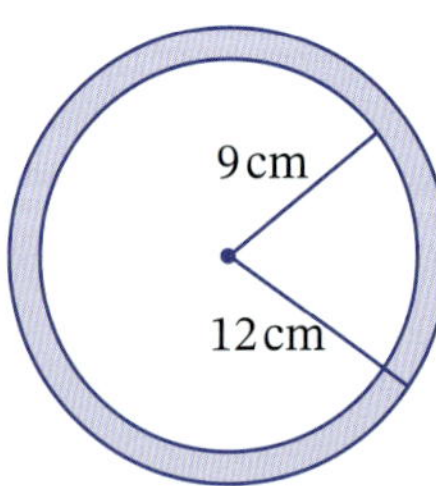


**k** Measurements are in metres

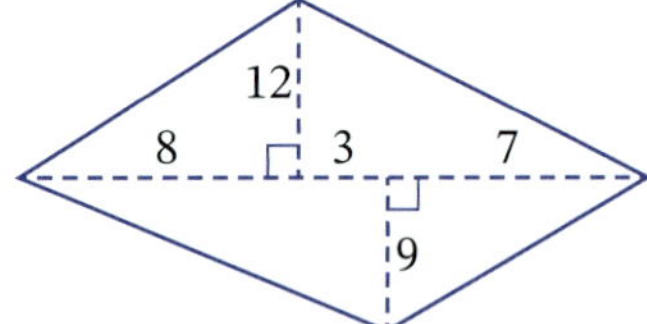


□ Foundation ○ Mastery ⬡ Complex

**9** Find each shaded area, correct to the nearest square unit where necessary.

**a**

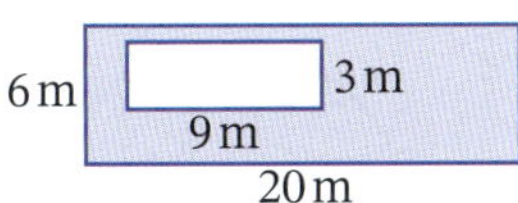


**b**

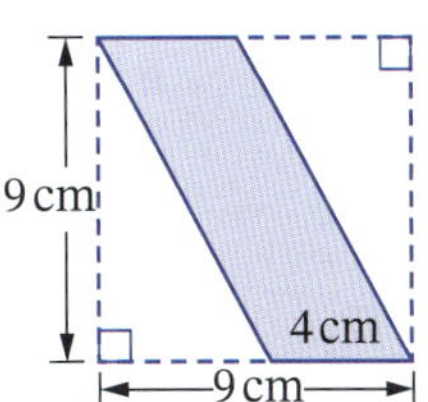


**c**

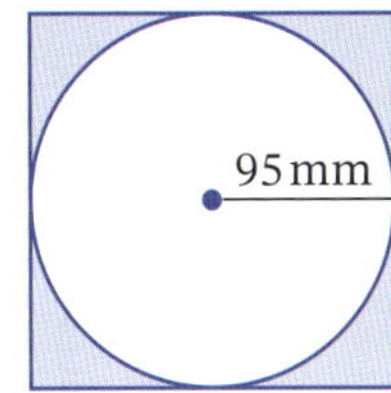


**d**

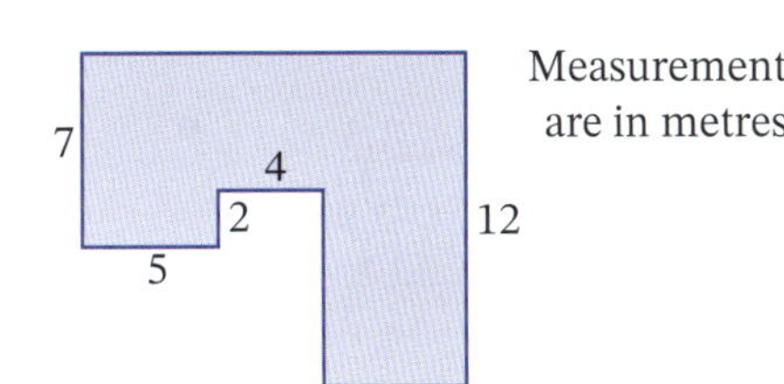


**e**

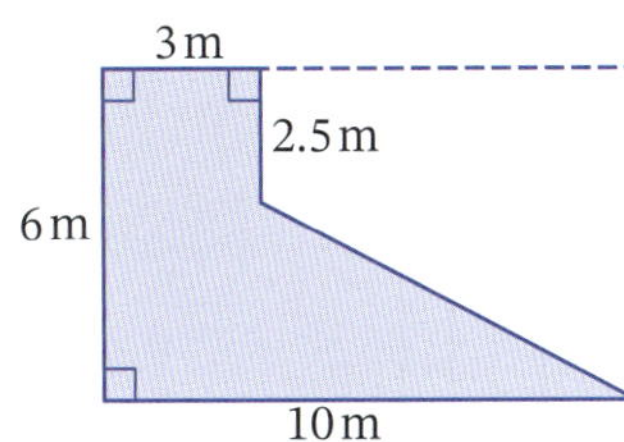


**f**

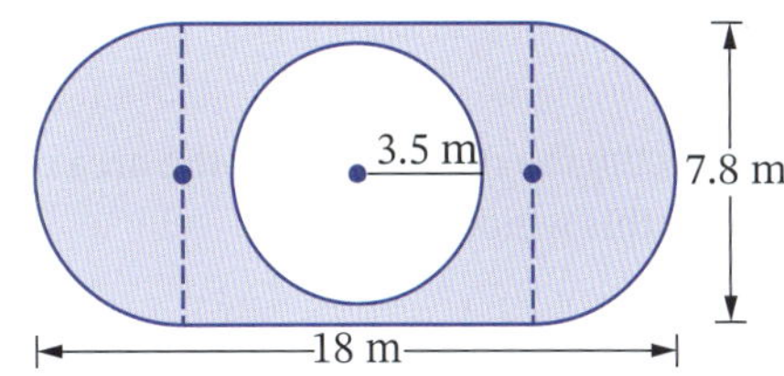


**10** Find the area of each shaded sector, correct to the nearest square unit.

**a**

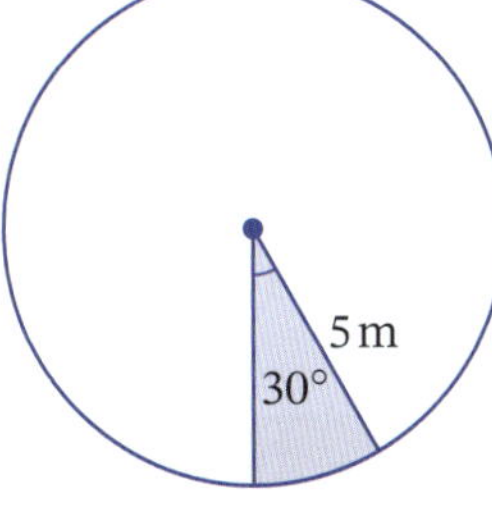


**b**

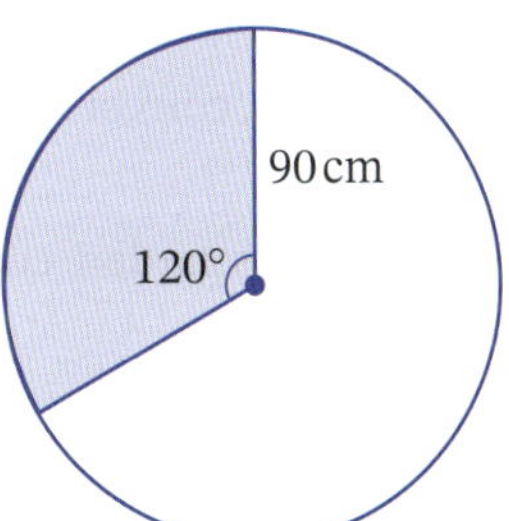


**c**

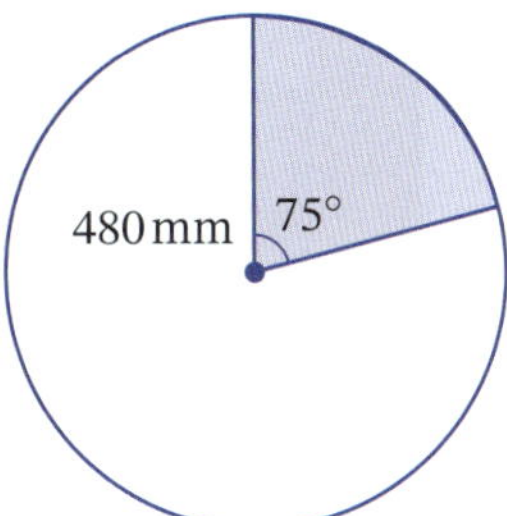


☐ Foundation ○ Mastery ⬡ Complex

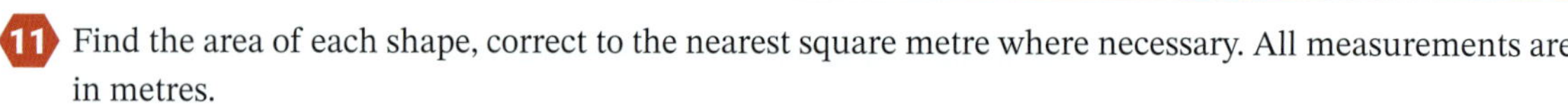

**11** Find the area of each shape, correct to the nearest square metre where necessary. All measurements are in metres.

**a**

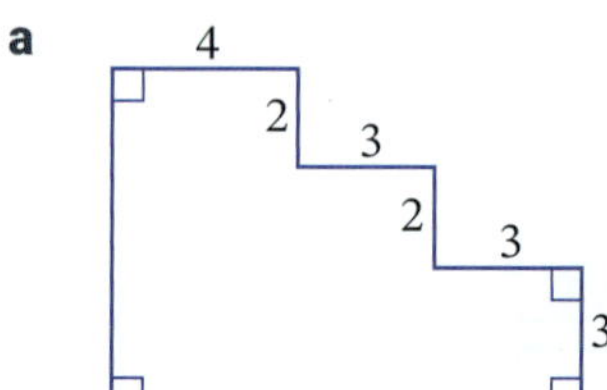


**b**

8
8

**c**

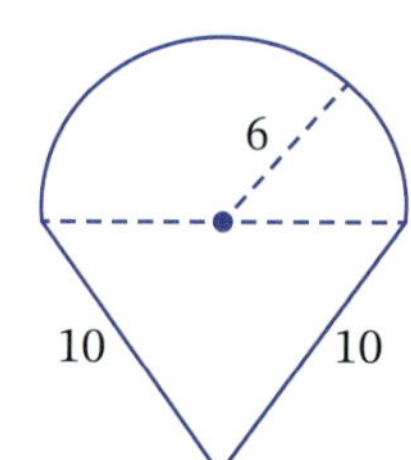


**d**

1.3
2.8
9.6

**e**

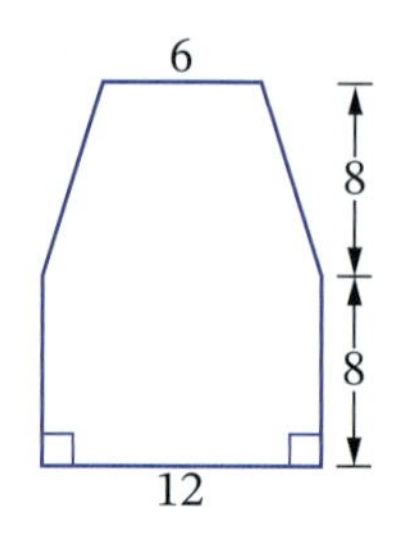


**f**

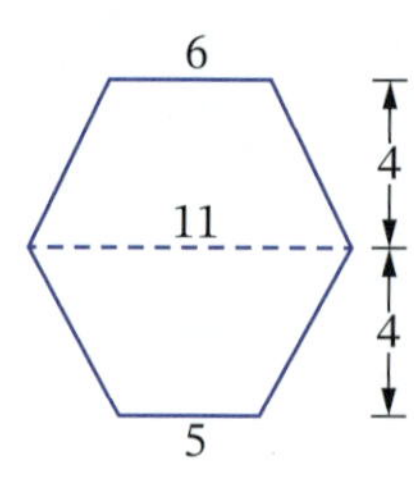


**12** Find, correct to the nearest 0.1 m$^2$, the area of the glass in this window.

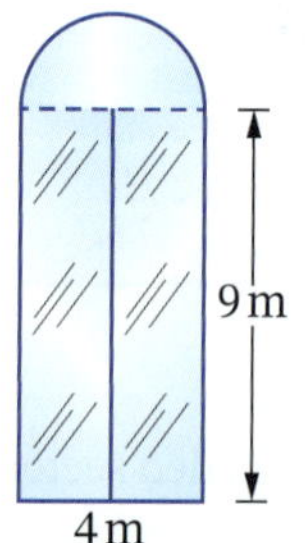

**13** **a** Find the area of this garden, correct to 2 significant figures.

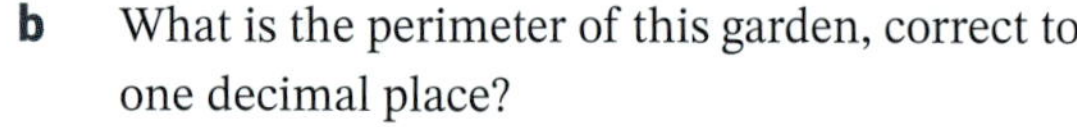

**b** What is the perimeter of this garden, correct to one decimal place?

60 m
200 m

**c** What is the cost of fencing the garden if fencing costs $124 per metre?

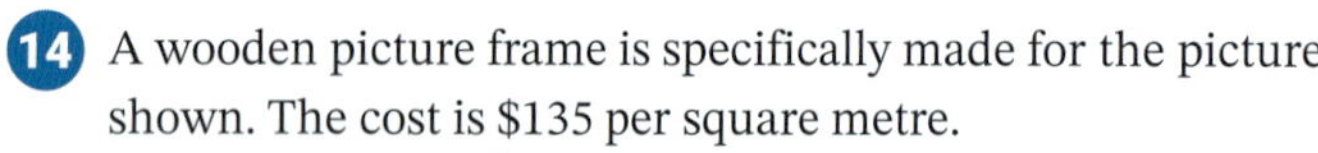

**14** A wooden picture frame is specifically made for the picture shown. The cost is $135 per square metre.

Calculate:

**a** the area of the frame, in square metres

**b** the cost of the frame, to the nearest cent.

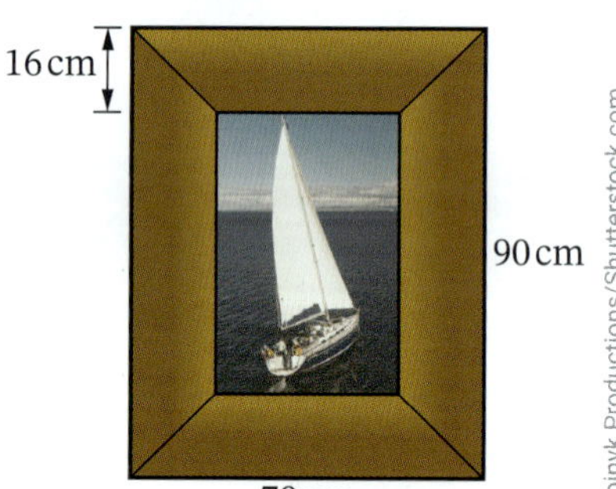


Goinyk Productions/Shutterstock.com

**15** **a** Find the area of each plastic throwing toy, correct to one decimal place.

**i**

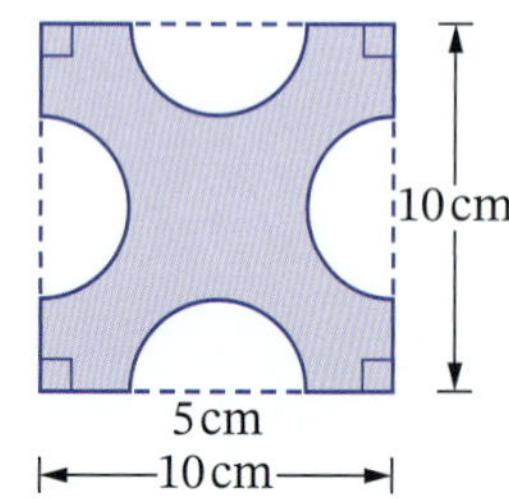


**ii**

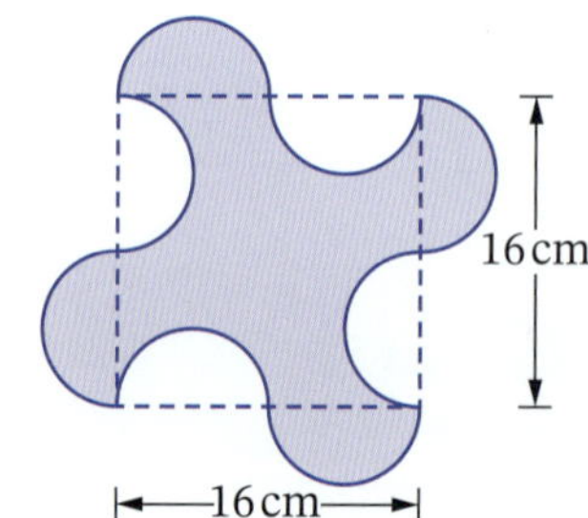


**b** Can you see an easy way of determining the area of the toy in part **ii** above? Explain.

□ Foundation ○ Mastery ⬡ Complex

**16** Harvey is building a path around a rectangular garden in the backyard of his house. The path is 80 cm wide. A diagram is shown below.

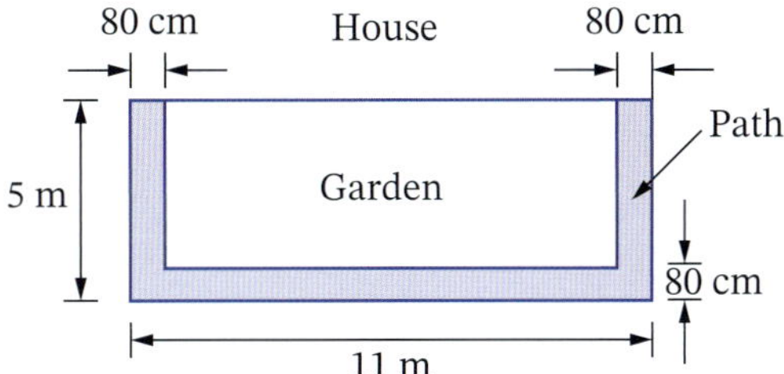


Calculate the area of the path around the garden.

# Land surveying and the trapezoidal rule 5.07

## Offset surveys

One method that surveyors use to measure the area of an irregular block of land is an **offset survey**, also called a **traverse survey**. The surveyor chooses a diagonal of the field to be the **traverse line** (traverse means 'to cross'), then measures the offsets (perpendicular distances) to each corner.

### Example 17

An offset survey was conducted on an irregular block of land, as shown on the **field diagram**, using $LF$ as the traverse line. All measurements are in metres. Find the area of the block.

As most measurements in the diagram are correct to 2 significant figures, write the final answer correct to 2 significant figures.

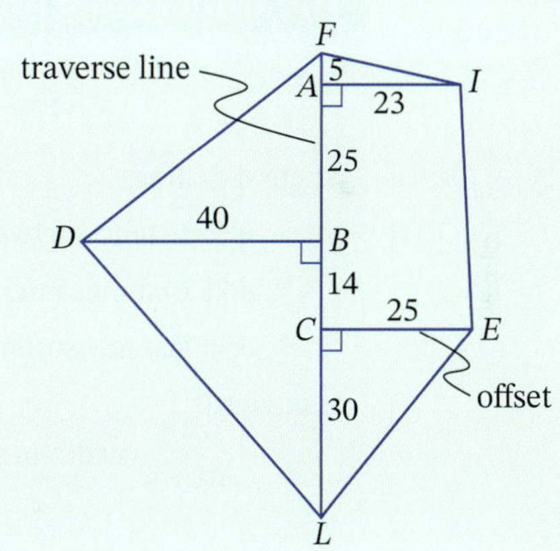


#### Solution

The block is made up of one big triangle on the left, two small triangles and a trapezium on the right.

Area of $\triangle FLD = \frac{1}{2} \times 74 \times 40$ $\qquad LF = 5 + 25 + 14 + 30 = 74$

$= 1480 \text{ m}^2$

Area of $\triangle FIA = \frac{1}{2} \times 23 \times 5$

$= 57.5 \text{ m}^2$

Area of trapezium $IECA = \frac{39}{2} \times (23 + 25)$ $\qquad AC = 25 + 14 = 39$

$= 936 \text{ m}^2$

Foundation Mastery Complex

Area of $\triangle CEL = \frac{1}{2} \times 25 \times 30$

$= 375 \text{ m}^2$

Total area of block $= 1480 + 57.5 + 936 + 375$

$= 2848.5$

$\approx 2800 \text{ m}^2$ correct to 2 significant figures

When conducting an offset survey, the surveyor marks the measurements in a notebook first. The **notebook entries** for the field in Example 17 look like those shown on the right.

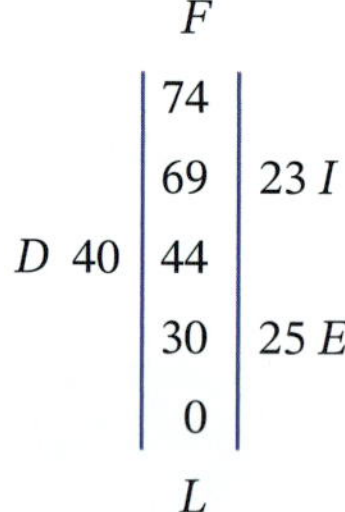


The surveyor uses a tape measure to measure the increasing distances along the traverse line from $L$ to $F$. For example, 74 m is the **entire** distance from $L$ to $F$, whereas the distance $CB = 44 - 30 = 14$ m.

## The trapezoidal rule

Worksheet
Trapezoidal rule

The **trapezoidal rule** is an approximation method for calculating the area of an irregular shape, such as one with a curved side. Approximate the shape as a trapezium with width, $h$, and take 2 vertical measurements, $d_f$ and $d_l$, as shown.

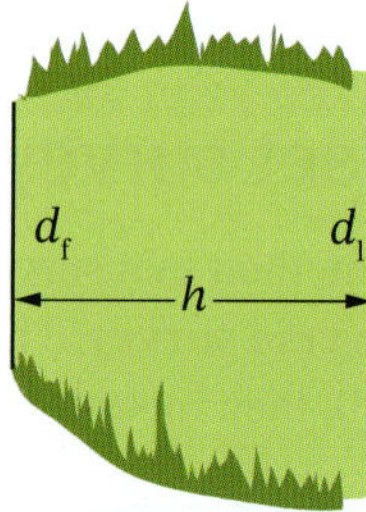


### The trapezoidal rule

$$A \approx \frac{h}{2}(d_f + d_l)$$

where $A$ = area

$h$ = distance between successive measurements

$d_f$ = first measurement

$d_l$ = last measurement

or in words:

$$\text{Area} \approx \frac{\text{width of strip}}{2}(\text{first measurement} + \text{last measurement})$$

## Example 18

A surveyor took 3 vertical measurements across a pond at 10 m intervals.

Find the approximate area of the pond using:

**a** one application of the trapezoidal rule

**b** 2 applications of the trapezoidal rule.

Which answer is more accurate?

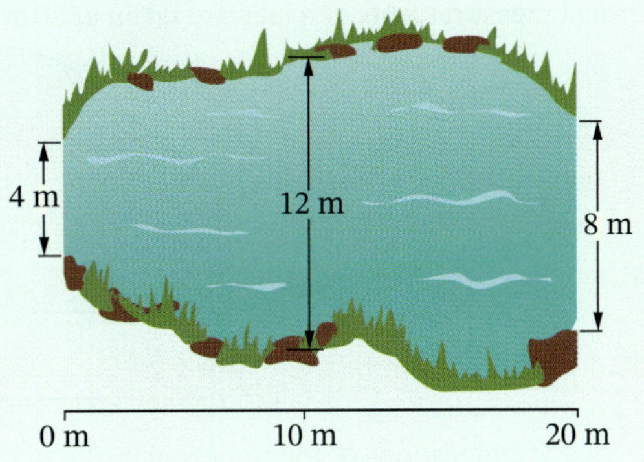


### Solution

**a** For one application, we treat the pond as a trapezium with parallel sides 4 m and 8 m, so $h = 20$, $d_f = 4$, $d_l = 8$.

$$A \approx \frac{h}{2}(d_f + d_l)$$

Ignore the 12 m in the middle of the pond.

$$= \frac{20}{2}(4 + 8)$$

$$= 120\,\text{m}^2$$

The approximate area of the pond is $120\,\text{m}^2$.

**b** For 2 applications, treat the pond as 2 trapeziums side by side, joined by the 12 m side.

For the left trapezium, $h = 10$, $d_f = 4$, $d_l = 12$.

$$A \approx \frac{h}{2}(d_f + d_l)$$

$$= \frac{10}{2}(4 + 12)$$

$$= 80\,\text{m}^2$$

For the right trapezium, $h = 10$, $d_f = 12$, $d_l = 8$.

$$A \approx \frac{h}{2}(d_f + d_l)$$

$$= \frac{10}{2}(12 + 8)$$

$$= 100\,\text{m}^2$$

$$\text{Total area} = 80 + 100$$

$$= 180\,\text{m}^2$$

The approximate area of the pond is $180\,\text{m}^2$.

The $180\,\text{m}^2$ is more accurate, because the more applications (trapeziums) we use, the closer the answer is to the exact area of the pond.

Video
The trapezoidal rule

## Example 19

Vertical measurements of a lake are taken at 20 m intervals.

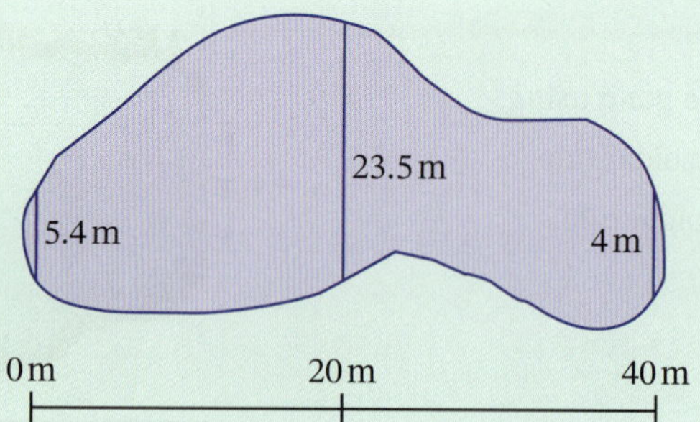


**a** Use 2 applications of the trapezoidal rule to estimate the area of the lake, correct to one decimal place.

**b** Find the volume of water in the lake if it is a constant depth of 2.5 m.

## Solution

**a** Left trapezium: $d_f = 5.4, d_l = 23.5, h = 20$

$$A = \frac{h}{2}(d_f + d_l)$$

$$= \frac{20}{2}(5.4 + 23.5)$$

$$= 289$$

Right trapezium: $d_f = 23.5, d_l = 4, h = 20$

$$A = \frac{h}{2}(d_f + d_l)$$

$$= \frac{20}{2}(23.5 + 4)$$

$$= 275$$

Total area of the lake $= 289 + 275$

$= 564\text{ m}^2$

**b** For the volume of the lake, multiply the above area by the depth.

$$V = Ah$$

$$= 564 \times 2.5$$

$$= 1410\text{ m}^3$$

The volume of the lake is $1410\text{ m}^3$.

**EXERCISE 5.07** Answers on p. 468

## Land surveying and the trapezoidal rule

In this exercise, all measurements are in metres and diagrams are not to scale.

**1** Find the area of each field, correct to 2 significant figures.

**a**

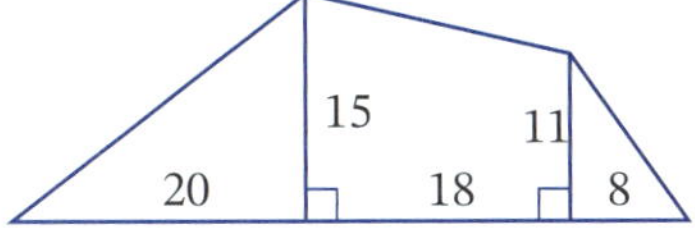


**b**

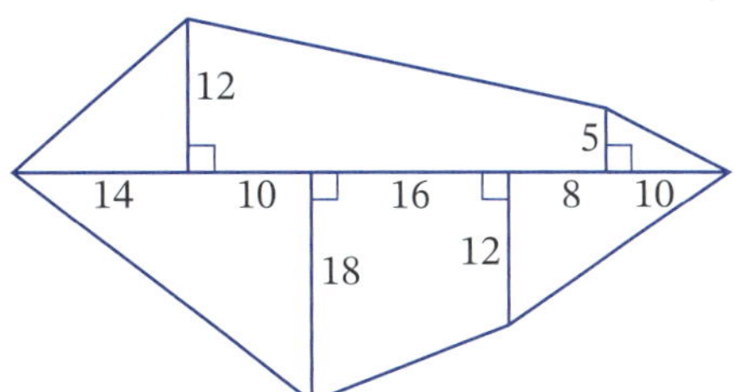


**c**

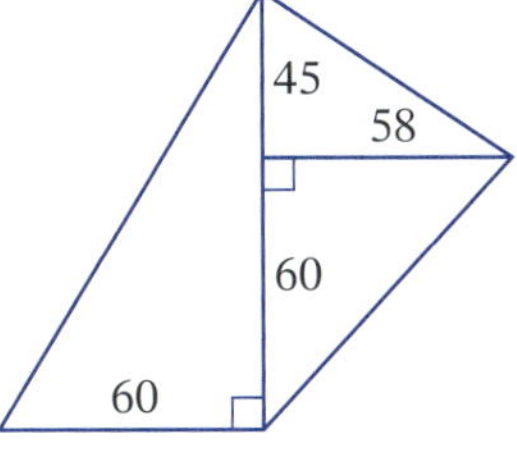


**d**

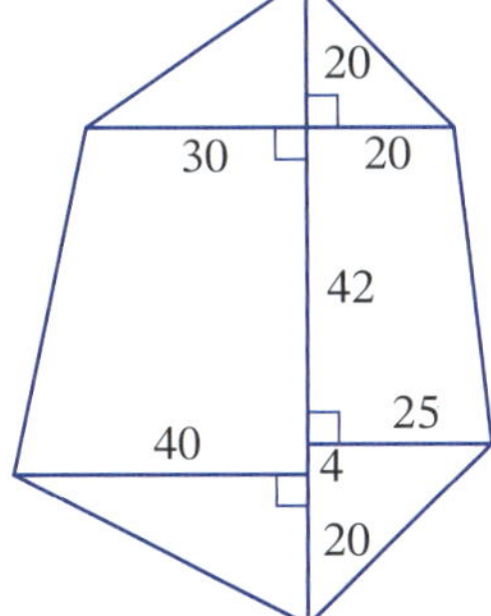


**e**

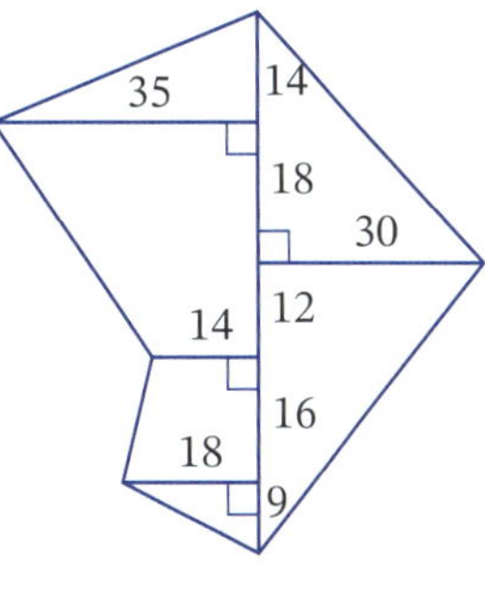


**2** For this surveyor's notebook entry, sketch a field diagram and then calculate its area, correct to the nearest square metre.

| | *F* | |
|---|---|---|
| | 52 | |
| | 34 | 36 *G* |
| | 20 | 15 *H* |
| *J* 40 | 12 | |
| | 0 | |
| | *I* | |

**3** **a** Find the area of this farm, correct to the nearest square metre.

**b** The farm is to be fenced around its perimeter. What length of fence is needed? Answer to the nearest metre.

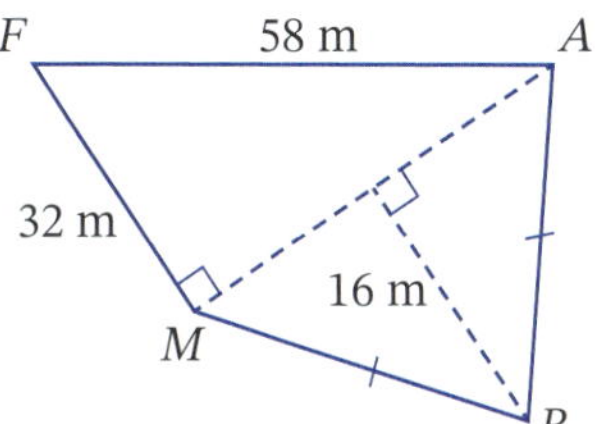


EXAMPLE 18

**4** A garden has the measurements shown in the figure (in metres). Use the trapezoidal rule to approximate its area to the nearest square metre. Select **A**, **B**, **C** or **D**.

**A** 7 **B** 17 **C** 18 **D** 45

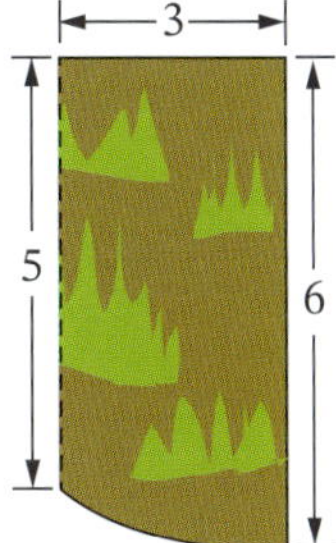


□ Foundation ○ Mastery ⬡ Complex

**5** Approximate the area of each field using one application of the trapezoidal rule.

**a**

21 m
12 m
18 m

**b**

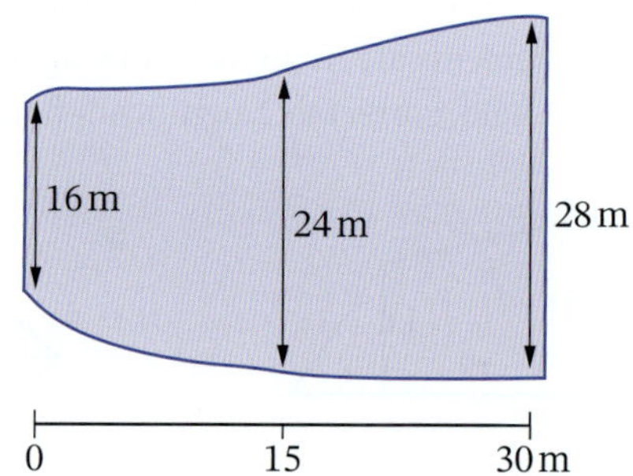


**6** To calculate an approximate area for her driveway, Olivia took 3 measurements at 5 m intervals.

**a** Use 2 applications of the trapezoidal rule to find the area of the driveway, correct to the nearest m$^2$.

**b** What is the cost of sealing the driveway if sealer costs \$45/L and 1 L covers 12 m$^2$?

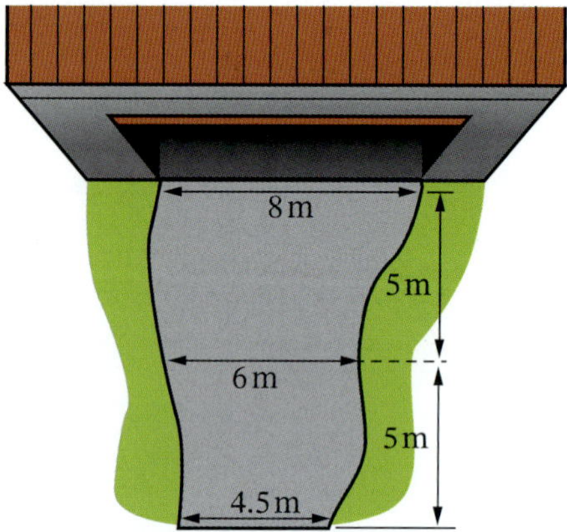


**7** The **cross-section** of a river is shown, with measurements in metres. Use 4 applications of the trapezoidal rule to approximate the area of the cross-section.

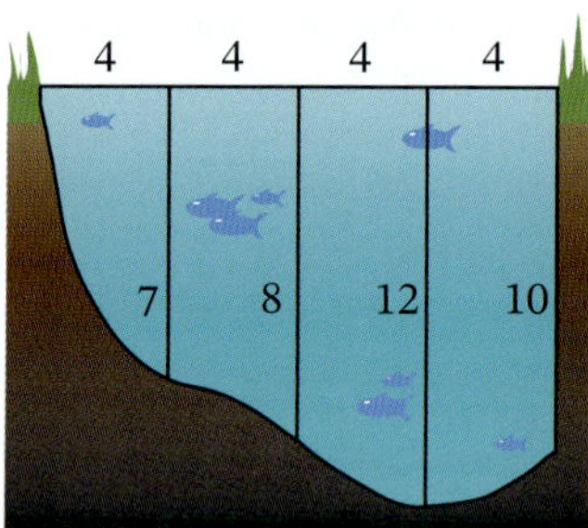


**8** The shape of a new swimming pool is marked out on the lawn and 5 vertical measurements are taken at 6 m intervals. Use 4 applications of the trapezoidal rule to find the area of the pool.

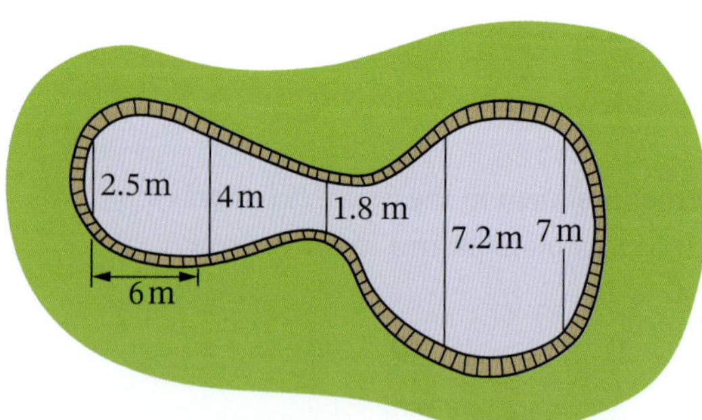


☐ Foundation ○ Mastery ⬡ Complex

**9** This diagram shows the outline of a lake in a botanical garden.

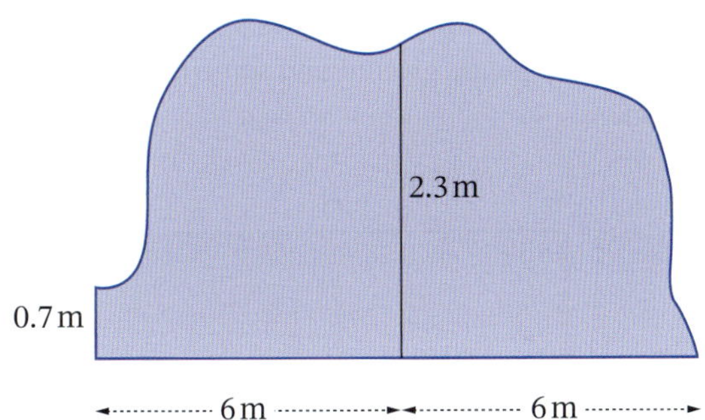


**a** Use 2 applications of the trapezoidal rule to find the approximate surface area of the lake.

**b** The average water depth in the lake is 1.2 metres. How many whole litres of water can the lake hold when full? ($1\,m^3 = 1\,kL$)

**10** An artificial pond is designed for a new shopping centre.

**a** Calculate the area of the cross-section using 4 applications of the trapezoidal rule.

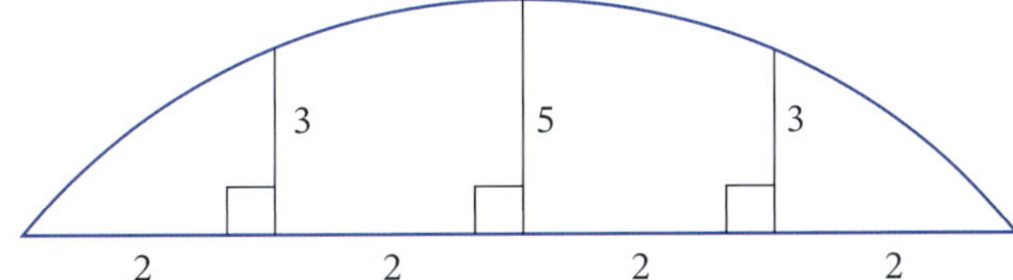


**b** If the pond is 1.6 m deep, calculate its capacity in litres.

**11** Chris has surveyed the property shown by this diagram and written the measurements for its lengths.

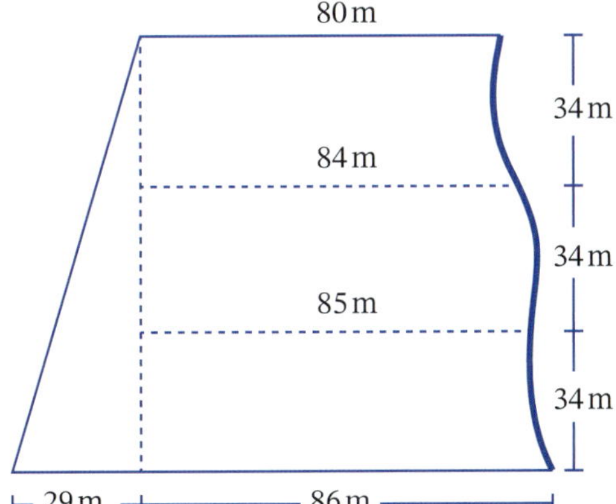


**a** Approximate the total area of this property by including 3 applications of the trapezoidal rule.

**b** Chris discovered that the average monthly rainfall on this property is 12.6 cm. What volume of rain can Chris expect to fall over this property in one month? Round your answer to the nearest cubic metre.

☐ Foundation ○ Mastery ○ Complex

**12** This diagram shows an area in a national park consisting of a rectangle with dimensions 200 m by 150 m and a semicircle with radius 75 m. A duck pond occupies a section of this area in the park, as shown in the diagram. The rest of the area is a picnic ground and some measurements from the edge of the picnic ground to the pond are shown.

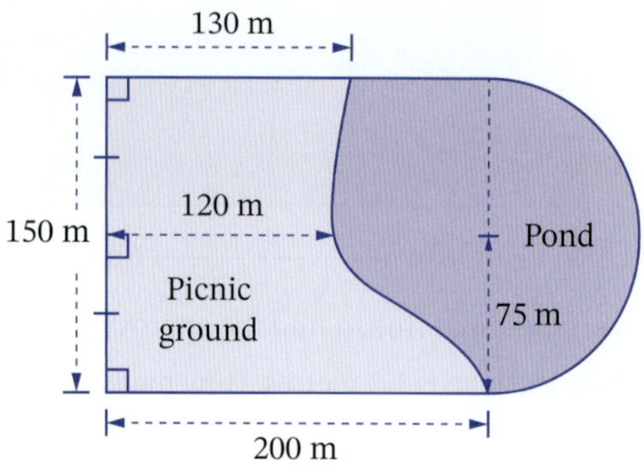


**a** Use 2 applications of the trapezoidal rule to calculate the approximate area of the picnic ground.

**b** Hence, calculate the approximate area of the pond, correct to 2 significant figures.

# 5.08 Volume and surface area of a prism

**Worksheet**
Estimating area and volume

The **volume** of a solid is the amount of space it occupies, measured in **cubic units**.

| Unit | Relationship | Size |
|---|---|---|
| cubic millimetre ($mm^3$) | $1\text{ mm}^3 = 1\text{ mm} \times 1\text{ mm} \times 1\text{ mm}$ | The volume of a 1 mm cube (about the size of a grain of raw sugar) |
| cubic centimetre ($cm^3$) | $1\text{ cm}^3 = 1\text{ cm} \times 1\text{ cm} \times 1\text{ cm}$<br>$= 10\text{ mm} \times 10\text{ mm} \times 10\text{ mm}$<br>$= 1000\text{ mm}^3$ | The volume of a 1 cm cube (about the size of a peanut) |
| cubic metre ($m^3$) | $1\text{ m}^3 = 1\text{ m} \times 1\text{ m} \times 1\text{ m}$<br>$= 100\text{ cm} \times 100\text{ cm} \times 100\text{ cm}$<br>$= 1\,000\,000\text{ cm}^3$ | The volume of a 1 m cube (about the size of two washing machines)<br>1 m<br>1 m<br>1 m |

$$\begin{aligned}1\text{ km}^3 &= 1\text{ km} \times 1\text{ km} \times 1\text{ km}\\ &= 1000\text{ m} \times 1000\text{ m} \times 1000\text{ m}\\ &= 1\,000\,000\,000\text{ m}^3\end{aligned}$$

When converting units of volume, we need to multiply or divide by 3 lots of units.
For example, if $1\text{ m} = 100\text{ cm}$, then $1\text{ m}^3 = (100 \times 100 \times 100)\text{ cm}^3 = 1\,000\,000\text{ cm}^3$.

## Example 20

Convert:

**a** $74\,m^3$ to $cm^3$ **b** $4600\,mm^3$ to $cm^3$

### Solution

**a** $74\,m^3 = (74 \times 100 \times 100 \times 100)\,cm^3$ because $1\,m^3 = (100 \times 100 \times 100)\,cm^3$

$= 74\,000\,000\,cm^3$

**b** $4600\,mm^3 = (4600 \div 10 \div 10 \div 10)\,cm^3$ because $1\,cm^3 = (10 \times 10 \times 10)\,mm^3$

$= 4.6\,cm^3$

## Volume and capacity

While volume is the amount of space taken up by a solid, **capacity** is the amount of material (usually liquid) that a solid or container can hold.

### Volume and capacity

$1\,cm^3 = 1\,mL$

$1\,m^3 = 1\,kL = 1000\,L$

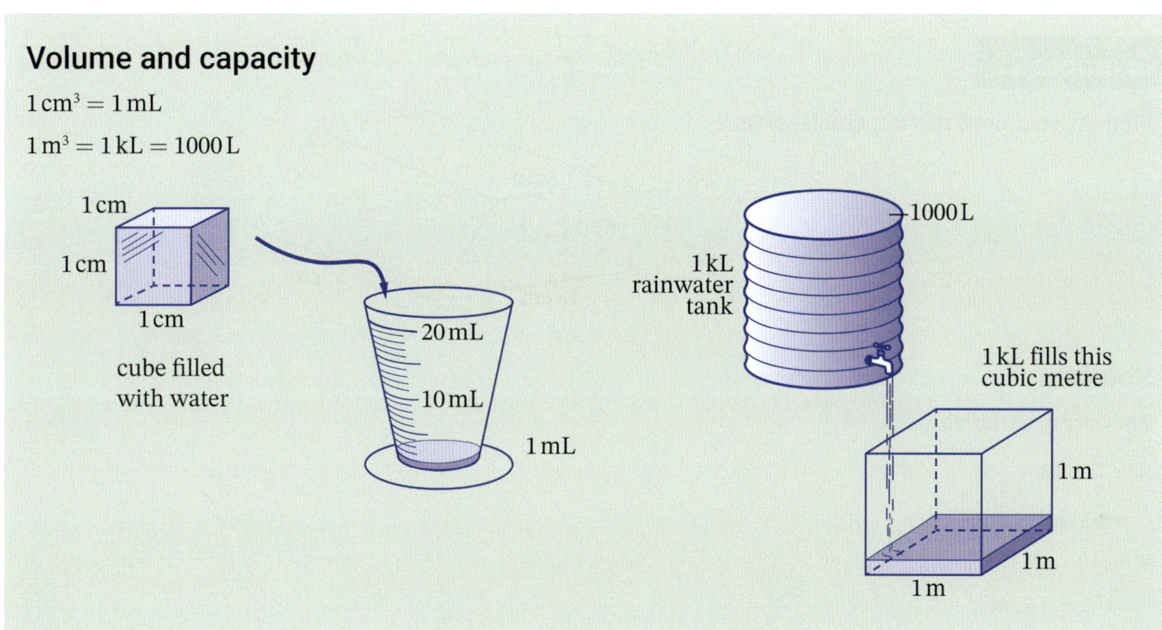


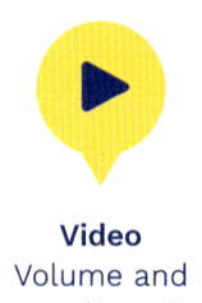

**Video** Volume and capacity units

## Example 21

Convert:

**a** $46\,L$ to $cm^3$ **b** $10\,400\,cm^3$ to $L$ **c** $3500\,m^3$ to $ML$

### Solution

**a** $46\,L = 46\,000\,mL$ $1\,L = 1000\,mL$

$= 46\,000\,cm^3$ $1\,cm^3 = 1\,mL$

**b** $10\,400\,cm^3 = 10\,400\,mL$ $1\,cm^3 = 1\,mL$

$= 10.4\,L$

**c** $3500\,m^3 = 3500\,kL$ $1\,m^3 = 1\,kL$

$= 3\,500\,000\,L$

$= 3.5\,ML$ $1\,ML = 1$ million litres

Video
Volumes of prisms and cylinders

A **prism** is a solid shape with identical cross-sections along its length. For example, a triangular prism has a **base** and cross-sections that are identical triangles.

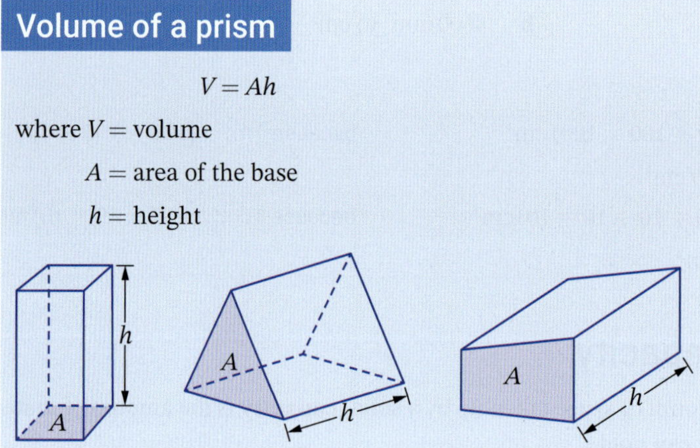

### Volume of a prism

$$V = Ah$$

where $V$ = volume

$A$ = area of the base

$h$ = height



### Example 22

Find the volume of this trapezoidal prism.

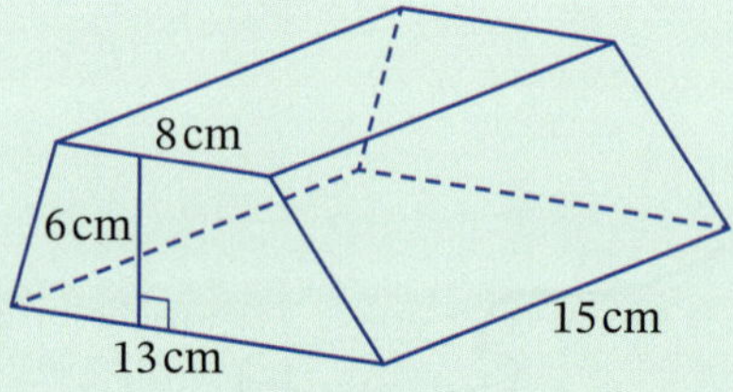


#### Solution

The base is a trapezium.

$A = \frac{6}{2} \times (8 + 13)$

$= 63\text{ cm}^2$

$V = Ah$

$= 63 \times 15\text{ cm}$

$= 945\text{ cm}^3$

Video
Surface area of a prism

Worksheet
Nets of solids

### Surface area of a prism

The surface area of a prism is the sum of the areas of its faces, measured in square units.

The solid can be closed or open.

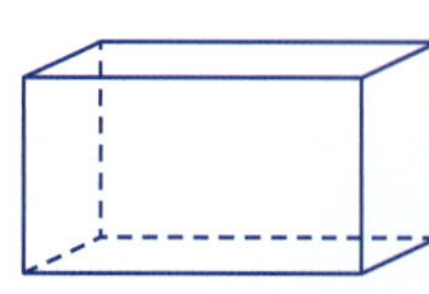

A closed rectangular prism has 6 faces.

An open triangular prism with no end faces has 3 faces.

## Example 23

This triangular prism is open at one end.
Calculate its surface area.

'Surface area' here means the external (outside) surface area.

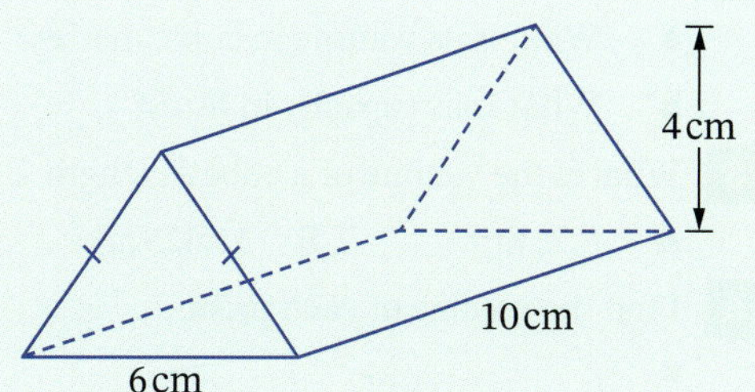


### Solution

The open prism has one triangular face and 3 rectangular faces.
To find the areas of the side faces, first find the equal lengths ($x$ cm) of the triangular face using Pythagoras' theorem.

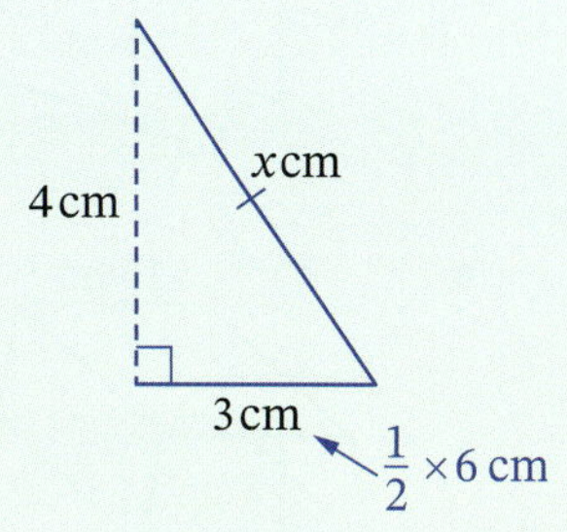


$x^2 = 3^2 + 4^2$

$= 25$

$x = \sqrt{25}$

$= 5$

Surface area = area of triangle + area of base rectangle + 2 × area of side rectangles

$= \frac{1}{2} \times 6 \times 4 + (10 \times 6) + 2 \times (10 \times 5)$

$= 172$

The surface area of this triangular prism is $172\,\text{cm}^2$.

**EXERCISE 5.08** Answers on p. 468

## Volume and surface area of a prism

**1** Convert:

| | | | | | |
|---|---|---|---|---|---|
| **a** | $7\,\text{m}^3$ to $\text{cm}^3$ | **b** | $50\,\text{cm}^3$ to $\text{mm}^3$ | **c** | $89\,000\,\text{cm}^3$ to $\text{m}^3$ |
| **d** | $0.468\,\text{m}^3$ to $\text{cm}^3$ | **e** | $2400\,\text{mm}^3$ to $\text{cm}^3$ | **f** | $5\,600\,000\,\text{cm}^3$ to $\text{m}^3$ |
| **g** | $9\,100\,000\,\text{mm}^3$ to $\text{cm}^3$ | **h** | $12\,\text{m}^3$ to $\text{cm}^3$. | **i** | $2.45\,\text{m}^3$ to $\text{mm}^3$ |

**2** Convert:

| | | | | | |
|---|---|---|---|---|---|
| **a** | $680\,\text{cm}^3$ to mL | **b** | $8500\,\text{cm}^3$ to L | **c** | $22\,\text{m}^3$ to L |
| **d** | 8000 L to $\text{m}^3$ | **e** | $3.5\,\text{m}^3$ to mL | **f** | 690 L to $\text{cm}^3$ |
| **g** | $55\,\text{m}^3$ to L | **h** | $4300\,\text{m}^3$ to kL | **i** | 9500 L to $\text{m}^3$ |
| **j** | $8.5 \times 10^4\,\text{cm}^3$ to L | **k** | $4.3 \times 10^{-3}$ kL to $\text{cm}^3$ | **l** | $10^6\,\text{m}^3$ to ML. |

**3** Find the volume of this rectangular prism. Select **A**, **B**, **C** or **D**.

**A** $1.2\,\text{m}^3$

**B** $7.6\,\text{m}^3$

**C** $7.2\,\text{m}^3$

**D** $2.4\,\text{m}^3$

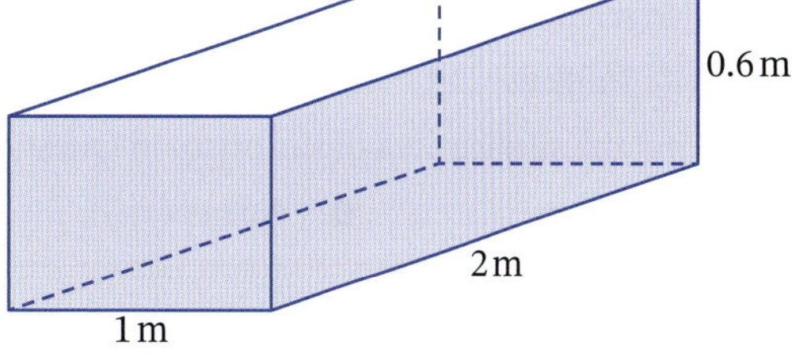


□ Foundation ○ Mastery ⬡ Complex

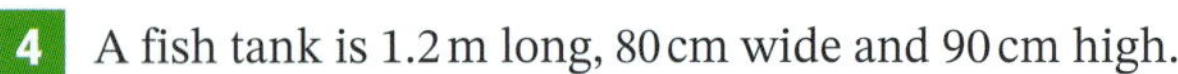

**4** A fish tank is 1.2 m long, 80 cm wide and 90 cm high.

**a** What is its volume in cubic metres?

**b** What is its capacity in litres?

**5** What is the volume of a cube of length 2.5 m? Select **A**, **B**, **C** or **D**.

**A** 6.25 mm$^3$ **B** 15.625 m$^3$ **C** 30 m$^3$ **D** 37.5 m$^3$

**6** Find the volume of each prism.

**a**

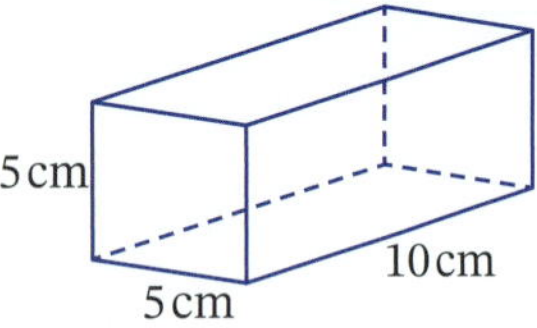


**b**

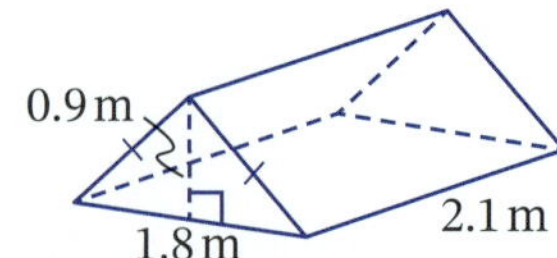


**c**

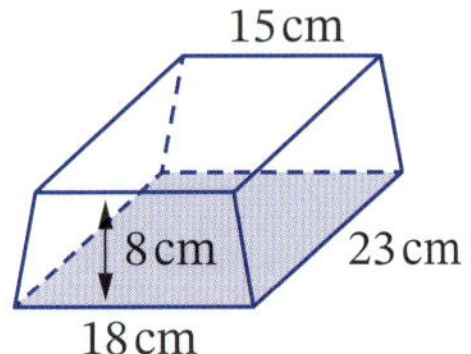


**d**

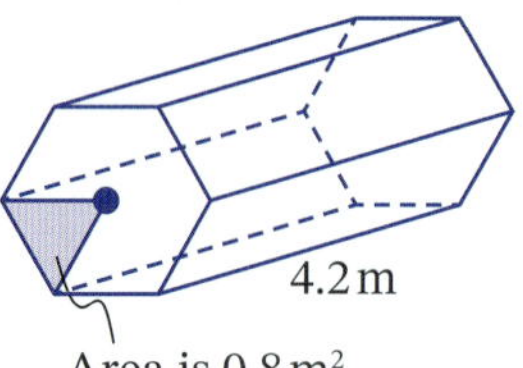


**7** What is the capacity of this bar fridge, to the nearest litre?

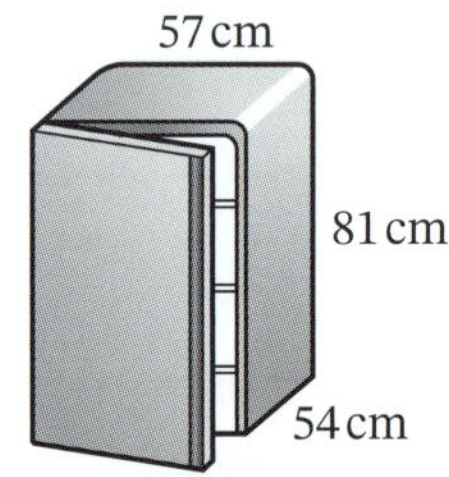


**8** The cross-section of a kidney-shaped swimming pool has an area of 11.5 m$^2$. The pool has a constant depth of 1.4 m. What volume of water does it hold:

**a** in cubic metres? **b** in litres?

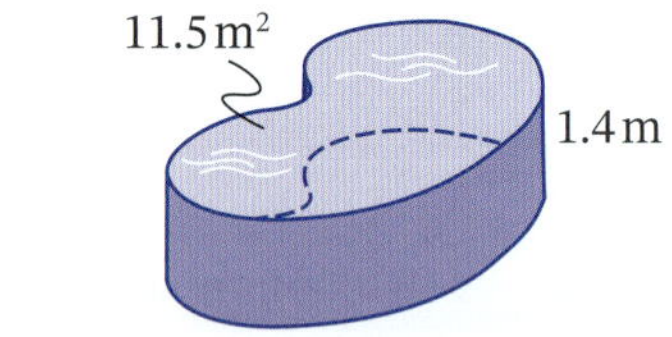


**9** **a** This skip has the shape of a trapezoidal prism. What volume of rubbish will it hold?

**b** If building rubbish costs $16.50 per cubic metre to dump at the local tip, how much will it cost to dump 4 full skips?

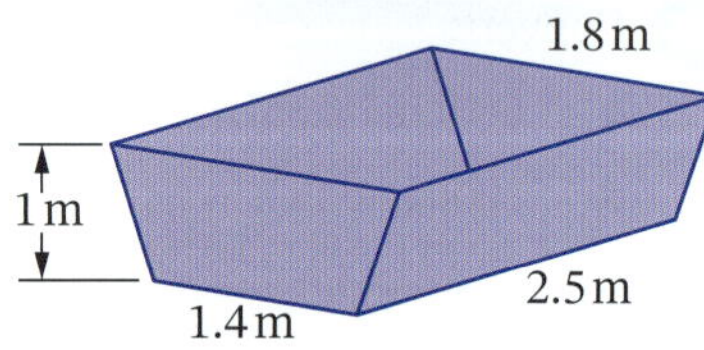


**10** Warragamba Dam supplies Sydney with water and has a capacity of $2.031 \times 10^6$ ML. Use scientific notation to express the dam's capacity:

**a** in litres **b** in cubic metres.

**11** Find the surface area of the prism in Question **3** if it is open at the top. Select **A**, **B**, **C** or **D**.

**A** 172 m$^2$ **B** 3.6 m$^2$ **C** 5.6 m$^2$ **D** 6.6 m$^2$

Foundation Mastery Complex

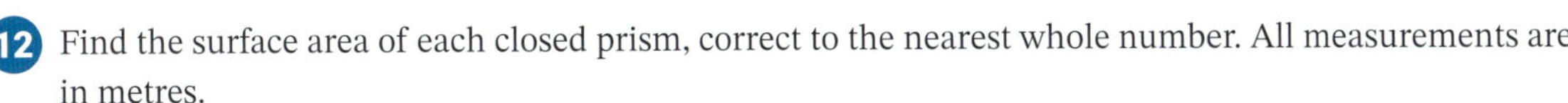

**12** Find the surface area of each closed prism, correct to the nearest whole number. All measurements are in metres.

**a**

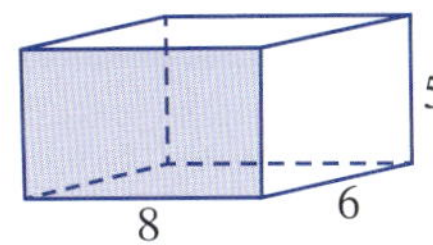


**b**

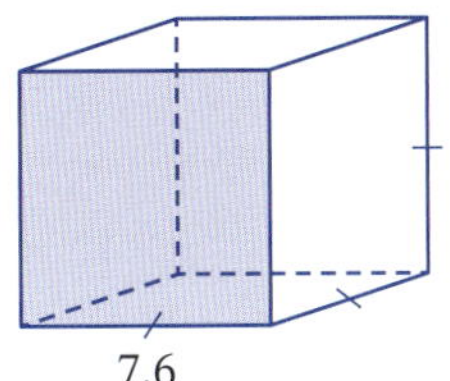


**c**

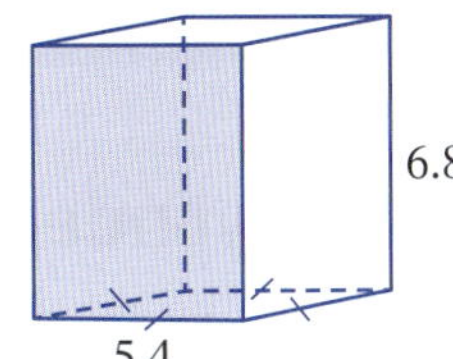


**d**

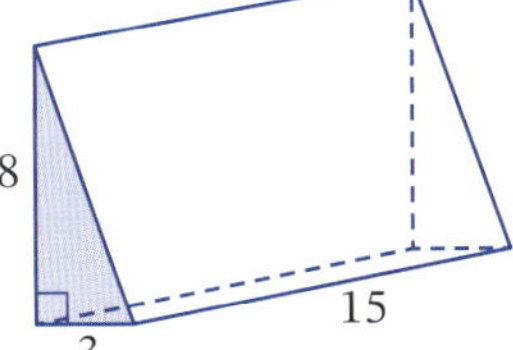


**e**

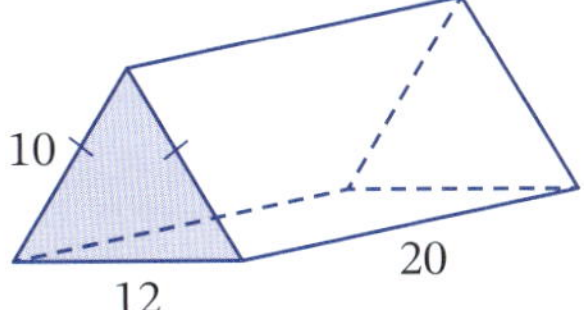


**13** **a** Find the external surface area of this open water trough, correct to 2 significant figures.

**b** What is the capacity of the trough, to the nearest litre?

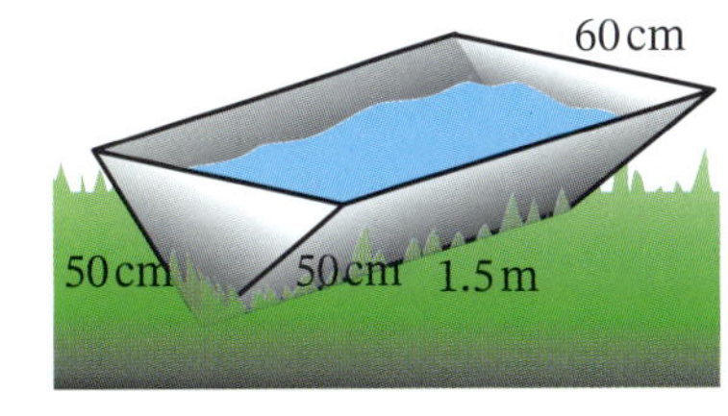


**14** If the cross-section of a prism has area, $A$, and perimeter, $P$, and the length of the prism is $h$, then its surface area can be calculated using the formula $S = 2A + Ph$. The $2A$ represents the area of the 2 identical end faces, while $Ph$ is the total area of the rectangular side faces, which combine to make one big rectangle.

**a** A portable set of stairs has dimensions as shown. Use the formula to calculate its surface area in square metres.

**b** Calculate its volume in cubic metres.

**15** **a** Find the volume of this tent.

**b** How many square metres (correct to one decimal place) of canvas were used to make this tent, including the floor?

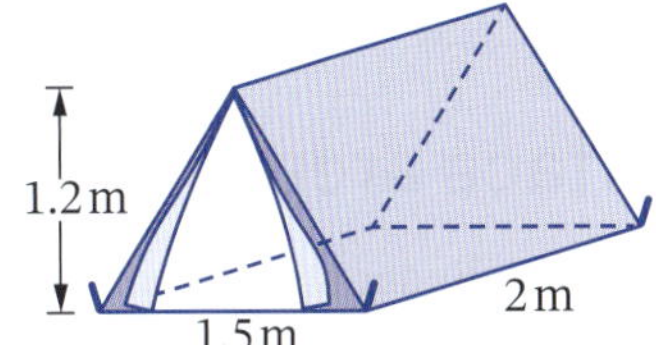


**16** This timber wedge is used to prop open a door.

**a** Calculate its volume.

**b** Kobi applies varnish to a box of wedges. One 250 mL pot of varnish covers 1 m² of surface area. How many wedges can he varnish with 5 pots?

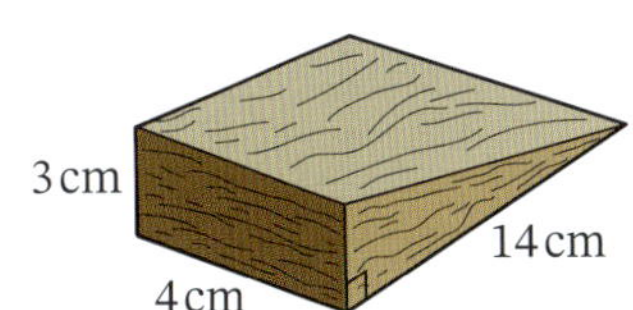


## Did you know?

### Mass and capacity

One litre of water has a mass of one kilogram. For water, 1 L is equivalent to 1 kg.

What is the mass of:

**a** 1 m³ of water?

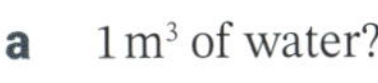

**b** 1 cm³ of water?

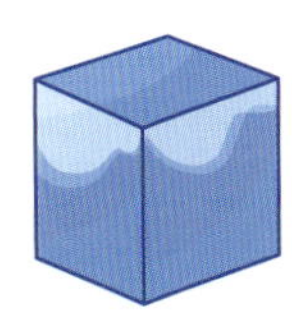

1 L weighs 1 kg

Foundation Mastery Complex

# 5.09 Volume and surface area of a cylinder

**Videos**
Volumes of prisms and cylinders

Capacity of a cylinder

**Worksheets**
Measurement in the home

Sweet areas and volumes

Surface areas of solids

## Volume of a cylinder

A cylinder has a circular base and its volume can also be found using the formula $V = Ah$, where $A = \pi r^2$.

**Volume of a cylinder**

$$V = \pi r^2 h$$

where $V$ = volume

$r$ = radius of circular base

$h$ = perpendicular height

**Example 24**

A can of soup has base radius 3.7 cm and height 11 cm.
What is its volume, correct to the nearest millilitre?

### Solution

$V = \pi r^2 h$

$= \pi \times 3.7^2 \times 11$

$= 473.0924 \ldots \text{ cm}^3$

$= 473.0924 \ldots \text{ mL}$ 　　 $1 \text{ cm}^3 = \text{mL}$

$\approx 473 \text{ mL}$

Volume of the can is 473 mL.

## Surface area of a cylinder

A closed cylinder has 2 circular ends and a curved surface. The curved surface is a rectangle whose length is equal to the circumference ($2\pi r$) of the circular end.

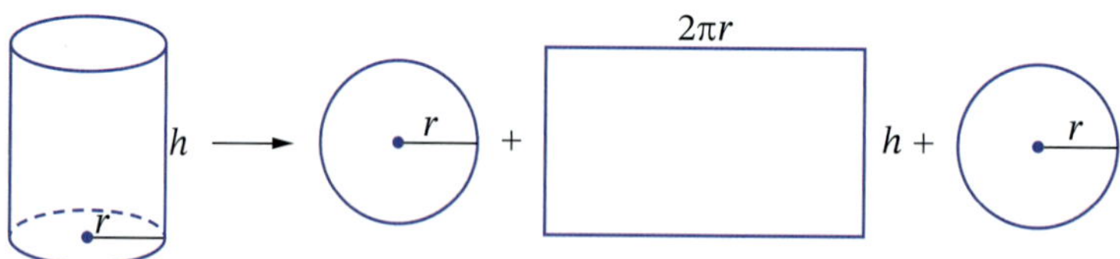


Surface area = area of 2 circles + area of rectangle

$= 2 \times \pi r^2 + 2\pi r \times h$

$= 2\pi r^2 + 2\pi rh$

## Surface area of a closed cylinder

$$A = 2\pi rh + 2\pi r^2$$

where $A$ = surface area

$r$ = radius of circular base

$h$ = perpendicular height

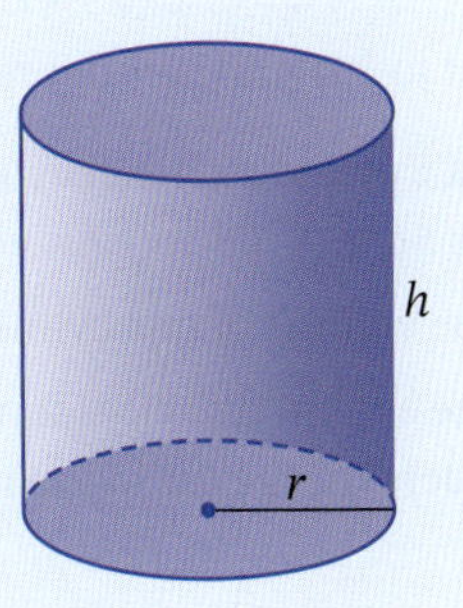

**Video**
Surface area of a cylinder

## Example 25

A can of tomatoes has base diameter 12 cm and height 15 cm. Calculate to the nearest square centimetre:

**a** the surface area of the can

**b** the area of the label wrapped around the can if there is a 2 cm overlap.

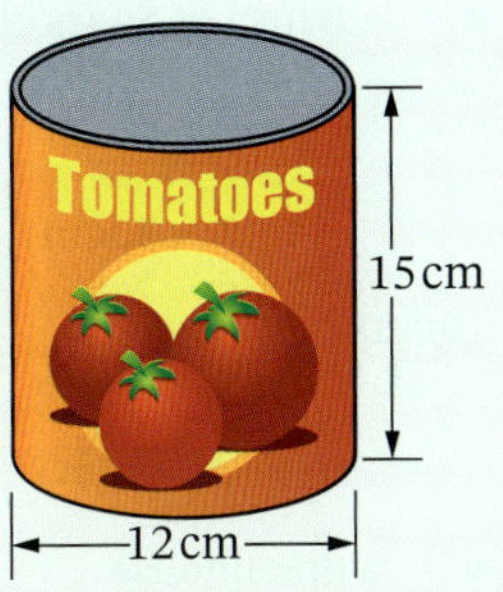


### Solution

**a** $A = 2\pi rh + 2\pi r^2$

$= 2 \times \pi \times 6 \times 15 + 2 \times \pi \times 6^2$

$= 791.6813 \ldots$

$\approx 792\text{ cm}^2$

$r = \frac{1}{2} \times 12\text{ cm} = 6\text{ cm}$

The surface area of the can is $792\text{ cm}^2$.

**b** Length of label $= 2\pi r + 2$, height of label $= h$

Area of label $= (2\pi r + 2) \times h$

$= (2 \times \pi \times 6 + 2) \times 15$

$= 595.4866 \ldots$

$\approx 595\text{ cm}^2$

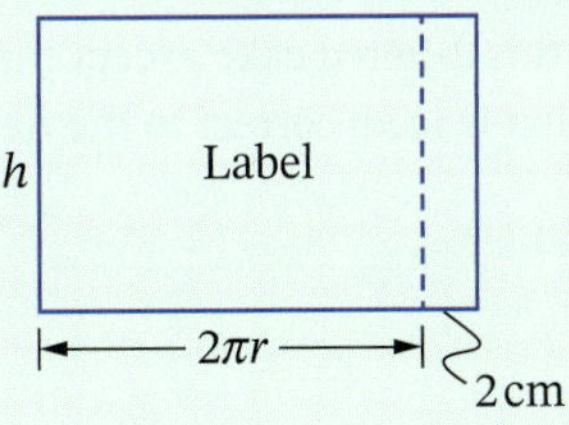


The area of the label is $595\text{ cm}^2$.

**EXERCISE 5.09** Answers on p. 468

## Volume and surface area of a cylinder

**1** Find, correct to 2 significant figures, the volume of each cylinder. Answer parts **b** and **d** in $\text{m}^2$.

EXAMPLE 24

**a**

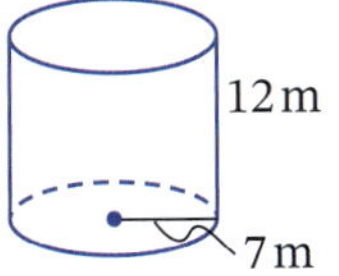


**b**

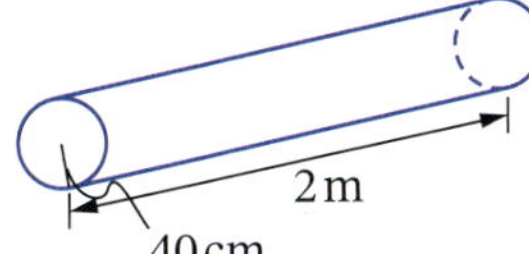


**c**

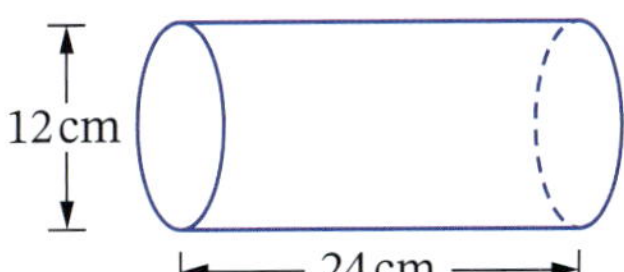


**d**

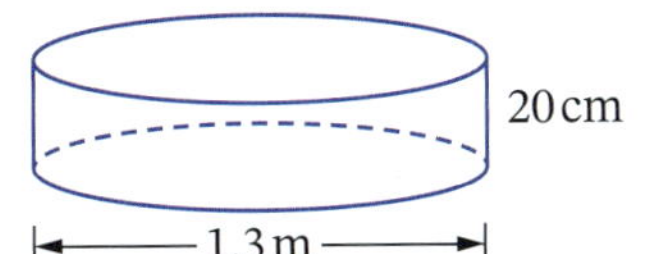


**e**

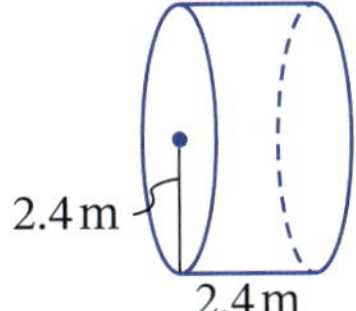


**f**

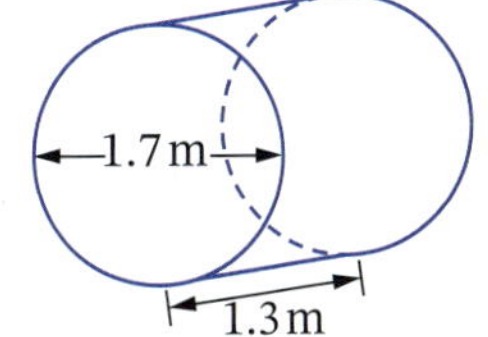

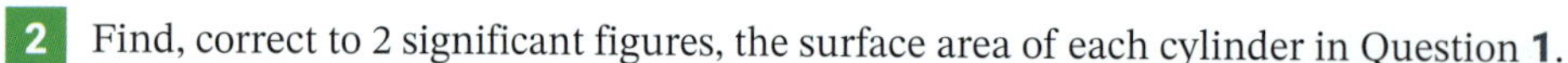

**2** Find, correct to 2 significant figures, the surface area of each cylinder in Question **1**.

**3** A V6 car has 6 cylinders. Each cylinder has a bore (diameter) of 8 cm and a stroke (height) of 6.5 cm.

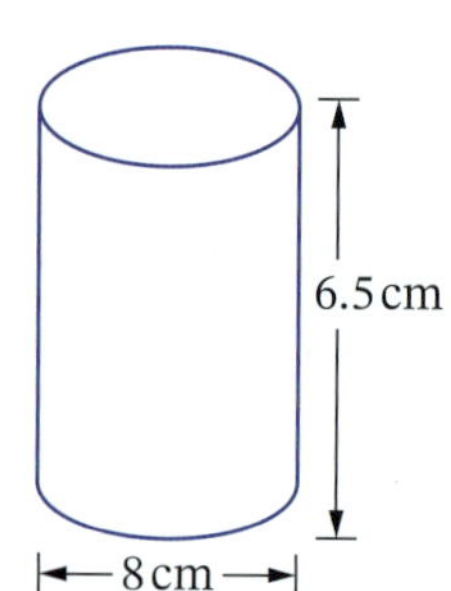


- **a** Find the volume of one cylinder, to the nearest cm$^3$.
- **b** Find the engine capacity (sum of the capacities of the 6 cylinders) of the car to the nearest litre.

**4** Find, correct to 2 significant figures, the volume of each solid.

- **a** a wheel of Swiss cheese

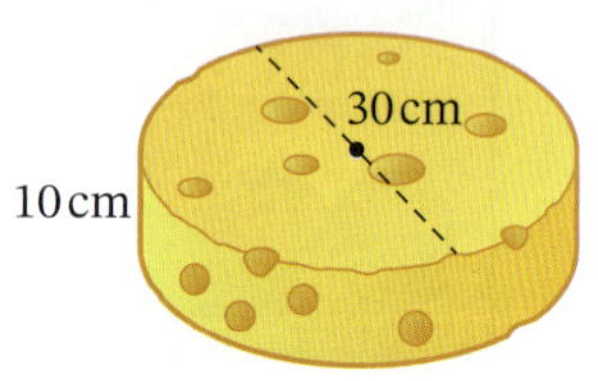

- **b** a water trough in the shape of a half-cylinder

1.5 m
1.2 m

- **c** a greenhouse

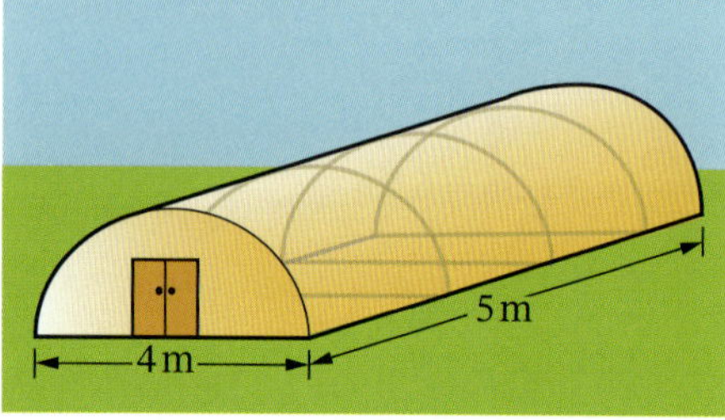


- **d** a piece of plumber's pipe

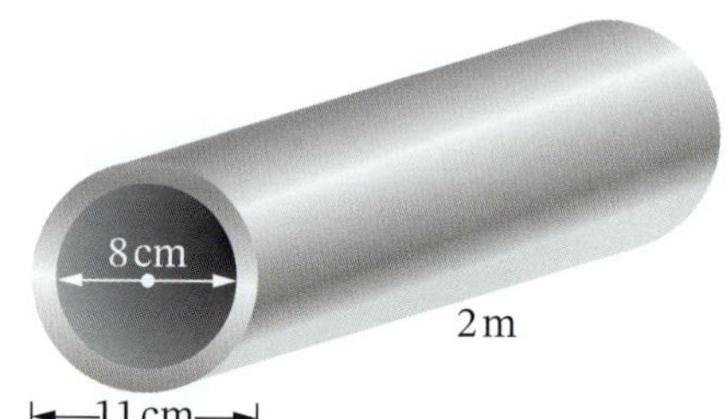


**5** All surfaces of this 2-tiered cake except the base are iced.
Find the area that is iced, correct to the nearest square centimetre.

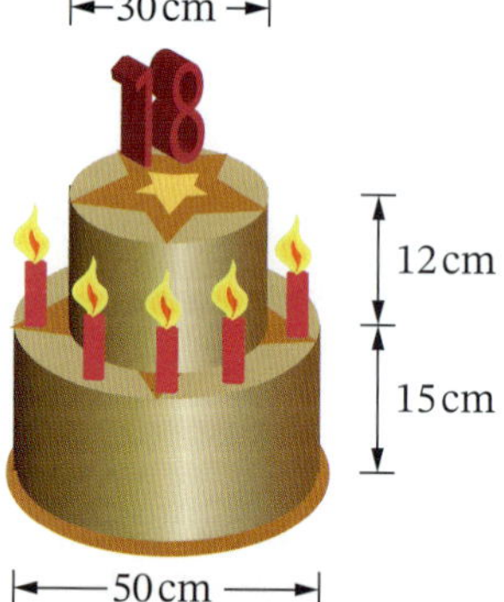


**6** A tent has the shape of a half-cylinder. The floor area is 2 m by 2 m.
Calculate, correct to one decimal place:

- **a** its volume
- **b** its surface area, including the floor.

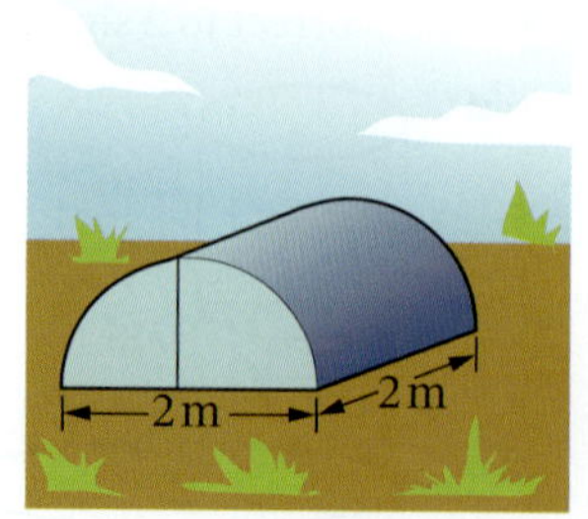


☐ Foundation ○ Mastery ⬡ Complex

**7** A cylindrical tank with diameter 4 m is placed in a 2 m deep circular hole so that there is a gap of 50 cm between the side of the tank and the hole. The top of the tank is level with the ground.

**a** What volume of dirt was removed to make the hole? Answer to the nearest $m^3$.

**b** What is the capacity of the tank, to the nearest litre?

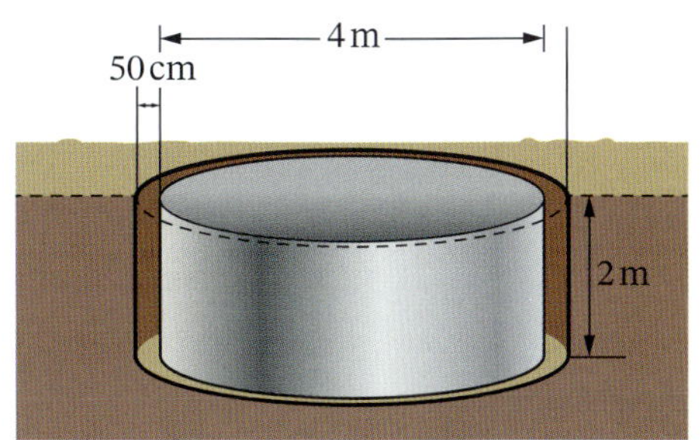


**8** A cylinder has a base radius of 5 cm. If the radius is doubled, what happens to the volume? Select **A**, **B**, **C** or **D**.

**A** doubles **B** triples **C** 4 times larger **D** halved

## Investigation

### Square or round peg?

Have you ever heard the saying that 'you can't put a square peg in a round hole'?

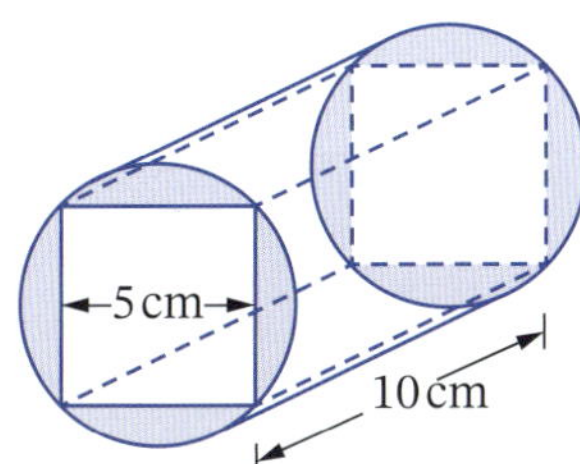


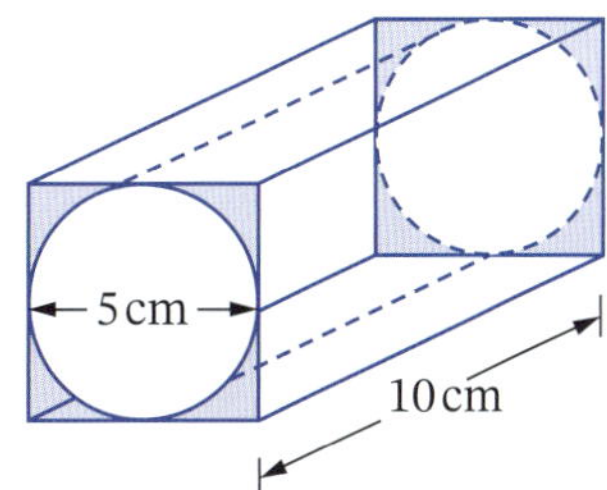


**1** Compare the volumes occupied by a square peg of side 5 cm in a round hole of depth 10 cm and a round peg of diameter 5 cm in a square hole of depth 10 cm.

**2** Which peg occupies more space?

**3** What percentage of the hole does each peg occupy?

**4** Is it better to put a square peg in a round hole or a round peg in a square hole? Justify your answer.

# Volumes of pyramids and cones 5.10

The volume of a **pyramid** is $\frac{1}{3}$ of the volume of a prism with the same base and height.

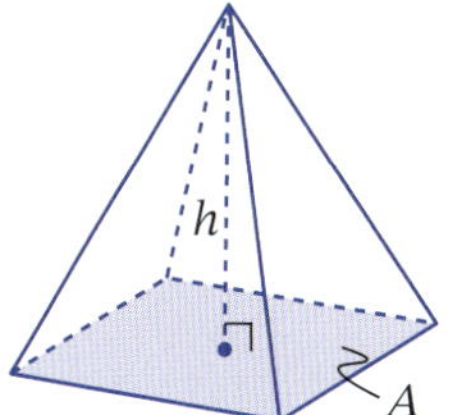


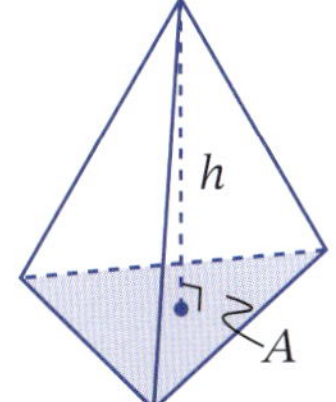


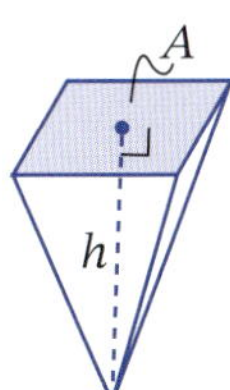


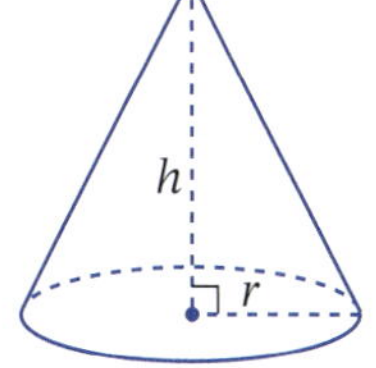


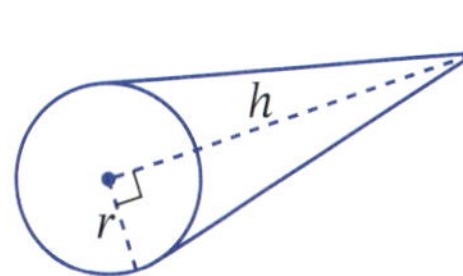


A cone is like a pyramid with a circular base.

The volume of a cone is $\frac{1}{3}$ of the volume of a cylinder with the same circular base and height.

## Volume of a pyramid

$$V = \frac{1}{3}Ah$$

where $A$ = area of base

$h$ = perpendicular height

## Volume of a cone

$$V = \frac{1}{3}Ah = \frac{1}{3}\pi r^2 h$$

where $r$ = radius of circular base

$h$ = perpendicular height

## Example 26

This grain bin has the shape of an open square pyramid.
How many cubic metres of grain will it hold?

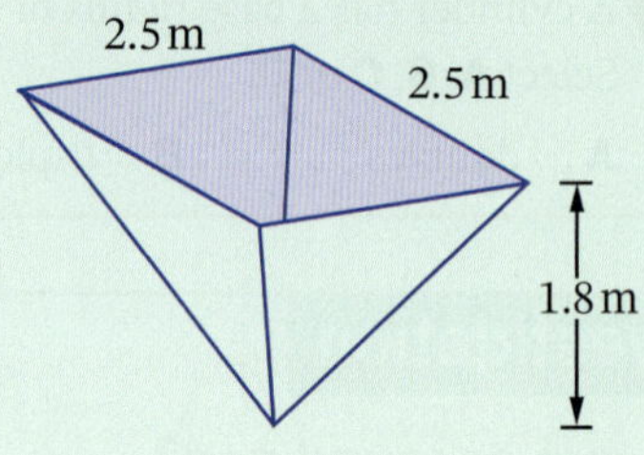


### Solution

$$V = \frac{1}{3}Ah$$

$$= \frac{1}{3} \times 2.5^2 \times 1.8 \quad \text{the base is a square of side 2.5}$$

$$= 3.75\,\text{m}^3$$

The bin will hold $3.75\,\text{m}^3$ of grain.

**Video**
Volume of a cone

## Example 27

Billy the plumber dug a rectangular trench, 1 m by 0.8 m, to lay gas pipes. He ended up with a conical pile of dirt, 1.2 m high with a base radius of 0.5 m.

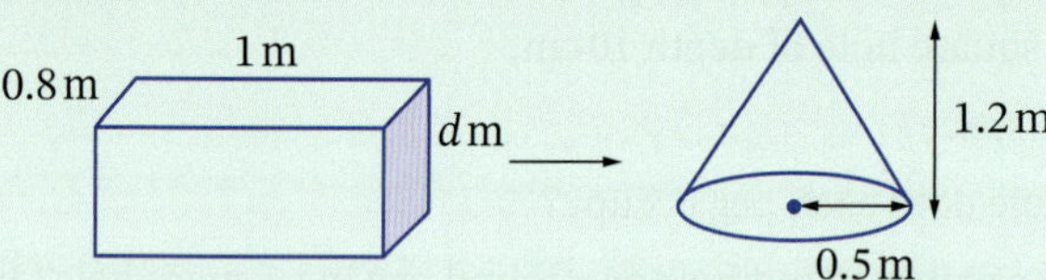


**a** How many square metres of dirt, correct to three decimal places, did Billy dig out?

**b** What was the depth of the hole, $d$ metres, correct to one decimal place?

### Solution

**a** $V = \frac{1}{3}\pi r^2 h$ — formula for the volume of a cone

$$= \frac{1}{3} \times \pi \times 0.5^2 \times 1.2$$

$$= 0.31415\ldots$$

$$\approx 0.314\,\text{m}^3$$

Billy dug out $0.314\,\text{m}^3$ of dirt.

**b** $V = Ah$ formula for the volume of a prism

$= 1 \times 0.8 \times d$

$= 0.8d$

But the volume is also $0.314\,\text{m}^3$ from part **a**.

$0.314 = 0.8d$

$0.8d = 0.314$

$d = \dfrac{0.314}{0.8}$

$= 0.3925$

$\approx 0.4$ m

The depth of the hole was 0.4 m.

**EXERCISE 5.10** Answers on p. 468

## Volumes of pyramids and cones

**1** Find the volume of each pyramid, correct to the nearest whole number.

**a**

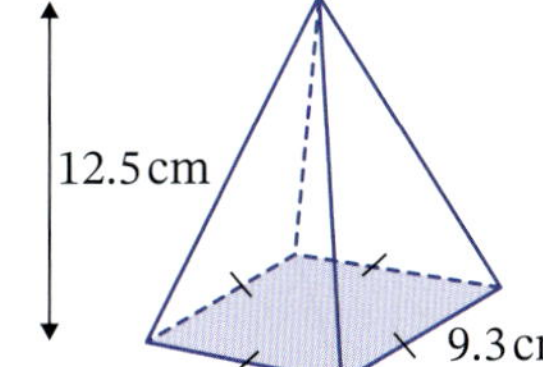


**b**

8.3 m
4.7 m
16 m

**c**

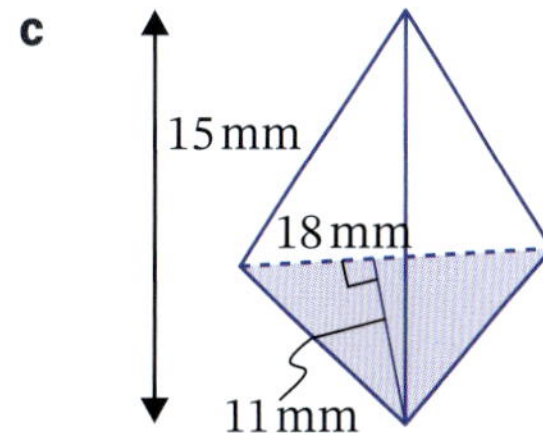


**d**

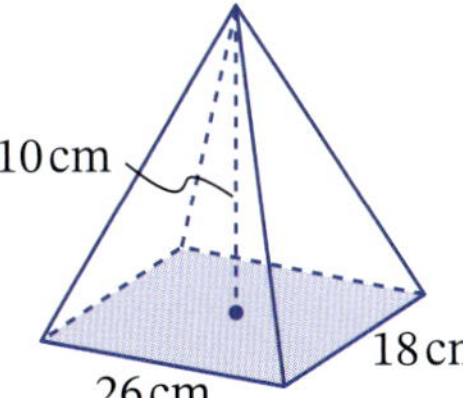


**e**

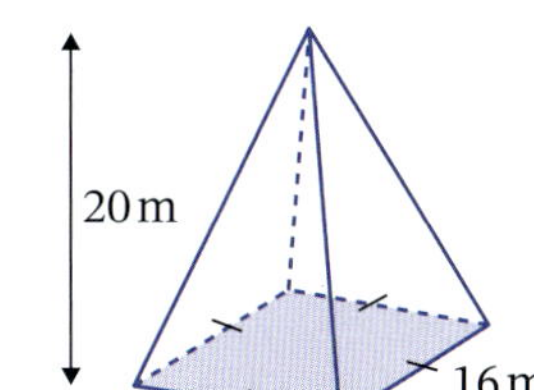


**f**

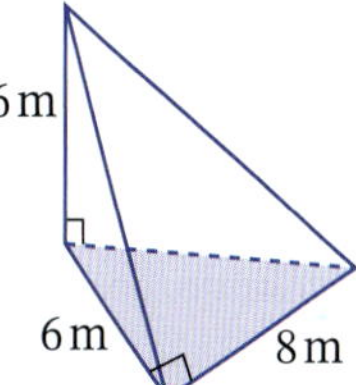


EXAMPLE 27

**2** Find the volume of each cone, correct to 2 decimal places.

**a**

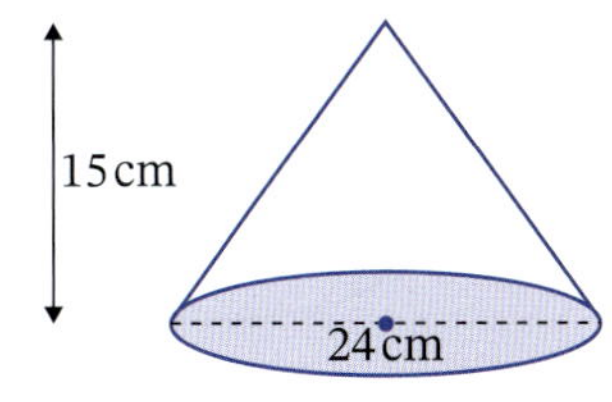


**b**

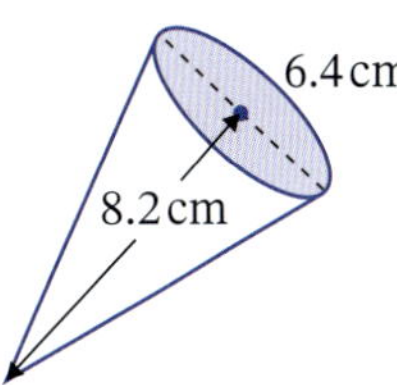


**c**

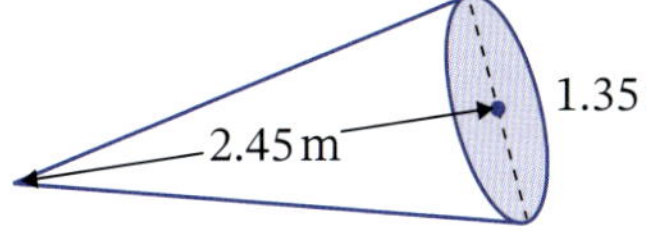


**d**

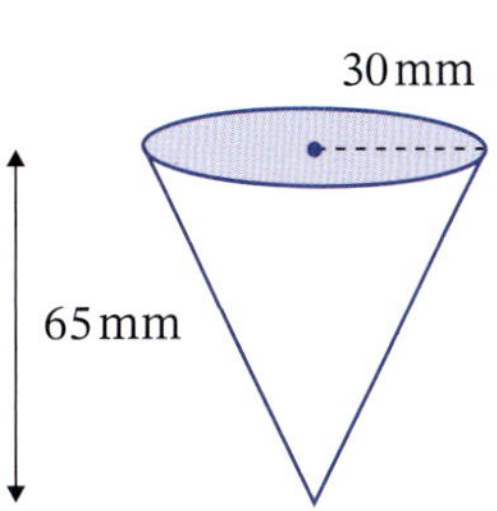


☐ Foundation ◯ Mastery ⬡ Complex

**3** This crystal is in the shape of an octahedron. Calculate its volume.

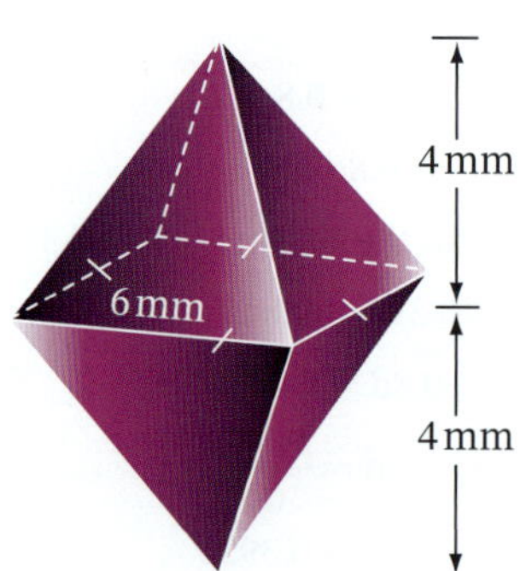


**4** A large paper cup in the shape of a cone is used to hold popcorn. It has a base diameter of 164 mm and a height of 205 mm.

Calculate, correct to 3 significant figures, the volume of the cone:

**a** in cubic centimetres

**b** in litres.

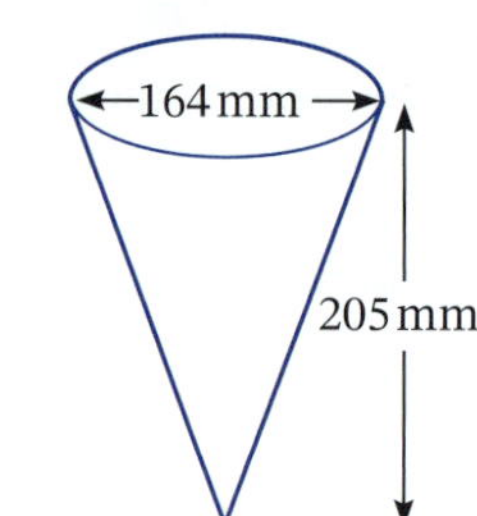


**5** A cone of height 25 cm and base diameter 20 cm has its top section removed to leave a frustum of height 15 cm. Find the volume of this frustum.

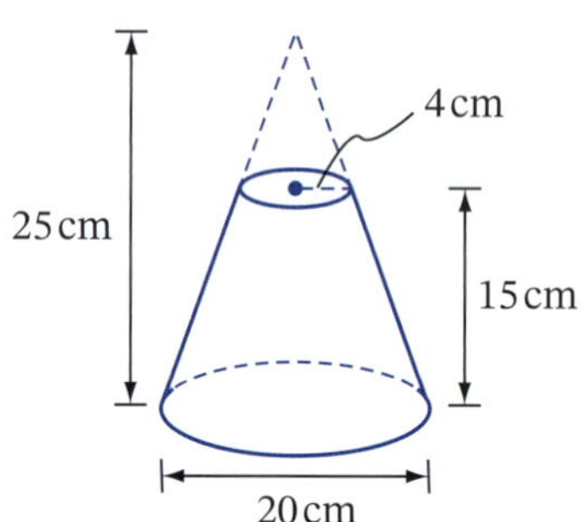


**6** The largest pyramid in Egypt is the Great Pyramid of Giza. It has a square base of side 230 m and a perpendicular height of 137 m. What is its volume, correct to 3 significant figures?

**7** A wheat silo is made up of a cylinder with a cone on top. What volume of wheat will it hold when full? Answer to the nearest cubic metre.

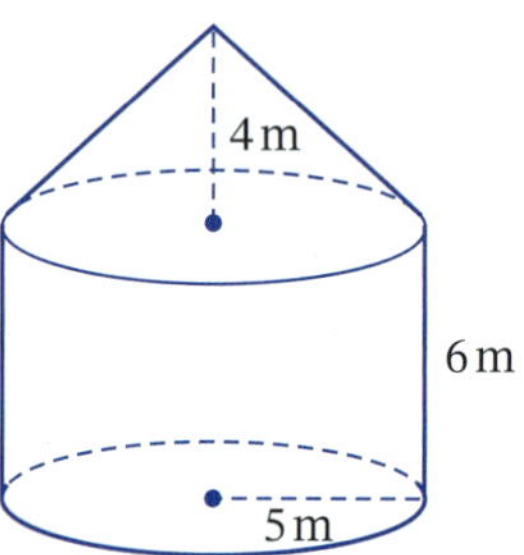


☐ Foundation ○ Mastery ○ Complex

# Volume and surface area of a sphere 5.11

A **sphere** is a perfectly round solid for which every point on its surface is the same distance from its centre. This distance is the sphere's **radius**.

Video
Volume of a sphere

Worksheets
Measurement formulas chart

Volumes of solids

Puzzles
Formula matching game

Surface area riddle

## Volume of a sphere

$$V = \frac{4}{3}\pi r^3$$

where $V$ = volume

$r$ = radius

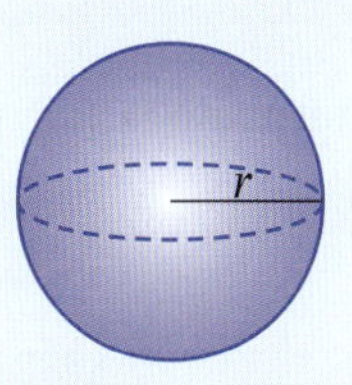


## Surface area of a sphere

$$A = 4\pi r^2$$

where $A$ = area

$r$ = radius

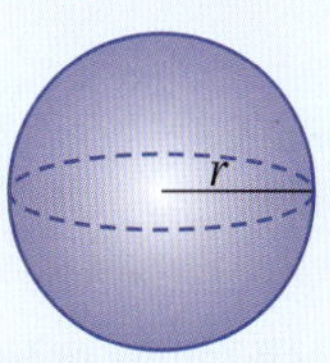


### Example 28

An asteroid is in the shape of a sphere with diameter 324 m.

**a** What is its volume in scientific notation, correct to 3 significant figures?

**b** What is its surface area, correct to the nearest hectare?

### Solution

**a** $V = \frac{4}{3}\pi r^3$

$= \frac{4}{3} \times \pi \times 162^3$ $\quad$ $r = \frac{1}{2} \times 324 = 162\text{ m}$

$= 17\,808\,758.84$

$\approx 1.78 \times 10^7\text{ m}^3$

The volume of the asteroid is approximately $1.78 \times 10^7\text{ m}^3$.

**b** $A = 4\pi r^2$

$= 4 \times \pi \times 162^2$

$= 329\,791.8304\text{ m}^2$

$= 32.9791\ldots\text{ ha}$ $\quad$ $1\text{ ha} = 10\,000\text{ m}^2$

$\approx 33\text{ ha}$

The surface area of the asteroid is approximately 33 ha.

## Example 29

Find the surface area of a closed **hemisphere** with radius 30 mm.
Give your answer correct to the nearest $10\,\text{mm}^2$.

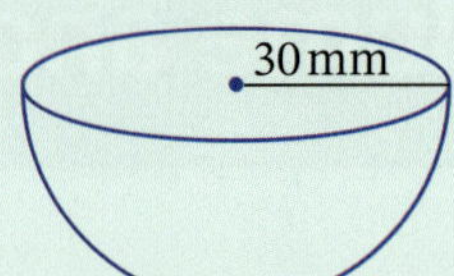


### Solution

$$\text{Area of curved surface} = \frac{1}{2} \times 4\pi r^2$$
$$= 2\pi r^2$$

Area of circular top $= \pi r^2$

$$\text{Surface area of closed hemisphere} = 2\pi r^2 + \pi r^2$$
$$= 3\pi r^2$$
$$= 3 \times \pi \times 30^2$$
$$= 8482.3001\ \ldots$$
$$\approx 8480\,\text{mm}^2$$

The surface area of the hemisphere is approximately $8480\,\text{mm}^2$.

**EXERCISE 5.11** Answers on p. 468

## Volume and surface area of a sphere

EXAMPLE 28

**1** Find the volume of each sphere or hemisphere, correct to 3 significant figures.

**a**

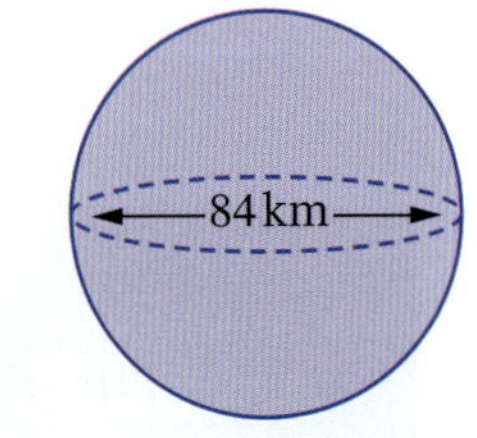


**b**

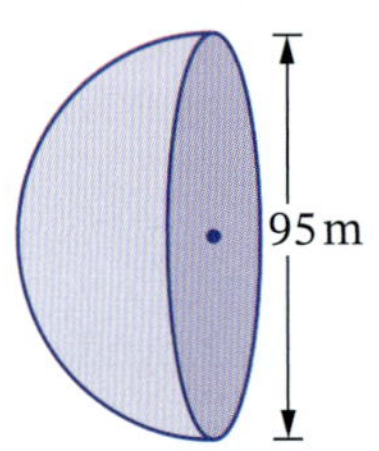


**c**

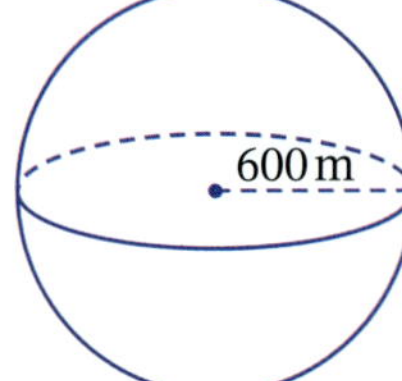


**d**

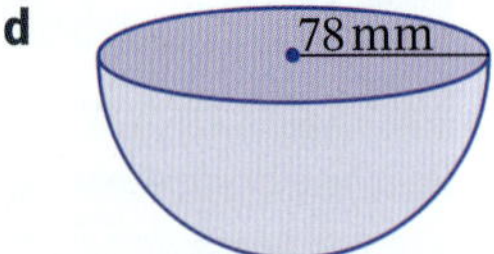


**e**

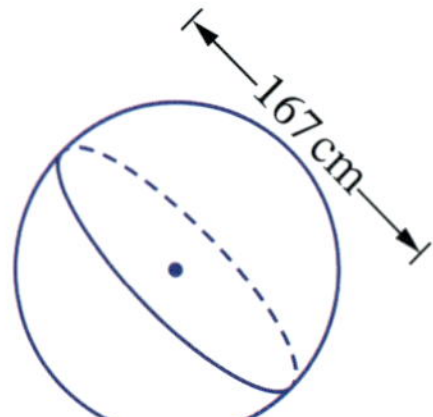


**f**

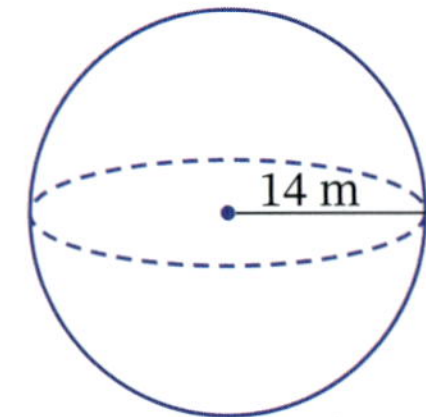


**2** Earth is a sphere with a diameter of approximately 12 683 km. 80% of Earth's surface is covered by water.

**a** Calculate in scientific notation, correct to 4 significant figures:

**i** the area of Earth's surface that is covered by land

**ii** Earth's volume in cubic kilometres.

**b** If Earth's mass is $5.974 \times 10^{21}$ tonnes, what is its mass per volume, or density, in $\text{t/km}^3$?

☐ Foundation ○ Mastery ⬡ Complex

EXAMPLE 29

3 A hemispherical metal bird bath has an internal radius of 0.5 m.

a What is its capacity, correct to the nearest litre?

b What is the internal surface area, correct to 2 decimal places?

4 Christmas baubles have the shape of a sphere with a diameter of 6 cm. They are packed in individual boxes so they just touch each side of the box.

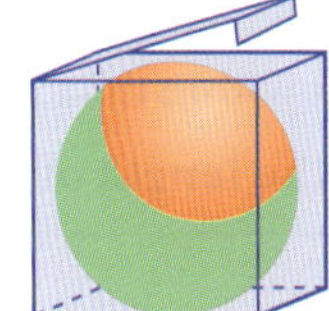

a What is the volume of each bauble, to the nearest $cm^3$?

b What volume of air, to the nearest $cm^3$, is in a box containing one bauble?

c Each bauble is covered in fabric costing $15 per $m^2$. How much does it cost to cover 100 baubles (assume no wastage)?

5 The volume of a sphere is 400 $cm^3$. What is its radius in cm, correct to one decimal place? Select **A**, **B**, **C** or **D**.

**A** 4.6 **B** 3.2 **C** 5.6 **D** 7.3

6 A mirror ball with a radius of 0.28 m hangs over the dance floor at a disco. Its surface is covered with 1 $cm^2$ mirror tiles.

a What is the volume of the mirror ball in $cm^3$, correct to 3 significant figures?

b How many whole mirror tiles cover the surface of the ball?

7 A very powerful computer-operated light is in the shape of 2 hemispheres hinged together. The curved surface of the light is solid steel and the front of each hemisphere is special glass. If the radius of the light is 0.48 m, find, to 2 significant figures:

a the area of steel covering the light

b the area of glass

c the volume of the light.

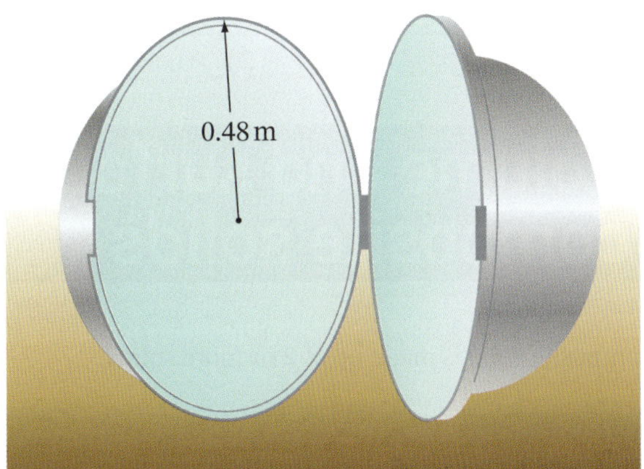


8 A solid capsule to hold dodge balls is designed with 2 hemispheres joined to each end of an open-ended cylinder.

Exactly 3 dodge balls fit into this capsule so that they just touch each other, and each ball has a diameter of 20 cm, as shown in the diagram.

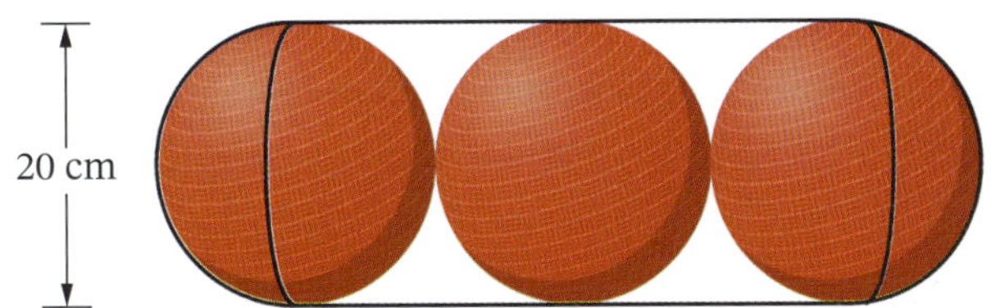


Calculate, correct to one decimal place, the total surface area of the capsule.

## Did you know?

### Circles and spheres in sport

The shotput and hammer throw are 2 athletics field events requiring the participant to start in a throwing circle of diameter 2.135 m and launch into a 40° sector.

The **shot** is a spherical iron ball weighing at least 7 kg for a man and 4 kg for a woman. The shot is 'put' (not thrown) from the throwing circle into the marked sector.

The **hammer** has a spherical head attached to a handle by a length of steel about 1 m long and is launched from the throwing circle using a circular wind-up action.

The shot and hammer can be launched from any position in the circle but must land within the marked sector outside the circle.

**What is the length of arc *AB*?**

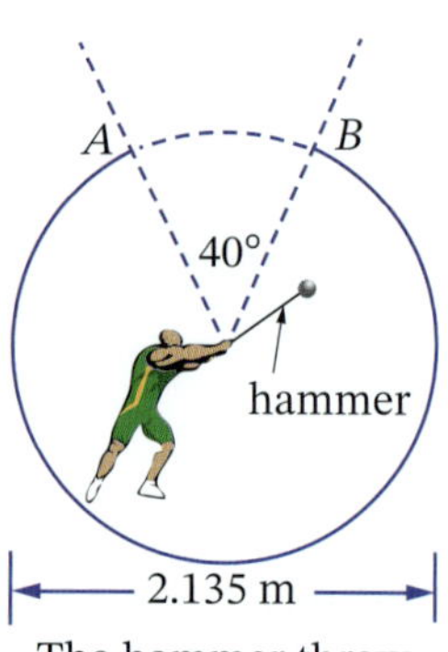


The hammer throw

## Investigation

### Sphere inside a cylinder

The sphere shown just fits into a cylinder of diameter and height both 20 cm.

**1** Calculate the volume of the sphere and the cylinder. Do not round your answers.

**2** What fraction of the cylinder's volume is taken up by the sphere?

**3** Is this fraction the same for any sphere and cylinder with equal diameters? Experiment with different diameters.

**4** Is there any way of proving this result?

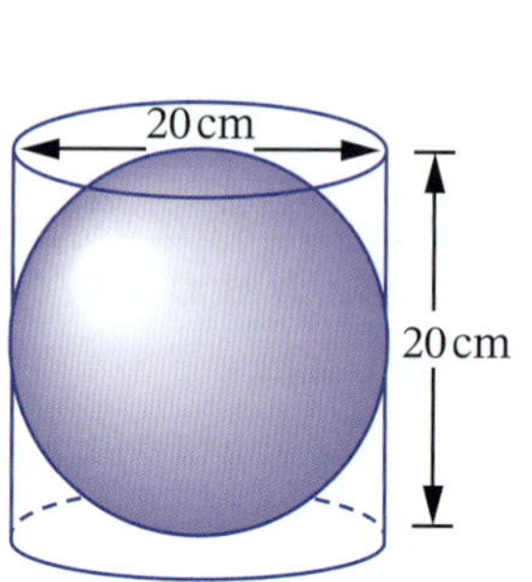


# 5.12 Volume and surface area of composite solids

**Video**
Volumes of composite shapes

**Worksheets**
A page of solid shapes

Volumes of water

A composite solid is made up of 2 or more solids.

### Example 30

The concrete block shown is a square prism with a cylindrical hole of diameter 10 cm drilled through it.

**a** What was the volume of the concrete block before drilling?

**b** What is the volume of the concrete block after drilling, to the nearest $cm^3$?

**c** What percentage, correct to one decimal place, of the concrete was removed to make the hole?

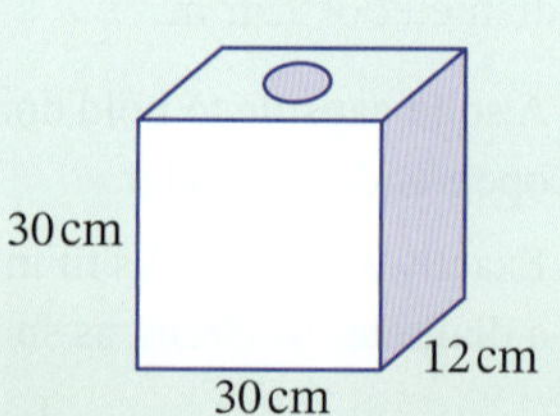

### Solution

**a** Volume of concrete block $= 30 \times 30 \times 12$

$= 10\,800\,\text{cm}^3$

**b** First find the volume of the cylindrical hole.

$V = \pi r^2 h$, where $r = \frac{1}{2} \times 10 = 5$ and $h = 30$

$= \pi \times 5^2 \times 30$

$= 2356.19\ ...\ \text{cm}^3$

Volume remaining $= 10\,800 - 2356.19...$

$= 8443.80\ ...$

$\approx 8444\,\text{cm}^3$

**c** Percentage removed $= \frac{2356.19\ ...}{10\,800} \times 100\%$

$= 21.816\ ...\%$

$\approx 21.8\%$

## Investigation

### Volume and packaging

**1** **a** Design a cardboard package to hold six golf balls (each with a diameter of 37 mm) so that the minimum amount of cardboard is used.

**b** Would the same packaging be suitable to hold:

**i** 6 billiard balls?

**ii** 6 ping pong balls?

Why?

**2** **a** Design a cardboard carton to hold 12 cans of soup:

**i** in one layer

**ii** in 2 or more layers.

**b** Which carton uses the least amount of cardboard?

**c** Which carton is the most appropriate for this product? Why?

**EXERCISE 5.12** Answers on p. 468

## Volume and surface area of composite solids

**1** A cylindrical can contains 3 tightly-packed tennis balls, each with a diameter of 75 mm.

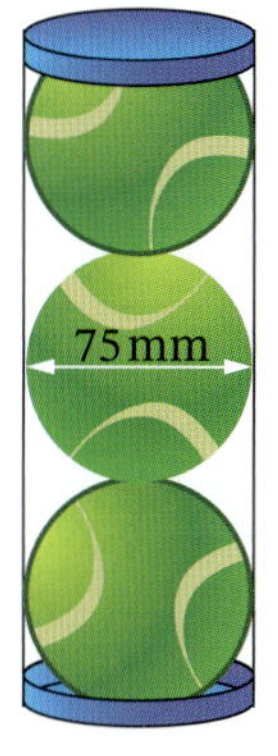


**a** What is the volume of the can (to the nearest cubic millimetre)?

**b** What percentage (correct to one decimal place) of the can's volume is taken up by the tennis balls?

**c** If the 3 balls were packed tightly into a square prism instead of a cylinder, what would be the volume of the prism?

**d** Why do manufacturers pack tennis balls into cylindrical cans? Justify your answer.

Foundation | Mastery | Complex

EXAMPLE 30

**2** A metal-fabricating company makes a machine part by drilling 3 holes of diameter 20 mm through a block of metal.

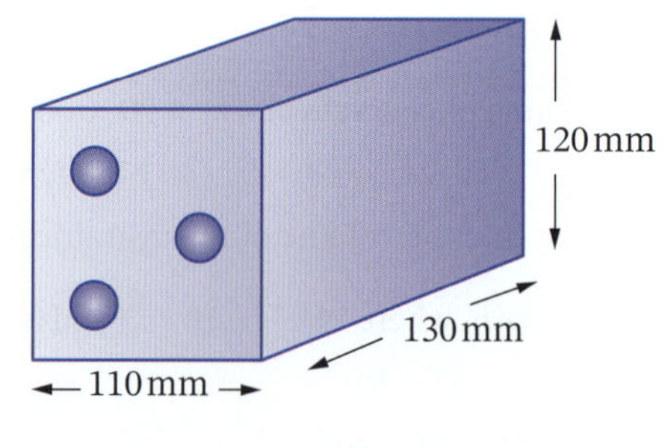


**a** What is the volume of metal before drilling?

**b** What is the volume of metal (to the nearest $cm^3$) after drilling?

**c** What percentage (correct to the nearest whole number) of metal was removed when the holes were drilled?

**d** What is the surface area of the metal before drilling?

**3** This cheese wheel is covered with red wax to keep it fresh. One of the sectors has been eaten.

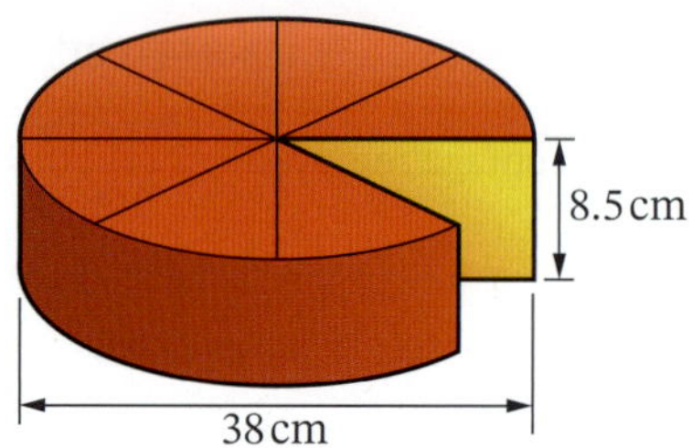


Find, correct to 2 significant figures:

**a** the volume of the cheese remaining

**b** the surface area of the cheese remaining

**c** the volume of one sector of cheese

**d** the area of red wax on one sector of cheese.

**4** Two neighbours are arguing over whose swimming pool holds more water.

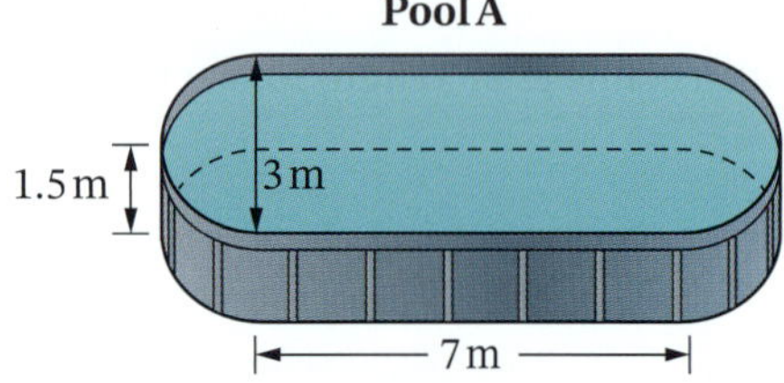


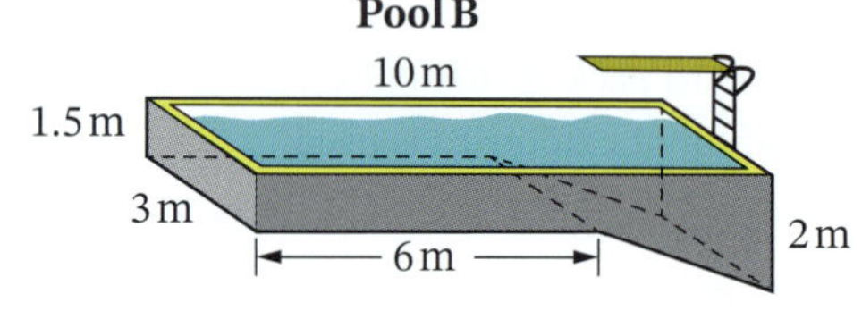


**a** What is the volume of water in each pool, correct to the nearest litre?

**b** Which pool holds more water, and by how much?

**c** If the base and sides of pool B are to be tiled, what area is to be tiled? Answer to the nearest square metre.

**5** This is a woodworker's bench, with measurements in centimetres.

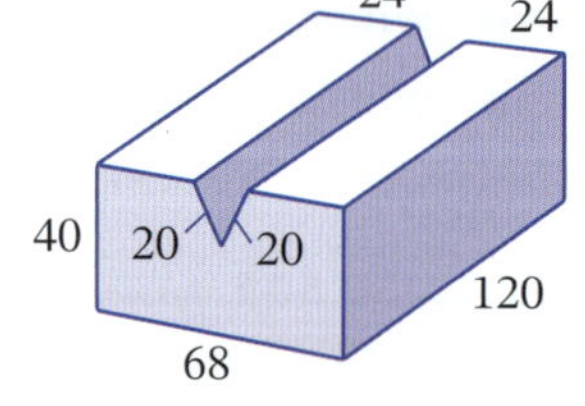


Find:

**a** its volume in cubic metres, correct to 3 significant figures

**b** its total surface area, correct to the nearest square centimetre.

**6** This marquee with a square floor was hired for a family celebration.

How much canvas (correct to the nearest 0.1 square metre) was used to make this marquee? (Do not include the floor.)

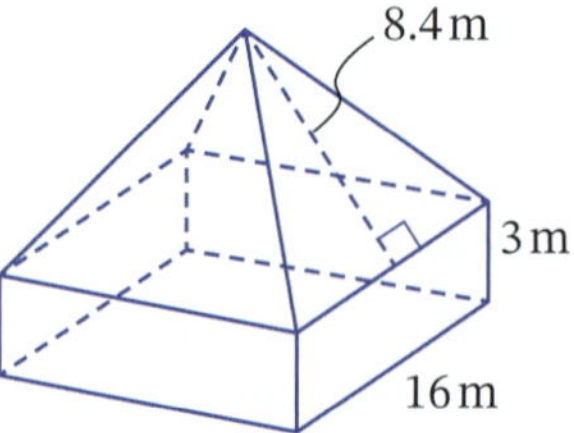


**7** For this time capsule, calculate correct to one decimal place:

**a** its surface area

**b** its volume

**c** its capacity in litres.

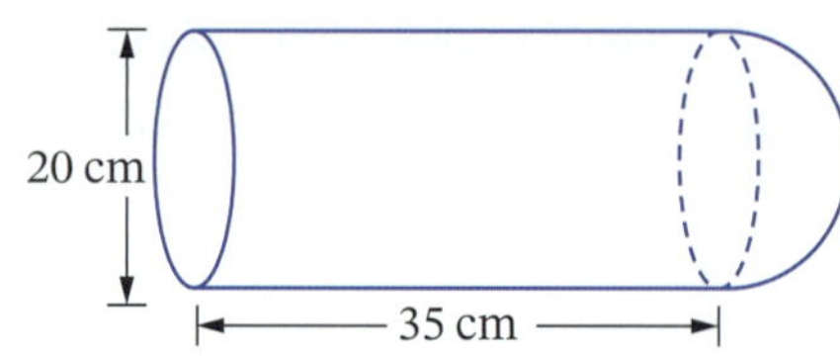


Foundation | Mastery | Complex

## Sample HSC problem Answers on p. 468

(3 marks) Sarah designed a cake for her daughter's birthday. Its shape is a cone covering the top of a cylinder, as shown in the diagram. The cylinder has a base diameter of 23 cm and a height of 9 cm.

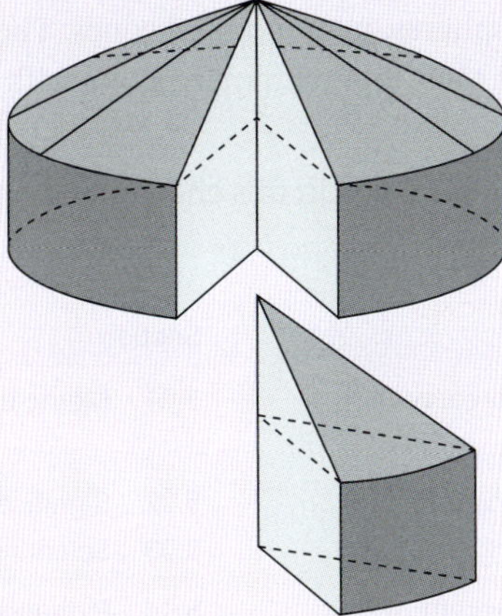

The cone is $\frac{1}{6}$ of the volume of the cylinder.

The cake is cut into slices, each with a volume of 218 cm$^3$.

How many whole slices are cut from the cake?

## Study Tip

### Making lists and managing your time

You cannot just make up your study routine as you go along. You need to have a plan, even if it is a rough one. Make 2 lists of the things you want to achieve:

- An 'A list' of urgent tasks, such as homework and reading
- A 'B list' of less pressing demands, such as research assignments and revising for an exam.

Write in your diary what you want to accomplish each day or each week, and include all important dates, deadlines and reminders.

Make sure your 'A list' is realistic, taking into account the time and effort required for each task. Remember that your time is both precious and limited, so it is not possible to do everything. Make the most of the time that you have and don't be afraid to cut corners. Set realistic goals and spend time only on what is important. Start studying straight away and avoid timewasters, such as cleaning your room, decorating title pages and even devising elaborate study timetables (if you do these, then you are really putting things off).

From your 'A list', choose the easiest task to begin with, so that you can ease yourself into the routine. Alternatively, some students like to start with a difficult task while their mind is fresh and alert. Once you have completed a significant amount of work, give yourself a break and a small reward.

# CHAPTER SUMMARY

This chapter, *Measurement*, builds on the measurement work covered in Years 9 and 10. You examined the perimeters, areas, surface areas, volumes and capacities of more complex and composite shapes, including sectors, prisms, cylinders, spheres, pyramids and cones. There are many formulas in this chapter, so be sure to include them in your topic summary along with appropriate diagrams and examples.

Before you move on, consider what you learned in this chapter and revisit any sections that may have been unclear.

| What you learned in this chapter... | Section |
|---|---|
| Convert between metric units of length, area, volume, capacity, mass and time | 5.01 Metric units |
| Express numbers in scientific notation and round to significant figures | 5.02 Significant figures<br>5.03 Scientific notation |
| Solve problems involving Pythagoras' theorem | 5.04 Pythagoras' theorem problems |
| Calculate perimeters and areas of shapes, including circular and composite shapes | 5.05 Perimeters of circular and composite shapes<br>5.06 Area |
| Calculate perimeters, areas and volumes of irregular shapes using dissection and the trapezoidal rule | 5.07 Land surveying and the trapezoidal rule |
| Calculate surface areas, volumes and capacities of prisms, cylinders, spheres and composite solids | 5.08 Volume and surface area of a prism<br>5.09 Volume and surface area of a cylinder<br>5.11 Volume and surface area of a sphere<br>5.12 Volume and surface area of composite solids |
| Calculate volumes of pyramids and cones | 5.10 Volumes of pyramids and cones |

To help master this topic, make a summary mind map. Use the chapter outline and the mind map below as a guide. Add your own words, symbols, diagrams, boxes and reminders. The summary should give you a 'whole picture' view of the topic and allow you to identify any weak areas to revisit in your revision.

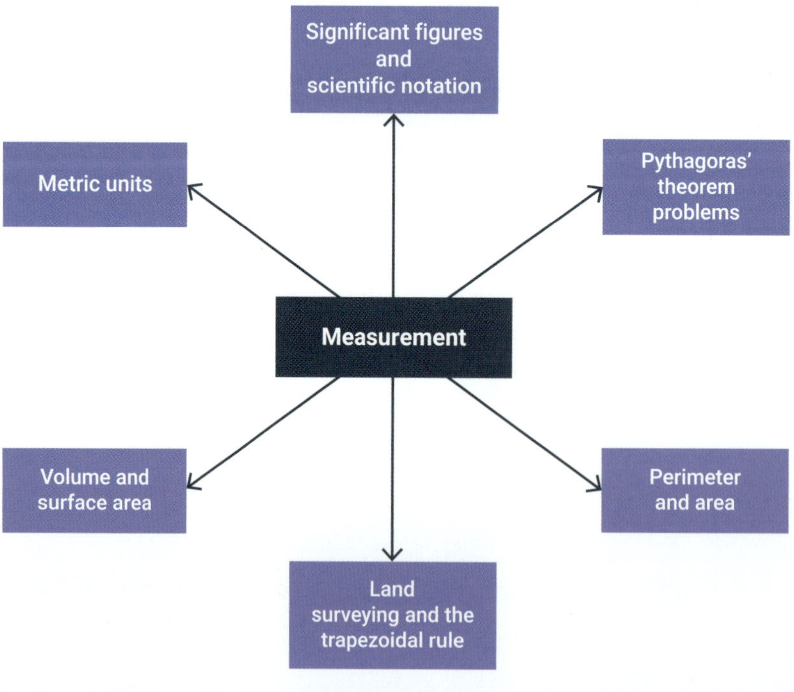

# Test yourself Answers on p. 469 5

**1** Convert: 5.01 

**a** 28.5 km to m **b** 6.4 t to kg **c** 340 mL to L.

**2** Farmer Joe's dam holds 43 ML of water. How many kilolitres does it hold? 5.01

**3** Write each value correct to 2 significant figures. 5.02

**a** 38.915 **b** 1036 **c** 0.00872 **d** 6 587 200

**4** A bank has \$8 350 000 000 in its vault. Write this value in scientific notation. 5.03 

**5** A rectangular dog park has a length that is twice its width. The park has a dividing wall to make 2 sections, one for large dogs and another for small dogs, as shown in the diagram. 5.04, 5.06

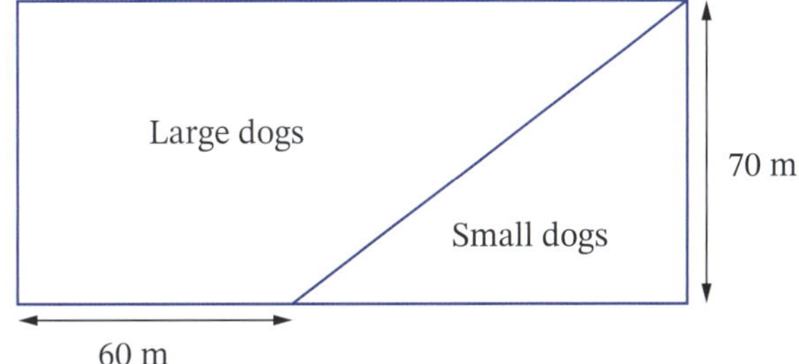


**a** Calculate, correct to one decimal place, the length of the dividing wall.

**b** Calculate the area of the section for large dogs.

**6** A cell is $4.6 \times 10^{-6}$ mm in diameter. Write this in normal decimal notation. 5.03

**7** Calculate the perimeter of each shape (correct to 2 decimal places for shape **c**). 5.05

**a**

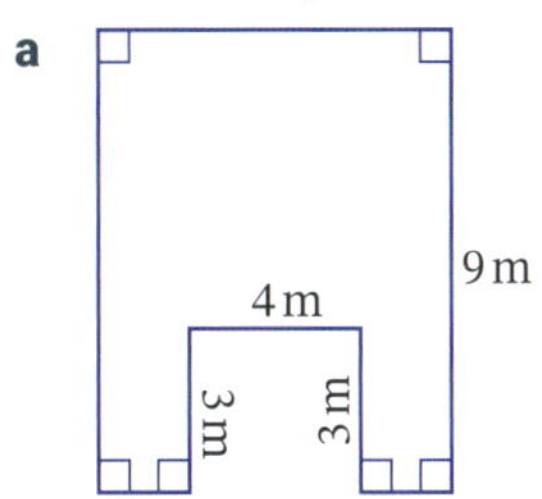


**b** **c**

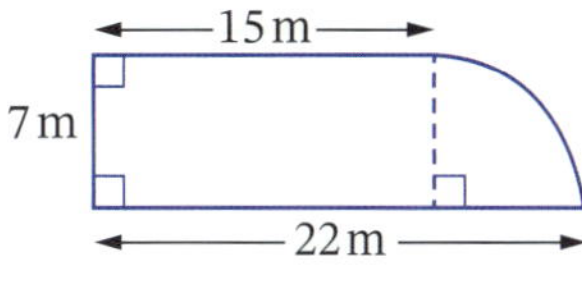


**8** Convert: 5.06

**a** 8400 mm$^2$ to cm$^2$ **b** 5.6 ha to m$^2$.

**9** Calculate the shaded area of each shape (correct to 2 significant figures for shape **c**). 5.06

**a**

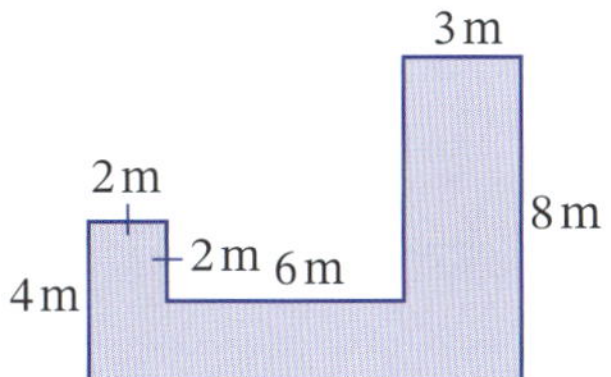


**b**

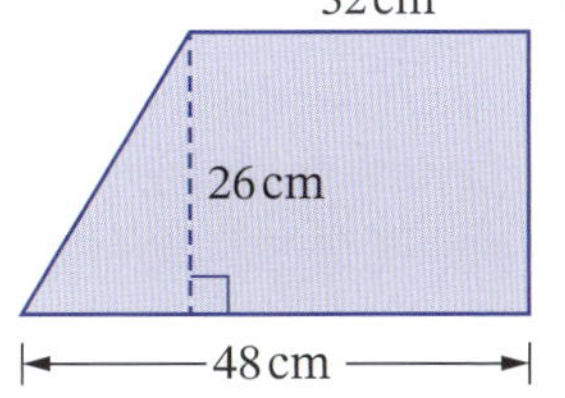


**c**

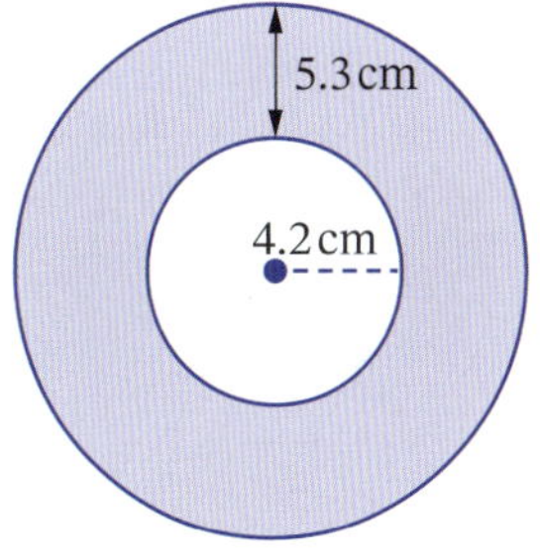


☐ Foundation ○ Mastery ○ Complex

5.06 **10** For this sector, calculate correct to one decimal place:

**a** its perimeter

**b** its area.

46 cm, 35°

5.07 **11** Felix completed an offset survey on a field, *PQRST*, and recorded the measurements in metres on this diagram. Calculate its area:

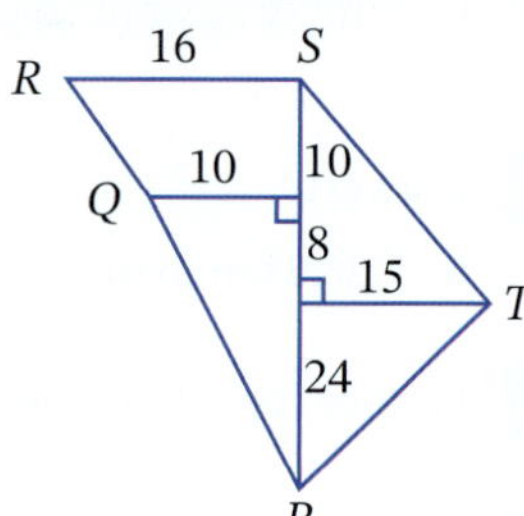


**a** in m$^2$ **b** in ha.

5.08 **12** Convert:

**a** 20.7 cm$^3$ to mm$^3$ **b** 1 650 000 cm$^3$ to m$^3$.

5.08, 5.09 **13** For each closed solid find, correct to 2 significant figures:

**i** the volume

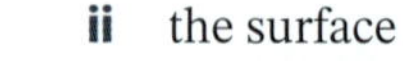
**ii** the surface area

**a**

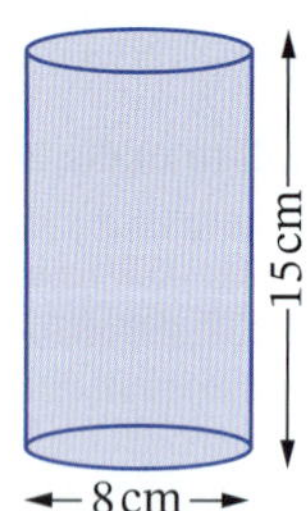


**b**

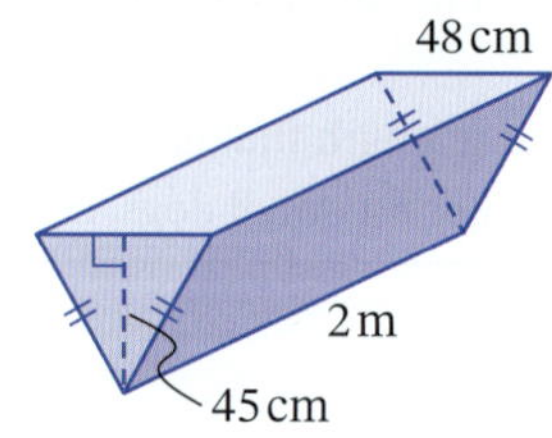


5.08 **14** **a** How many millilitres will a container of volume 894 cm$^3$ hold?

**b** How many litres will a container of volume 6.5 m$^3$ hold?

5.07 **15** Find the area of this field, correct to the nearest square metre, using:

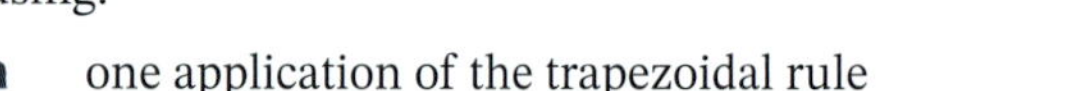

**a** one application of the trapezoidal rule

**b** 2 applications of the trapezoidal rule.

**c** Which answer is more accurate? Why?

6 m 12 m 4 m
11 m 11 m

5.08 **16** Blocks of concrete are used in the construction of a sports stadium. One of the blocks is shown in the diagram. Calculate, correct to one decimal place:

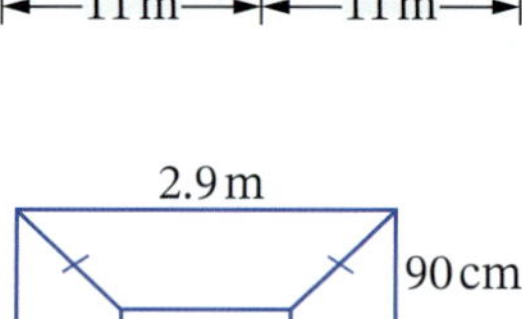

**a** its surface area **b** its volume.

5.09 **17** A large stainless steel cylinder for storing wine is 7 m high and has a diameter of 2.8 m. How many full 750 mL bottles of wine can be filled from this container?

5.10 **18** The Great Pyramid of Giza has a square base of length 230 m and had a perpendicular height of 146 m when first built. Calculate, correct to 4 significant figures:

**a** its original volume

**b** the length of the side edge from a corner of the base to the top of the pyramid.

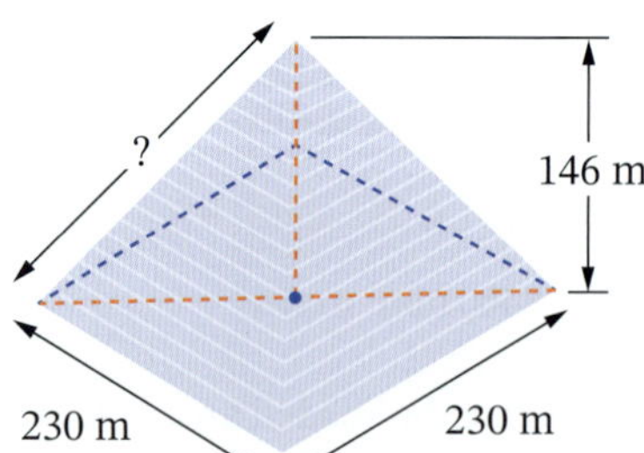


Foundation Mastery Complex

**19** Rosie is having a party and decides to make an 'ice bowl' to hold punch. She freezes water in the space between two hemispherical bowls of diameter 16 cm and 22 cm. Find, correct to 2 decimal places: 5.11

**a** the volume of ice used to make the bowl

**b** how many litres of punch the ice bowl will hold.

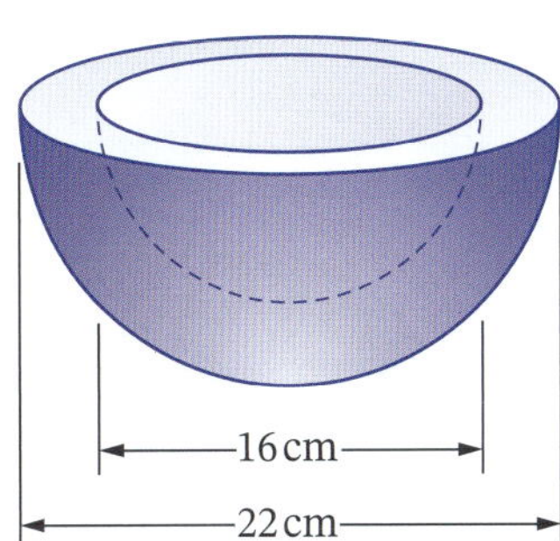


**20** **a** Find the volume of the barn shown. 5.12

**b** If the barn is to be painted (including the doors) and one litre of paint covers 14 m², how much paint (rounded up to the nearest whole litre) do you need, to put 2 coats on the barn?

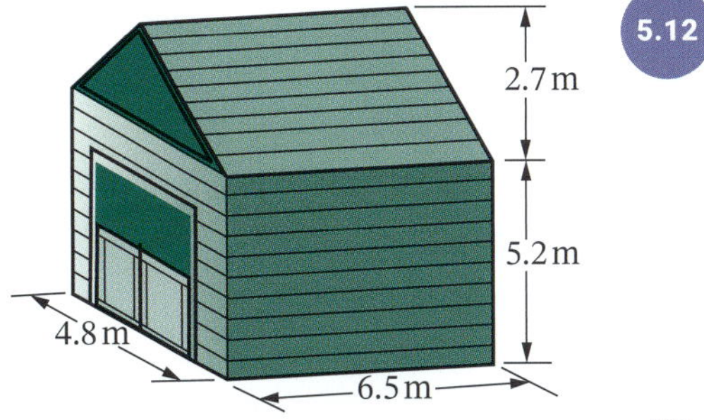


**21** A high-top loaf of bread is cut into thick slices in the shape of a rectangle combined with a semicircle. Find, correct to 2 significant figures: 5.12

**a** the volume of the loaf

**b** the length of crust around one slice of bread.

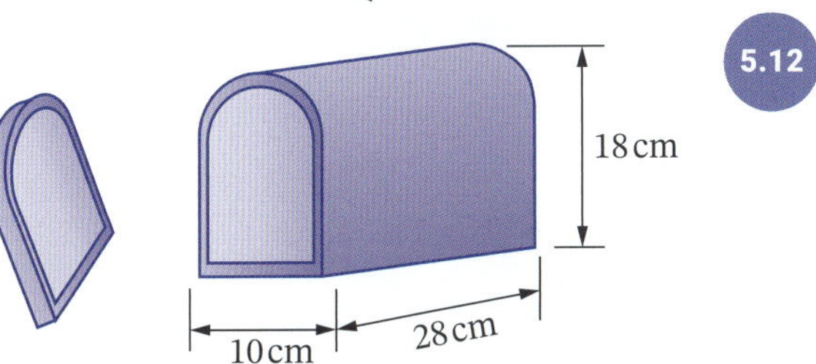


Foundation Mastery Complex

# MANAGING MONEY

Managing money effectively involves understanding the financial mathematics behind personal spending and budgeting. When buying items on credit, interest calculations and repayment schedules are necessary to avoid falling into debt. Household expenses like water, electricity and gas can vary, so it's important to find ways to reduce consumption and save money.

## Chapter outline

## In this chapter you will:

- apply percentage increase and decrease to problems involving markups, discounts, profit and loss
- calculate goods and services tax (GST) in Australia and value-added tax (VAT) internationally
- examine and compare 'buy now, pay later' credit plans and term payment plans, and the fees and financial calculations associated with both
- examine and make calculations with household water, electricity and gas bills
- create a personal budget, taking into account fixed and discretionary spending.

**Videos (3):**

**6.02** GST 1 • Discounts and GST • GST 2

**Worksheets (7):**

**SkillCheck** Assignment 6

**6.01** Profit and loss

**6.05** Water usage and costs • Water usage in your home

**6.06** Power problems

**6.07** Budget grid • Budgeting scenarios

Nelson MindTap

To access resources above, visit **cengage.com.au/nelsonmindtap**

## Terminology

| | | | |
|---|---|---|---|
| budget | buy now pay later | capacity | decrease |
| discount | discretionary spending | emergency funds | expense |
| fixed spending | goods and services tax (GST) | increase | kilolitre (kL) |
| kilowatt (kW) | kilowatt-hour (kWh) | lay-by | loss |
| markup | megajoule (MJ) | off-peak rate | peak rate |
| profit | term payments | usage charges | value-added tax (VAT) |

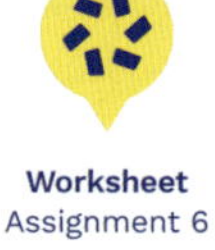

Worksheet
Assignment 6

## SkillCheck Answers on p. 469

**1** How many days from 5 May to 3 August of the same year?

**2** The electricity account for the period 5 August to 9 November was $486.24. What was the average daily cost for the electricity usage?

**3** If the cost of petrol is $1.91 per litre, what is the cost of 46 L of petrol?

**4** Express each percentage as a decimal.

**a** 2.5% **b** 84% **c** 106% **d** 18.75%

**5** Copy and complete each conversion.

**a** 10% p.a. = _____% per quarter

**b** 6.5% p.a. = _____% per week

**c** 5.4% p.a. = _____% per month

**d** 3.9% p.a. = _____% per fortnight

**6** Calculate each amount.

**a** 3% of $2780 **b** 2.5% of $45 800 **c** 1.1% of $10 500

**d** 4% of $9530 **e** 0.065% of $56 125 **f** 0.0075% of $150 600

**7** Calculate the simple interest earned on a principal of $5400 invested at 4.3% p.a. for 6 years.

**8** **a** Increase $28.50 by 3%.

**b** Decrease $2500 by 5%.

**c** Increase $88 by 12% and then decrease this amount by 10%.

# Markups, discounts, profit and loss

6.01

A retail store makes money by buying items and selling them at a higher price, to make a **profit**.

The **cost price** is how much they bought the item for, while the **selling price** is how much they sold the item for, after **marking up** (increasing) the cost price.

Sometimes, to encourage shoppers to buy, the store may **discount** (decrease) the selling price, but this may result in a smaller profit.

If the store sells the item at below the cost price, then it loses money and makes a **loss**.

**Worksheet**
Profit and loss

### Profit and loss

profit = selling price – cost price (when selling price is higher)

loss = cost price – selling price (when selling price is lower)

### Example 1

**a** A clothing store buys a jacket for \$90. They make a 25% profit by marking up the price. What is the selling price of the jacket?

**b** A TV priced at \$1600 is discounted by 15%. What is the sale price?

#### Solution

**a** Increasing an amount by 25% is the same as finding 125% of it, because 100% + 25% = 125%

$$\begin{aligned}\text{Selling price} &= 125\% \times \$90\\ &= 1.25 \times \$90\\ &= \$112.50\end{aligned}$$

**b** Decreasing an amount by 15% is the same as finding 85% of that amount, because 100% – 15% = 85%

$$\begin{aligned}\text{Sale price} &= 85\% \times \$1600\\ &= \$1360\end{aligned}$$

### Example 2

A furniture store sells a work desk for \$125, which is marked up by 30% from the wholesale price. How much did the store pay for the desk?

#### Solution

Let the wholesale price be $\$x$. The selling price is 30% more, so it is 130% of $x$.

$$\begin{aligned}\text{Selling price} = 1.3x &= \$125\\ x &= \frac{\$125}{1.3}\\ &= \$96.1538\ldots\\ &\approx \$96.15\end{aligned}$$

The furniture store paid approximately \$96.15 for the work desk.

## Percentage profit and loss

Percentage profit and percentage loss are usually calculated as a percentage of the cost price.

$$\text{percentage profit} = \frac{\text{profit}}{\text{cost price}} \times 100\%$$

$$\text{percentage loss} = \frac{\text{loss}}{\text{cost price}} \times 100\%$$

## Example 3

A company buys laptops for $450 each. After 2 years, due to a new model release, the company decides to sell the laptops for $360. Calculate the percentage loss.

### Solution

$$\begin{aligned}\text{Loss} &= \$450 - \$360 \\ &= \$90\end{aligned}$$

$$\begin{aligned}\text{Percentage loss} &= \frac{\text{loss}}{\text{cost price}} \times 100\% \\ &= \frac{\$90}{\$450} \times 100\% \\ &= 20\%\end{aligned}$$

## Example 4

A wholesaler buys boxes of chocolates for $3 each, marks up the price by 60%, and then sells 400 boxes to a retail store at a 15% discount on the marked-up price. How much profit or loss does the wholesaler make on the 400 boxes?

### Solution

For one box of chocolates:

$$\begin{aligned}\text{Marked-up price} &= 160\% \times \$3 \\ &= \$4.80\end{aligned}$$

A discount of 15% is the same as 85% of the marked-up price.

$$\begin{aligned}\text{Discounted price} &= 85\% \times \$4.80 \\ &= \$4.08\end{aligned}$$

The wholesaler still makes a profit because they bought the chocolates for $3 per box.

$$\begin{aligned}\text{Profit per box} &= \$4.08 - \$3 \\ &= \$1.08\end{aligned}$$

$$\begin{aligned}\text{Profit on 400 boxes} &= 400 \times \$1.08 \\ &= \$432\end{aligned}$$

**EXERCISE 6.01** Answers on p. 469

## Markups, discounts, profit and loss

1. A store buys a pair of shoes for $45 and marks up the price by 25%. What is the selling price of the shoes?

EXAMPLE 2

2. A toy store sells a doll that was marked up by 20%. What was its cost price if it was sold for $96?

□ Foundation ○ Mastery ○ Complex

3 A school purchased 250 calculators at $35 each and was given a 12% discount. Find the total paid for the calculators.

4 If Australia's population is 29 773 492, what will be the population next year if the expected growth for the year is 1.4%?

5 Adash's salary of $82 450 p.a. increases by 2.3%. What is his new salary?

6 The value of Elin's computer depreciated (decreased) by 33% this year. If its value was $2399, calculate its current value.

7 A gym charges a monthly membership fee of $114 and decides to increase it by 18%. What will the cost of the monthly membership fee be after the increase?

8 The Bargain Basement is offering a 15% discount on all items in their store and if payment is by cash, a further discount of 10% of the discounted price will be given. How much will Kim pay for a TV marked at $3299 if she pays in cash?

9 Harry's Hardware is having an end-of-financial-year (EOFY) sale. All items in the store are discounted by 18%.

**a** Allegra is a house painter. She buys ten 4 L tins of paint at the Harry's Hardware EOFY sale regularly priced at $67 per tin.

**i** Calculate the cost of the paint at its regular price.

**ii** Calculate the cost of the paint after the 18% discount.

**b** Allegra is also entitled to a trade discount of 10% on the discounted price. Calculate the final cost Allegra will pay for the paint.

10 The Toy Factory stores increased the price of all toys by 15% in July. In the following January, they discounted all prices by 8%. For a drone originally priced at $89, calculate: EXAMPLE 3

**a** the price after the 15% increase

**b** the sale price after discount

**c** the overall change in price and the percentage change on the original price.

11 During a sale, a lounge is sold for $599. The lounge cost the store $720. Calculate the percentage loss, correct to one decimal place.

12 Ant bought 200 keyrings for $6 each and sells them for $8.20 each.

**a** How much total profit did he make?

**b** Calculate, correct to one decimal place, the profit percentage on the cost price.

13 Maia bought 900 shares for $8.47 each. She paid a stockbroker's fee of 2.5% of the value of the shares. Three years later, she sold all the shares for $7065 and paid a stockbroker's fee of 2% of the value of the shares.

**a** Calculate the total profit or loss Maia made on the shares.

**b** Calculate, correct to one decimal place, the profit or loss percentage on the cost price.

14 Hudson bought 9 boxes of bananas for $56 each. He sells 6 of the boxes for $64 but discounted the remaining boxes by $10 before the bananas went bad. Calculate the percentage profit Hudson made on the sale of the bananas.

15 A company buys smartphones for $450 each and marks up the price by 20%. After one year, due to a new model release, the company sells all old models at a 20% discount on the selling price, rounded to the nearest dollar.

**a** Does the company make a profit or a loss? How much?

**b** What percentage (correct to one decimal place) of the cost price is the profit or loss?

16 A store buys bicycles for $150 each and marks up the price by 25%. However, due to a sudden market downturn, the store has to sell 40 of the bicycles at a 10% discount from the marked-up price. How much profit or loss does the store make on those 40 bicycles?

# 6.02 GST and VAT

Videos
GST 1

Discounts and GST

GST 2

While income tax is a tax on earnings, the GST and VAT are taxes on spending or consuming.

The **goods and services tax (GST)** in Australia is 10% of the cost of the goods or services.

New Zealand, Canada and Singapore also have a GST. In some countries, it is called the **value-added tax (VAT)**.

The consumer pays GST for goods and services, such as electrical appliances, lease of premises, plumbing work and IT support. The GST is not charged on essential items such as basic foods, childcare or renting a home.

Prices in Australia are usually printed with the GST included. Prices in some countries, for example, the USA, are printed 'before tax', with the tax added when the consumer pays.

The GST/VAT rate in some countries (correct in 2025) is shown in the table.

| Australia | 10% | Singapore | 9% |
|---|---|---|---|
| Canada | 5% | South Africa | 15% |
| Japan | 10% | Sweden | 25% |
| Mexico | 16% | Thailand | 7% |
| New Zealand | 15% | United Kingdom | 20% |

## Example 5

**a** A camera costs \$568 plus 10% GST. What is its selling price?

**b** A TV costs \$1995, including GST. What was the original price of the TV, to the nearest 5 cents, before GST was added?

### Solution

**a** Selling price = 110% × \$568 — increasing the price by 10% (100% + 10% = 110%)

= \$624.80

The selling price of the camera is \$624.80.

**b** 110% of original price = \$1995

1% of original price = \$1995 ÷ 110 — using the unitary method to find 1% first

= \$18.136 36 ...

Original price = \$18.136 36 ... × 100 — multiplying by 100 to find 100%

= \$1813.636

≈ \$1813.65 — rounding to the nearest 5 cents

The original price of the TV was \$1813.65.

## Example 6

A handbag in Norway costs 425 kroner before VAT and 531.25 kroner after.

What is the rate of VAT charged?

### Solution

VAT = 531.25 – 425

= 106.25 kroner

$$\text{VAT rate} = \frac{106.25}{425} \times 100\%$$

= 25%

The rate of VAT charged is 25%.

**EXERCISE 6.02** Answers on p. 469

## GST and VAT

**1** A ceiling light was purchased for a price of $750, plus 10% GST. What is the ceiling light's selling price after GST is added? Select **A**, **B**, **C** or **D**.

**A** $750.75 **B** $757.50 **C** $775.00 **D** $825.00

**2** A computer was purchased in Ireland for €1500 (Euro). What is the computer's selling price when 23% VAT is added?

**3** A watch in Singapore costs $142 plus 9% VAT. What is the watch's selling price?

**4** A car costs $42 000 including 10% GST. What was the car's price before GST, correct to the nearest dollar? Select **A**, **B**, **C** or **D**.

**A** $38 182 **B** $37 800 **C** $38 181 **D** $46 200

**5** A train ticket from Sydney to Bomaderry costs $35.20 including 10% GST. What is the price of the ticket before tax?

**6** A night's accommodation in London is £126.90 (pounds) including 20% VAT. What was the price before tax?

**7** An overseas telephone call from South Africa costs 53.01 rand after a VAT of 15% is imposed. What was its price before VAT, rounded to 2 decimal places?

**8** A box of chocolates from the Netherlands costs €6.50, plus €1.37 VAT. What percentage VAT, to the nearest whole number, has been imposed?

**9** The cost of a laundry service in Greece was €9.88 after a VAT of €1.91 was added.

**a** What was the pre-VAT price of the laundry service?

**b** What was the rate of VAT charged, to the nearest percentage?

**10** Sarah had her home bathroom renovated. A 10% GST of $3855 was added to the total price of the renovations paid to the builder.

**a** What was the cost of the bathroom renovations before the GST was added?

**b** Calculate the total cost of the renovations.

**11** Phoebe is holidaying in the United Kingdom and finds 2 shops selling the same sports shorts that she would like to purchase.

| Shop A | Shop B |
|---|---|
| £85.00 | £72.00 |
| Price includes VAT | Price does not include VAT |

The rate of VAT is 20%. Which shop has the cheaper shorts and by how much?

**12** Freyja lives in Sweden and bought a desk costing 1599 kronor. This includes VAT of 25%. Calculate the amount of VAT that she paid.

**13** Solange wants to buy 6 new chairs for her dining table. The furniture shop has a special offer:

*Buy 2 chairs, get 1 free*

If each chair costs $185 plus 10% GST, how much will Solange pay in total for the chairs?

**14** Nikko is in New Zealand and bought a new caravan. The cost of the caravan is $65 000 plus a VAT of 15%. Nikko paid a $9000 deposit, and the rest is paid in monthly repayments over 4 years.

**a** How much will the caravan cost, including the VAT?

**b** How much will Nikko repay in total over 4 years, after the deposit is deducted?

**c** What will he repay each month? Answer to the nearest cent.

**15** Part of Asritha's food shopping receipt is shown on the right. Determine the missing values, $A$ and $B$, to complete the receipt.

**16** A supermarket receipt shows a total amount of \$213.90. The GST shown on the receipt is \$14.70. If 10% GST is charged on only some of the items, what was the value of the items that did not have GST charged?

**SUPERMARKET**
RECEIPT

| Description | \$ |
|---|---|
| * Chocolate 300 g | $A$ |
| Tomatoes 1 kg | 7.50 |
| Natural almonds 400 g | $B$ |
| Cheese slices 500 g | 11.00 |
| Milk 2 L | 3.80 |
| Bananas 570 g | 3.55 |
| **Total for 6 items** | **47.80** |
| GST included in total | 1.20 |

*GST of 10% is included in the price of item.

# 6.03 Buy now, pay later

Many customers purchase expensive items using 'buy now, pay later' plans, a credit scheme that involves 'paying off' the item by regular instalments over time, usually after paying a deposit. This type of payment plan is often offered by a finance company linked to the retail store as an alternative to paying by credit card. This plan is a short-term loan that is becoming more popular, especially for online purchases, but the interest rate may be high.

You will learn more about credit cards in Year 12.

With a buy now, pay later plan, there may be an interest-free period at the start, such as 6 months, when you can pay off or reduce the loan without being charged interest. However, there could be administrative fees and charges added if the regular instalments are not paid on time.

**Lay-by** is a type of plan where you don't receive the item until it is paid off, so no interest is charged on the cost of the items. However, there may be service charges as the lay-by items have to be stored in a secure location within the store or warehouse. A deposit must be paid to lay-by items and the rest of the purchase price is repaid in regular instalments.

☐ Foundation ○ Mastery ⬡ Complex

## Example 7

Eve buys a camera for $1650 using a buy now, pay later plan. She is offered 2 different payment options:

**Option 1**: A $75 activation fee and a $10 monthly administrative fee for 18 months.

**Option 2**: A $100 activation fee and a $15 monthly administrative fee for 12 months.

**a** Calculate the total payment for each option.

**b** Which option should Eve choose to minimise her total payment, and why?

### Solution

**a** For Option 1:

Total payment = $1650 + $75 + 18 × $10

= $1905

For Option 2:

Total payment = $1650 + $100 + 12 × $15

= $1930

**b** Eve should choose Option 1 because it is the cheaper option by $25.

## Example 8

Ryan purchased a suit valued at $190 on lay-by. If he paid a deposit of 20% and repaid the balance off with weekly payments of $32, how many weeks until he takes possession of the suit? Round up your answer to the nearest week.

### Solution

Deposit = 20% × $190

= $38

Balance = $190 − $38

= $152

Number of weekly payments = $152 ÷ $32

= 4.75

= 5 rounding up

Ryan can take possession of the suit in 5 weeks.

**EXERCISE 6.03** Answers on p. 469

## Buy now, pay later

**1** Jenna purchases a laptop using a buy now, pay later option. She agrees to pay $650 in 6 months, within the interest-free period, plus $25 as a processing fee. How much will Jenna pay in total? Select **A**, **B**, **C** or **D**.

**A** $675 **B** $800 **C** $3925 **D** $4050

**2** Ethan buys a $600 phone using 12 monthly payments. Each payment is $60.

**a** How much is the total amount that Ethan will pay for the phone by the end of the 12 months?

**b** How much interest was charged?

**3** Manal purchases a $600 bicycle on a buy now, pay later plan that charges a single $30 establishment fee, an $8 monthly administrative fee and a $15 late fee if she does not make a payment on time. Manal made 4 monthly payments, one of them late. How much did she pay in total for the bicycle?

EXAMPLE 8

4 Before Christmas, Ashwin bought $235 worth of toys on lay-by for his children. He paid a $60 deposit and then made weekly payments of $15/week or 6.5% of the balance owing, whichever was greater. How many weeks did it take Ashwin to pay off the lay-by? Round your answer up to the nearest week.

5 Emilia buys a treadmill from a sports store and pays using the Longitude app, with interest-free terms for 12 months. The treadmill costs $590 and Emilia is required to pay 35% of the purchase price, followed by 4 equal monthly instalments. She is also charged an administration fee of $6.50 per month.

**a** How much is each instalment?

**b** Calculate the total amount Emilia paid for the treadmill.

6 Jasmine buys a dining table for $1000 on lay-by. The plan includes:

- a $50 processing fee
- a $20 delivery fee
- a $5 monthly maintenance fee for 12 months.

Jasmine has 2 options to pay for the dining table:

**Option 1:** Payments over 12 months

**Option 2:** Payments over 18 months with a reduced $4 monthly maintenance fee

**a** Calculate the total amount Jasmine will pay for the dining table under each option.

**b** Which option is cheaper?

**c** Which option would Jasmine choose if she prefers lower monthly repayments?

7 Virat is comparing 2 plans for purchasing a laptop. Both plans are interest-free.

**Plan A:** $50 administration fee and a $9.95 monthly account fee for 12 months

**Plan B:** $40 administration fee and a $12 monthly account fee for 12 months

How much more will Virat pay under Plan B compared to Plan A?

8 Olivia buys a sofa for $1850 using a plan that charges a $60 establishment fee. She has 2 options for repayment:

**Option 1:** Repay over 12 months with a $12 monthly account fee

**Option 2:** Repay over 6 months with a $18 monthly account fee

Which option should Olivia choose if she prefers to minimise her total payment?

9 Conduct an internet search of stores offering a lay-by or buy now, pay later plan and write down 3 fees stores charge for each type of plan.

10 Jake is purchasing a home entertainment system worth $2670. He is considering buy now, pay later options offered by 2 apps:

**ZAP:** Pay 5.556% of purchase price per month, with a $100 administrative fee and a $10 monthly account keeping fee for 18 months

**BUZZ:** Pay fortnightly interest-free instalments, with a $75 administrative fee and a $12 monthly account keeping fee for 15 months

Which plan should Jake choose if he wants to minimise his total payment?

☐ Foundation ○ Mastery ⬡ Complex

# Term payments

6.04

For term payment plans, it is important to read the terms and conditions, the 'fine print' that may indicate hidden costs where additional fees and charges may apply. If you fail to keep up with the payments, higher interest may be charged, or the item may be **repossessed** (taken back).

A **term payments plan** is sometimes called **hire purchase** because the customer actually hires the item until it is completely paid off.

## Example 9

Daniel is buying new outdoor furniture and will pay it off over 3 years, with no interest for the first 6 months, followed by a **simple interest** rate of 18% p.a. being applied for the remainder of the time. The cost of the furniture is \$4339. If Daniel pays monthly payments of \$180, how much will he owe after 6 months:

**a** before interest is applied?

**b** with interest included?

### Solution

**a** Payments after 6 months $= 6 \times \$180$

$= \$1080$

Balance owing $= \$4339 - \$1080$

$= \$3259$

**b** 3 years $= 3 \times 12 = 36$ months

Period of loan remaining $= 3$ years $- 6$ months

$= 2.5$ years

Interest on loan remaining $= Prn$ — simple interest formula

$= \$3259 \times 0.18 \times 2.5$

$= \$1466.55$

Total owing $= \$3259 + \$1466.25$ — including interest

$= \$4725.55$

## Example 10

Sophie buys a smartwatch for \$500 using a repayment plan for 9 months with simple interest charged at 6% p.a.

**a** How much interest will Sophie pay?

**b** How much will Sophie pay in total?

**c** Calculate Sophie's monthly repayment.

### Solution

**a** Interest $= Prn$ — simple interest formula

$= \$500 \times \frac{0.06}{12} \times 9 \text{ (months)} \quad \text{OR} \quad \$500 \times 0.06 \times \frac{9}{12} \text{ (years)}$

$= \$22.50$

Sophie will pay \$22.50 in interest.

The interest rate $r$ and number of periods $n$ must be in the same units, either months or years

**b** The total repayment amount is the principal plus interest:

Total repayment = \$500 + \$22.50

= \$522.50

Sophie will pay a total of \$522.50.

**c** The total repayment is \$522.50, over 9 months.

$$\text{Monthly repayment} = \frac{\$522.50}{9}$$

$$= \$58.0555\ldots$$

$$\approx \$58.06$$ always round up for repayments

Sophie's monthly repayment will be \$58.06.

With a **deferred payment plan**, you do not make any repayments until a later date, such as after 3 years.

'Deferred' means delayed.

## Example 11

Awesome Appliances and Krazy Kitchens offered deferred payment plans to customers wanting to buy a dishwasher for \$1350.

'Unpaid balance' and 'outstanding amounts' both mean the remaining part of the loan still owing.

**a** Which plan should Petro choose if he wants more time to save up enough to pay off the dishwasher before any interest is charged?

**b** Souki chose the Krazy Kitchens plan and had paid \$800 when the 300 days were up. How much does she owe now?

## Solution

**a** Awesome Appliances, because its deferred payment period of 2 years is longer than Krazy Kitchens' period of 300 days.

**b** Outstanding amount = $1350 – $800

= $550

For the simple interest formula $I = Prn$, $P = \$550$, $r = 0.0011$ per day, $n = 300$ days.

The outstanding amount is charged interest from the date of purchase.

$I = \$550 \times 0.0011 \times 300$

$= \$181.50$

Total owing = $550 + $181.50

= $731.50

**EXERCISE 6.04** Answers on p. 469

## Term payments

EXAMPLE 9

**1** Lily purchases a tablet device for $500 using a repayment plan charging simple interest at an annual rate of 8%. The term of this plan is 12 months. How much will Lily pay in interest over the course of the year? Select **A**, **B**, **C** or **D**.

**A** $20 **B** $40 **C** $60 **D** $80

**2** Marnus buys a bicycle for $750 and is charged simple interest at a rate of 6% p.a. If Marnus plans to pay off the loan in 8 months, how much interest will he have paid by the end of the loan?

**3** Nabilah buys a sofa valued at $800. She pays 0% interest for the first 6 months, after which an interest rate of 15% p.a. applies. If Nabilah pays monthly payments of $100, how much will she owe after 6 months, before interest is applied? Select **A**, **B**, **C** or **D**.

**A** $200 **B** $400 **C** $600 **D** $800

**4** Babu bought a second-hand tractor for $11 000 for his farm. He paid a 10% deposit and monthly repayments of $360 for 4 years. How much interest did he pay in total? Select **A**, **B**, **C** or **D**.

**A** $18 380 **B** $17 280 **C** $7380 **D** $6280

**5** Scott purchased an engagement ring for $7560, paying $100 deposit and 7.2% p.a. interest for 18 months. What was the total amount paid for the ring? Select **A**, **B**, **C** or **D**.

**A** $8365.68 **B** $8265.68 **C** $7460 **D** $8056.80

**6** Madison bought a new pair of shoes for $450 with a payment plan that charges simple interest at an annual rate of 9%. She plans to pay off the loan in 8 months.

**a** What is the amount of interest Madison will pay by the end of the loan?

**b** How much will Madison pay in total?

**c** How much will her monthly repayment be?

**7** David and Sarah bought a motorhome for $19 500 to travel around Australia. There were 2 options for paying for the motorhome:

**Plan A:** 20% deposit and 36 monthly instalments of $595

**Plan B:** No deposit and monthly instalments over 4 years at 8.9% p.a.

By calculating and comparing, find which plan:

**a** charges more interest

**b** has lower monthly instalments.

**8** Zoe bought a backyard pool to be installed in her backyard for \$45 500. She paid a 10% deposit and the balance is to be repaid in equal monthly instalments over 5 years, with flat-rate interest charged at 9.5% p.a.

iStock.com/StockSeller_ukr

Calculate Zoe's:

- **a** deposit
- **b** balance owing
- **c** interest charged
- **d** total to be repaid
- **e** monthly instalment
- **f** total price paid for the pool.

**9** Autocar is a car dealership selling new cars on terms.

**AUTOCAR weekly specials**

Mark Scheuern/Alamy Stock Photo

Grzegorz Czapski/Alamy Stock Photo

Heritage Images/Getty Images

- **a** How much would you pay for the Toyota hybrid on Autocar's terms?
- **b** How much would you save on the Kia hatch by paying cash rather than term payments?
- **c** What is the flat rate of interest per annum, correct to one decimal place, charged on the Mazda 2 hatch if making term payments?
- **d** Give a possible reason Autocar charges different rates of interest for different cars.

**10** The Rosa Finance company had the following agreement for purchasing a yacht.

Rosa finance

**Hire Purchase agreement**

| | |
|---|---|
| Cash price of yacht | \$47 500 |
| Deposit | \$6500 |
| Trade-in allowance on old yacht | \$5400 |
| Stamp duty | \$475 |
| Registration | \$1203 |

*Flat interest rate of 15.8% p.a.*
*Equal monthly repayments over 5 years*

Find:

- **a** the total cost of the yacht, including charges and allowing for trade-in if paying by cash.
- **b** the monthly repayment if paying on terms (including stamp duty and registration).

**11** Two finance companies offer deferred payment plans and charge interest of 0.2% per day on outstanding amounts from the purchase date.

Local Loans: 10% deposit then nothing to pay for 12 months

Feelgood Finance: Buy now and pay nothing for 100 days

**a** Miguel bought a $1600 desk through Local Loans. Find:

**i** the amount owing 12 months after purchase

**ii** the amount owing after 396 days if only the deposit had been paid.

**b** Renate bought a computer for $1350 using Feelgood Finance. Calculate the amount if she repaid the debt in full after:

**i** 90 days

**ii** 101 days

**c** Write down one positive and one negative feature of each plan.

EXAMPLE 11

**12** Boris bought a drum kit for $1600 and was deciding between buying with term repayments or with the deferred payment plan.

How much will Boris pay for the drum kit if he:

**a** buys using term repayments?

**b** uses the deferred payment plan and pays the amount owing 2 years after purchase?

**c** uses the deferred payment plan, pays an $800 deposit upfront and fees only for 2 years, and the rest after the 2 years is up?

**13** Samanvi bought outdoor furniture priced at $3299. She paid a deposit of $900 and took out a loan for the balance, which was repaid in 24 monthly instalments of $130.18. What flat interest rate per annum was charged on Samanvi's loan? Select **A**, **B**, **C** or **D**.

**A** 11% **B** 15% **C** 28% **D** 47%

14 A store offers terms of one-third deposit on the purchase price with the balance plus interest paid in equal, monthly instalments over 18 months. The interest is charged at 9% p.a. Dami buys a bedroom package priced at \$6000.

- **a** Calculate her deposit.
- **b** Determine the balance owing.
- **c** Calculate the interest to be paid.
- **d** Determine the total amount Dami owes.
- **e** Hence, calculate Dami's monthly repayment.

15 Taylor purchased a new car valued at \$32 000. She paid a 10% deposit and was told that one option was to repay the balance owing on the purchase price of the car plus interest over 4 years. Alternatively, she could pay off the balance of the purchase price of the car plus interest in 8 years. If Taylor chooses the second option, she would pay \$19 584 more in comparison to taking the first option. The interest rate on the 8-year option is 1% greater than for the 4-year option.

- **a** How much was Taylor's deposit?
- **b** What was the balance owing when she purchased the car?
- **c** Find the interest rate for each option.
- **d** Determine the total amount that would be repaid for the car for each option.
- **e** Calculate the monthly repayment Taylor would pay for each option.

# 6.05 Water bills

**Worksheets**
Water usage and costs

Water usage in your home

If your home is connected to mains water, it has a **water meter** to measure the amount of water used in your home. A water bill is usually sent to the resident every quarter, listing the costs of the water service and **sewerage** (which are fixed), and the cost of the water used in the home (which is variable).

## Example 12

Yitong's water bill shows the following charges:

| Water service | \$41.39 |
|---|---|
| Wastewater (sewerage) | \$162.88 |
| Water usage 33.5 kL @ \$2.276 per kL | $A$ |
| **Amount due** | $B$ |

Calculate Yitong's total water usage ($A$) and the total amount due ($B$).

### Solution

Total water usage, $A = 33.5 \times \$2.276$

$= \$76.246$

$\approx \$76.25$

Total amount due, $B = \$41.39 + \$162.88 + \$76.25$

$= \$280.52$

□ Foundation  

**EXERCISE 6.05** Answers on p. 469

## Water bills

EXAMPLE 12

**1** Anil's water bill shows the following charges:

| Water service | \$41.39 |
|---|---|
| Wastewater (sewerage) | \$162.88 |
| Water usage 52.1 kL @ \$2.276 per kL | *A* |
| **Amount due** | *B* |

**a** Calculate Anil's total water usage (*A*).

**b** Find the total amount due (*B*).

**2** For the month of June, a household used 425 L of water per day. If water costs \$2.28/kL, what was the water bill amount for June?

**3** Below is a water bill from Riverstone Water.

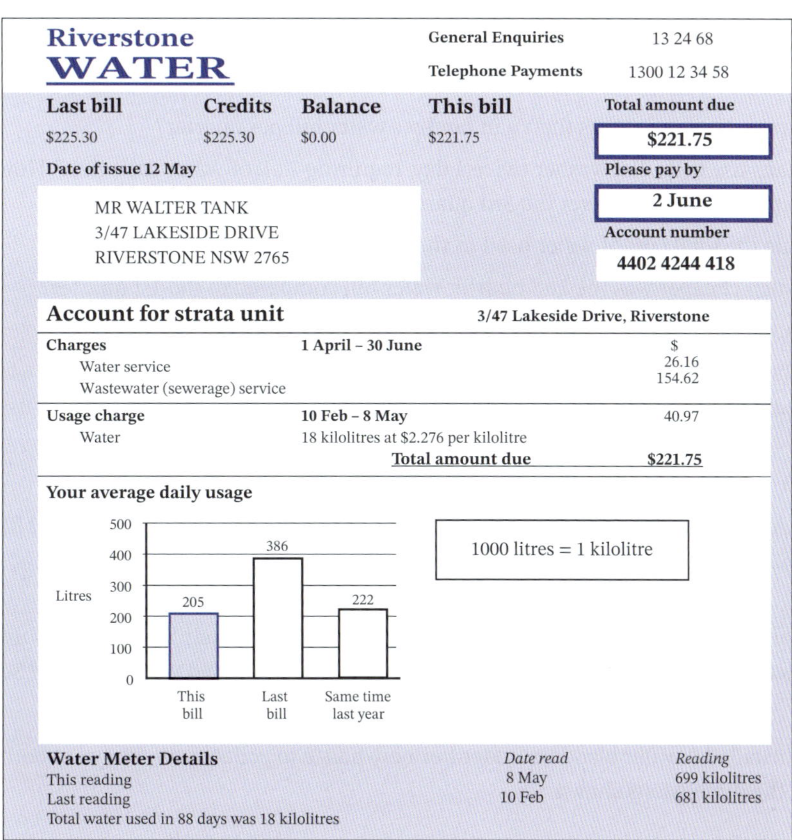

Riverstone WATER

General Enquiries 13 24 68
Telephone Payments 1300 12 34 58

| Last bill | Credits | Balance | This bill | Total amount due |
|---|---|---|---|---|
| \$225.30 | \$225.30 | \$0.00 | \$221.75 | \$221.75 |

Date of issue 12 May

Please pay by 2 June

MR WALTER TANK
3/47 LAKESIDE DRIVE
RIVERSTONE NSW 2765

Account number 4402 4244 418

**Account for strata unit** 3/47 Lakeside Drive, Riverstone

| | | \$ |
|---|---|---|
| **Charges** | **1 April – 30 June** | |
| Water service | | 26.16 |
| Wastewater (sewerage) service | | 154.62 |
| **Usage charge** | **10 Feb – 8 May** | 40.97 |
| Water | 18 kilolitres at \$2.276 per kilolitre | |
| | **Total amount due** | **\$221.75** |

**Your average daily usage**



**Water Meter Details**

| | *Date read* | *Reading* |
|---|---|---|
| This reading | 8 May | 699 kilolitres |
| Last reading | 10 Feb | 681 kilolitres |

Total water used in 88 days was 18 kilolitres

**a** How many days are there from the bill's day of issue to the payment's due date?

**b** Is this bill paid monthly, 2-monthly or quarterly?

**c** What are the fixed charges for, and how much are they in total?

**d** How many kilolitres of water were used in the period covered by this bill?

> **Quarterly** means 4 times per year or every 3 months.

**e** Was more water used each day for this bill compared to the last bill? Give a possible reason for this.

**f** Why is it more useful to compare average daily usage with that of the same time last year?

**g** Verify the average daily usage of 205 L by dividing the volume of water used by the number of days covered by this bill.

**h** What is the cost per kilolitre for water usage?

**i** Verify the water usage charge of \$40.97 using this rate.

**j** A backyard swimming pool contains 61 kL of water. Calculate the cost of filling this pool with water.

**4** Hannah's water bill shows that water usage is charged at \$2.276/kL.

**a** If Hannah used 24 kL of water, how much will she need to pay?

**b** Hannah paid \$35.05 in water usage charges on her latest bill. How many kilolitres of water did she use?

**c** Hannah was also charged \$29.18 for the water service and \$165.87 for the wastewater (sewerage) service. What was the total of Hannah's bill?

**5** The Yuan family has 2 adults and 2 teenagers. In one year, their water usage for each quarter was as shown below:

| Quarter | Date | Water usage (kL) |
|---|---|---|
| 1 | 1 Jan–31 Mar | 66.0 |
| 2 | 1 Apr–30 Jun | 59.3 |
| 3 | 1 July–30 Sept | 58.5 |
| 4 | 1 Oct–31 Dec | 71.8 |

**a** What was the Yuan family's total water usage for the year?

**b** Water costs \$2.67/kL. How much did the Yuan family spend on water usage for the whole year?

**c** On average, how much was the Yuan family's water bill per quarter?

**d** The Yuans use their dishwasher twice a day, requiring 21 L of water each time. How much water did their dishwasher use over the 3rd quarter?

**e** Calculate the total cost of water used in the dishwasher in the 3rd quarter.

**f** How much cheaper was the 2nd quarter water bill compared to the 1st quarter?

**g** How much more was the 4th quarter water bill compared to the 3rd quarter?

**6** The Chand family are calculating their water usage for the quarter. This reading is 8764 kL and the previous reading was 8112 kL. What was their water usage bill for the quarter? Water usage is charged at \$2.276/kL. Select **A**, **B**, **C** or **D**.

**A** \$652.00 **B** \$985.60 **C** \$1483.95 **D** \$2967.90

**7** Jeremy and Annie pay a recycled water charge of \$4.27 per quarter. If they also use 12 kL of recycled water at \$2.62/kL and 20.2 kL of drinking water at \$2.276 kL, calculate their charges for water usage for the quarter.

**8** The Green family pay a recycled water charge of \$4.22 per bill. On their latest bill they used 39 kL of recycled water at \$2.52/kL. How much did they pay in total for recycled water charges? Select **A**, **B**, **C** or **D**.

**A** \$66.30 **B** \$102.50 **C** \$168.80 **D** \$262.86

**9** Biljana has installed 2 water bladders under her new house to collect rain. They are both rectangular prisms with the dimensions shown.

Not to scale

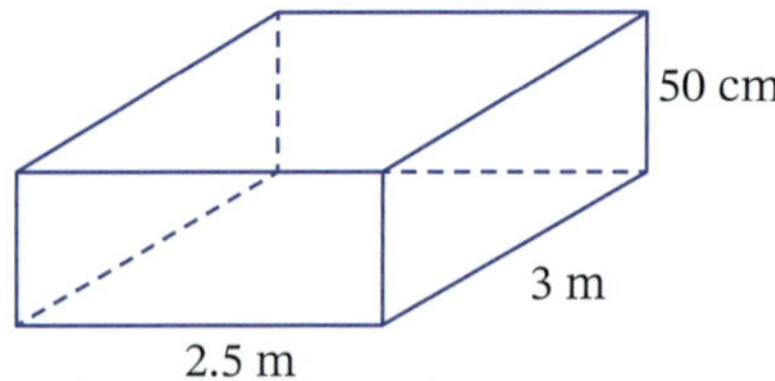


**a** Calculate the combined capacity of the water bladders, in litres.

**b** In April, Biljana's family uses an average of 570 L/day. If the water bladders are full, how long will it take to use all the water in the bladders (assuming no rainfall). Answer to the nearest day.

**c** If water charges are \$2.276/kL, how much money will Biljana's family save by using the water in these bladders?

☐ Foundation ○ Mastery ○ Complex

**10** The average daily water usage for 2 families, for the quarter 1 January–31 March, is:

- the Baskaran family (4 people), 560 L/day
- the Zhong family (6 people), 710 L/day

**a** Calculate the water usage costs for each family for the quarter if water costs $2.276 per kL and state which family pays more.

**b** Calculate the average daily water use per person for each family and state which family uses more.

## Did you know?

### Water usage in the home

- Washing your hands/face — 5 L
- Brushing your teeth (tap running) — 5 L
- Brushing your teeth (tap not running) — 1 L
- Cooking and making coffee/tea — 8 L per day
- Flushing the toilet — 9 L to 13 L
- Flushing the toilet (half flush) — 4.5 L to 6 L
- Household tap — 18 L per minute
- Washing the dishes (by hand) — 18 L
- Washing the dishes (dishwasher) — 25 L per cycle
- Bath use — 85 L to 150 L
- Shower (8 minutes) — 80 L to 120 L
- Washing machine (front loading) — 120 L per cycle
- Washing machine (top loading) — 180 L per cycle
- Washing the car (with hose) — 100 L to 300 L
- Garden sprinkler — 1 kL to 1.5 kL per hour
- Garden hose — 1.8 kL per hour
- Swimming pool (backyard) — 20 kL to 55 kL

On average, a 4-person Sydney house (with garden) uses 936 litres of water per day. Outside taps and toilet flushing account for half of this usage.

**How much water does your household use each day? Find out by investigating your water bill.**

## Investigation

### Water costs and savings

Investigate the cost of water usage in your local area by visiting the website of your water supplier and examining the following water usage costs:

1. Water service charges
2. Cost of drinking water per kilolitre
3. Cost of recycled water
4. Wastewater (sewerage) charges
5. Stormwater charges

Investigate strategies for making savings on your water bill by minimising and recycling water use. Investigate water-rating websites to find out more about Water Efficient Targets and compare your household use with the daily water consumption targets. Find a recent water bill for your home and interpret the different costs.

**Can you recommend any changes to your family's usage to help save water in your home?**

# 6.06 Power bills

Worksheet
Power problems

## Units of power and electrical usage

The **watt** (abbreviation W) is a unit of **power** and is equal to one joule of energy per second.

| Unit | Relationships |
|---|---|
| watt (W) | |
| kilowatt (kW) | 1 kW = 1000 W |
| megawatt (MW) | 1 MW = 1000 kW = 1 000 000 W or $10^6$ W |
| gigawatt (GW) | 1 GW = 1000 MW = 1 000 000 000 W or $10^9$ W |

## Units of energy usage

**Electrical energy usage** is measured in **watt-hours (Wh),** which is the amount of electrical energy used by a one-watt load (such as a light globe) drawing power for one hour.

The electricity usage for households is measured in **kilowatt-hours (kWh)**, which is 1000 watt-hours. The **cost** of electricity usage is given in **cents/kilowatt-hour** (c/kWh).

## Domestic and off-peak electricity

**Domestic** electricity refers to electricity that is used during the day.

**Off-peak** electricity refers to electricity that is used during the late evening and early morning, such as between 10 pm and 7 am, when the demand from households and businesses is much lower. Off-peak electricity is charged at a cheaper rate than domestic electricity, and is often used to heat water in homes.

### Example 13

Calculate the cost of electricity usage for a household where the supply charge is 101.9 c/day for 92 days (one quarter), domestic usage for the quarter is 954 kWh at \$0.594 89/kWh and off-peak usage is 748 kWh at \$0.247 65/kWh.

#### Solution

Supply charge cost = 92 × \$1.019
= \$93.75

Cost of domestic usage = 954 × \$0.594 89
= \$567.53

Cost of off-peak usage = 748 × \$0.247 65
= \$185.24

Total cost = \$567.53 + \$185.24 + \$93.75
= \$846.52

## Gas usage

Gas is a reliable and readily available power source, mainly used for cooking and heating. Many homes use natural gas as a cost-effective supplement to electricity.

The gas usage for households is measured in **megajoules** (MJ), and the cost of gas usage is given in cents/megajoule (c/MJ). Gas appliances show their energy consumption in terms of megajoules per hour (MJ/h).

### Example 14

A portion of Nate's gas bill is shown.

| **Billing period** | Start date: 1 June | End date: 31 August |
|---|---|---|
| **USAGE** | First 7560 MJ | \$0.0483/MJ |
| **CHARGE** | Additional usage over 7560 MJ | \$0.0364/MJ |
| Supply charge is \$0.6572 per day. | | |

If Nate uses 9425 MJ in the quarter covered by the bill, how much is his gas bill?

### Solution

First 7560 MJ: Charge = 7560 × \$0.0483

= \$365.148

Additional amount: 9425 − 7560 = 1865 MJ

Charge = 1865 × \$0.0364

= \$67.886

Number of days from 1 June to 31 August = 30 + 31 + 31

= 92

Supply charge = 92 × \$0.6572

= \$60.4624

Total gas bill = \$365.148 + \$67.886 + \$60.4624

= \$493.4964

≈ \$493.50

Fabrizio Maffei/Shutterstock.com

**EXERCISE 6.06** Answers on p. 470

## Power bills

EXAMPLE 13

**1** A section of an electricity bill for a 3-month period is shown below. This bill shows 3 rates: peak, shoulder and off-peak.

***Energy Used and Costs***

| METER ID | THIS READING | LAST READING | ENERGY USED | RATE (per kWh) | COST |
|---|---|---|---|---|---|
| Peak Energy Charge | | | | | |
| AUEN0812/01 | 531.2 | 274.8 | 256.4 kWh | 48.2400c | $123.69 |
| Shoulder Energy Charge | | | | | |
| AUEN0812/02 | *A* | 560.9 | 523.5 kWh | 29.8700c | *B* |
| Off-peak (Night) Energy Charge | | | | | |
| AUEN0812/03 | 242.5 | 0.0 | 242.5 kWh | 20.1500c | *C* |

**a** Calculate the values at *A*, *B* and *C*.

**b** Calculate the total of this electricity bill for 91 days, if there is also a supply charge of 87.45 c/day.

**2** Justin's electricity meter reading (in kWh) for June is shown below.

Justin is required to pay the following:

Supply charge: 100.97 c/day

First 190 kWh used: 36.94 c/kWh

For usage over 190 kWh: 38.78 c/kWh

Calculate Justin's total bill to be paid for June.

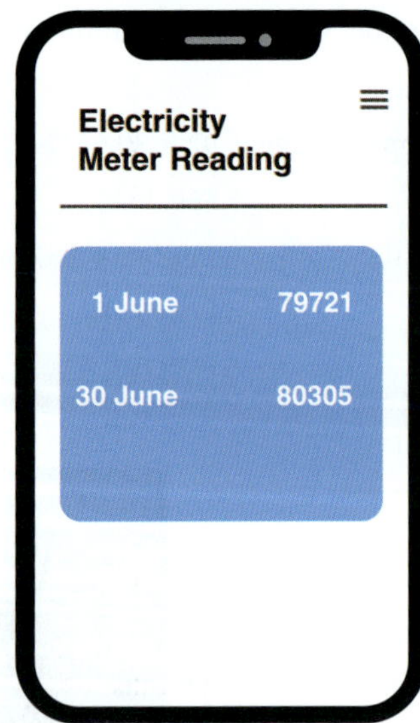

EXAMPLE 14

**3** Vasa's gas bill has usage charges as shown below.

| USAGE CHARGE | First 7560 MJ | $0.0475 |
|---|---|---|
| | Additional usage over 7560 MJ | $0.0353 |

If the supply charge is $0.7163 per day, how much is Vasa's gas bill if he used 11 430 MJ and the bill is for the March, April and May quarter?

☐ Foundation ○ Mastery ⬡ Complex

**4** An electricity bill from Lightning Energy is shown below.

## Lightning Energy

**Account Number:** 71433973
**Date Due:** 6 December
**Amount Due:** $847.84

Ms B. SPARKS
13 AMPERE CRESCENT
LIGHTNING RIDGE, NSW 2834

ELECTRICITY ACCOUNT

### Account Summary

| Description | Amount |
|---|---|
| Previous Balance | $764.92 |
| Less payment(s) received to 05/09/18 | $764.92 CR |
| Subtotal | $0.00 |
| Bill period: 16 Aug to 15 Nov (91 days) | $847.84 |
| **Total Amount Payable (Incl. GST)** | **$847.84** |

### How much energy are you using?

Average daily cost and usage.

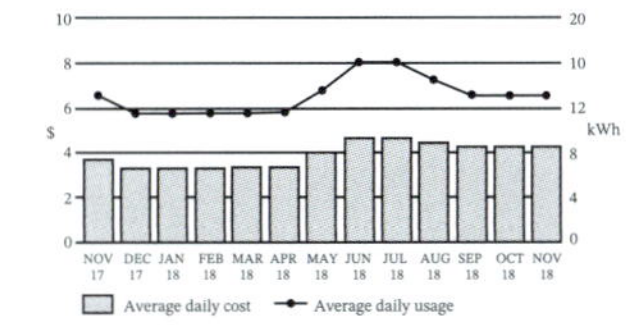

NOV 17, DEC 17, JAN 18, FEB 18, MAR 18, APR 18, MAY 18, JUN 18, JUL 18, AUG 18, SEP 18, OCT 18, NOV 18

Average daily cost — Average daily usage

Never use more than one adaptor on the one power outlet. Better still, use a power board to plug in additional low current applications as they have their own safety overload cut out switch.

### Compare with other homes in your area.

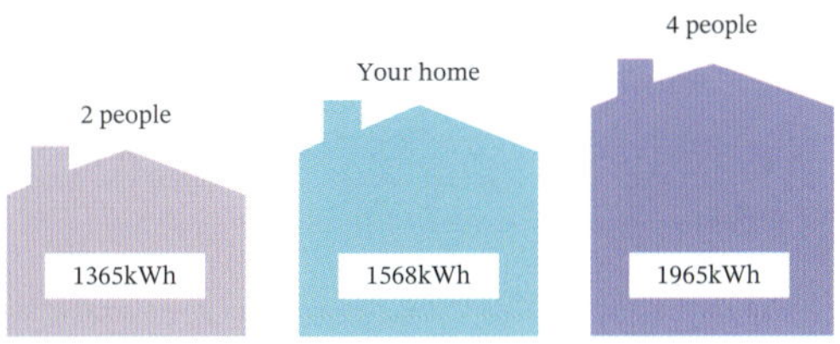

2 people: 1365kWh
Your home: 1568kWh
4 people: 1965kWh

Snapshot.
Average daily cost: **$9.32**
Average daily usage: **17.23 kWh**
Same time last year: **14.38 kWh**

### Your electricity supply details.

| Meter No. | Read Date | Days | Start Read | End Read | kWh |
|---|---|---|---|---|---|
| 0874221 | 14 Nov | 91 | 57144 | 58712 | 1568 |

Your next meter read is between **13 Feb** and **17 Feb.**
Please ensure easy access to your meter on these days.

### New charges and credits.

| **Usage and supply charges** | **Units** | **Price** | **Amount** |
|---|---|---|---|
| Peak (2:00-7:59 pm) | 647kWh | $0.6544 | $423.40 |
| Shoulder (7 am-1:59 pm, 8:00-9:59 pm, and weekends 7 am-9:59 pm) | 415kWh | $0.3442 | $142.84 |
| Off Peak (10 pm-6:59 am) | 506kWh | $0.2724 | $137.83 |
| Supply charge | 91 days | $1.1209 | $102.00 |
| **Other Charges** | | | |
| Payment processing fee | | | $5.24 |
| Total charges | | | $811.31 |
| **Credits** | | | |
| 5% Direct Debit Discount | | | $40.57CR |
| TOTAL NEW CHARGES AND CREDITS | | | $770.74 |
| Total GST | | | $77.10 |
| TOTAL AMOUNT PAYABLE (incl. GST) | | | $847.84 |

**a** Is this bill paid monthly, 2-monthly or quarterly?

**b** What was the average daily electricity usage over this period?

**c** From the graph, find the time of the year when the electricity usage was greatest. Why do you think this is so?

**d** How much electricity was used in the period covered by this bill?

**e** How many days of usage were covered by this bill?

**f** When was the electricity meter read? What was the reading?

**g** Has the electricity bill increased or decreased since the last bill? By how much?

**h** Consider the average usage data under the heading 'Compare with other homes in your area'. This bill is for a household of 2 adults and 2 children. Do you think this house is running efficiently? Justify your answer.

**i** Explain the 4 different cost rates of electricity usage.

**5** A gas bill is shown below.

## Bogas Energy

PAYMENT REFERENCE NUMBER 0150 1626 8000 305
INVOICE NUMBER 00030
DATE OF ISSUE 29 October

**Customer number**
**5016338**

MR LUKE WARM
6 BOYLE AVENUE
SUMMER BAY NSW 2284

| | |
|---|---|
| Accounts and Services – 24 hours | 131 666 |
| Card Payments | 1300 658 999 |
| Emegency – 24 hours | 131 999 |

**Payment Due**
15 November

| Last Account | Payments Received | Balance | New Charge |
|---|---|---|---|
| $285.76 | $285.76 | $0.00 CR | $277.14 |

**Total Due**
$277.144

Average daily usage

For this bill 55.76 MJ

Same time last year 44.78 MJ

Understand your bill

Gas charges are based on an actual meter reading
**Bill period: 26 Jul to 24 Oct (91 days)**

| **Previous balance and payments** | **Amount** |
|---|---|
| Previous balance | $285.76 |
| 16 Aug payment | $285.76cr |
| **Balance brought forward** | **$0.00** |

**New charges and credits**

| **Usage and supply charges** | **Time of use** | **Units** | **Price** | | **Amount** |
|---|---|---|---|---|---|
| General usage | At all times | 1,885 MJ | $0.04465 | | *A* |
| General usage next | At all times | 1,855 MJ | $0.03318 | | *B* |
| General usage next | At all times | 1,334 MJ | $0.03218 | | *C* |
| Supply charge | Daily | 91 days | $0.67332 | | *D* |
| **Other charges** | | | | | |
| Credit card payment fee | | | | | $2.03 |
| **Total charges** | | | | + | **$251.95** |
| **Total new charges and credits (excluding GST)** | | | | = | **$251.95** |
| Total GST | | | | + | $25.19 |
| **Total new charges and credits (including GST)** | | | | = | **$277.14** |

Understand your usage

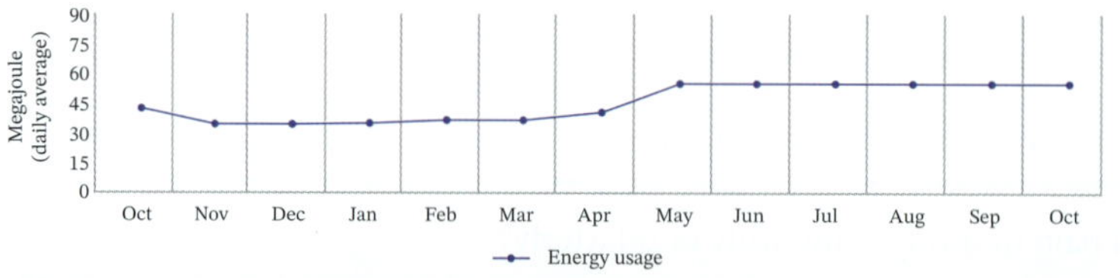

Megajoule (daily average): 0, 15, 30, 45, 60, 75, 90
Oct, Nov, Dec, Jan, Feb, Mar, Apr, May, Jun, Jul, Aug, Sep, Oct
Energy usage

Meter details

| Meter number | Read date | Read type | Start read | End read | Heating value | Conversion factor | Usage MJ |
|---|---|---|---|---|---|---|---|
| MR075043 | 24 Oct | Actual | 3,903 | 4,036.01 | 38.26000 | 0.997100 | 5,074 |

**a** Calculate the difference between this bill and the last one.

**b** How many days does this gas bill cover?

c How many megajoules were consumed during the period for this bill?

d What is the difference in usage in MJ per day for this bill compared to the same time last year?

e On which 2 dates was the gas meter read at this property?

f Calculate the costs for the gas usage and the total supply charge, as shown at $A$, $B$, $C$ and $D$.

g Consider the graph showing average daily gas usage in MJ at this property. How has the usage changed in 12 months? Also, give a reason to explain this change.

**6** The following questions refer to the electricity bill shown below.

## MIF Power and Gas

**Account Number: 56007899**
**Date Due: 23 August**
**Amount Due: $602.87**

FOR ACCOUNT ENQUIRIES: PHONE 131 002
FOR 24HR EMERGENCY ASSISSTANCE: PHONE 131 002

MR B POWER
125 FARADAY DR
PICTON 2571

ELECTRICITY ACCOUNT

Issue Date: 3 August

### Summary of Your Account

| **Description** | | **Amount** |
|---|---|---|
| Balance of last account | | $492.16 |
| Less payments and rounding | | $492.16 CR |
| Subtotal | | $0.00 |
| **Account Summary** – 5 May to 3 August | | |
| Electricity charges (see next page for details)* | $548.06 | |
| Plus GST payable | $54.81 | |
| Total amount of this account (including GST) | | $602.87 |
| **Total Amount Due (including GST)** | | **$602.87** |

*GST applies to this item

### Your Electricity Usage

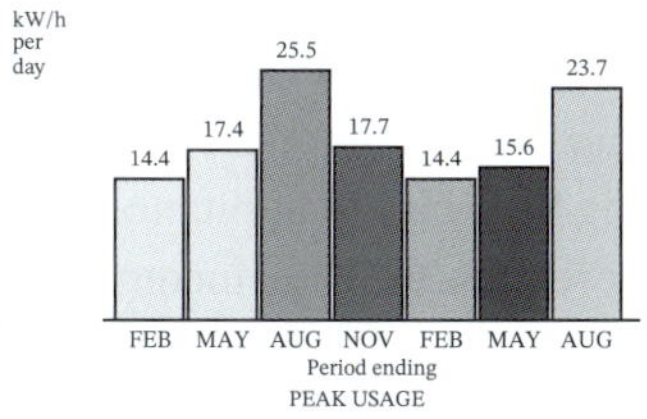

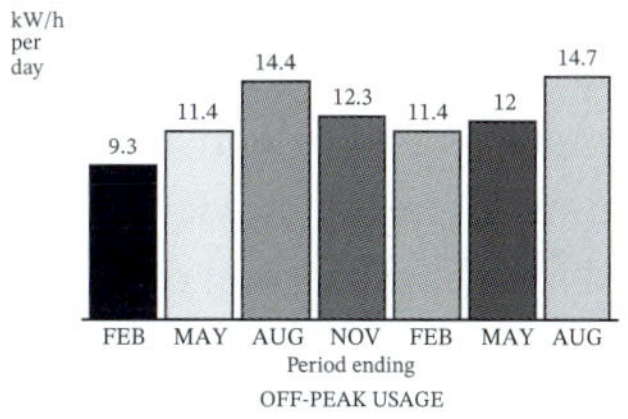

kW/h per day

| Period ending | FEB | MAY | AUG | NOV | FEB | MAY | AUG |
|---|---|---|---|---|---|---|---|
| PEAK USAGE | 14.4 | 17.4 | 25.5 | 17.7 | 14.4 | 15.6 | 23.7 |
| OFF-PEAK USAGE | 9.3 | 11.4 | 14.4 | 12.3 | 11.4 | 12 | 14.7 |

### Itemised Details

**Meter information for Period 5 May to 3 August – 90 days**

| Meter No. | Const | Days | Usage (kWh) |
|---|---|---|---|
| 1234530 | 1 | 90 | 437 |
| OFF PEAK – Total Usage | | | 437 |
| 1234532 | 1 | 90 | 718 |
| DOMESTIC – Total Usage | | | 718 |

**Your Account Calculations**

| Pricing Option | GST | Usage Applies | Rate | Amount |
|---|---|---|---|---|
| **DOMESTIC OFF PEAK(10 pm–7 am Mon–Fri)** | | | | |
| OFF PEAK<br>Off Peak 1 | Yes | 369.000 | $0.1805 | $66.60 |
| **DOMESTIC OFF PEAK (SHOULDER USAGE 7 am–2 pm, 8 pm–10 pm)** | | | | |
| Off Peak 2 | Yes | 68.000 | $0.3125 | $21.25 |
| **DOMESTIC PEAK(2 pm–8 pm Mon–Fri)** | | | | |
| DOMESTIC<br>Domestic | Yes | 718.00 | $0.5924 | $425.34 |
| Daily Supply Charge $0.9964 | Yes | 90 days | | $89.68 |
| **Electricity subject to GST** | | | | **$602.87** |

a What was the average daily electricity usage over the period of this bill? Answer separately for peak and off-peak.

b From the graph, find the time of the year when electricity usage was the highest. Why do you think this is so?

c An extra fee of $89.68 is charged in addition to off-peak and peak usage. What is this extra fee for and how is it calculated?

d Explain why the off-peak shoulder rate is more expensive than the off-peak rate?

e At what time of day are peak rates charged?

f Did this household use more peak or off-peak power?

g By how much has electricity usage increased or decreased when compared to the same period the previous year? Answer separately for peak and off-peak.

**7** The items for an electricity bill were:

| | |
|---|---|
| General usage | 1230 kWh at $0.372 900/kWh |
| Daily supply | 91 days at $0.897 820 |
| Solar meter charge | 91 days at $0.000 000 |

The solar feed-in credit was 815 kWh at $0.050 000.

Calculate the total for this bill.

# 6.07 Personal budgeting

Worksheets
Budget grid

Budgeting scenarios

A **budget** is an estimate of income and spending for a certain period of time. It is a plan to help manage your income wisely. Whether you are running a household or a business or just managing your pocket money, it is essential that you do not spend more than you earn. A budget is divided into **income** and **expenses** and a balanced budget is one in which the total expenses equal the total income.

When preparing a budget, it is important to identify items that qualify as **fixed spending** and **discretionary spending**. **Fixed spending** is money spent on necessities such as food, clothes, fuel and household bills. **Discretionary spending** is money spent by consumers on items other than necessities. These purchases could involve technology or entertainment items, movie tickets or upgrading furniture or appliances. Overspending on discretionary items could lead to debt if the total of all expenses is more than the total income.

## Example 15

Lucy was saving for an overseas trip and accepted some extra casual work. She set up a weekly budget, as shown in the table, and aimed to save $170 per week.

| Income | | Expenses | |
|---|---|---|---|
| Wages | $1120 | Rent | $425 |
| Casual wage | ______ | Groceries | $163 |
| | | Fares | $103 |
| | | Car loan | $188 |
| | | Petrol/car maintenance | $93 |
| | | Entertainment | $185 |
| | | Savings | ______ |
| Total: | $1305 | Total: | $______ |

a How much did Lucy earn from her casual job?

b What should be Lucy's total expenses?

c To balance her budget, how much can Lucy save per week?

d Name an expense that represents discretionary spending.

e Suggest a way Lucy could increase her savings to $170 per week.

☐ Foundation ○ Mastery ○ Complex

## Solution

**a** Casual wage = $1305 – $1120

= $185

**b** Total expenses = $1305 equal to total income

**c** Savings = $1305 – ($425 + $163 + $103 + $188 + $93 + $185)

= $148

**d** Sample responses: entertainment or possibly the car loan (in addition to fares listed). Perhaps they live near public transport and don't need a car.

**e** Suggestions: cut down on entertainment, rent with a friend, sell the car, or look for weekly grocery specials.

Some people set aside some of their savings and put it in an **emergency fund**, a savings account to cover unexpected expenses, emergencies or urgent situations that require immediate money. The fund acts as a financial safety net to help you avoid going into debt, borrowing money or using credit cards when something unexpected happens. As a guide, an emergency fund should be established with enough money to cover 3 to 6 months of essential living expenses, such as food, housing and bills and transportation. Building an emergency fund is an important part of budgeting because it helps you stay financially secure and confident during difficult times.

Examples of where an emergency fund might be used for unplanned expenses:

- **Medical or vet bills:** if you or your pet get sick or injured and need to pay for a doctor/vet, hospital treatment and/or medications.
- **Car repairs:** if your car breaks down or is involved in an accident.
- **Home repairs:** if something in your house, like a broken hot water system or a leaking roof, needs to be fixed.
- **Job loss**: if you or another family member loses their job and needs money to live on until they find new work.
- **Emergency travel:** unplanned travel, for example, for a funeral or to care for a family member.

Some of these expenses could be covered by insurance (another expense), but it's financially wise to deposit a small amount regularly into an emergency fund by automatic transfer.

**EXERCISE 6.07** Answers on p. 470

## Personal budgeting

**1** Categorise each item as either income or expense.

**a** a car insurance payment

**b** an inheritance from your grandmother

**c** a birthday present for a friend

**d** the cost of a plane ticket to Fiji

**e** a prize from a Lotto win

**f** the monthly rental of an office space

**g** a rent payment to a real estate agent

**h** the admission price to the cinema

**i** the rent from an investment unit

**j** pay to an electrician for installing new power points

**k** your weekly wage

**l** the cost of a gym class

**m** pocket money

**n** a Christmas present from your uncle

**o** the cost of a Friday night pizza

**2** Identify each expense item from Question **1** as either fixed spending or discretionary spending.

**3** Leah has monthly expenses of $1800. She wants to set up an emergency fund that will cover 6 months' worth of expenses. If she saves $300 per month, how many months will it take her to reach her goal? Select **A**, **B**, **C** or **D**.

**A** 14 **B** 18 **C** 24 **D** 36

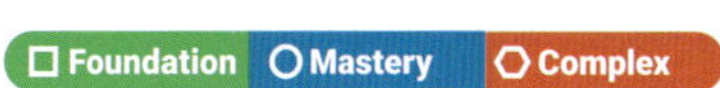

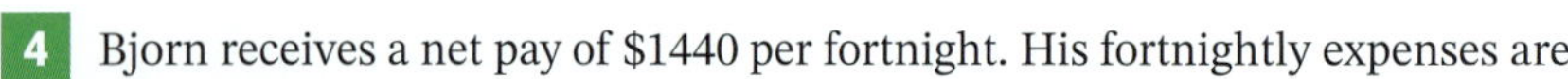

**4** Bjorn receives a net pay of $1440 per fortnight. His fortnightly expenses are:

Rent: $800 Transportation: $100

Groceries: $250 Insurance: $50

Utilities: $150

How much money can he save each fortnight? Select **A**, **B**, **C** or **D**.

**A** $140 **B** $90 **C** $40 **D** $240

EXAMPLE 15

**5** Mariella's budget allowed her to save $100 each week. Her budget is shown below.

| Income | | Expenses | |
|---|---|---|---|
| Youth allowance | $663.30 | Rent | $290 |
| Casual job | ________ | Fares | $76 |
| | | Groceries | $124 |
| | | Clothes | $85 |
| | | Entertainment | ________ |
| | | Bills | $88 |
| | | Savings | $100 |
| **Total:** | $921.00 | **Total:** | $ |

**a** How much did Mariella earn for her casual job?

**b** Is entertainment a fixed or discretionary item? What amount did Mariella spend on entertainment?

**c** Suggest 2 ways that Mariella could increase her weekly savings by $80.

**6** Louis has an emergency fund of $3000, but his monthly expenses have increased to $1500 due to a new car payment and higher rent. He wants his emergency fund to cover 4 months of expenses. How much more does Louis need to save to reach his new emergency fund goal?

**7** Rhian lives at home and is preparing her annual budget for 2028. Her expected income is:

$1050 each week in wages

Simple interest from her $5200 investment at 3.5% p.a.

Rhian has planned expenses of:

$55/week on public transport

$35/week on work lunches and snacks

$90/month on entertainment and going out

$60/month on phone and streaming services.

She uses this spreadsheet for her annual budget.

| | A | B | C | D |
|---|---|---|---|---|
| 1 | Rhian's Annual Budget for 2028 | | | |
| 2 | **Income** | | **Expenses** | |
| 3 | Wages | 54 600 | Transport | *B* |
| 4 | Interest | *A* | Work lunches | 1820 |
| 5 | | | Entertainment | 1080 |
| 6 | | | Phone/Streaming | 720 |
| 7 | | | Savings | *C* |



Each question below refers to the above spreadsheet. Select **A**, **B**, **C** or **D**.

**a** Find the amount of interest, *A*.

**A** $3.50 **B** $103.50 **C** $182 **D** $218.75

**b** Find the cost of transport, *B*.

**A** $2860 **B** $1430 **C** $1120 **D** 550

c Find the maximum possible savings for the year, *C*.

**A** $28 502 **B** $48 302 **C** $52 180 **D** $55 700

d Calculate Rhian's total expenses (excluding savings) for the year.

**A** $3802 **B** $4740 **C** $5050 **D** $6480

**8** Hollie created this spreadsheet to plan her annual car and bus expenses.

| | A | B | C | D | E |
|---|---|---|---|---|---|
| 1 | **Transport Budget - 1 year period** | | | | |
| 2 | | | | | |
| 3 | **Transport costs** | **Amount** | **Frequency** | **Payment** | **Total Cost** |
| 4 | Car Repayments | $230.50 | Monthly | 12 | $2,766.00 |
| 5 | Bus fares | $42.60 | Weekly | 52 | $2,215.20 |
| 6 | Car Insurance | $158.38 | Monthly | 12 | $1,900.56 |
| 7 | Car Registration | $497 | Annual | 1 | $497.00 |
| 8 | Fuel Costs | $85 | Weekly | 52 | $4,420.00 |
| 9 | Car Maintenance | *A* | Quarterly | 4 | $950.00 |
| 10 | **TOTAL** | | | | *B* |



a Determine the values in the cells labelled *A* and *B*.

b Hollie wants to put a weekly amount into an account so that over one year, she can cover all of her transport costs. Calculate this weekly amount if she allows for a 10% increase on all costs listed. Write your answer to the nearest 10 cents.

**9** Jess wants to save $4200 for an end of year trip. She earns $1312.90 net a week and has the following weekly expenses: meals $140, rent $350, groceries $180, fares $57, tennis $40, entertainment $80, car loan $185, bills $164 and magazines $20.

a Calculate the amount Jess has left for savings each week.

b Will Jess have enough saved to pay for her trip in one year?

c If the trip price increases by 20%, would Jess have enough saved in one year to pay the new price?

**10** The Abdelal family has the following net weekly incomes: wages $2089.77, part-time job wages $589.53 and parenting allowance $357.55. The family also has the following weekly expenses: bills $190.50, school fees $185.80, entertainment costs $180, health fund payment $154.96, clothes $165, home maintenance $200, groceries $340, petrol $130, Friday night takeaway $96, newspapers and magazines $46, home loan repayments $1012.84, and car loan repayment $195.

a Use a table or spreadsheet to display this data as a budget, adding an entry for savings.

b What amount does the Abdelal family save per week?

**11** Zac shares a flat and pays $320 a week towards the rent. He spends $158 on food and $96 on public transport. His share of the quarterly electricity bill is $260 and he has a monthly phone bill of $56. He earns $1084.90 net per week, spends about $95 per week on entertainment and visits a chiropractor once a month at a cost of $125 a visit. Convert Zac's expenses to weekly amounts and draw up a weekly budget, including an amount for savings.

a What is Zac's weekly income?

b What is the total of Zac's weekly expenses?

c How much can Zac save per week?

☐ Foundation ○ Mastery ○ Complex

## Technology

### A budget spreadsheet

Use a spreadsheet to model a personal budget for each of the 3 scenarios described.

**1** Every week, Rebecca spends about \$120 on groceries, \$40 on fruit and vegetables, \$35 on meat, \$50 on transport, \$20 for drum lessons, \$20 on streaming services, \$30 on clothing and \$60 on entertainment. She makes mortgage (home loan) payments of \$527 each month. In one year, she pays \$769.15 for electricity, \$890 for council rates, \$685.95 for water rates and \$540 for her phone bills.

- **a** How much does Rebecca spend on mortgage payments:
  - **i** each year?
  - **ii** each week?
- **b** How much does she spend each week on electricity?
- **c** If Rebecca earns \$783 a week, create a spreadsheet like the one below for her weekly budget.

| | A | B | C | D |
|---|---|---|---|---|
| 1 | Rebecca's Weekly Budget | | | |
| 2 | **Income** | **\$/week** | **Expenses** | **\$/week** |
| 3 | Wages | | Groceries | |
| 4 | Other | | Transport | |
| 5 | | | Drum Lessons | |
| 6 | | | Streaming services | |
| 7 | | | Clothing | |
| 8 | | | Entertainment | |
| 9 | | | Mortgage | |
| 10 | | | Electricity | |
| 11 | | | Council rates | |
| 12 | | | Water rates | |
| 13 | | | Phone | |
| 14 | | | Savings | |
| 15 | **Total** | | **Total** | |



**2** Brady earns \$484.50 a week at his regular job. He also earns \$120 a week working part-time at a café. Every week, he pays \$55 for car loan repayments, \$30 for leisure, \$80 for clothing and \$50 for other expenses. Brady wants to save \$20 per week towards a holiday and to also save some money generally. Draw up a budget for Brady in a spreadsheet.

**3** Create a weekly budget for Zac's expenses from the last question of the previous exercise, using the spreadsheet below, which converts all his expenses to weekly amounts first. Ensure that all cells that have money amounts are formatted in currency: highlight column, right click, **format cells** and choose **currency** with 2 decimal places and \$.

| | A | B | C | D |
|---|---|---|---|---|
| 1 | | 551.4 | Take home pay/ week | |
| 2 | | | | **Weekly** |
| 3 | | | **Expenses** | **Budget** |
| 4 | 110 | week | Rent | =A4 |
| 5 | 90 | week | Food | =A5 |
| 6 | 46 | week | Public Transport | =A6 |
| 7 | 60 | quarter | Electricity | =A7/12 |
| 8 | 55 | bimonthly | Phone costs/access to streaming services | =A8/2 |
| 9 | 70 | week | Entertainment/going out | =A9 |
| 10 | 42 | month | Chiropractor | =A10/4 |
| 11 | | | **Total expenses** | =SUM(D4:D10) |
| 12 | | | **Savings** | =B1-D11 |

## Investigation

### Preparing my budget

Imagine that you are 25 years old with a net monthly income of $5340.

| | Amount | Per | Weekly amount |
|---|---|---|---|
| **Income** | | | |
| | $5340.00 | month | |
| **Expenses** | | | |
| Rent | $185.00 | week | |
| Mobile phone | | month | |
| Gym membership | $584.00 | year | |
| Internet | $49.95 | month | |
| Water | $114.80 | quarter | |
| Electricity | $222.50 | quarter | |
| Health fund | $26.40 | week | |
| Groceries | | week | |
| Petrol/car maintenance | | week | |
| Fares | | week | |
| Clothes | | week | |
| Entertainment | | week | |
| Sport | | week | |
| Gifts/donations | | week | |
| Savings | | week | |
| Other | | week | |

1. Use a calculator or spreadsheet to copy this weekly budget and complete it after converting the fixed expenses to weekly amounts.
2. Choose the amount you wish to spend per week on each variable expense and enter these amounts.
3. Suggest 3 ways that you could save more per week.

The **MoneySmart** website (moneysmart.gov.au) has helpful advice about preparing budgets and creating savings goals and emergency funds.

## Sample HSC problem Answers on p. 470

(3 marks) This is Jordan's quarterly water bill for his 4-person family.

***Catchment Water Account*** **Quarterly Account**

| *Fixed Service Charges for period 1st June to 30th Aug* | | | |
|---|---|---|---|
| | **Number of Services** | **Charge** | **Amount** |
| Water Service - Residential | 1 | \$17.60 | \$17.60 |
| Wastewater (sewerage) service - Residential | 1 | \$155.89 | \$155.89 |
| | **Total Service Charges** | | **\$173.49** |

| *Usage Charges* | | | |
|---|---|---|---|
| | **Usage (kL)** | **Charge** | **Amount** |
| Water Usage (per kL) | 55 | \$2.6900 | \$147.95 |
| **Total Usage Charges** | | | **\$147.95** |

| | **Total Amount Due** | **\$321.44** |
|---|---|---|

Jordan wants to budget an amount from each weekly pay to cover his family's future water accounts. He allows for 65 kL of water usage per quarter in case there are extra usages or price rises.

How much should Jordan put aside each week?

## Study Tip

### Organising your notes

- One feature of effective study is to keep an organised collection of class notes. This allows you to quickly and easily locate your work on any topic. The key to maintaining a neat set of notes is to label everything: the names of the topics, the theory, the class examples, the exercises, your folders.
- Your maths class follows a program of topics, which may correspond roughly to the chapters of this textbook. Ask your teacher for a copy of the program or syllabus and organise your folder according to these topics. Use coloured dividers or title pages to mark the start of each new topic.
- Start each maths lesson with the title of the lesson and the date. Pay attention to your teacher's explanations and copy the class examples into your book. Make sure you write the question as well as each step in the solution. You can add any personal notes, comments or reminders to help you later, when revising the work. If you don't understand something, ask your teacher to explain again.
- If using a loose-leaf folder, make sure your notes are filed in the correct order. Number the pages of each topic so that you can refile if they get out of order. If your folder becomes too bulky to manage, start a new one or remove some of the older topics and file them at home. Realise that you do not need to bring to school all your notes for the entire maths course, just what you are currently working on in class.

# CHAPTER SUMMARY

This chapter, *Managing money*, covered the mathematics of personal finance, from profit, loss and GST to term payment plans, household bills and personal budgeting. Make sure you develop a deep understanding of the terminology and calculations involved in spending money, budgeting and managing household expenses.

Before you move on, consider what you learned in this chapter and revisit any sections that may have been unclear.

| What you learned in this chapter... | Section |
|---|---|
| apply percentage increase and decrease to problems involving markups, discounts, profit and loss | 6.01 Markups, discounts, profit and loss |
| calculate goods and services tax (GST) in Australia and value-added tax (VAT) internationally | 6.02 GST and VAT |
| examine and compare 'buy now, pay later' credit plans and term payment plans, and the fees and financial calculations associated with both | 6.03 Buy now, pay later<br>6.04 Term payments |
| examine and make calculations with household water, electricity and gas bills | 6.05 Water bills<br>6.06 Power bills |
| create a personal budget, taking into account fixed and discretionary spending | 6.07 Personal budgeting |

To help master this topic, make a summary mind map. Use the chapter outline and the mind map below as a guide. Add your own words, symbols, diagrams, boxes and reminders. The summary should give you a 'whole picture' view of the topic and allow you to identify any weak areas to revisit in your revision.

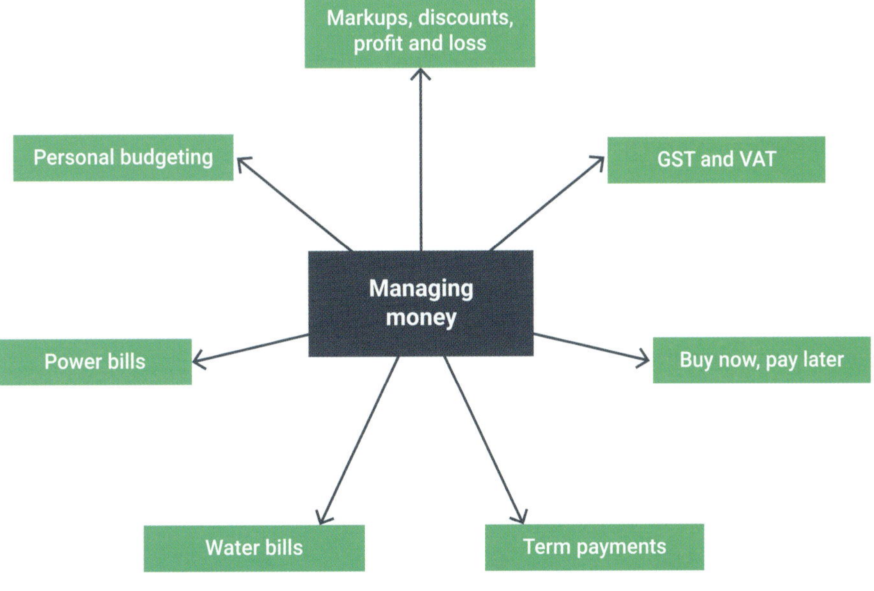

# 6 Test yourself Answers on p. 470

6.01 **1** A café adds an 11% surcharge on Sundays. Calculate the total to be paid if the bill for lunch for 2 people on a Sunday was $92.40.

6.01 **2** The population of a small town decreased from 12 500 to 10 200 over 3 years due to people moving away for employment opportunities. What was the percentage decrease in the town's population? Select **A**, **B**, **C** or **D**.

**A** 18% **B** 15% **C** 12% **D** 20%

6.01 **3** A farmer buys 500 kg of apples for $2.50 per kilogram. He wants to make a 30% profit on the total cost price. However, due to a sudden oversupply of apples in the market, he sells half of the apples at a 20% discount of the original selling price. What is the percentage profit or loss on the cost price that the farmer makes on all of the apples?

6.02 **4** The cost of an air fryer is $169, including GST. What was the cost before GST was added? Select **A**, **B**, **C** or **D**.

**A** $152.10 **B** $153.64 **C** $159.00 **D** $185.90

6.02 **5** **a** A book originally costs $25. Calculate its final price after VAT of 15% is added.

**b** A set of kitchen appliances has a total price of $550, including VAT. If the VAT rate is 8%, how much is the VAT amount?

6.03 **6** Jack buys a TV priced at $1200 by lay-by. He pays a deposit of $160 and makes 4 fortnightly payments of $270 each. The store charges a lay-by fee of $40.

**a** How much does Jack still owe after making the deposit?

**b** What is the total amount Jack pays for the TV?

6.03 **7** Elsa buys a gaming console for $500 using a 'buy now, pay later' no-interest plan that charges a $30 processing fee and a $15 monthly maintenance fee. The repayment term is 10 months.

**a** How much will Elsa pay in total for the gaming console?

**b** What will Elsa's monthly repayment be?

6.03 **8** Grant buys a bicycle for $750 using a payment plan with the following conditions:

term of 12 months

$30 processing fee

$10 insurance fee

$15 monthly maintenance fee

**a** How much will Grant pay in total for the bicycle?

**b** What will his monthly repayment be?

6.04 **9** Charissa buys a laptop valued at $1800 and is charged simple interest at 10% p.a. with the loan over 15 months. How much will she owe in interest after 15 months? Select **A**, **B**, **C** or **D**.

**A** $145 **B** $180 **C** $225 **D** $270

6.04 **10** Brad wants to buy a refrigerator priced at $1200. He makes a deposit of $250, and the balance is to be repaid by monthly payments at 8% p.a. simple interest. The store charges a $50 processing fee which must be paid when the deposit is made. Brad repays the loan over 24 months.

**a** What is the balance owing after Brad makes the deposit?

**b** How much interest does he pay on the loan?

**c** Calculate the total amount Brad will repay.

**d** Determine his monthly repayment.

6.05 **11** The Hamper family is calculating their water usage for the quarter. This reading is 9845 kL and the previous reading was 9423 kL. If water usage is charged at $2.2760/kL, calculate the cost of their water usage for the quarter.

☐ Foundation ○ Mastery ○ Complex

12 Jed's water bill shows that his water usage is charged at \$2.2760/kL. He used 34.75 kL and was also charged \$18.35 for the water service and \$174.68 for the wastewater service. What was the total of Jed's bill? 6.05

13 The average daily water usage for 2 families, for a 90-day quarter is: 6.05

The Noh family (4 people), 560 L/day

The Waddock family (6 people), 710 L/day.

**a** Calculate the water usage costs for each family for the quarter if water costs \$2.2760 per kL and state which family pays more.

**b** Calculate the average daily water use per person for each family and average quarterly cost per person and state which family uses more water per person.

14 Calculate the cost of electricity usage for a household where the supply charge is 87.45 c/day for a 92-day quarter and the readings were as shown below. 6.06

| Usage type | Usage (kWh) | Cost (c/kWh) |
|---|---|---|
| Peak | 1188.2 | 48.24 |
| Off-peak | 494.9 | 20.15 |
| Off-peak shoulder (weekends) | 230.5 | 29.87 |

15 The gas bill charges for the April to June quarter one year, are shown below. 6.06

| | | |
|---|---|---|
| **USAGE** | First 7560 MJ | \$0.044 65 |
| | Next 7440 MJ | \$0.033 18 |
| | Next 18 000 MJ | \$0.032 18 |
| **CHARGE** | Supply charge per day | \$0.712 20 |

How much is the gas bill if 19 765 MJ was the usage for this quarter, and credit card simple interest of 0.65% p.a. is also charged on the total amount (including the supply charge)?

16 Chloe earns \$1654 per fortnight as a graphic designer, as well as collecting \$1270 monthly rent from an apartment she owns. 6.07

**a** Calculate Chloe's total *weekly* income.

**b** Prepare a balanced weekly budget for Chloe using the items below.

| | | |
|---|---|---|
| Groceries \$142 | Home loan repayment \$235 | Health fund \$28.20 |
| Household bills \$43 | Coffee \$18 | Take-away meals \$32 |
| Petrol \$70 | Phone plan \$56.70 | Entertainment \$? |
| Clothes \$? | Savings \$? | |

**c** Name one way Chloe could reduce her discretionary spending and save \$40 more per week.

17 Darius wants to create a budget and save for an emergency fund. His monthly expenses are: 6.07

Rent: \$1200 Utilities (Electricity and Water): \$200

Groceries: \$400 Transportation: \$150

Insurance: \$100

Darius wants to build an emergency fund that will cover 5 months of his expenses.

**a** How much does he need to save for his emergency fund?

**b** Darius saves \$500 per month. How many months will it take him to reach his goal?

18 Emily wants to save \$6000 for a trip to South America next year. She has already saved \$1500. How much more does Emily need to save each month to reach her goal in 12 months? 6.07

☐ Foundation ◯ Mastery ◯ Complex

# PRACTICE EXAM 2

Answers on p. 470

**Recommended time: One hour**

**Section 1**
**10 multiple-choice questions: 1 mark each**
**Select the correct answer A, B, C or D.**

**1** How many millimetres in 9.2 km?

**A** $9.2 \times 10^7$ **B** $9.2 \times 10^6$ **C** $9.2 \times 10^5$ **D** $9.2 \times 10^4$

**2** Find the value of $x$ in the diagram.

**A** 5
**B** 10
**C** 12
**D** 13

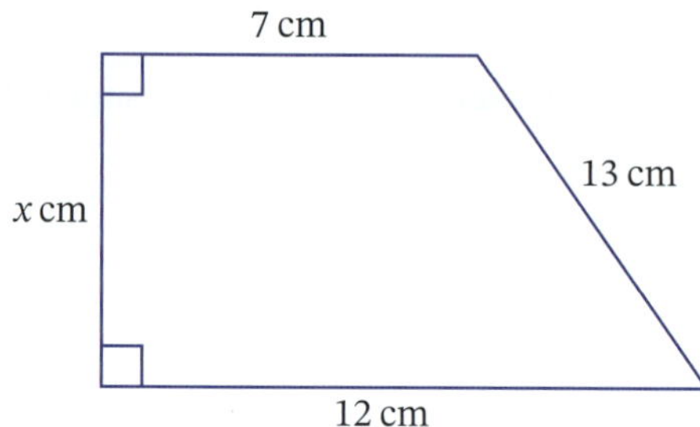


5.10

**3** The volume of this cone is 150 cm³. Its perpendicular height is 9 cm.
Find the radius of the cone's base, correct to one decimal place.

**A** 5.3 cm
**B** 4.6 cm
**C** 4.0 cm
**D** 2.3 cm

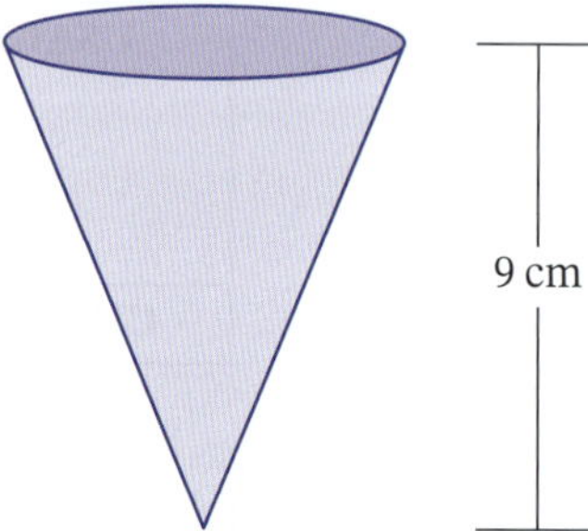


6.04

**4** Manon wants to buy a lounge suite package for \$2948. She makes a deposit of \$300, and the balance is paid off in monthly repayments over 2 years. The balance is charged at a simple interest rate charged at 12% p.a. Calculate Manon's monthly repayment.

**A** \$122.83 **B** \$136.81 **C** \$142.65 **D** \$151.57

6.02

**5** A portable speaker has a price of \$137.50 after a GST of 10% is added. What was the speaker's price before GST?

**A** \$123.75 **B** \$125.15 **C** \$127.50 **D** \$125.00

6.01

**6** Ben buys a slow cooker marked at \$259 for a discounted price of \$199.
Danil buys a laptop marked at \$449 and gets a discount of \$90 off this price.
Which statement is correct about their percentage discounts?

**A** Ben's percentage discount is higher than Danil's.
**B** Ben's percentage discount is lower than Danil's.
**C** Ben's percentage discount is the same as Danil's.
**D** There is not enough information provided to calculate Ben and Danil's percentage discounts.

**7** Felipe's gas bill has usage charges as shown below. 6.06

| Usage Charge Per MJ | First 7560 MJ | \$0.0475 |
|---|---|---|
| | Additional usage over 7560 MJ | \$0.0353 |

What is Felipe's gas bill if the supply charge is \$0.7163 per day, he used 10 098 MJ and the bill is for the October, November and December quarter?

**A** \$806.93 **B** \$602.24 **C** \$514.59 **D** \$513.87

**8** Jameil's cookbook sells for \$29.00 per copy. He earns a royalty of 12% of the selling price of the book. How much money will Jameil receive if 7200 books are sold in total? 4.02

**A** \$25 056 **B** \$11 232 **C** \$13 824 **D** \$38 880

**9** Evaluate $(4.2 \times 10^5) \times (3.5 \times 10^{-2})$, correct to 2 significant figures. 5.03

**A** $1.47 \times 10^4$ **B** $1.4 \times 10^3$ **C** $1.5 \times 10^3$ **D** $1.5 \times 10^4$

**10** Vanita's car service bill consists of: 6.02

\$429 for parts, including 10% GST

\$250 for the mechanic's labour, excluding 10% GST

If the mechanic needs to add on GST to the labour cost, how much GST will the customer pay in total for the car service?

**A** \$67.90 **B** \$64.00 **C** \$42.90 **D** \$39.00

## Section 2
**3 questions: 10 marks each**

### Question 11 (10 marks)

**a** Soojung's water bill shows that water usage is charged at \$2.276/kL. 6.05

**i** On her latest bill, she paid \$40.75 in water usage charges. How many kilolitres of water did she use? Answer correct to one decimal place. (1 mark)

**ii** Soojung was also charged \$26.16 for the water service, \$154.62 for the wastewater service and a recycled water charge of \$2.62. She also used 21 kL of recycled water at \$2.62/kL. What was the total of Soojung's water bill? (2 marks)

**b** A dam has 3 vertical measurements taken at 24 m intervals. 5.07

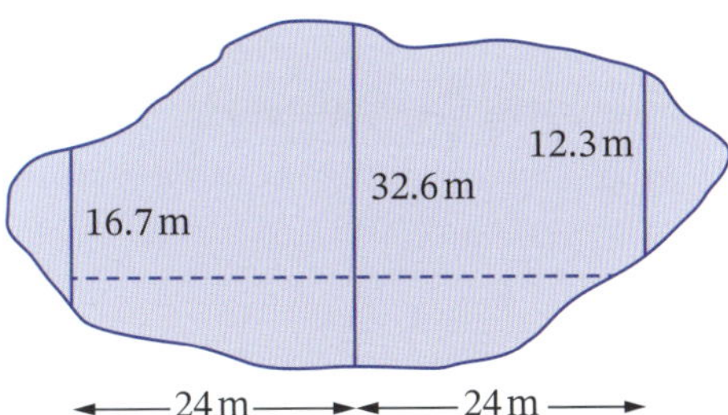


Use 2 applications of the trapezoidal rule to approximate the area of the lake, correct to 3 significant figures. (3 marks)

$$\left[A \approx \frac{h}{2}(d_f + d_l)\right]$$

4.01 **c** The table below shows the weekly pay of 4 employees at Wonka's chocolate factory. Overtime is paid at time-and-a-half. Calculate the missing values of **i**, **ii**, **iii** and **iv** in the table. 4 marks

| Name | Pay rate (per hour) | Normal hours | Overtime hours | Weekly pay |
|---|---|---|---|---|
| Ilaria | $25.00 | 38 | 2 | **i** |
| Kai | $33.50 | 12 | **ii** | $452.25 |
| Lauren | **iii** | 38 | 4 | $1204.72 |
| Harry | $31.00 | **iv** | 3 | $976.50 |

## Question 12 10 marks

4.06 **a** In the last financial year, Jannik was paid a total income of $67 200 from his main job working in retail. He also earned $12 500 from sports coaching twice a week and umpiring matches on weekends. Jannik has deductions totalling $725 for work-related items.

**i** Calculate Jannik's taxable income. 1 mark

**ii** Use the tax table below to calculate how much tax he should pay. 2 marks

**iii** Calculate the 2% Medicare levy Jannik will have to pay. 1 mark

| Taxable income | Tax on this income |
|---|---|
| 0 – $18 200 | Nil |
| $18 201 – $45 000 | 16c for each $1 over $18 200 |
| $45 001 – $135 000 | $4288 plus 30c for each $1 over $45 000 |
| $135 001 – $190 000 | $31 288 plus 37c for each $1 over $135 000 |
| $190 001 and over | $51 638 plus 45c for each $1 over $190 000 |

Source: Tax rates – Australian resident, Australian Taxation Office, 2025

5.12 **b** Rainwater is collected in a water tank from the roof of a block of classrooms at a school. The area of the roof is 300 m$^2$. During a day of heavy rainfall, 15 mm of rain falls on the roof and is collected in the water tank. How many litres of water were collected? 2 marks

5.10 **c** The planet Venus has a diameter of 12 104 km. Assuming that it is a sphere, calculate in scientific notation, correct to 4 significant figures:

**i** its volume 2 marks

**ii** its surface area in hectares. 2 marks

$$\left[V = \frac{4}{3}\pi r^3, A = 4\pi r^2\right]$$

## Question 13 10 marks

**a** Leo's most recent quarterly water bill for his 4-person household is shown. 6.05

| *North West Water Account* | Quarterly Account | | |
|---|---|---|---|
| Fixed Service Charges for period 1st October to 30th November | | | |
| | Number of Services | Charge | Amount |
| Water Service - Residential | 1 | $17.60 | $17.60 |
| Wastewater (sewerage) service - Residential | 1 | $155.89 | $155.89 |
| **Total Service Charges** | | | **$173.49** |
| Usage Charges – Water Meter readings - see over the page | | | |
| | Usage (kL) | Charge | Amount |
| Water Usage (per kL) | 57 | $2.6900 | $153.33 |
| **Total Usage Charges** | | | **$153.33** |
| | Total Amount Due | | **$326.82** |

Leo wants to lower the total cost of his water bill each quarter by 10%. What would his new water usage, per kL, need to be, if fixed service charges remain unchanged? Answer to the nearest kL. 3 marks

**b** At the animal park, newspaper is cut in the shape of a sector and rolled to hold animal food. For one piece of newspaper, find:

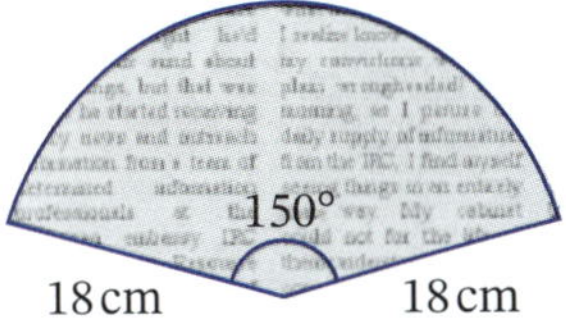


**i** its area, correct to the nearest square centimetre 1 mark 5.05

**ii** its perimeter, correct to the nearest centimetre. 2 marks 5.06

$$\left[\text{arc length} = \frac{\theta}{360} \times 2\pi r, \text{ area of a sector} = \frac{\theta}{360} \times \pi r^2\right]$$

**c** Bree earns $925 per week. Calculate her holiday loading if it is 17.5% of 4 weeks' pay. 2 marks 4.03

**d** Jarryd earns 2% commission from each property he sells. How much would he earn in a month in which he sells 3 units for $810 000, $675 000 and $408 600? 2 marks 4.02

**END OF EXAMINATION**

# OWNING A CAR

When buying a car, as well as the initial purchase price (or loan repayments), you need to consider other on-road costs such as insurance, registration and stamp duty. Running costs of a car include fuel consumption, servicing, repairs and tyres.

## Chapter outline

## In this chapter you will:

- identify the different types of car insurance available, including compulsory third-party (CTP)
- calculate on-road costs for buying a car, including finance, stamp duty, registration and insurance
- calculate the running costs of owning a car, including service and repairs
- identify fuel consumption and prices as rates and use them to calculate the amount and cost of fuel used on a trip
- compare the costs of owning different types of car.

**Video (1):**
**7.04** Fuel consumption

**Worksheets (4):**
**SkillCheck** Assignment 7
**7.01** Investigating car insurance
**7.02** Stamp duty
**7.04** How much petrol?

Nelson MindTap

To access resources above, visit **cengage.com.au/nelsonmindtap**

## Terminology

| | | |
|---|---|---|
| comprehensive insurance | Compulsory Third Party (CTP) insurance | excess |
| fuel consumption | green slip | on-road costs |
| premium | stamp duty | Third Party Property insurance |
| vehicle registration | | |

**Worksheet**
Assignment 7

## SkillCheck Answers on p. 470

**1** Copy and complete:

**a** \$148 per month = \$________ per year

**b** \$350 per fortnight = \$________ per year

**c** \$276 per week = \$________ per year

**d** \$2754 per year = \$________ per month

**e** \$182 per month = \$________ per week [Hint: find amount per year first]

**2** A loan repayment is \$128 per fortnight. To the nearest cent, how much is this:

**a** per week? **b** per year? **c** per month?

**3** Find these percentages, correct to the nearest cent:

**a** 3% of \$25 680 **b** 2.5% of \$64 320 **c** 11% of \$12 700

**d** 4% of \$2564 **e** 6.5% of \$68 530 **f** 1.9% of \$72 350

**4** If a car is stolen in Australia every 10 minutes, how many cars are stolen:

**a** per day? **b** per week? **c** per month? **d** per year?

**5** A Range Rover SUV has a 90 L fuel tank. What is the cost to fill the car with fuel at \$2.19 per litre?

**6** If the buyer of a \$65 000 car received a 10% discount for paying cash and a further 3% discount in the sale, how much did he pay for the car?

# Car insurance

7.01

Worksheet
Investigating car insurance

Car **insurance** is a financial scheme to protect you in the event of a car accident. In return for paying a regular fee, called the **premium**, the insurance company agrees to pay you a large amount to cover the costs of repair or replacement if you have an accident.

## Types of car insurance

There are 3 main types of car insurance:

- **Compulsory Third Party (CTP) insurance** protects vehicle owners and drivers from legal liability for personal injury or death to any other person ('third party') including drivers, passengers, cyclists and pedestrians. It is also called '**green slip**' insurance because of the colour of the insurance document and it must be obtained before a vehicle can be registered.
- **Third Party Property insurance** is optional and covers damage to other vehicles and property in an accident where you are at fault. It does not cover damage to your vehicle.
- **Comprehensive insurance** is optional and covers damage to all vehicles and property, including your own. It also protects against theft, attempted theft, accidental loss, fire, vandalism and hail damage to your vehicle.

### Example 1

Mia bought a second-hand utility vehicle (ute) for her farm. Her comprehensive insurance was \$843.06 annually upfront or \$71.78 per month. How much extra would Mia pay per year if she chose to pay her insurance per month?

#### Solution

Cost when paid upfront $= \$843.06$

Cost when paid monthly $= \$71.78 \times 12$

$= \$861.36$

Difference $= \$861.36 - \$843.06$

$= \$18.30$

Mia pays an extra \$18.30 per year if she pays per month.

iStock.com/jodie777

9780170497886

## Insurance premiums and discounts

A motor vehicle insurance premium depends on factors such as vehicle type, driver age and driving record, where and how your vehicle is stored (garage, driveway, street parking) and whether the vehicle is financed or fully paid off. Some insurance companies give a **loyalty discount** if you have been with them for several years, a **multi-policy discount** for having more than one type of insurance policy with them, and a **no claim discount** if you have not made a claim for several years.

If you have an accident and it is your fault, an assessor will view the damage and assess the costs involved. You will be asked to complete a **claim form** and pay an **excess**, that is, some money upfront before the company pays the rest. The excess is stated in your policy document and some customers agree to pay a larger excess in return for a cheaper policy.

### Example 2

This table shows 5 examples of comprehensive car insurance premiums.

| Driver gender | Driver age | Town where vehicle is garaged | Car make | Car year | Basic premium |
|---|---|---|---|---|---|
| male | 55 | Goulburn | Ford | 2014 | \$1016 |
| female | 33 | Wollongong | Mazda | 2023 | \$1438 |
| female | 24 | Dubbo | Mitsubishi | 2019 | \$2174 |
| male | 21 | Newcastle | Holden | 2018 | \$2435 |
| female | 45 | Bathurst | Hyundai | 2024 | \$926 |

Calculate the premium for:

**a** the female from Bathurst who gets a 10% loyalty discount

**b** the 55-year-old Ford driver who pays a 5% surcharge for an older car

**c** the owner of the Mitsubishi who pays a 13% levy as the car is financed

**d** the Mazda driver from Wollongong who gets a discount of 2.5% for a good driving record, as well as a further discount of 1.7% for loyalty.

### Solution

**a** Premium = \$926

Discount = 10%

Discounted premium = 90% × \$926

= \$833.40

**b** Premium = \$1016

Surcharge = 5%

Actual premium = 105% × \$1016

= \$1066.80

**c** Basic premium = \$2174

Levy = 13%

Actual premium = 113% × \$2174

= \$2456.62

**d** Basic premium = \$1438

Discount = 2.5% followed by 1.7%

Discounted premium = \$97.5% × 98.3% × \$1438 [100 − 2.5 = 97.5, 100 − 1.7 = 98.3]

= \$1378.21515

≈ \$1378.22

**EXERCISE 7.01** Answers on p. 470

## Car insurance

**1** Third Party Property insurance costs $38.17 per month. What is the cost per week? Select **A**, **B**, **C** or **D**.

**A** $8.80 **B** $9.54 **C** $8.81 **D** $7.63

**2** Mia paid her comprehensive car insurance at $150 per month. How much would she save if she paid a lump sum of $1265? Select **A**, **B**, **C** or **D**.

**A** $685 **B** $1115 **C** $0 **D** $535

**3** Jack bought a new van for $28 350. His CTP insurance is $626 per year and his comprehensive insurance is $75.78 per month. What is the total cost of insurance for the van per year?

**4** Which insurance is the cheapest option? Select **A**, **B** or **C**.

**A** $985 per year paid in a lump sum

**B** $83.10 per month

**C** $38.25 per fortnight

**5** The following table gives the Third Party Property insurance premiums for the RIGHT4U insurance company.

| Car make | Driver gender | Driver age | Suburb where vehicle is garaged | Place where car is usually kept | Third Party Property premium |
|---|---|---|---|---|---|
| Suzuki | male | 19 | Campbelltown | on street | $345 |
| Honda | female | 56 | Campbelltown | carport | $198 |
| Mitsubishi | male | 42 | Parramatta | driveway | $297 |
| Ford | female | 23 | Parramatta | locked garage | $243 |
| Toyota | male | 83 | Bondi | on street | $312 |
| Nissan | female | 65 | Bondi | security compound | $302 |

What is the Third Party Property insurance premium, to the nearest cent, for:

**a** the female from Campbelltown who gets a 20% no claim discount?

**b** the driver who has the car in a locked garage and gets a 3.5% discount?

**c** the driver under 20 years of age who must pay an 11.6% surcharge as he lives in a high-theft area?

**d** the Toyota driver from Bondi who pays a 24% age levy but also gets a 4.2% loyalty discount?

**6** Fahaad paid $1235.20 for a new comprehensive insurance policy after receiving a 15% discount for buying online. What discount did Fahaad receive?

**7** The base premium for the green slip on Emily's car is $379.67 (GST excluded) and the MCIS Levy (a tax to fund Medical Care and Injury Services) is 24.42% of the base premium. Find the total cost of the green slip, including the MCIS levy and the 10% GST on the base premium.

**8** Colin bought a motor scooter. His annual CTP green slip was $238 and his Third Party Property insurance was $19.35 per month. He also agreed to pay the first 8.5% of a claim if he had an accident that was his fault.

**a** What was his yearly insurance cost?

**b** He had an accident that was his fault and the damage bill to the other car was $7320. How much did he contribute to the claim?

**c** In the same accident, his scooter suffered $1360 worth of damage. How much did Colin pay towards this claim?

▷

□ Foundation ○ Mastery ○ Complex

**9** The NO FRILLS insurance company advertises a fixed insurance premium of \$500 for a car garaged in a metropolitan area and \$550 otherwise. The premium is adjusted for age and gender of the driver as shown in the table.

| NO FRILLS insurance | | |
|---|---|---|
| **Driver gender** | **Driver age** | **% adjustment** |
| male | <29 | +14.6% |
| female | <29 | +13.5% |
| male | 30–69 | −4.6% |
| female | 30–69 | −6.3% |
| male | >70 | +17.4% |
| female | >70 | +9.8% |

Calculate the insurance premium for each driver.

| | Driver gender | Driver age | Vehicle garaged |
|---|---|---|---|
| **a** | male | 24 | metropolitan area |
| **b** | female | 76 | country town |
| **c** | male | 32 | on a farm |
| **d** | female | 47 | metropolitan area |
| **e** | male | 28 | metropolitan area |
| **f** | female | 18 | country town |

**10** Ellie bought a small second-hand car for \$12 350. Her Compulsory Third Party insurance was \$286 and she took out comprehensive insurance at \$53.95 per month.

**a** What was her yearly insurance cost?

**b** She ran into the back of a car that had stopped at the traffic lights and did \$6320 damage to the car, as well as \$4280 to her own car. If Ellie had to pay an excess of 8.5% of the repair costs, how much did the insurance company pay?

**c** Ellie's comprehensive insurance premium will increase by 13.5% next year. How much will she pay annually, to the nearest cent?

## Did you know?

### Car theft

- About 59% of vehicles stolen in Australia are stolen from outside the home of the owner or a friend of the owner, so park your car in a garage or in a secure parking area.
- Most vehicles are stolen on a weekday at night-time.
- Most cars are stolen for short-term use (such as joyriding, transport) or for resale (either as a whole vehicle or as stripped parts).
- A car is stolen in Australia every 10 minutes.
- About $\frac{2}{3}$ of stolen vehicles in NSW are recovered.

**Cheaper cars are more likely to be stolen, and account for about 40% of all short-term vehicle thefts. Give a possible reason for this happening.**

☐ Foundation ○ Mastery ⬡ Complex

# On-road costs

7.02

Worksheet
Stamp duty

**Stamp duty** is a tax paid to the Office of State Revenue when you buy a new or used vehicle.

It is calculated on the market value of the vehicle or the price actually paid, whichever is the greater. Stamp duty is usually included in the on-road price of a new vehicle.

In NSW, the rates for stamp duty on a motor vehicle are shown in the table below.

| Vehicle value | Rate |
|---|---|
| \$0–\$44 999 | \$3 for every \$100 (or part of \$100) |
| \$45 000 and over | \$1350 plus \$5 for every \$100 (or part of \$100) over \$45 000 |

The price of a motor vehicle needs to be rounded up to the nearest \$100 when calculating stamp duty.

## Example 3

Lexie purchased a vehicle for \$56 240. The stamp duty is \$3 for every \$100 (or part of \$100) of the purchase price up to \$44 999, and \$1350 plus \$5 for every \$100 (or part of \$100) for \$45 000 and over.

How much stamp duty did Lexie pay?

### Solution

Purchase price $= \$56\,240$

$\approx \$56\,300$     for stamp duty, round purchase price up to the nearest \$100

Stamp duty $= \$1350 + 5\% \times (\$56\,300 - \$45\,000)$

$= \$1915$

Lexie paid \$1915 stamp duty.

These are the costs of purchasing a car:

- purchase price
- stamp duty
- **vehicle registration**: yearly process of paying and obtaining permission for using a vehicle on public roads
- CTP (Compulsory Third Party) insurance – a 'green slip'
- Third Party Property or comprehensive insurance (optional but advisable)
- loan interest and fees.

The cost of registration and CTP insurance is often included in the purchase price of a car from a dealership, and stamp duty if the car is new. These are called **on-road costs**.

iStock.com/sturti

### Example 4

Kayne purchased a van for $40 850 to use in his landscaping business. He set up a table of costs associated with his purchase.

| | |
|---|---|
| Purchase price | $40 850 |
| GST | 10% of purchase price |
| Stamp duty | $3 for every $100 (or part of $100) of GST price |
| Registration (per year) | $764 |
| CTP insurance (green slip) | $412 |
| Comprehensive insurance (per year) | $1425 |

What was the total purchase cost of this vehicle?

### Solution

| | | Cost |
|---|---|---|
| GST price | 110% × $40 850 | $44 935 |
| Stamp duty on GST price | 3% × $45 000 (rounding $44 935 up to the nearest $100) | $1350 |

Total cost = $44 935 + $1350 + $764 + $412 + $1425
= $48 886

## Technology

### Stamp duty online calculator

Search for 'vehicle stamp duty calculator' on the internet to find one that calculates the duty for different states of Australia, such as paycal.com.au.

The stamp duty in NSW for a vehicle with a purchase price of $40 000 is $1200.

**1** For a 4-cylinder car with a purchase price of $40 000, find:

- **a** which state has the highest stamp duty
- **b** which state has the lowest fee to transfer the car registration to a new owner
- **c** the total payable in Victoria
- **d** the range of stamp duties
- **e** the average stamp duty for the 6 states and the 2 territories.

**2** Find the current stamp duty payable on:

- **a** a car worth $22 000 in NSW
- **b** a van with a value of $32 500 in Tasmania
- **c** a hybrid car in Queensland valued at $31 480
- **d** an SUV in Western Australia worth $102 450.

## Technology

### Purchasing your first car

Research the following questions using an internet search and visiting websites such as **NSW Government and Services NSW**.

1 What things should be on your vehicle checklist before you buy a car?

2 What should you consider before buying an electric vehicle?

3 What are the payment options when buying a car?

4 What should you consider before buying a car by private sale?

Investigate other websites that may assist you in the purchase of a vehicle.

**EXERCISE 7.02** Answers on p. 471

## On-road costs

1 Stamp duty is \$3 for every \$100 (or part of \$100) of the purchase price of a small car. What is the stamp duty on a small Hyundai with a purchase price of \$22 120? Select **A**, **B**, **C** or **D**.

**A** \$660 **B** \$664 **C** \$662 **D** \$666

2 Stamp duty on a vehicle is \$3 for every \$100 (or part of \$100) for the first \$45 000 and \$5 for every \$100 (or part of \$100) on the remainder. What is the stamp duty on a \$55 000 ute? Select **A**, **B**, **C** or **D**.

**A** \$1350 **B** \$1650 **C** \$1850 **D** \$2750

3 Vehicle stamp duty is 3% of the purchase price up to \$30 000 plus 11% for every dollar between \$30 000 and \$40 000 plus 4% for every dollar over \$40 000. EXAMPLE 3

Find the stamp duty on a vehicle costing:

**a** \$30 000 **b** \$45 000 **c** \$60 000

4 If stamp duty on a new vehicle is 2.75% of the purchase price up to \$20 000 plus 5.75% for every dollar between \$20 000 and \$45 000 plus 6.5% for every dollar over \$45 000, find the stamp duty will you pay on a car worth:

**a** \$20 000 **b** \$45 000 **c** \$78 500

5 In one state of Australia, there is no stamp duty on a car valued at less than \$6000. The stamp duty on a car valued at \$6000 or more attracts a stamp duty charge of \$10 for every \$200 or part thereof of its value. To the nearest dollar, what is the stamp duty on a car valued at:

**a** \$6350? **b** \$68 980? **c** \$146 000?

6 If stamp duty on a vehicle is \$60 for the first \$3000 of the purchase price plus 4% for every dollar over \$3000, calculate the stamp duty payable on a vehicle with a purchase price of:

**a** \$8000 **b** \$12 500 **c** \$69 750

7 Angela paid cash for a second-hand motor scooter with these on-road costs.

| Purchase price (including GST) | \$15 000 |
|---|---|
| Stamp duty | 2% of purchase price |
| Transfer of registration | \$200 |
| CTP insurance (green slip) | \$150 |
| Third party insurance | \$240 |

What was the total cost of purchasing this scooter? Select the correct answer **A**, **B**, **C** or **D**.

**A** \$15 590 **B** \$20 090 **C** \$15 620 **D** \$15 890

Foundation Mastery Complex

**8** Naazmi bought a new SUV with on-road costs as shown.

| Purchase price | $125 000 |
|---|---|
| GST | 10% |
| Stamp duty | 5% of GST price |
| Registration | $860 |
| Green slip | $620 |

Calculate the total amount paid. Select **A**, **B**, **C** or **D**.

**A** $145 230 **B** $145 855 **C** $138 530 **D** $126 480

EXAMPLE 4

**9** Libby bought a sports car for $59 600 plus costs as shown.

| Purchase price | $59 600 |
|---|---|
| GST | 10% of purchase price |
| Stamp duty on GST price | $1350 plus $5 for every $100 (or part of $100) over $45 000 |
| Registration | $615 |
| CTP insurance (green slip) | $419 |
| Comprehensive insurance | $1625 |

What was the total purchase cost of her sports car?

**10** Demir bought his second-hand station wagon with the following upfront costs.

| Purchase price | $31 280 |
|---|---|
| Stamp duty | 1.5% of purchase price |
| Transfer of registration | $39 |
| CTP Insurance (green slip) | $427 |
| Third Party Property insurance | $317 |

How much did he pay for this vehicle?

**11** A car dealer was offering purchasers a 15% discount on the marked price of all vehicles. The costs of GST, stamp duty, registration and CTP insurance are then added to the discounted price to give the total cost.

| Type of car | Registration | CTP insurance (compulsory) | Third Party Property insurance (optional) | Comprehensive insurance (optional) |
|---|---|---|---|---|
| Kia | $325 | $614 | $250 | $1268 |
| Ford | $318 | $598 | $264 | $1095 |
| Toyota | $336 | $437 | $206 | $999 |
| Nissan | $384 | $503 | $295 | $1134 |
| Honda | $396 | $602 | $301 | $1003 |
| Subaru | $401 | $599 | $278 | $1236 |

GST: 10% of discounted price

Stamp duty: $3 for every $100 (or part of $100) up to $45 000 or

$1350 plus $5 for every $100 (or part of $100) over $45 000

Dealer special: No stamp duty on GST prices of $6500 or less.

 Foundation  Mastery  Complex

Use the table to find the total cost of:

- **a** Kia with a marked price of $31 468 and Third Party Property insurance
- **b** Ford with a marked price of $6500 and no optional insurance
- **c** Toyota with a marked price of $65 489 and comprehensive insurance
- **d** Nissan with a marked price of $27 650 and no optional insurance
- **e** Honda with a marked price of $18 639 and comprehensive insurance
- **f** Subaru with a marked price of $58 790 and Third Party Property insurance.

**12** Sascha bought a minivan with the following upfront costs.

| | |
|---|---|
| **Purchase price** | $43 860 |
| **Stamp duty** | 1.5% for first $25 000 plus 3% for each $ over $25 000 |
| **Registration** | $726 |
| **CTP Insurance (green slip)** | $854 |
| **Comprehensive insurance** | $1352 |

What was the total cost of his van?

## Did you know?

### Best-selling car brands

The table shows the 10 best-selling car brands in Australia in 2016 and 2024.

| | 2016 | 2024 |
|---|---|---|
| 1 | Toyota HiLux | Ford Ranger |
| 2 | Toyota Corolla | Toyota RAV4 |
| 3 | Hyundai i30 | Toyota HiLux |
| 4 | Ford Ranger | Isuzu D-Max |
| 5 | Mazda 3 | Mitsubishi Outlander |
| 6 | Toyota Camry | Ford Everest |
| 7 | Holden Commodore | Toyota Corolla |
| 8 | Mazda CX-5 | Mazda CX-5 |
| 9 | Mitsubishi Triton | MG ZS |
| 10 | Hyundai Tucson | Kia Sportage |

Data source: racv.com.au

1. What notable changes have occurred from 2016 to 2024?
2. Find out what the list looks like this year.

# 7.03 Running costs

The running costs of a car include fuel, servicing, repairs, tyres, renewal of car registration and insurance premiums, and depreciation. Motorway tolls, parking fees and fines for traffic infringements also need to be considered.

## Example 5

Rabiba owns a Hyundai Sonata and drives about 15 000 km per year. Her car was serviced twice during the year, with each service capped at $250. She also needed 2 new tyres at $215 per tyre. Rabiba had to pay $422 for car registration, $385.21 for the green slip and $991.88 for comprehensive insurance. Her weekly fuel bill is $85. Parking fees and tolls amounted to $48 per week.

**a** How much did Rabiba spend on her car during the year?

**b** What was the weekly cost of the car?

iStock.com/Minerva Studio

### Solution

**a** $\text{Total cost} = \$250 \times 2 + \$215 \times 2 + \$422 + \$385.21 + \$991.88 + \$85 \times 52 + \$48 \times 52$

$= \$9645.09$

**b** $\text{Weekly cost} = \dfrac{\$9645.09}{52}$

$= \$185.4825$

$\approx \$185.48$

**EXERCISE 7.03** Answers on p. 471

## Running costs

**1** The average weekly cost of running a Mazda 3 5-door hatchback has been estimated to be $202.67. What is the yearly cost of running this car?

**2** The weekly fuel bill of a car is $87.69 based on the car travelling 15 000 km per year. If the average fuel price is $2.05 per litre, find:

**a** the amount of fuel (to the nearest litre) used per week

**b** the annual fuel cost.

**3** Jackson estimates that he travels approximately 500 km per week, using a car with a fuel consumption of 11 km per litre . The average fuel price is $1.98 per litre. Car registration and Compulsory Third Party insurance cost $785 p.a. and the comprehensive insurance on his car is $725. Jackson calculates that $1100 should cover servicing and maintenance of the car. He has paid $180 for membership to the NRMA road service. He also pays $50 per week on motorway tolls and his parking fees amount to $500 for the year.

**a** Calculate Jackson's fuel bill for the year.

**b** How much will Jackson pay for the annual running costs of his car?

**c** How much will Jackson need to budget per fortnight to cover the cost of running his car?

□ Foundation ○ Mastery ○ Complex

4 The list price for a car is $32 990 and the on-road price is $36 528. The car depreciates by $74.86 per week. What is the value of the car after one year?

5 The Tesla Model Y was the highest-selling electrical vehicle in 2024. The average total weekly running cost is $235.87. Find the annual running cost of a Tesla Model Y.

6 Sarita bought a new car for $21 000. She is told that the running cost of the car will be 49.29 cents per km. Sarita travels 20 000 km per year. What is her average weekly running cost for the car?

7 Ben's annual fuel cost is $3744. His registration, CTP insurance and comprehensive car insurance amount to $1432. His car has had 2 services costing $267 and $321. At the last service, Ben was informed he needed 4 new tyres, which cost $212 each. The NRMA fee for gold membership was $180, and tolls and parking fees amounted to $890 for the year. Parking fines added $320 to his costs.

What is the weekly running cost of Ben's car?

# Fuel consumption and prices 7.04

Worksheet
How much petrol?

The types of fuel used to power vehicles include petrol, diesel, liquefied petroleum gas (LPG), ethanol and electricity. Petrol is the most common fuel and most cars use unleaded petrol (ULP).

**Fuel consumption** is the rate at which a car uses fuel and is usually measured in litres per 100 km (L/100 km). For example, a large car may use 12.3 L of fuel to travel 100 km, whereas a small car may use only 5.8 L. The small car has a better fuel consumption and is more fuel-efficient. The table shows the fuel consumption for some of Australia's best-selling cars.

| Vehicle | Fuel consumption (L/100 km) | | |
|---|---|---|---|
| | Average | City | Highway |
| Ford Ranger (diesel) | 6.6 | 7.8 | 5.8 |
| Honda Civic | 6.5 | 8.6 | 5.3 |
| Hyundai i30 | 7.3 | 10.3 | 5.5 |
| Mazda 3 | 5.8 | 7.6 | 4.8 |
| Nissan X-Trail | 8.1 | 11.0 | 6.4 |
| Toyota Camry Hybrid | 5.2 | 5.7 | 4.9 |
| Toyota Corolla | 4.1 | 3.9 | 4.1 |
| Toyota HiLux (diesel) | 9.0 | 10.9 | 7.8 |
| Toyota Prado | 10.0 | 11.1 | 8.7 |
| Toyota Yaris | 5.8 | 7.1 | 5.1 |

Data source: www.greenvehicleguide.gov.au © Copyright 2024, Commonwealth of Australia

The fuel consumption of a vehicle depends on several factors. You use more fuel when:

- stop–start driving in traffic (city cycle)
- driving on unsealed roads
- using a badly tuned engine
- using a driving style involving hard acceleration and/or braking
- using under-inflated tyres
- towing a trailer or caravan
- using the air conditioner.

Foundation Mastery Complex

Video
Fuel consumption

## Example 6

A MINI Cooper has an advertised fuel consumption of 5.8 L/100 km.

**a** How much fuel, to 2 decimal places, does it use for a trip of 1530 km?

**b** If fuel is \$2.04 per litre, what is the fuel cost, to the nearest dollar, for this trip?

**c** How far, to the nearest kilometre, can the MINI Cooper travel on a full tank of 44 L at the advertised fuel consumption?

**d** If the MINI Cooper used 25 L to go 400 km on an outback trip, how far could it go on 42 L at the same rate?

### Solution

**a** Fuel for 100 km = 5.8 L

Fuel for 1 km = $\frac{5.8}{100}$ L    using the unitary method

Fuel for 1530 km = $1530 \times \frac{5.8}{100}$ L

= 88.74 L

**b** Fuel cost = $88.74 \times \$2.04$

= \$181.0296

≈ \$181

**c** 5.8 L is used for 100 km

44 L is used for $\frac{100}{5.8} \times 44$ km = 758.6206 ... km

≈ 759 km

At the advertised fuel consumption, the MINI Cooper can travel 759 km (to the nearest km) on a full tank of fuel.

**d** 25 L takes the car 400 km

1 L takes the car $\frac{400}{25}$ km

42 L takes the car $\frac{400}{25} \times 42$ km = 672 km

The MINI Cooper can travel 672 km on 42 L.

**EXERCISE 7.04** Answers on p. 471

## Fuel consumption and prices

**1** A motorcycle has a fuel consumption of 3.2 L/100 km. How far can it travel on 1 L of petrol? Select **A**, **B**, **C** or **D**.

**A** 32 km  **B** 16.1 km  **C** 31.25 km  **D** 3.2 km

**2** A car travels 160 km on 14 L of petrol. What is its fuel consumption in L/100 km? Select **A**, **B**, **C** or **D**.

**A** 11.4  **B** 8.7  **C** 22.4  **D** 8.8

EXAMPLE 6

**3** A Mazda has a fuel consumption of 11.5 L/100 km.

**a** What is the fuel consumption in L/km?

**b** How much fuel, to the nearest litre, does it use to travel 980 km?

**c** How far could it travel on a 55 L tank of fuel? Answer to the nearest kilometre.

**d** If the Mazda used 48 L to go 695 km, how far (to the nearest km) could it go on 102 L at the same rate?

**4** A Hyundai i30 has a fuel consumption of 5.5 L/100 km. How much fuel, in litres correct to 2 decimal places, does it use to travel:

**a** 10 km?  **b** 2347 km?

☐ Foundation ○ Mastery ○ Complex

**5** If a car uses 8 L of fuel to travel 100 km, how much will it use to travel:

a 1 km? b 150 km?

Syda Productions/Shutterstock.com

**6** Refer to the table of fuel consumption of some of Australia's best-selling cars on page 267.

a Which car has the best average fuel consumption?

b How far (to the nearest kilometre) can this car travel on a 45 L tank of fuel?

c How much petrol does the Mazda 3 use for a 2500 km trip at the rate shown?

d What is the cost of 1348 km of city driving in the Toyota HiLux if the price of diesel is $2.02/L?

**7** Tyson filled up his 77 L Ford Ranger fuel tank with petrol in Melbourne at $1.81/L.

a How much did he pay?

b If his fuel consumption is 10.7 L/100 km, how many whole kilometres would he go before running out of petrol?

**8** What does it cost Anastacia to fill her 50 L diesel Golf fuel tank at $2.25 per litre?

**9** A Volkswagen 'beetle' car travelled 440 km on a 40 L tank of fuel. What was the fuel consumption in L/100 km?

**10** Lachlan, Brodie and Mitchell each did a 3-day car trip and recorded their fuel usage.

| Brodie | Distance travelled (km) | Fuel used (L) |
|---|---|---|
| Friday | 497 | 47 |
| Saturday | 364 | 43 |
| Sunday | 579 | 56 |

| Lachlan | Distance travelled (km) | Fuel used (L) |
|---|---|---|
| Friday | 543 | 68 |
| Saturday | 297 | 43 |
| Sunday | 410 | 56 |

| Mitchell | Distance travelled (km) | Fuel used (L) |
|---|---|---|
| Friday | 624 | 56 |
| Saturday | 208 | 39 |
| Sunday | 524 | 45 |

Whose car had the best average fuel consumption?

Foundation Mastery Complex

9780170497886

**11** Copy and complete the table.

| Distance travelled (km) | Fuel used (L) | Fuel consumption (L/100 km) |
|---|---|---|
| 863 | 62 | |
| | 43 | 11.2 |
| 1304 | | 8.4 |
| | 86 | 9.7 |
| 2356 | | 12.9 |
| 1730 | 167 | |

## Technology

### Calculating fuel consumption and costs

**1** Search for an online fuel consumption calculator similar to this one.

Fuel spent in litres (L): 60

Kilometres (km): 725

Select units type: Metric units (L, km)

Rounding options: 2 digits after decimal point

Calculate the fuel consumption in L/100 km for:

**a** a Jeep that travels 3200 km and uses 380 L of fuel

**b** a Volkswagen that travels 1850 km and uses 160 L of fuel

**c** a Smart car that travels 790 km and uses 60 L of fuel.

**2** Search for an online calculator that finds annual fuel costs, such as this one.

| Calculate your annual fuel costs | | | | |
|---|---|---|---|---|
| | Fuel price ($/litre) | Fuel consumption (L/100 km) | Annual distance travelled (km) | Annual fuel cost ($) |
| | CAR 1 | | | |
| | 1.95 | 8.28 | 15000 | CALCULATE |

**a** Calculate the annual fuel costs for Brad, who travels 550 km each week, when fuel costs $2.17 per litre. His car can travel 600 km on a 50-litre tank of fuel.

**b** Survey 5 friends or family members to see how far each travels per year and the fuel consumption of their cars in L/100 km. Calculate their annual fuel costs at the current petrol price.

## Did you know?

### Ten ways of saving fuel

1. Drive at a moderate speed: not too slow, not too fast. Driving at 110 km/h and beyond uses much more fuel than driving at 80–100 km/h.
2. Service your car regularly and check the tyre pressure.
3. Avoid driving in stop–start traffic during peak times and on busy roads.
4. Avoid excessive braking. Allow the car to come to a natural stop.
5. If waiting in traffic for a long time, switch off the engine.
6. Close the windows when driving to reduce drag.
7. Switch off the air conditioner.
8. Reduce the load on your car by removing heavy objects from the boot and removing roof racks if they are not being used.
9. Plan to do all your small jobs in one trip rather than lots of short trips.
10. Use public transport, walk or ride a bike.

**Choose 2 of the ways above and explain how or why they save fuel.**

# Choosing the best car 7.05

Choosing a car to buy will depend on your personal tastes and needs, and most importantly, on your budget. Initial costs that need to be considered are registration, CTP insurance, car insurance, maintenance, repairs, and cost of fuel. If a loan is required to buy the car, then the fortnightly/monthly payments also need to be included in your budget.

Other essential factors that may influence your choice of car are:

- the size of the motor (for example, 1.8 L, 2.0 L, 3.6 L)
- the type of fuel the car uses (petrol, diesel, electric, hybrids)
- fuel consumption
- warranty periods for a car purchased from a dealership
- security and safety features
- whether to buy a new car or a used car.

## Example 7

Jasmine decides to buy a new car and apart from the purchase price and on-road costs, she is allowing \$6500 per year for running costs. Jasmine borrows \$25 000 over 3 years to pay for the car. Her fortnightly repayments are \$355.77.

How much will Jasmine have to budget for the car each week (to the nearest dollar)?

### Solution

Running costs/year = \$6500

Loan repayments/year = $\$355.77 \times 26 = \$9250.02$

Total = $\$6500 + \$9250.02$

$= \$15\,750.02$

Weekly amount $= \dfrac{\$15\,750.02}{52}$

$= \$302.89$

$\approx \$303$

Jasmine will have to budget \$303 per week.

## Example 8

Samin bought a car for \$39 044, which included on-road costs. Samin paid 40% of the total of his car in cash and borrowed the rest at 8.5% p.a. flat (simple) interest over 4 years.

**a** How much did Samin borrow, rounded up to the nearest dollar?

**b** How much will he have to budget per week to cover the cost of repayments?

### Solution

**a** Amount borrowed $= 60\% \times \$39\,044$     $100\% - 40\% = 60\%$

$= \$23\,426.40$

$\approx \$23\,427$

**b** Interest $= \$23\,427 \times 0.085 \times 4$

$= \$7965.18$

Total to repay $= \$23\,427 + \$7965.18$

$= \$31\,392.18$

Weekly amount $= \dfrac{\$31\,392.18}{208}$     $52 \times 4 = 208$

$= \$150.923...$

$\approx \$150.92$

**EXERCISE 7.05** Answers on p. 471

## Choosing the best car

1. The annual running costs for Nicole's car include \$780 for registration and green slip, \$589 for comprehensive insurance, \$1800 for fuel, \$1100 for maintenance, servicing and tyres. She also has to pay \$610 per month for the car loan. Calculate Nicole's weekly budget for the car.

2. Dylan is purchasing a new car. He has selected 2 cars, which are similar, but one uses diesel and the other uses petrol.

   **Car 1**: Diesel – cost \$44 430, average weekly running cost \$235.87

   **Car 2**: Unleaded petrol – cost \$42 700, average weekly running cost \$241.58

   Calculate the cost of each car over 5 years to determine which car is the better option.

Evannovostro/Shutterstock.com

3. At a petrol station, the price of regular unleaded petrol is 179.9c/L and the price of diesel is 191.9c/L. Two cars, one petrol and the other diesel, both fill their tanks with 65 L of fuel. What is the difference between their fuel bills for one tank of fuel?

☐ Foundation ○ Mastery ⬡ Complex

EXAMPLE 8

4 Mariah paid $32 700 for a second-hand station wagon. She borrowed 75% of the total cost at 9.4% p.a. flat interest over 3 years.

**a** How much did she borrow, rounded up to the nearest dollar?

**b** How much will she need to budget per week to cover the cost of loan repayments?

5 Daniel wants to purchase a new car costing $29 950. He needs finance to pay for the car and has 2 options:

**A:** monthly payments of $731.17 for 4 years

**B:** fortnightly payments of $441.98 for 3 years.

Which is the better option? Justify your answer.

6 Many car dealerships offer a 5-year warranty or 100 000 km. What does this mean?

7 Copy this table to compare the cost of buying 2 different cars.

| | Car 1 | Car 2 |
|---|---|---|
| Make | | |
| Model | | |
| Year | | |
| Cost of car | $ | $ |
| Cash price of car | $ | $ |
| Less trade-in or deposit | $ | $ |
| Stamp duty and registration transfer | $ | $ |
| Comprehensive and third party insurance | $ | $ |
| Fees and extras | $ | $ |
| | $ | $ |
| | $ | $ |
| Other costs | $ | $ |
| Extended warranty | $ | $ |
| **Total amount needed for car** | **$** | **$** |

Complete this table for 2 cars, using the internet to obtain information such as price, stamp duty, insurance, cost of registration and so on, to calculate the total amount needed to buy a car.

vectorfusionart/Shutterstock.com

Foundation Mastery Complex

## Sample HSC problem Answers on p. 471

(6 marks) Nitara is buying a used car that has a sale price of \$18 700. She will have to pay a registration transfer fee of \$39 and stamp duty, which is calculated at \$3 for every \$100 or part thereof of the sale price.

**a** Calculate the stamp duty Nitara has to pay. 1 mark

**b** Nitara borrows the total amount of the cost of the car including the transfer of registration and stamp duty. She pays \$540 per month over 4 years to repay the loan. Calculate:

- **i** the total amount Nitara borrows 2 marks
- **ii** the total amount Nitara pays for the car 1 mark
- **iii** how much Nitara needs to allow for the loan repayments if she works on a weekly budget. 2 marks

## Study Tip

### Your study routine

- Your study routine will be more effective if you maintain a healthy lifestyle: eat, sleep, rest, exercise, socialise and generally be positive!
- Try to study at the same time and place each day.
- When studying, alternate between easy tasks and hard tasks.
- Stop and reflect on what you have achieved before moving on.
- Use your time wisely and take regular breaks.
- Ask yourself: 'Am I really studying effectively or am I wasting time? Do I feel a sense of achievement?'
- Take a 1-minute break after completing each study or homework task.
- Take a longer break after working for more than 45 minutes continuously.
- Reward yourself when you feel you have completed a significant amount of work: listen to some music, have a snack, play some sport or a game, or go for a walk.

# CHAPTER SUMMARY

This chapter, *Owning a car*, looked at the costs of owning a car, from on-road costs such as the purchase price, insurance, registration and stamp duty, to running costs such as fuel consumption, service, maintenance and repairs.

Before you move on, consider what you learned in this chapter and revisit any sections that may have been unclear.

| What you learned in this chapter... | Section | |
|---|---|---|
| Identify the different types of car insurance available, including compulsory third-party (CTP) | 7.01 | Car insurance |
| Calculate on-road costs for buying a car, including finance, stamp duty, registration and insurance | 7.02 | On-road costs |
| Calculate running costs of owning a car, including service and repairs | 7.03 | Running costs |
| Identify fuel consumption and prices as rates and use them to calculate the amount and cost of fuel used on a trip | 7.04 | Fuel consumption and prices |
| Compare the costs of owning different types of car | 7.05 | Choosing the best car |

To help master this topic, make a summary mind map. Use the chapter outline and the mind map below as a guide. Add your own words, symbols, diagrams, boxes and reminders. The summary should give you a 'whole picture' view of the topic and allow you to identify any weak areas to revisit in your revision.

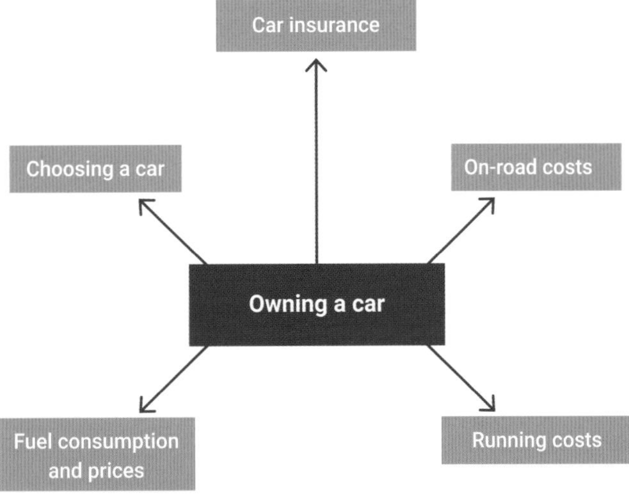

# 7 Test yourself Answers on p. 471

**1** This table gives 2 examples of Third Party Property insurance premiums.

| Car make | Driver gender | Driver age | Town where vehicle is garaged | Place where car is usually kept | Third Party Property premium |
|---|---|---|---|---|---|
| Mitsubishi | male | 42 | Parramatta | driveway | $297 |
| Nissan | female | 65 | Bondi | security compound | $302 |

What is the insurance premium for:

**a** the Nissan driver who gets a 5% discount for secure garaging?

**b** the male driver who pays an 8% levy for non-secure garaging in the Parramatta area?

**2** Ben had Third Party Property insurance, paid by a direct debit of $15.80 per fortnight.

**a** How much would he save if he paid a yearly amount of $402.50 upfront?

**b** He hit a tree and wrote off his car, worth $15 600. What does his insurance cover?

7.02 **3** The rates for stamp duty in NSW are shown in this table.

| Vehicle value | Rate |
|---|---|
| $0–$44 999 | $3 for every $100 (or part of $100) |
| $45 000 and over | $1350 plus $5 for every $100 (or part of $100) over $45 000 |

Calculate the stamp duty on a car valued at:

**a** $21 370 **b** $44 020 **c** $58 700 **d** $89 220

7.02 **4** Raphael bought a new car with the following upfront costs.

| Purchase price | $24 999 |
|---|---|
| Stamp duty | 1.5% up to $5000<br>plus 2.5% for each dollar between $5000 and $20 000<br>plus 4.5% for each dollar over $20 000 |
| Registration | $328 |
| CTP green slip | $503 |
| Comprehensive insurance | $1204 |

What was the total upfront cost of his car?

7.03 **5** Nadia spends $42 on fuel per week and her comprehensive car insurance is $70.81 per month. She also pays $812.68 for car registration and CTP insurance, and services cost her $590 annually. Other repairs and replacement of tyres cost $450. Nadia uses the motorways 5 days a week at an average cost of $62 per week in tolls.

Find the total running cost of Nadia's car for:

**a** one year **b** one week.

☐ Foundation ○ Mastery ○ Complex

**6** The fuel consumption of a four-wheel drive vehicle was measured under various conditions. 7.04

| Conditions | Fuel consumption (L/100 km) | Trip distance (km) | Fuel used (L) |
|---|---|---|---|
| City driving non-peak hour | 14.5 | 35 | |
| City driving in peak hour | 12.7 | 26 | |
| Highways | 10.8 | 460 | |
| Outback dirt roads | 8.4 | 2635 | |
| Towing a caravan | 7.3 | 9870 | |

**a** What is the average fuel consumption (in L/100 km) for these 5 conditions?

**b** Copy and complete the table by calculating, correct to one decimal place, the fuel used for the trip distances for each condition.

**c** List 3 ways to improve fuel consumption.

**7** The table shows the fuel consumption for some vehicles. 7.04

| Car type | Fuel consumption (L/100 km) |
|---|---|
| Ford Everest | 8.5 |
| Haval H6 | 7.4 |
| Honda Jazz | 5.7 |
| Hyundai Sonata SLX | 8.0 |
| Lexus ES250 | 6.6 |
| Mini Cooper | 5.8 |
| Mitsubishi Colt | 5.6 |
| Nissan Patrol | 17.2 |
| Peugeot 308 | 6.7 |
| Toyota Aurion | 9.9 |

**a** How many kilometres can the Ford Everest travel on a 80 L tank of fuel?

**b** How much fuel does the Lexus use for a trip of 2500 km?

**8** Unleaded petrol at a service station is 209.0 cents per litre. Diesel fuel is priced at 198.6 cents per litre. What is the difference in cost of filling up tanks that hold 70 L? 7.04

☐ Foundation ○ Mastery ○ Complex

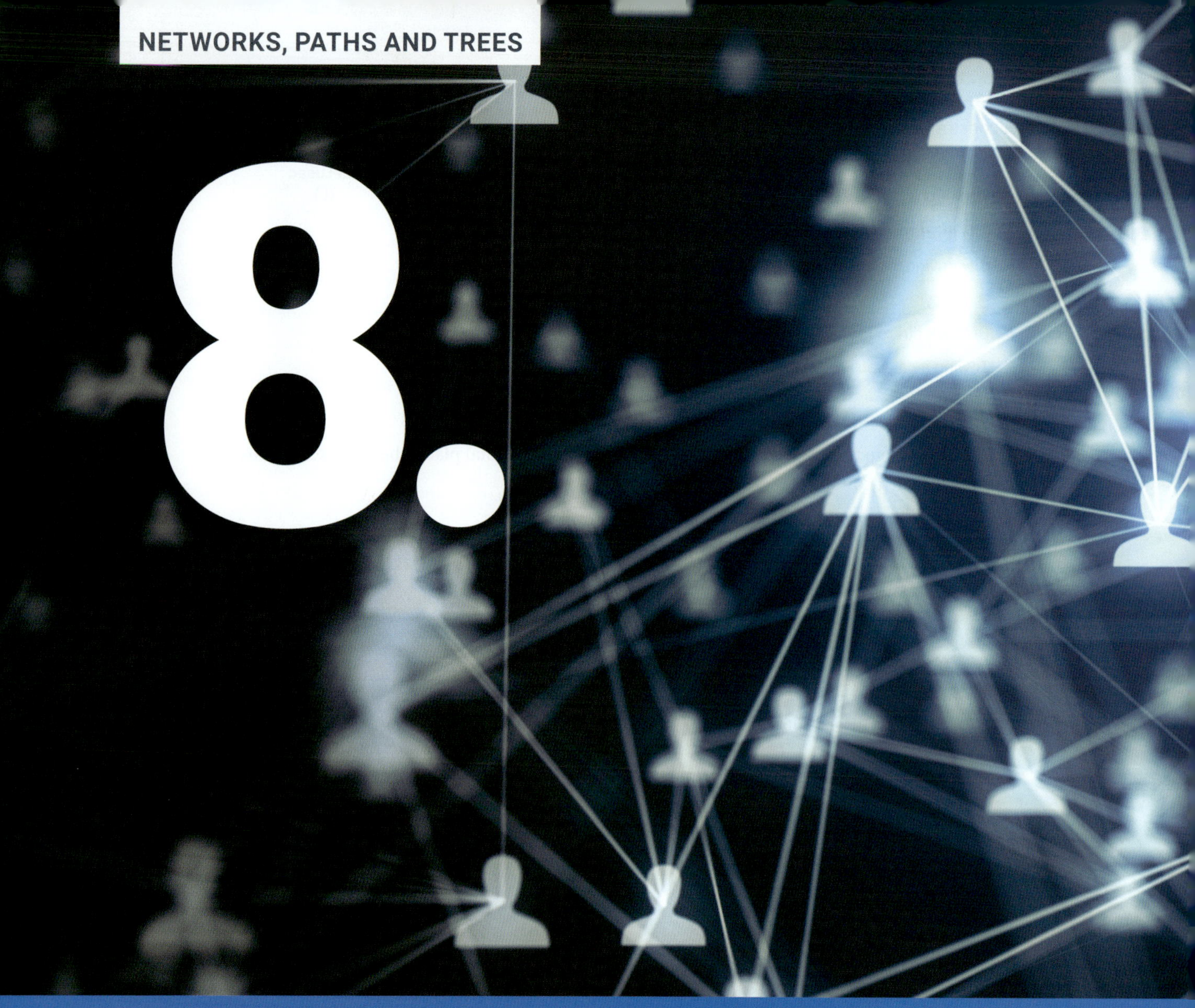

# 8. NETWORKS

A **network** is a system or structure of things that are related. Bus, rail and air routes are examples of networks. Mathematicians study networks and use graph theory to try to improve them: this can lead to faster process times, shorter distances and lower costs. Networks are used to represent computer systems, electrical and plumbing systems for housing, GPS road navigation, garbage bin collection routes and social media connections, for such platforms as Instagram.

## Chapter outline

flashmovie/Adobe Stock Photos

## In this chapter you will:

- learn the parts of a network and associated terminology, including vertex, edge and degree
- represent situations using tables and by drawing network diagrams
- draw network diagrams from information given in a table or map
- find the minimum spanning tree of a network by inspection and by using Kruskal's or Prim's algorithms, and apply this skill to solve minimal connector problems
- find the shortest path between 2 vertices in a network, and apply this skill to solve problems.

**Videos (6):**

**8.01** Networks • Degree of a vertex

**8.02** Networks and maps

**8.03** Kruskal's algorithm • Prim's algorithm

**8.04** Shortest path

**Worksheets (6):**

**SkillCheck** Assignment 8

**8.01** Network diagrams

**8.03** Minimum spanning trees • Spanning tree diagrams

**8.04** Shortest path problems

**Chapter summary** Network summary cards 1

**Puzzles (2):**

**8.01** Networks terminology

**8.04** Networks find-a-word

Nelson MindTap

To access resources above, visit **cengage.com.au/nelsonmindtap**

## Terminology

| | | | |
|---|---|---|---|
| degree | directed network | edge | Kruskal's algorithm |
| minimum spanning tree | network | path | Prim's algorithm |
| shortest path | spanning tree | tree | vertex |
| vertices | weight | weighted edges | weighted network |

## SkillCheck Answers on p. 471

**1** Copy each diagram and using a pencil, trace the diagram along the edges, using each edge only once and without lifting your pencil.

**a** 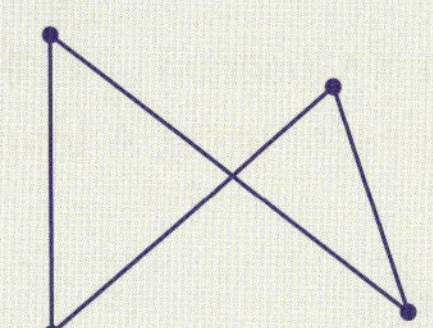

**b** 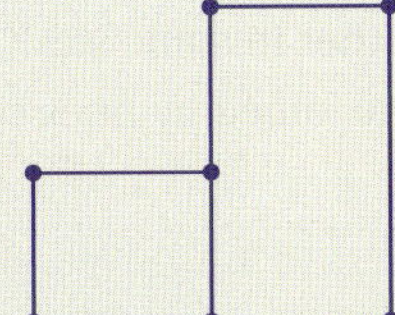

**c** 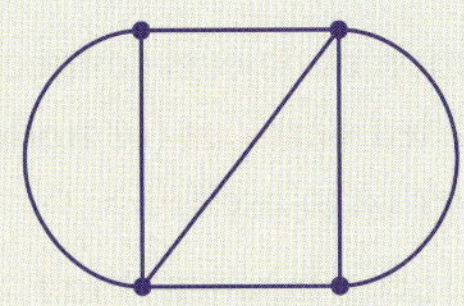

**2** This diagram shows the roads linking 5 towns, labelled $A$, $B$, $C$, $D$ and $E$. Mitch starts from town $A$, visits each town once and returns to town $A$.

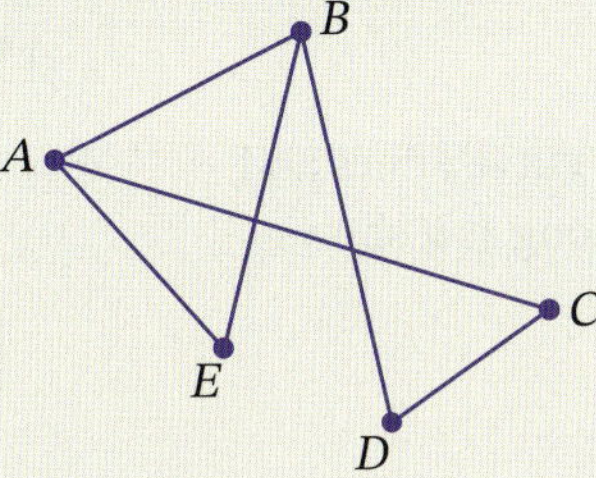


**a** Describe a route that Mitch could take.

**b** List any roads that Mitch does not travel on.

**3** This diagram shows the distances, in kilometres, along roads between the towns of Mayberry and Vincent.

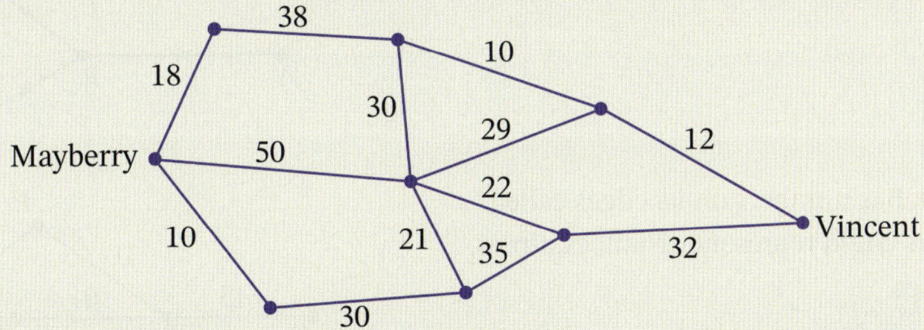


Find the shortest distance between Mayberry and Vincent.

**4** This diagram shows a network of friends who went to school together, where the lines show people who still have direct contact with each other.

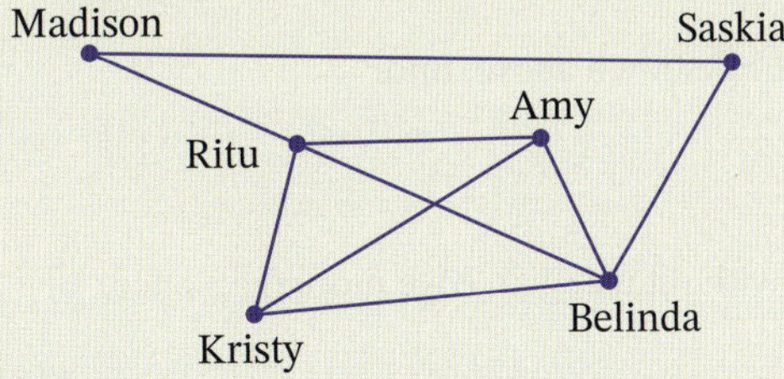


**a** Who has direct contact with

**i** Amy? **ii** Saskia?

**b** Who does not have direct contact with Belinda?

**c** How could Ritu get in contact with Saskia?

# 8.01 Networks

Video
Networks

Puzzle
Networks terminology

A **network** is a collection of objects connected to each other in some way. A **graph** is a diagram of a network and contains points called **vertices** (or **nodes**), which are joined by lines called **edges** (or **arcs**). A **loop** is an edge that connects a vertex to itself.

This network has 6 vertices, labelled $A$ to $F$.

It has 8 edges, including one loop at $D$.

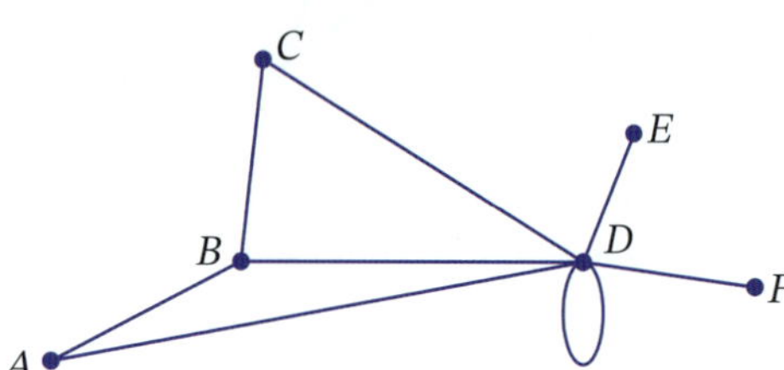


The same network can also be drawn with curved lines and with the vertices in different positions, as long as it shows the same links.

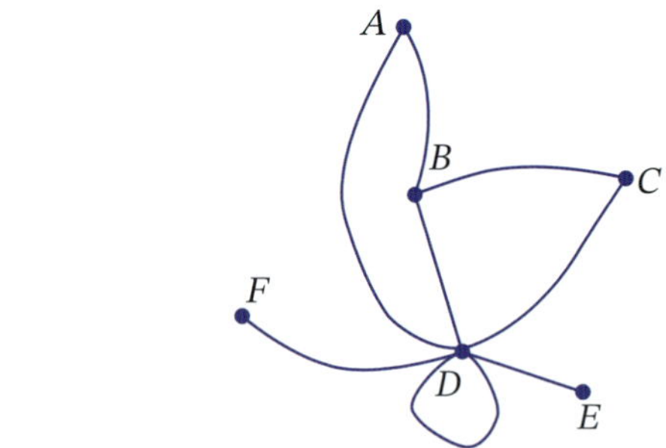


A **directed network** has arrows on its edges that indicate direction.

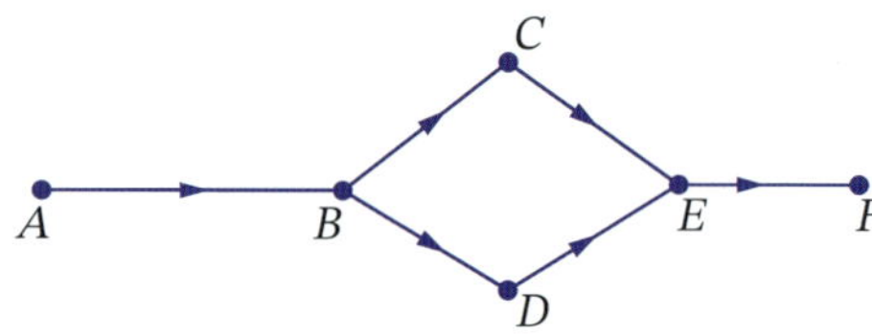


A **weighted network** has numbers on its edges called **weights**. The numbers could represent distances, times, costs or other quantities.

The **weight** of a network is the sum of the weights of its edges.

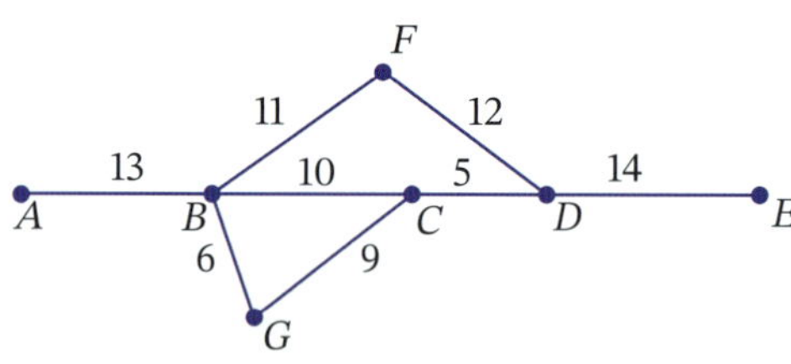


## Example 1

For the weighted network $ABCDEFG$ shown above, find:

**a** the weight of the network

**b** the shortest path from $A$ to $C$

**c** the longest path from $A$ to $C$ without using an edge more than once.

### Solution

**a** Weight of network $= 13 + 10 + 5 + 14 + 11 + 12 + 6 + 9$

$= 80$

**b** Shortest path from $A$ to $C = 13 + 10$ — the path $ABC$

$= 23$

**c** Longest path $= 13 + 11 + 12 + 5$ — the path $ABFDC$

$= 41$

## Degree of a vertex

The **degree** (or **order**) of a vertex is the number of edges connected to it. An **even vertex** has an even number of edges connected to it. An **odd vertex** has an odd number of edges.

The **sum of the degrees** of all vertices in a network is equal to **twice the number of edges**, because each edge belongs to 2 vertices. This means that the sum of degrees in a network is always even.

### Example 2

**a** Find the degree of each vertex in this network.

**b** How many vertices are even?

**c** Show that the sum of the degrees is equal to twice the number of edges.

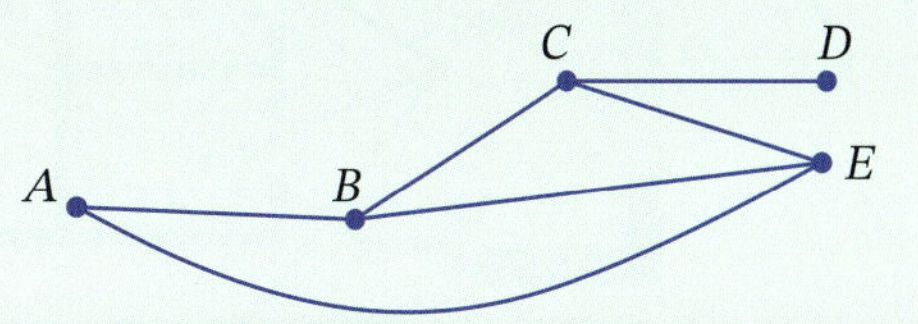


### Solution

**a**

| Vertex | *A* | *B* | *C* | *D* | *E* |
|---|---|---|---|---|---|
| Degree | 2 | 3 | 3 | 1 | 3 |

**b** There is one even vertex. *A* has degree 2, which is even.

**c** Number of edges = 6

Sum of degrees of vertices = 12, which is twice the number of edges.

### Example 3

Model this house plan as a network, showing doors as edges and the rooms (including entry and hall) as vertices.

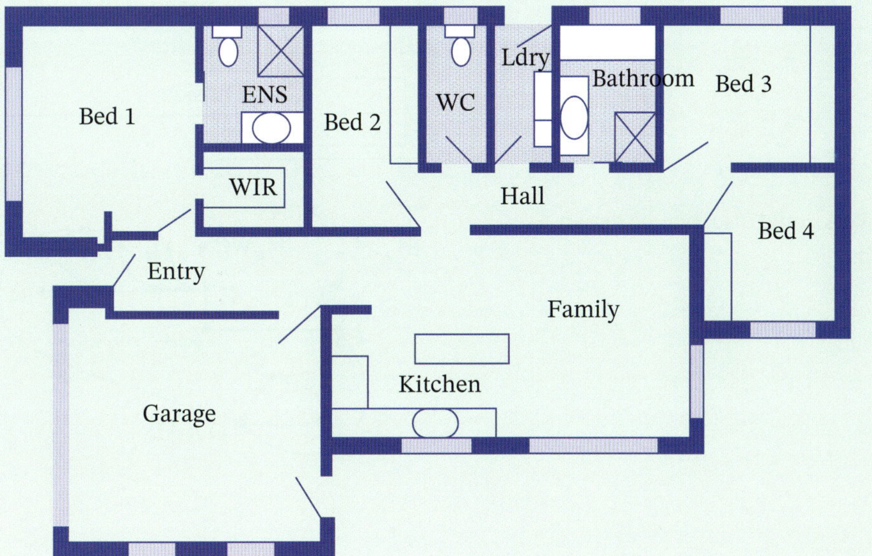

### Solution

Put a dot over each room (including the entry and hall). Also put one dot outside.

Then connect the dots through the doorways.

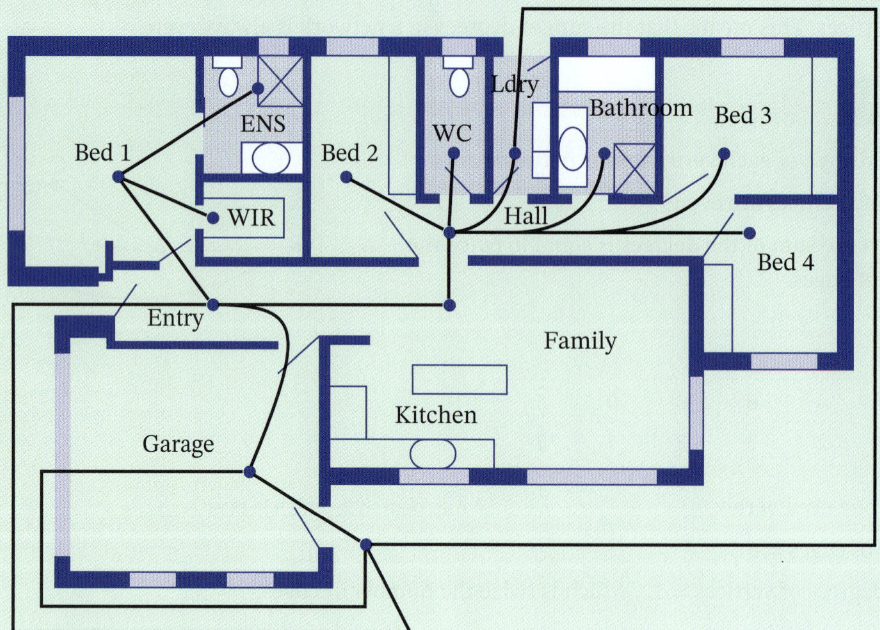


1 Redraw the network without the plan.

2 Show connections clearly and label each vertex.

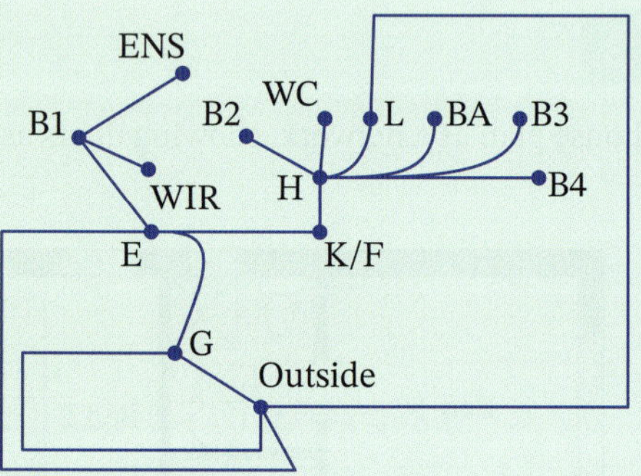


The network could also then be drawn as shown.

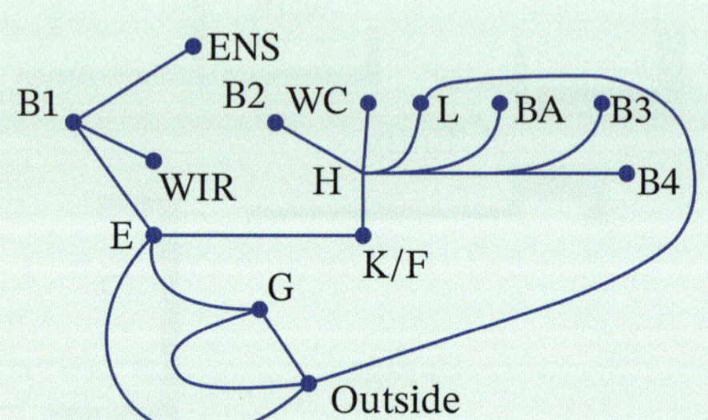

**EXERCISE 8.01** Answers on p. 471

## Networks

**1** For each network, count the number of:

**i** vertices **ii** edges.

**a**

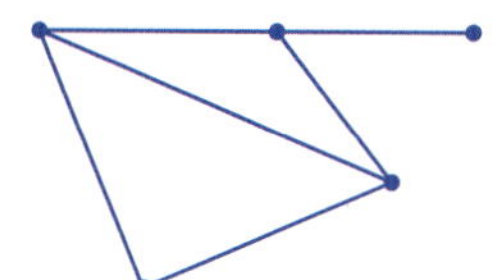

**b**

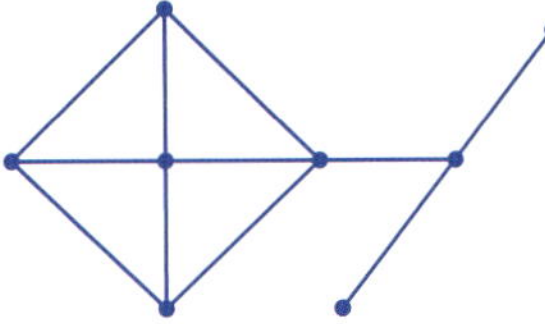

**c**

*D*, *E*, *C*, *B*, *F*

**d**

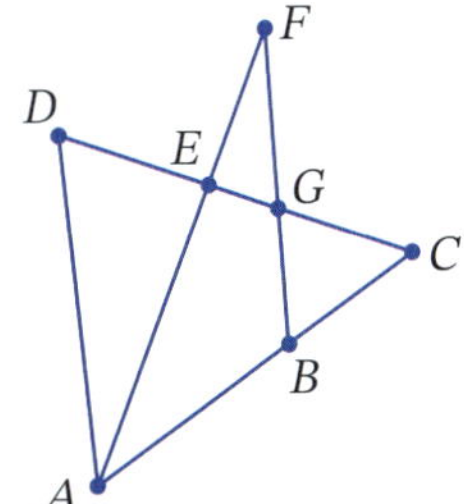


**e**

**f**

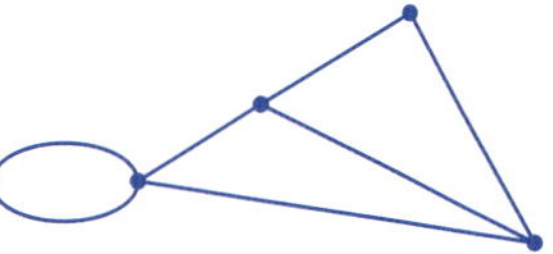

**2** This directed network shows water pipes connecting facilities in a camping ground. Between which 2 facilities can water flow? Select **A**, **B**, **C** or **D**.

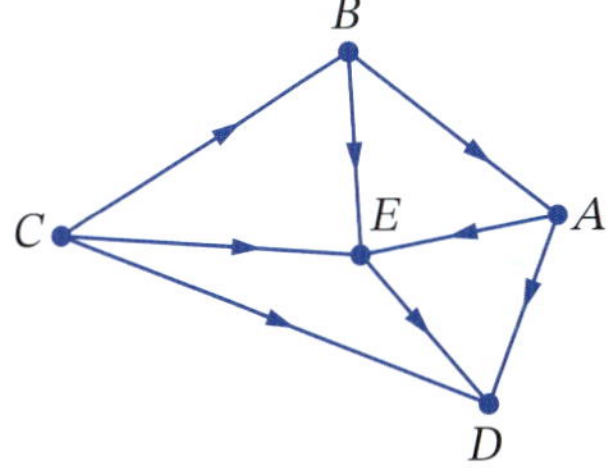


**A** *A* to *C* **B** *B* to *D*

**C** *B* to *C* **D** *A* to *B*

**3** This directed network shows places surrounding a central place, *K*.

How many places cannot be reached from *K*? Select **A**, **B**, **C** or **D**.

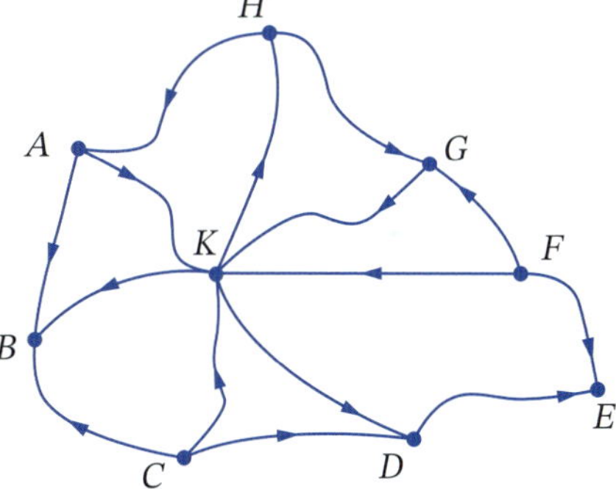


**A** 1 **B** 2

**C** 3 **D** 4

**4** This network diagram shows social connections between Alex (*A*), Mariah (*M*), Georgia (*G*) and Rabibah (*R*). An edge represents a friendship between 2 people.

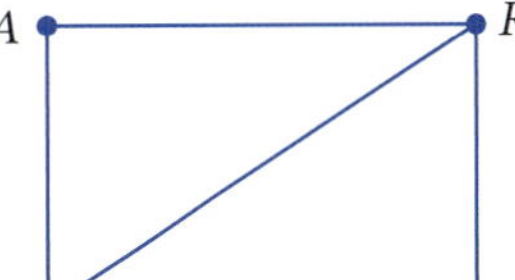


Which network diagram does *not* show the same information?

**A**

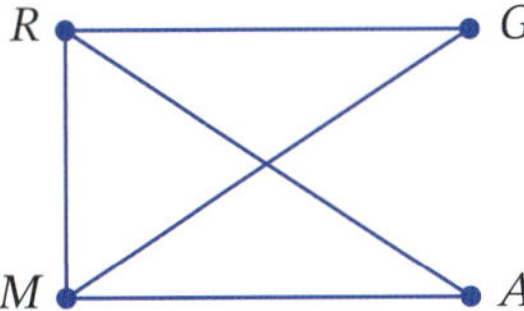


**B**

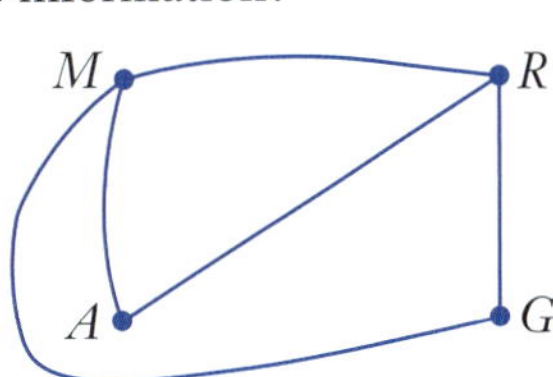
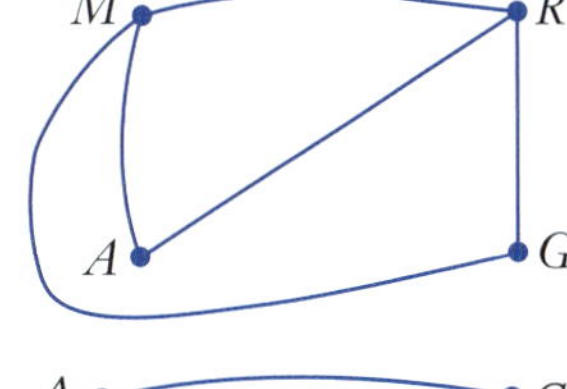


**C**

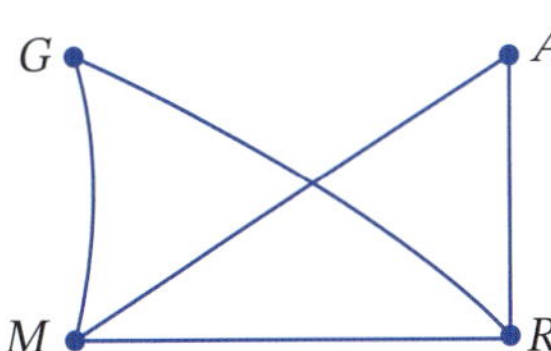


**D**

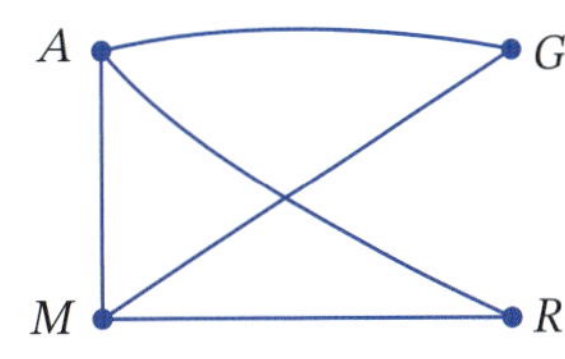


Foundation Mastery Complex

**5** Which network diagram represents this map, which shows the roads linking 4 towns? Select **A**, **B**, **C** or **D**.

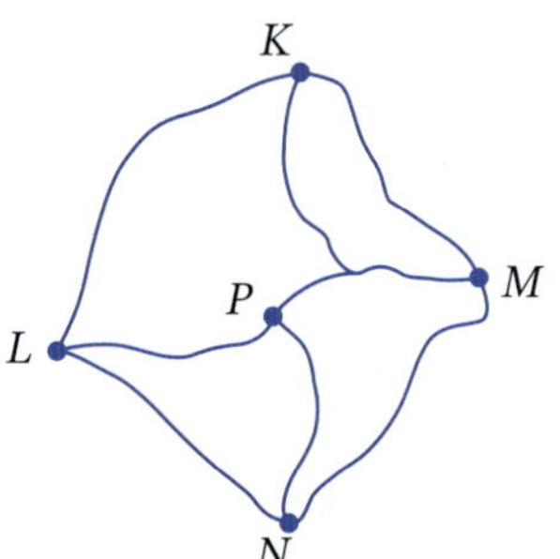


**A**

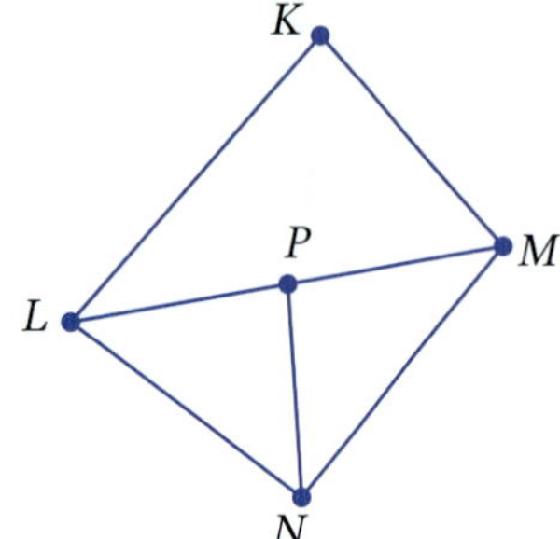


**B**

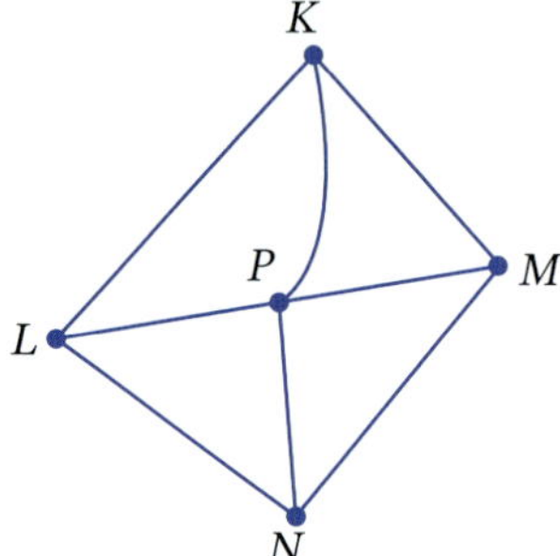


**C**

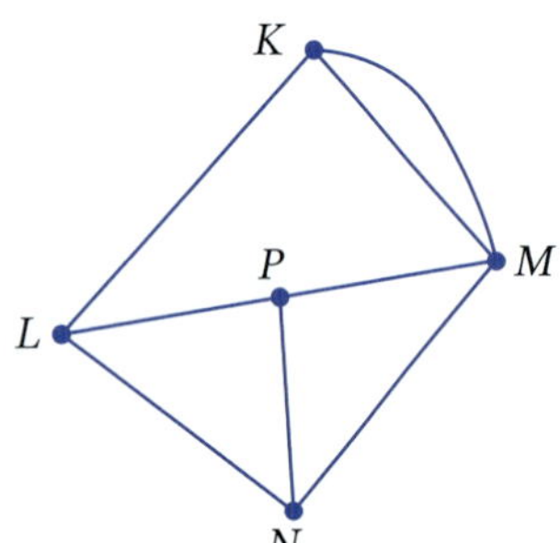


**D**

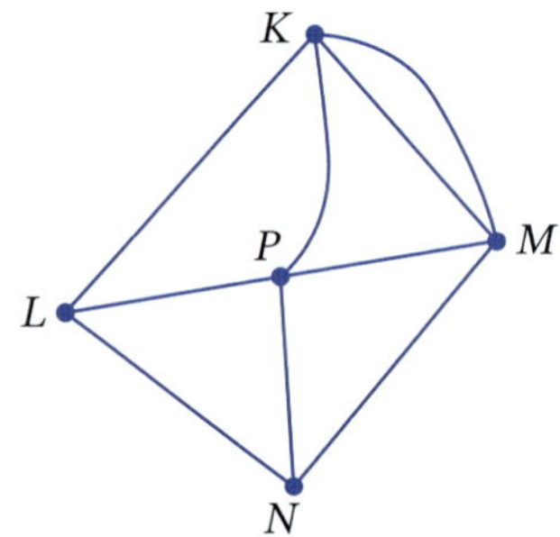

**Worksheet** Network diagrams

The worksheet 'Network diagrams' contains the diagrams from Questions **6–10** and **14** for you to write on.

**6** For each network, find:

**i** the weight of the network

**ii** the shortest path from $A$ to $F$

**iii** the longest path from $A$ to $F$ without using an edge more than once.

**a**

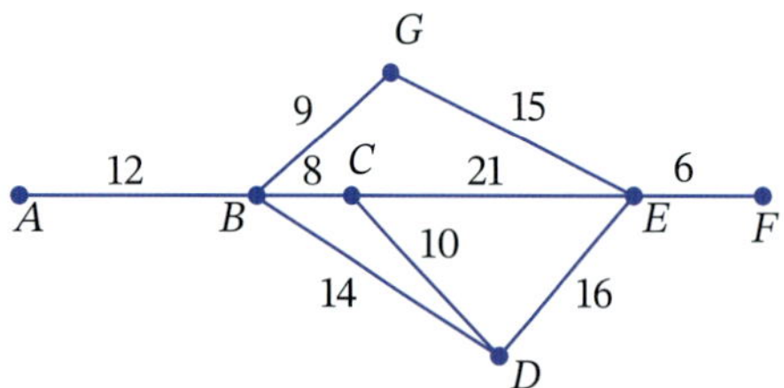


**b**

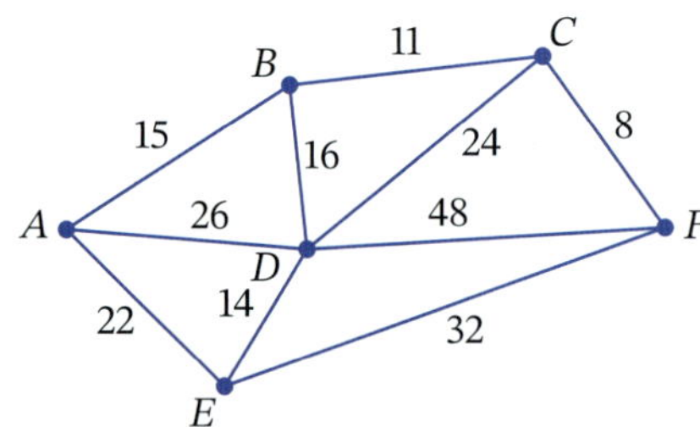


**7** What is the length of the longest path from $L$ to $W$ that doesn't use an edge more than once? Select **A**, **B**, **C** or **D**.

**A** 55 **B** 58

**C** 73 **D** 75

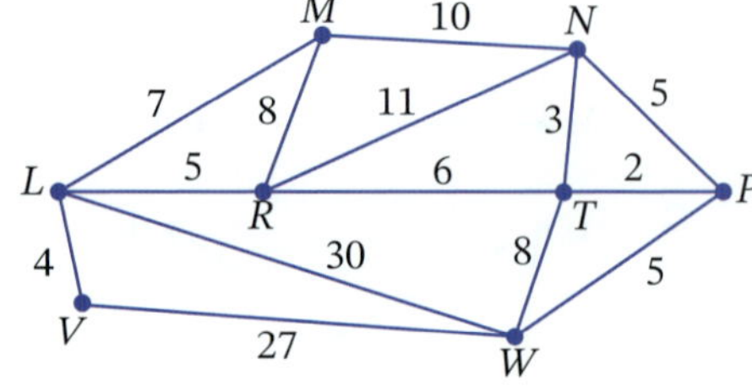


☐ Foundation ○ Mastery ⬡ Complex

**8** The network shows different suburbs and the distance between them in kilometres. What is the shortest path from $D$ to $X$?

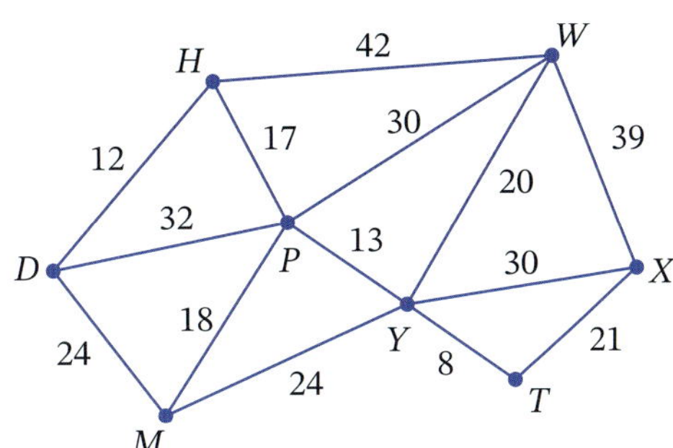


**9** The network represents a large park with walking paths. The vertices are the rest areas and the numbers are the lengths of the paths in metres. Saieesha goes to the park and walks from $G$ to $C$.

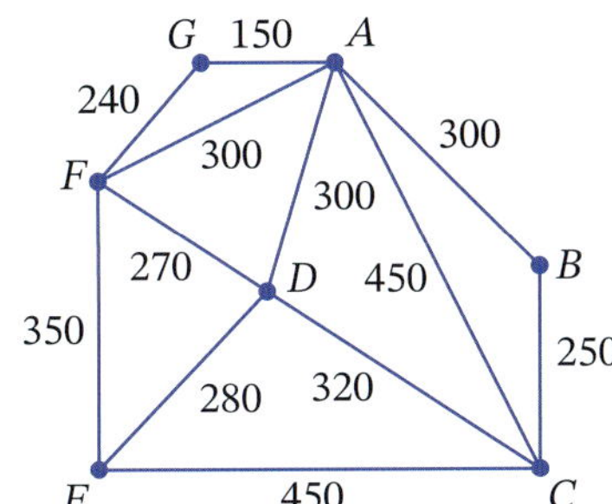


What is the longest walk Saieesha could do while only using any path once?

**10** For each network:

EXAMPLE 2

- **i** find the degree of each vertex
- **ii** count how many vertices are even
- **iii** show that the sum of the degrees is equal to twice the number of edges.

**a**

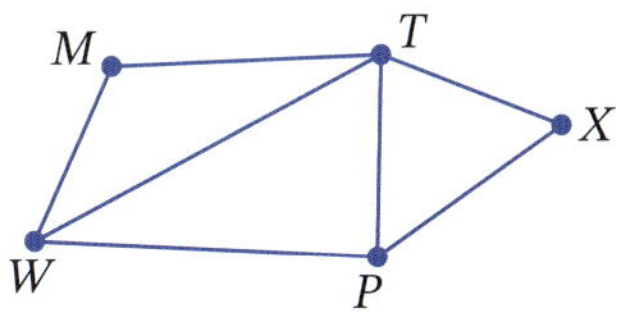


**b**

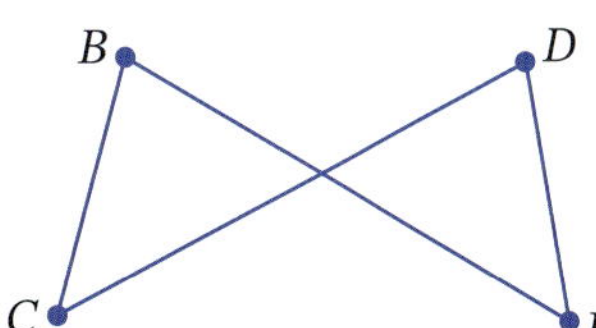


**c**

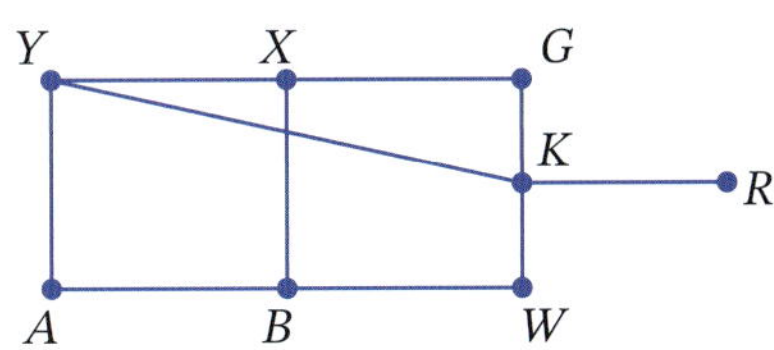


**d**

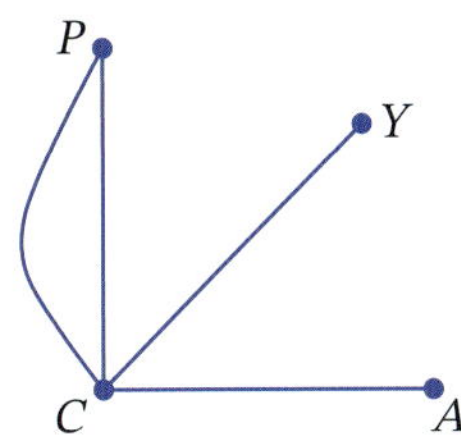


**11** How many odd vertices does this network have?

**A** 1 **B** 2

**C** 3 **D** 4

□ Foundation ○ Mastery ⬡ Complex

**12** How many even vertices does this network have?

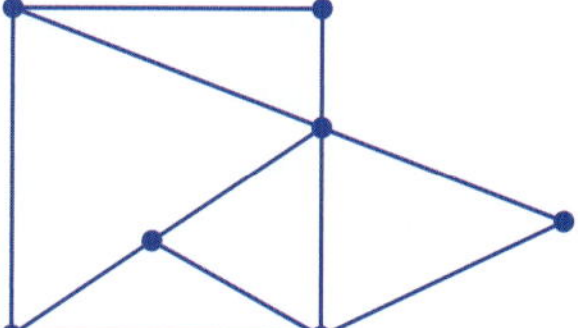

**A** 2 **B** 3

**C** 4 **D** 5

**13** What is the sum of the degrees of all vertices in this network?

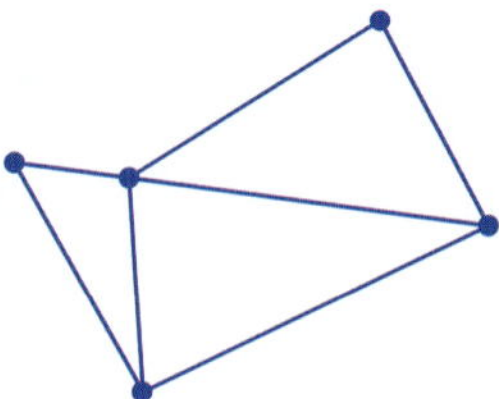

**A** 10 **B** 12

**C** 14 **D** 16

EXAMPLE 3

**14** Model each house plan as a network, showing doors as edges and the rooms (including entry and hall) as vertices.

**a**

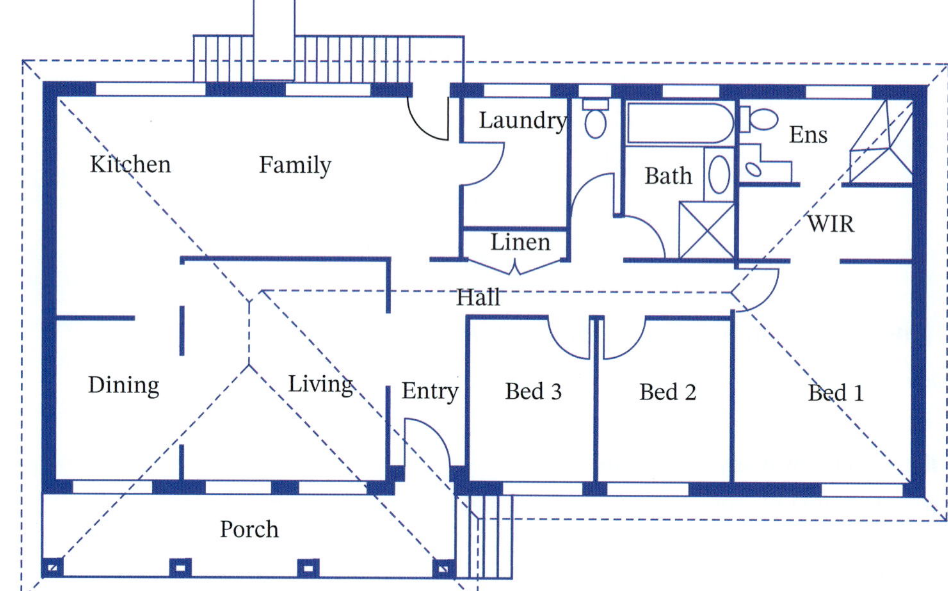


**b**

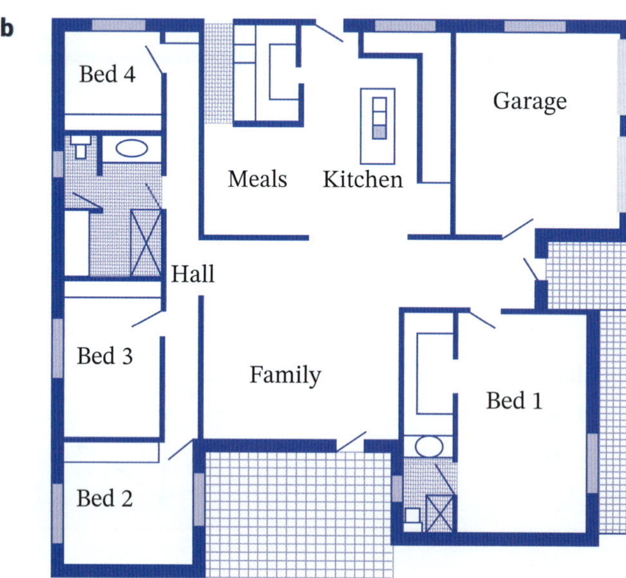

Foundation Mastery Complex

15 On the Sydney rail network, trains travelling from Campbelltown to Central go along 2 possible routes. They both stop at Glenfield, and then one branch passes through Liverpool, Lidcombe, Redfern and Central, while the other branch passes through East Hills, Wolli Creek, International Airport and Domestic Airport, before reaching Central. From Central, all trains go around the City Circle, stopping at Town Hall, Wynyard, Circular Quay, St James, Museum and returning to Central.

Draw a network diagram that represents these routes.

16 If 2 people shake hands, a network diagram can be drawn where the vertices represent the people and an edge represents the handshake, as shown below.

So for 2 people there is one handshake.

Draw a network diagram to represent each number of people shaking hands with everybody else and count:

- **i** the number of handshakes for each person
- **ii** the total number of handshakes.

**a** 3 people shake hands

**b** 4 people shake hands

**c** 5 people shake hands

**d** 8 people shake hands

H_Ko/Shutterstock.com

# Constructing network diagrams 8.02

Networks can be drawn from maps and represented by a table of numbers (or matrix).

The network diagram below is represented by the table on the right, where a 1 means the 2 vertices are joined by an edge and 0 (or '–') means the 2 vertices are not joined by an edge.

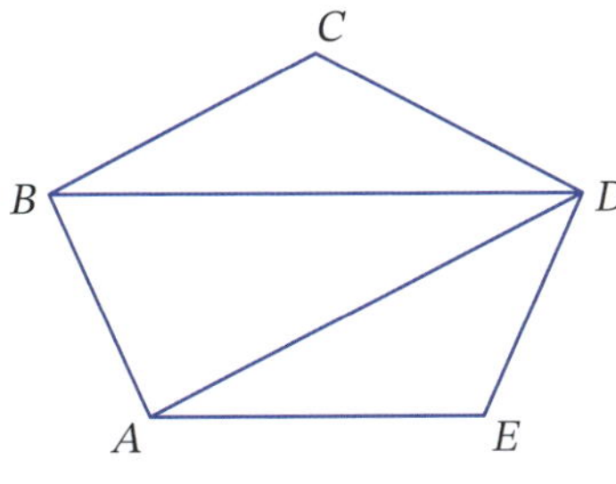


**Diagram/graph**

| | *A* | *B* | *C* | *D* | *E* |
|---|---|---|---|---|---|
| ***A*** | 0 | 1 | 0 | 1 | 1 |
| ***B*** | 1 | 0 | 1 | 1 | 0 |
| ***C*** | 0 | 1 | 0 | 1 | 0 |
| ***D*** | 1 | 1 | 1 | 0 | 1 |
| ***E*** | 1 | 0 | 0 | 1 | 0 |

**Table**

Foundation Mastery Complex

**Directed networks** can be represented in a similar way.

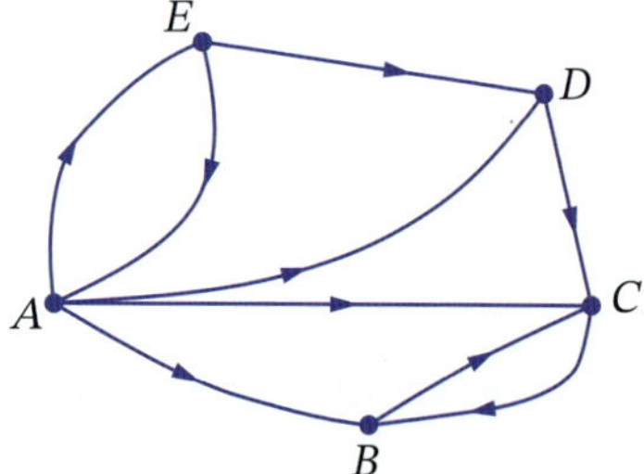


| | | To | | | | |
|---|---|---|---|---|---|---|
| | | *A* | *B* | *C* | *D* | *E* |
| From | *A* | 0 | 1 | 1 | 1 | 1 |
| | *B* | 0 | 0 | 1 | 0 | 0 |
| | *C* | 0 | 1 | 0 | 0 | 0 |
| | *D* | 0 | 0 | 1 | 0 | 0 |
| | *E* | 1 | 0 | 0 | 1 | 0 |

**Weighted networks** can also be shown by this method.

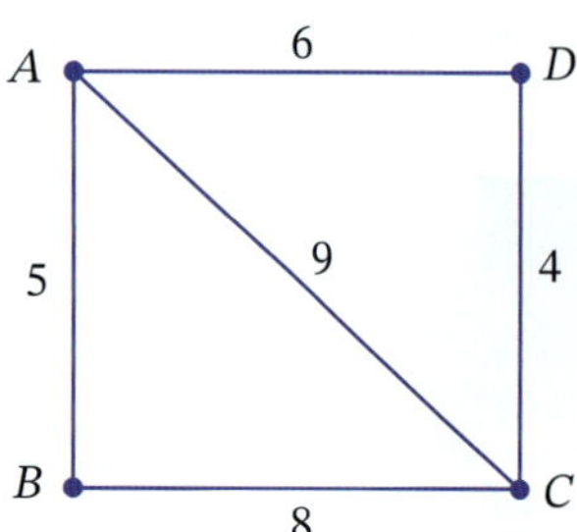


| | *A* | *B* | *C* | *D* |
|---|---|---|---|---|
| *A* | 0 | 5 | 9 | 6 |
| *B* | 5 | 0 | 8 | 0 |
| *C* | 9 | 8 | 0 | 4 |
| *D* | 6 | 0 | 4 | 0 |

## Example 4

Draw the directed weighted network represented by this table.

| | | To | | | |
|---|---|---|---|---|---|
| | | *A* | *B* | *C* | *D* |
| From | *A* | – | 4 | 8 | 3 |
| | *B* | – | – | 6 | 4 |
| | *C* | 5 | – | – | 7 |
| | *D* | – | – | 7 | – |

### Solution

Two possible network diagrams are:

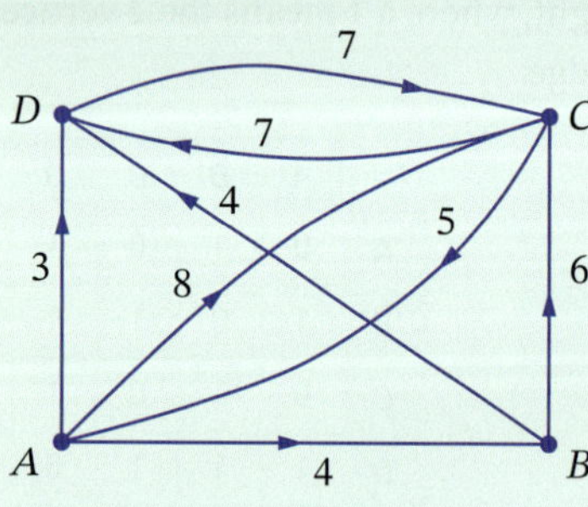


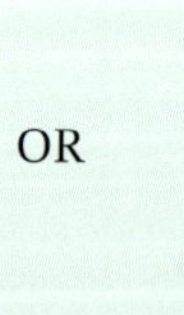


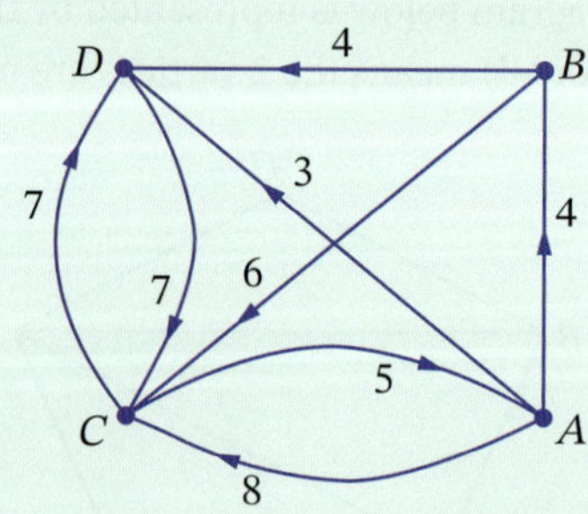

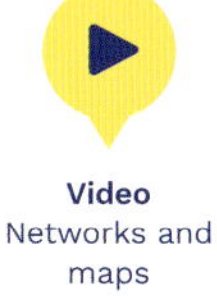


## Example 5

**a** Use the map of Australia to draw a network diagram representing its states and territories as vertices and the borders shared between them as edges.

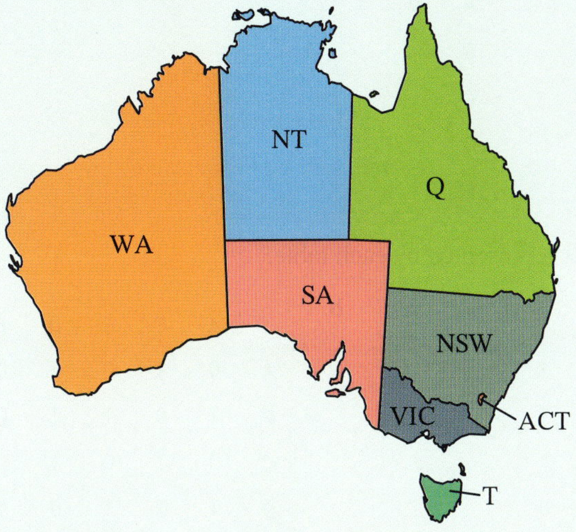


**b** Which vertex has the highest degree? Explain why.

**c** Explain why vertex T has a zero degree.

### Solution

**a**

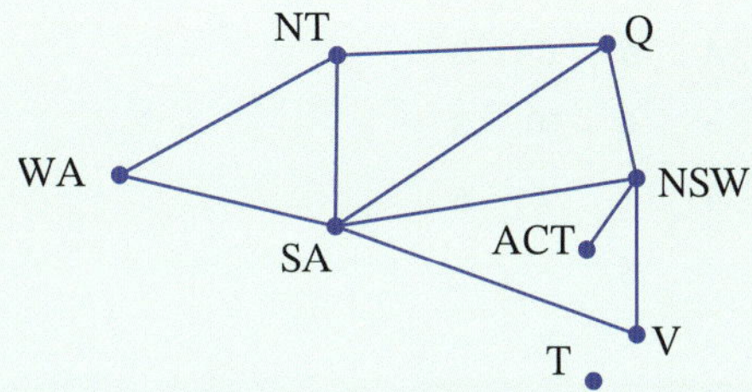


**b** SA has degree 5 since South Australia shares borders with 5 states or territories.

**c** Tasmania shares no land borders with any state or territory.

**EXERCISE 8.02** Answers on p. 472

## Constructing network diagrams

**1** Five towns, $L$, $M$, $P$, $Q$ and $R$, are linked by roads as shown by the network diagram and table.

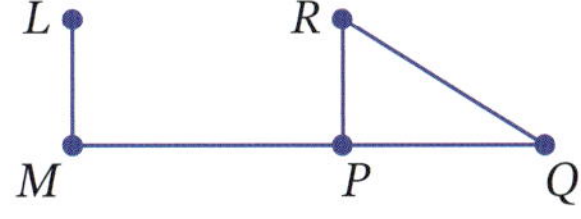


| | L | M | P | Q | R |
|---|---|---|---|---|---|
| **L** | 0 | 1 | 0 | 0 | 0 |
| **M** | 1 | 0 | 1 | 0 | 0 |
| **P** | 0 | $a$ | 0 | 1 | 1 |
| **Q** | 0 | 0 | 1 | 0 | 1 |
| **R** | 0 | $b$ | 1 | 1 | 0 |

**a** What is the meaning of a 0 in the table?

**b** Write the values of $a$ and $b$.

Foundation Mastery Complex

**2** Copy and complete the table below for this network.

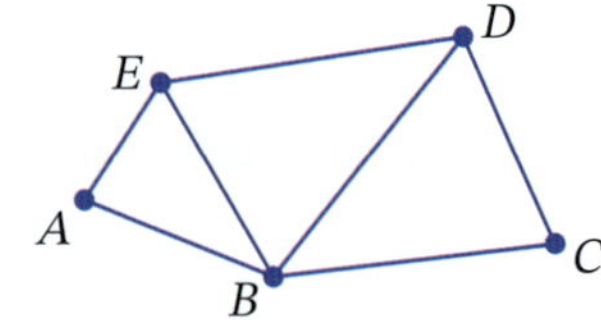


| | A | B | C | D | E |
|---|---|---|---|---|---|
| A | 0 | 1 | | | |
| B | | | 1 | | |
| C | | | | | |
| D | | | | | |
| E | | | | | |

**3** Draw the network that is represented by each table.

**a**

| | A | B | C | D |
|---|---|---|---|---|
| A | 0 | 1 | 1 | 1 |
| B | 1 | 0 | 1 | 1 |
| C | 1 | 1 | 0 | 1 |
| D | 1 | 1 | 1 | 0 |

**b**

| | A | B | C | D | E |
|---|---|---|---|---|---|
| A | 0 | 1 | 0 | 1 | 0 |
| B | 1 | 0 | 1 | 1 | 0 |
| C | 0 | 1 | 0 | 1 | 0 |
| D | 1 | 1 | 1 | 0 | 1 |
| E | 0 | 0 | 0 | 1 | 0 |

**4** Draw the directed weighted network that is represented by each table.

**a**

| | | To | | | | |
|---|---|---|---|---|---|---|
| | | J | K | L | M | N |
| From | J | – | – | – | – | – |
| | K | 5 | – | 5 | – | 4 |
| | L | – | 4 | – | 8 | 4 |
| | M | – | – | – | – | 6 |
| | N | – | – | 6 | – | – |

**b**

| | | To | | | |
|---|---|---|---|---|---|
| | | A | B | C | D |
| From | A | – | 6 | 10 | 3 |
| | B | 4 | – | 7 | 6 |
| | C | – | 5 | – | – |
| | D | – | – | 5 | – |

**5** The network diagram shows bus routes between 4 suburbs, labelled *M*, *P*, *R* and *T*. The directed arrows represent direct routes between the suburbs. Copy and complete the table to show the number of routes between pairs of suburbs.

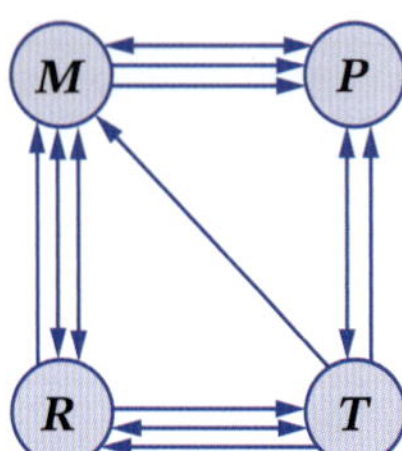


| | | To | | | |
|---|---|---|---|---|---|
| | | M | P | T | R |
| From | M | | | | |
| | P | | | | |
| | T | | | | |
| | R | | | | |

☐ Foundation ○ Mastery ⬡ Complex

6 The time it takes to drive between 6 locations can be shown using a network diagram or table. Use the network diagram to copy and complete the table.

| | A | B | G | H | P | W |
|---|---|---|---|---|---|---|
| A | 0 | | | | | |
| B | | 0 | | | | |
| G | | | 0 | | | |
| H | | | | 0 | | |
| P | | | | | 0 | |
| W | | | | | | 0 |

7 The table shows the road distances (in km) between 9 towns in northern NSW.

| | A | B | C | GI | Gr | T | U | W | Y |
|---|---|---|---|---|---|---|---|---|---|
| A | – | – | – | 98 | – | – | 172 | – | – |
| B | – | – | 64 | – | 131 | – | – | – | – |
| C | – | 64 | – | – | 100 | 126 | – | – | – |
| GI | 98 | – | – | – | 164 | 93 | – | 129 | – |
| Gr | – | 131 | 100 | 164 | – | – | 110 | – | – |
| T | – | – | 126 | 93 | – | – | – | – | 193 |
| U | 172 | – | – | – | 110 | – | – | – | – |
| W | – | – | – | 129 | – | – | – | – | 85 |
| Y | – | – | – | – | – | 193 | – | 85 | – |

*A* – Armidale
*B* – Ballina
*C* – Casino
*GI* – Glen Innes
*Gr* – Grafton
*T* – Tenterfield
*U* – Urunga
*W* – Warialda
*Y* – Yetman

Draw a network diagram to represent the information in the table.

8 This map shows some regions of Sydney.

EXAMPLE 5

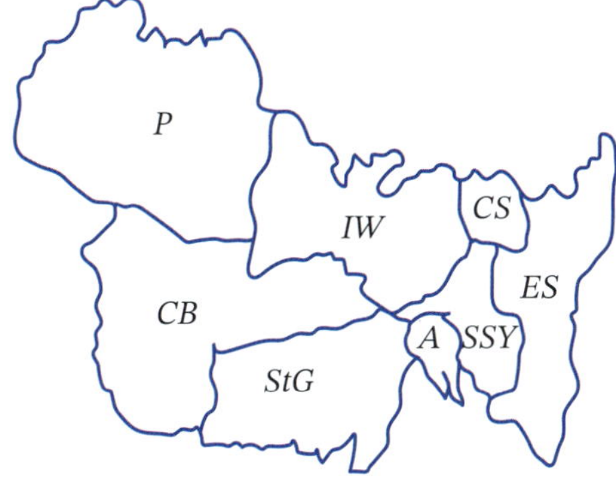


| | |
|---|---|
| A | Sydney Airport |
| CB | Canterbury-Bankstown |
| CS | Central Sydney |
| ES | Eastern Suburbs |
| IW | Inner West |
| P | Parramatta |
| SSY | South Sydney |
| StG | St George |

**a** Draw a network diagram for the map where the vertices represent the regions and the edges represent the borders.

**b** Which vertex has the highest degree?

**c** Which region shares the most borders?

Foundation | Mastery | Complex

9 A park has 6 different areas, $W$, $P$, $R$, $T$, $U$ and $V$.

A network diagram, with the different areas represented by vertices, has been drawn. The boundaries between the areas are represented by the edges.

**a** Find the degree of each vertex.

**b** Copy this map of the park and use the network diagram to represent the areas of the park.

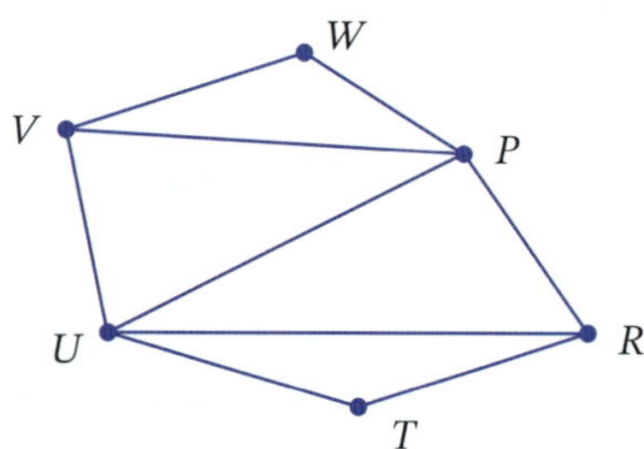


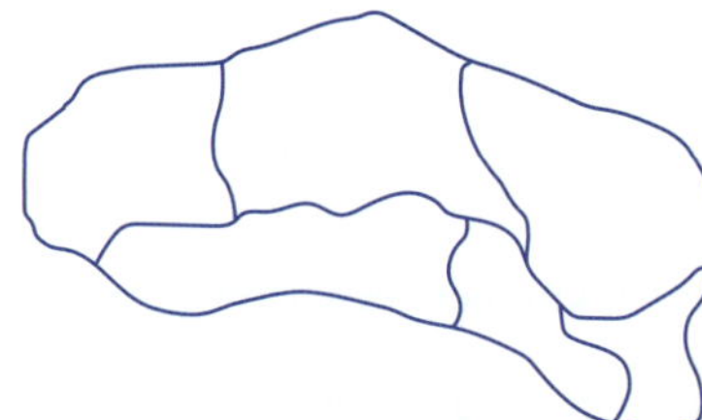

10 This map shows the number of ways you can travel between 4 towns. Copy and complete the table.

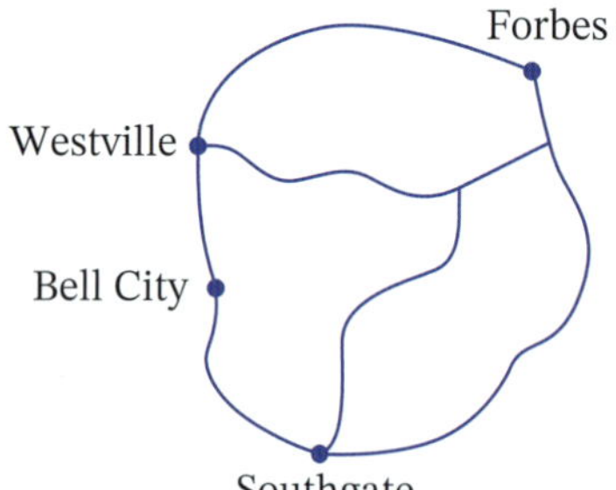


| | B | W | F | S |
|---|---|---|---|---|
| B | | | | |
| W | | | | |
| F | | | | |
| S | | | | |

11 This table shows the number of different walking trails between rest areas in a nature reserve. Draw a network diagram to show the information in the table.

| | A | B | C | D | E | F |
|---|---|---|---|---|---|---|
| A | 0 | 1 | 0 | 0 | 1 | 0 |
| B | 1 | 0 | 3 | 1 | 1 | 1 |
| C | 0 | 3 | 0 | 1 | 0 | 1 |
| D | 0 | 1 | 1 | 1 | 2 | 1 |
| E | 1 | 1 | 0 | 2 | 0 | 0 |
| F | 0 | 1 | 1 | 1 | 1 | 0 |

12 A company has several departments and the lines of communication between them are shown by this network diagram.

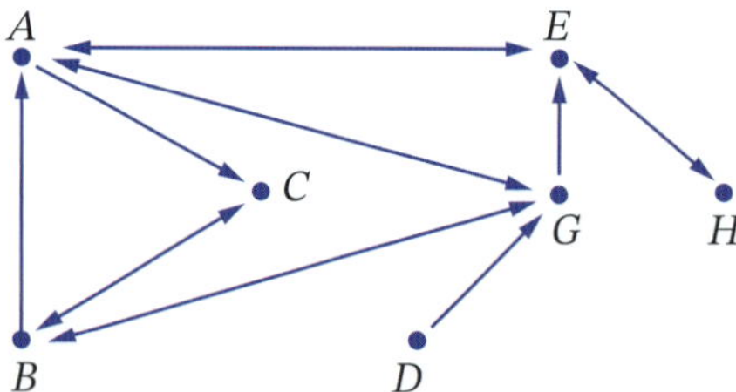


**a** How can department $H$ communicate with department $B$?

**b** Create a table for the information shown in the network diagram.

Foundation Mastery Complex

## Investigation

### Using networks

1 Go to the **Sydney Water** website and access the water network.

Obtain a copy of the water network diagram and show how water is supplied to:

a Sydney

b Blacktown.

2 Visit the **qantas.com** website and search for its Australian domestic flight network map.

a From which state and territory capital cities can you fly directly to every other capital city?

b From which capital cities can you fly directly to:

i Uluru (Ayers Rock)?

ii Lord Howe Island?

iii Broome?

iv Cairns?

3 Network diagrams can also be used to represent connections between people, such as contacts on social media.

With a group of about 10 people in your class or Year 12, draw a diagram showing contacts through a social media platform.

4 What is meant by the idea of '6 degrees of separation'? Write a brief report on your findings and decide whether this could apply to you and any randomly selected student at your school.

# 8.03 Minimum spanning trees

Worksheet
Minimum spanning trees

A **tree** is a network in which any 2 vertices are connected by exactly one path. This means that a tree is connected and has no cycles.

Here are some examples of trees.

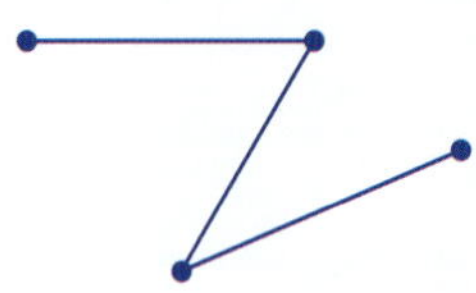

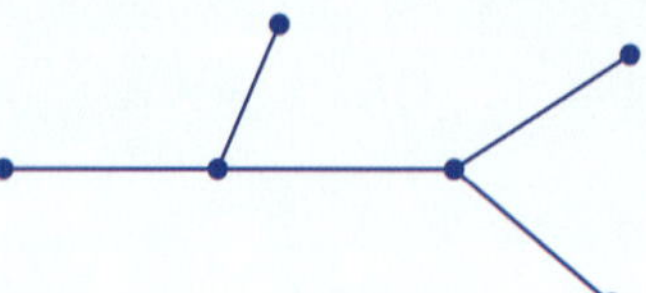

These are not trees. The left one is not connected and the right one has a cycle.

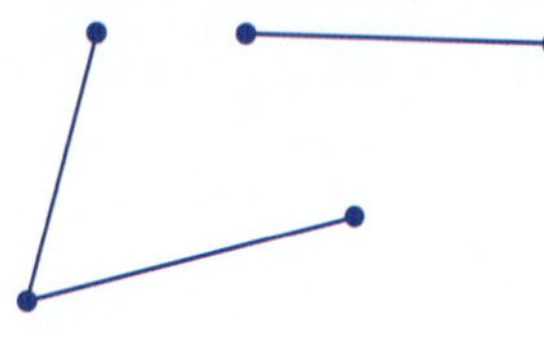

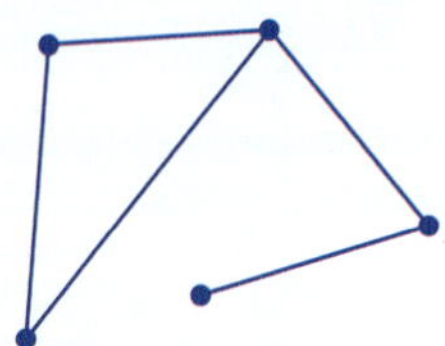

A **spanning tree** in a network is a tree in the network that connects all of the vertices. A network can have many different spanning trees.

For example, for this network:

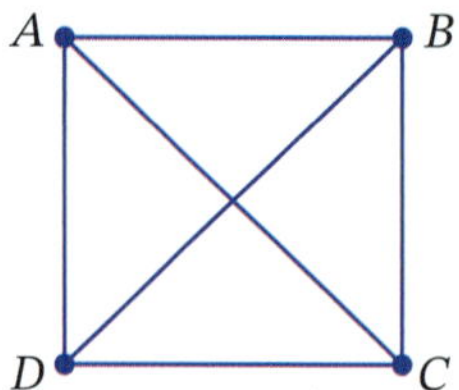


some of the spanning trees are:

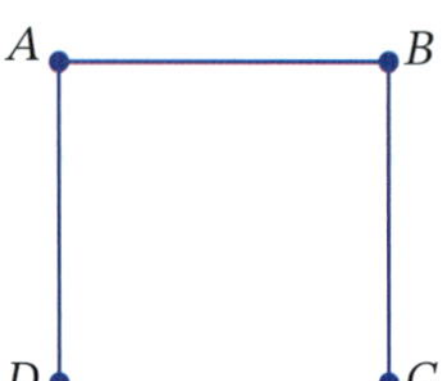


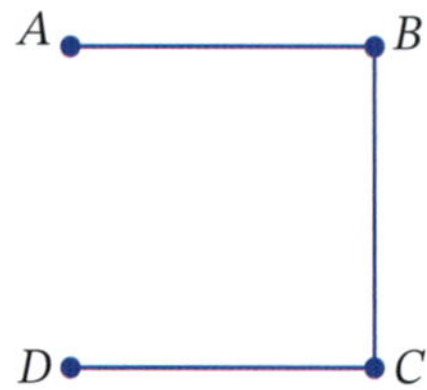


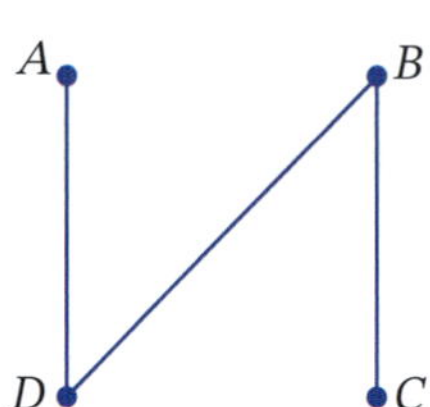


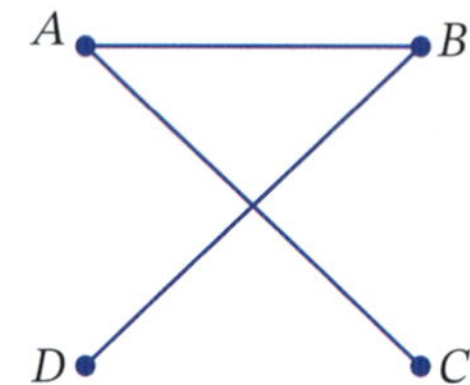


Note that each spanning tree connects the 4 vertices with 3 edges. Generally, if there are $n$ vertices in the network, then there will be $(n - 1)$ edges in a spanning tree. Each edge can be paired with a vertex, except for one vertex.

## Example 6

Determine whether each graph is a spanning tree of the network shown on the right.

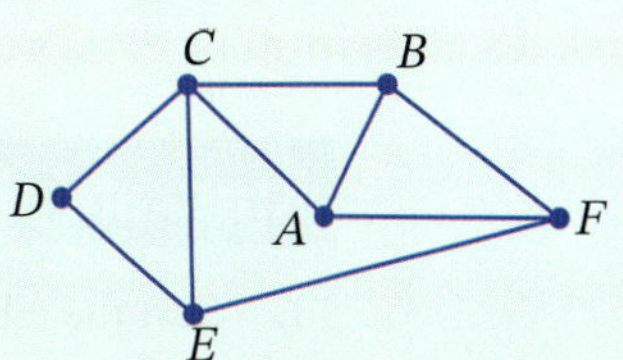

**a**

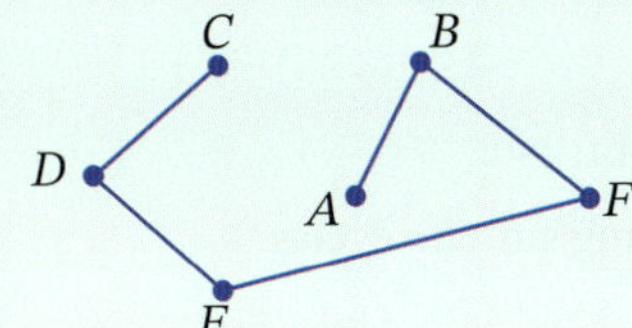

**b**

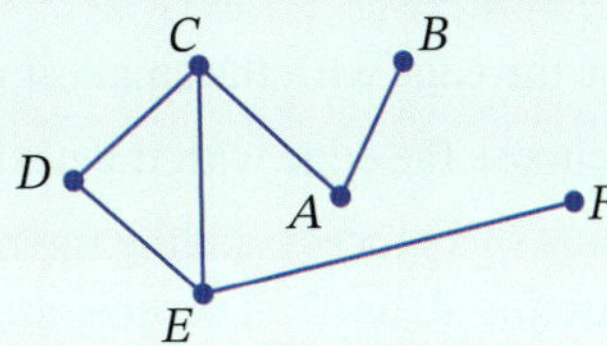

**c**

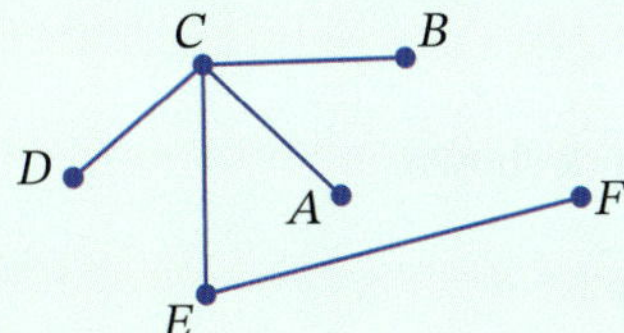

**d**

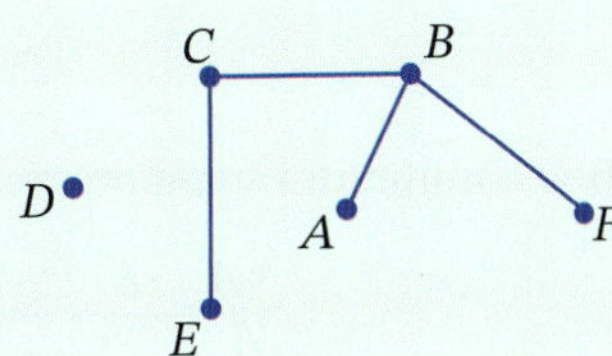

### Solution

**a** This is a spanning tree as it connects all the vertices and has no cycles.

**b** This is not a spanning tree as there is a cycle ($CDEC$).

**c** This is a spanning tree.

**d** This is not a spanning tree because it does not connect vertex $D$.

## Example 7

This network diagram represents 8 computers that are to be connected with optical fibre cables. The numbers show the costs (in $100s) of the cables.

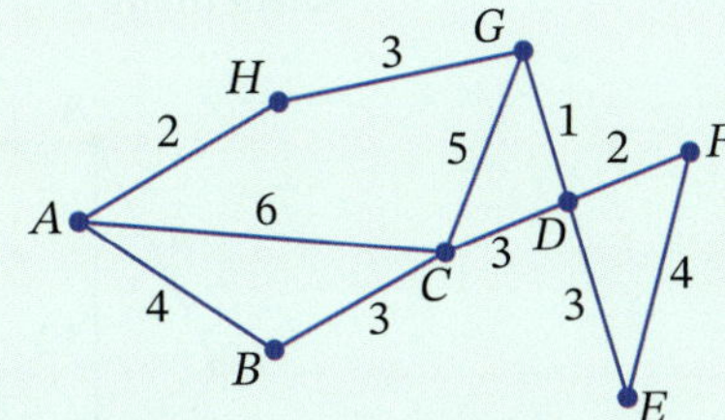

Draw 2 spanning trees for the network and determine which is the cheaper option.

### Solution

Two possible spanning trees are:

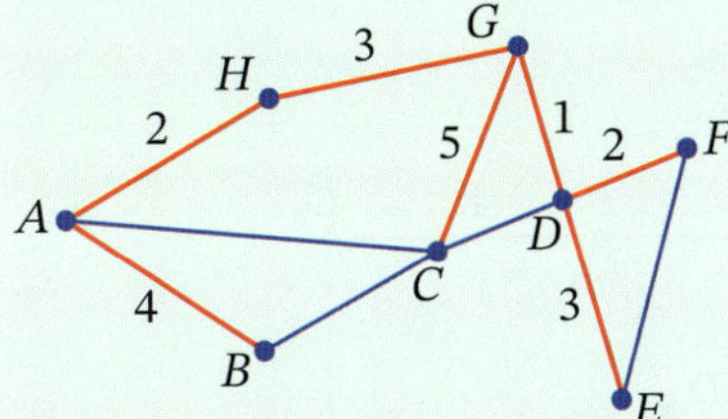

$$\text{Weight} = 4 + 2 + 3 + 5 + 1 + 2 + 3 = 20$$

Cost = $2000

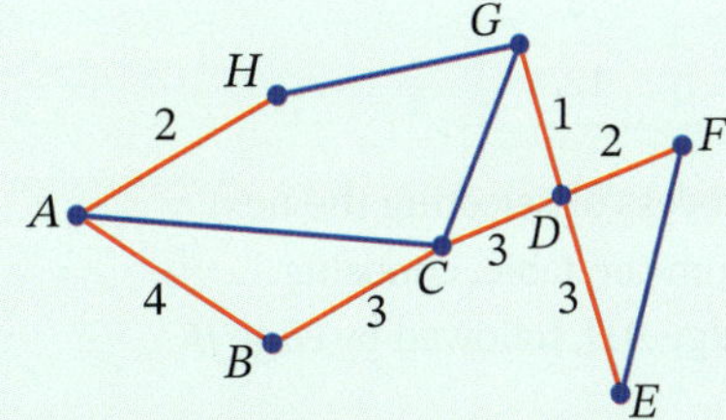

$$\text{Weight} = 2 + 4 + 3 + 3 + 3 + 1 + 2 = 18$$

Cost = $1800

This option is cheaper.

A **minimum spanning tree** in a weighted network is the spanning tree with the smallest weight. It is possible to find minimum spanning trees by trial and error but this can be time-consuming. Instead, we use 2 methods, **Kruskal's algorithm** or **Prim's algorithm**, where an algorithm is a process of precise steps, often repeated.

## Kruskal's algorithm for minimum spanning trees

1. Sort the edges in order of increasing weight.
2. Make a diagram of the network with all vertices, but no edges.
3. Choose the edge with the smallest weight.
4. Then choose the edge with the next smallest weight.
5. Continue this process, adding the next edge, making sure no cycles are introduced, until all vertices are connected.

Video
Kruskal's algorithm

## Example 8

Use Kruskal's algorithm to find a minimum spanning tree for the network shown.

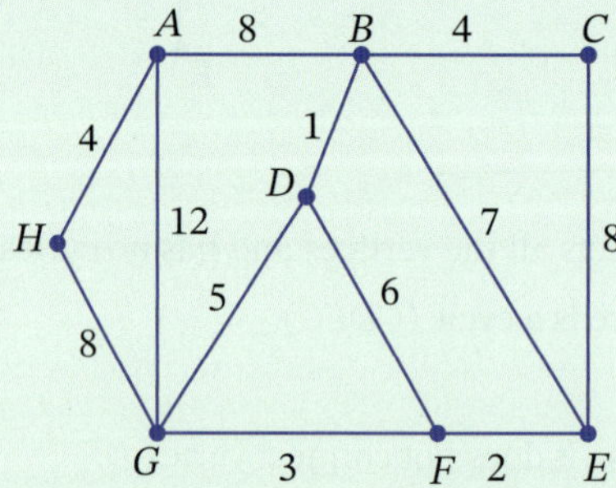


### Solution

**1** Sort the edges in increasing order.

| Edge | Weight | |
|---|---|---|
| *BD* | 1 | ✓ |
| *EF* | 2 | ✓ |
| *FG* | 3 | ✓ |
| *AH* | 4 | ✓ |
| *BC* | 4 | ✓ |
| *DG* | 5 | ✓ |
| *DF* | 6 | × |
| *BE* | 7 | × |
| *AB* | 8 | ✓ |
| *HG* | 8 | |
| *CE* | 8 | |
| *AG* | 12 | |

**2** Choose the edge with the smallest weight, *BD*, then the edge with the next smallest weight, *EF*. Tick off each edge from the table as you select them.

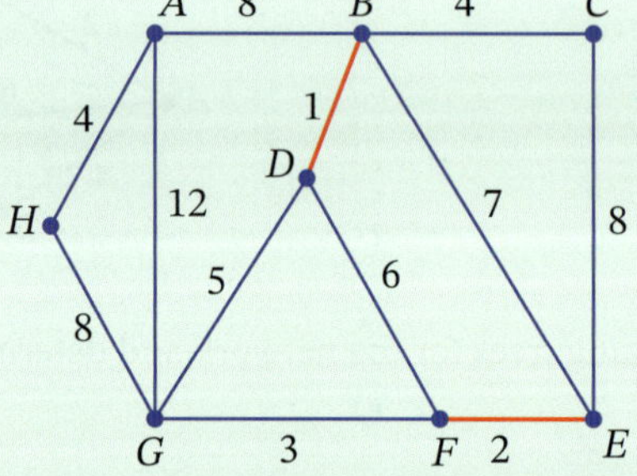


**3** Continue the process of selecting the next smallest edge from the table, choosing edge *FG*, then edge *AH*, followed by edge *BC*.

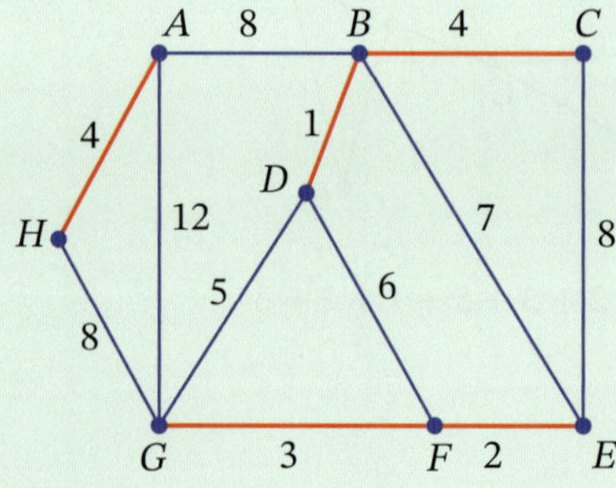

4 Choose $DG$, but **not** $DF$ as it will make a cycle. Put a cross next to it in the table. For the same reason, don't choose $BE$. Choose $AB$ next.

All vertices are now connected (8 vertices connected by 7 edges) so the minimum spanning tree is complete.

Weight of minimum spanning tree
$= 1 + 2 + 3 + 4 + 4 + 5 + 8$
$= 27$

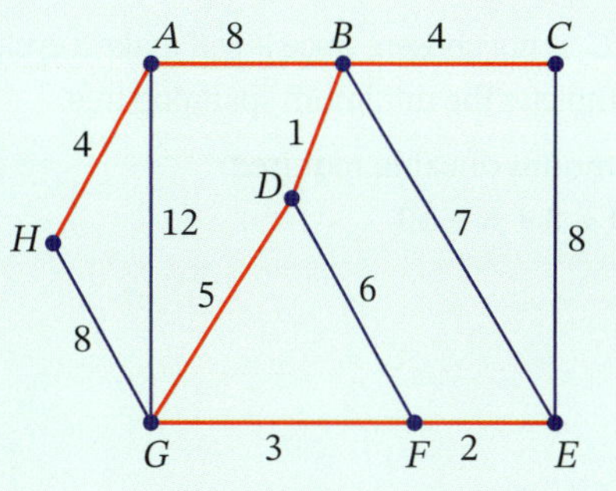


Note: This is not the only answer. A network can have more than one minimum spanning tree, but the minimum spanning trees will all have the same minimum weight.

## Example 9

William needed to connect electricity to 5 cabins at a camping site. The weighted graph shows the distances between the cabins (in metres) and the location of the electricity source ($M$).

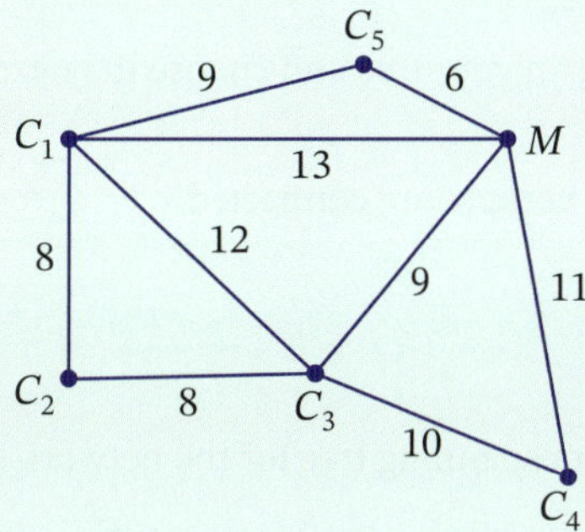


Use Kruskal's algorithm to find the least amount of cable needed to connect electricity to all sites.

### Solution

1 Sort the edges in increasing order.

| Edge | Weight | |
|---|---|---|
| $MC_5$ | 6 | ✓ |
| $C_1C_2$ | 8 | ✓ |
| $C_2C_3$ | 8 | ✓ |
| $C_1C_5$ | 9 | ✓ |
| $MC_3$ | 9 | × |
| $C_3C_4$ | 10 | ✓ |
| $MC_4$ | 11 | |
| $C_1C_3$ | 12 | |
| $MC_1$ | 13 | |

2 Choose the first 4 edges from the table, starting with $MC_5$.

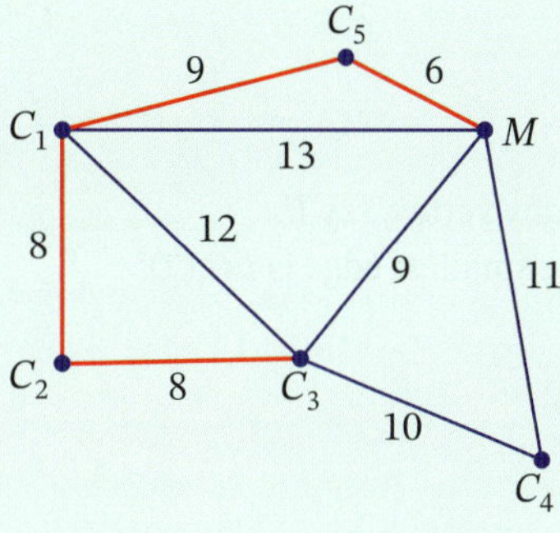

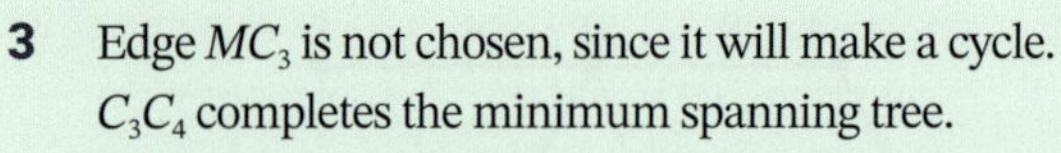

**3** Edge $MC_3$ is not chosen, since it will make a cycle.
$C_3C_4$ completes the minimum spanning tree.

Least amount of cable required
$= 6 + 9 + 8 + 8 + 10$
$= 41$ m

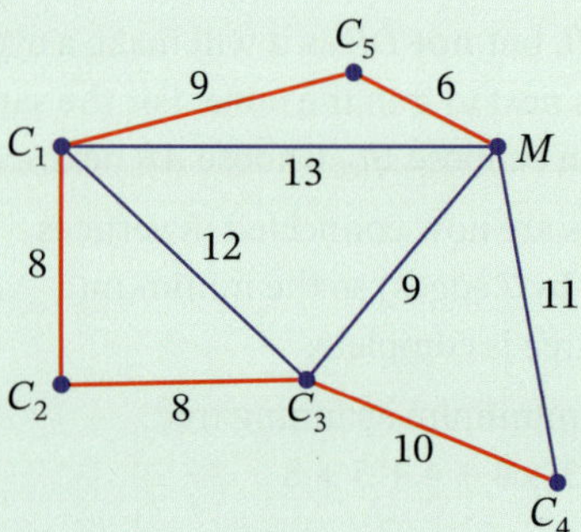

## Prim's algorithm for minimum spanning trees

While Kruskal's algorithm looks at the edges, Prim's algorithm focuses on the vertices.

1. Choose any vertex within the network.
2. Choose the edge with the least weight connecting this vertex to another vertex.
3. Look at all edges connecting with these 2 vertices and select the next edge with the least weight to add to the tree.
4. Look at all the vertices in your tree and choose the next edge with the least weight to add to the tree.
5. Repeat step 4 until all vertices are connected.

## Example 10

Use Prim's algorithm to find a minimum spanning tree for the network shown.

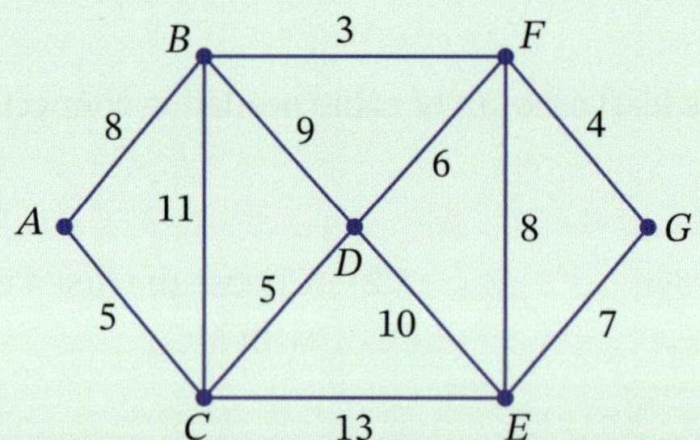


### Solution

**1** Start with any vertex, say $E$.
From $E$, the smallest edge is $EG$ (7).

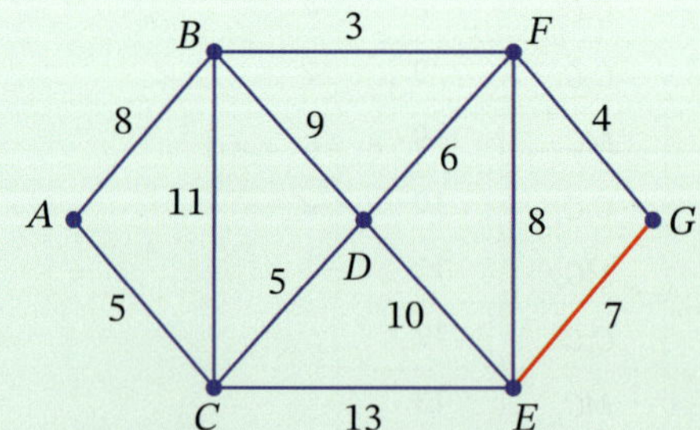


**2** The next smallest edge from $E$ or $G$ is $GF$ (4).

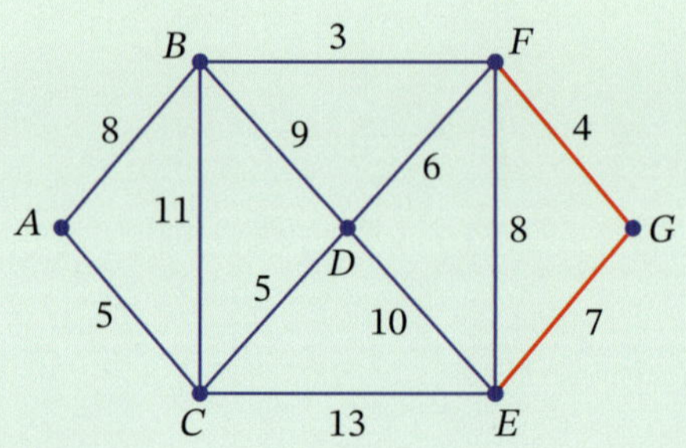

**3** The next smallest edge from $F$ or $E$ is $FB$ (3), which is added to the tree.

Then the the next smallest edge from $B$, $E$ or $F$ is $FD$ (6).

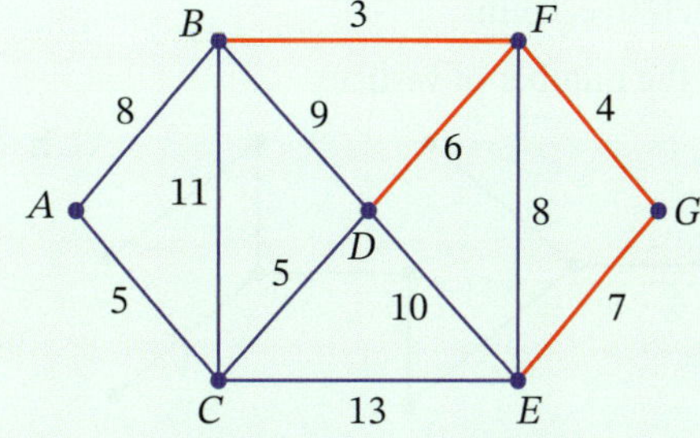


**4** The next smallest edge to connect to the tree is $DC$ (5), then edge $CA$ (5).

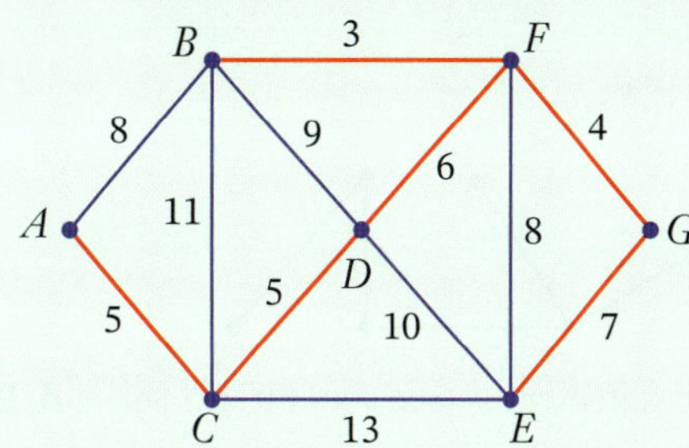


The minimum spanning tree is complete (7 vertices, 6 edges).

Weight of minimum spanning tree = $7 + 4 + 3 + 6 + 5 + 5 = 30$

**EXERCISE 8.03** Answers on p. 473

## Minimum spanning trees

**1** For each network, state whether or not it is a tree. Give reasons.

**a**

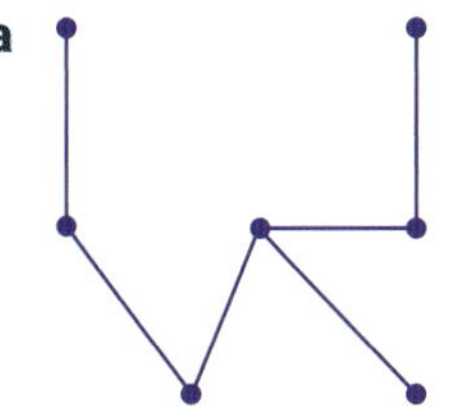

**b**

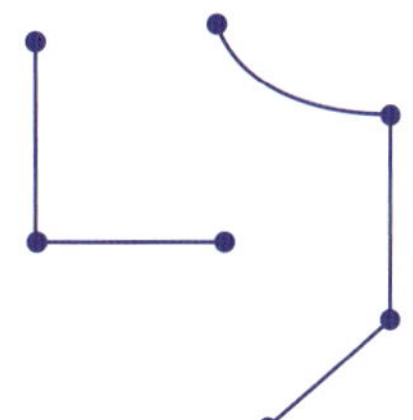

**c**

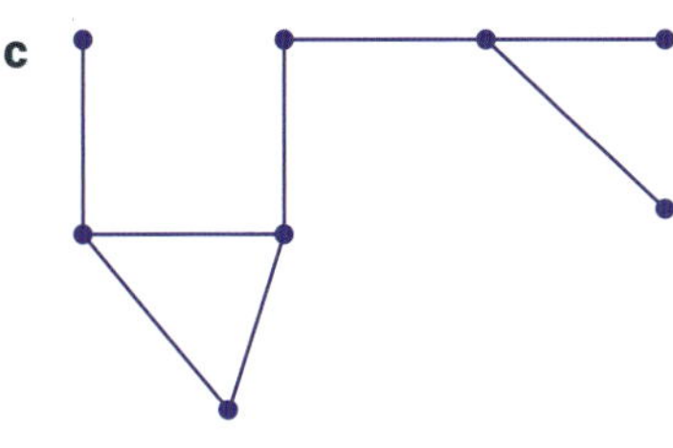

**d**

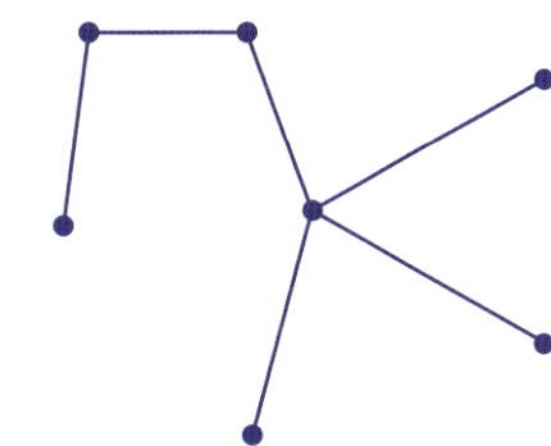

**e**

**2** For each tree, count:

**i** the number of vertices **ii** the number of edges.

**a**

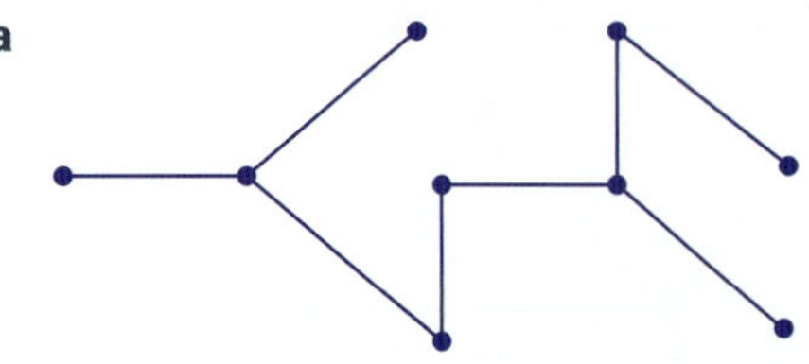

**b**

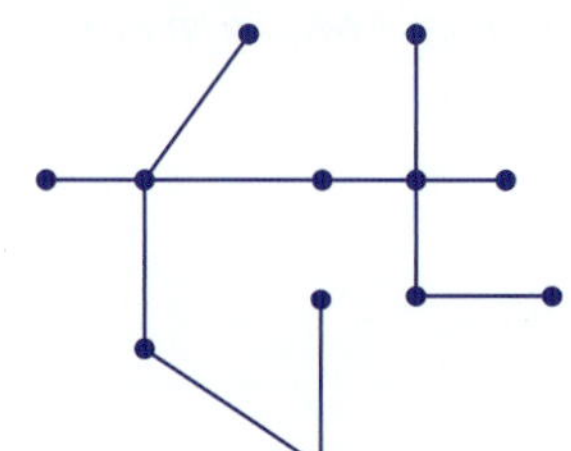

**c**

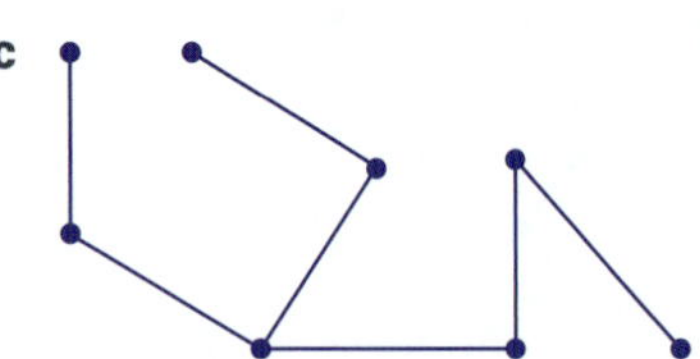

**3** Which statement is true about trees? Select **A**, **B**, **C** or **D**.

**A** The number of vertices is equal to the number of edges.

**B** A tree with $n$ vertices has $(n - 1)$ edges.

**C** A tree with $n$ vertices has $(n + 1)$ edges.

**D** A tree with $n$ vertices has $2n$ edges.

EXAMPLE 6

**4** Determine which of the following are spanning trees of the network shown.

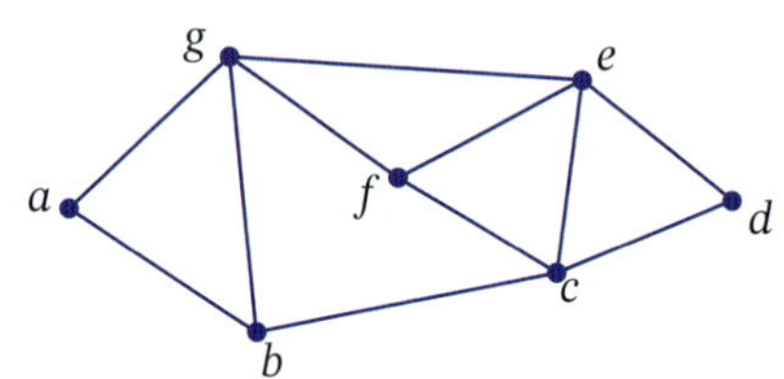


**a**

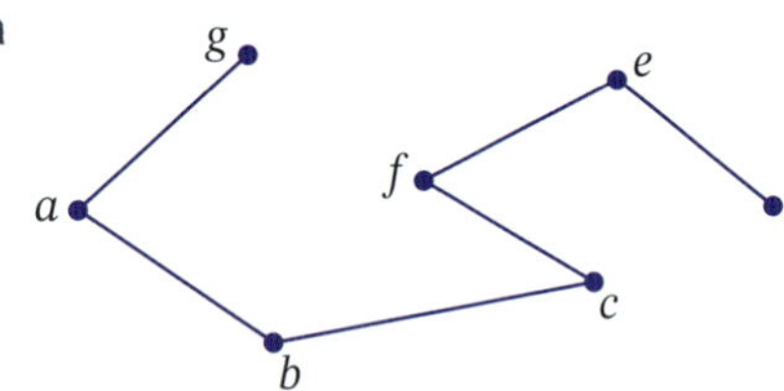


**b**

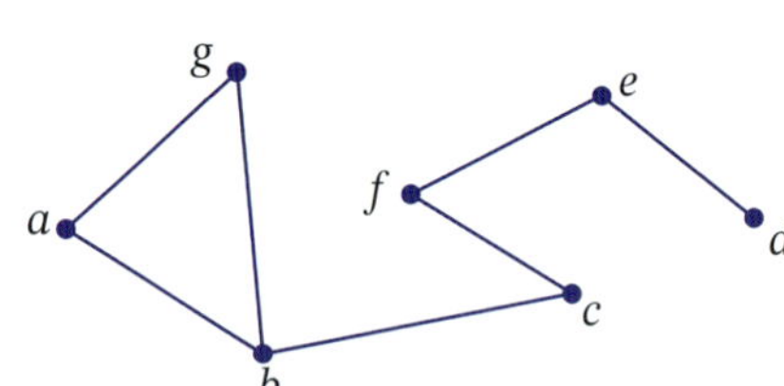


**c**

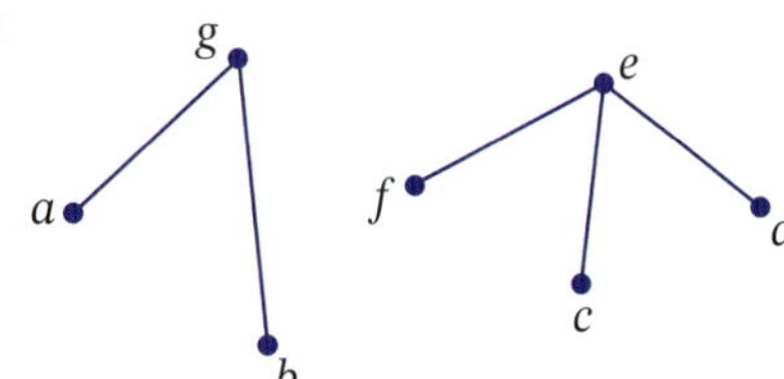


**d**

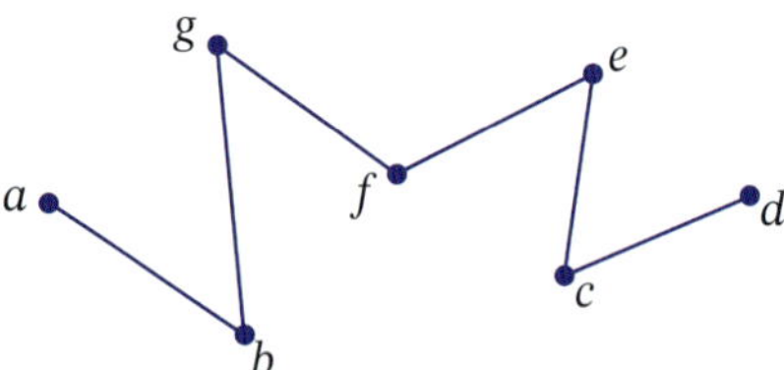


**e**

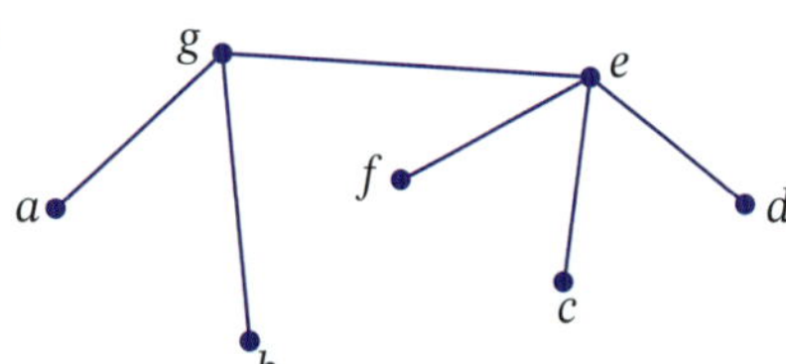


**f**

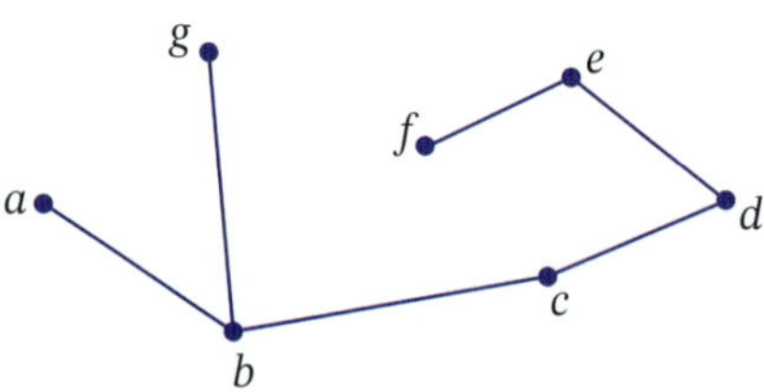


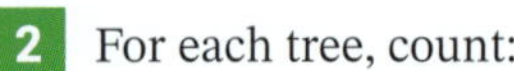

Foundation Mastery Complex

**5** Find 3 spanning trees for the network below.

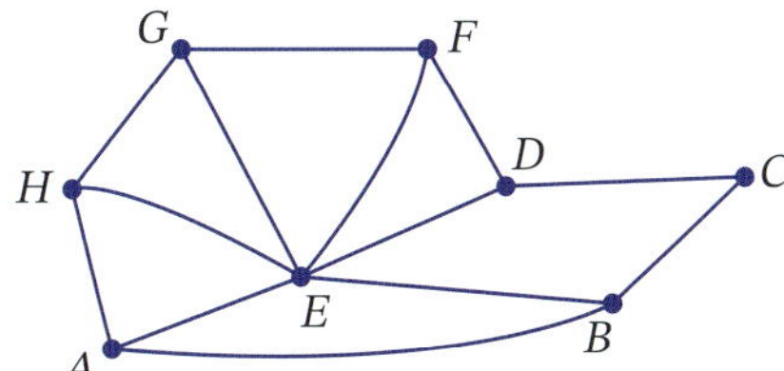


**6** Consider this network, which shows the cost (in dollars) of laying power cables from a mains power supply ($M$) to different areas of a camping and caravan park.

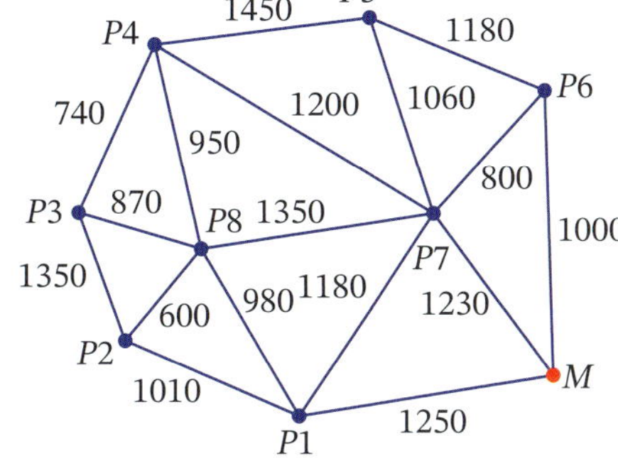


**a** Spanning trees for the network are shown below. Find the weight of each tree.

**i**

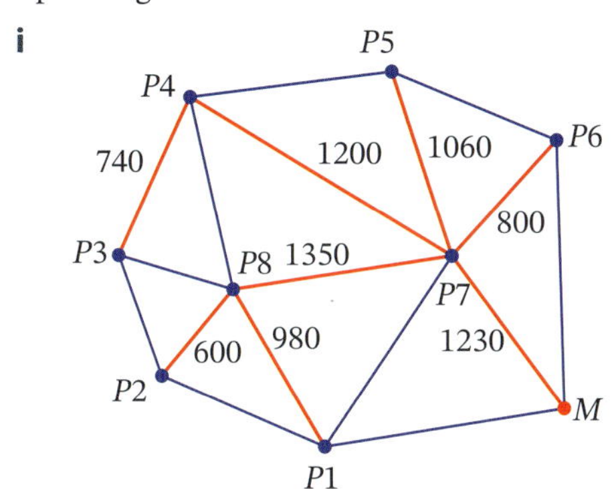


**ii**

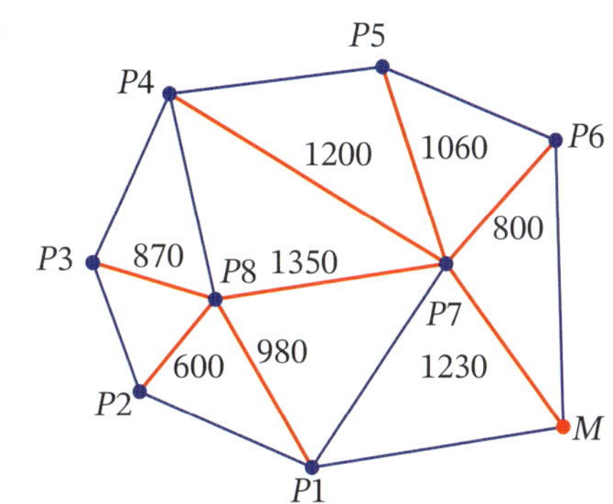


**iii**

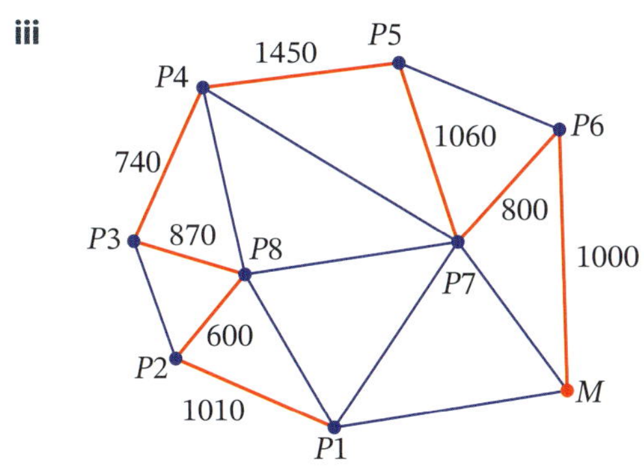


**iv**

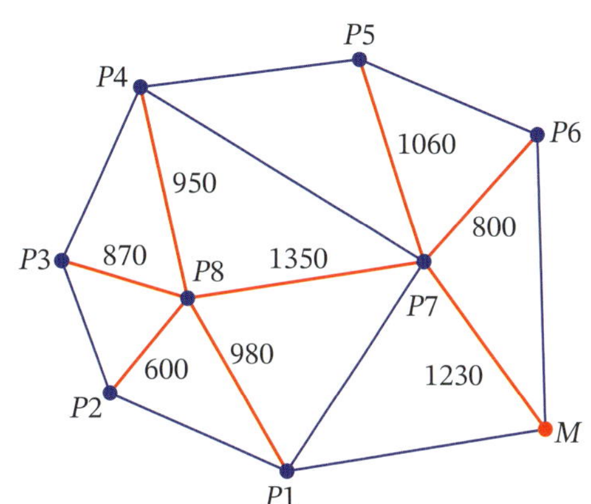


**b** Which spanning tree in part **a** is the best option? Is this tree the minimum spanning tree?

Foundation Mastery Complex

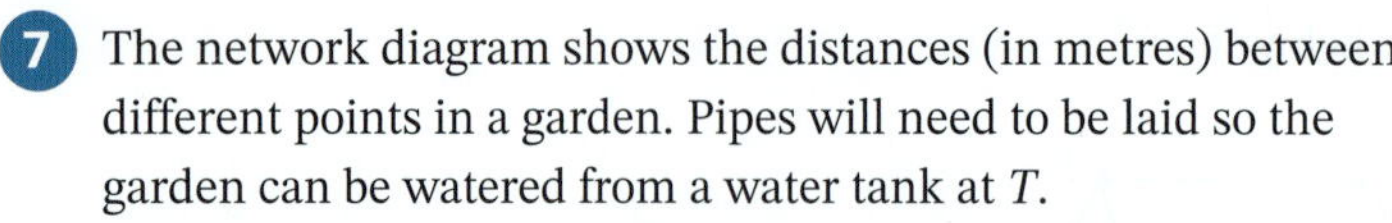

**7** The network diagram shows the distances (in metres) between different points in a garden. Pipes will need to be laid so the garden can be watered from a water tank at $T$.

Draw spanning trees for the network and by finding their weight, determine the minimum length of water pipe required to supply water to all parts of the garden.

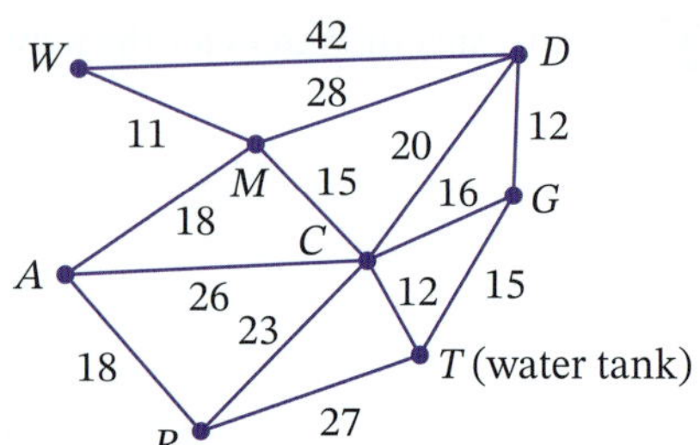

**8** **a** List the edges of this network in increasing order.

**b** Use Kruskal's algorithm to find the minimum spanning tree.

**c** What is the weight of the minimum spanning tree?

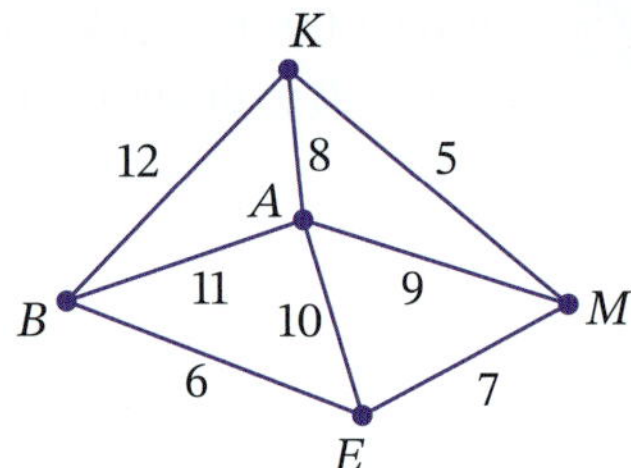


Worksheet
Spanning tree diagrams

The worksheet 'Spanning tree diagrams' contains the diagrams from Questions **9**, **13**, **14** and **15**.

**9** For each network below:

**i** use Kruskal's algorithm and showing the order in which the edges are selected, find the minimum spanning tree for each network and calculate its weight

**ii** verify that a spanning tree with $n$ vertices has $(n - 1)$ edges.

**a**

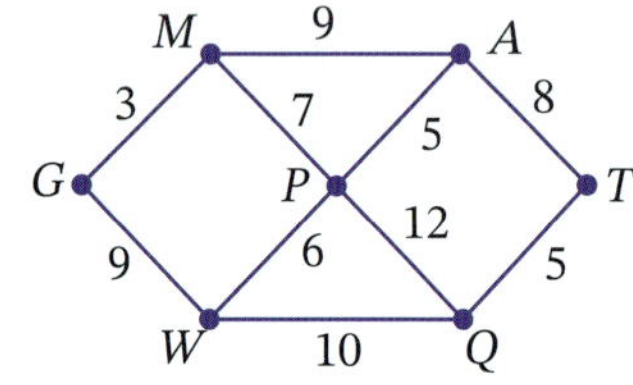


**b**

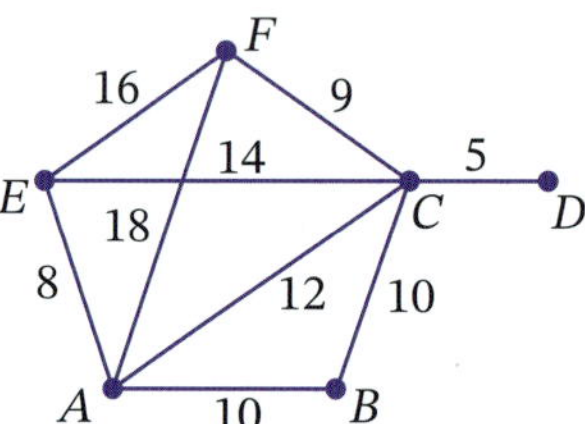


**c**

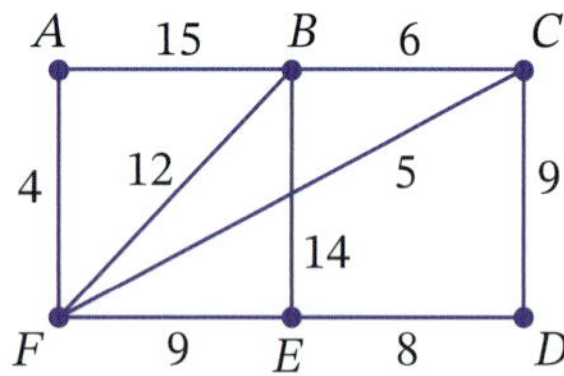


**d**

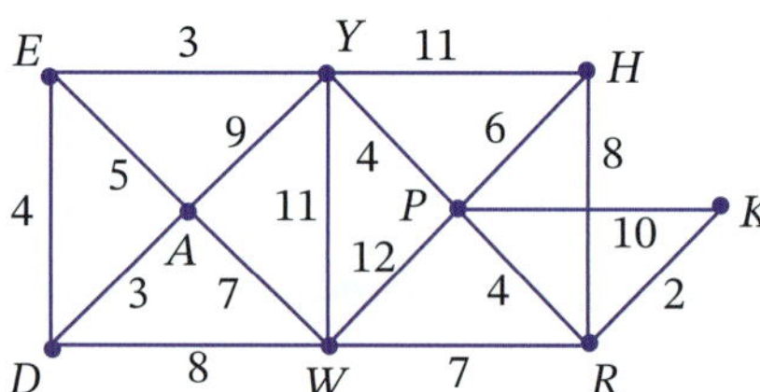

**10** The cost involved in connecting a school's computer network is shown in the network diagram.

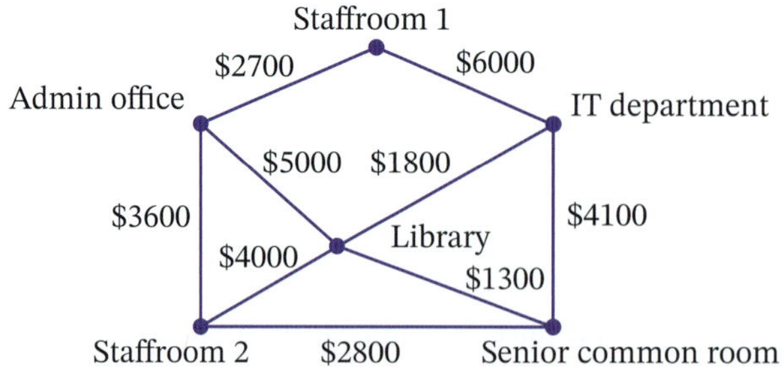


**a** Use Kruskal's algorithm to find the minimum spanning tree for the network.

**b** Calculate the minimum cost of connecting the network.

☐ Foundation ○ Mastery ○ Complex

**11** The minimum spanning tree for the network includes 2 edges with weighting $a$ and $b$.

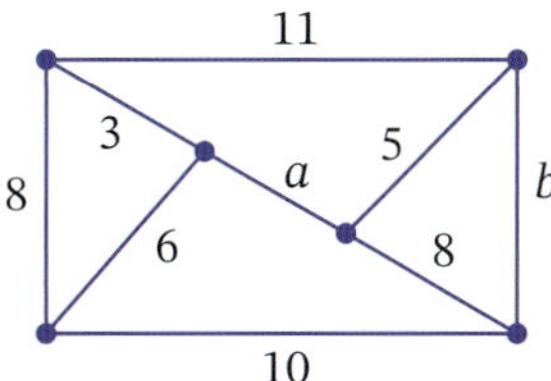


The length of the minimum spanning tree is 25.

What are the values of $a$ and $b$? Select **A**, **B**, **C** or **D**.

**A** $a = 6, b = 5$ **B** $a = 8, b = 4$

**C** $a = 9, b = 3$ **D** $a = 7, b = 10$

**12**

EXAMPLE 10

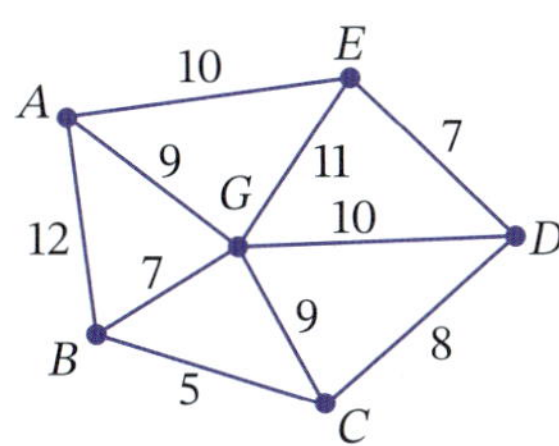


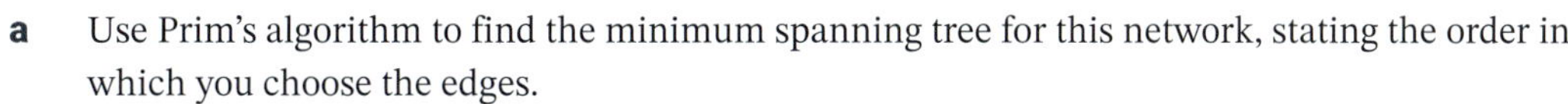

**a** Use Prim's algorithm to find the minimum spanning tree for this network, stating the order in which you choose the edges.

**b** What is the total weight of the tree?

**c** Is there another minimum spanning tree for this network?

**13** For each network below:

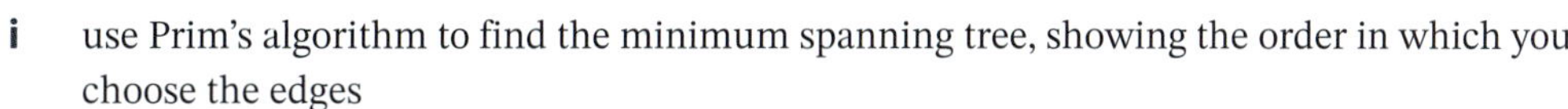

**i** use Prim's algorithm to find the minimum spanning tree, showing the order in which you choose the edges

**ii** calculate the weight of the tree.

**a**

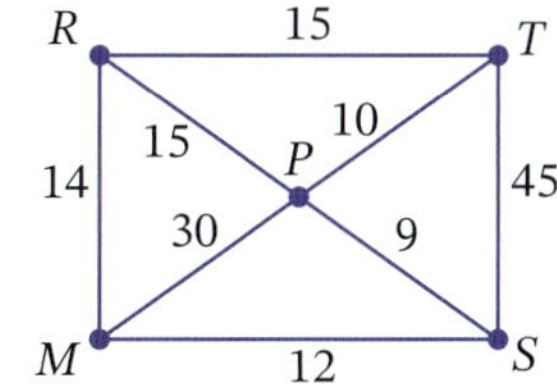


**b**

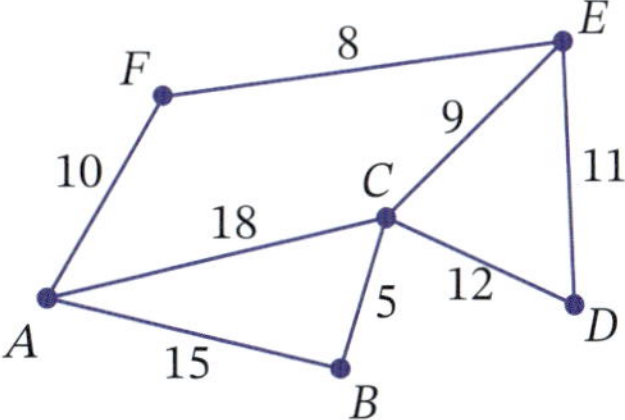


**c**

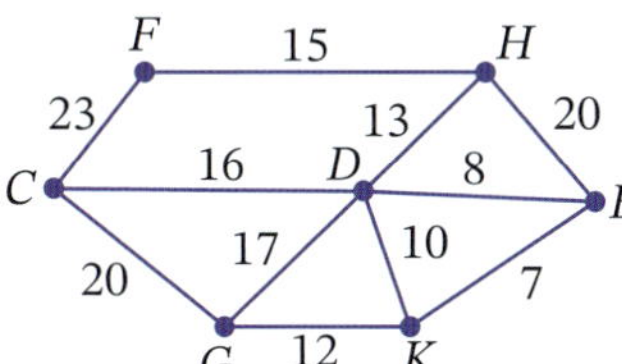


**d**

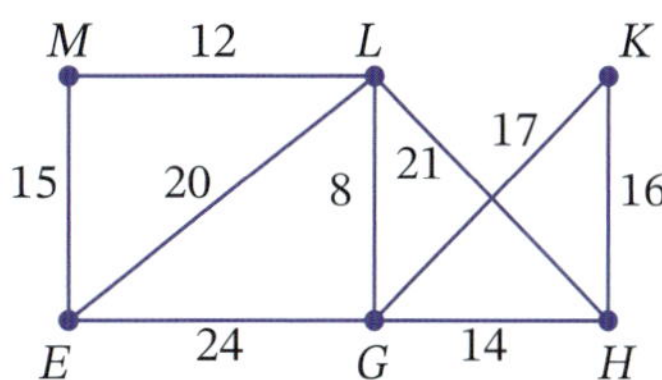

Foundation Mastery Complex

**14** The network diagram shows the distances, in kilometres, between Adelaide ($A$), Brisbane ($B$), Canberra ($C$), Melbourne ($M$) and Sydney ($S$).

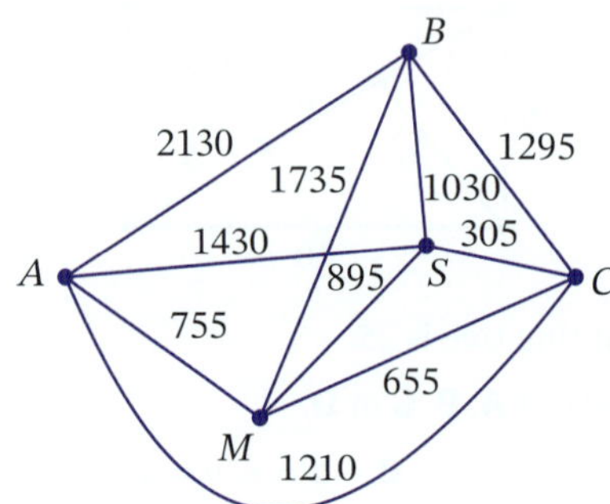


**a** Use Prim's algorithm to find a minimum spanning tree for the 5 cities, showing the order in which the edges are chosen.

**b** Calculate the length of your spanning tree.

**15** **a** The graph shows the road distances between 9 suburbs of a city, in kilometres.

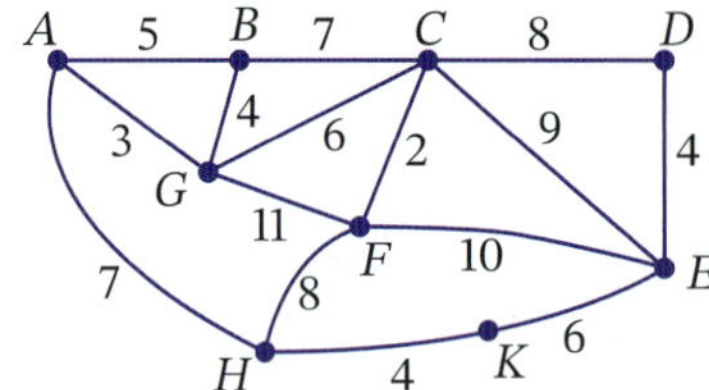


Find a minimum spanning tree for the road network, using:

**i** Kruskal's algorithm

**ii** Prim's algorithm.

**b** Are the minimum spanning trees the same?

**c** Find the length of the minimum spanning tree.

**d** Find the length of the shortest route from $A$ to $D$.

**16** The table shows the road distances between 6 towns in kilometres. A dash (–) indicates the towns are not directly connected.

| | A | B | C | D | E | F |
|---|---|---|---|---|---|---|
| **B** | 22 | | | | | |
| **C** | 30 | 8 | | | | |
| **D** | 15 | 42 | – | | | |
| **E** | – | 10 | 15 | – | | |
| **F** | 40 | 9 | 12 | 24 | – | |

**a** Draw a network diagram showing the information in the table.

**b** Find the minimum spanning tree for the network using:

**i** Kruskal's algorithm

**ii** Prim's algorithm.

**c** Find the length of the minimum spanning tree.

☐ Foundation ◯ Mastery ◯ Complex

**17** This network diagram shows the lengths of water pipes to different locations in a garden. From the options, which tree marked by red edges is a minimum spanning tree? Select **A**, **B**, **C** or **D**.

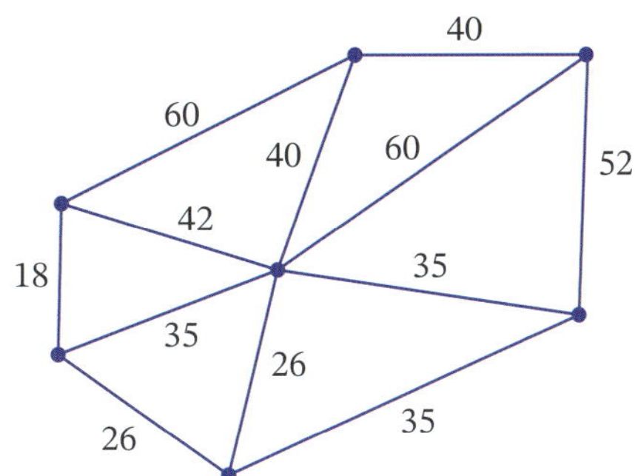


**A**

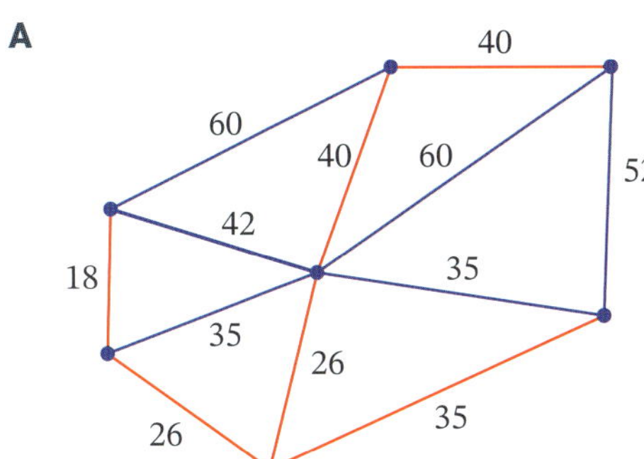


**B**

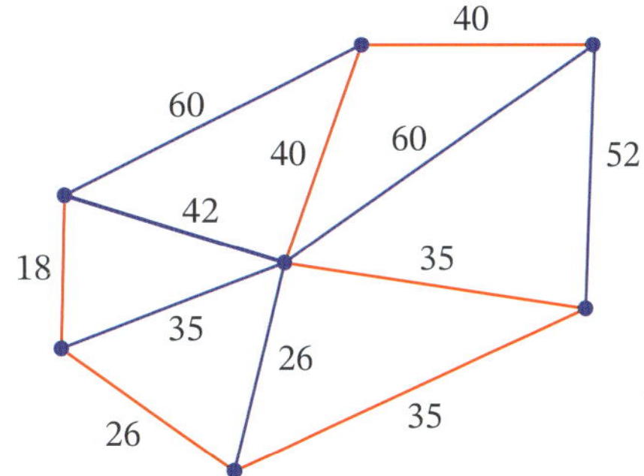


**C**

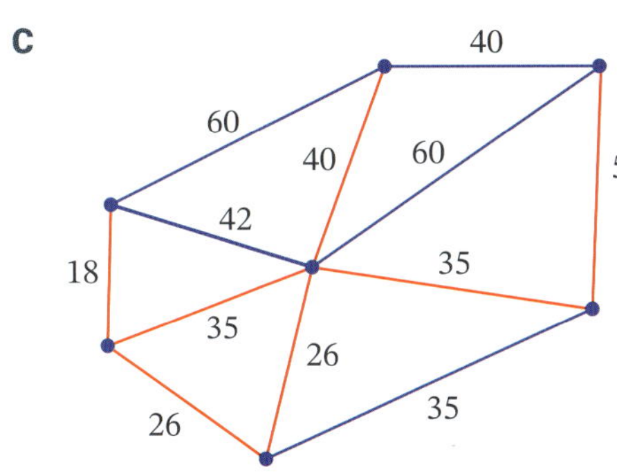


**D**

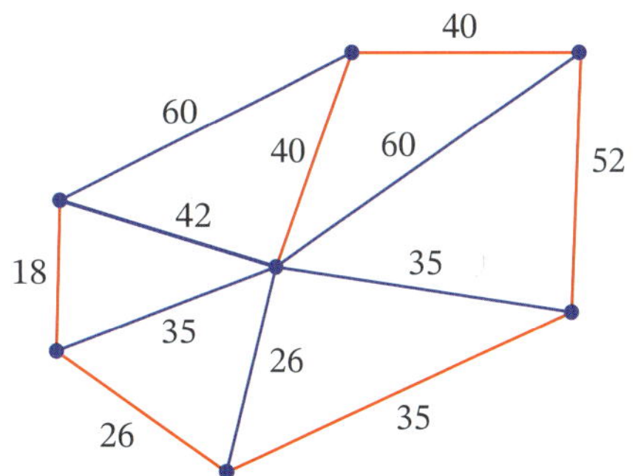


**18** This network diagram shows a minimum spanning tree (in red).

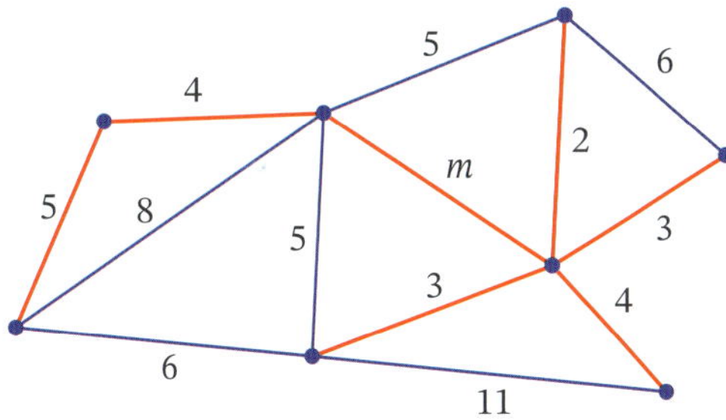


Find the possible values for $m$ if this is the only minimum spanning tree.

**19** This table shows the distances in kilometres between towns in NSW.

| | *D* | *B* | *C* | *WaW* | *N* | *H* | *WeW* |
|---|---|---|---|---|---|---|---|
| Dubbo (*D*) | – | 206 | – | – | – | – | 259 |
| Bathurst (*B*) | 206 | – | 106 | – | – | – | – |
| Cowra (*C*) | – | 106 | – | 208 | – | – | 159 |
| Wagga Wagga (*WaW*) | – | – | 208 | – | 98 | – | – |
| Narrandera (*N*) | – | – | – | 98 | – | 173 | 135 |
| Hay (*H*) | – | – | – | – | 173 | – | 256 |
| West Wyalong (*WeW*) | 259 | – | 159 | – | 135 | 256 | – |

**a** Draw a network diagram for the information given in the table.

**b** Find the minimum spanning tree for the network.

**c** Find the length of the minimum spanning tree.

**20** The costs of connecting 6 locations in a school with optical fibre cable are given in this table.

| | *B* | *C* | *D* | *E* | *F* |
|---|---|---|---|---|---|
| *A* | 7000 | – | 8000 | – | 10 000 |
| *B* | – | 3000 | 4200 | 1500 | – |
| *C* | – | – | 5100 | 4000 | 6000 |
| *D* | – | – | – | 6000 | 3600 |

**a** Draw a network diagram showing the information in the table.

**b** Find the minimum spanning tree for the network (showing the order in which edges are added).

**c** Use the minimum spanning tree to find the lowest cost of connecting the cable.

## Technology

### Spanning tree practice

Search for 'minimum spanning tree practice—GeoGebra' on the internet to practise finding the minimum spanning tree for a network.

Foundation Mastery Complex

# Shortest path problems

## 8.04

For small weighted networks, the **shortest path** between 2 vertices can be found by looking at the network (by inspection) and using trial and error.

**Video**
Shortest path

**Puzzle**
Networks find-a-word

### Example 11

This network shows the travel times (in minutes) along roads from Dilnoor's house to Vanessa's house. Find the shortest time it will take Dilnoor to travel to Vanessa's house.

#### Solution

By inspection, the shortest trip will be along the roads numbered 13, 8, 3, 2.

$13 + 8 + 3 + 2 = 26$

The shortest time for Dilnoor to get to Vanessa's house is 26 minutes.

### Example 12

**a** Find a shortest path from $B$ to every other vertex in the network.

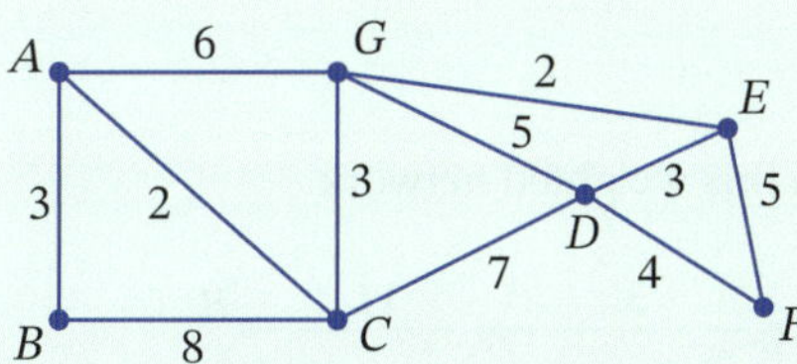


**b** Use the shortest paths from vertex $B$ to obtain a spanning tree for the network and find its weight.

**c** Find the minimum spanning tree for the network and its weight.

**d** Compare the shortest path spanning tree with the minimum spanning tree.

#### Solution

**a** The shortest paths from $B$ are shown in the table below.

| Vertex | $A$ | $C$ | $D$ | $E$ | $F$ | $G$ |
|---|---|---|---|---|---|---|
| Path | $BA$ | $BAC$ | $BACD$ | $BACGE$ | $BACGEF$ | $BACG$ |
| Weight | 3 | 5 | 12 | 10 | 15 | 8 |

**b** When the shortest paths from $B$ are used, this spanning tree is obtained.

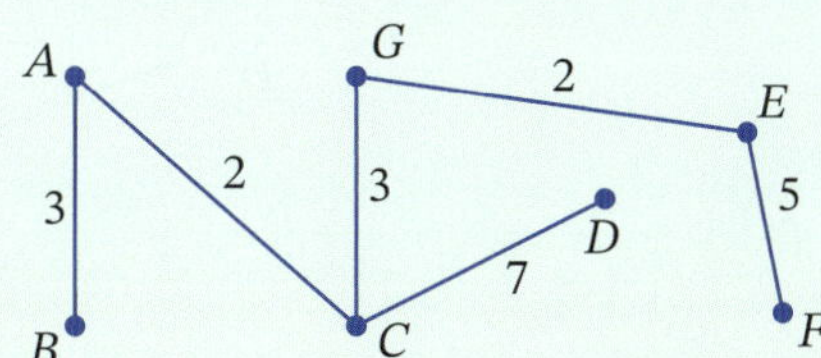


Weight of spanning tree $= 3 + 2 + 3 + 2 + 5 + 7 = 22$

**c** The minimum spanning tree is:

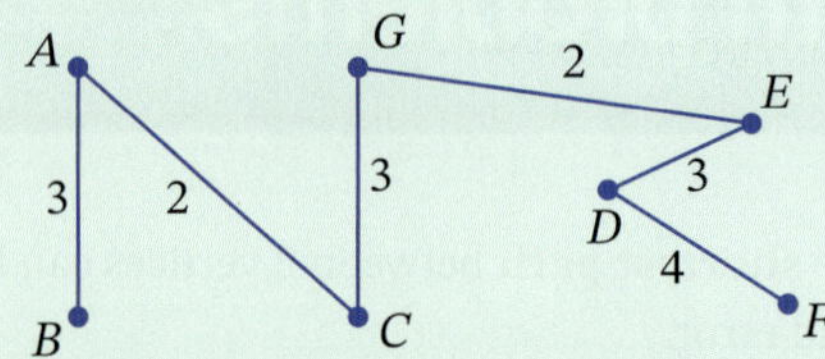


Weight of minimum spanning tree $= 3 + 2 + 3 + 2 + 3 + 4 = 17$

**d** The weight of the shortest path spanning tree is greater than the weight of the minimum spanning tree. Also, the shortest paths between the vertices $C$ and $D$ and between $E$ and $F$ are not contained in the minimum spanning tree.

Note that the shortest path between 2 vertices is a path connecting those vertices, whereas a minimum spanning tree is a tree that connects all vertices. For some networks, there may be a shortest path between 2 vertices that is not contained in any minimum spanning tree.

## Finding the shortest path

1. Redraw the network with circles at each vertex, except at the starting vertex, $S$.
2. For all vertices one edge away from $S$, write down the shortest distance inside the circle.
3. For all vertices 2 edges away from $S$, write down the shortest distance inside the circle.
4. Continue this process until the finish vertex, $F$, is reached.
5. The shortest path is then identified by starting at $F$ and moving backwards to the vertex from which the shortest distance at $F$ was obtained, and continuing until $S$ is reached.

## Example 13

Find the shortest path from $X$ to $Y$ in this weighted network.

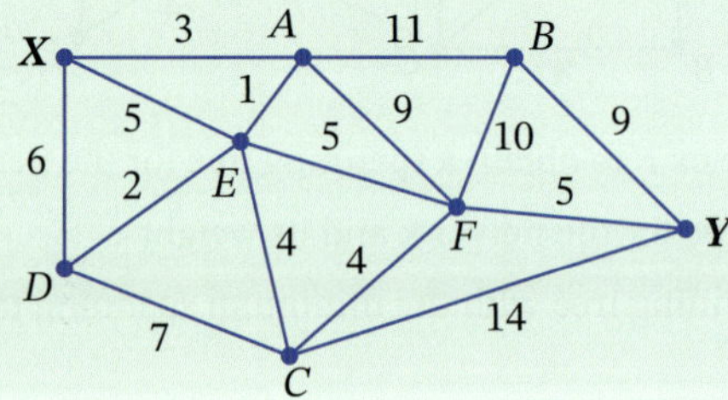


### Solution

**1** Redraw the network with circles at each vertex, except at the starting vertex, $X$.

**2** For all vertices **one** edge away from $X$, write down the shortest distance inside the circle.

The shortest distance to $E$ is $3 + 1 = 4$ (path $XAE$).

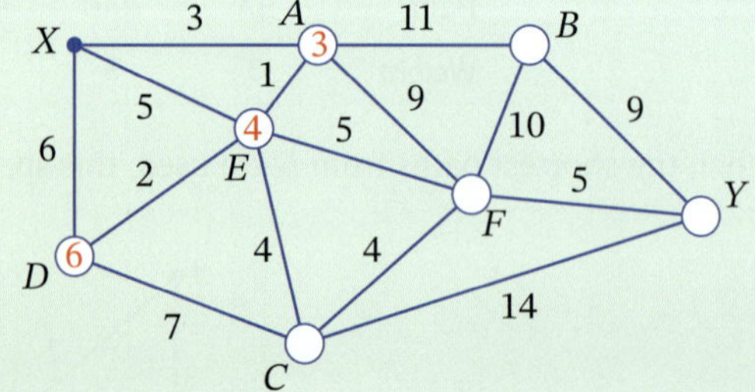

**3** For all vertices **2** edges away from $X$, write down the shortest distance inside the circle with the help of the previous circle values.

The shortest distance to $C$ is $4 + 4 = 8$, using the path $XAEC$.

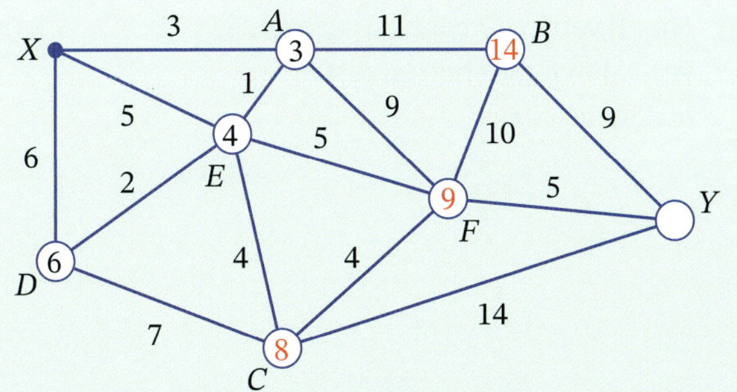

**4** Continuing the process, the shortest distance from $X$ to $Y$ is $9 + 5 = 14$ (via $F$).

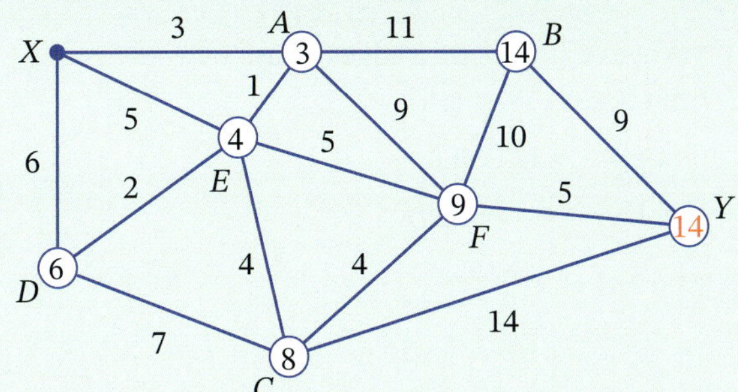

**5** To identify the shortest path, start at $Y$, move back to the vertex from which the shortest distance to $X$ was obtained, continuing this until $X$ is reached.

$14 - 5 = 9$

$9 - 5 = 4$

$4 - 1 = 3$

$3 - 3 = 0$

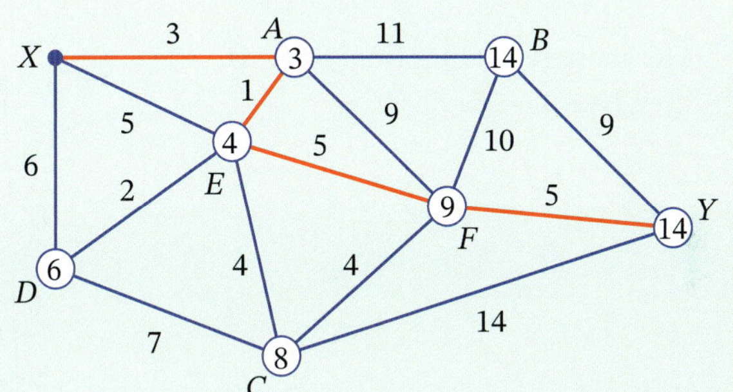

The shortest path is $XAEFY$ and has weight 14.

## Example 14

Rohan wanted to run the shortest distance, starting at $S$, through a large nature reserve, finishing at $F$. The reserve is shown by the graph, where the distances are in kilometres.

Along which path did Rohan run and how far did he run?

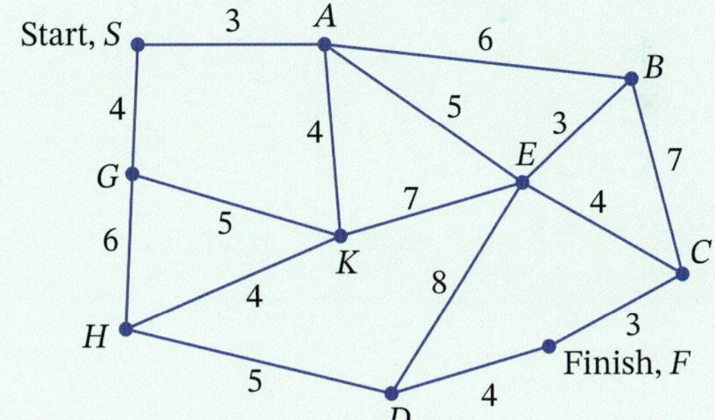

### Solution

**1** For all vertices **one** edge away from $S$, write down the shortest distance inside the circle.

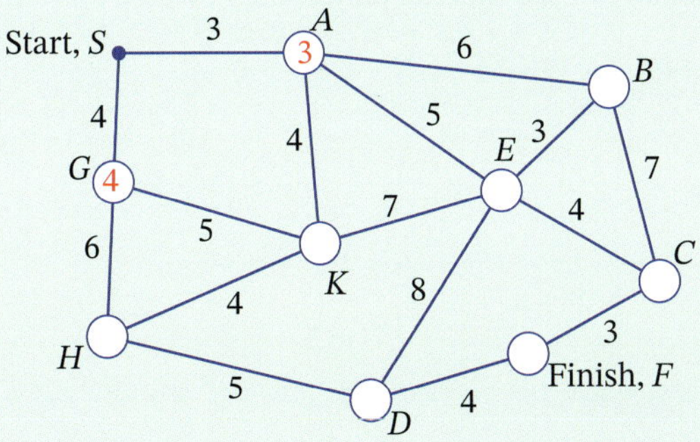

**2** For all vertices **2** edges away from $S$, write down the shortest distance inside the circle.

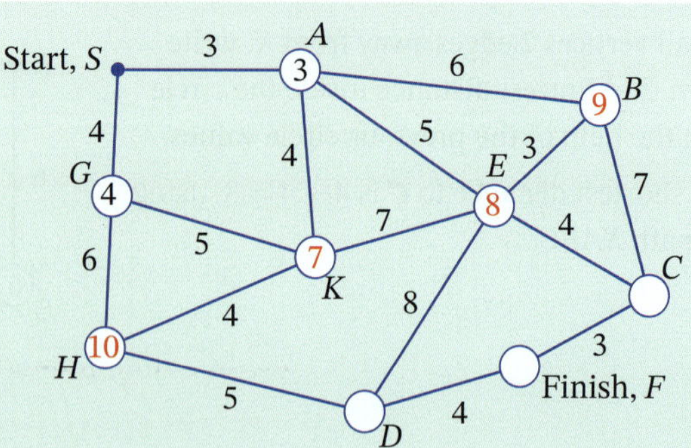


**3** Vertices $C$ and $D$ are **3** edges away from $S$, so:

$C$ is $8 + 4 = 12$ (via $E$)

$D$ is $10 + 5 = 15$ (via $H$)

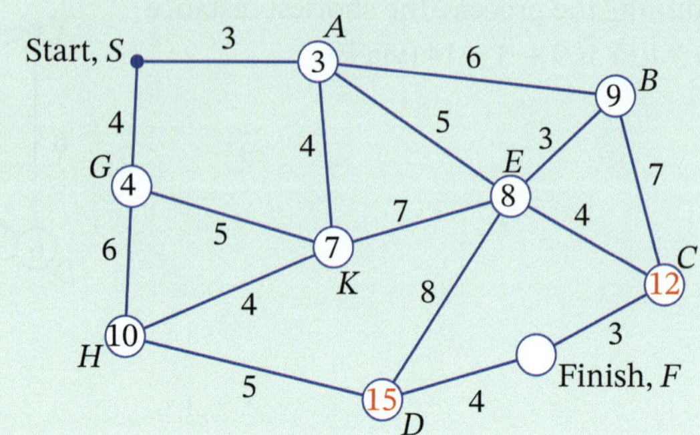


**4** The shortest distance from $S$ to $F$ is then $12 + 3 = 15$ (via $C$).

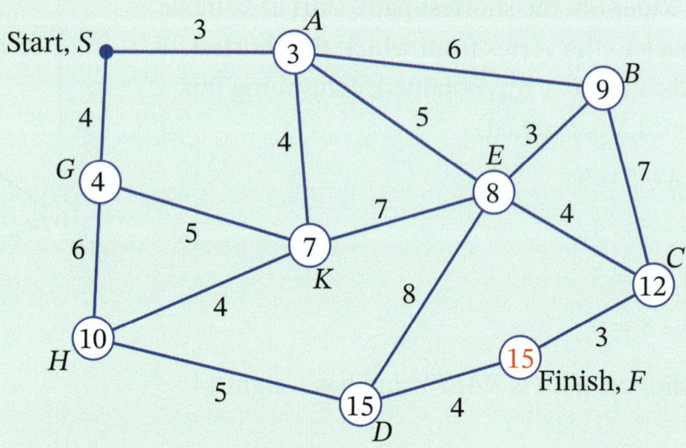


**5** To identify the shortest path, start at $F$, move backwards until $S$ is reached.

$15 - 3 = 12$

$12 - 4 = 8$

$8 - 5 = 3$

$3 - 3 = 0$

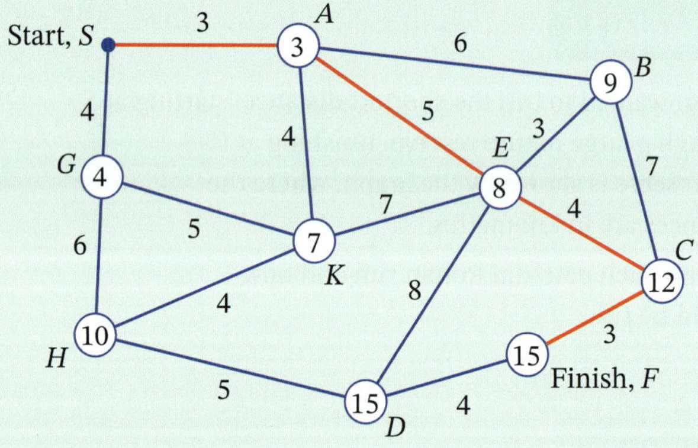


**6** So Rohan ran the shortest path $SAECF$, which has a distance of 15 km.

## EXERCISE 8.04 Answers on p. 475

# Shortest path problems

**1** The network shows walking trails in a national park and the times in minutes it takes to walk the trails.

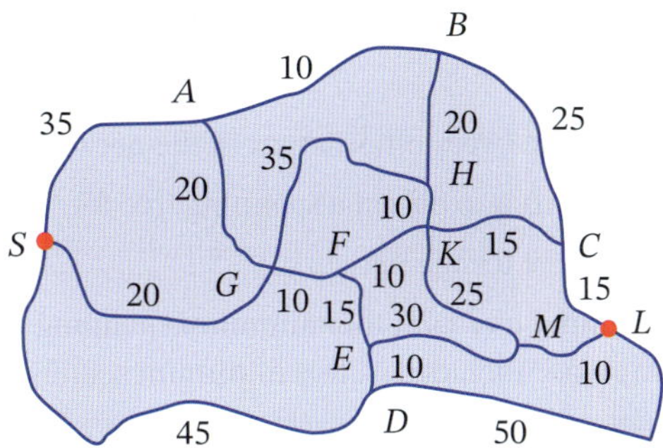


Find the shortest time it will take to walk from the start, $S$, to a lookout, $L$. Select **A**, **B**, **C** or **D**.

**A** 55 min **B** 70 min **C** 85 min **D** 90 min

**2** The graphs show the network of roads between 6 suburbs of south-western Sydney.

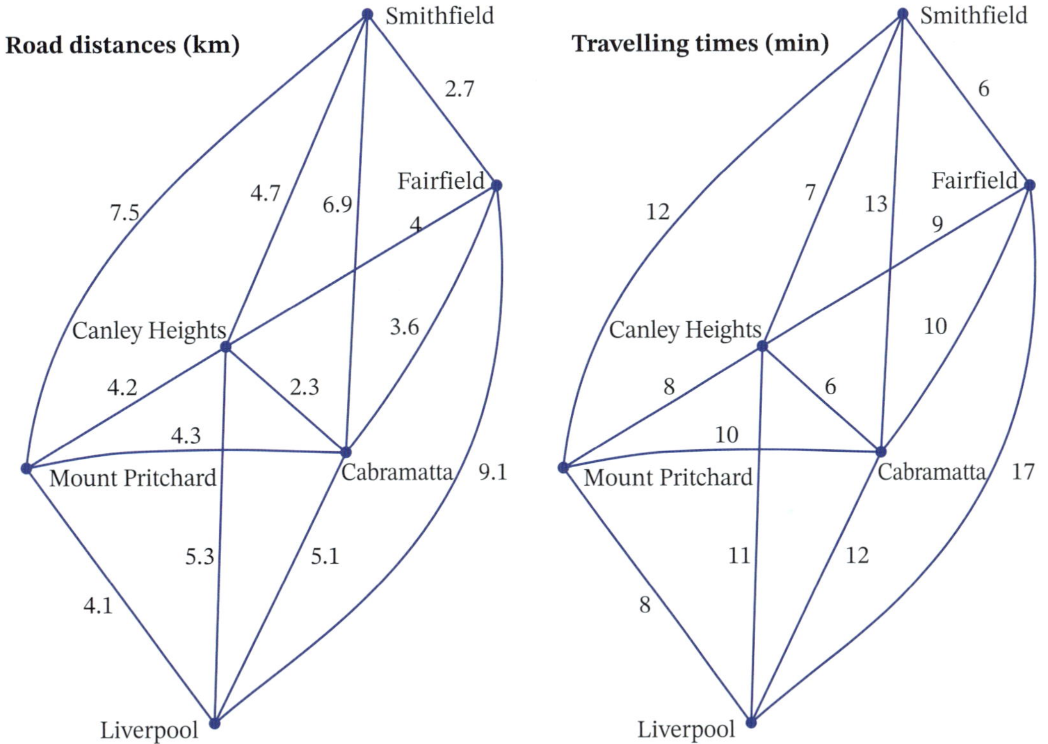


**a** If travelling from Mount Pritchard to Fairfield, find:

**i** the route that is the shortest distance

**ii** the route that takes the least amount of time.

**b** Are the 2 routes the same? Provide a possible reason for your answer.

□ Foundation ○ Mastery ⬡ Complex

EXAMPLE 12

3

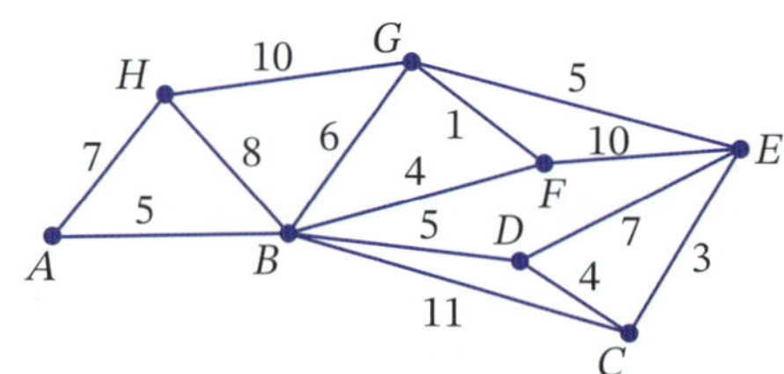


a Find the shortest path from $A$ to every other vertex in the network and their weights.

b Use the shortest paths from part **a** to obtain a spanning tree for the network and find its weight.

c Draw a minimum spanning tree for the network.

d Compare the shortest spanning tree with the minimum spanning tree. Are they the same? Is the shortest path from $A$ to $E$ included in this minimum spanning tree?

4 a Use the first graph from Question **2** to:

i draw the spanning tree that shows the shortest path from Liverpool to every other suburb and find its weight

ii draw the minimum spanning tree for the network and find its weight.

b Is the shortest path from Liverpool to Smithfield contained in the minimum spanning tree?

c Which other shortest distances from Liverpool are not contained in the minimum spanning tree?

5 a Use the second graph from Question **2** to:

i draw the spanning tree that shows the shortest time to travel from Liverpool to every other suburb and find its weight

ii draw the minimum spanning tree for the network and find its weight.

b Is the shortest time to travel from Liverpool to Smithfield contained in the minimum spanning tree?

c Which other shortest travel times from Liverpool are not contained in the minimum spanning tree?

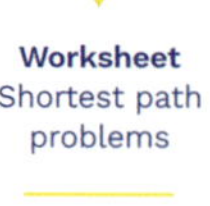

**Worksheet** Shortest path problems

The worksheet 'Shortest path problems' contains the diagrams from Questions **6**, **7**, **9** and **10**.

EXAMPLE 13

6 Copy this network diagram to find the shortest path from $S$ to $F$, or download the worksheet 'Shortest path problems' from Nelson MindTap.

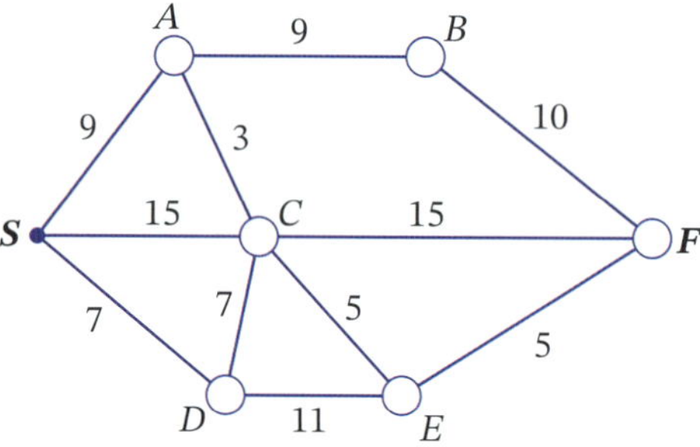


a For all vertices one edge away from $S$, write down the shortest distance inside the circle.

b Do the same for vertices 2 edges away, and continue the process until $F$ is reached.

c Then start at $F$ and move back to the vertex from which the shortest distance to $F$ was obtained, continuing this until $S$ is reached. This path will give the shortest distance.

☐ Foundation ○ Mastery ○ Complex

**7** Find the shortest path from $S$ to $F$ in each network.

**a**

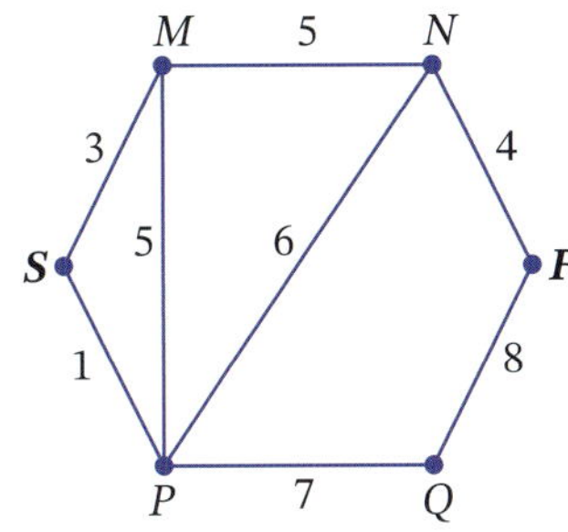


**b**

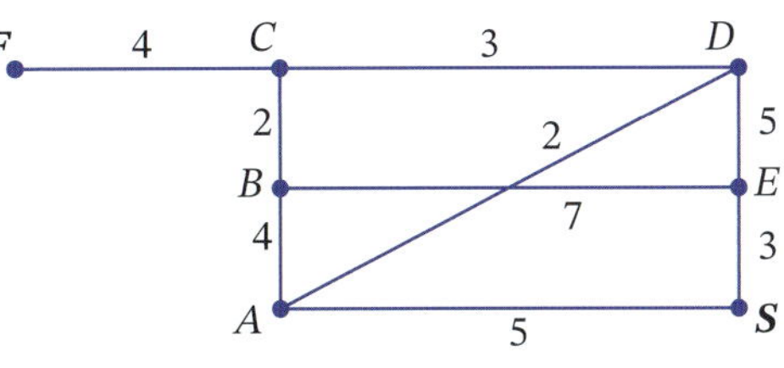


**c**

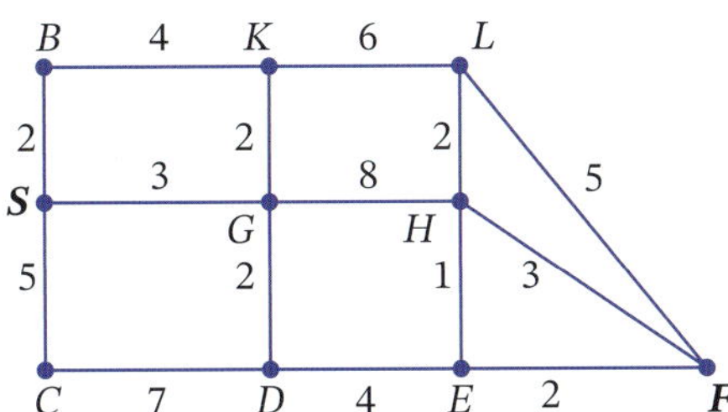


**d**

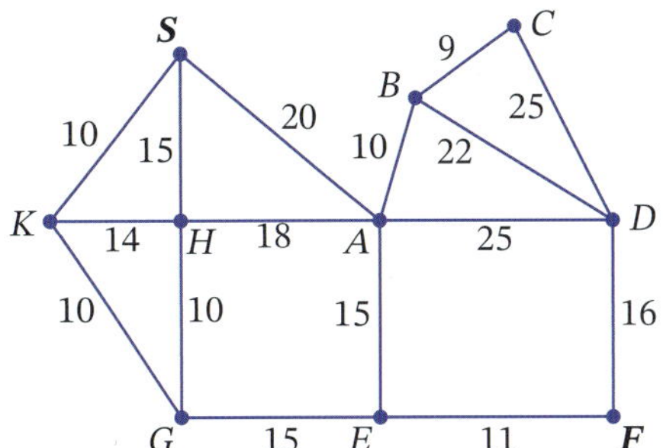


**8** The network shows Emily's bus travel times (in minutes) along roads from her home to school. What is the shortest time, in minutes, for Emily to travel from home to school? Select **A**, **B**, **C** or **D**.

EXAMPLE 14

**A** 20 **B** 24 **C** 26 **D** 27

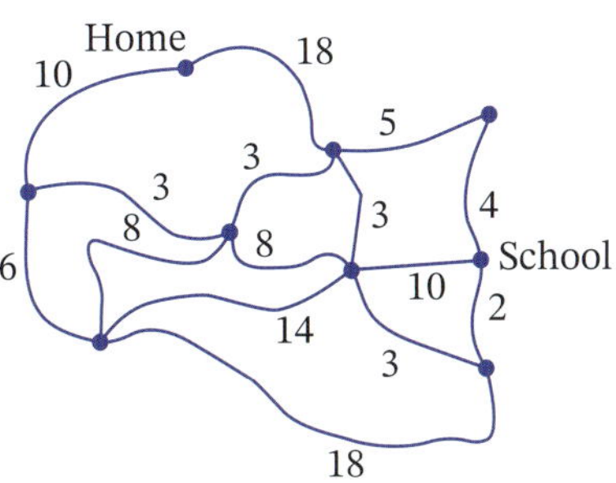


**9** The network diagram shows walking trails in a nature reserve. The numbers on the edges are in kilometres.

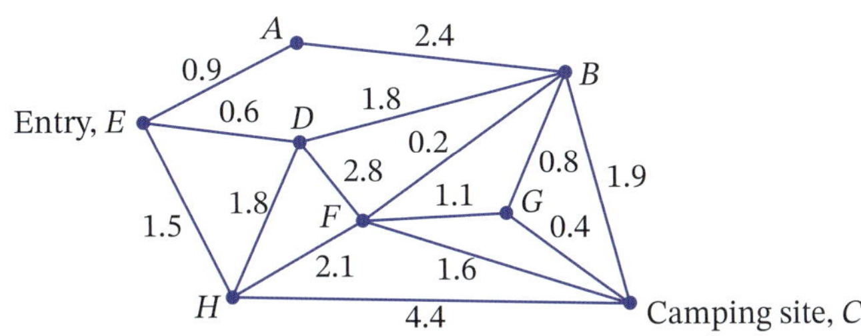


Find the length of the shortest path from the entry, $E$, to the camping site, $C$.

**10** **a** Draw a minimum spanning tree for the network shown using either Kruskal's or Prim's algorithm and find its weight.

**b** Is this spanning tree unique?

**c** **i** Find the shortest path from $Q$ to $N$.

**ii** Is this path contained in the minimum spanning tree?

**d** Find the spanning tree that gives the shortest distance from $Q$ to every other vertex.

Foundation Mastery Complex

**11** The table shows the travelling times (in minutes) between 8 different locations in the Sydney metropolitan area, as shown on the map.

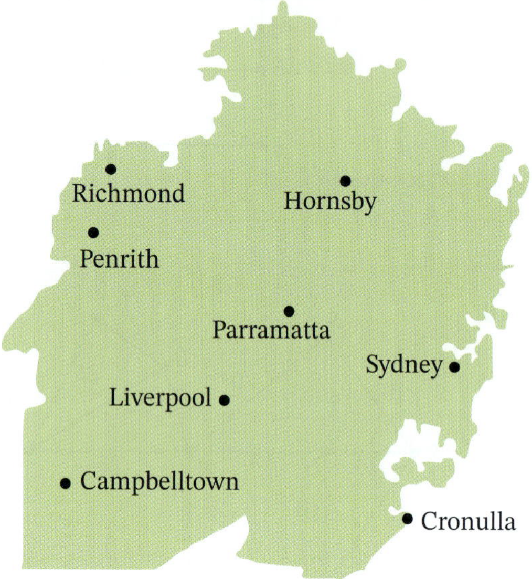


| | Cron | Cam | Liv | Par | Horn | Pen | Rich |
|---|---|---|---|---|---|---|---|
| **Syd** | 38 | – | 38 | 39 | 41 | 60 | 55 |
| | **Cron** | – | 47 | 59 | – | – | – |
| | | **Cam** | 26 | – | – | 45 | – |
| | | | **Liv** | 31 | – | 42 | 47 |
| | | | | **Par** | 34 | 35 | 40 |
| | | | | | **Horn** | 56 | 50 |
| | | | | | | **Pen** | 22 |

**a** Draw a network diagram showing the information in the table.

**b** What is the length of the shortest path from Cronulla to Richmond?

**c** Draw a spanning tree to show the quickest travelling time from Sydney to each location and find its weight.

**d** **i** Draw a minimum spanning tree for the network. Write down the order in which the edges were added.

**ii** What is the weight of the minimum spanning tree?

**e** Is the path for the quickest travelling time from Cronulla to Richmond contained in the minimum spanning tree? Give a possible reason.

## Sample HSC problem Answers on p. 477

6 marks The Robinson family is going to a national park and there are 5 towns, *A*, *B*, *C*, *D* and *E*, that they can pass through on the way.

**a** Use this table to draw a network diagram. A dash (–) in the table means there is no direct road between the 2 towns. 3 marks

| Town | *A* | *B* | *C* | *D* | *E* | Destination |
|---|---|---|---|---|---|---|
| Start | 50 | 70 | 60 | – | – | – |
| *A* | | 15 | – | 65 | – | – |
| *B* | | | 25 | 55 | 40 | – |
| *C* | | | | – | 55 | – |
| *D* | | | | | 20 | 60 |
| *E* | | | | | | 85 |

**b** Find a minimum spanning tree for the network. 1 mark

**c** Determine the shortest path from the start to the destination. 1 mark

**d** Is the shortest path contained in the minimum spanning tree? 1 mark

☐ Foundation ○ Mastery ⬡ Complex

## Study Tip

### Useful exam tips

- Make yourself familiar with the format of the exam: the number of sections and questions, the types of questions (for example, multiple-choice, short answer), the time allowed and the number of marks allocated.
- Don't worry if you are a bit nervous. This is normal and will help you perform better.
- However, being too casual or too anxious can harm your performance.
- Calculate the average amount of time you should spend on each question or section.
- Spend the reading time of the exam browsing through the paper to see what work lies ahead of you.
- Easier questions are usually at the beginning, with harder ones at the end. Do an easy question first to boost your confidence. It will also save time.
- Put a mark next to the harder questions and allow more time for working on them. Leave them if you get stuck and come back to them later.
- Show all working. Even if your answer is incorrect, you may be awarded some marks for correct working.
- Attempt every question. It is better to do most of every question and score some marks than to ignore some questions completely and score zero for them.
- Don't leave multiple-choice questions unanswered. Even if you guess, there is a chance of being correct. Some students like to leave multiple-choice questions till last so that, if they run out of time, they can make quick guesses. However, some multiple-choice questions can be quite difficult.

# CHAPTER SUMMARY

**Worksheet**
Network summary cards 1

This chapter, *Networks*, looked at ways of modelling situations using graphs called networks. You became familiar with network terminology such as vertex, edge, tree, weight and degree, and used algorithms to find minimum spanning trees and shortest paths.

Before you move on, consider what you learned in this chapter and revisit any sections that may have been unclear.

| What you learned in this chapter... | Section |
|---|---|
| Learn the parts of a network and associated terminology, including vertex, edge and degree | 8.01 Networks |
| Represent situations using tables and by drawing network diagrams | 8.01 Networks |
| Draw network diagrams from information given in a table or map | 8.02 Constructing network diagrams |
| Find the minimum spanning tree of a network by inspection and by using Kruskal's or Prim's algorithms, and apply this skill to solve minimal connector problems | 8.03 Minimum spanning trees |
| Find the shortest path between 2 vertices in a network, and apply this skill to solve problems | 8.04 Shortest path problems |

To help master this topic, make a summary mind map. Use the chapter outline and the mind map below as a guide. Add your own words, symbols, diagrams, boxes and reminders. The summary should give you a 'whole picture' view of the topic and allow you to identify any weak areas to revisit in your revision.

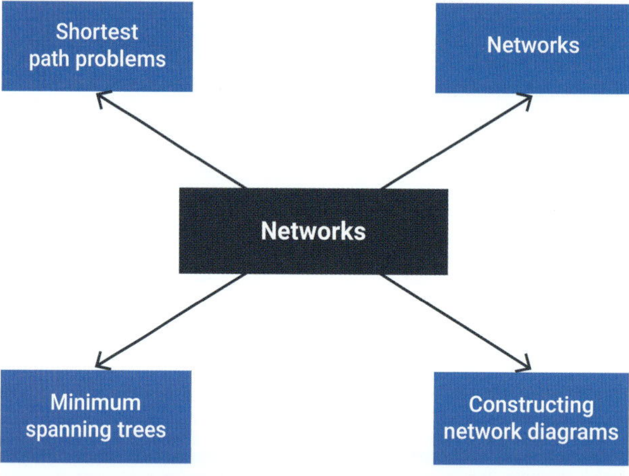

# Test yourself 8

Answers on p. 477

**1** For each network: 8.01

**i** find the degree of each vertex

**ii** show that the sum of the degrees of the vertices is equal to twice the number of edges.

**a**

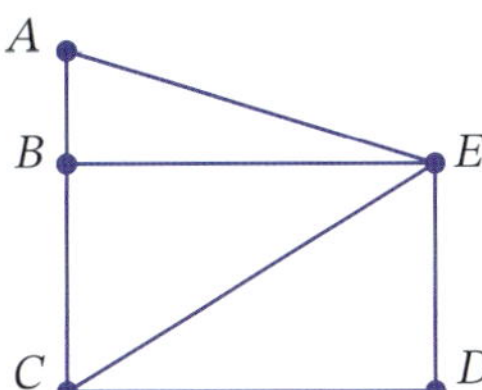


**b**

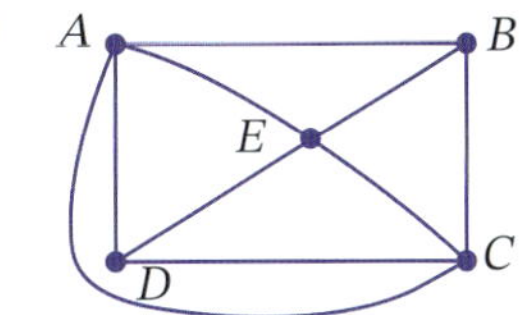


**2** For this network, find: 8.01

**a** the weight of the network

**b** the shortest path from $L$ to $W$

**c** the longest path from $L$ to $W$ without using an edge more than once.

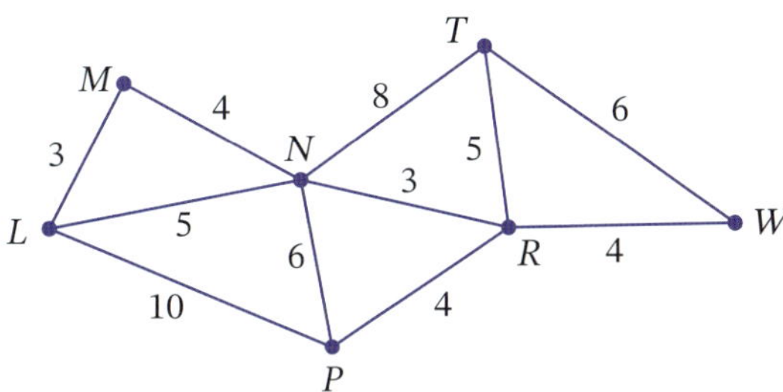


**3** Draw a network diagram for this map where the vertices represent the regions and the edges represent the borders between regions. 8.02

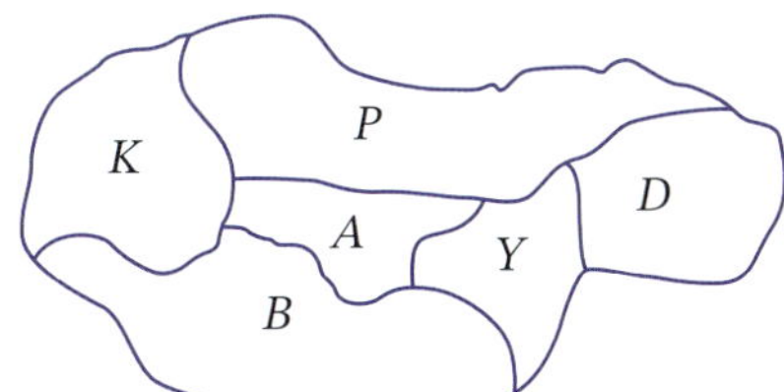


**4** Copy and complete the table for this network. 8.02

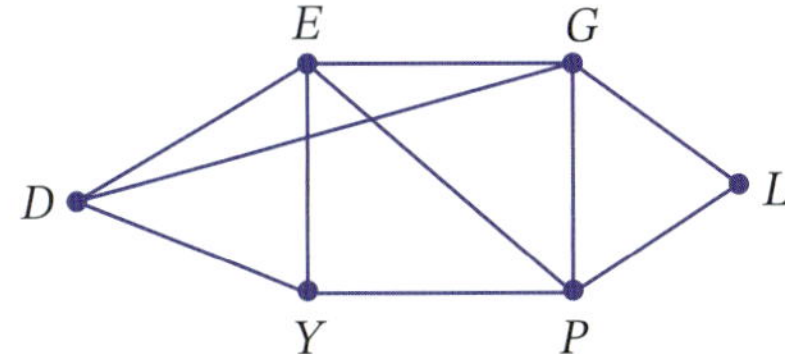


| | D | E | G | L | P | Y |
|---|---|---|---|---|---|---|
| D | | | | | | |
| E | | | | | | |
| G | | | | | | |
| L | | | | | | |
| P | | | | | | |
| Y | | | | | | |

☐ Foundation ○ Mastery ⬡ Complex

8.02 **5** Represent the information in the table as a weighted network.

| | *A* | *B* | *C* | *D* | *E* | *F* |
|---|---|---|---|---|---|---|
| *A* | – | 7 | 6 | 8 | – | – |
| *B* | 7 | – | 5 | 2 | – | – |
| *C* | 6 | 5 | – | 4 | – | – |
| *D* | 8 | 2 | 4 | – | 3 | 5 |
| *E* | – | – | – | 3 | – | 4 |
| *F* | – | – | – | 5 | 4 | – |

8.02 **6** Draw the directed network that is represented by this table.

| | | To | | | | |
|---|---|---|---|---|---|---|
| | | *D* | *E* | *F* | *G* | *H* |
| From | *D* | 1 | 1 | 2 | 1 | 1 |
| | *E* | 1 | – | 2 | – | – |
| | *F* | – | 1 | – | 1 | – |
| | *G* | – | – | 1 | – | 1 |
| | *H* | 1 | – | 1 | 1 | – |

8.03 **7** Draw all the spanning trees for the network below.

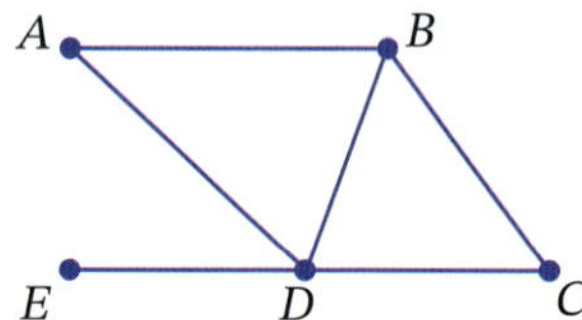


8.03 **8** Use Kruskal's algorithm to find a minimum spanning tree for the network below.

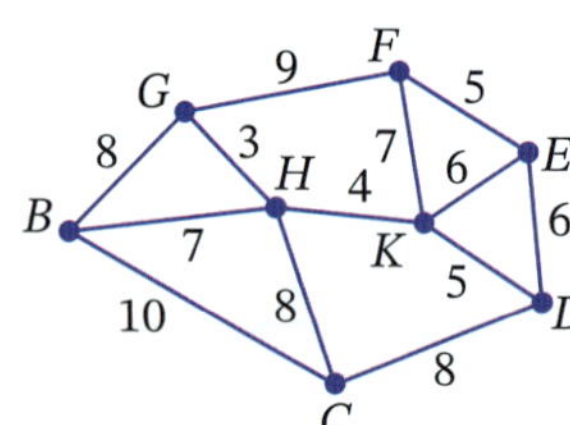


Show the order in which the edges for the spanning tree are selected and calculate its weight.

8.03 **9** Use Prim's algorithm to find the minimum spanning tree for the network in Question **8**, showing the order in which you choose the edges.

8.03 **10**

**a** Use Prim's algorithm to find the minimum spanning tree, showing the order in which you choose the edges.

**b** Can you find more than one minimum spanning tree for this network?

☐ Foundation ○ Mastery ⬡ Complex

**11** The minimum spanning tree for this network has a length of 23. 8.03

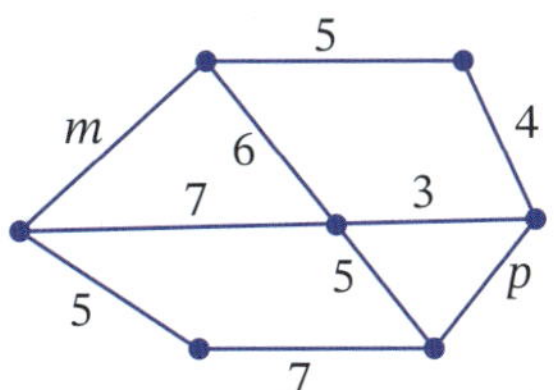


What are the values of $m$ and $p$?

**12** The network diagram shows the distances between locations in a camping ground. The distances are given in hundreds of metres. 

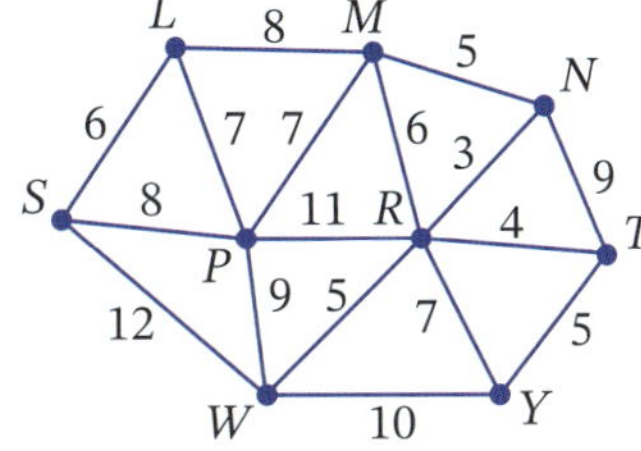


**a** Draw the minimum spanning tree for the network and find its length.

**b** Find the shortest path from $S$ to $T$. Is this path contained in the minimum spanning tree?

**13** Find the shortest distance from $S$ to $F$ in this network. 8.04

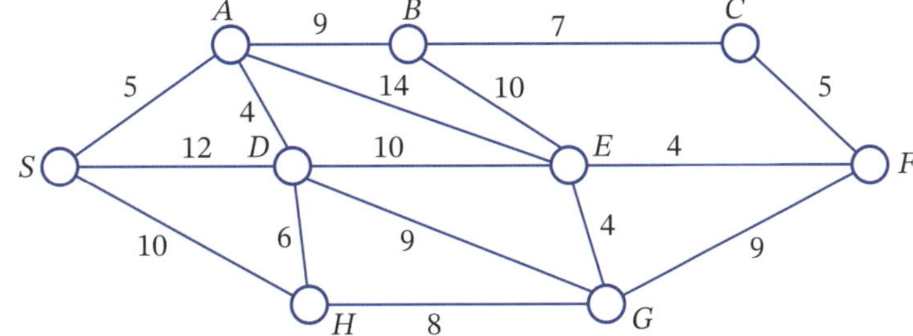


**14** This network shows the distances in kilometres between several towns. 8.04

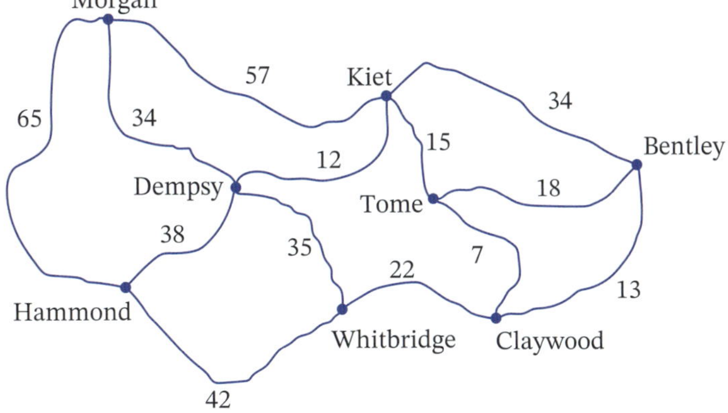


**a** Find a minimum spanning tree for this network and determine its length.

**b** Find the shortest path from Morgan to Bentley and calculate its length.

**c** Is the shortest path contained in the minimum spanning tree?

Foundation Mastery Complex

# LINEAR FUNCTIONS

A criminologist discovered that, in a big city, if the number of police on street patrol was increased, the number of crimes committed would decrease. She noticed the following pattern:

| Number of police | 50 | 150 | 200 | 250 | 300 |
|---|---|---|---|---|---|
| Number of crimes (per month) | 3100 | 2800 | 2650 | 2500 | 2350 |

Is it possible for the criminologist to find an algebraic formula to fit this pattern? Algebraic modelling is the study of relationships and the formulation of a mathematical rule or model to describe such relationships.

## Chapter outline

myphotobank.com.au/Adobe Stock Photos

## In this chapter you will:

- graph linear functions of the form $y = mx + c$
- find the gradient and $y$-intercept of a line from its graph
- use linear functions to model practical situations and interpret the meaning of the gradient and $y$-intercept of the line in context
- graph and solve problems involving direct linear variation.

**Videos (5):**

**9.01** Gradient and $y$-intercept of a line

**9.03** Applying linear functions • Linear modelling 1 • Linear modelling 2

**9.04** Direct linear variation

**Skillsheet (1):**

**9.01** Graphing linear equations

**Worksheets (10):**

**SkillCheck** Assignment 9

**9.01** Drawing gradients • Graphing linear functions 1 • Graphing linear functions 2 • A page of number planes • Gradient and $y$-intercept • $y = mx + c$ • A page of number planes • Number plane grid paper

**9.02** Finding the equation of a line • Finding the gradient between two points on a line

**Puzzle (1):**

**9.03** Linear functions code puzzle

**Spreadsheet (1):**

**9.02** Graphing $y = mx + c$

Nelson MindTap

To access resources above, visit **cengage.com.au/nelsonmindtap**

## Terminology

| | | | |
|---|---|---|---|
| constant of variation | dependent variable | direct linear variation | extrapolate |
| gradient | independent variable | interpolate | linear function |
| linear modelling | proportional to | rise | run |
| vertical intercept | $y$-intercept | | |

## SkillCheck

Answers on p. 479

Worksheet Assignment 9

1 Graph this table of values on a number plane and rule a line through the points.

| $x$ | −1 | 0 | 1 | 2 |
|---|---|---|---|---|
| $y$ | −4 | −1 | 2 | 5 |

2 Copy and complete each of these tables of values using the given formula.

a $y = 2x + 7$

| $x$ | −1 | 0 | 1 | 2 |
|---|---|---|---|---|
| $y$ | | | | |

b $y = -x - 4$

| $x$ | −1 | 0 | 1 | 2 |
|---|---|---|---|---|
| $y$ | | | | |

3 If $y = mx + c$, find the value of $c$ if $x = 2$, $y = 5$ and $m = 3$.

4 If $y = kx$:

a find the value of $y$ if $k = 0.7$ and $x = 4$

b find the value of $k$ if $x = 2.5$ and $y = 10$.

5 Given the formula $V = -450t + 2575$:

a find the value of $V$ if:

i $t = 3$ ii $t = 4.2$

b find the value of $t$ if:

i $V = 1450$ ii $V = 10$

# Graphing linear functions 9.01

Skillsheet Graphing linear equations

Worksheets Drawing gradients

Graphing linear functions 1

Graphing linear functions 2

A page of number planes

The formula or equation $y = 3x - 2$ is called a **linear function** because its graph is a straight line. **Linear** means 'of a line', while a **function** is an algebraic rule similar to the 'number machine' shown below, which changes an input value, $x$, into an output value, $y$.

Function: $y = 3x - 2$

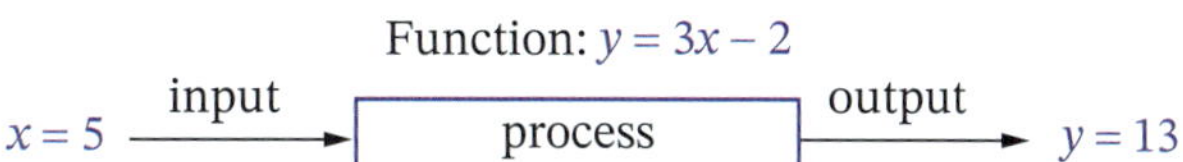


A line has a **gradient** (the measure of its steepness) and a ***y*-intercept** (the value where the line crosses the $y$-axis).

### Gradient of a line

The gradient of a line is given by the formula:

$$m = \frac{\text{rise } (\uparrow)}{\text{run } (\rightarrow)}$$

- The **rise** is the vertical change in position (going up).
- The **run** is the horizontal change in position (going across to the right).

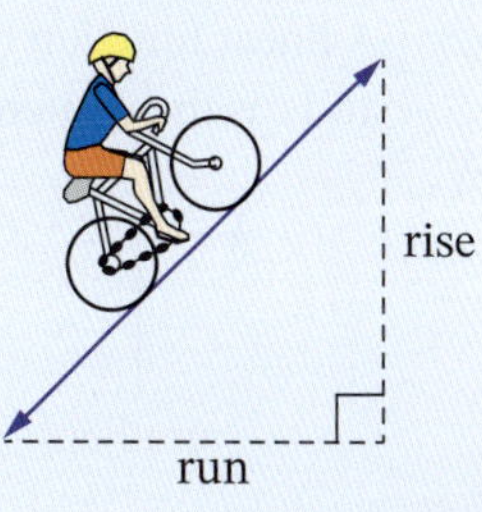

**Worksheets**
Gradient and $y$-intercept
$y = mx + c$

## Example 1

**a** Copy and complete this table of values for the linear function $y = 3x - 2$.

| $x$ | −1 | 0 | 1 | 2 | 3 |
|---|---|---|---|---|---|
| $y$ | | | | | |

**b** What pattern do you notice in the $y$ values in the bottom row of the completed table?

**c** Graph $y = 3x - 2$ on a number plane.

**d** Find the gradient and $y$-intercept of the line you drew in part **c**.

### Solution

**a**

| $x$ | −1 | 0 | 1 | 2 | 3 |
|---|---|---|---|---|---|
| $y$ | −5 | −2 | 1 | 4 | 7 |

**b** The $y$ values increase by 3 each time.

**c** Use the $(x, y)$ ordered pairs from the table to graph the points: $(-1,-5)$, $(0,-2)$, $(1,1)$, $(2,4)$, $(3,7)$. Then rule a line through the points and label the line with the equation.

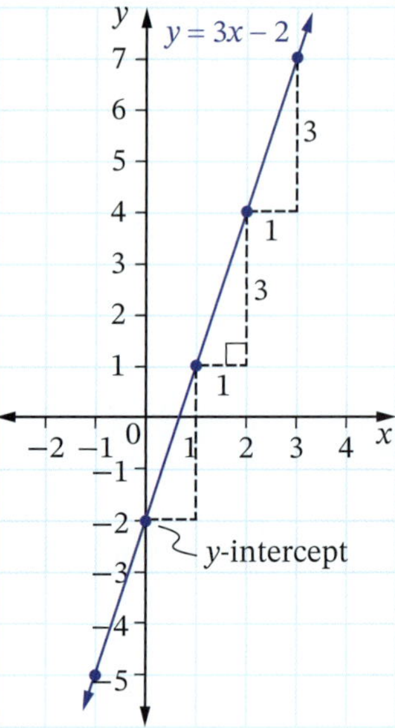


**d** Choose 2 points on the line, say $(1,1)$ and $(2,4)$.

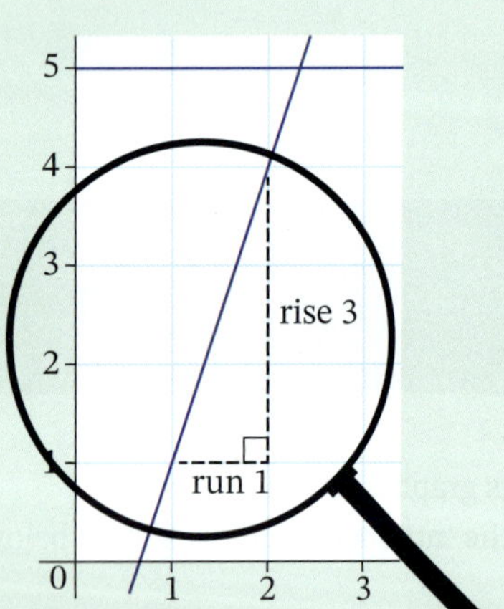


$$\text{Gradient } (m) = \frac{\text{rise}}{\text{run}} = \frac{3}{1} = 3$$

The $y$-intercept is −2.

**Video**
Gradient and $y$-intercept of a line

### Linear function

A linear function has the form $\boldsymbol{y = mx + c}$, where $m$ is the gradient and $c$ is the $y$-intercept.

$y = mx + c$ — $m$: gradient; $c$: $y$-intercept

$y = 3x - 2$ — 3: gradient 3; −2: $y$-intercept −2

## Example 2

**a** Complete this table of values for the linear function $y = -2x + 4$.

| $x$ | −1 | 0 | 1 | 2 | 3 |
|---|---|---|---|---|---|
| $y$ | | | | | |

**b** What pattern do you notice in the $y$ values in the completed table?

**c** Graph $y = -2x + 4$ on a number plane.

**d** Find the gradient and $y$-intercept of the line you drew in part **c**.

### Solution

**a**

| $x$ | −1 | 0 | 1 | 2 | 3 |
|---|---|---|---|---|---|
| $y$ | 6 | 4 | 2 | 0 | −2 |

**b** The $y$ values decrease by 2 each time.

**c**

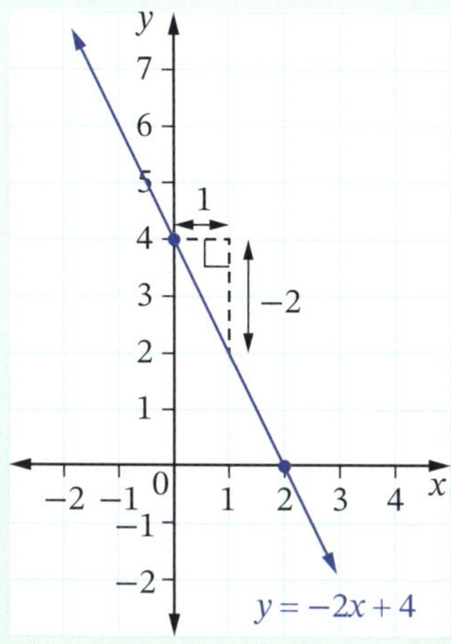


**d** Using points $(0, 4)$ and $(1, 2)$:

$$\text{Gradient } (m) = \frac{\text{rise}}{\text{run}} = \frac{-2}{1} = -2$$

The $y$-intercept is 4.

## Positive and negative gradients

A line with a **positive gradient** slopes upwards (from left to right) because its $y$ values are *increasing*. A gradient of 3 means that, for each 1 unit to the right, the line goes up 3 units or, as the $x$ values increase by 1, the $y$ values *increase* by 3.

Positive gradient

**Graph sloping upwards**

A line with a **negative gradient** slopes downwards (from left to right) because its $y$ values are *decreasing*. A negative gradient has a 'negative rise', or a drop. A gradient of −2 means that, for each unit to the right, the line goes down 2 units, or as the $x$ values increase by 1, the $y$ values *decrease* by 2.

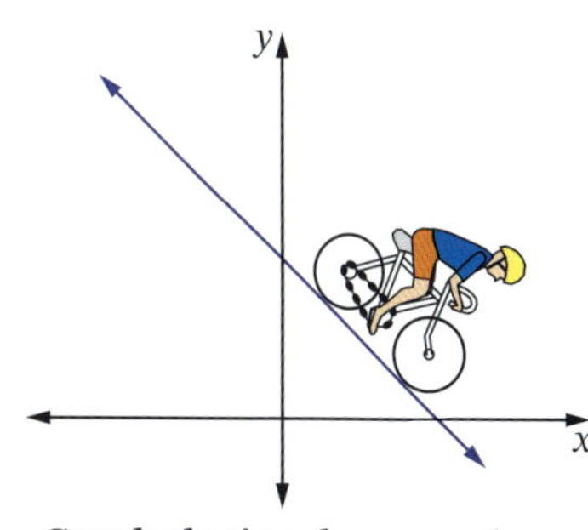


**Graph sloping downwards**

## Example 3

Graph each linear function after finding its gradient and $y$-intercept.

**a** $y = 2x + 1$ **b** $y = -4x$ **c** $y = \frac{2}{3}x - 4$

### Solution

**a** gradient = 2, $y$-intercept = 1

To graph the line:

1. plot the $y$-intercept, 1, on the $y$-axis
2. indicate a gradient of 2 from this point, by moving across 1 unit and up 2 units and marking that point
3. rule a line through both points you marked
4. draw arrows at the ends and label the line with its equation.

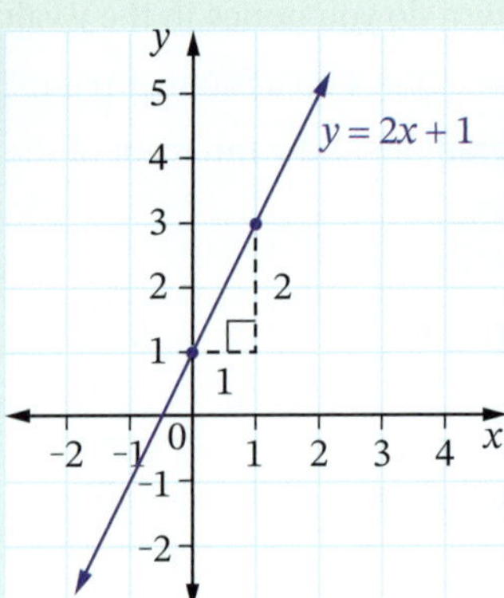


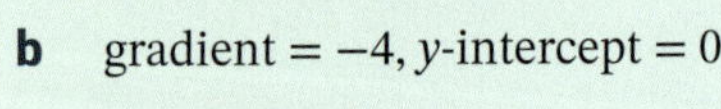

**b** gradient = −4, $y$-intercept = 0

To graph the line:

1. plot the $y$-intercept, 0, at the origin
2. indicate a gradient of −4 from this point, by moving across 1 unit and down 4 units and marking that point
3. rule a line through both points you marked, add arrows and label.

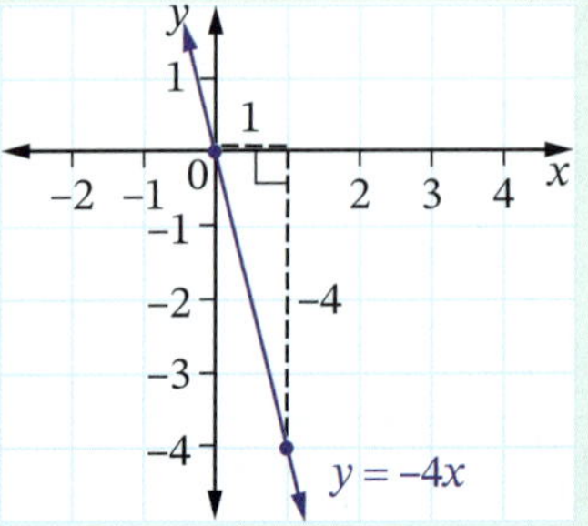


**c** gradient = $\frac{2}{3}$, $y$-intercept = −4

To graph the line:

1. plot the $y$-intercept −4 on the $y$-axis
2. indicate a gradient of $\frac{2}{3}$ from this point, by moving across 3 units and up 2 units and marking that point
3. rule a line through both points you marked, add arrows and label.

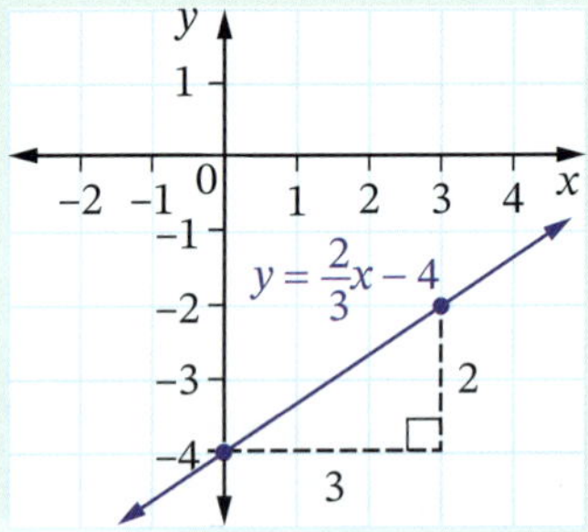

Worksheets
A page of number planes

Number plane grid paper

**EXERCISE 9.01** Answers on p. 479

## Graphing linear functions

**1** Graph each table of values, joining the points to form a straight line.

**a**

| $x$ | −2 | −1 | 0 | 1 | 2 |
|---|---|---|---|---|---|
| $y$ | −7 | −5 | −3 | −1 | 1 |

**b**

| $x$ | −2 | −1 | 0 | 1 | 2 |
|---|---|---|---|---|---|
| $y$ | 5 | 2 | −1 | −4 | −7 |

**2** Graph each linear function after copying and completing the table of values, then find the gradient and $y$-intercept of each line.

**a** $y = x + 2$

| $x$ | −2 | −1 | 0 | 1 | 2 |
|---|---|---|---|---|---|
| $y$ | | | | | |

**b** 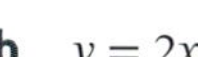 $y = 2x - 2$

| $x$ | −2 | −1 | 0 | 1 | 2 |
|---|---|---|---|---|---|
| $y$ | | | | | |

**c** $y = 3x + 1$

| $x$ | −2 | −1 | 0 | 1 | 2 |
|---|---|---|---|---|---|
| $y$ | | | | | |

**d** $y = \frac{x}{2} - 1$

| $x$ | −4 | −2 | 0 | 2 | 4 |
|---|---|---|---|---|---|
| $y$ | | | | | |

**e** $N = 1 - M$

| $M$ | −2 | −1 | 0 | 1 | 2 |
|---|---|---|---|---|---|
| $N$ | | | | | |

**f** $P = \frac{5t - 1}{2}$

| $t$ | −3 | −1 | 0 | 1 | 3 |
|---|---|---|---|---|---|
| $P$ | | | | | |

EXAMPLE 2

**3** Write the equation of the line with:

**a** gradient 3, $y$-intercept 7

**b** gradient −2, $y$-intercept 1

**c** gradient 1, $y$-intercept −1

**d** gradient $\frac{1}{3}$, $y$-intercept $\frac{1}{2}$

**e** gradient $-\frac{5}{4}$, $y$-intercept 0

**f** gradient 0, $y$-intercept 5

**4** Graph each linear function.

**a** $y = 4x - 3$

**b** $y = \frac{1}{2}x + 2$

**c** $y = x - 2$

**d** $y = -2x$

**e** $y = 2x - 5$

**f** $y = \frac{3}{5}x - 1$

**g** $y = -x + 4$

**h** $y = \frac{4}{3}x + 1$

**i** $y = -\frac{1}{4}x + 3$

**5** What is the gradient of this line? Select **A**, **B**, **C** or **D**.

**A** $\frac{1}{3}$

**B** $-\frac{1}{3}$

**C** 3

**D** −3

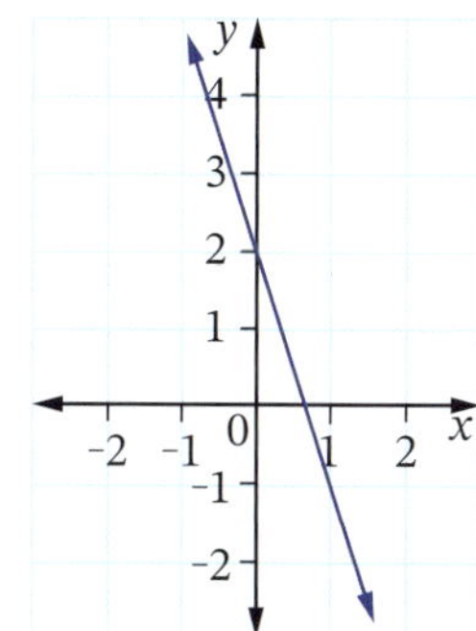


□ Foundation ○ Mastery ⬡ Complex

**6** Find the gradient and $y$-intercept of each line below, and write its equation.

**a**

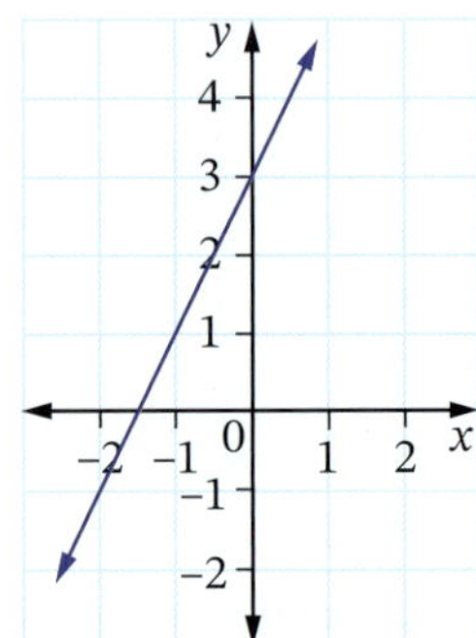

**b**

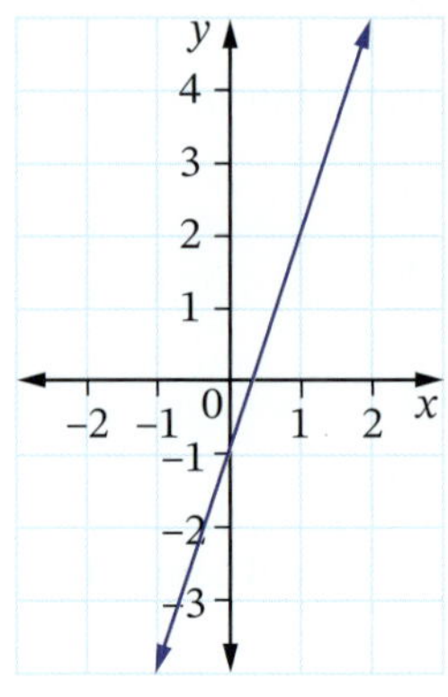

**c**

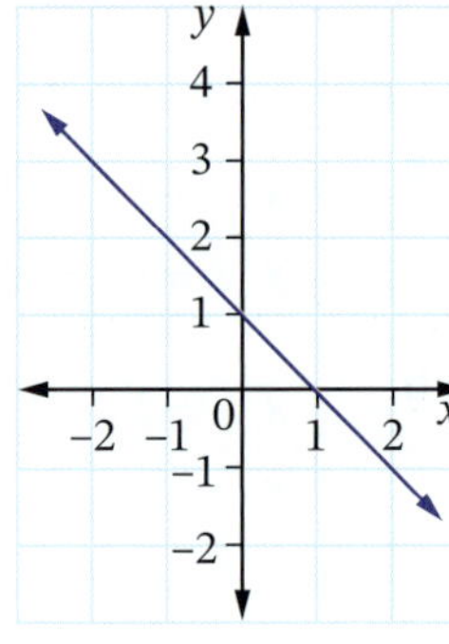

**d**

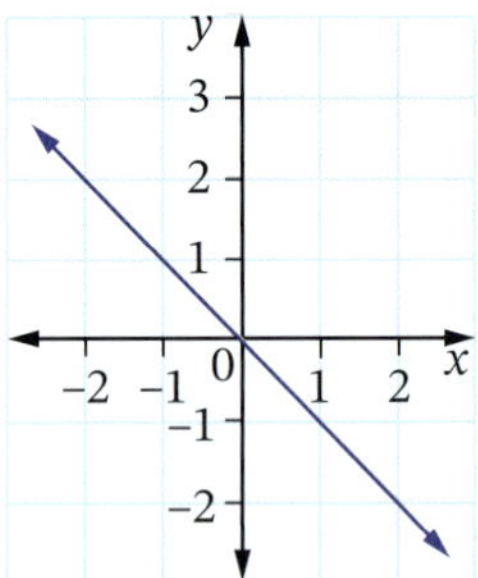

**e**

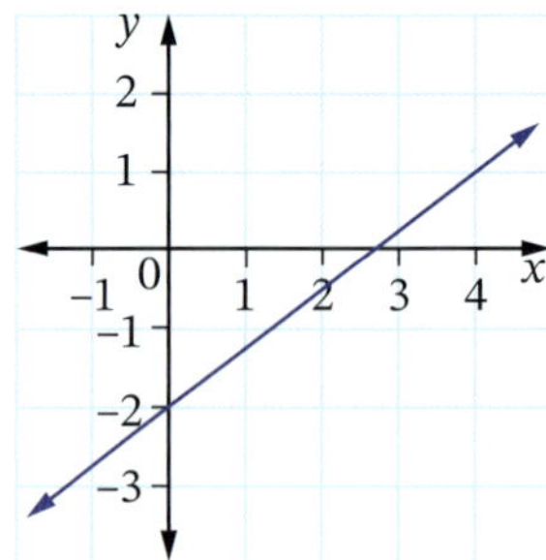

**f**

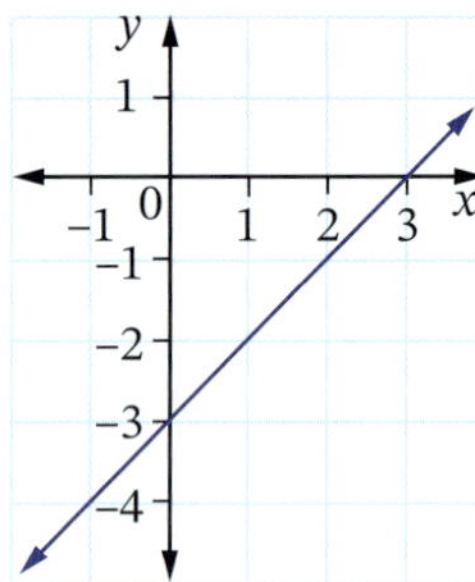

**7** What is the equation of the line $k$? Select **A**, **B**, **C** or **D**.

**A** $y = -3x + 3$

**B** $y = 3x + 3$

**C** $y = -\frac{x}{3} + 3$

**D** $y = \frac{x}{3} + 3$

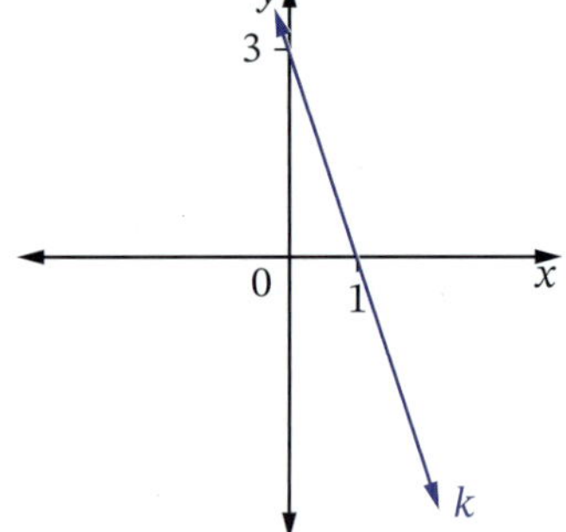

**8** This line has equation $y = mx + c$.

Which statement is true? Select **A**, **B**, **C** or **D**.

**A** $m$ is positive and $c$ is negative

**B** $m$ is negative and $c$ is positive

**C** $m$ and $c$ are both positive

**D** $m$ and $c$ are both negative

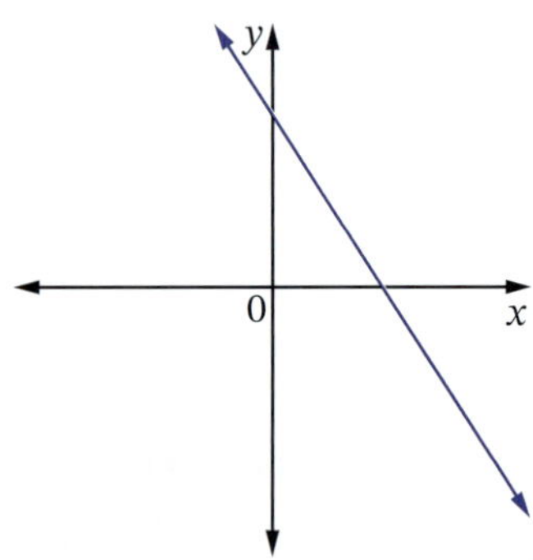

☐ Foundation ◯ Mastery ⬡ Complex

**9** Which is the graph of $y = 3x - 3$? Select **A**, **B**, **C** or **D**.

**A**

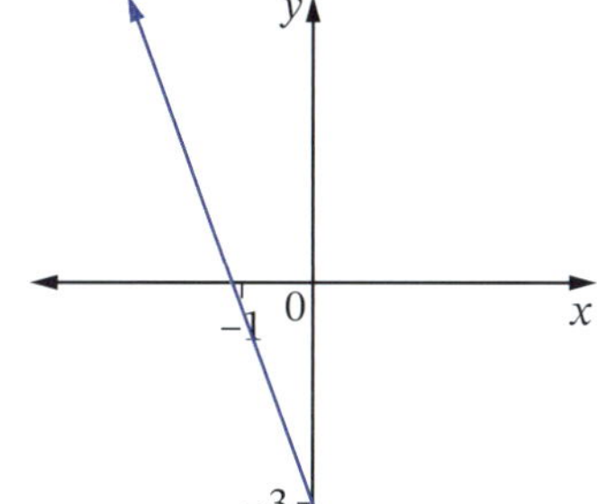


**B**

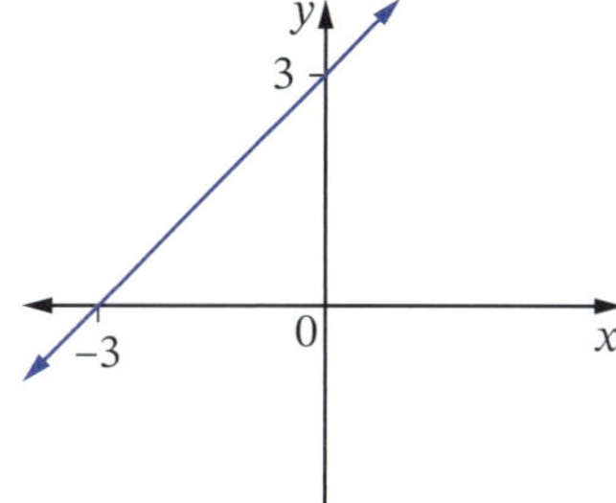


**C**

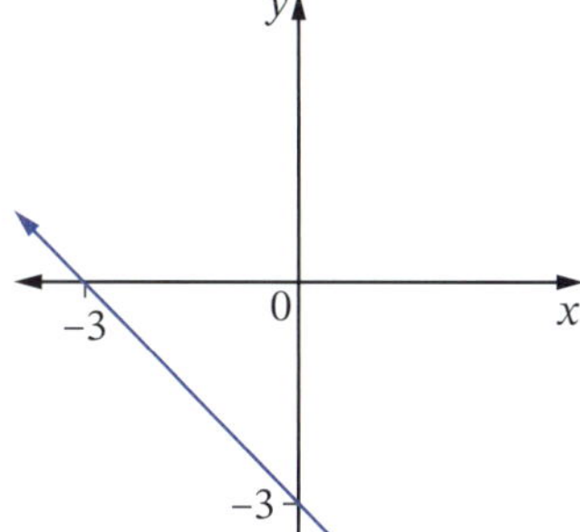


**D**

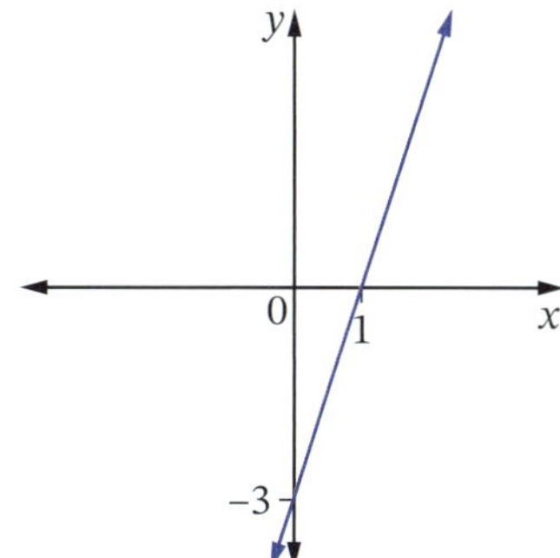


## Investigation

### Gradients

1. What does a line with a gradient of 0 look like?
2. What does a line with a gradient of 1 look like?
3. What does a line with a gradient of −1 look like?
4. What is the highest gradient a line can have?

## Did you know?

### Steep roads and hills

Australia is a fairly flat continent. You don't often see road gradient signs here, but in Europe, road signs often display the steepness of a road for drivers (or the steepness of a hill for hikers).

Consider this example:

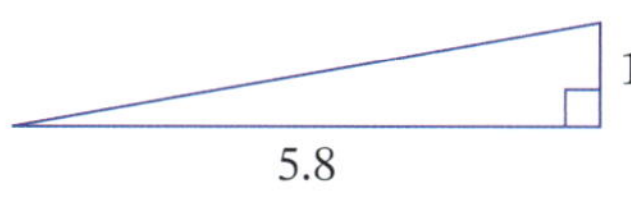


A gradient of 8% $\left(\text{that is, } \frac{8}{100}\right)$ means a rise of 8 units for every 100 units of run.

Bennian/Shutterstock.com

For this example, the steepness is $\frac{1}{5.8}$, also written as '1 : 5.8' or '1 in 5.8'.

**Bulli Pass, north of Wollongong, has a gradient of $\frac{1}{7}$ and Victoria Pass, east of Lithgow, has a gradient of $\frac{1}{8}$. Which pass is steeper?**

Foundation Mastery Complex

# 9.02 The gradient formula

**Worksheets**
Finding the equation of a line

Finding the gradient between two points on a line

**Spreadsheet**
Graphing $y = mx + c$

The formula for the gradient of a line is useful if:

- the **coordinates** of 2 points on the line are given, or
- a table of values is given.

**Gradient formula**

$$m = \frac{\text{rise } (\uparrow)}{\text{run } (\rightarrow)} = \frac{\text{change in } y}{\text{change in } x}$$

## Example 4

Find the gradient of each line.

**a**

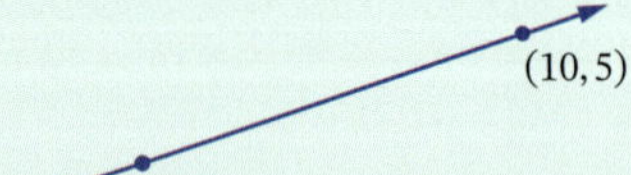


**b**

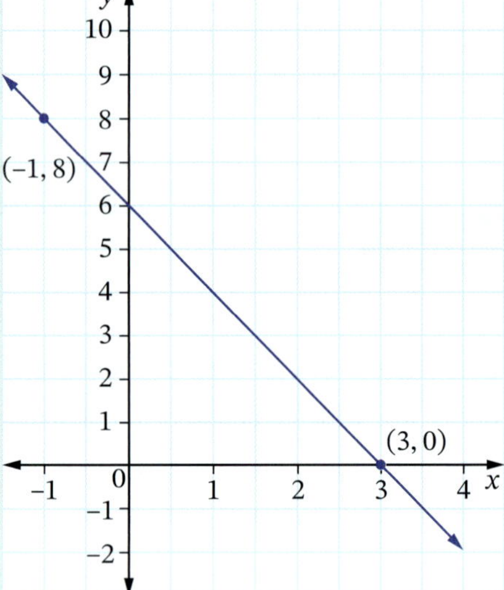


**c**

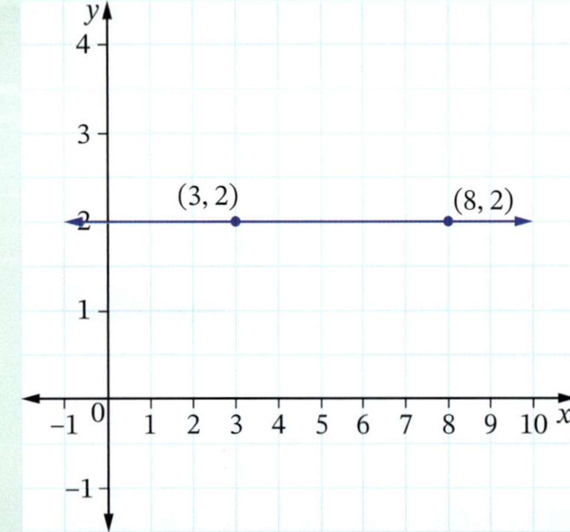


**d**

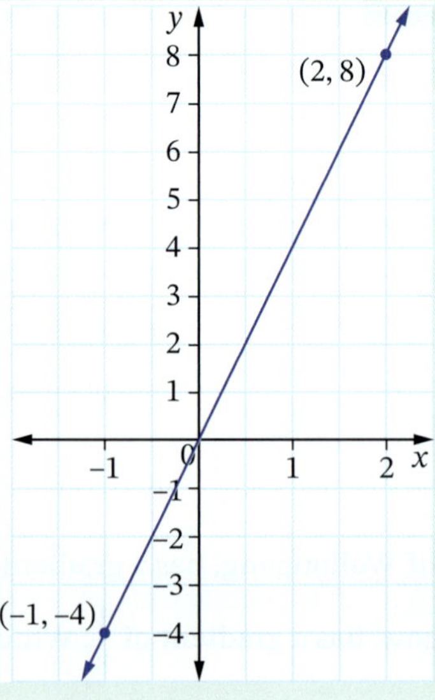

## Solution

**a** $m = \frac{\text{change in } y}{\text{change in } x}$

$= \frac{5-3}{10-4}$

$= \frac{2}{6}$

$= \frac{1}{3}$ a gradient of $\frac{1}{3}$ is not steep

**b** $m = \frac{0-8}{3-(-1)}$

$= \frac{-8}{4}$

$= -2$ negative gradient, the line slopes downward

**c** $m = \frac{2-2}{8-3}$

$= \frac{0}{5}$

$= 0$ a zero-gradient line is flat

**d** $m = \frac{\text{change in } y}{\text{change in } x}$

$= \frac{\text{rise}}{\text{run}}$

$= \frac{8-(-4)}{2-(-1)}$

$= \frac{12}{3}$

$= 4$ a gradient of 4 is steep

## Example 5

Find the gradient of the line represented by each table of values.

**a**

| *x* | 0 | 4 | 8 | 20 |
|---|---|---|---|---|
| *y* | 5 | 11 | 17 | 35 |

**b**

| *x* | 2 | 6 | 9 | 12 |
|---|---|---|---|---|
| *y* | 4 | −8 | −17 | −26 |

## Solution

**a** Choose any 2 points from the table, say $(0, 5)$ and $(4, 11)$:

$m = \frac{\text{change in } y}{\text{change in } x}$

$= \frac{11-5}{4-0}$

$= \frac{6}{4}$

$= \frac{3}{2}$

The gradient is $\frac{3}{2}$.

**b** Using (2, 4) and (6, −8) from the table:

$$m = \frac{-8-4}{6-2}$$

$$= \frac{-12}{4}$$

$$= -3$$

negative gradient ($y$ decreases as $x$ increases)

**EXERCISE 9.02** Answers on p. 481

## The gradient formula

EXAMPLE 4

**1** Find the gradient of each line.

**a**

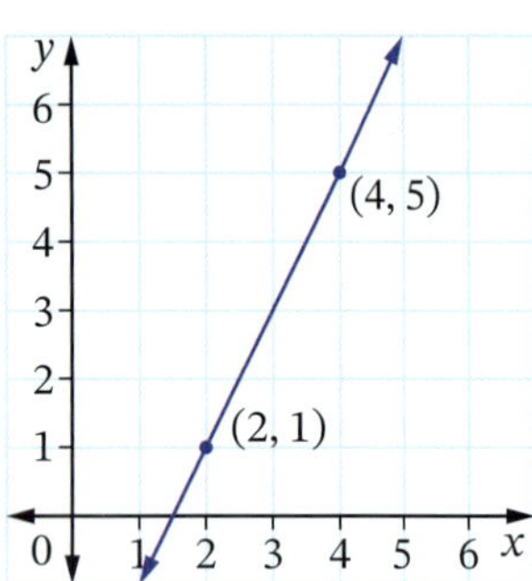


**b**

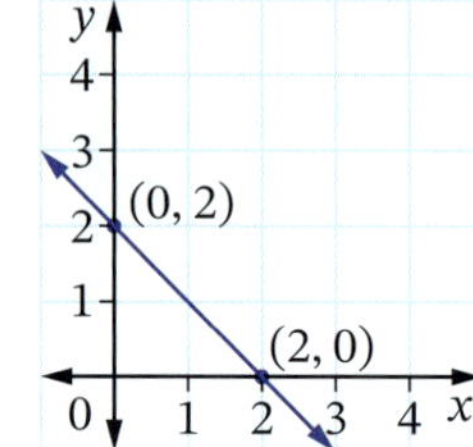


**c**

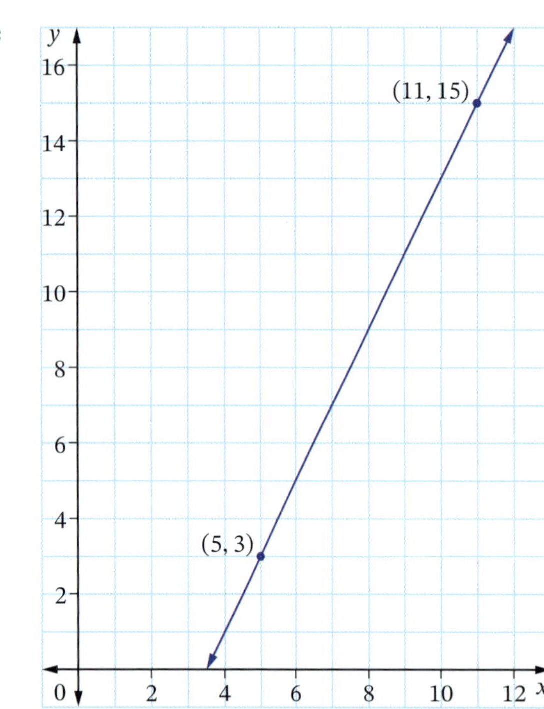


**d**

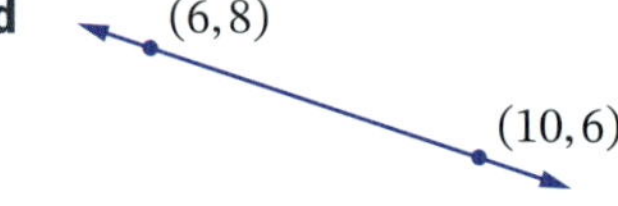


**e**

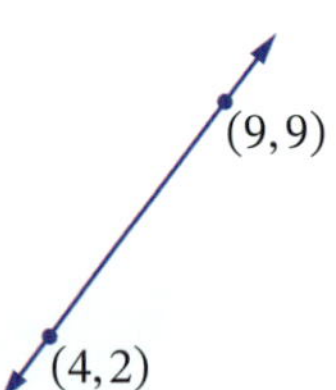


**f**

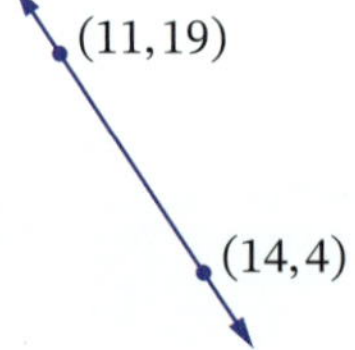


□ Foundation ○ Mastery ⬡ Complex

**g**

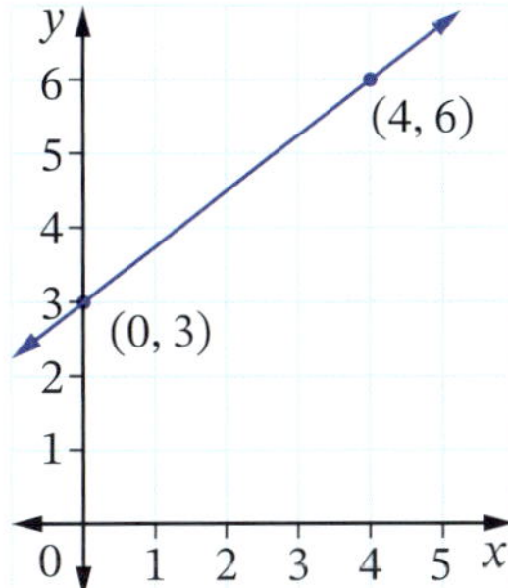


**h**

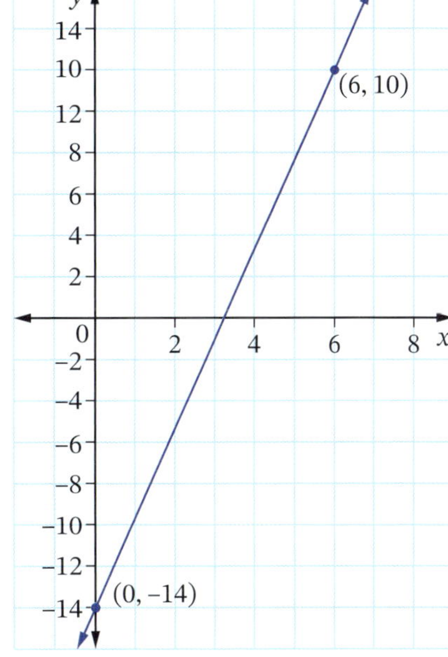


**i**

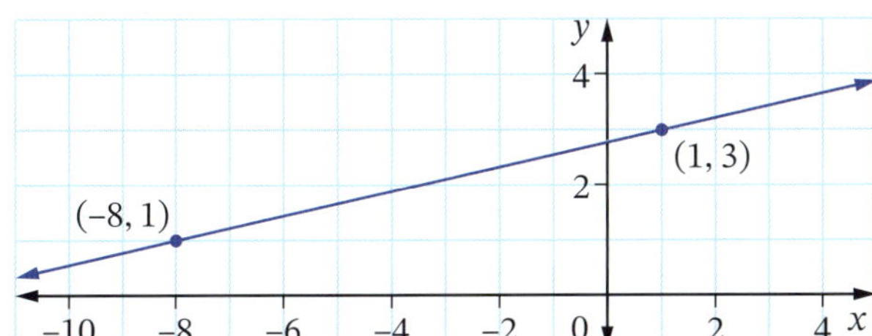


**2** Find the gradient of the line represented by each table of values.

EXAMPLE 5

**a**

| x | 0 | 4 | 7 | 10 |
|---|---|---|---|---|
| y | –1 | 11 | 20 | 29 |

**b**

| x | 1 | 5 | 7 | 12 |
|---|---|---|---|---|
| y | 1 | 21 | 31 | 56 |

**c**

| x | 0 | 8 | 12 | 20 |
|---|---|---|---|---|
| y | 3 | 5 | 6 | 8 |

**d**

| x | 1 | 3 | 6 | 8 |
|---|---|---|---|---|
| y | 7 | 3 | –3 | –7 |

**e**

| x | 0 | 10 | 25 | 40 |
|---|---|---|---|---|
| y | –4 | 2 | 11 | 20 |

**f**

| x | 2 | 8 | 14 | 16 |
|---|---|---|---|---|
| y | 4 | 1 | –2 | –3 |

**3** Find the equation of each line.

**a**

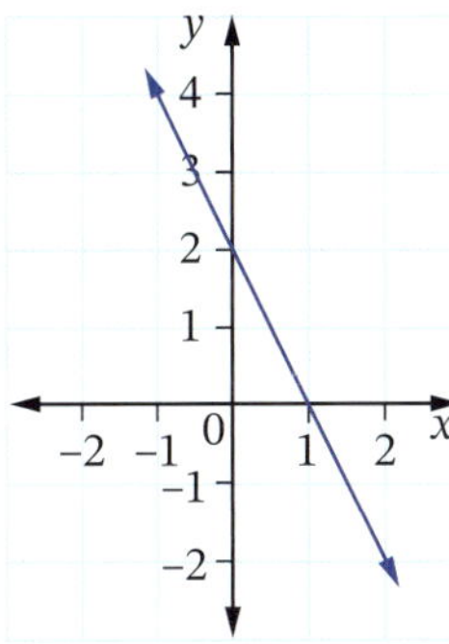

**b**

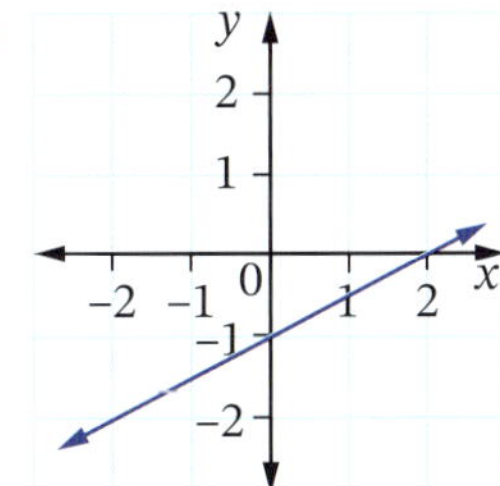

**c**

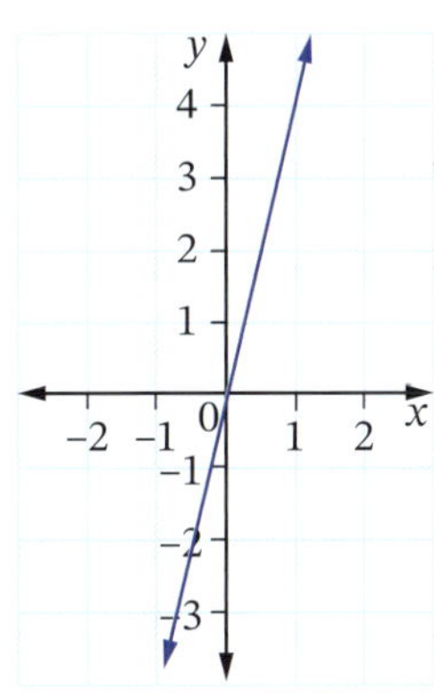

Foundation Mastery Complex

## Technology

### Graphing linear functions of the form $y = mx + c$

In this activity, use graphing technology to graph linear functions.

**1** Graph each set of linear functions. Identify and describe any similarities and differences for each set.

**a** $y = 2x$ $\quad y = 2x - 1$ $\quad y = 2x + 3$

**b** $y = -x$ $\quad y = 5 - x$ $\quad y = -x + 2$

**c** $y = \frac{1}{4}x + 3$ $\quad y = \frac{1}{4}x$ $\quad y = \frac{x-1}{4}$

**2** Graph each linear function and find the gradient and $y$-intercept of the line graphically.

**a** $y = 2x + 5$ **b** $y = -2x - 4$

**c** $y = 3x - 3$ **d** $y = 7 - x$

**e** $y = 2 - \frac{1}{2}x$ **f** $y = \frac{5x-2}{2}$

# 9.03 Linear modelling

**Video**
Applying linear functions

**Puzzle**
Linear functions code puzzle

Scientists and researchers observing patterns in nature and society often use a mathematical formula to represent a real-life situation. This is called **algebraic modelling** and if the relationship uses the linear function $y = mx + c$, it is called **linear modelling**.

## Dependent and independent variables

Because the value of $y$ depends on the value of $x$, $y$ is called the **dependent variable** and $x$ is called the **independent variable**.

When a function is graphed, the independent variable is shown on the horizontal axis, while the dependent variable is shown on the vertical axis.

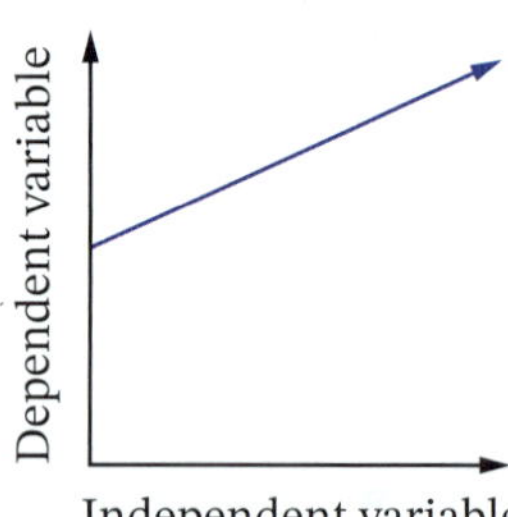


## Interpreting the gradient and the $y$-intercept

The **gradient** measures the steepness of a line, but it also shows how quickly the $y$ values are changing.

Look at this table of values for $y = 5x - 4$:

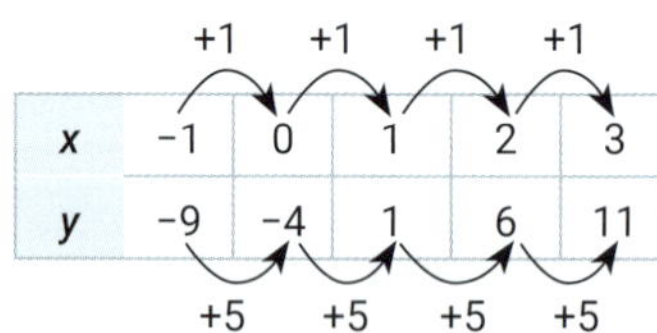


| $x$ | -1 | 0 | 1 | 2 | 3 |
|---|---|---|---|---|---|
| $y$ | -9 | -4 | 1 | 6 | 11 |

As the $x$ values increase by 1, the $y$ values increase by 5. If this linear function is graphed, for every 'run' of 1 unit, there is a 'rise' of 5 units, giving a gradient of $\frac{5}{1} = 5$.

### The gradient and $y$-intercept

- The **gradient** of a linear function is the **rate of change** of $y$.
- The higher the gradient, the steeper the line, and the faster $y$ increases relative to $x$.
- The **$y$-intercept** is the value of $y$ when $x = 0$.

## Interpolation and extrapolation

In mathematics, interpolation and extrapolation are used to predict values in relation to known data.

**Interpolation** (pronounced 'in-TERP-o-lay-shon') is making predictions for data values within the known data set.

**Extrapolation** (pronounced 'ex-TRAP-o-lay-shon') is making predictions for data values beyond or outside the known data set. There can be limitations involved when extrapolating because the pattern or trend may not continue beyond the known data.

For example, consider the linear relationship between volume and mass of a liquid shown in the graph, for masses from 0 to 5 grams.

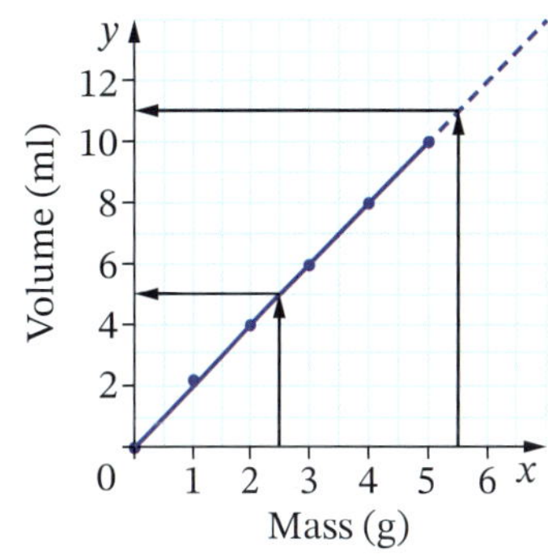


On the graph, if we consider the mass of 2.5 grams, we can **interpolate** that its volume is 5 mL. For a mass of 5.5 grams, *outside* the known values, we could **extrapolate** that its volume is about 10.9 mL.

### Example 6

Videos
Linear modelling 1
Linear modelling 2

Brett is organising a 21st birthday party. The total cost of the party will include \$200 for hiring the hall, \$120 for the music and \$14 per person for catering. The table below shows the total cost, \$$C$, for different numbers of guests, $n$, attending the party.

| **Number of guests, $n$** | 80 | 100 | 120 | 130 | 150 |
|---|---|---|---|---|---|
| **Party cost, \$$C$** | 1440 | 1720 | 2000 | 2140 | 2420 |

**a** Which variable, $n$ or $C$, is the dependent variable?

**b** Find the gradient of the linear relationship between $n$ and $C$.

**c** Find the vertical intercept of this linear relationship.

**d** Write the linear function for $C$ in terms of $n$.

**Vertical intercept** is the more general name for the $y$-intercept, because linear functions can use variables other than $x$ and $y$.

**e** If this function was graphed, which variable would be shown on the horizontal axis?

**f** What does the vertical intercept of this function represent?

**g** Find the cost of the party if there were 95 guests.

#### Solution

**a** $C$, because $C$ depends on $n$.

**b** Choosing the ordered pairs (80, 1440) and (100, 1720) from the table:

$$m = \frac{1720 - 1440}{100 - 80} = \frac{280}{20} = 14$$

The gradient is 14.

**c** The linear function is $C = mn + c$.

The gradient is $m = 14$. — from the answer to part **b**

So $C = 14n + c$.

To find the value of $c$, the vertical intercept, substitute an ordered pair into the function.

Substitute (100, 1720):

$1720 = 14(100) + c$

$1720 = 1400 + c$

$c = 320$

The vertical intercept is 320.

**d** The function is $C = 14n + 320$.

This formula should work for the other ordered pairs in the table.

**e** $n$, because it is the independent variable.

**f** The vertical intercept, 320, is the value of $C$ when $n = 0$, which is the cost of the party if there are no guests attending. This makes sense because it is the total fixed cost of hiring the hall and the music ($200 + $120).

**g** Substitute $n = 95$ into the function:

$C = 14n + 320$

$= 14(95) + 320$

$= 1650$

The cost for the party with 95 guests is $1650.

## Example 7

A criminologist discovered that the number of crimes committed per month, $C$, in a big city decreased as the number of police officers, $P$, patrolling the city increased. After graphing her data on a number plane, she found the linear relationship to be $C = -3P + 3250$.

**a** What is the independent variable in this relationship?

**b** Copy and complete this table for the equation $C = -3P + 3250$.

| Number of police, $P$ | 50 | 150 | 200 | 250 | 300 |
|---|---|---|---|---|---|
| Crimes per month, $C$ | | | | | |

**c** Graph $C = -3P + 3250$ on the number plane.

**d** What is the gradient of the graph you drew in part **c** and what does it represent?

**e** What is the vertical intercept of the graph you drew in part **c** and what does it represent?

**f** Calculate how many crimes are committed when 100 police officers are on patrol. Is this interpolation or extrapolation?

**g** Calculate how many police officers are required to reduce the number of crimes to 1900. Is this interpolation or extrapolation?

**h** Describe any limitations to this linear model for large values of $P$.

## Solution

**a** $P$, because $C$ depends on $P$.

**b**

| Number of police, $P$ | 50 | 150 | 200 | 250 | 300 |
|---|---|---|---|---|---|
| Crimes per month, $C$ | 3100 | 2800 | 2650 | 2500 | 2350 |

**c**

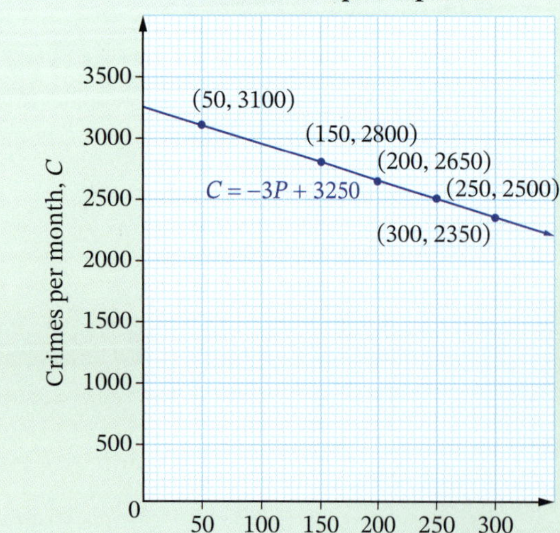

**d** From the equation, the gradient is −3 and it represents the rate of change in crime as the number of police increases. As the number of police increases by 1, the number of crimes decreases by 3.

**e** The vertical intercept is 3250, and this represents the number of crimes if no police officers were on patrol. It is the value of $C$ when $P = 0$.

**f** Substitute $P = 100$:

$$C = -3P + 3250$$
$$C = -3(100) + 3250$$
$$= 2950$$

2950 crimes are committed when 100 police are on patrol.

This is interpolation because $P = 100$ is within the data values in the table and graph.

**g** Substitute $C = 1900$:

$$C = -3P + 3250$$
$$1900 = -3P + 3250$$
$$-3P = -1350$$
$$P = \frac{-1350}{-3}$$
$$= 450$$

450 police officers are required to reduce the number of crimes to 1900.

This is extrapolation because $C = 1900$ and $P = 450$ are beyond or outside the data values in the table and graph.

**h** It may be impractical to keep increasing the number of police officers indefinitely, and the effect on reducing crime may become less (for example, it is impossible to have zero crime).

**EXERCISE 9.03** Answers on p. 481

## Linear modelling

**1** This table shows the cost of coffee $\$C$ by weight $W$ kg.

| Weight ($W$ kg) | 1 | 2 | 3 | 4 | 5 |
|---|---|---|---|---|---|
| Cost ($\$C$) | 12 | 24 | 36 | 48 | 60 |

**a** What is the independent variable?

**b** Graph the table of values.

**c** Use the graph to find the cost of 1.5 kg of coffee.

**d** Use the graph to find the weight of \$30 worth of coffee, correct to one decimal place.

**2** A children's outdoor party business charges \$125 to set up a party, \$75 to clean up after the party and \$110 per hour, to hire staff to run the party. Which equation represents the model for the total cost, $\$C$, and number of hours for a party, $h$, of using this service? Select **A**, **B**, **C** or **D**.

**A** $C = 200h + 110$ **B** $C = 185h + 125$

**C** $C = 110h + 200$ **D** $C = 90h + 200$

Foundation Mastery Complex

**3** **a** For this graph, determine the value of $y$:

**i** when $x = 3$

**ii** when $x = -1$.

**b** Find the gradient, $m$, and $y$-intercept, $c$, and the equation of the line.

**c** Hence, calculate the value of $x$ when $y = 10$ and determine whether this is interpolated or extrapolated data.

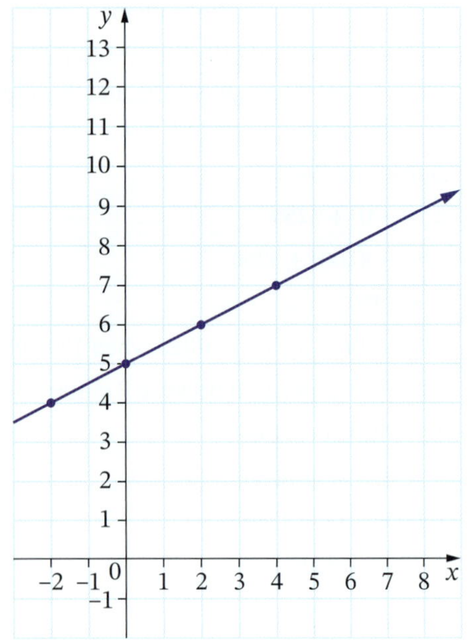


**4** Scott installs windows and charges a fixed cost of \$20 plus \$80 for every window he installs.

**a** If $n$ represents the number of windows installed, write an equation for the total amount earned by Scott, $\$w$, for installing windows, and copy and complete this table.

| $n$ | 0 | 1 | 2 | 3 | 4 | 5 | 6 |
|---|---|---|---|---|---|---|---|
| $\$w$ | | | | | | | |

**b** Graph the table of values.

**c** **i** What is the vertical intercept of the line?

**ii** What does this value represent in terms of the model?

**d** Using the graph or otherwise, solve each problem and determine whether it is interpolation or extrapolation.

**i** If Scott installs 10 windows, how much will a customer be charged?

**ii** Scott earns \$580. How many windows were installed?

**5** A cricket team's progressive score during a one-day cricket match can be approximated by the linear function graphed here.

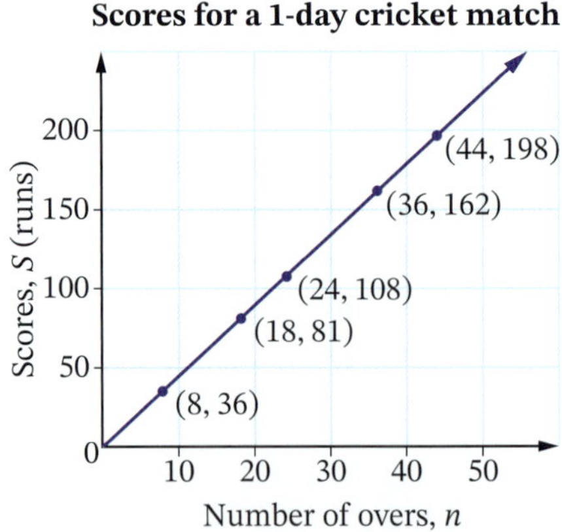


A one-day match has 50 overs, where an over is a set of 6 balls bowled by the same bowler. The variable, $n$, represents the number of overs bowled, while $S$ represents the total number of runs scored by the batting team.

□ Foundation ○ Mastery ○ Complex

a Is $S$ the dependent variable or the independent variable?

b Find the formula for $S$ in terms of $n$.

c The gradient of this linear function is also the team's run rate. What is the gradient and in what units is this run rate measured?

d What is the vertical intercept and what does it represent?

e What was the score after:

i the 21st over? ii the 50th over?

f At the end of which over had the score reached:

i 54 runs? ii 180 runs?

g The graph for an actual cricket match would not be a straight line but would normally flatten out later in the innings. Why?

**6** The value of a computer depreciates according to the formula $V = -420t + 1900$, where $V$ is the value in dollars and $t$ is the time in years. EXAMPLE 7

a Copy and complete this table for the formula $V = -420t + 1900$.

| Time, $t$ (years) | 1 | 2 | 3 | 4 |
|---|---|---|---|---|
| Value of computer, $V$ ($) | | | | |

b Graph this linear relationship on a number plane.

c What does the gradient of this linear function represent?

d What was the value of the computer after $2\frac{1}{2}$ years?

e What was the original value of the computer?

f This linear model does not work when $t = 5$ and beyond. Why not?

g Find, correct to 1 decimal place, the time when the computer has zero value.

**7** Yasmin works for a pizza shop and, each day, she earns a base pay of $50, plus $3 for every pizza she delivers.

a If $n$ is the number of pizzas Yasmin delivers in a day and $P$ is her total pay in dollars, write a formula for $P$ in terms of $n$.

b If this linear function was graphed, which variable would be represented on the horizontal axis?

c What is the vertical intercept of this function and what does it represent?

d How much will Yasmin earn for delivering 28 pizzas in a day?

e If Yasmin earned $98 today, how many pizzas did she deliver?

**8** This table shows the linear relationship between distances measured in miles and distances measured in kilometres.

| Miles, $M$ | 15 | 25 | 30 | 45 |
|---|---|---|---|---|
| Kilometres, $K$ | 24 | 40 | 48 | 72 |

a Graph this linear relationship on a number plane.

b Find the equation of the line.

c What can you say about the value of the vertical intercept? Why?

d What is the gradient and what does it represent?

e Use your equation from part **b** to convert:

i 100 miles to kilometres ii 100 km to miles.

f Use your graph from part **a** to convert:

i 12 miles to kilometres ii 20 kilometres to miles.

☐ Foundation ○ Mastery ○ Complex

**9** This graph shows the linear relationship between the distance, $d$ km, travelled by a truck and the running costs, $\$C$, for the trip.

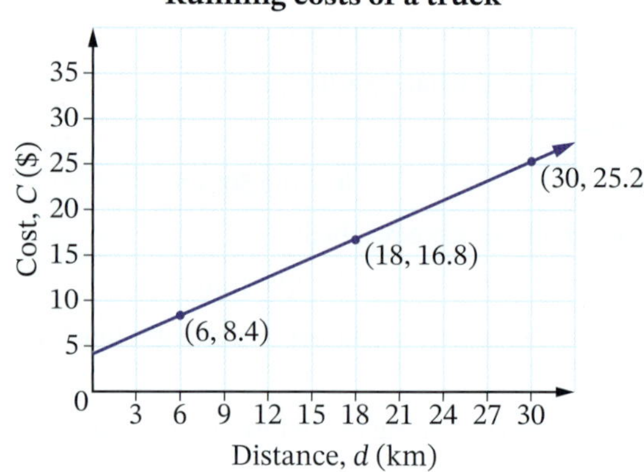


**a** What is the dependent variable?

**b** Find the gradient and the vertical intercept of this linear relationship.

**c** Write the formula for this linear relationship.

**d** If the length of a trip is extended by 5 km, by how much will the charge increase?

**e** Calculate the running costs for a trip of length:

**i** 20 km **ii** 0 km

**f** Calculate the distance travelled if the running costs were \$37.80.

**10** During summer, crickets chirp faster at night if the temperature is higher. There is a linear relationship between the temperature and a cricket's chirping rate, shown in the table below.

| Temperature, $T$ (°C) | 12 | 15 | 19 | 22 | 28 |
|---|---|---|---|---|---|
| Chirp rate, $n$ (chirps/min) | 72 | 96 | 128 | 152 | 200 |

**a** Is $T$ the dependent variable or the independent variable?

**b** Find the linear function for $n$ in terms of $T$.

**c** Graph the linear function you found in part **b**.

**d** If the temperature increases by 2°C, what happens to the cricket's chirp rate?

**e** Find the chirp rate of a cricket when the temperature is 26°C. Is this interpolation or extrapolation?

**f** At what temperature does a cricket chirp 144 times per minute?

**g** What is the vertical intercept of this function? Why doesn't the linear model work for this value?



**11** A plumber charges a callout fee of \$100 and \$1.50 per minute during his visit. If the plumber works for $t$ hours, which equation represents the amount charged by the plumber, $\$C$, as a function of time, $t$ hours? Select **A**, **B**, **C** or **D**.

**A** $C = 1.5 + 100t$ **B** $C = 1.5t + 100$

**C** $C = 90 + 100t$ **D** $C = 90t + 100$

Foundation Mastery Complex

**12** This graph represents the costs to hire an electrician for a job, excluding the cost of materials and equipment.

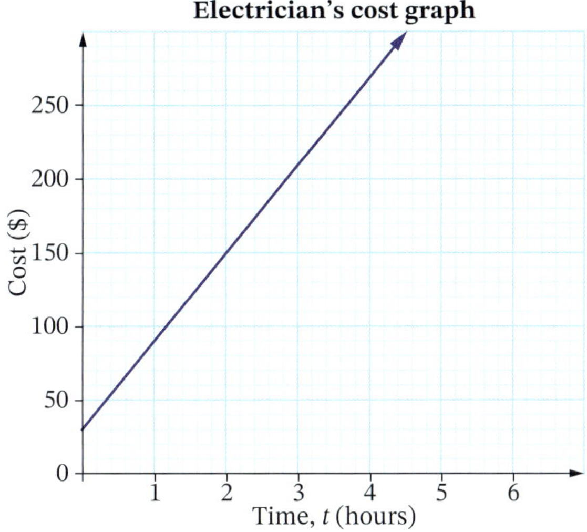


- **a** What is the vertical intercept of the graph and what does it represent?
- **b** Estimate how much it will cost for the electrician to work for 3.5 hours.
- **c** Determine how long the electrician would need to work to earn $120.
- **d** Describe one possible limitation of this model if the data was extrapolated beyond 4 hours.

**13** This height–weight graph shows the healthy weight range (shaded) for Australian adults of different heights.

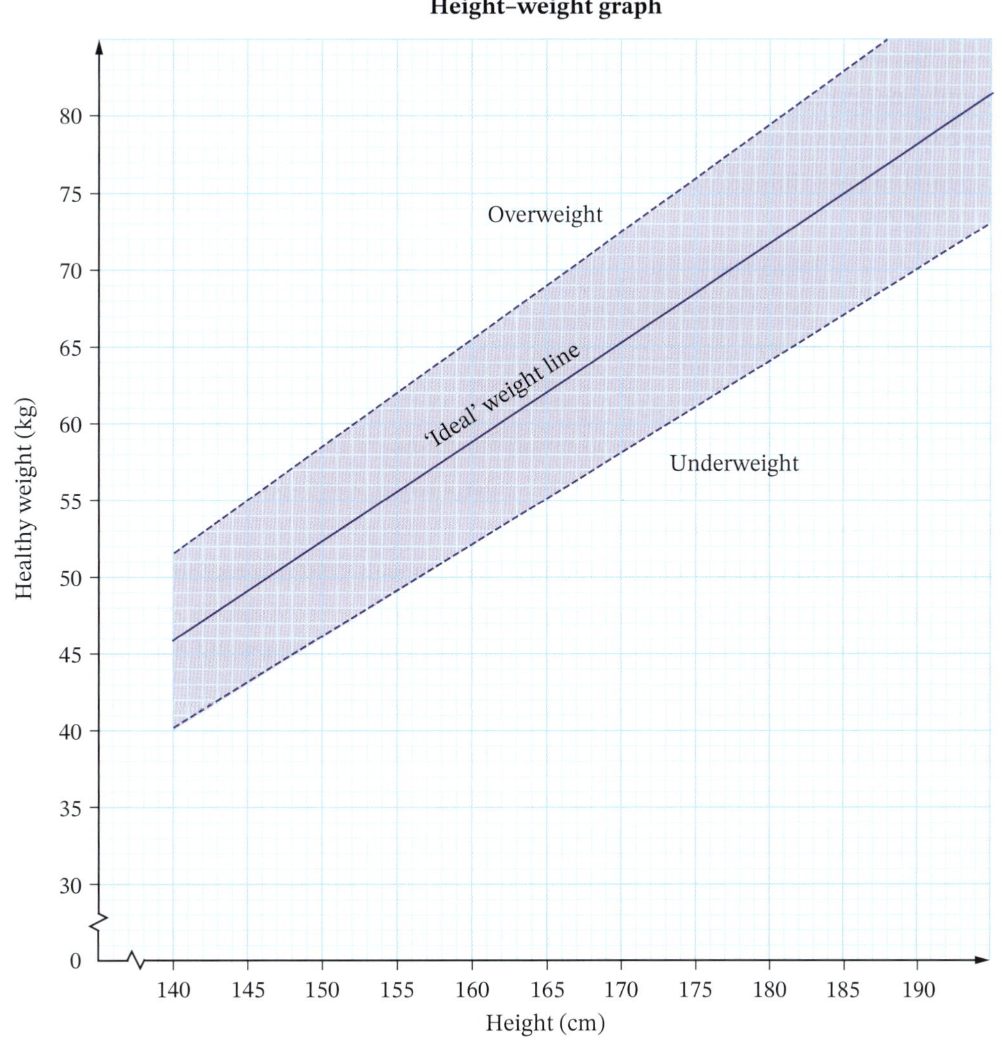

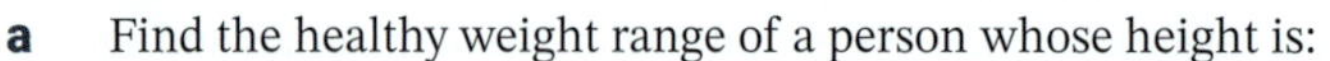

**a** Find the healthy weight range of a person whose height is:

**i** 180 cm **ii** 155 cm **iii** 192 cm

**b** Find the height range of a person with a healthy weight of:

**i** 72 kg **ii** 58 kg **iii** 80 kg

**c** Calculate the gradient of the 'ideal' weight line. What rate does this gradient value represent?

## Technology

### Linear modelling

Spreadsheets and graphing technology can be used to represent and model applications of linear functions.

**1** **Taxi fares:** The day rate (9 am – 5 pm) for hiring a taxi is $4.85 flagfall and $3.90 per kilometre travelled.

**a** Copy and complete the table of values using $d$ to represent distance travelled and $\$C$ for the cost of a taxi ride.

| Distance travelled $d$ (km) | 0 | 10 | 20 | 30 | 40 | 50 | 60 |
|---|---|---|---|---|---|---|---|
| Cost, $C$ ($) | | | | | | | |

**b** Use a spreadsheet or graphing technology to graph the table of values.

**c** Find the gradient, $m$, and $y$-intercept, $c$, and determine the equation representing this model.

**d** Use the graph to determine the cost of a taxi ride of:

**i** 24 km **ii** 75 km

**e** In part **d** above, is this interpolation or extrapolation? Justify your answer.

**f** Using your graph, determine the maximum number of kilometres travelled if the cost to hire a taxi is $85.

**g** Identify one limitation of this linear model.

**2** **Hiking trip:** A hiker walks at a constant rate of 3 km/h for 4 hours.

**a** Copy and complete this table of values using $t$ to represent time in hours and $d$ for distance travelled in kilometres.

| Time, $t$ (h) | 1 | 2 | 3 | 4 |
|---|---|---|---|---|
| Distance travelled, $d$ (km) | | | | |

**b** Use technology to graph the information in the table.

**c** Find the gradient, $m$, and $y$-intercept, $c$, and hence write the formula representing this model.

**d** Determine the distance travelled if the hiker walks for 2.5 hours. Is this interpolation or extrapolation?

**e** Determine the time taken to travel 7 km. Is this interpolation or extrapolation?

**f** Identify a limitation of this model.

**3** **Ashley's bank account:** Ashley is a professional tennis player with $25 000 in his bank account. Every month he spends $1500 on coaching and other costs and he currently does not add any money to the account.

**a** Copy and complete the table of values using $t$ to represent time in months and $A$ to represent the amount remaining in his account.

| Time, $t$ (months) | 1 | 2 | 3 | 4 |
|---|---|---|---|---|
| Amount remaining, $A$ ($) | | | | |

**b** Use technology to graph the table of values.

**c** If Ashley maintains this pattern of spending for one year, extend the table and determine:

**i** how much money he has left in the account after 7 months

**ii** how many months until he has $10 000 remaining in the account.

**d** Identify a limitation of this model that would affect extrapolation of the data as $t$ increases indefinitely and include any calculations for specific values for $t$ in your response.

Foundation Mastery Complex

# Direct linear variation



Ann-Marie noticed that, with her new car, she could drive a distance of 567 km on a full tank of petrol (42 litres). She also discovered the results shown in the table below.

Video
Direct linear variation

| Amount of petrol used, $p$ (L) | 10 | 14 | 22 | 32 | 36 | 42 |
|---|---|---|---|---|---|---|
| Distance travelled, $d$ (km) | 135 | 189 | 297 | 432 | 486 | 567 |

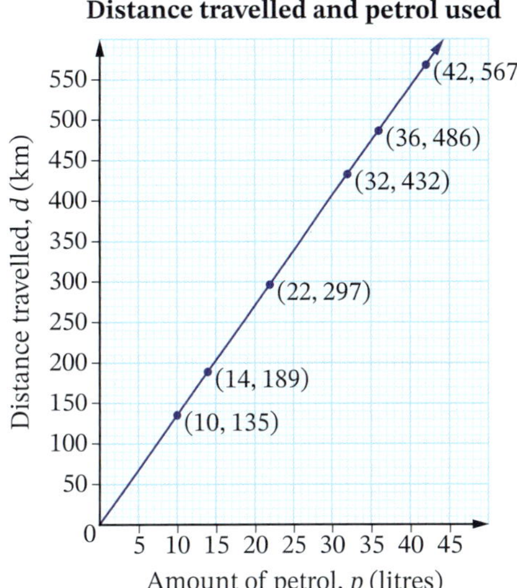


When she graphed these values, Ann-Marie found a linear relationship.

This relationship must be of the form $d = mp + c$, but the vertical intercept is 0 because, when $p = 0$, $d = 0$ (no petrol, no distance). So the linear function is $d = mp$.

To find the value of $m$, substitute any point from the table into the equation, for example (10, 135).

$$d = mp$$
$$135 = m(10)$$
$$m = 10$$
$$= 13.5$$

Choose another point from the table to check that this formula is correct.

$\therefore$ the linear function is $d = 13.5p$.

Because $d = 13.5p$, the distance, $d$, is found by multiplying the amount of petrol, $p$, by a constant amount, 13.5. This is an example of **direct linear variation**, and we can say that '$d$ varies directly as $p$', or that '$d$ is directly **proportional to** $p$'.

### Direct linear variation

If $y$ **varies as** $x$, or $y$ is **directly proportional to** $x$, then $y = kx$, where $k$ is a **constant**. $k$ is called the **constant of variation** or **constant of proportionality**.

A direct linear relationship exists between $x$ and $y$. If $x$ increases (or decreases), $y$ increases (or decreases). If $x$ is doubled (or halved), $y$ is doubled (or halved).

### Example 8

The mass, $M$ (in kilograms), of a metal varies directly as its volume, $V$ (in cubic centimetres).

| Volume, $V$ (cm$^3$) | 0 | 48 | 60 | 80 | 116 | 140 |
|---|---|---|---|---|---|---|
| Mass, $M$ (kg) | 0 | 14.4 | 18 | 24 | 34.8 | 42 |

**a** Graph the relationship between $M$ and $V$.

**b** Find the equation for $M$ in terms of $V$.

**c** Find the mass of 212 cm$^3$ of the metal.

## Solution

**a**

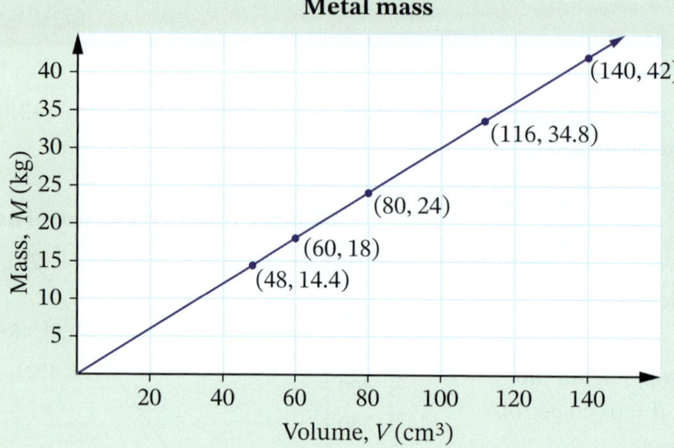


**b** $M = kV$

Substituting (60, 18) from the table to find $k$:

$18 = k(60)$

$k = \frac{18}{60}$

$= \frac{3}{10}$

$\therefore M = \frac{3V}{10}$ (or $M = 0.3V$)

Choose another point from the table to check that this formula is correct.

**c** Substitute $V = 212$ into the formula.

$M = \frac{3(212)}{10}$

$= 63.6$

The mass of the metal is 63.6 kg.

## Example 9

The stretch, $S$ (in centimetres), of a spring varies as the mass, $M$ (in kilograms), of the load pulling it. A load of 24 kg causes the spring to stretch 15 cm.

**a** Find the variation equation relating $S$ to $M$.

**b** What is the stretch caused by a load of 13 kg?

**c** What is the mass of the load that will cause a stretch of 25 cm?

**d** What is a limitation of this linear model?

## Solution

**a** $S = kM$

Substitute $M = 24$ and $S = 15$ to find $k$.

$15 = k(24)$

$k = \frac{15}{24}$

$= \frac{5}{8}$ (or 0.625)

$\therefore S = \frac{5M}{8}$ (or $S = 0.625M$)

**b** When $M = 13$:

$$S = \frac{5(13)}{8}$$

$$= 8.125$$

A stretch of 8.125 cm is caused by a load of 13 kg.

**c** When $S = 25$:

$$25 = \frac{5M}{8}$$

$$5M = 200$$

$$M = 40$$

A 40 kg load will cause a stretch of 25 cm.

**d** For a very heavy load, the spring will become permanently stretched and will not return to its original length when the load is removed.

## Linear variation problems

To solve a linear variation problem:

1. Identify the 2 variables (say $x$ and $y$) and form a variation equation, $y = kx$.
2. Substitute values for $x$ and $y$ to find $k$, the constant of variation.
3. Rewrite $y = kx$ using the value of $k$.
4. Substitute a value for $x$ or $y$ into $y = kx$ to solve the problem.

**EXERCISE 9.04** Answers on p. 482

## Direct linear variation

**1** Copy and complete this table after finding the value of $k$ in $y = kx$.

| x | 1 | 4 | 6 | 10 | 11 | 15 |
|---|---|---|---|---|---|---|
| y | | 14 | 21 | | | 52.5 |

**2** The distance travelled by a bicycle varies directly with the number of revolutions made by the pedals.

**a** Form a variation equation and find the constant of variation, given that the bicycle travels 55 metres for 20 revolutions of the pedals.

**b** Calculate the distance travelled for 33 revolutions of the pedals.

**c** Calculate the number of revolutions of the pedals required for the bike to travel 99 metres.

**3** The increase in pressure experienced by a scuba diver is directly proportional to her depth under the water. The increase in pressure at 25 m is 147 kilopascals (kPa).

**a** Form a variation equation.

**b** What increase in pressure is experienced at 40 m?

**c** At what depth is the increase in pressure 833 kPa?

☐ Foundation ◯ Mastery ⬡ Complex

**4** The graph illustrates the fact that the mass of fuel an aeroplane requires varies directly with the distance it flies.

**a** Form a variation equation, using a point on the graph to find the constant of variation.

**b** Use the equation to calculate the amount of fuel required to fly 3250 km.

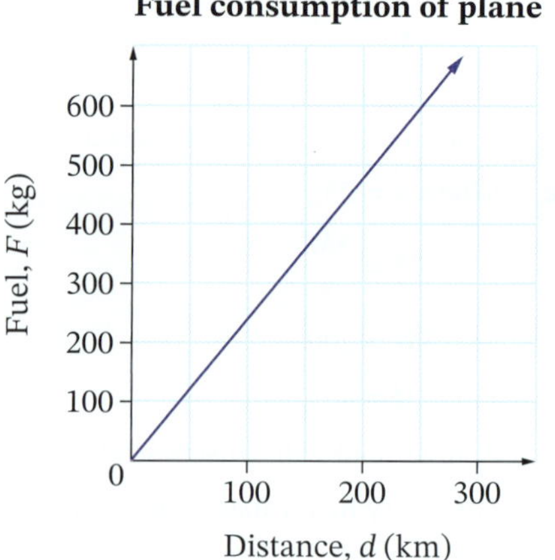


**5** A speed expressed in km/h is directly proportional to the same speed expressed in miles/h.

| Speed in miles/h, $x$ | 30 | 36 | 45 | 50 | 62.5 | 74 |
|---|---|---|---|---|---|---|
| Speed in km/h, $y$ | 48 | 57.6 | 72 | 80 | 100 | 118.4 |

**a** Find the equation for this table of values.

**b** Convert 60 miles/h to km/h.

**c** Convert 120 km/h to miles/h.

**6** For an object that is cooling, the drop in temperature varies directly with time. If the temperature drops 8°C in 5 minutes, which of the following is the amount of time it would take for the temperature to drop 10°C? Select **A**, **B**, **C** or **D**.

**A** 6.25 min **B** 7 min **C** 12.8 min **D** 16 min

**7** The volume, $V$ litres, of water in a tank as it is filled is proportional to the time taken, $t$ minutes, as shown on the graph.

**a** Find the constant of variation.

**b** What does the constant you found in part **a** represent?

**c** Calculate the amount of water in the tank after 8 minutes.

**d** Calculate how long it takes to pump 960 L of water into the tank.

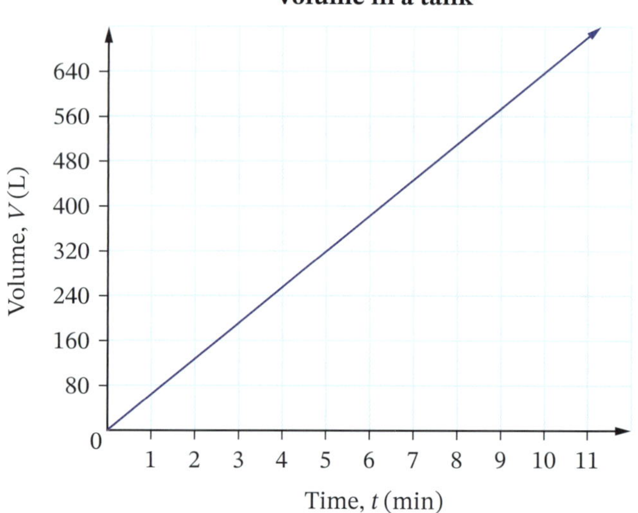


Foundation Mastery Complex

8. The distance travelled by a marathon runner varies directly with time.
   - **a** If he covers 9.75 km in 45 minutes, calculate how long it will take him to run 26 km.
   - **b** What is a limitation of this linear model?

9. The weight of an astronaut on Mars is proportional to his weight on Earth.
   A 72 kg astronaut weighs 27.4 kg on Mars.
   - **a** Calculate how much a 60 kg astronaut weighs on Mars, correct to one decimal place.
   - **b** If an astronaut weighs 32 kg on Mars, calculate his weight on Earth, correct to one decimal place.

10. Nick noticed that, during his road trip, the amount of petrol used by his car varied with the amount of time he drove. It consumed 45 L of petrol in 4 hours. How long, to the nearest minute, will Nick's car take to consume 100 L of petrol?

11. The download time of a computer file is directly proportional to the size of the file.
    If a file of 100 megabytes requires 36 seconds to download, calculate:
    - **a** how long it will take to download a 180 megabyte file
    - **b** the size of a file that requires 80 seconds to download.

12. The compression of a car spring is proportional to the force applied. A force of 200 N (newtons) causes a compression of 1.6 cm. Which of the following amounts of compression is caused by a 5000 N force? Select **A**, **B**, **C** or **D**.

    **A** 25 cm **B** 40 cm **C** 62.5 cm **D** 200 cm

13. At 3:30 pm one day, the shadows of objects of different heights were measured, then graphed.

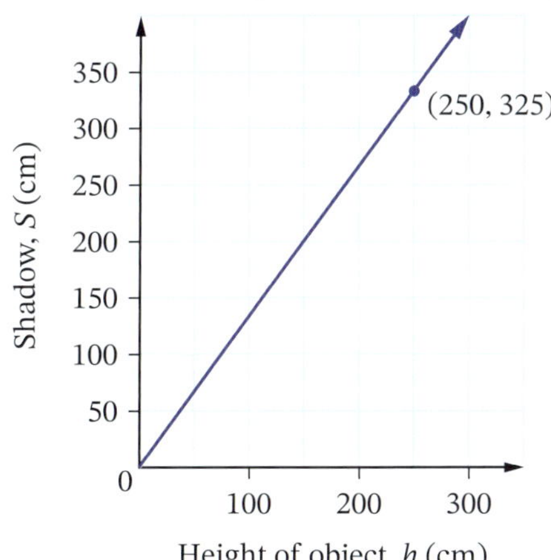


   - **a** Find the equation of the line.
   - **b** Calculate the length of the shadow of a lamppost of height 244 cm.
   - **c** What is the height of a letterbox, correct to the nearest centimetre, if its shadow is 120 cm long?

14. When an object is dropped under gravity, its speed varies with time. Its speed after 5 seconds is 49 m/s.
    - **a** Find its speed after 12 seconds.
    - **b** Calculate the time taken to reach a speed of 175 m/s (to the nearest second).
    - **c** What is a limitation of this linear model?

Foundation Mastery Complex

**15** This graph converts between the euro (€) and the Australian dollar (A$). Use this conversion graph to answer the questions on the following page.

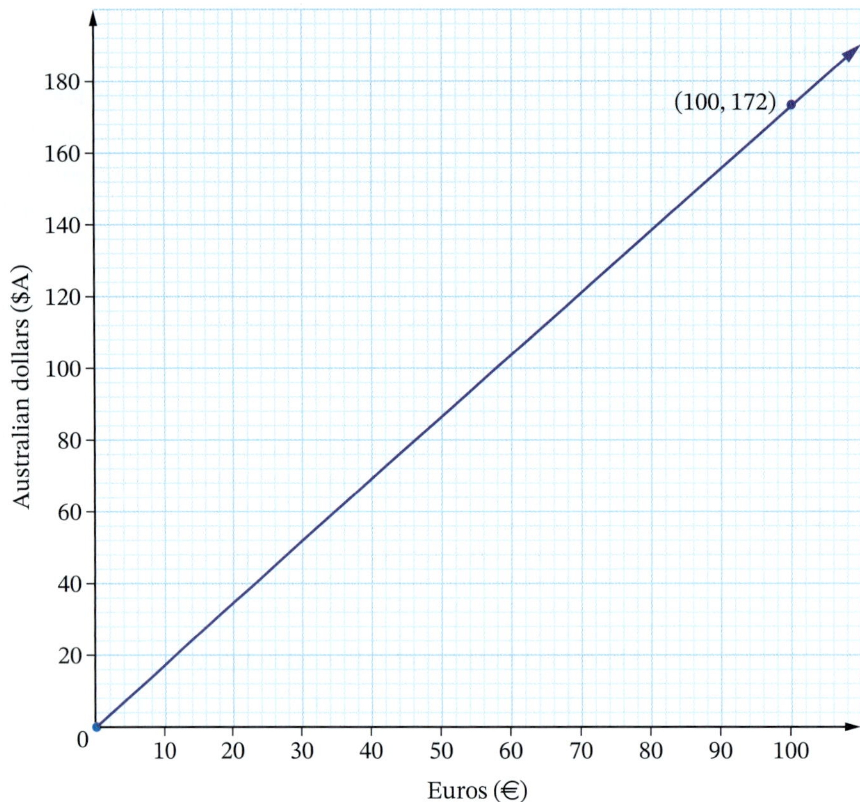


**a** Convert each European price to Australian dollars, to the nearest dollar

| | | | | | |
|---|---|---|---|---|---|
| **i** | a smart watch | €88 | **ii** | theme park entry | €63 |
| **iii** | a mobile phone | €76 | **iv** | a camera | €52 |
| **v** | a city bus tour | €25 | **vi** | car hire per day | €42 |

**b** Calculate the gradient of the graph above, correct to 2 decimal places. What does this value represent?

Foundation Mastery Complex

**16** This graph converts from pounds (an imperial unit of mass, with abbreviation lb) to kilograms (the metric unit).

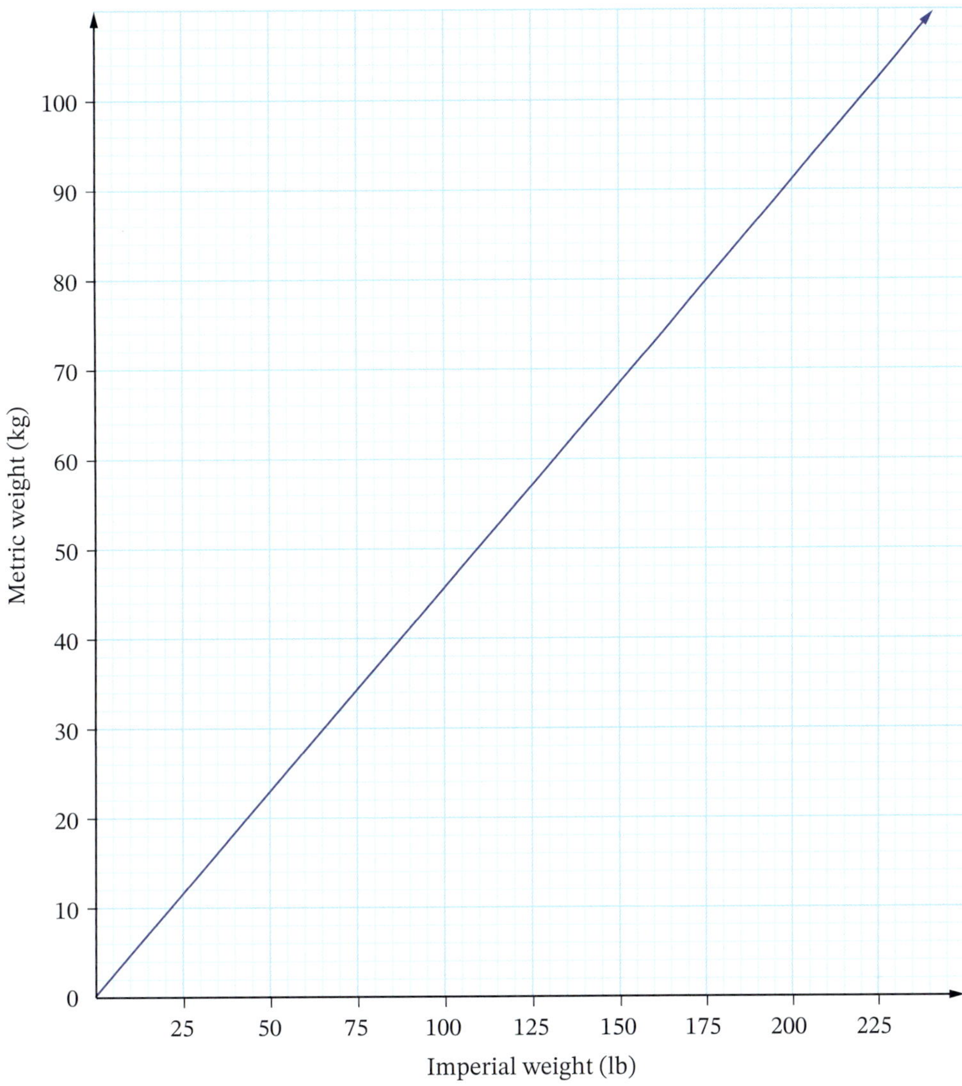


**a** Convert each mass to kilograms.

**i** 90 lb **ii** 135 lb **iii** 160 lb

**b** Convert each mass to pounds.

**i** 40 kg **ii** 64 kg **iii** 90 kg

**c** What is the conversion rate in kg/lb, correct to 2 decimal places?

**d** Hence, write the direct variation equation for converting $P$ pounds to $K$ kilograms.

**e** Use the equation to convert (to one decimal place):

**i** 135 lb to kg **ii** 64 kg to lb.

□ Foundation ○ Mastery ○ Complex

## Sample HSC problem

Answers on p. 483

10 marks A bucket contains 750 mL of water. Under the sun, the water evaporates at a rate of 60 mL per hour.

**a** Copy and complete the table of values, using $t$ to represent time in hours and $V$ for volume in millilitres. 2 marks

| Time, $t$ (h) | 0 | 1 | 2 | 3 | 4 | 5 | 6 |
|---|---|---|---|---|---|---|---|
| Volume, $V$ (mL) | | | | | | | |

**b** Graph the table of values. 2 marks

**c** Find the equation of the graph. 2 marks

**d** Use the equation to determine the volume of water left in the bucket after 4.2 hours in the sun. 1 mark

**e** Calculate the time taken for the volume to reach 150 mL and explain why this is extrapolation. 2 marks

**f** What is a limitation of this linear model? Include any relevant calculations in your explanation. 1 mark

## Study Tip

### Attacking your weak areas

Most of your study time should be spent on attacking your weak areas to fill in any gaps in your Maths knowledge. Don't spend too much time on work you already know well, unless you need a confidence boost! Ask your teacher, use study guides or other textbooks to improve the understanding of your weak areas and to practise Maths skills. Use your topic summaries for general revision, and spend longer study periods on overcoming any difficulties in your mastery of the course.

# CHAPTER SUMMARY

This chapter, *Linear functions*, introduced the concept of the linear function and its graph and applications. Learn the meanings of the gradient and vertical intercept of a linear function and practise solving problems involving linear modelling and direct linear variation.

Before you move on, consider what you learned in this chapter and revisit any sections that may have been unclear.

| What you learned in this chapter... | Section | |
|---|---|---|
| Graph linear functions of the form $y = mx + c$ | 9.01 | Graphing linear functions |
| Find the gradient and $y$-intercept of a line from its graph | 9.01 | Graphing linear functions |
| | 9.02 | The gradient formula |
| Use linear functions to model practical situations and interpret the meaning of the gradient and $y$-intercept of the line in context | 9.03 | Linear modelling |
| Graph and solve problems involving direct linear variation | 9.04 | Direct linear variation |

To help master this topic, make a summary mind map. Use the chapter outline and the mind map below as a guide. Add your own words, symbols, diagrams, boxes and reminders. The summary should give you a 'whole picture' view of the topic and allow you to identify any weak areas to revisit in your revision.

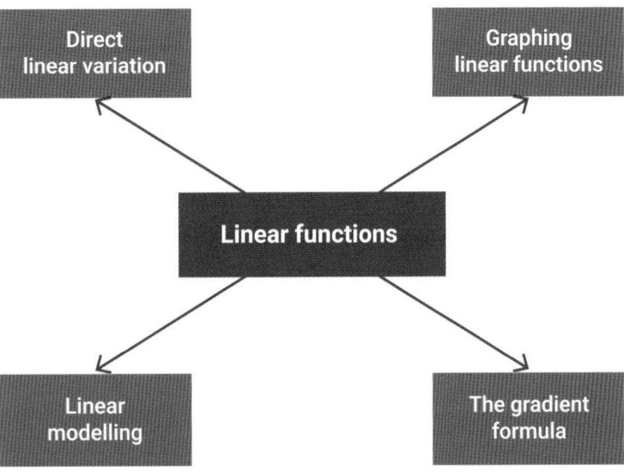

# 9 Test yourself Answers on p. 483

9.01 **1** Graph each linear function on a number plane.

**a** $y = -2x + 1$ **b** $y = 4x$

9.01 **2** Find the equation of each line.

**a**

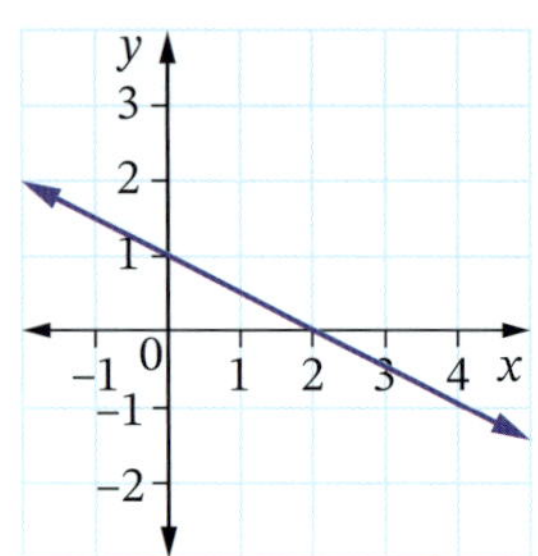

**b**

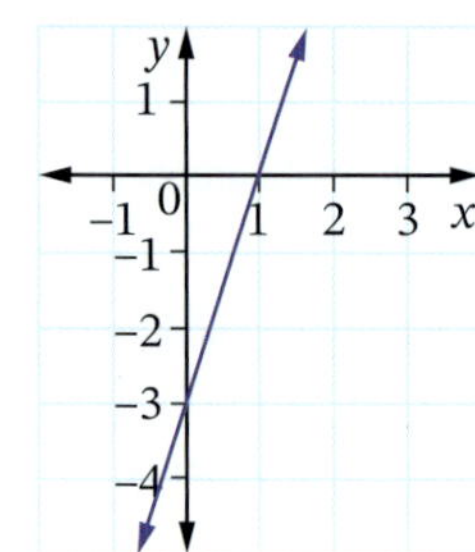

9.02 **3** Find the gradient of each line.

**a**

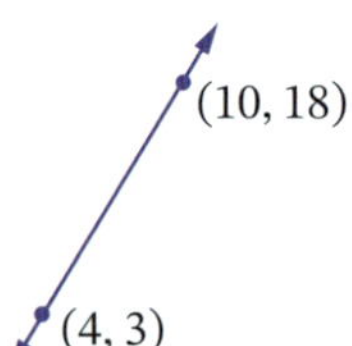


**b**

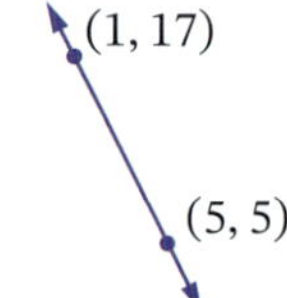


9.02 **4** Find the gradient of the function for each table of values.

**a**

| *x* | 0 | 4 | 6 | 12 |
|---|---|---|---|---|
| *y* | 12 | 14 | 15 | 18 |

**b**

| *x* | 2 | 5 | 9 | 11 |
|---|---|---|---|---|
| *y* | 42 | 30 | 14 | 6 |

9.03 **5** The length, $L$ inches, of a shoe has a linear relationship with the size, $S$, of the shoe.

| Shoe size, *S* | 2 | 5 | 7 | 8 | 12 |
|---|---|---|---|---|---|
| Length, *L* (inches) | 9 | 10 | $10\frac{2}{3}$ | 11 | $12\frac{1}{3}$ |

**a** Which is the independent variable?

**b** Graph this linear relationship.

**c** What is the gradient, and what does this represent?

**d** How long is a shoe of size 0? How is this shown on the graph?

**e** What is the formula for $L$ in terms of $S$?

**f** What is the length of a shoe of size $7\frac{1}{2}$?

**g** What size is a shoe of length 13 inches?

Foundation Mastery Complex

**6** Mr and Mrs Dillon give their children weekly pocket money according to a function based on the age of the child, as shown in the table below. 9.03

| Age, *n* years | 7 | 9 | 10 | 13 | 15 | 18 |
|---|---|---|---|---|---|---|
| Pocket money, \$*P* | 4 | 12 | 16 | 28 | 36 | 48 |

**a** Find the formula for $P$ in terms of $n$.

**b** What is the gradient for this function and what does it stand for?

**c** What is the pocket money for a child aged 16?

**d** What is the age of the child who receives \$20 per week?

**e** According to Mr and Mrs Dillon, this linear model holds true only for values of $n$ from 7 to 18. Why do you think the formula is not valid for values of $n$:

**i** below 7? **ii** above 18?

**7** The distance travelled by a car is directly proportional to the number of rotations of its tyres. If 950 metres are travelled after 540 rotations, calculate how much distance, to the nearest kilometre, is covered after 10 000 rotations. 

**8** This graph converts between Thai baht (the currency of Thailand) and Australian dollars (A\$). 9.04

**a** Convert each of these Thai prices to Australian dollars (to the nearest dollar).

**i** 800 baht **ii** 2050 baht

**b** Convert each of these Australian prices to Thai baht (to the nearest 50 baht).

**i** A\$65 **ii** A\$82

**c** Calculate the gradient of this graph, correct to 3 decimal places. What does this value represent?

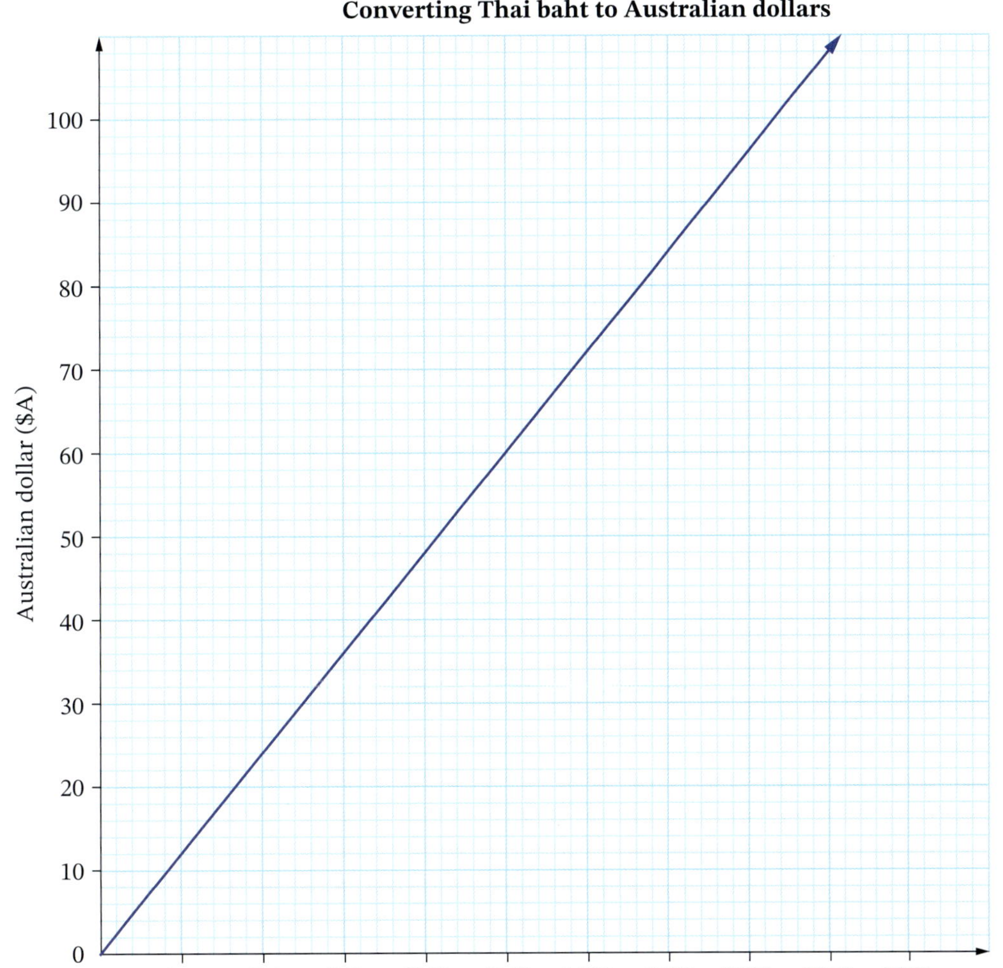


Foundation Mastery Complex

# PRACTICE EXAM 3

Answers on p. 483

**Recommended time: One hour**

**Section 1**
**10 multiple-choice questions: 1 mark each**
**Select the correct answer A, B, C or D.**

7.03 **1** Elena drives from Sydney to Brisbane, a distance of 920 km. She must charge her car every 300 km, at a cost of 25 cents per minute for 2 hours. What is Elena's total charge cost for the trip?

**A** \$30 **B** \$60 **C** \$90 **D** \$100

7.04 **2** A car can travel 465 km on a 50 L tank of petrol. What is its fuel consumption in L/100 km?

**A** 0.9 **B** 1.1 **C** 9.3 **D** 10.8

7.01 **3** Sienna pays for Third Party Property motor insurance by a direct debit of \$41.40 per month. How much would she save by paying a yearly amount of \$399 instead?

**A** nothing **B** \$97.80 **C** \$169.80 **D** \$185.20

9.04 **4** $T$ is directly proportional to $M$. When $M = 4$, $T = 10$. What is the value of $T$ when $M = 5$?

**A** 20 **B** 12 **C** 12.5 **D** 2.5

9.01 **5** Find the equation of the graphed line.

**A** $y = 3 + x$

**B** $y = x - 3$

**C** $y = 3 - 3x$

**D** $y = 3x + 3$

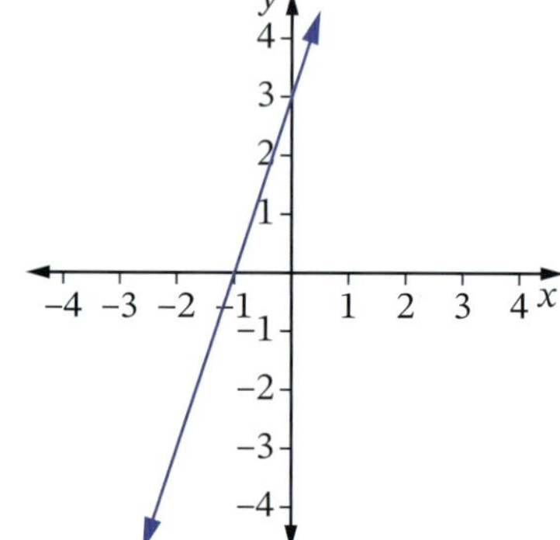


7.02 **6** The stamp duty for purchasing a car is \$3 per \$100 or part thereof up to \$45 000 and \$1350 plus \$5 for every \$100 or part thereof for purchases over \$45 000. How much stamp duty did Taylor pay when buying a car for \$39 450?

**A** \$1185 **B** \$1183.50 **C** \$1182 **D** \$1975

**7** How many edges does a spanning tree of this network have? 8.03

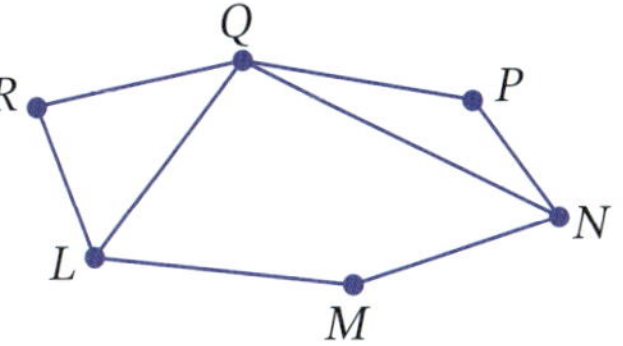


**A** 4

**B** 5

**C** 6

**D** 7

**8** Find the sum of the degrees of all the vertices in the network from Question **7** above. 8.01

**A** 15 **B** 16 **C** 17 **D** 18

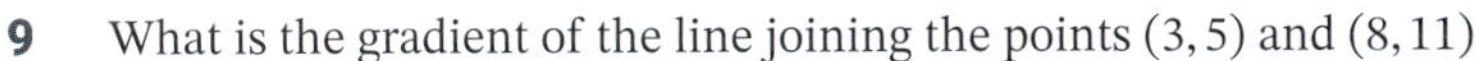

**9** What is the gradient of the line joining the points $(3, 5)$ and $(8, 11)$? 9.01

**A** $\frac{2}{3}$ **B** $-\frac{5}{6}$ **C** $\frac{6}{5}$ **D** $-\frac{6}{5}$

**10** Which of the following is a spanning tree of this network?

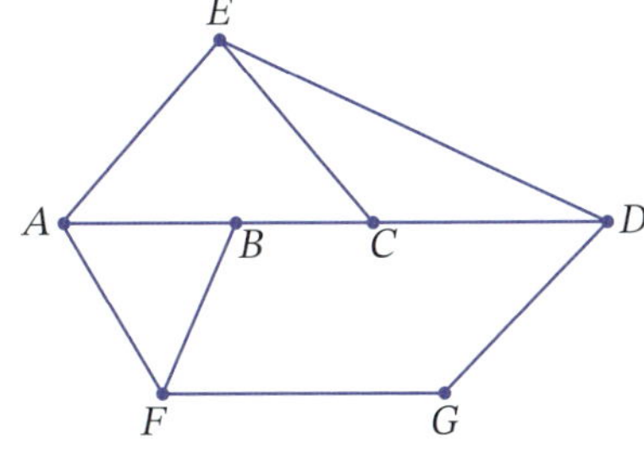


**A**

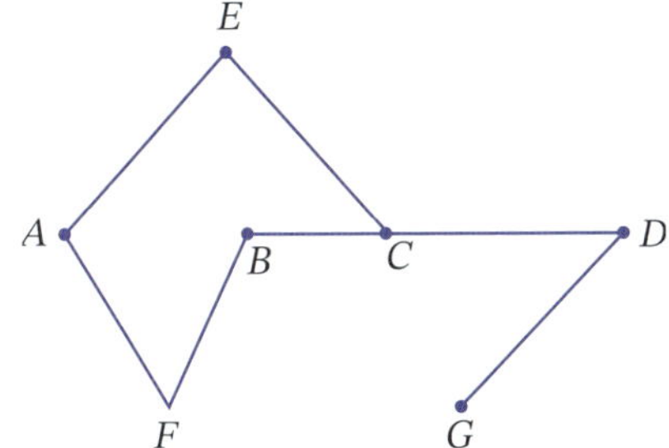


**B**

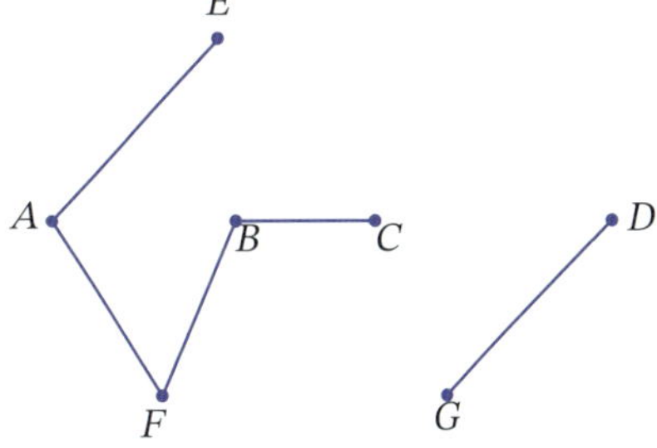


**C**

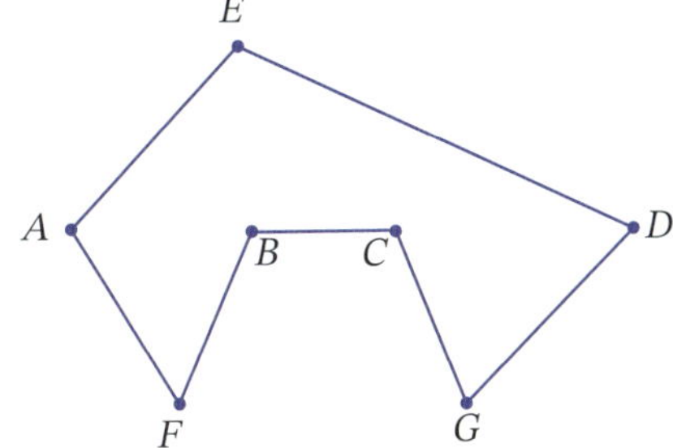


**D**

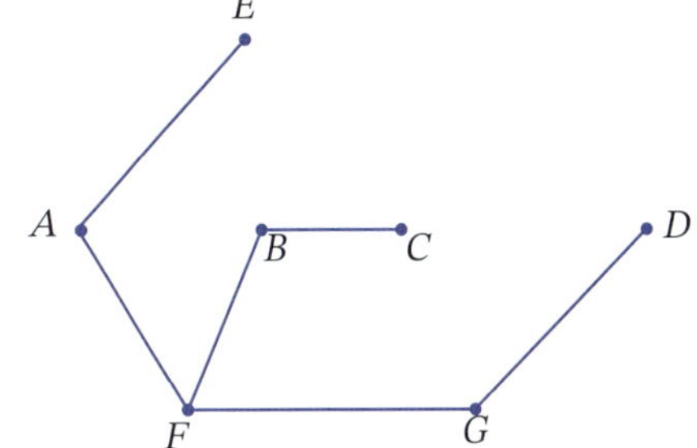


PRACTICE EXAM 3

## Section 2

**3 questions: 10 marks each**

8.02

### Question 11 10 marks

**a** Draw the directed network that is represented by the table. 2 marks

| | | To | | | | |
|---|---|---|---|---|---|---|
| | | *D* | *E* | *F* | *G* | *H* |
| From | *D* | – | 2 | – | 1 | 1 |
| | *E* | 1 | – | 3 | 2 | – |
| | *F* | – | 1 | – | 1 | 2 |
| | *G* | 1 | 1 | 1 | – | 2 |
| | *H* | 1 | – | 1 | 1 | 1 |

9.01

**b** **i** What is the gradient and *y*-intercept of the line $y = -\frac{1}{2}x + 4$? 2 marks

**ii** Graph the line on a number plane. 2 marks

7.02

**c** The stamp duty for purchasing a car is 3% of the purchase price up to \$45 000.

9.03

**i** Copy and complete the table below. 2 marks

| Price \$ | 0 | 10 000 | 20 000 | 30 000 | 40 000 | 45 000 |
|---|---|---|---|---|---|---|
| Stamp duty \$ | 0 | 300 | | | | |

**ii** Graph the table of values. 1 mark

**iii** Use the graph to estimate the stamp duty on a vehicle purchased for \$35 000. 1 mark

### Question 12 10 marks

7.01

**a** Briefly explain the main differences between Compulsory Third Party (CTP) insurance and Third Party Property insurance. 2 marks

9.03

**b** The table below shows the volume of water in kilolitres in a pool as it is drained.

| Time, *t* hours | 2 | 3 | 6 | 8 | 9 |
|---|---|---|---|---|---|
| Volume, *V* kL | 73 | 67.5 | 51 | 40 | 34.5 |

**i** Find the linear function for $V$ in terms of $t$. 1 mark

**ii** What does the gradient of this function represent? 1 mark

**iii** What does the vertical intercept of this function represent? 1 mark

**iv** After how many hours, to one decimal place, is the pool drained completely? 1 mark

8.03, 8.04

**c** The diagram shows the distances in metres between the office and cabins in a coastal holiday resort. Pathways are to be laid between the cabins and the office.

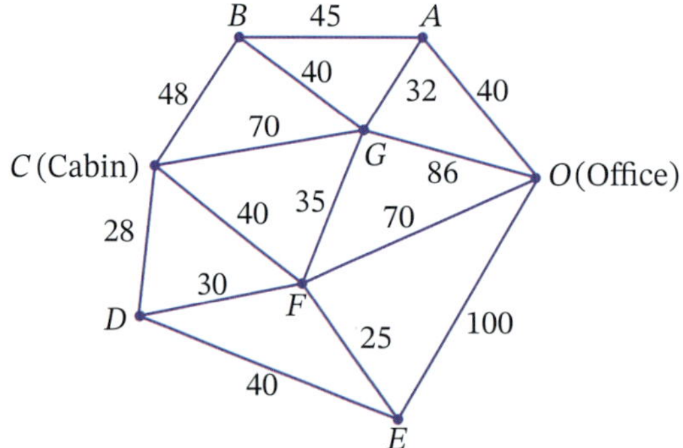


**i** Name the 2 vertices with degree 4. 1 mark

**ii** Find the shortest path from the office ($O$) to the cabin ($C$). 1 mark

**iii** Draw a minimum spanning tree for the resort. 1 mark

**iv** If the cost of the pathways is \$140/metre, what is the lowest cost of laying the pathways? 1 mark

**Question 13** 10 marks

9.04

**a** The graph shows that the distance covered varies directly with the amount of fuel used.

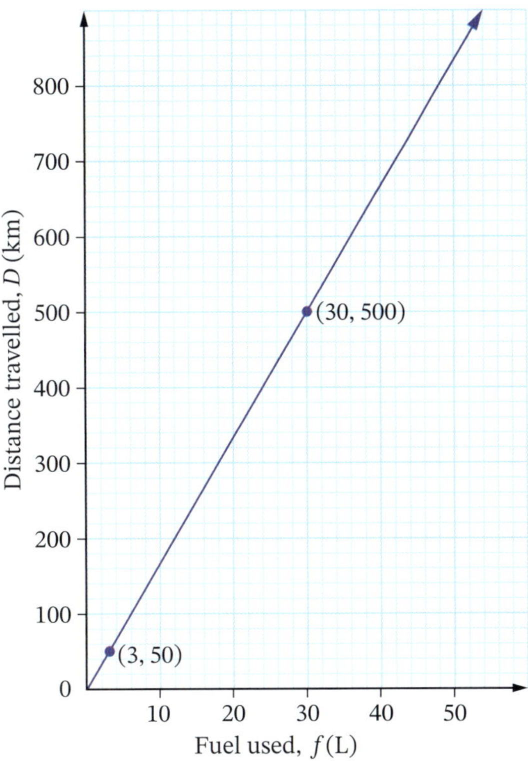


**i** What is the gradient of the line? 1 mark

**ii** What is the equation of the line? 1 mark

**iii** Use the equation to find the amount of fuel used to cover a distance of 900 km. 1 mark

**b** Tamim purchased a new ute for \$48 250 to use in his lawn-mowing business. 7.02

The costs associated with his purchase are shown in the table.

| | |
|---|---|
| **Purchase price** | \$48 250 |
| **GST** | 10% of purchase price |
| **Stamp duty on GST price** | \$1350 plus \$5 for every \$100 (or part of \$100) over \$45 000 |
| **Registration** | \$736 |
| **CTP insurance** | \$387 |
| **Comprehensive insurance** | \$682 |

Calculate the on-road costs that Tamim has to pay for the ute. 3 marks

**c** The fuel consumption of Regan's car is 8.4 L/100 km. 7.04

**i** Regan drives 47 km each way to work 5 days per week.

Calculate how much petrol she would use going to work each week. 2 marks

**ii** She also uses the car on weekends, covering a distance of about 150 km.

If Regan pays 215.9 c/L for petrol, calculate her weekly fuel bill. 2 marks

**END OF EXAMINATION**

# 10.

# ANALYSING DATA

The meteorologists at the Bureau of Meteorology measure and record weather data from over 1000 sites in Australia and Antarctica, and calculate statistics about temperature, rainfall and humidity. Climate averages, such as the median monthly rainfall or the mean number of rainy days per month, are calculated from weather data gathered over many years and can assist farmers to decide on the best times to plant crops.

## Chapter outline

## In this chapter you will:

- use the statistical investigation process
- design a survey or questionnaire that is clear and free from bias
- calculate and interpret the mean, median and mode of sets of data, including ungrouped data
- calculate and interpret the range, interquartile range and standard deviation of sets of data, including ungrouped data
- identify outliers in a set of data and examine their effects on statistical measures
- calculate cumulative frequency and construct cumulative frequency histograms and polygons
- use a cumulative frequency polygon to find the median, quartiles and interquartile range of a data set
- use a five-number summary to construct box plots
- describe the shape of a distribution using its graph or display
- compare data sets displayed on back-to-back stem-and-leaf plots and parallel box plots.

**Videos (20):**

**10.02** The mode, median and mean • The mean, mode, median and range • Analysing dot plots 1 • Analysing dot plots 2 • Statistics from a frequency table

**10.03** Interquartile range 1 • Interquartile range 2

**10.04** Standard deviation on a Casio calculator

**10.05** Outliers • Identifying outliers • Testing for outliers

**10.06** IQR from cumulative frequency polygon • Cumulative frequency and the median • Cumulative frequency graphs

**10.07** Box-and-whisker plots • Double box plots

**10.08** The shape of a frequency distribution • Shape and modality of distributions

**10.09** Comparing data sets • Box plots and outliers

**Skillsheet (1):**

**10.02** Statistical measures

**Worksheets (23):**

**SkillCheck** Assignment 10

**10.02** Mean, median, mode 1 • Mean, median, mode 2 • Mode, median and mean • Mean, median, mode and range

**10.03** Interquartile range

**10.04** Standard deviation • Statistical calculations

**10.05** Calculating and interpreting summary statistics

**10.06** Cumulative frequency graphs

**10.07** Five-number summaries 1 • Five-number summaries 2 • Box-and-whisker plots • Box plots • Box plots 1 • Box plots 2

**10.08** Shapes of distributions

**10.09** Comparing group measures • Comparing sports scores • Comparing city temperatures • Comparing word lengths • Investigating young drivers

**Chapter summary** Statistics review

**Puzzles (5):**

**10.02** Mode, median and mean • Mean, median, mode 2

**10.03** Statistical match-up • Statistical measures puzzle

**Chapter summary** Statistics crossword

Nelson MindTap

To access resources above, visit **cengage.com.au/nelsonmindtap**

## Terminology

| | | | |
|---|---|---|---|
| box plot | class centre | class interval | cumulative frequency |
| distribution | five-number summary | interquartile range | mean |
| measure of centre | measure of spread | median | median class |
| modal class | mode | ogive | outlier |
| peak | quartile | questionnaire | range |
| standard deviation | survey | symmetrical | |

Worksheet
Assignment 10

## SkillCheck Answers on p. 484

**1** This stem-and-leaf plot shows the ages of visitors entering the Royal Easter Show in a 5-minute period.

| Stem | Leaf |
|---|---|
| 0 | 3 8 9 |
| 1 | 0 2 2 2 5 6 7 9 |
| 2 | 0 2 3 4 6 7 |
| 3 | 1 3 3 4 9 |
| 4 | 3 4 7 8 |
| 5 | 5 5 8 |

**a** How many visitors entered the show during the 5-minute period?

**b** What was the age of the oldest visitor?

**c** What was the most common age?

**d** How many visitors were under 16 years old?

**e** What was the middle age?

**2** Is a frequency histogram a line graph or a column graph?

**3** The dot plot shows the shoe sizes of a sample of Year 11 students.

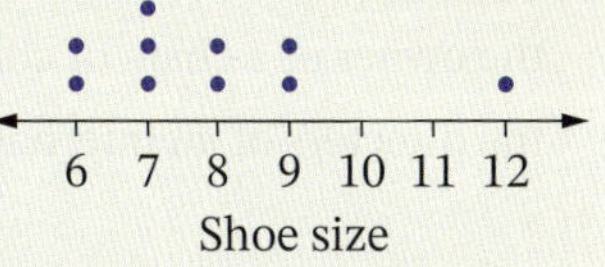


**a** How many students in the sample?

**b** What is the most common shoe size for these students?

**c** Find the outlier and describe the student that has this outlier.

**d** How many students had a shoe size of 10?

**e** What percentage of students had a shoe size over 8?

**4** A sample of students was surveyed about the number of cars owned by each of their families. The results are shown in the table.

| Number of cars | Frequency |
|---|---|
| 0 | 4 |
| 1 | 16 |
| 2 | 11 |
| 3 | 0 |
| 4 | 1 |

**a** How many families did not own a car?

**b** What was the most common number of cars owned?

**c** What was the highest number of cars owned?

**d** How many students were surveyed?

**e** What was the total number of cars owned?

**5** The masses (in kilograms) of 40 skydivers were recorded. The results are shown below.

| | | | | | | | | | |
|---|---|---|---|---|---|---|---|---|---|
| 58 | 63 | 77 | 82 | 53 | 69 | 65 | 80 | 96 | 105 |
| 79 | 63 | 52 | 90 | 104 | 85 | 65 | 87 | 68 | 105 |
| 65 | 87 | 109 | 84 | 62 | 75 | 102 | 78 | 93 | 84 |
| 68 | 105 | 74 | 59 | 68 | 74 | 88 | 66 | 70 | 62 |

Copy and complete the frequency table below using the data about the mass of the skydivers.

| Mass (kg) | Class centre | Frequency |
|---|---|---|
| 50–<60 | | |
| 60–<70 | | |
| 70–<80 | | |
| 80–<90 | | |
| 90–<100 | | |
| 100–<110 | | |

# Statistical investigations and surveys

The word **statistics** originally meant 'information about the state' (nation). Today, statistics is the science of collecting, classifying and analysing any kind of information, while the singular word 'statistic' means a numerical value (such as the mean or range) that is calculated from a set of data.

Governments, businesses and other organisations collect and use statistics to plan and make informed decisions for the future. For example:

- government ministers want to know how many people use public transport
- a baby goods manufacturer wants to test the safety of a new highchair
- the owner of a chain of coffee shops wants to decide where to open the next shop
- the directors of an international relief agency need to find out which countries require aid most urgently.

## Stages of a statistical investigation

There are 6 stages in a statistical investigation.

**Stage 1: Posing questions**

Before an investigation can begin, the researcher needs to determine clearly the key questions of the study: the main aims, the ideas and the topics of concern.

**Stage 2: Collecting data**

Collecting data involves the researcher going out 'into the field' to find the necessary facts, through surveys, interviews, questionnaires, observations, tests, experiments and existing records. This usually requires a great deal of work and time.

**Stage 3: Organising data**

Once the raw data have been collected, they are converted into meaningful information for further study. This involves grouping, ordering or classifying the data in different ways, such as into categories, lists, tables or spreadsheets.

**Stage 4: Summarising and displaying data**

Any patterns or trends are clearer when displayed graphically, such as on a line graph or stem-and-leaf plot, or summarised by measures such as the mean, median, mode or range. These calculated values are called summary statistics.

**Stage 5: Analysing data and drawing conclusions**

'Analyse' means to examine in closer detail, and analysing the data is the most important stage of a statistical investigation. This involves comparing statistics and noting any significant results, answering the questions posed in Stage 1 and making important decisions and recommendations.

**Stage 6: Writing a report**

Finally, the researcher needs to write a report about the investigation, summarising clearly the project's aims, methods, findings and conclusions.

## Example 1

Suppose the state government is investigating the best location to build a new hospital.

Describe what might happen at each of the 6 stages of the investigation.

### Solution

**Stage 1: Posing questions**

- 'Which location is central to most areas?'
- 'Which location has the fastest-growing population?'
- 'Which location has the facilities and road networks to support a new hospital?'

**Stage 2: Collecting data**

The researcher searches for data by looking at existing records or conducting surveys and interviews.

**Stage 3: Organising data**

The researcher groups the data into a table showing the features of each possible location for the hospital.

**Stage 4: Summarising and displaying data**

The researcher calculates statistics and/or uses graphs to compare the locations and look for a pattern.

**Stage 5: Analysing data and drawing conclusions**

The researcher examines the data and patterns in more detail, aiming to make a definite decision on the best location for the hospital.

**Stage 6: Writing a report**

The researcher reports on the investigation, clearly describing what data about hospital locations were collected, how they were collected, and the patterns, results and recommendations.

## Designing a questionnaire

As we learned in Chapter 1, a survey is often conducted using a questionnaire, which asks for details about a person's lifestyle, beliefs, attitudes and behaviours. For example:

- a market researcher may use a questionnaire to collect data about online shopping habits
- a software company may gather information about customer preferences before introducing a new app
- voters are surveyed before an election to find out which political party they prefer.

This table illustrates the features of an effective questionnaire.

| Feature | Poor example | Good example |
|---|---|---|
| **1 Simple language**<br>All instructions and questions should be easily understood. | 'Describe the frequency with which you consume takeaway food on a weekly basis.' | 'How many times per week do you eat takeaway food?' |
| **2 Unambiguous questions**<br>Questions must not have more than one meaning or interpretation. | 'What type of car do you own?' Does the question mean:<br>• the make of car (for example, Toyota)?<br>• the style of car (for example, hatchback)?<br>• the size of car (for example, small)? | 'What make of car do you own? Tick one box from the following list: ...' (This is followed by a list of different makes of cars.) |
| **3 Respect for privacy**<br>A survey should be anonymous and not require personal details. Questions that are too personal should be optional. | 'What is your email address?' People are concerned about receiving junk email if they give out their email address. | 'In case we need to contact you further regarding this survey, please supply your email address. This is optional and your details will not be forwarded to other parties or used for promotional purposes.' |

| Feature | Poor example | Good example |
|---|---|---|
| **4 Freedom from bias**<br>A question should not suggest or influence a person's response. | 'What do you think of changing the Australian flag, given that it is part of our history and our soldiers have died for it in the past?' Too emotional, not objective, influencing people to vote against a new flag. | 'Do you think the current Australian flag should be changed?' |
| **5 Consideration of a number of choices**<br>A question may have many different answers, so either make the question very specific or provide a list of answers from which to choose. | 'How did you enjoy your meal?' Too open-ended. There would be a wide range of possible answers that would be difficult to organise and analyse. | 'How satisfied were you with your meal? Choose from one of the following:<br>Very satisfied<br>Satisfied<br>Not satisfied<br>Very unsatisfied<br>No opinion' |

Here are some steps for designing a questionnaire.

*Step 1*: Decide what information you want to gather.

*Step 2*: Carefully word each question to ensure that you collect the data you decided upon in Step 1.

*Step 3*: Select an appropriate format for each answer: open-ended, words, sentences or numbers, tick boxes, or a scale of 1 to 10.

*Step 4*: Trial your questionnaire on a few people and revise the questions if necessary.

- Select from given word options.

| What type of vehicle do you own? | Car Motorbike Truck Other None |
|---|---|

- Rating on a scale.

| How did you enjoy your meal? | 0 ______________________ 10 |
|---|---|

- Put a tick (✓) or cross (×) in the appropriate box.

| Gender | Male ☐ Female ☐ Non-binary ☐ |
|---|---|

- Write a numeral.

| How many children in your family? | ________ |
|---|---|

- Write a sentence.

| Why did you decide to buy a computer? | ________<br>________<br>________ |
|---|---|

**EXERCISE 10.01** Answers on p. 484

## Statistical investigations and surveys

**1** By researching on the internet, match each organisation that collects data to its field of activity or business.

| Organisation | Activity |
|---|---|
| **a** CSIRO | A National intelligence and security |
| **b** RMS | B Administration of the rugby league competition |
| **c** UNICEF | C Scientific research |
| **d** NRMA | D International aid for underprivileged children |
| **e** BOM | E Weather forecasting and record-keeping |
| **f** ASX | F Administration of road transport and maritime services |
| **g** ASIO | G International health authority |
| **h** WHO | H Association for motorists, road service and insurance |
| **i** NRL | I Buying and selling shares |

2 Research what each set of initials used to name an organisation in Question **1** stands for.

3 Select 4 of the organisations from Question **1** and, for each one, state:

**a** an example of the data it collects

**b** how it uses the data.

4 A medical researcher is investigating whether a new drug improves a person's memory. There are 50 patients in the test group who take the memory pill, while 50 patients in a control group take a placebo (a fake pill that has no effect but the patients think it's the memory pill). Both groups then complete a memory test and their scores are compared.

Describe what might happen at each stage of the investigation.

**Stage 1:** Posing questions

**Stage 2:** Collecting data

**Stage 3:** Organising data

**Stage 4:** Summarising and displaying data

**Stage 5:** Analysing data and drawing conclusions

**Stage 6:** Writing a report

5 At what stage of a statistical investigation does the researcher:

**a** make a decision or recommendation after the results of the study become clear?

**b** determine the key goals and issues of the investigation?

**c** calculate statistics from the data and look for some relationship between them?

**d** gather the raw data required for the study?

6 Visit the **Australian Bureau of Statistics** website, www.abs.gov.au, to collect the following data from the last Australian census:

**a** the population of Australia

**b** the actual date the census was taken

**c** the number of females in Australia

**d** the percentage of Australians who were born in New Zealand

**e** the most popular language spoken at home after English

**f** the percentage of Australians who are married.

7 You should always be aware of privacy issues and the need to keep your personal details confidential. What data can be collected about yourself when:

**a** you use an ATM?

**b** you respond to an online survey?

**c** you send a text message on a phone?

**d** you purchase items with a credit card?

**e** you tap on and off with your Opal card when using public transport?

**f** you post on social media?

8 Identify 2 dangers of revealing too much information about yourself.

9 Here are some questions from the 2021 Census.

**i**

**What is the person's date of birth and age?**

- If date of birth is not known, please give age.
- Example: Day Month Year

1 8 0 3 1 9 9 3

AND Age 2 8 Years

Day Month Year

AND Age Years

**ii**

**What is the person's current marital status?**

- 'Married' refers to registered marriages.
- Mark one box, like this: ▬

Never married

Widowed

Divorced

Separated but not divorced

Married

**iii**

**In what year did the person first arrive in Australia to live for one year or more?**

- For example, for arrival in 1987 write: 1 9 8 7

Year

Will be in Australia for less than one year

**iv**

**How well does the person speak *English*?**

- Mark one box, like this: ▬

Very well

Well

Not well

Not at all

**v**

**What is the person's religion?**

- Answering this question is **OPTIONAL.**
- Examples of 'Other': LUTHERAN, SALVATION ARMY, JUDAISM, TAOISM, ATHEISM.
- Mark one box, like this: ▬

No religion

Catholic

Anglican (Church of England)

Uniting Church

Islam

Buddhism

Presbyterian

Hinduism

Greek Orthodox

Baptist

Other (please specify)

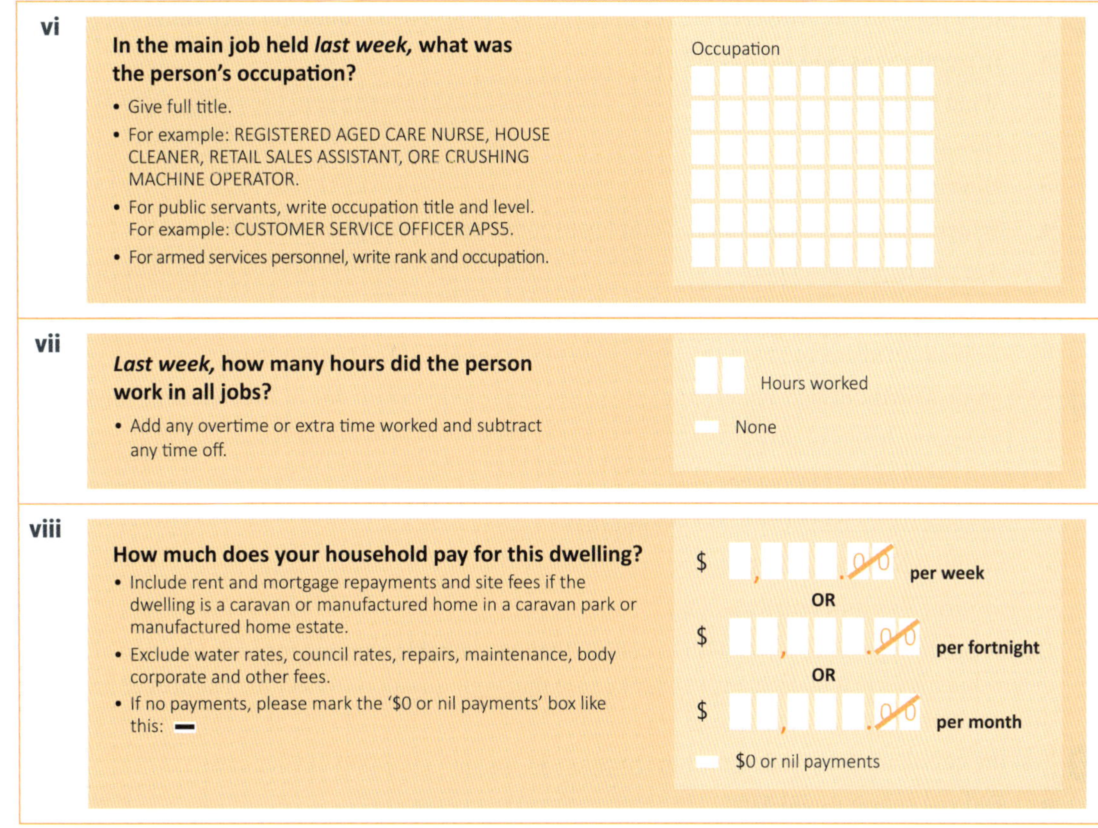

**vi**

**In the main job held *last week*, what was the person's occupation?**

- Give full title.
- For example: REGISTERED AGED CARE NURSE, HOUSE CLEANER, RETAIL SALES ASSISTANT, ORE CRUSHING MACHINE OPERATOR.
- For public servants, write occupation title and level. For example: CUSTOMER SERVICE OFFICER APS5.
- For armed services personnel, write rank and occupation.

Occupation

**vii**

***Last week*, how many hours did the person work in all jobs?**

- Add any overtime or extra time worked and subtract any time off.

Hours worked

None

**viii**

**How much does your household pay for this dwelling?**

- Include rent and mortgage repayments and site fees if the dwelling is a caravan or manufactured home in a caravan park or manufactured home estate.
- Exclude water rates, council rates, repairs, maintenance, body corporate and other fees.
- If no payments, please mark the '$0 or nil payments' box like this: ▬

$ , . 00 **per week**

**OR**

$ , . 00 **per fortnight**

**OR**

$ , . 00 **per month**

$0 or nil payments

Source: Australian Bureau of Statistics (ABS), 2021 Census

**a** In Question part **i**, how could asking for 'age' be ambiguous?

**b** Which questions have multiple-choice answers?

**c** Which questions do not have multiple-choice answers, but still give choices about the format of the expected answer?

**d** What would happen if Question part **iv** about speaking English did not have multiple-choice answers?

**e** Why is Question part **v** about religion optional?

**f** Why are there 3 different ways of answering Question part **viii**?

**g** Why is Question part **viii** about how much a person pays for their dwelling (home) so wordy? Select **A**, **B**, **C** or **D**.

- **A** To protect a person's privacy.
- **B** To make the question simpler to read.
- **C** To make the question optional.
- **D** To make the question clearer, more precise and unambiguous.

10 Explain what is wrong with each survey question and write a better question each time.

**a** Are politicians paid too much?

**b** What is your favourite food?

**c** How often do you vape?

**d** How many TVs do you own?

**e** Which mode of transport do you employ to commute to your place of work?

**f** Are you happy at school?

**g** How much do you earn?

**h** How satisfied are you with the level of customer service at the Nelson Bank?

11 As a group activity, design a questionnaire of 10–12 questions about one of the following topics.

- TV viewing habits of people
- Spending habits of teenagers
- Students' use of computers and the internet
- Family and household details
- Favourite sports to watch and play

Decide on the type of facts you wish to collect, then use suitable wording for each question and determine the format of each answer. Make sure that your questions reflect the features of an effective questionnaire (see pages 367–8).

12 Test the questionnaire you designed in Question **11** on another group in your class.

**a** Did you collect useful data?

**b** Were there too many questions or too few questions?

**c** Were any questions ambiguous, misleading or unnecessary?

**d** How could the questionnaire be improved?

**e** Was the survey successful? Why or why not?

☐ Foundation ○ Mastery ○ Complex

# The mean, median and mode

10.02

The **mean**, **median** and **mode** are 3 **summary statistics** that represent the centre or average of a set of data. They are called the **measures of centre** (or **measures of central tendency**).

The **mean** (or average) has the symbol $\bar{x}$, and is the sum of all data values divided by the number of data values.

### The mean

$$\text{mean, } \bar{x} = \frac{\text{sum of data values}}{\text{number of data values}}$$

$$= \frac{\sum x}{n}$$

$\Sigma$ means 'the sum of'.
$x$ represents a data value.
$n$ is the total number of values.

### The median and mode

When the data are ordered from lowest to highest, the **median** is:

- the middle value, for an odd number of data values
- the average of the 2 middle values for an even number of data values.

The **mode** is the most common value or category.

A set of data can have more than 1 mode, or no mode at all.

**Videos**
The mode, median and mean
The mean, mode, median and range
Analysing dot plots 1
Analysing dot plots 2

**Skillsheet**
Statistical measures

**Worksheets**
Mean, median, mode 1
Mean, median, mode 2
Mode, median and mean
Mean, median, mode and range

**Puzzles**
Mode, median and mean
Mean, median, mode 2

### Example 2

For each data set below, find:

**i** the mean (correct to one decimal place)

**ii** the median

**iii** the mode.

**a** The maximum daily temperature (in °C) in Mudgee for the first 2 weeks in January:

30 28 26 31 34 35 32 33 21 25 28 32 32 35

**b** A stem-and-leaf plot of the marks (out of 100) in a maths test for a class of students:

| Stem | Leaf |
|---|---|
| 4 | 4 7 7 8 |
| 5 | 2 6 8 9 9 |
| 6 | 1 3 5 5 7 8 |
| 7 | 0 2 3 4 5 |
| 8 | 3 7 8 |
| 9 | 2 8 |

## Solution

**a** **i** sum of data values $(\Sigma x) = 30 + 28 + 26 + 31 + 34 + 35 + 32 + 33 + 21 + 25 + 28 + 32 + 32 + 35$
$= 422$

$$\text{mean, } \bar{x} = \frac{\text{sum of data values}}{\text{number of data values}}$$
$$= \frac{422}{14}$$
$$= 30.142\,85\,...$$
$$\approx 30.1$$

Note that the mean temperature of 30.1°C is at the centre of all 14 temperatures.

**ii** Placing the data in order:

21 25 26 28 28 30 31 32 32 32 33 34 35 35

$$\text{median} = \frac{31+32}{2}$$
$$= 31.5$$

For 14 values, the middle values are the 7th and 8th values.

**iii** mode = 32

the most common value (it occurred 3 times)

Note that the mean (30.1), median (31.5) and mode (32) are all around the same central value.

**b** **i** $\text{mean, } \bar{x} = \frac{1671}{25}$
$= 66.84$
$\approx 66.8$

The sum of the 25 marks is 1671.

**ii**

| Stem | Leaf |
|---|---|
| 4 | 4 7 7 8 |
| 5 | 2 6 8 9 9 |
| 6 | 1 3 5 5 7 8 |
| 7 | 0 2 3 4 5 |
| 8 | 3 7 8 |
| 9 | 2 8 |

For 25 values, the middle value is the 13th value.

median = 65

**iii** mode = 47, 59 and 65

## The statistics mode on a calculator

Scientific calculators have a statistics mode (SD or STAT). Follow the instructions in the table below to calculate the mean of the temperatures from Example **2a** using your calculator's statistics mode.

| Operation | Casio *fx*-82 | Sharp |
|---|---|---|
| Start statistics mode. | MODE STAT 1-VAR | MODE STAT = |
| Clear the statistical memory. | SHIFT 1 Edit, Del-A | 2nd F DEL |
| Enter data. | SHIFT 1 Data to get table<br>30 = 28 =, etc. to enter in column<br>AC to leave table | 30 M+ 28 M+, etc. |
| Calculate the mean.<br>($\bar{x}$ = 30.142 85 ...) | SHIFT 1 Var $\bar{x}$ = | RCL $\bar{x}$ |
| Check the number of values.<br>($n$ = 14) | SHIFT 1 Var $n$ = | RCL $n$ |
| Return to normal (COMP) mode. | MODE COMP | MODE 0 |

| Operation | Casio *fx*-8200 |
|---|---|
| Start statistics mode. | HOME Statistics EXE 1-Variable EXE |
| Clear the statistical memory. | TOOLS Edit EXE Delete All EXE |
| Enter data. | 30 EXE 28 EXE, etc. to enter in the X column |
| Calculate the mean.<br>($\bar{x}$ = 30.142 85 ...)<br>Calculate the sum of values.<br>($\Sigma x$ = 422)<br>Check the number of values.<br>($n$ = 14) | EXE 1-Var Results EXE to calculate many statistics (scroll down for more) |

## The mean, median and mode from a frequency table

If the values in a data set are presented in a **frequency distribution table** then, by adding an $fx$ column, the mean can be calculated using the formula shown below.

Video
Statistics from a frequency table

### Calculating the mean from a frequency table

$$\text{mean, } \bar{x} = \frac{\text{sum of } fx}{\text{sum of } f} = \frac{\sum fx}{\sum f}$$

### Example 3

The scores for the players in a 9-hole golf competition were sorted into the frequency table.

**a** How many players were there?

**b** For this data, find:

**i** the mean (correct to one decimal place)

**ii** the mode

**iii** the median.

| Score ($x$) | Frequency ($f$) |
|---|---|
| 37 | 2 |
| 38 | 4 |
| 39 | 7 |
| 40 | 4 |
| 41 | 1 |

### Solution

**a** 18 players $\quad \Sigma f = 18$

**b** **i**

| Score ($x$) | Frequency ($f$) | $fx$ |
|---|---|---|
| 37 | 2 | 74 |
| 38 | 4 | 152 |
| 39 | 7 | 273 |
| 40 | 4 | 160 |
| 41 | 1 | 41 |
| **Totals** | $\Sigma f = 18$ | $\Sigma fx = 700$ |

$fx$ means $f \times x$
$2 \times 37 = 74$
$4 \times 38 = 152$, etc.

This means that there were 2 scores of 37, 4 scores of 38, etc. The $fx$ column groups equal scores and adds them together.

$$\text{mean, } \bar{x} = \frac{\Sigma fx}{\Sigma f} = \frac{700}{18}$$

sum of all 18 scores = 700

$$= 38.8888\ldots$$
$$\approx 38.9$$

The mean of these golf scores can also be found using the calculator's statistics mode: see instructions next page.

**ii** mode = 39

39 has the highest frequency, 7

**iii** Add a cumulative frequency column to the table, which keeps a running total of the frequencies.

| Score ($x$) | Frequency ($f$) | Cumulative frequency | |
|---|---|---|---|
| 37 | 2 | 2 | |
| 38 | 4 | 6 | $2 + 4 = 6$ |
| 39 | 7 | 13 | $6 + 7 = 13$, etc. |
| 40 | 4 | 17 | |
| 41 | 1 | 18 | |

Because there are 18 scores, the 2 middle scores are the 9th and 10th values. Reading from the cumulative frequency column, the 6th score is 38, and the 13th score is a 39, so the 9th and 10th values must both be 39.

$$\text{median} = \frac{39 + 39}{2} = 39$$

Note that the mean (38.9), median (39) and mode (39) are all at the centre of the data set.

Follow the instructions below to use a calculator's statistics mode to calculate the mean of the golf scores.

| Operation | Casio *fx*-82 | Sharp |
|---|---|---|
| Start statistics mode. | MODE STAT 1-VAR<br>SHIFT MODE scroll down to STAT<br>Frequency? ON | MODE STAT = |
| Clear the statistical memory. | SHIFT 1 Edit, Del-A | 2nd F DEL |
| Enter data. | SHIFT 1 Data to get table<br>37 = 38 =, etc. to enter in the $X$ column<br>2 = 4 =, etc. to enter in the FREQ column<br>AC to leave table | 37 2nd F STO<br>2 M+<br>38 2nd F STO<br>4 M+<br>etc. |
| Calculate the mean.<br>($\bar{x} = 38.8888\ldots$) | SHIFT 1 Var $\bar{x}$ = | RCL $\bar{x}$ |
| Check the number of values.<br>($n = 18$) | SHIFT 1 Var $n$ = | RCL n |
| Return to normal (COMP) mode. | MODE COMP | MODE 0 |

| Operation | Casio *fx*-8200 |
|---|---|
| Start statistics mode. | HOME Statistics EXE 1-Variable EXE TOOLS Frequency EXE On AC |
| Clear the statistical memory. | TOOLS Edit EXE Delete All EXE |
| Enter data. | 37 EXE 38 EXE, etc. to enter in the X column<br>2 EXE 4 EXE, etc. in the Freq column |
| Calculate the mean.<br>($\bar{x} = 38.8888$ ...)<br>Calculate the sum of values.<br>($\Sigma x = 700$)<br>Check the number of values.<br>($n = 18$) | EXE 1-Var Results EXE to calculate many statistics (scroll down for more) |

## The mean of grouped data

For data grouped into **class intervals**, an **estimate** of the mean can be calculated using the **class centres**. It is only an estimate because, with class intervals, we do not know the exact value of all data values.

### Example 4

The ages of the patients at a medical centre in one afternoon were recorded and grouped into this frequency table.

| Age | Frequency |
|---|---|
| 0–9 | 8 |
| 10–19 | 7 |
| 20–29 | 6 |
| 30–39 | 8 |
| 40–49 | 5 |
| 50–59 | 4 |
| 60–69 | 3 |
| 70–79 | 1 |

**a** Calculate, correct to one decimal place, the estimated mean age of the patients.

**b** How many patients went to the medical centre?

### Solution

**a**

| Age | Class centre, $x$ | Frequency, $f$ | $fx$ |
|---|---|---|---|
| 0–9 | 4.5 | 8 | 36 |
| 10–19 | 14.5 | 7 | 101.5 |
| 20–29 | 24.5 | 6 | 147 |
| 30–39 | 34.5 | 8 | 276 |
| 40–49 | 44.5 | 5 | 222.5 |
| 50–59 | 54.5 | 4 | 218 |
| 60–69 | 64.5 | 3 | 193.5 |
| 70–79 | 74.5 | 1 | 74.5 |
| | **Totals** | $\Sigma f = 42$ | $\Sigma f = 1269$ |

Estimate of the mean, $\bar{x} = \dfrac{\Sigma fx}{\Sigma f}$

$$= \frac{1269}{42}$$

$$= 30.2142\ldots$$

$$\approx 30.2$$

Note that the estimated mean age of 30.2 is a central value of the data set.

**b** 42 patients; $\Sigma f = 42$

## The median class and modal class of grouped data

### Median class and modal class

The **median class** is the class interval that contains the median value.

The **modal class** is the most common class interval(s).

### Example 5

The monthly costs of a sample of mobile phone users were grouped in the cumulative frequency table.

For this data, find:

**a** the median class

**b** the modal class.

| Monthly cost ($) | Frequency | Cumulative frequency |
|---|---|---|
| 0–<20 | 6 | 6 |
| 20–<40 | 8 | 14 |
| 40–<60 | 13 | 27 |
| 60–<80 | 17 | 44 |
| 80–<100 | 23 | 67 |
| 100–<120 | 20 | 87 |
| 120–<140 | 16 | 103 |
| 140–<160 | 10 | 113 |
| 160–<180 | 4 | 117 |
| 180–<200 | 3 | 120 |

#### Solution

**a** There are 120 values. The 2 middle values are the 60th and 61st values.

From the cumulative frequency column, the 60th and 61st values are in the 80–<100 class.

The median class is 80–<100.

**b** The modal class is 80–<100. class with the highest frequency (23)

## Comparing measures of centre

| Measure of centre | Features | When it is most appropriate |
|---|---|---|
| **Mean**<br>$\bar{x} = \frac{\text{sum of values}}{\text{number of values}}$<br>$\bar{x} = \frac{\Sigma x}{n}$<br>$\bar{x} = \frac{\Sigma fx}{\Sigma x}$ | Depends on all values in the data set<br>Is affected by outliers | When the data set does not have many outliers |
| **Median**<br>Middle value or average of 2 middle values | Not affected by outliers | When the data set has many outliers; for example, house prices, salaries |
| **Mode**<br>Most popular value(s) | Not affected by outliers | When the most common value or category is needed (for example, dress size); also useful for categorical data |

## Example 6

Which measure of centre is most appropriate for describing each average?

**a** the average price of a new car

**b** the most common number of bedrooms in a house

**c** a cricket player's batting average

**d** average weekly income

### Solution

**a** Median, because there would be many outliers (the prices of expensive cars).

**b** Mode, because the most frequent value is needed.

**c** Mean, because all data values are required in the calculation.

**d** Median, because there would be outliers (the incomes of very rich people).

## Example 7

Ten houses were sold this week at Nelson Lakes for the following prices.

| | | | | |
|---|---|---|---|---|
| \$976 000 | \$1 800 000 | \$870 000 | \$908 000 | \$972 000 |
| \$1 009 000 | \$987 000 | \$1 182 000 | \$1 060 000 | \$838 000 |

**a** Calculate the mean house price.

**b** Calculate the median house price.

**c** Which measure of centre is higher, the mean or the median?

**d** Which measure is more appropriate to describe the average house price?

### Solution

**a** $\text{mean, } \bar{x} = \frac{10\,602\,000}{10}$

$= \$1\,060\,200$

**b** Prices in order:

| | | | | |
|---|---|---|---|---|
| \$838 000 | \$870 000 | \$908 000 | \$972 000 | \$976 000 |
| \$987 000 | \$1 009 000 | \$1 060 000 | \$1 182 000 | \$1 800 000 |

$\text{median} = \frac{\$976000 + \$987000}{2}$

$= \$981500$

**c** The mean is higher.

Note that 8 of the 10 house prices are below the mean (1 060 200).

**d** The median, because it is not distorted by the outlier of \$1 800 000.

**EXERCISE 10.02** Answers on p. 485

## The mean, median and mode

**1** For each set of data, find:

**i** the mean **ii** the median **iii** the mode(s).

| | | | | | | | | | | | |
|---|---|---|---|---|---|---|---|---|---|---|---|
| **a** | 1 | 1 | 2 | 5 | 5 | 7 | 9 | 10 | | | |
| **b** | 37 | 31 | 35 | 39 | 31 | 32 | 34 | 32 | 35 | 38 | |
| **c** | 28 | 40 | 38 | 42 | 45 | 29 | 31 | 41 | 30 | | |
| **d** | 5 | 8 | 14 | 9 | 10 | 7 | 11 | 15 | 8 | 7 | 5 |

**2** The stem-and-leaf plot on the right represents the number of points scored by the Roosters in every round of the football season.

**a** How many rounds were played in the season?

**b** Calculate the mean score (correct to the nearest whole number).

**c** Find the median number of points scored.

**d** What is the mode?

| Stem | Leaf |
|---|---|
| 0 | 6 6 |
| 1 | 2 3 4 4 4 8 8 9 |
| 2 | 0 0 0 5 6 |
| 3 | 0 0 2 4 4 6 7 |
| 4 | 0 |
| 5 | |
| 6 | 2 |

**3** Ngaire is training for a triathlon. She swam the following times, in minutes, in her last 10 races.

28 34 22 24 25 24 26 26 24 27

**a** Which of the following is Ngaire's mean swim time? Select **A**, **B**, **C** or **D**.

**A** 24 **B** 25 **C** 25.5 **D** 26

**b** Which of the following was her median swim time in minutes? Select **A**, **B**, **C** or **D**.

**A** 24 **B** 25 **C** 25.5 **D** 26

**c** Which of the following was Ngaire's modal swim time for the 10 races? Select **A**, **B**, **C** or **D**.

**A** 24 **B** 25 **C** 25.5 **D** 26

EXAMPLE 3

**4** 'Average contents 50' is printed on each box of Meg's Matches. A quality controller counted the contents of a sample of 160 matchboxes from the production line and tabulated the results, as shown.

| Number of matches ($x$) | Frequency ($f$) |
|---|---|
| 48 | 10 |
| 49 | 45 |
| 50 | 52 |
| 51 | 39 |
| 52 | 9 |
| 53 | 5 |

**a** Use an $fx$ column, or your calculator's statistics mode, to calculate the mean number of matches per box, correct to one decimal place.

**b** Is the claim 'Average contents 50' justified? Give a reason for your answer.

**c** Find the mode.

**d** Find the median.

☐ Foundation ○ Mastery ⬡ Complex

**5** This dot plot shows the number of children in each family living on Willard Crescent.

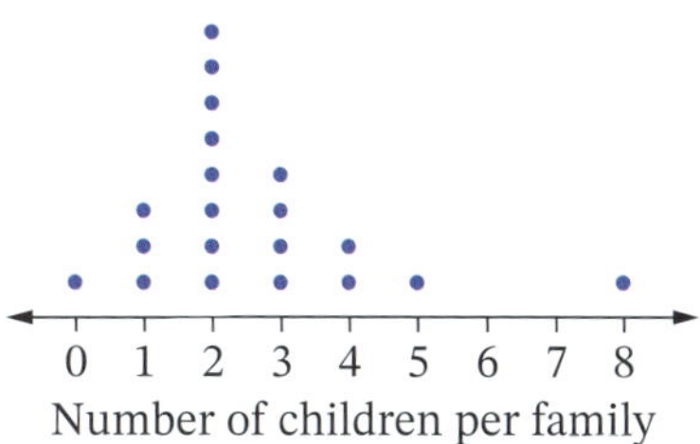


**a** How many families live on Willard Crescent?

**b** Use a frequency table, or your calculator's statistics mode, to calculate the mean number of children per family.

**c** What is the median?

**d** What is the mode?

**e** What is the outlier?

**f** If the outlier is removed from the data set, how does this affect:

**i** the mean? **ii** the median? **iii** the mode?

**6** This frequency histogram shows the number of mobile phone calls made by Elena each day over a number of days.

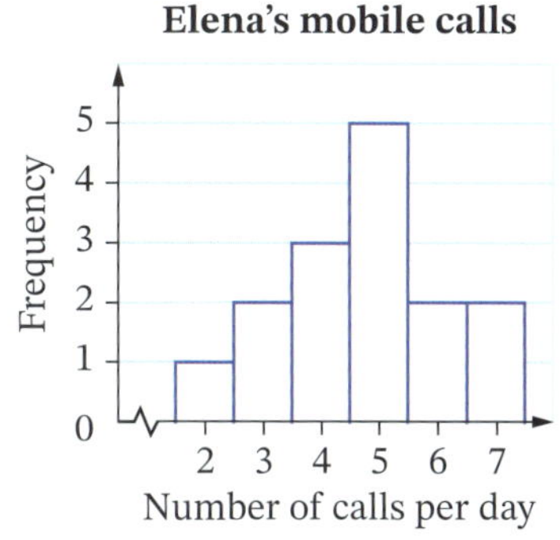


**a** Draw a frequency table for this data, including an $fx$ column.

**b** Over how many days was the number of calls Elena made recorded?

**c** Find the mode of this data.

**d** Find the median of this data.

**e** Calculate the mean number of phone calls made by Elena per day, correct to one decimal place.

**7** The police used radar to check the speeds of motor vehicles driving in a 40 km/h zone outside a local primary school one morning. They recorded the results in the table.

EXAMPLE 4

| Speed (km/h) | Number of cars, $f$ |
|---|---|
| 36–40 | 64 |
| 41–45 | 36 |
| 46–50 | 18 |
| 51–55 | 15 |
| 56–60 | 11 |
| 61–65 | 5 |

**a** Add a column of class centres to the table and calculate an estimate for the mean speed of the vehicles, correct to 2 decimal places.

**b** How many motor vehicles had their speeds checked?

EXAMPLE 5

**8** The heights of young trees in a section of a nursery were measured before planting. The results are shown in the table.

| Height (cm) | Number of trees |
|---|---|
| 20–29 | 28 |
| 30–39 | 45 |
| 40–49 | 74 |
| 50–59 | 63 |
| 60–69 | 24 |

For this data, find:

**a** the median class

**b** the modal class.

**9** This dot plot shows the minimum daily temperatures (in °C) in Camden over a 3-week period.

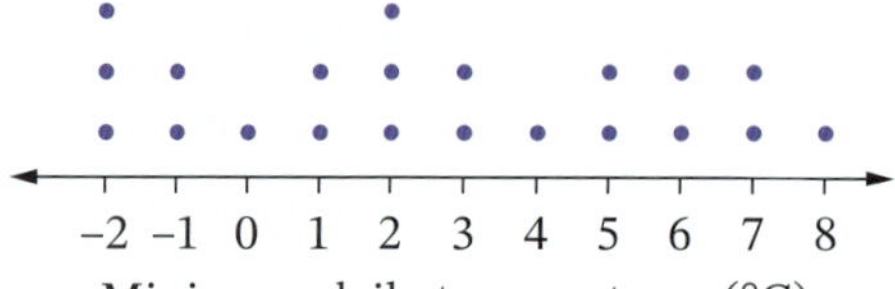

Minimum daily temperatures (°C)

**a** What is the mode?

**b** What is the median?

**c** Calculate the mean, correct to one decimal place.

**10** The weekly wages of the staff at Yen's restaurant are shown in the frequency table.

| Wage ($) | Number of employees |
|---|---|
| 800–<900 | 5 |
| 900–<1000 | 11 |
| 1000–<1100 | 20 |
| 1100–<1200 | 4 |
| 1200–<1300 | 3 |
| 1300–<1400 | 1 |

**a** What is the modal class for the wages?

**b** What is the median class?

**11** Decide which $M$ (mean, median or mode) is correct for each sentence.

**a** This $M$ takes all values in the data set into account.

**b** This $M$ is one of the data values if there is an odd number of values.

**c** Half of the data values are above this $M$, the other half are below.

**d** There can be more than one of this $M$ in a set of data.

**e** This $M$ often needs to be rounded to decimal places.

**f** This $M$ can also be used for categorical data.

**g** This $M$ can be distorted by many outliers.

**h** This $M$ must be one of the values in the data set.

☐ Foundation ○ Mastery ⬡ Complex

**12** Which measure of centre is most appropriate for describing each average?

**a** The average exam mark for the class

**b** The average shirt size for teenage girls

**c** The average rent paid for a house in Sydney

**d** The average screen size of a notebook computer

**e** The average mass of football players in a team

**f** The average brand of mobile phone

**13** A small business employs staff with the following salaries:

| General manager | $168 300 |
|---|---|
| 3 factory workers | $74 300 each |
| Supervisor | $95 600 |
| 2 office workers | $78 500 each |

**a** How many people are there on staff?

**b** Calculate the mean salary of the staff, correct to the nearest dollar.

**c** Calculate the median salary of the staff.

**d** Which measure of centre is higher, the mean or median? Why?

**e** Which measure of centre best describes the average salary at this business?

**14** The ages of the mathematics teachers at Westvale Christian College are:

49 32 37 32 25 41 39 50

**a** For this data, find:

**i** the mean **ii** the median **iii** the mode.

**b** The 39-year-old teacher is replaced by a new teacher, aged 22. Describe how this will affect:

**i** the mean **ii** the median **iii** the mode.

**15** The colours of the new cars sold last week at Huxley Motors were recorded. The results are shown in the table below.

| **Colour** | Black | Blue | Red | Silver | White |
|---|---|---|---|---|---|
| **Frequency** | 4 | 7 | 7 | 9 | 12 |

**a** How many new cars were sold?

**b** What is the mode for this data?

**c** Why is the mode the only valid measure of centre here?

**16** The weekly mortgage repayments (in dollars) of 11 home owners are:

370 628 299 417 354 1027 585 435 509 652 481

**a** For this data, find:

**i** the mean, correct to the nearest dollar

**ii** the median

**iii** the mode.

**b** Why isn't the mean or mode an appropriate measure of centre for this set of data?

**c** If the outlier is removed from the data, check whether the new mean will be closer to the new median than the mean was to the median for the original set of data.

**17** The dot plot shows the shoe sizes of a sample of Year 11 students.

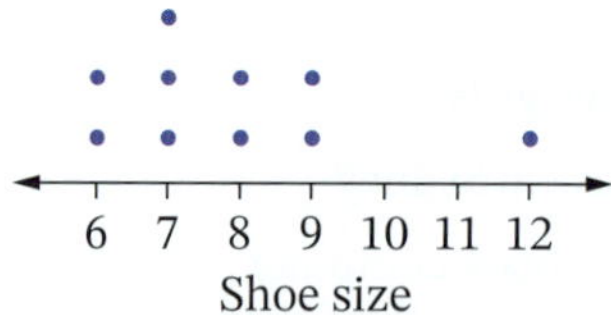


**a** For this data, find:

**i** the mean **ii** the median **iii** the mode.

**b** If the outlier is removed, state what will happen to:

**i** the mean? **ii** the mode?

**c** A shoe store needs to buy more shoes for a back-to-school sale. Which measure of centre is most appropriate for the store to use in this situation?

**18** The stem-and-leaf plot on the right shows the maximum daily temperatures (in °C) in Port Macquarie for the last 2 weeks in December.

| Stem | Leaf |
|---|---|
| 2 | 2 4 4 5 6 6 7 7 7 8 8 9 |
| 3 | 1 4 |

**a** For this data, find:

**i** the mean **ii** the median **iii** the mode.

**b** Which measure of centre is the most appropriate for describing the average maximum daily temperature?

# Technology

## Calculating measures of centre

*Step 1*: Open a blank spreadsheet to enter the following temperature data about Mudgee from Example 2 on page 371.

| | A | B | C | D | E | F | G |
|---|---|---|---|---|---|---|---|
| 1 | **Maximum daily temperatures in Mudgee** | | | | | | |
| 2 | 30 | 28 | 26 | 31 | 34 | 35 | 32 |
| 3 | 33 | 21 | 25 | 28 | 32 | 32 | 35 |
| 4 | | | | | | | |
| 5 | | | | Mean | | | |
| 6 | | | | Median | | | |
| 7 | | | | Mode | | | |
| 8 | | | | | | | |



*Step 2*: In cell E5, enter the formula =**AVERAGE(A2:G3)** to calculate the mean (30.142 85 ...).
*Step 3*: In cell E6, enter the formula =**MEDIAN(A2:G3)** to calculate the median (31.5).
*Step 4*: In cell E7, enter the formula =**MODE(A2:G3)** to calculate the mode (32).

If there is more than one mode in a data set, the spreadsheet displays *only one* of the modes.

Foundation Mastery Complex

# The range and interquartile range

10.03

While the mean, median and mode describe the centre of a data set, there are 3 summary statistics that describe the **spread** of data: the **range**, the **interquartile range** and the **standard deviation**. These are called **measures of spread**.

**Worksheet**
Interquartile range

**Puzzles**
Statistical match-up

Statistical measures puzzle

**Videos**
Interquartile range 1

Interquartile range 2

## The median and quartiles

### Quartiles

The 3 **quartiles** of a data set are those values that separate the data into quarters.

- The **lower** or **1st quartile**, $Q_1$, separates the bottom quarter (25%) of values from the rest of the data.

| Lowest value | Lower quartile, $Q_1$ | Median, $Q_2$ | Upper quartile, $Q_3$ | Highest value |
|---|---|---|---|---|

- The **upper** or **3rd quartile**, $Q_3$, separates the top quarter (25%) of values from the rest of the data.
- The **middle** or **2nd quartile**, $Q_2$, is the median, and separates the 2 middle quarters.

These speeds (in km/h) were recorded for 11 cars driving along a major country road:

104 86 95 100 81 120 84 78 93 92 107

When we sort the data in ascending order, we can find the quartiles.

A speed of 81 km/h is in the bottom quarter of data.

A speed of 100 km/h is in the 2nd top quarter of data.

A speed of 107 km/h is in the top quarter of data.

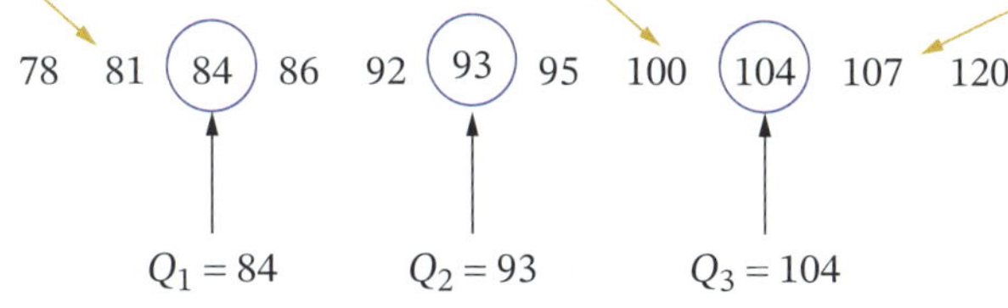


### Quartiles of a data set

To find the quartiles of a data set:

*Step 1*: Sort the data in order, find the median and call it $Q_2$.

*Step 2*: Find the median of the bottom half of the data and call it $Q_1$.

*Step 3*: Find the median of the top half of the data and call it $Q_3$.

### Example 8

Find the quartiles for each data set.

**a** The marks obtained by a class of students for an art project are:

51 41 60 38 46 57 39 61 43 64

**b** The scores obtained by a golfer for the first 9 holes of a golf course are:

4 3 5 6 4 3 8 6 6

## Solution

**a** First, sort the marks and place them in order:

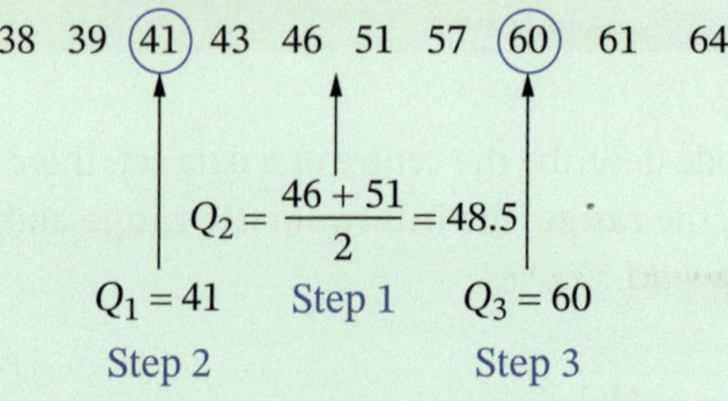


**b**

3 3 4 4 (5) 6 6 6 8

$Q_1 = \frac{3+4}{2} = 3.5$ $\quad Q_3 = \frac{6+6}{2} = 6$

$Q_2 = 5$

### Range and interquartile range

range = highest value − lowest value

interquartile range (IQR) = upper quartile − lower quartile = $Q_3 - Q_1$

**Standard deviation** will be explained in the next section.

## Example 9

For each data set, find:

**i** the range **ii** the interquartile range.

**a** The maximum daily temperature (in °C) in Mudgee for the first 2 weeks in January:

30 28 26 31 34 35 32 33 21 25 28 32 32 35

**b** The body temperatures (in °C) of a sample of hospital patients, as shown in the dot plot.

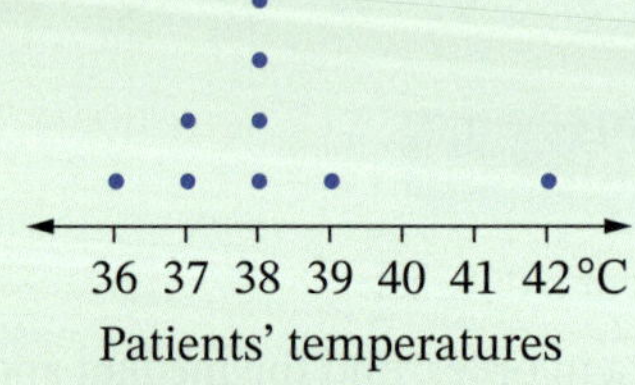


Patients' temperatures

## Solution

**a** **i** range = 35 − 21 = 14

**ii** Sorting the data in order:

21 25 26 (28) 28 30 31 ↑ 32 32 32 (33) 34 35 35

$Q_1 = 28$ $\quad Q_2$ (Step 1) $\quad Q_3 = 33$

$$\begin{aligned} \text{IQR} &= Q_3 - Q_1 \\ &= 33 - 28 \\ &= 5 \end{aligned}$$

**b** **i** range = 42 − 36 = 6

**ii**

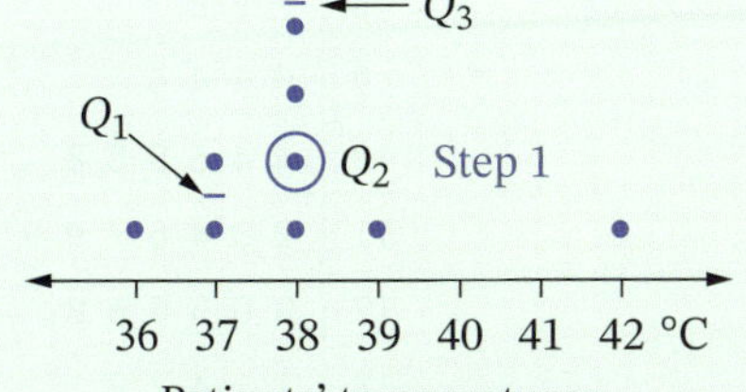


Of 9 data values, the median, $Q_2$, is the 5th value, counting upwards from the left.

There are 4 values below the median and 4 values above.

The median of the bottom half is the average of the 2nd and 3rd values.

The median of the top half is the average of the 7th and 8th values.

$$Q_1 = \frac{37+37}{2} = 37,\ Q_3 = \frac{38+39}{2} = 38.5$$

$$\begin{aligned}\text{IQR} &= Q_3 - Q_1\\ &= 38.5 - 37\\ &= 1.5\end{aligned}$$

The **range** represents the total spread of data but it is not a good measure if there are outliers.
The **interquartile range** is not affected by outliers, because it measures the range of the middle 2 quarters only.

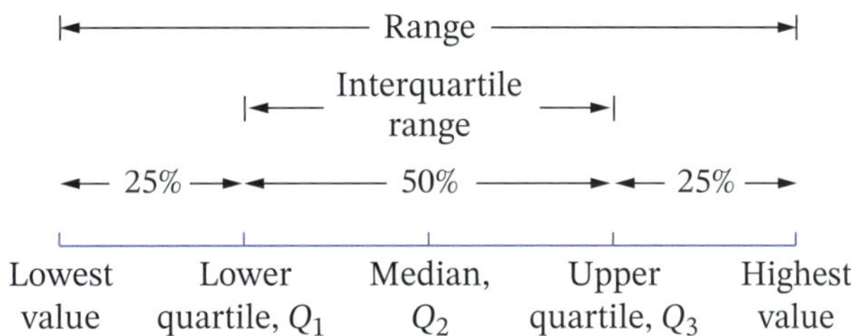


**EXERCISE 10.03** Answers on p. 485

## The range and interquartile range

**1** Find the quartiles $Q_1$, $Q_2$ and $Q_3$ for each data set.

**a** The times, in seconds, to run 100 metres:

8.7 9.1 11.0 13.5 10.6 8.9 10.1 9.6
12.3 9.9 9.0 10.8 9.2 13.1 10.6

**b** The number of matches in a box:

49 50 52 48 50 51 49 50 52 51 50 50

**c** The prices, in dollars, of a bag of potatoes:

6.50 6.20 6.50 7.10 6.00 6.50 6.90 7.80 6.40 6.00

**d** The weekly rainfall, in millimetres, over 3 months:

16 24 18 26 21 27 5 7 17 21 9 0 22

Foundation Mastery Complex

**2** The stem-and-leaf plot on the right shows the game scores of a group of 10-pin bowlers.

For this data, find:

**a** the median

**b** the lower quartile

**c** the upper quartile.

| Stem | Leaf |
|---|---|
| 8 | 2 7 8 |
| 9 | 0 3 4 6 9 |
| 10 | 4 4 5 8 8 8 |
| 11 | 2 3 4 6 7 9 9 |
| 12 | 0 0 5 6 6 8 |
| 13 | 1 1 4 7 9 |

**3** The dot plot shows the number of vehicles driving past Westvale High School per minute in a 20-minute period.

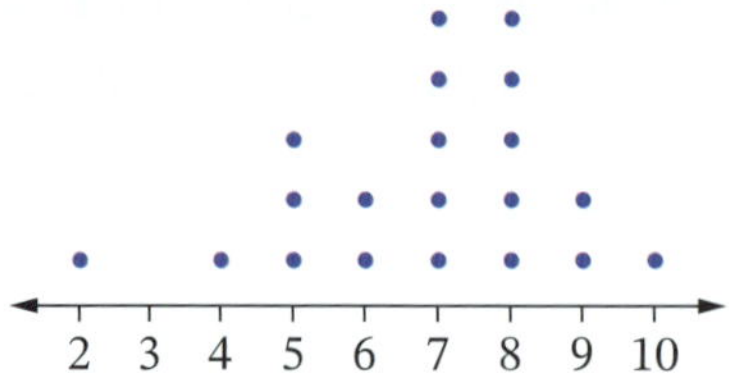


Which of the following is the upper quartile, $Q_3$? Select **A**, **B**, **C** or **D**.

**A** 7.5 **B** 7 **C** 8.5 **D** 8

EXAMPLE 9

**4** Calculate the range of each data set.

**a** Number of accidents per month in a factory:

3 0 0 1 2 1 6 0 0 2 1 0

**b** A golfer's scores for the last 9 holes of a golf course:

4 3 5 6 4 3 8 6 6

**c** Weekly mortgage repayments, in dollars:

370 628 299 417 354 1027 585 435 509 652 481

**d** Times, in minutes, for the swim-leg of a triathlon:

28 34 22 24 25 24 26 26 24 27

**5** Calculate the interquartile range of each data set in Question **4**.

**6** Which of the following is the interquartile range of the dot plot in Question **3**? Select **A**, **B**, **C** or **D**.

**A** 2.5 **B** 3 **C** 5 **D** 8

**7** This stem-and-leaf plot shows the marks out of 100 for a class of students in a maths test.

For this data, find:

**a** the range

**b** the interquartile range.

| Stem | Leaf |
|---|---|
| 3 | 0 7 |
| 4 | 2 3 4 6 8 |
| 5 | 0 1 4 5 7 7 |
| 6 | 2 3 5 7 8 8 |
| 7 | 4 5 6 9 |
| 8 | 2 2 7 |
| 9 | 3 |

**8** 15 job applicants took a short general knowledge multiple-choice quiz. Their times (in seconds) to complete this test were:

| 45 | 37 | 46 | 34 | 26 | 15 | 35 | 61 |
|---|---|---|---|---|---|---|---|
| 43 | 48 | 52 | 38 | 30 | 44 | 37 | |

**a** What was the range of times?

**b** What was the interquartile range?

**c** Give a possible reason for the outlier.

**9** The stem-and-leaf plot below represents the number of points per match scored by the GWS Giants in a football season.

What is the interquartile range of this data set? Select **A**, **B**, **C** or **D**.

**A** 80

**B** 38

**C** 44

**D** 42

| Stem | Leaf |
|---|---|
| 7 | 8 9 |
| 8 | 3 5 8 9 |
| 9 | 1 2 3 5 8 |
| 10 | 0 5 |
| 11 | 1 7 |
| 12 | 6 7 9 |
| 13 | |
| 14 | 6 9 |
| 15 | 1 8 |

**10** Calculate the range of the data set from Question **9**.

Matt King/AFL Photos/Getty Images

Foundation Mastery Complex

# 10.04 Standard deviation

**Video**
Standard deviation on a Casio calculator

**Worksheets**
Standard deviation

Statistical calculations

**Standard deviation** is a better **measure of spread** than the range and interquartile range because, like the mean, its value depends on every value in the data set.

Standard deviation measures how different each value in a data set is from the mean.

The symbol for standard deviation is $\sigma$, the lowercase Greek letter sigma, and it is more precisely called the *population standard deviation*.

The formula for calculating standard deviation is quite complicated, and does not need to be learned. Instead, you can use your calculator's statistics mode.

## Example 10

Calculate, correct to one decimal place, the standard deviation of each data set.

**a** The maximum daily temperature (in °C) in Mudgee for the first 2 weeks in January:

| 30 | 28 | 26 | 31 | 34 | 35 | 32 |
|---|---|---|---|---|---|---|
| 33 | 21 | 25 | 28 | 32 | 32 | 35 |

**b** The body temperatures (in °C) of a group of hospital patients:

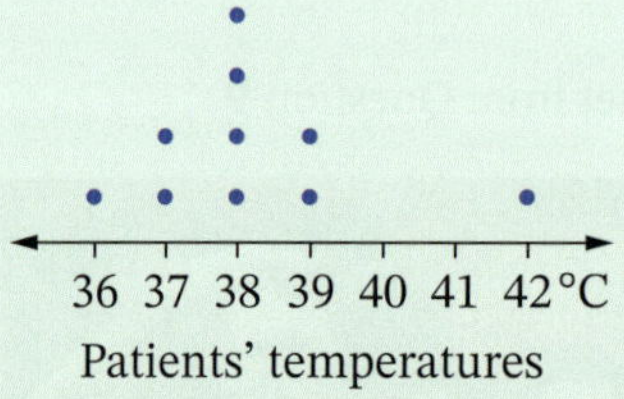


Patients' temperatures

### Solution

**a** $\sigma = 3.9434 \ldots \approx 3.9$

On the calculator, the symbol for standard deviation includes an '$x$' to indicate that it is the standard deviation of $x$ values.

| Operation | Casio *fx*-8200 | Sharp |
|---|---|---|
| Refer to page 373 to enter the data. | | |
| Calculate the population standard deviation ($\sigma_x = 3.9434 \ldots$) | SHIFT 1 Var $\sigma_x$ = | RCL $\sigma X$ |
| | EXE 1-Var Results EXE to calculate many statistics (scroll down for more) | |

**b** $\sigma = 1.5362 \ldots$

$\approx 1.5$

To calculate the standard deviation of data presented in a frequency table, refer to the table of calculator instructions on page 374, then follow the instructions from part **a** above.

## Example 11

36 people were given a concentration task and the time taken (in seconds) to complete the exercise are shown below.

| **Males** | 32 | 44 | 44 | 29 | 40 | 26 | 64 | 21 | 65 | 32 | 42 | 30 | 66 | 51 | 53 | 30 | 55 | 42 |
|---|---|---|---|---|---|---|---|---|---|---|---|---|---|---|---|---|---|---|
| **Females** | 35 | 35 | 41 | 41 | 49 | 38 | 33 | 44 | 36 | 53 | 28 | 42 | 37 | 35 | 28 | 54 | 60 | 61 |

**a** Find the mean and standard deviation of each group.

**b** Is there a significant difference between the times it took to complete the exercise for males and females? How much variation is there in each group? Give reasons for your answer.

### Solution

**a** Using the calculator's statistics mode:

Males: $\bar{x} = 42.56$, $\sigma = 13.58$

Females: $\bar{x} = 41.67$, $\sigma = 9.72$

**b** The difference in mean time for males and females to complete the task was only 0.89 seconds, so there was not a significant difference. However, the standard deviation for females was 3.88 seconds lower than the times for males, showing that the times for females were more consistent than for males.

**EXERCISE 10.04** Answers on p. 485

## Standard deviation

**1** The number of monthly accidents at a construction site over 8 months was:

3 0 4 2 3 0 2 2

**a** Calculate the mean number of accidents per month.

**b** Find the standard deviation for the data, correct to one decimal place.

**2** An express train from Central Station was late in arriving at Homebush by the following times (in minutes):

6 0 3 −2 5 −1 0 3 −1 6 7 1

**a** Find the mean, $\bar{x}$.

**b** Calculate the standard deviation, $\sigma$, correct to 2 decimal places.

**c** Evaluate $\bar{x} + \sigma$ and $\bar{x} - \sigma$, the values that are, respectively, one standard deviation below and one standard deviation above the mean.

**d** How many of the given scores lie within one standard deviation of the mean, that is, between the 2 values you calculated in part **c**?

**e** What percentage, correct to one decimal place, of scores were within 1 standard deviation from the mean?

**3** A sample of mobile phone batteries was tested for charge life (in hours).

60 73 65 84 77 64 66 73 88 90 79 81

Find, correct to 2 decimal places:

**a** the mean

**b** the standard deviation.

**4** Bani's weekly wages (in dollars) for providing online tech support were:

540 510 1100 1350 780 650 920 590 1080

Calculate for this data, correct to the nearest dollar:

**a** the mean **b** the standard deviation.

Foundation Mastery Complex

**5** Students were surveyed on the number of movies they had streamed in the last fortnight, with the results shown in the frequency table.

| Score ($x$) | Frequency ($f$) |
|---|---|
| 0 | 6 |
| 1 | 7 |
| 2 | 8 |
| 3 | 10 |
| 4 | 9 |
| 5 | 5 |
| 6 | 5 |

**a** For this data, find the mean, $\bar{x}$.

**b** Calculate the standard deviation, correct to one decimal place.

**c** How many data values were within one standard deviation of the mean?

**d** What percentage of values were within one standard deviation of the mean?

For many large sets of data, approximately 68% $\left(\text{slightly more than } \frac{2}{3}\right)$ of the scores lie within one standard deviation of the mean.

**6** This dot plot shows the number of vehicles driving past Westvale High School every minute for a 20-minute period.

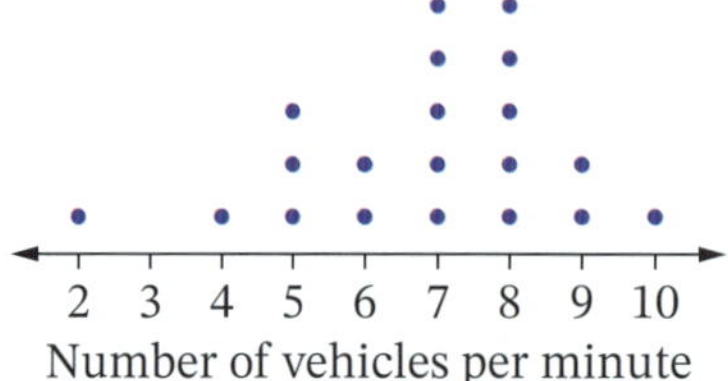


**a** Find the mean.

**b** Calculate, correct to 2 decimal places, the standard deviation.

**c** How many values were within one standard deviation of the mean?

**d** What percentage of values were within one standard deviation of the mean?

**7** This table shows the weekly wages of employees at Great Gals electrical store, grouped in classes of $100.

| Weekly wage ($) | Class centre | Frequency |
|---|---|---|
| $900–<$1000 | | 7 |
| $1000–<$1100 | | 20 |
| $1100–<$1200 | | 36 |
| $1200–<$1300 | | 17 |
| $1300–<$1400 | | 11 |
| $1400–<$1500 | | 3 |

**a** Copy and complete the table.

**b** Find, to the nearest cent, an estimate for:

**i** the mean

**ii** the standard deviation.

☐ Foundation ○ Mastery ⬡ Complex

EXAMPLE 11

**8** The heights (in cm) of male and female students in a Year 11 PDHPE class are shown.

| **Males** | 183 | 160 | 178 | 179 | 171 | 175 | 184 | 172 | 173 | 187 | 179 | 165 |
|---|---|---|---|---|---|---|---|---|---|---|---|---|
| **Females** | 172 | 160 | 162 | 160 | 173 | 165 | 165 | 163 | 168 | 150 | 160 | 177 |

**a** Find the mean and standard deviation for males and for females.

**b** Is there a significant difference between the heights of males and females? Give reasons for your answer.

**9** The results of 2 Maths tests given to a Year 11 class are displayed in this back-to-back stem-and-leaf plot.

| Test 1 | | Test 2 |
|---:|:---:|:---|
| 4 | 3 | 2 |
| 4 3 | 4 | 9 |
| 9 8 0 | 5 | 2 7 9 |
| 9 8 7 4 0 | 6 | |
| 9 7 5 5 5 3 1 | 7 | 0 1 1 2 4 4 8 |
| 9 9 | 8 | 0 1 2 4 5 5 7 8 |

**a** Find the mean mark and standard deviation for each test.

**b** Are there significant differences between the means and standard deviations of the 2 tests?

**c** In which test did the students perform better? Justify your answer.

**10** A group of men and women were timed on the length of time (in seconds) of the last call they made on their mobile phone.

| **Men** | 292 | 360 | 840 | 60 | 60 | 900 | 60 | 328 | 217 | 16 |
|---|---|---|---|---|---|---|---|---|---|---|
| | 1565 | 58 | 22 | 98 | 73 | 537 | 51 | 49 | 1210 | 15 |
| **Women** | 653 | 73 | 202 | 58 | 74 | 75 | 58 | 168 | 354 | 600 |
| | 1560 | 2220 | 56 | 900 | 481 | 60 | 139 | 80 | 72 | 110 |

**a** Find the mean and standard deviation for each group.

**b** Calculate the mean and standard deviation of the times for men and women if the outliers (1565 s and 1210 s for men, 1560 s and 2220 s for women) are excluded.

**c** Do the men or the women of this group make longer calls? Justify your answer.

Foundation Mastery Complex

## Technology

### Calculating measures of spread

**1** Open a blank spreadsheet and enter the temperature data about Mudgee from Example 10 on page 388.

| | A | B | C | D | E | F | G |
|---|---|---|---|---|---|---|---|
| 1 | **Maximum daily temperatures in Mudgee** | | | | | | |
| 2 | 30 | 28 | 26 | 31 | 34 | 35 | 32 |
| 3 | 33 | 21 | 25 | 28 | 32 | 32 | 35 |
| 4 | | | | | | | |
| 5 | | | Highest value | | | | |
| 6 | | | Lowest value | | | | |
| 7 | | | Q3 | | | | |
| 8 | | | Q1 | | | | |
| 9 | | | | | | | |
| 10 | | | Range | | | | |
| 11 | | | Interquartile range | | | | |
| 12 | | | Standard deviation | | | | |
| 13 | | | | | | | |



**2** In cell E5, enter the formula =**MAX(A2:G3)** to calculate the highest value (35).

**3** In cell E6, enter =**MIN(A2:G3)** to calculate the lowest value (21).

**4** In cell E7, enter =**QUARTILE(A2:G3,3)** to calculate the upper quartile, $Q_3$ (32.75).

**5** In cell E8, enter =**QUARTILE(A2:G3,1)** to calculate the lower quartile, $Q_1$ (28).

*Note:* A spreadsheet calculates quartiles using a slightly different method to the method we have described, so its answers for the interquartile range may not be exactly the same as ours, but they should be close.

**6** In cell E10, enter =**E5-E6** to calculate the range (14).

**7** In cell E11, enter =**E7-E8** to calculate the interquartile range (4.75).

**8** In cell E12, enter =**STDEV.P(A2:G3)** to calculate the population standard deviation.

# The effect of outliers

10.05

An **outlier** is a very high or very low value in a data set that is clearly apart from the other data. It can occur for a variety of reasons and should be investigated. If it was obtained through incorrect measurement, it should be excluded.

**Worksheet**
Calculating and interpreting summary statistics

## Outliers

An **outlier** is a value in a data set that is either:

- less than $Q_1 - 1.5 \times \text{IQR}$ or
- greater than $Q_3 + 1.5 \times \text{IQR}$

This is only one of several ways of determining whether a data value is an outlier.

## Example 12

**Videos**
Outliers
Identifiying outliers
Testing for outliers

The following scores are marks achieved by students in a test.

| 11 | 8 | 12 | 12 | 15 | 13 | 10 | 25 | 12 | 11 | 7 |
|---|---|---|---|---|---|---|---|---|---|---|
| 10 | 13 | 16 | 10 | 12 | 16 | 11 | 12 | 16 | 17 | 20 |

Test which scores are outliers.

### Solution

The scores arranged in order are:

7 8 10 10 10 11 11 11 12 12 (12 12) 12 13 13 15 16 16 16 17 20 25

$Q_1$ = 11, $Q_2$ = 12 (between the two circled 12s), $Q_3$ = 16

$$\begin{aligned}\text{IQR} &= Q_3 - Q_1\\ &= 16 - 11\\ &= 5\end{aligned}$$

$$\begin{aligned}\therefore 1.5 \times \text{IQR} &= 1.5 \times 5\\ &= 7.5\end{aligned}$$

$$\begin{aligned}\therefore Q_1 - 1.5 \times \text{IQR} &= 11 - 7.5\\ &= 3.5\end{aligned}$$

$$\begin{aligned}Q_3 + 1.5 \times \text{IQR} &= 16 + 7.5\\ &= 23.5\end{aligned}$$

$\therefore$ a score is an outlier if it is less than 3.5 or greater than 23.5.

$\therefore$ 25 is an outlier.

## Outliers and measures of centre and spread

Outliers can affect the measures of centre and spread of a data set.

- The mean and standard deviation are most affected by outliers (because their value depends on every value in the data set).
- The median can be affected, but not by much.
- The mode is not affected at all.
- The range is affected because its value depends on the highest and lowest values of the data set.

## Example 13

The dot plot shows the temperatures of patients in a hospital ward.

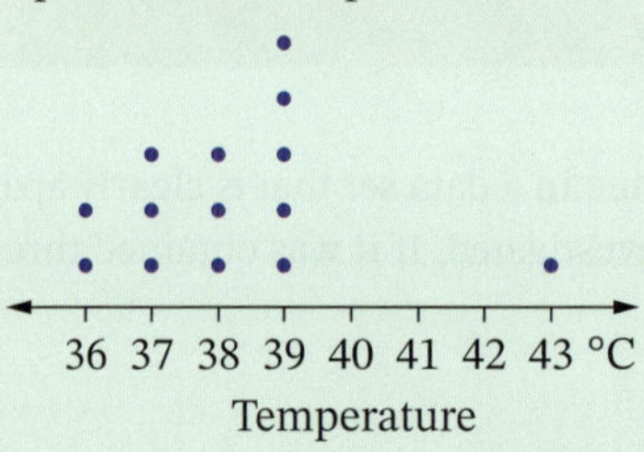


**a** Calculate the mean, mode, median, range, interquartile range and standard deviation of this data set.

**b** What is the outlier temperature?

**c** Recalculate the measures of centre and spread of this data set if the outlier is excluded.

**d** Describe the effect the outlier has on the measures of centre and spread of the distribution.

### Solution

**a** mean $= \dfrac{535}{14} \approx 38.2$

mode $= 39$

median $= \dfrac{38+38}{2} = 38$ — the average of the 7th and 8th scores

range $= 43 - 36$

$= 7$

From the dot plot, $Q_1 = 37$, $Q_3 = 39$.

$\text{IQR} = Q_3 - Q_1$

$= 39 - 37$

$= 2$

standard deviation $= 1.6978\ ...$

$\approx 1.70$

**b** $1.5 \times \text{IQR} = 1.5 \times 2$

$= 3$

If 43 is an outlier, it must be greater than $Q_3 + 1.5 \times \text{IQR}$.

$\therefore Q_3 + 1.5 \times \text{IQR} = 39 + 3$

$= 42$

$\therefore$ the outlier is 43.

**c** mean $= \dfrac{492}{13} \approx 37.8$

mode $= 39$

median $= 38$

range $= 39 - 36$

$= 3$

$\text{IQR} = Q_3 - Q_1$

$= 39 - 37$

$= 2$

standard deviation $= 1.10$

**d** The high outlier does not affect the mode, median and interquartile range but it increases the mean, range and standard deviation.

**EXERCISE 10.05** Answers on p. 485

## The effect of outliers

EXAMPLE 12

**1** The following scores are the number of goals scored by a hockey team during a season.

| 3 | 2 | 0 | 0 | 1 | 2 | 3 | 2 | 4 | 8 |
|---|---|---|---|---|---|---|---|---|---|
| 2 | 3 | 5 | 2 | 1 | 3 | 4 | 4 | 2 | 3 |

**a** Find the interquartile range.

**b** Find the value of:

**i** $Q_1 - 1.5 \times \text{IQR}$ **ii** $Q_3 + 1.5 \times \text{IQR}$

**c** Is the score of 8 goals an outlier? Give reasons for your answer.

**2** Determine whether each data set has outliers.

**a** 2 5 6 6 7 8 10 10 15

**b** 9 13 13 14 14 15 15 15 15
16 16 16 16 16 17 17 18

**c**

| Stem | Leaf |
|---|---|
| 1 | 2 9 |
| 2 | 0 3 4 4 8 |
| 3 | 4 5 6 7 |
| 4 | 1 4 9 |
| 5 | 0 2 |
| 6 | 8 |

**d**

| Score ($x$) | Frequency ($f$) |
|---|---|
| 4 | 3 |
| 5 | 12 |
| 6 | 4 |
| 7 | 3 |
| 8 | 0 |
| 9 | 1 |

**3** The employees at the Bread and Butter Cafe earned the following wages in a week.

\$750 \$820 \$910 \$530 \$1200 \$720 \$890

**a** What is the mean wage?

**b** What is the median wage?

**c** Find the interquartile range.

**d** The manager's wage is an outlier. What is this wage and how do we verify that it is an outlier?

**e** If the manager's wage is not included, how does this affect the mean, median and interquartile range?

**f** If each employee receives a 10% pay rise, what will be the new mean, median and interquartile range? Is it 10% more than the old mean, median and interquartile range?

**4** The cups of coffee drunk by a sample of HSC exam markers in one night is shown in the table.

| Cups of coffee | No. of markers |
|---|---|
| 2 | 1 |
| 3 | 4 |
| 4 | 5 |
| 5 | 9 |
| 6 | 0 |
| 7 | 0 |
| 8 | 1 |

**a** How many markers were surveyed?

**b** What is the outlier?

**c** What is the mean, range and standard deviation if the outlier:

**i** is included? **ii** is not included?

**d** If the outlier is included, what effect does this have on the mean, range and standard deviation?

**5** A group of friends go to the cinema. The ages of the group are:

13 12 11 14 12 15 14 13

If Kait brings her 5-year-old sister as well, what will happen? Select **A**, **B**, **C** or **D**.

**A** The range of ages decrease. **B** The median age decreases.

**C** The mode age decreases. **D** The mean age decreases.

Foundation Mastery Complex

6 In a netball tournament of 5 matches, the points scored by 3 teams are:

| | | | | | |
|---|---|---|---|---|---|
| **The Wombats** | 24 | 18 | 14 | 6 | 22 |
| **The Possums** | 16 | 16 | 15 | 18 | 15 |
| **The Koalas** | 36 | 8 | 14 | 16 | 12 |

**a** What are the mean, median and range for each team?

**b** Which team is the most consistent? Why?

**c** An error was made in the scoring for the Wombats – the score of 6 should have been 16. What are the new mean, median and range for the Wombats?

**d** Which team is most consistent now? Why?

7 Sam and Terri sell copiers. The numbers of copiers that they sell each week are sorted in ascending order.

| | | | | | | | | | | |
|---|---|---|---|---|---|---|---|---|---|---|
| **Sam** | 1 | 2 | 3 | 3 | 5 | 6 | 7 | 8 | 12 | 25 |
| **Terri** | 3 | 3 | 3 | 14 | 16 | 18 | 18 | 24 | 32 | 35 |

**a** What is the modal number of copiers sold by each person?

**b** What could you say about each person if you only knew the mode?

**c** What is the median number of copiers sold by each person?

**d** What is the mean number of copiers sold by each person?

**e** Which measure of centre is best for comparing their sales performances?

**f** Who is the better salesperson? Justify your answer.

8 This dot plot represents the number of accidents at a factory each month over a year.

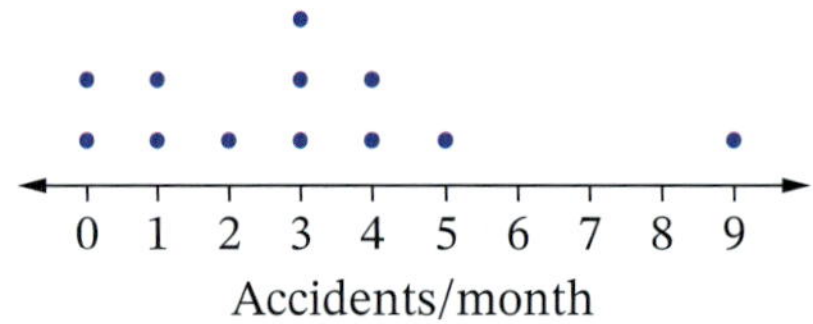


**a** Calculate the 6 measures of centre and spread of this data set.

**b** What is the outlier number of accidents? Explain why.

**c** Calculate the measures of centre and spread of this data set if the outlier is excluded.

**d** Describe the effect the outlier has on the measures of centre and spread.

9 Mandy's bookstore employs the following people with annual wages as shown.

| | |
|---|---|
| 1 store manager | \$173 800 |
| 2 cashiers | \$64 200 each |
| 2 part-time clerical staff | \$28 500 each |
| 3 salespeople | \$66 500 each |
| 2 part-time cleaners | \$23 500 each |

**a** Find the mean, median and modal annual salary for the 10 employees.

**b** Which measure of centre could Mandy use to make the salaries appear higher? Why?

**c** Which measure best represents the average wage for an employee at Mandy's bookstore? Why?

☐ Foundation ○ Mastery ⬡ Complex

# Cumulative frequency graphs

10.06

A **cumulative frequency histogram** is a column graph of **cumulative frequency**.

A **cumulative frequency polygon**, also called an **ogive** (pronounced 'oh-jive') is drawn by joining the **top right-hand corner** of each column of a cumulative frequency histogram.

**Worksheet**
Cumulative frequency graphs

**Videos**
IQR from cumulative frequency polygon

Cumulative frequency and the median

Cumulative frequency graphs

## Example 14

The maximum daily temperatures (in °C) in Campbelltown in June were recorded and grouped into the frequency table.

| Temperature (°C) | Frequency | Cumulative frequency |
|---|---|---|
| 12 | 1 | 1 |
| 13 | 2 | 3 |
| 14 | 6 | 9 |
| 15 | 2 | 11 |
| 16 | 6 | 17 |
| 17 | 3 | 20 |
| 18 | 6 | 26 |
| 19 | 1 | 27 |
| 20 | 2 | 29 |
| 21 | 1 | 30 |

**a** Draw a cumulative frequency histogram and polygon for the data.

**b** Use the frequency polygon to find the median and calculate the interquartile range.

### Solution

**a**

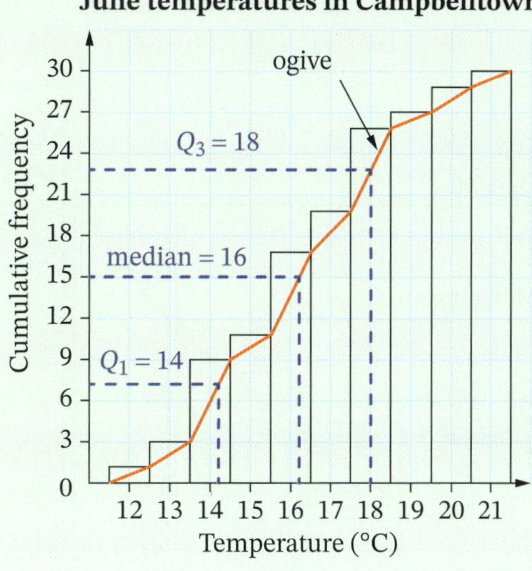


The ogive (polygon) is contained inside the columns.

**b** Draw a horizontal line from the halfway mark (15) on the Cumulative frequency axis to where it meets the ogive. The median is the corresponding value on the Temperature axis.

median = 16

To find $Q_1$, draw a horizontal line from the quarter mark $\left(\frac{1}{4} \times 30 = 7.5\right)$ on the Cumulative frequency axis to where it meets the ogive, then read the temperature value.

$Q_1 = 14$

To find $Q_3$, draw a horizontal line from the 3-quarter mark $\left(\frac{3}{4} \times 30 = 22.5\right)$ on the Cumulative frequency axis.

$Q_3 = 18$

$$\begin{aligned} \text{interquartile range} &= Q_3 - Q_1 \\ &= 18 - 14 \\ &= 4 \end{aligned}$$

## Example 15

The number of cases of ovarian cancer in women from various age groups is shown below.

| Age (years) | Class centre | Frequency | Cumulative frequency |
|---|---|---|---|
| 35–<45 | 40 | 28 | 28 |
| 45–<55 | 50 | 61 | 89 |
| 55–<65 | 60 | 65 | 154 |
| 65–<75 | 70 | 92 | 246 |
| 75–<85 | 80 | 74 | 320 |

Draw an ogive for this data and use it to find an estimate for:

**a** the median **b** the interquartile range.

### Solution

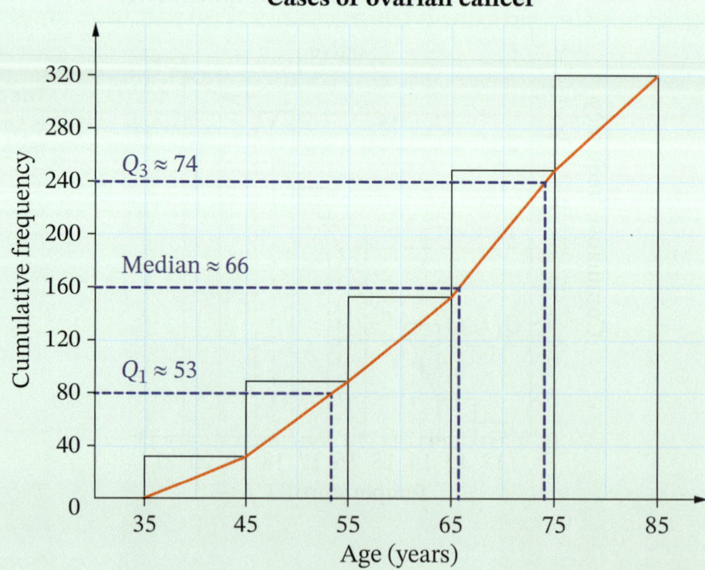


**a** Halfway point on the cumulative frequency axis: 160

median ≈ 66 estimating from the Age axis

All these values are estimates because the data has been grouped into class intervals.

**b** The three-quarter point on the cumulative frequency axis:

$\frac{3}{4} \times 320 = 240$

$Q_3 \approx 74$

Quarter point on the cumulative frequency axis:

$\frac{1}{4} \times 320 = 80$

$Q_1 \approx 53$

interquartile range $= 74 - 53$

$= 21$

**EXERCISE 10.06** Answers on p. 486

## Cumulative frequency graphs

**1** A sample of households was surveyed on the number of computers owned.

| Number of computers owned | Frequency | Cumulative frequency |
|---|---|---|
| 1 | 1 | |
| 2 | 7 | |
| 3 | 9 | |
| 4 | 6 | |
| 5 | 0 | |
| 6 | 1 | |

**a** Copy the table and complete the cumulative frequency column to find the median.

**b** Construct a cumulative frequency histogram and polygon.

**c** Use the graphs you drew in part **b** to find:

**i** the median

**ii** the interquartile range.

**2** This ogive shows the speeds of motor vehicles travelling along the main street of a town.

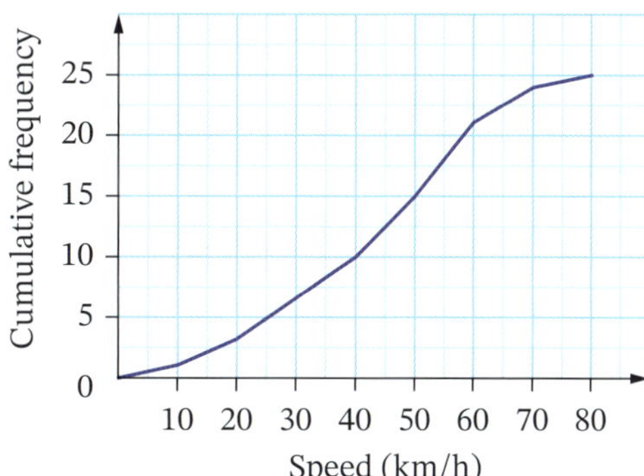


**a** How many vehicles were in the survey?

**b** Estimate the median speed of the vehicles.

**c** Estimate the interquartile range.

Foundation Mastery Complex

3 A packet of jelly beans is labelled 'Contents 30' but a quality control check found the results shown in the table.

| Number of jelly beans | Frequency | Cumulative frequency |
|---|---|---|
| 28 | 6 | |
| 29 | 34 | |
| 30 | 56 | |
| 31 | 28 | |
| 32 | 5 | |
| 33 | 1 | |

**a** Copy the table and complete the cumulative frequency column.

**b** Construct an ogive and use it to find an estimate of:

**i** the median

**ii** the interquartile range.

EXAMPLE 15

4 The heights of 50 students were measured and grouped into class intervals.

| Height (cm) | Class centre | Frequency | Cumulative frequency |
|---|---|---|---|
| 134–<141 | | 2 | |
| 141–<148 | | 3 | |
| 148–<155 | | 4 | |
| 155–<162 | | 13 | |
| 162–<169 | | 15 | |
| 169–<176 | | 11 | |
| 176–<183 | | 2 | |

**a** Copy and complete the table.

**b** What is the modal class?

**c** What is the median class?

**d** Construct an ogive and use it to estimate:

**i** the median **ii** the interquartile range.

## Did you know?

### The *Challenger* space shuttle disaster

In 1986, an engineer working on the space shuttle program at NASA predicted that at low air temperatures, the potential for damage to the shuttle would be extremely high. For a temperature of 12°C, he calculated a damage index of 11. He compared this to data from previous flights (as shown in the table below) and recommended that the *Challenger* flight be delayed until a warmer day.

| Year | Data from previous flights | | | | 1986 |
|---|---|---|---|---|---|
| Air temperature (°C) | 26 | 14 | 19 | 23 | 12 |
| Damage index | 0 | 4 | 0 | 0 | 11 |

However, his advice was ignored and the outlier was not considered important enough to delay the flight. The *Challenger* exploded just after takeoff, killing all 7 crew members, including a high school teacher. Later it was found that 2 rubber O-rings had failed to seal a joint at low temperatures, causing the shuttle to disintegrate.

**Research why teacher Christa McAuliffe was on board the *Challenger*.**

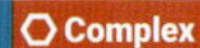

# Box plots

## 10.07

A **box plot** (or **box-and-whisker plot**) displays the quartiles of a set of data and the lowest and highest scores. The 'box' represents the middle 50% of scores and the interquartile range, while the 'whiskers' represent the lowest and highest 25% of scores.

**Video**
Box-and-whisker plots

**Worksheets**
Five-number summaries 1
Five-number summaries 2
Box-and-whisker plots
Box plots
Box plots 1
Box plots 2

Interquartile range
Whisker
Box
Lowest value
$Q_1$
$Q_3$
Highest value
Median

A box plot gives a **five-number summary** of a data set:

- the lowest value
- the lower quartile, $Q_1$
- the median, $Q_2$
- the upper quartile, $Q_3$
- the highest value.

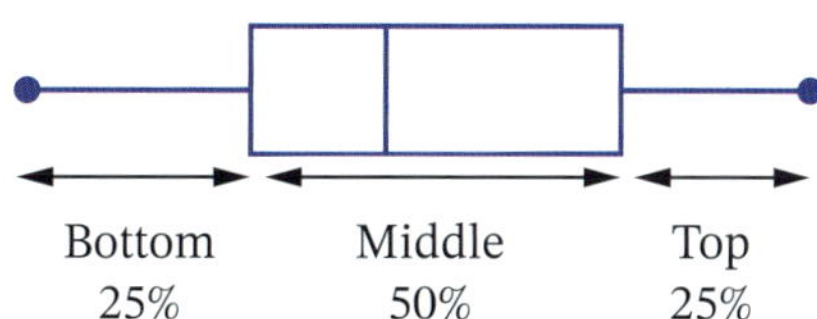


### Example 16

The ages of 10 people at a park were:

21 13 64 75 35 83 7 71 18 29

**a** Find the five-number summary for this data.

**b** Represent this data on a box plot.

### Solution

**a** In order:

7 13 (18) 21 29 35 64 (71) 75 83

$Q_1$ $Q_2$ $Q_3$

lowest value = 7 $\qquad Q_1 = 18 \qquad \text{median} = \dfrac{29+35}{2} = 32$

$Q_3 = 71 \qquad$ highest value = 83

The five-number summary for the ages is:
7, 18, 32, 71, 83

**b**

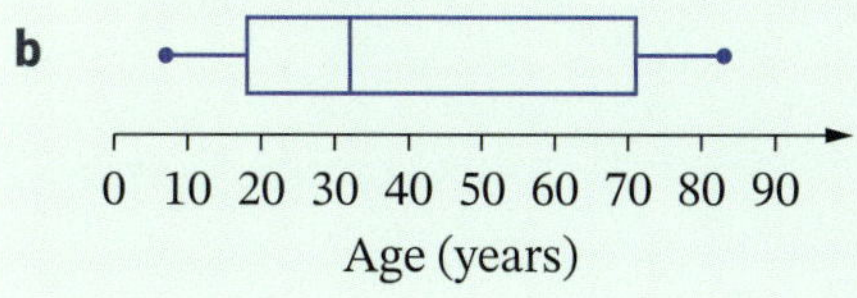


This box plot shows that, roughly:

- the bottom 25% of values lie from 7 to 18
- the next 25% of values lie from 18 to 32
- the median is 32
- the top 25% of values lie from 71 to 83.

## Example 17

This box plot represents the amount of pocket money in dollars earned by a sample of 48 children.

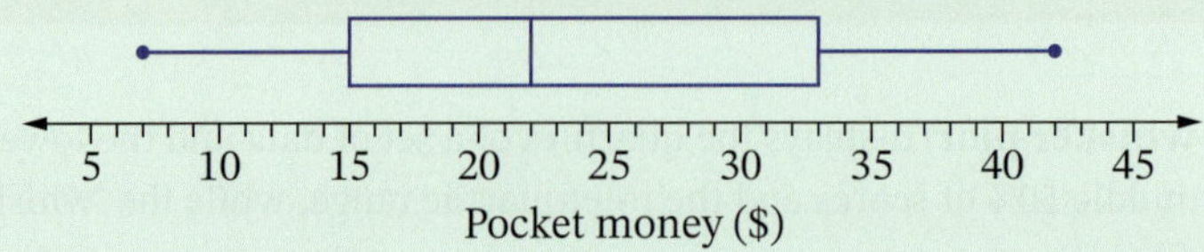


**a** Find the median.

**b** Find the range.

**c** How many children earned between:

**i** \$33 and \$42? **ii** \$15 and \$42?

**d** Find the interquartile range.

### Solution

**a** median = \$22

**b** range = \$42 − \$7 = \$35

**c** **i** $\frac{1}{4} \times 48$ children = 12 children top 25%

**ii** $\frac{3}{4} \times 48$ children = 36 children top 75%

**d** interquartile range = \$33 − \$15 = \$18

## Parallel box plots

**Video**
Double box plots

Parallel box plots can be used to represent 2 or more sets of data. They are drawn on the same scale, one above the other.

## Example 18

The mean maximum monthly temperatures for Sydney and Melbourne are shown in this table.

| Month | Jan | Feb | Mar | Apr | May | Jun | Jul | Aug | Sep | Oct | Nov | Dec |
|---|---|---|---|---|---|---|---|---|---|---|---|---|
| Sydney | 25.9 | 25.8 | 24.8 | 22.5 | 19.5 | 17.0 | 16.4 | 17.9 | 20.1 | 22.2 | 23.7 | 25.2 |
| Melbourne | 26.0 | 25.8 | 23.9 | 20.3 | 16.7 | 14.1 | 13.5 | 15.0 | 17.3 | 19.7 | 22.0 | 24.2 |

**a** Find the five-number summary for each city.

**b** Draw a parallel box plot to display the data.

**c** For each city, find:

**i** the range **ii** the interquartile range.

**d** Compare the temperatures for both cities. Are there significant differences between the spread of the temperatures for Sydney and Melbourne?

## Solution

**a** In order:

**Sydney**

16.4 17.0 17.9 19.5 20.1 22.2 22.5 23.7 24.8 25.2 25.8 25.9

(arrows: $Q_1$ between 17.9 and 19.5; $Q_2$ between 22.2 and 22.5; $Q_3$ between 24.8 and 25.2)

lowest value = 16.4 $\quad Q_1 = \frac{17.9 + 19.5}{2} = 18.7 \quad$ median $= \frac{22.2 + 22.5}{2} = 22.35$

$Q_3 = \frac{24.8 + 25.2}{2} = 25.0 \quad$ highest value = 25.9

The five-number summary for Sydney is:

16.4, 18.7, 22.35, 25.0, 25.9

**Melbourne**

13.5 14.1 15.0 16.7 17.3 19.7 20.3 22.0 23.9 24.2 25.8 26.0

(arrows: $Q_1$ between 15.0 and 16.7; $Q_2$ between 19.7 and 20.3; $Q_3$ between 23.9 and 24.2)

lowest value = 13.5 $\quad Q_1 = \frac{15.0 + 16.7}{2} = 15.85 \quad$ median $= \frac{19.7 + 20.3}{2} = 20.0$

$Q_3 = \frac{23.9 + 24.2}{2} = 24.05 \quad$ highest value = 26.0

The five-number summary for Melbourne is:

13.5, 15.85, 20.0, 24.05, 26.0

**b**

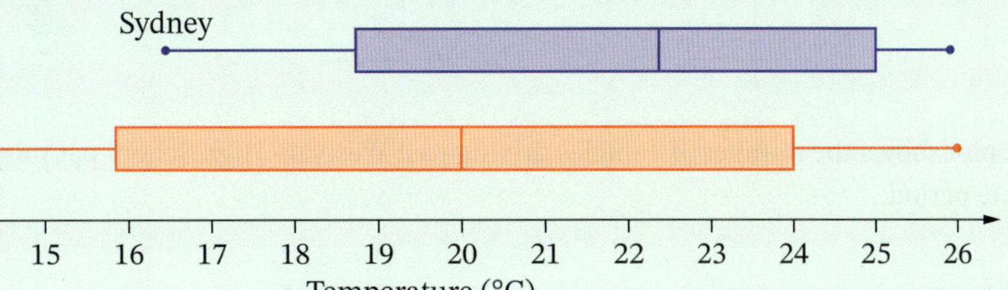


**c** **i** Sydney: range = 25.9 − 16.4 = 9.5

Melbourne: range = 26.0 − 13.5 = 12.5

**ii** Sydney: interquartile range = 25.0 − 18.7 = 6.3

Melbourne: interquartile range = 24.05 − 15.85 = 8.2

**d** The range of temperatures in Melbourne is 3° more than that of Sydney and the IQR is 1.9° more, so there is a significant difference. Sydney's mean maximum monthly temperatures are more consistent than Melbourne's.

**EXERCISE 10.07** Answers on p. 486

## Box plots

**1** Tom's scores for the 18 holes of a golf course were:

| 3 | 4 | 6 | 8 | 7 | 9 | 5 | 9 | 11 |
|---|---|---|---|---|---|---|---|---|
| 5 | 7 | 4 | 5 | 8 | 6 | 9 | 10 | 5 |

**a** Find a five-number summary for this data.

**b** Represent this data on a box plot.

**2** 15 job applicants took a short general knowledge multiple-choice quiz. Their times, in seconds, to complete this test were as shown below. Show this data on a box plot.

| 45 | 37 | 46 | 34 | 26 | 15 | 35 | 61 |
|---|---|---|---|---|---|---|---|
| 43 | 48 | 52 | 38 | 30 | 44 | 37 | |

**3** Find a five-number summary for the data in this stem-and-leaf plot of ages of people at the cinema, then draw a box plot for them.

| Stem | Leaf |
|---|---|
| 1 | 4 7 7 8 |
| 2 | 6 8 9 9 |
| 3 | 1 3 5 5 7 8 |
| 4 | 0 2 2 4 5 |
| 5 | 3 7 8 |
| 6 | 2 9 |

**4** This box plot illustrates the number of cigarettes smoked per day by a sample of 60 smokers who are trying to quit.

**a** What is the median number of cigarettes smoked per day?

**b** What is the interquartile range?

**c** What is the lowest value?

**d** How many people smoked between 20 and 25 cigarettes per day?

**e** How many people smoked fewer than 20 cigarettes per day?

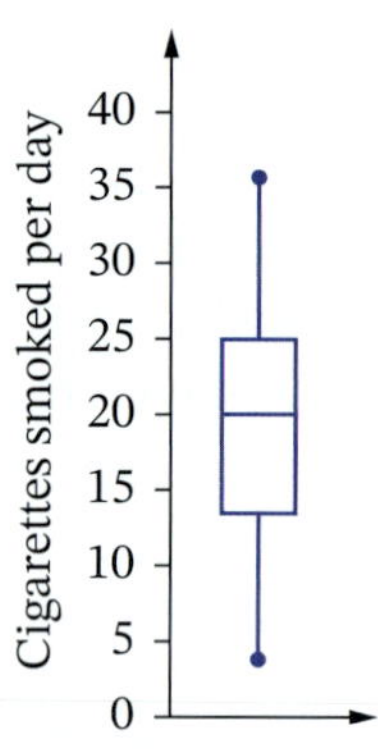


**5** This dot plot shows the number of vehicles driving past Westvale High School per minute in a 20-minute period.

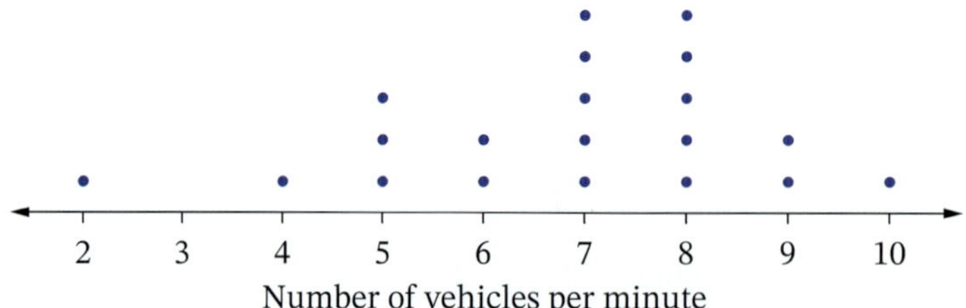


**a** Find the five-number summary for this data and draw a box plot.

**b** Compare the box plot you drew in part **a** with the original dot plot.
Which one do you prefer? Why?

☐ Foundation ○ Mastery ⬡ Complex

6 This box plot represents the annual wages (× \$1000) of the workers at a call centre.

**Annual wages of call centre workers**

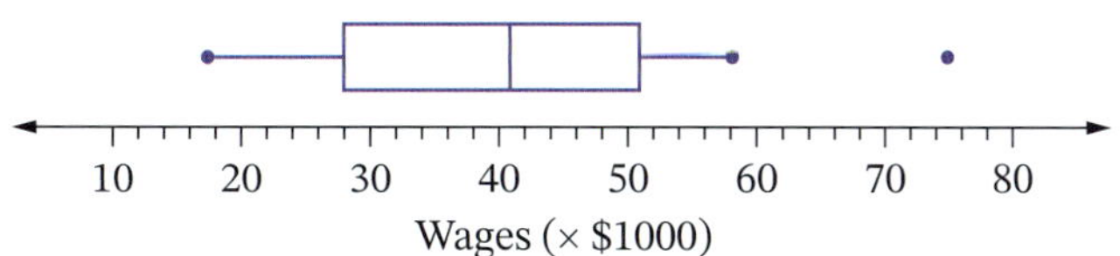


- **a** One of the wages is an outlier and it was not included in the box plot. What is the outlier?
- **b** What is the median wage?
- **c** Excluding the outlier, what is the range of wages?
- **d** Including the outlier, what is the range of wages?
- **e** Between what 2 amounts are the middle 50% of staff wages?
- **f** What percentage of the staff earn less than \$28 000?

7 In Year 11, the results of the first assessment task of 40 students who do both Modern History and Geography are displayed on the parallel box plot below.

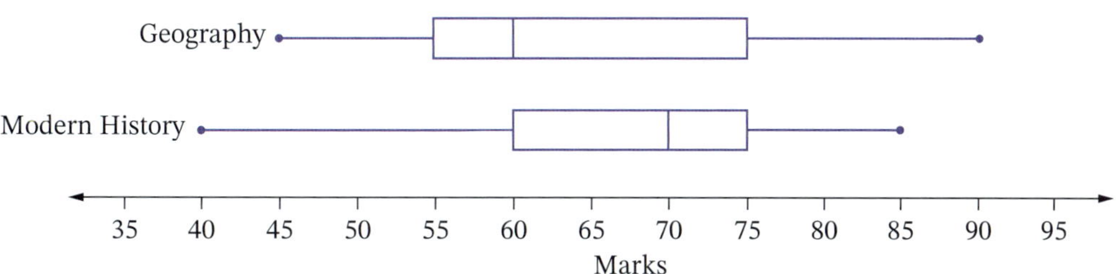


- **a** Find the five-number summary for each subject.
- **b** For each subject, find:
  - **i** the range
  - **ii** the interquartile range.
- **c** What is the median for each subject?
- **d** Which subject has the least spread? Give reasons for your answer.
- **e** How many students scored between 60 and 75 in:
  - **i** Geography?
  - **ii** Modern History?
- **f** In which subject did the Year 11 students perform better? Give reasons for your answer.

8 Year 12 students at Baramvale High had their pulse taken. The results are as follows.

| | | | | | | | | | | | | | | | |
|---|---|---|---|---|---|---|---|---|---|---|---|---|---|---|---|
| **Male** | 106 | 70 | 69 | 58 | 60 | 68 | 64 | 63 | 75 | 70 | 84 | 88 | 59 | 60 | 66 |
| **Female** | 68 | 74 | 59 | 75 | 74 | 82 | 82 | 71 | 120 | 55 | 77 | 91 | 73 | 60 | 79 |

- **a** Find the five-number summary for each group and draw a parallel box plot to show the information.
- **b** Find the range and interquartile range for each group.
- **c** Compare the spread between the 2 groups. Are there significant differences between the pulse rates for these males and females?
- **d** Which group had the lower pulse rates? Give reasons for your answer.

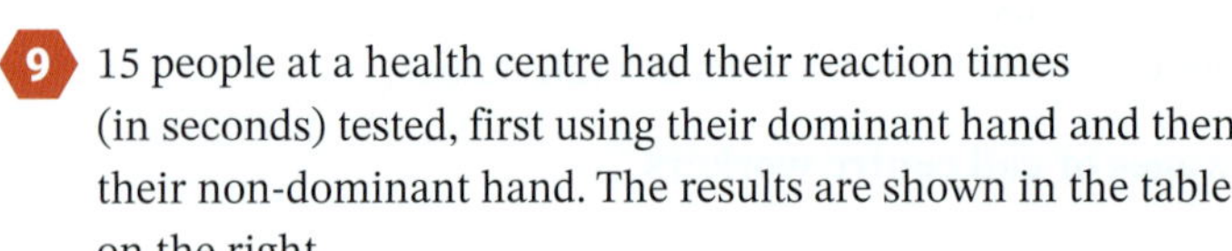

**9** 15 people at a health centre had their reaction times (in seconds) tested, first using their dominant hand and then their non-dominant hand. The results are shown in the table on the right.

**a** Find the five-number summary for both sets of results and draw a parallel box plot to display the data.

**b** Find the range and interquartile range for the dominant hand and the non-dominant hand.

**c** Are there significant differences between the 2 sets of results?

| Dominant hand | Non-dominant hand |
|---|---|
| 0.41 | 0.48 |
| 0.31 | 0.34 |
| 0.38 | 0.38 |
| 0.50 | 0.45 |
| 0.38 | 0.38 |
| 0.33 | 0.35 |
| 0.36 | 0.30 |
| 0.46 | 0.45 |
| 0.29 | 0.9 |
| 0.44 | 0.41 |
| 0.52 | 0.50 |
| 0.43 | 0.41 |
| 0.37 | 0.40 |
| 0.31 | 0.34 |
| 0.32 | 0.35 |

**10** The box plot shows the results of tests in Physics and Chemistry.

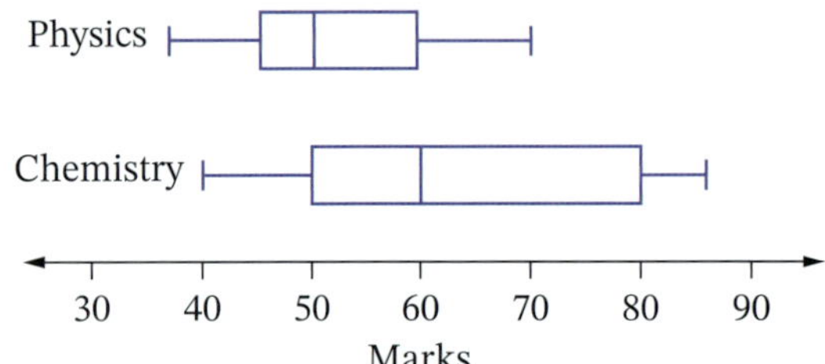


In Chemistry, 48 students completed the yearly exam and the number of students who scored above 50 or more was the same for both subjects.

How many students completed the Physics exam? Select **A**, **B**, **C** or **D**.

**A** 24 **B** 12 **C** 54 **D** 72

# 10.08 The shape of a distribution

**Video**
The shape of a frequency distribution

**Worksheet**
Shapes of distributions

The shape of a statistical **distribution** (data set) shows how the data are spread, and can be seen by drawing a curve around its graph or display.

A distribution is **symmetrical** if the data are balanced or evenly spread about the centre of the distribution, with the mean, median and mode being equal. One example of a symmetrical distribution is the heights of Year 11 students.

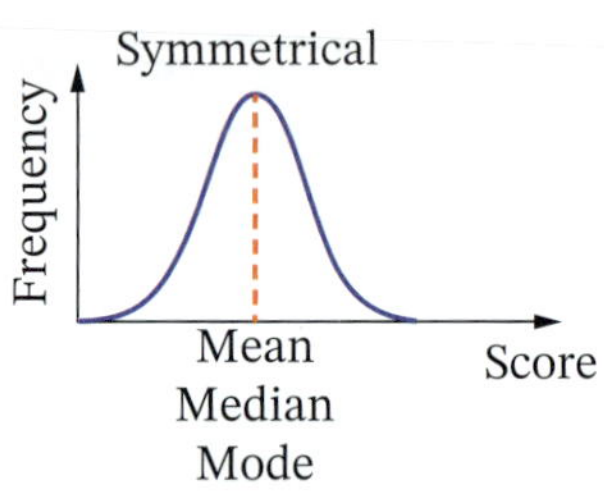


A distribution is **positively skewed** if its **tail** points to the right (the positive direction), because the mean is above the mode and median. This means that there are *more* higher data values, so more data is concentrated below the mean.

The word '**skewed**' means twisted.

One example of a positively skewed distribution is house prices in a town, where the mean is skewed up by a small number of very expensive houses.

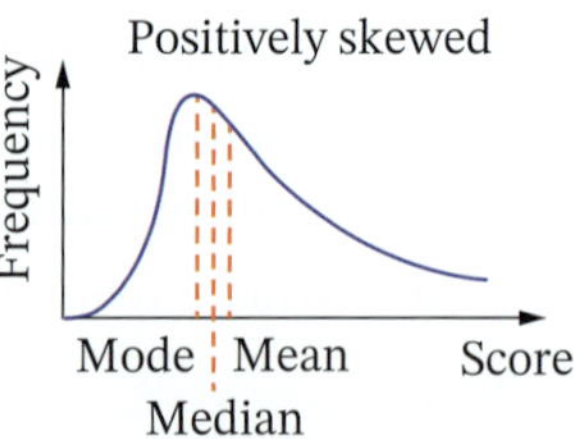

If a distribution is **negatively skewed**, then its tail points to the left (the negative direction) because the mean is below the mode and median. This means that the lower data values are overrepresented, so more data is concentrated above the mean.

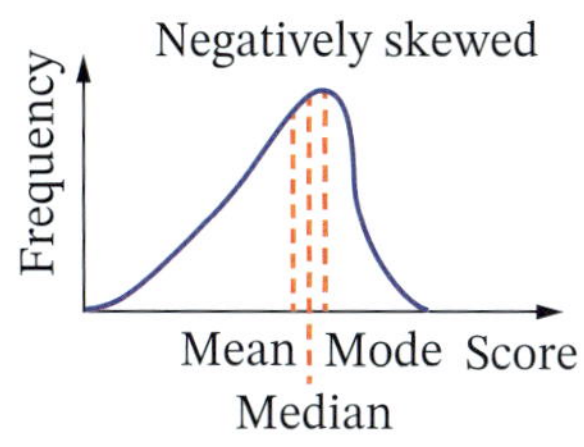


One example of a negatively skewed distribution is the retirement age of a person, where the mean is skewed down by a small number of people who retire early.

**Peaks** are the high points of the distribution and represent the more frequent scores. The highest peak is the mode.

On these graphs, the median divides the *area* under the curve exactly in half.

The **modality** is the number of peaks occurring in a distribution. A distribution can have one peak only (**unimodal**) or have more than one peak (**multimodal**).

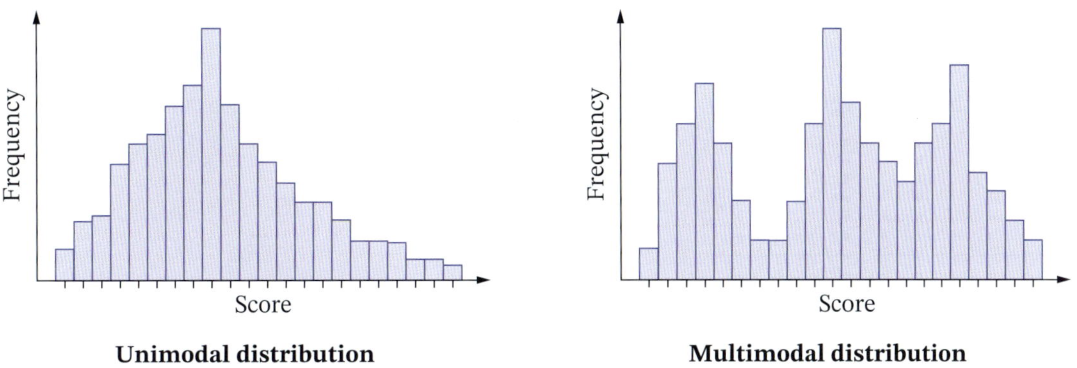


**Unimodal distribution** **Multimodal distribution**

If a distribution is **bimodal**, it has 2 peaks. For example, this frequency histogram is bimodal, having 2 peaks at 2 and 7. The mode, however, is 7.

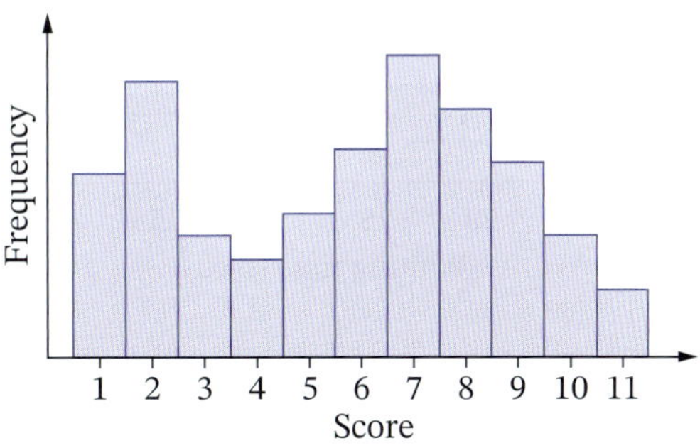


**Bimodal distribution**

If a distribution is fairly even and has no peaks, then it is a **uniform** distribution.

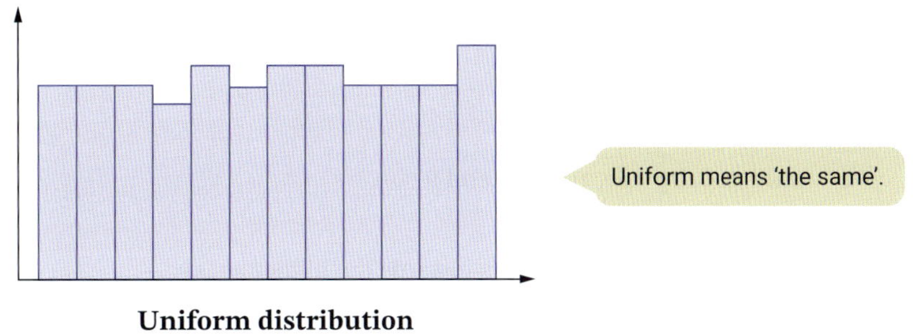


**Uniform distribution**

**Clusters** are values in the data set that are bunched or close together.

Video
Shape and modality of distributions

## Example 19

For each distribution shown below:

**i** describe its shape **ii** state the modality **iii** identify any clusters.

**a** Marks in a Japanese test

| Stem | Leaf |
|---|---|
| 3 | 1 2 |
| 4 | 3 5 9 |
| 5 | 0 2 6 |
| 6 | 4 5 |
| 7 | 3 5 6 7 7 8 9 |
| 8 | 0 2 4 6 6 8 8 9 |
| 9 | 1 2 4 8 9 |

**b** Amount of traffic on Sydney's roads

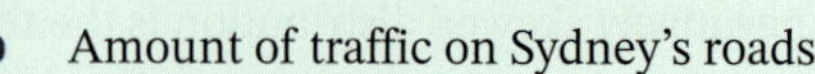

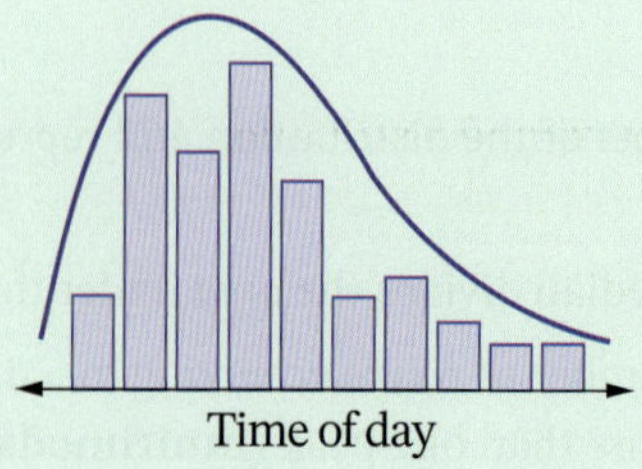


**c** Ages of children at a cinema

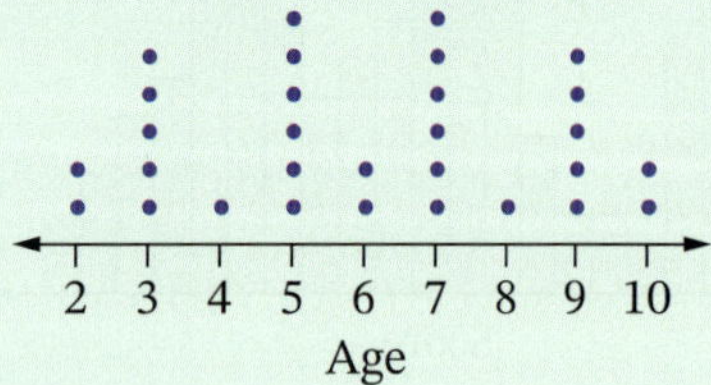


**d** Ages of people in a small coastal town

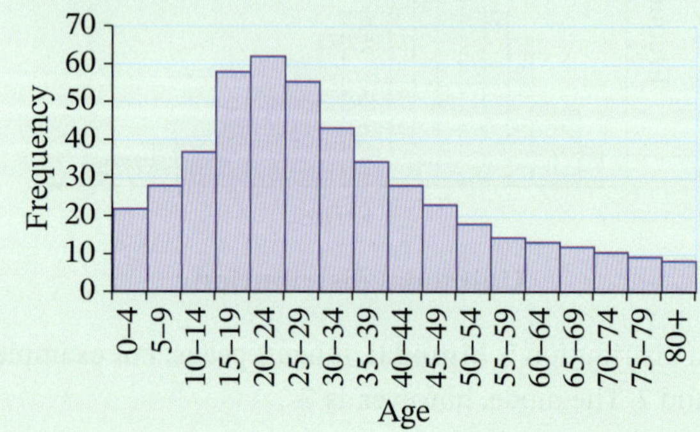


**e** Waiting time in a doctor's surgery

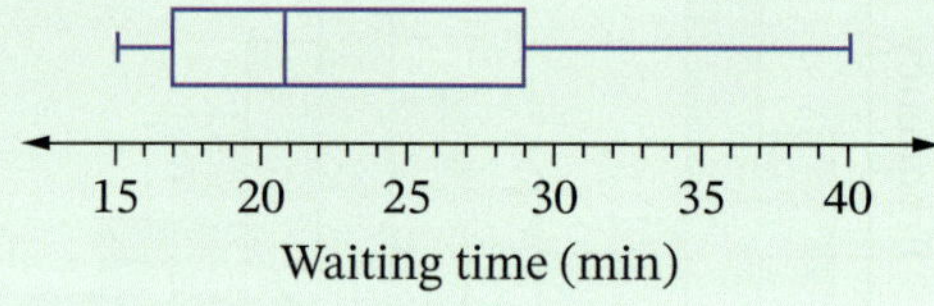


### Solution

**a** **i** negatively skewed (tail points towards the left)
**ii** multimodal, peaks at 77, 86, 88
**iii** clusters in the 70–90s

**b** **i** positively skewed (tail points towards the right)
**ii** bimodal, 2 peaks
**iii** clusters at earlier hours

**c** **i** symmetrical
**ii** multimodal, peaks at 3, 5, 7 and 9
**iii** no clusters

**d** **i** positively skewed (tail points towards the higher ages)
**ii** unimodal class, 1 peak
**iii** cluster from 15 to 29

**e** **i** positively skewed (tail points towards the right)
**ii** unable to determine since individual data values are not known
**iii** cluster from 15–17 min (25% of patients)

## The shape of a box plot

Remember that a box plot represents how the values in a data set are spread out into quarters, with the middle bar indicating the median.

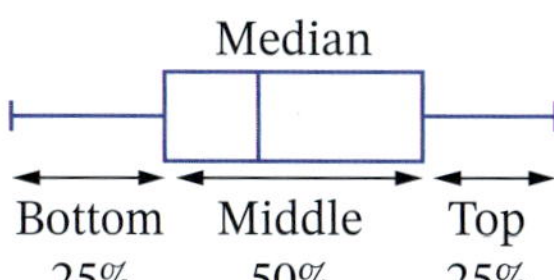


This table shows the shapes of a distribution represented on a curve and a box plot.

**Symmetrical distribution:** Even spread of data

**Example:** Heights of Year 11 students

| | |
|---|---|
| 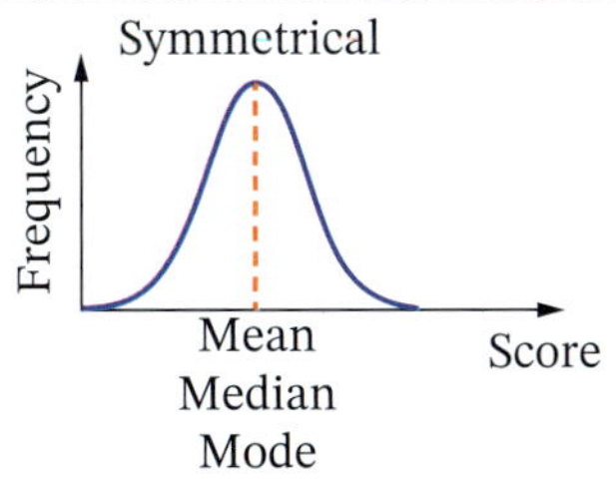  |  |

**Positively-skewed distribution:** Higher values skew the mean up, more data below the mean

**Example:** House prices in a town

| | |
|---|---|
| 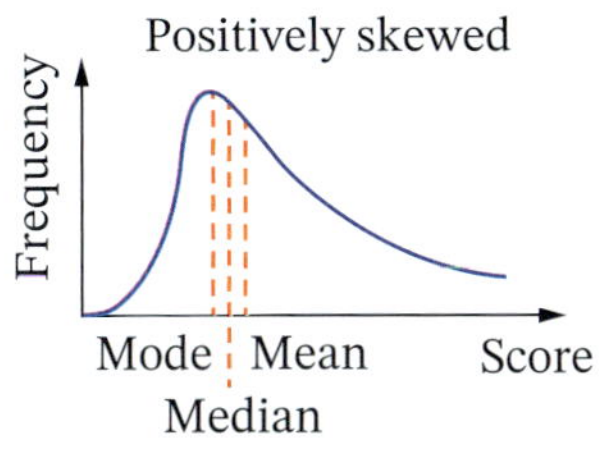  Tail points to the right (higher values) | 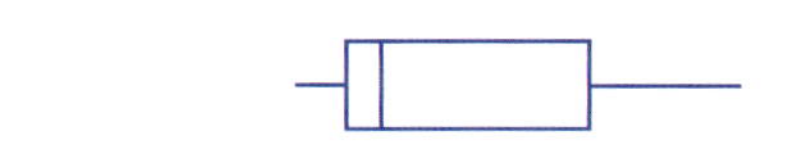 Right half longer (more data concentrated on the left) |

**Negatively-skewed distribution:** Lower values skew the mean down, more data above the mean

**Example:** Age of retirement of a person

| | |
|---|---|
| 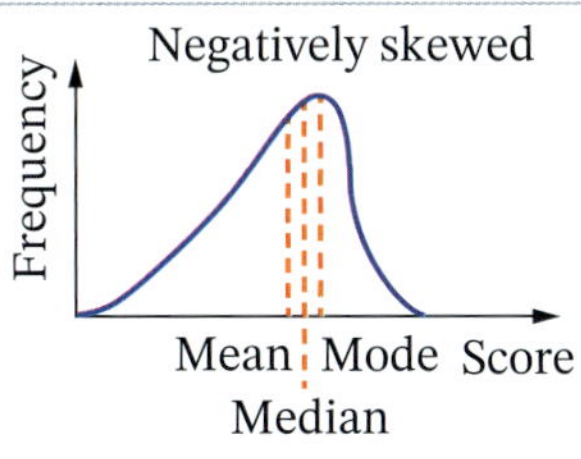  Tail points to the left (lower values) | Left half longer (more data concentrated on the right) |

**EXERCISE 10.08** Answers on p. 487

### The shape of a distribution

**1** This dot plot shows the judges' scores in a diving competition. Which of the following statements is true about the distribution? Select **A**, **B**, **C** or **D**.

**A** The data is positively skewed with a cluster around 6 to 8.

**B** The data is symmetrical with no modes.

**C** The data is negatively skewed with one mode.

**D** The data is positively skewed with a cluster around 0 to 4.

Foundation Mastery Complex

2 For each distribution:

i describe its shape  ii state the modality  iii identify any clusters.

**a**

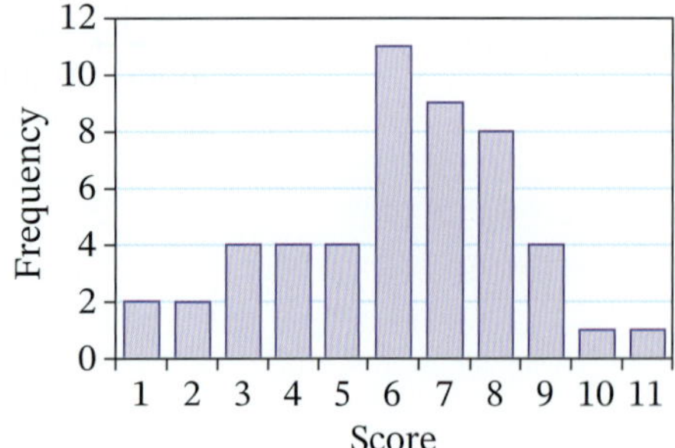

**b**

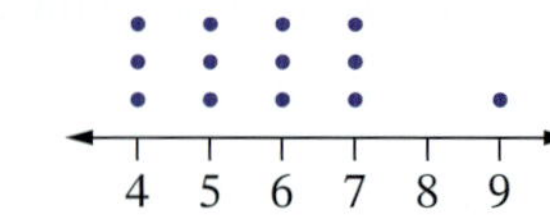


**c**

| Stem | Leaf |
|---|---|
| 1 | 3 4 6 6 6 7 8 9 9 |
| 2 | 0 7 |
| 3 | 1 2 2 5 7 8 8 9 |
| 4 | 0 2 3 |
| 5 | 2 9 |

**d**

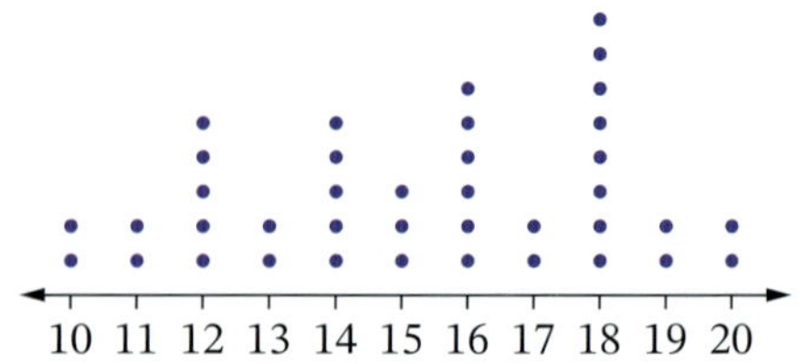


**e**

| Stem | Leaf |
|---|---|
| 4 | 1 3 |
| 5 | 5 5 6 |
| 6 | 0 3 5 5 6 8 |
| 7 | 2 6 |
| 8 | 5 5 8 |

**f**

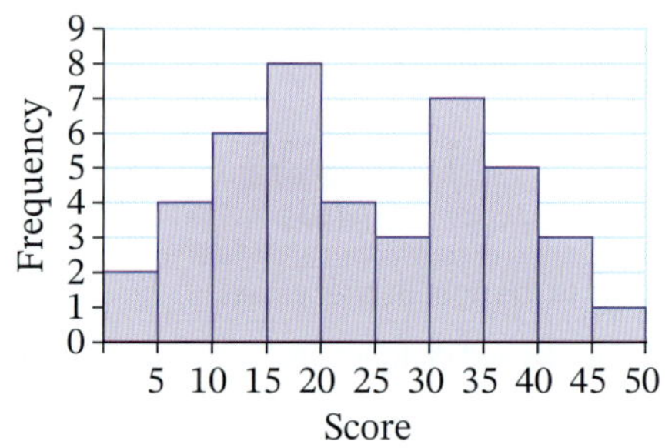


3 This stem-and-leaf plot shows the number of mobile phones sold in January across various OzTel stores in Australia.

**a** How many OzTel stores were surveyed?

**b** Describe the shape of the data.

**c** Where does the clustering occur?

**d** What is the mode?

| Stem | Leaf |
|---|---|
| 2 | 2 6 |
| 3 | 0 1 |
| 4 | 4 8 |
| 5 | 2 6 9 |
| 6 | 1 3 4 5 |
| 7 | 0 2 3 4 4 5 5 7 7 7 8 9 |
| 8 | 3 5 7 7 8 8 8 8 9 |
| 9 | 2 8 |

4 The number of visits to the **InstaChat** website was recorded between 1200 (noon) and 2100 (9 pm) one day.

| **Hour** | 1201–1300 | 1301–1400 | 1401–1500 | 1501–1600 | 1601–1700 | 1701–1800 | 1801–1900 | 1901–2000 | 2001–2100 |
|---|---|---|---|---|---|---|---|---|---|
| **Hits** | 1300 | 800 | 400 | 2100 | 2500 | 4500 | 3900 | 5300 | 2300 |

**a** Draw a histogram to represent this data.

**b** Comment on the shape of your histogram, also referring to modality and clusters.

**c** Suggest a possible reason for the skewness of this data.

☐ Foundation ○ Mastery ○ Complex

**5** Which statement is true about the data sets below? Select **A**, **B**, **C** or **D**.

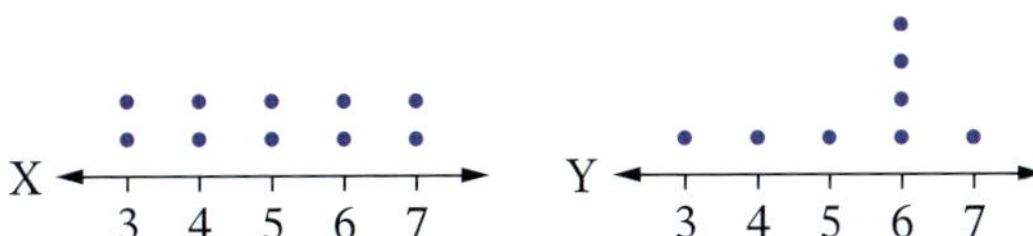


**A** Y is positively skewed.

**B** X does not have a mode.

**C** The mean of Y is 5.

**D** X and Y are both uniform.

**6** These are the ages of employees at the Berry Good Biscuit factory.

16 36 15 16 15 19 55 59 18 20 50 22 21 35 22 19 15 17 43 49

**a** Draw a stem-and-leaf plot for this data.

**b** Comment on the shape of the distribution, mentioning skewness, peaks and clusters.

**7** This dot plot represents the number of accidents per month at a factory over a year.

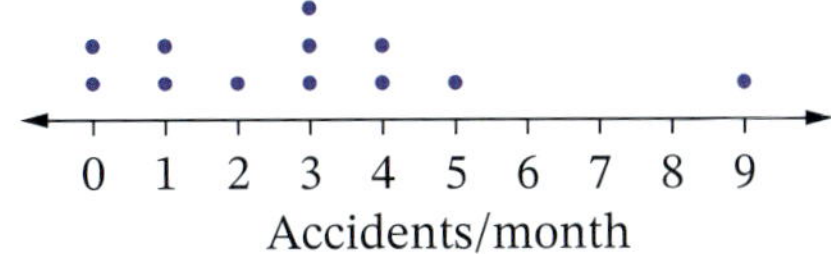


**a** Comment on the shape of the dot plot.

**b** What is the mode?

**c** Calculate the mean (correct to one decimal place) and compare it to the mode.

**8** The results of a Year 12 Maths exam are shown on the parallel box plot.

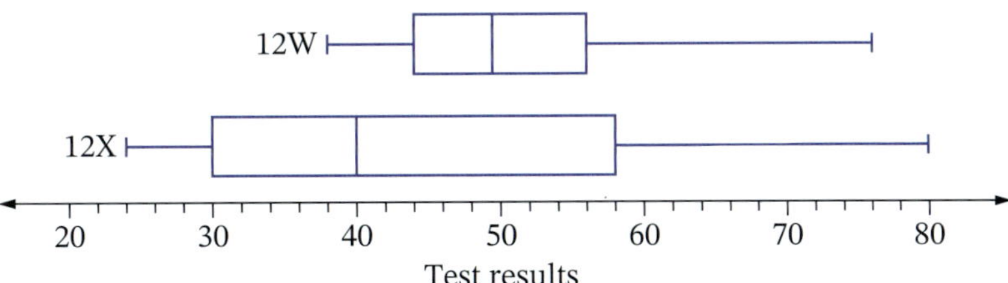


**a** What is the median result for each class?

**b** Find the range and interquartile range for each class.

**c** Describe the shape of the results for each class.

**d** Which class had the better test results? Give reasons.

□ Foundation ○ Mastery ○ Complex

# 10.09 Comparing data sets

**Videos**
Comparing data sets
Box plots and outliers

**Worksheets**
Comparing group measures
Comparing sports scores
Comparing city temperatures
Comparing word lengths
Investigating young drivers

## Example 20

The daily maximum temperatures for Sydney and Brisbane for December are shown below.

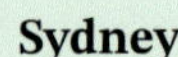


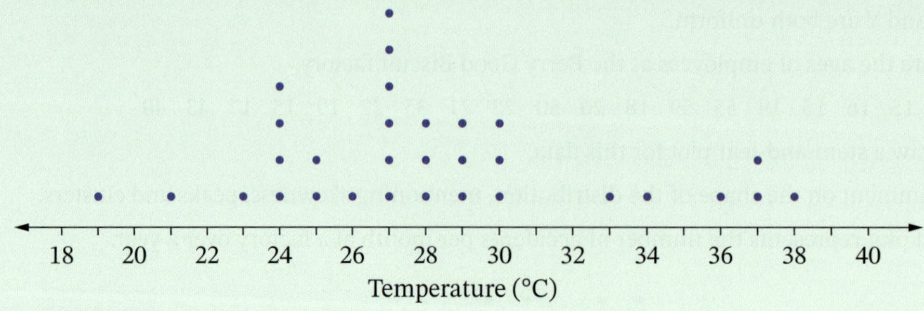


**Brisbane**

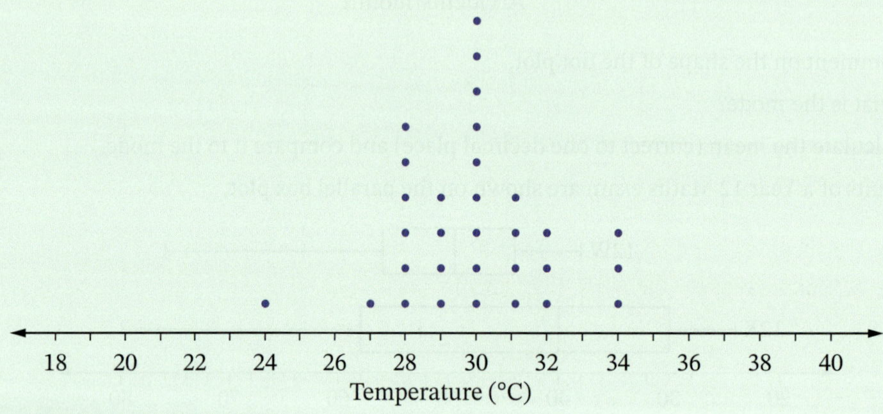


**a** Find the mean, the median and modal temperatures for each city.

**b** Find the range, interquartile range and standard deviation for each city.

**c** Describe the shape of the distribution of temperatures for each city and identify any outliers and clusters.

**d** Compare the temperatures in Sydney and Brisbane. Comment on measures of centre and spread.

### Solution

**a** Sydney: mean = 28.2°C
median = 27°C
mode = 27°C

Brisbane: mean = 29.9°C
median = 30°C
mode = 30°C

**b** Sydney: range = 38° − 19°
= 19°
$\text{IQR} = Q_3 - Q_1$
$= 30 - 25$
$= 5$
standard deviation = 4.6

Brisbane: range = 34° − 24°
= 10°
$\text{IQR} = Q_3 - Q_1$
$= 31 - 28$
$= 3$
standard deviation = 2.1

**c** Sydney's distribution of temperatures is positively skewed and 38°C is just an outlier:
$(Q_3 + 1.5 \times \text{IQR} = 30 + 1.5 \times 5 = 37.5)$.
Brisbane's temperatures are fairly symmetrical with no outliers.
Sydney's temperatures are bimodal, with peaks at 24°C and at 27°C, and are clustered at 27–30°C.
Brisbane's temperatures are also bimodal, with peaks at 28°C and at 30°C and are clustered at 30°C.

**d** Brisbane is the warmer city as shown by the mean, median and mode, which are 2–3°C above those of Sydney.
The spread of Sydney's temperatures is significantly greater than Brisbane's, as shown by larger values of the range, interquartile range and standard deviation. Sydney also had the lowest and the highest temperatures in December.

## Example 21

This back-to-back stem-and-leaf plot compares the amounts spent (in dollars) by a group of children visiting the Royal Easter show.

| Boys | | Girls |
|---:|:---:|:---|
| 2 1 | 3 | 1 2 3 8 9 |
| 3 0 | 4 | 0 2 4 4 6 7 8 |
| 8 6 6 5 4 4 2 0 | 5 | 4 |
| 6 6 6 5 5 3 | 6 | 9 |

**a** What is the modal amount spent for the whole group?

**b** What is the mean amount, to the nearest dollar, spent by:

**i** boys? **ii** girls? **iii** the whole group?

**c** Did boys or girls have the greater range of amounts spent?

**d** Compare the shapes of the 2 data sets (boys and girls).

**e** Who, on average, spent more – boys or girls? Justify your answer.

### Solution

**a** \$66

**b** **i** mean for boys $= \dfrac{\$972}{18}$
$= \$54$

**ii** mean for girls $= \dfrac{\$607}{14}$
$= \$43.357\,14\ldots$
$\approx \$43$

**iii** mean for group $= \dfrac{\$1579}{32}$
$= \$49.343\,775$
$\approx \$49$

**c** range for boys = \$66 − \$31 = \$35
range for girls = \$69 − \$31 = \$38
Girls had the greater range.

**d** The boys' spending was negatively skewed, while the girls' spending was positively skewed.

**e** The boys spent more, as shown by the greater mean of the boys, and in the shape of the stem-and-leaf plot.

## Example 22

Liz and Gaurav deliver parcels to addresses. The number of parcels delivered per day is shown.

| Liz | 24 | 25 | 26 | 27 | 28 | 28 | 31 | 32 | 32 | 32 | 35 | 35 |
|---|---|---|---|---|---|---|---|---|---|---|---|---|
| Gaurav | 15 | 18 | 21 | 24 | 25 | 29 | 31 | 31 | 32 | 38 | 38 | 45 |

**a** Find the five-number summary for each data set and draw a parallel box plot for the data.

**b** What is the interquartile range of the number of parcels delivered by each person?

**c** What is the median number of parcels delivered by each person?

**d** Compare the 2 data sets, describing their shapes and measures of centre and spread.

**e** Is Liz is a better worker than Gaurav? Give reasons for your answer.

### Solution

**a**

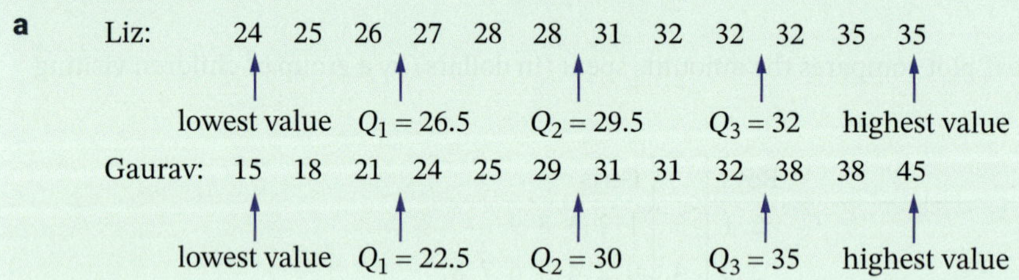


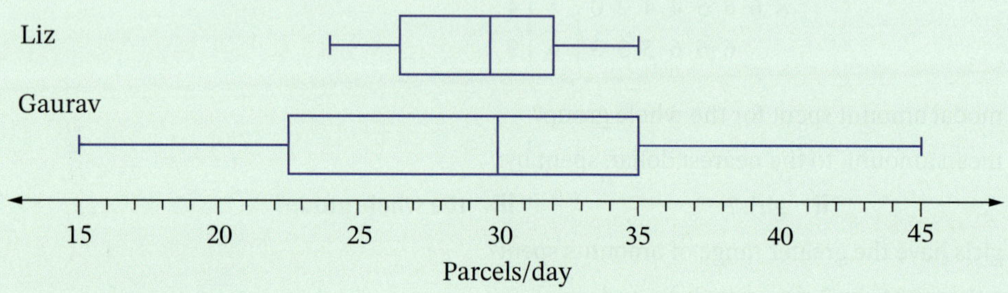


**b** interquartile range for Liz = 32 − 26.5 = 5.5

interquartile range for Gaurav = 35 − 22.5 = 12.5

**c** median for Liz = 29.5

median for Gaurav = 30

**d** Both distributions are fairly symmetrical and have similar medians. The data set for Liz has a much smaller range and interquartile range, so its spread is generally smaller.

**e** Liz is not necessarily the better worker because the medians of both data sets were similar. Liz is, however, more consistent, as there is less spread in her delivery rate.

**EXERCISE 10.09** Answers on p. 487

## Comparing data sets

**1** The dot plots show the reading scores for a group of kindergarten students at the start and finish of the year.

Start of kindergarten

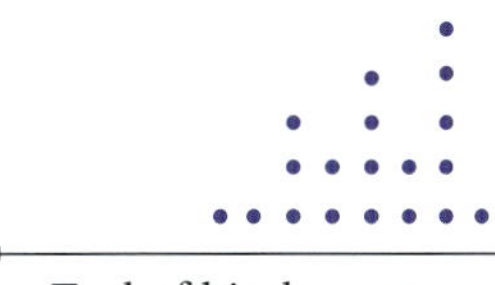

End of kindergarten

Which statement is true about the reading scores of the kindergarten students at the end of the year? Select **A**, **B**, **C** or **D**.

**A** The mean has increased and the standard deviation has decreased.

**B** The mean and standard deviation have both increased.

**C** The mean and standard deviation have both decreased.

**D** The mean has decreased and the standard deviation has increased.

**2** A class was given 2 different Science tests and their marks are displayed in the back-to-back stem-and-leaf plot.

EXAMPLE 21

**a** Find the range, interquartile range and median for each test.

**b** Which Science test was easier? Justify your answer.

| Test A | | Test B |
|---|---|---|
| 2 | 0 | 3 |
| 5 | 1 | |
| 5 | 2 | |
| | 3 | 8 |
| 8 | 4 | 5 |
| | 5 | 2 4 7 |
| 5 | 6 | 2 |
| 7 | 7 | |
| 0 | 8 | |
| 2 | 9 | 4 |

**3** The weights (in kilograms) of 2 sport teams were recorded.

| Netball players | 77 | 72 | 80 | 77 | 62 | 72 | 62 | 79 | 58 | 75 | | |
|---|---|---|---|---|---|---|---|---|---|---|---|---|
| Hockey players | 81 | 86 | 64 | 74 | 92 | 75 | 73 | 81 | 64 | 52 | 82 | 79 |

**a** Display the data in a back-to-back stem-and-leaf plot.

**b** Comment on the shape of each set of data, mentioning skewness, clusters and peaks.

**c** What is the mean weight of each team?

**d** What is the median weight of each team?

**e** Which is the better measure of centre, the mean or median? State a reason for your answer.

sergio_kumer/iStock

4 What is the shape of the data shown in this box-and-whisker plot? Select **A**, **B**, **C** or **D**.

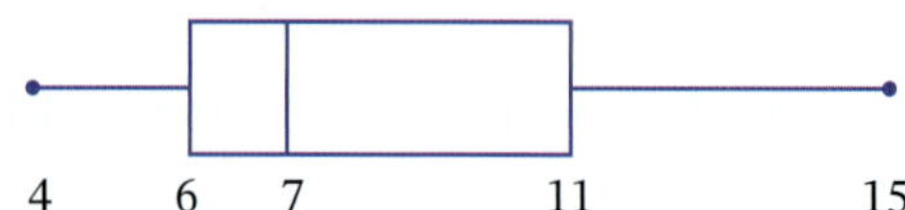


**A** uniform **B** positively skewed **C** negatively skewed **D** multimodal

EXAMPLE 22

5 Yashneel decided to investigate his theory that 'People always underestimate the length of a piece of string'. He asked students to estimate the lengths of several pieces of string, measure the actual lengths and display the results in box-and-whisker plots.

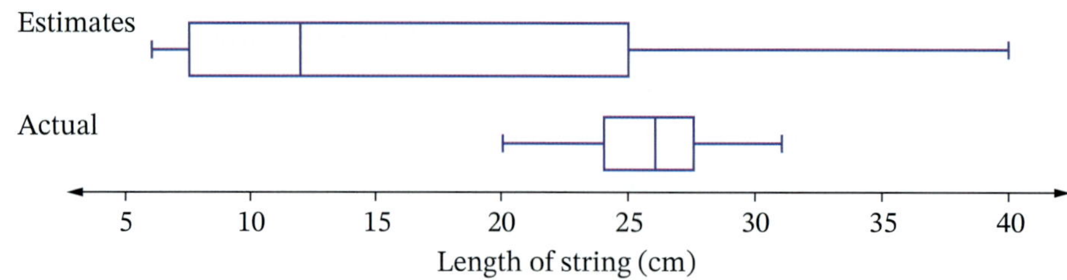


**a** Write down the median of:

**i** the estimated lengths **ii** the actual lengths.

**b** Compare the spreads of each data set.

**c** Do you agree with Yashneel's theory? Justify your answer.

6 **a** Use the data from the back-to-back stem-and-leaf plot in Example 21 on page 414 to create 2 five-number summaries and draw their box-and-whisker plots.

**b** What information is seen more easily in the box plots than in the stem-and-leaf plots?

7 This parallel box plot show the distribution of petrol prices in 6 Australian capital cities.

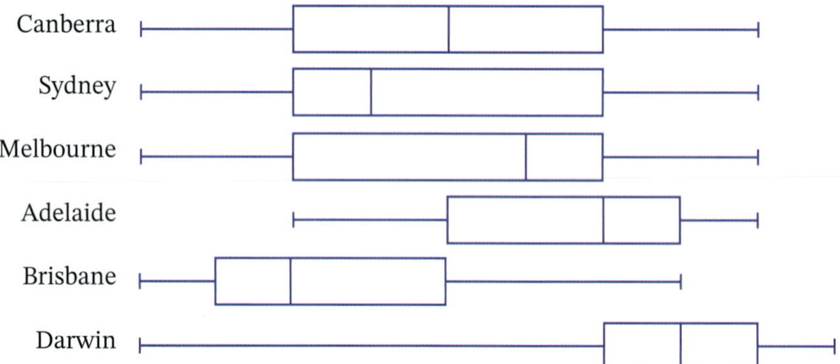


**a** Which city has:

**i** the greatest range of prices?

**ii** the lowest median price?

**iii** the smallest interquartile range of prices?

**iv** a symmetrical distribution of prices?

**b** Is petrol cheaper in Sydney than Melbourne? How can you tell?

☐ Foundation ◯ Mastery ◯ Complex

**8** A group of 20 people had their pulse rates taken before and after an exercise class and the results displayed. Two outlier pulse rates (shown as +) were excluded from the 'After exercise' data.

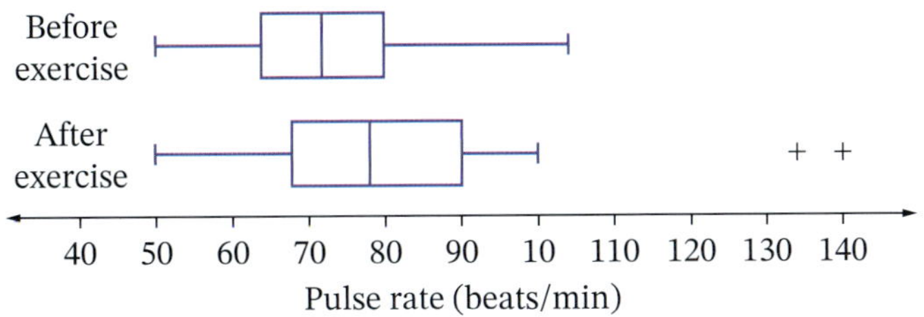


**a** How many people had a pulse rate between 64 and 72 before the exercise class?

**b** The lowest value before and after the class did not change. Give a possible reason for this.

**c** Give a possible reason for the 2 outlier pulse rates.

**d** What was the interquartile range of pulse rates after the class?

**e** By how much did the median pulse rate increase?

**9** A group of 12 people took part in the QUIT smoking program. The number of cigarettes smoked per day was recorded before the start of the program and 6 weeks later in the following table.

| **Before** | 21 | 10 | 36 | 16 | 32 | 42 | 9 | 14 | 21 | 18 | 34 | 45 |
|---|---|---|---|---|---|---|---|---|---|---|---|---|
| **6 weeks later** | 6 | 24 | 31 | 21 | 16 | 19 | 16 | 18 | 28 | 32 | 8 | 13 |

**a** Find the five-number summary for each data set and draw a parallel box plot.

**b** What is the interquartile range of each data set?

**c** Is the QUIT program generally working for this group of people? Justify your answer.

**10** This table shows the average number of rainy days per month for Sydney and Melbourne.

| **Month** | J | F | M | A | M | J | J | A | S | O | N | D |
|---|---|---|---|---|---|---|---|---|---|---|---|---|
| **Sydney** | 12 | 12 | 13 | 12 | 12 | 12 | 10 | 10 | 10 | 12 | 11 | 12 |
| **Melbourne** | 8 | 7 | 9 | 12 | 14 | 14 | 15 | 16 | 15 | 14 | 12 | 11 |

**a** Use a back-to-back stem-and-leaf plot to display the data.

**b** Draw a parallel box plot representing the data.

**c** Write down one fact that is clearly shown in each display.

**d** Does Melbourne generally have more rainy days than Sydney? Justify your answer.

**11** This back-to-back stem-and-leaf plot compares the ages of the people in 2 tour groups.

EXAMPLE 20

| Australians | | New Zealanders |
|---:|:---:|:---|
| 8 | 3 | 4 |
| 9 7 7 | 4 | 5 8 |
| 9 7 7 6 6 3 | 5 | 2 7 9 |
| 9 8 7 4 | 6 | 3 4 4 6 7 |
| 7 3 2 | 7 | 1 1 2 3 4 6 8 |
| 5 3 | 8 | 2 4 |

**a** Find the mean, median and mode for each group.

**b** Find the range, interquartile range and standard deviation of the ages of each group.

**c** Describe the shape of the distribution of ages of each group.

**d** Compare the ages for both groups and determine which group was older, commenting on shape, measures of centre and measures of spread.

Foundation Mastery Complex

**12** A Year 11 Biology class was asked to estimate their test results before completing the test. The estimates and actual test results are shown below.

| | | | | | | | | | | |
|---|---|---|---|---|---|---|---|---|---|---|
| **Estimates** | 87 | 80 | 83 | 65 | 82 | 82 | 92 | 73 | 82 | 89 |
| | 93 | 77 | 70 | 65 | 85 | 33 | 87 | 77 | 78 | 75 |
| | 88 | 89 | 86 | 58 | | | | | | |
| **Test results** | 80 | 73 | 86 | 52 | 91 | 91 | 72 | 64 | 91 | 87 |
| | 79 | 46 | 78 | 85 | 82 | 32 | 87 | 73 | 79 | 86 |
| | 95 | 79 | 49 | 73 | | | | | | |

**a** Display the data in a back-to-back stem-and-leaf plot.

**b** Comment on the shape of each set of data, mentioning skewness, modality and clusters.

**c** For each group of results, find:

- **i** the mean
- **ii** the median
- **iii** the mode.

**d** For each data set, find:

- **i** the range
- **ii** the interquartile range
- **iii** the standard deviation.

**e** Compare the 2 sets of results. Did the students overestimate their results? Justify your answer.

## Technology

### Box plots using GeoGebra

1. To create a box plot, click **View** and **Spreadsheet View**.
   Right-click in the **Graphics View** and choose **GRID**.
2. Use the rainfall data from Question **10** on page 417.
   Click on cell A1 in the spreadsheet view (right side of your screen).
3. Type: **Boxplot[]**
   Note the use of square brackets (outer brackets) and the 'curly' brackets inside. The entry below represents the following formula:
   **Boxplot[width of box, y-scale, {data set}]**
   Enter the Sydney rainfall data in cell A1 to create the box plot.
4. Find the five-number summary for this data set.

☐ Foundation ○ Mastery ○ Complex

## Investigation

### Self-serve supermarket checkouts

One of the main roles of a statistician is to critically analyse related data sets and report on the findings. Businesses often use the results of an analysis for promotional purposes and companies report to their shareholders.

To critically analyse data sets, you need to draw suitable displays, find measures of centre and spread and write a report on the relationship between the data sets, mentioning skewness, clusters, peaks and outliers as well as noting any patterns or unusual features. From this report you can draw conclusions and make recommendations.

At a supermarket, the times (in seconds) to check a basket of 20 grocery items at 15 self-serve and 15 cashier-assisted checkouts were recorded.

| **Self-serve** | 45 | 58 | 63 | 43 | 75 | 69 | 84 | 65 | 96 | 73 | 90 | 61 | 84 | 72 | 96 |
|---|---|---|---|---|---|---|---|---|---|---|---|---|---|---|---|
| **Cashier** | 95 | 105 | 82 | 110 | 125 | 148 | 136 | 137 | 86 | 99 | 145 | 119 | 101 | 97 | 124 |

Critically analyse the data and report back to the manager of the store on any benefits of self-serve checkouts.

## Sample HSC problem Answers on p. 488

(7 marks) The ages, in years, of a sample of patients at a hospital are shown in the stem-and-leaf plot.

| Stem | Leaf |
|---|---|
| 1 | 2 2 3 4 6 |
| 2 | 1 2 |
| 3 | 0 0 0 3 |
| 4 | 4 7 8 |
| 5 | 1 1 |
| 7 | 5 7 8 |
| 8 | 1 |

**a** Find the mean age of the patients. — 1 mark

**b** Find the median age of the patients. — 1 mark

**c** Is the mean or median more appropriate for describing the average age of the patients? Give a reason for your answer. — 2 marks

**d** Find the interquartile range of the patients' ages. — 1 mark

**e** Represent this data set on a box plot. — 2 marks

## Study Tip

### More exam tips

- Read all instructions carefully. Don't rush.
- Calculate the average time you should spend on each question.
- If one answer is taking too long, stop and ask yourself: 'Am I on the wrong track?' Don't get bogged down. You may need to retrace your steps, start again or come back later.
- When you have finished a question, make sure you have actually answered it. Do you need to write the answer in a sentence? Put a circle or box around the answer to highlight it.
- Make sure that your answer is reasonable, especially if it involves money or measurement. Did you round the answer correctly and include the correct units?
- If you have some time left at the end of the exam, double check your answers, especially for the more difficult or uncertain questions.

# CHAPTER SUMMARY

**Worksheet** Statistics review

**Puzzle** Statistics crossword

This chapter, *Analysing data*, examined the statistical measures of centre (mean, median, mode) and spread (range, interquartile range, standard deviation). You should be competent at making statistical calculations on sets of numerical data, including those represented in frequency tables, class intervals (grouped data), dot plots, box plots and stem-and-leaf plots.

Before you move on, consider what you learned in this chapter and revisit any sections that may have been unclear.

| What you learned in this chapter... | Section | |
|---|---|---|
| Use the statistical investigation process | 10.01 | Statistical investigations and surveys |
| Design a survey or questionnaire that is clear and free from bias | 10.01 | Statistical investigations and surveys |
| Calculate and interpret the mean, median and mode of sets of data, including ungrouped data | 10.02 | The mean, median and mode |
| Calculate and interpret the range, interquartile range and standard deviation of sets of data, including ungrouped data | 10.03<br>10.04 | The range and interquartile range<br>Standard deviation |
| Identify outliers in a set of data and examine their effects on statistical measures | 10.05 | The effect of outliers |
| Calculate cumulative frequency and construct cumulative frequency histograms and polygons | 10.06 | Cumulative frequency graphs |
| Use a cumulative frequency polygon to find the median, quartiles and interquartile range of a data set | 10.06 | Cumulative frequency graphs |
| Use a five-number summary to construct box plots | 10.07 | Box plots |
| Describe the shape of a distribution using its graph or display | 10.08 | The shape of a distribution |
| Compare data sets displayed on back-to-back stem-and-leaf plots and parallel box plots | 10.09 | Comparing data sets |

To help master this topic, make a summary mind map. Use the chapter outline and the mind map below as a guide. Add your own words, symbols, diagrams, boxes and reminders. The summary should give you a 'whole picture' view of the topic and allow you to identify any weak areas to revisit in your revision.

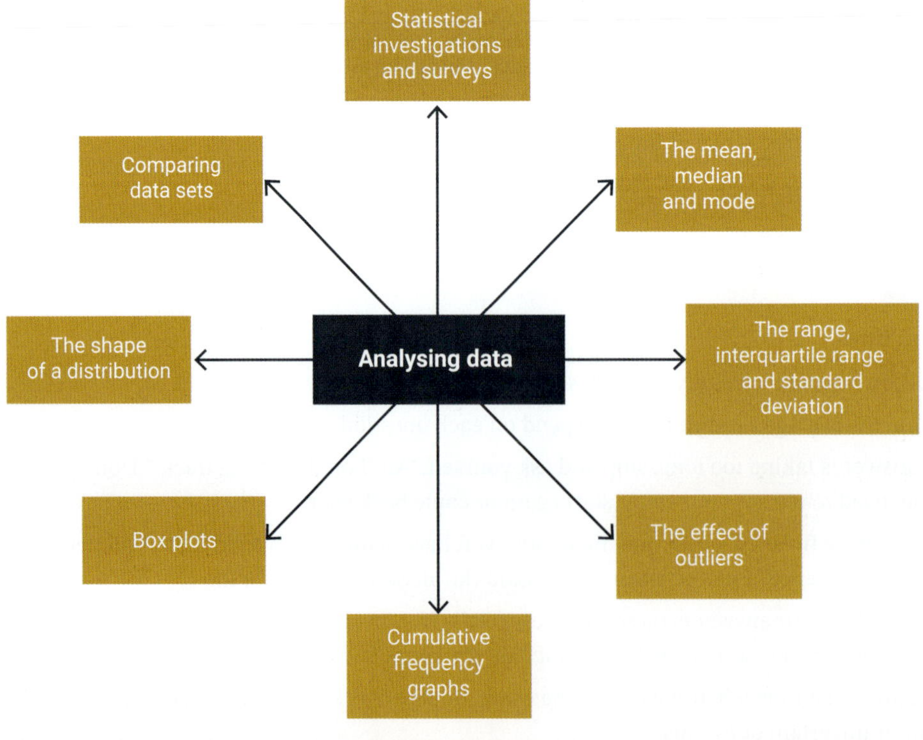

# Test yourself

Answers on p. 488

10

**1** **a** Name one feature of an effective questionnaire.  10.01

**b** Design 5 questions for a questionnaire about students' attitudes to school sport choices.

**2** Explain what is wrong with each survey question and write a better question.  10.01

**a** 'How often do you visit a doctor?'

**b** 'What did you like or dislike about the film?'

**c** 'Is it time that Australia grew up and became a republic?'

**3** The heights (in centimetres) of a group of ballet dancers are:  10.02

| | | | | | | | |
|---|---|---|---|---|---|---|---|
| 165 | 183 | 170 | 168 | 175 | 179 | 168 | 170 |
| 181 | 168 | 172 | 177 | 171 | 170 | 175 | 179 |

**a** Calculate the mean, correct to one decimal place.

**b** Find the median height.

**c** What is the mode?

**4** Motor vehicles were clocked, by police radar, travelling at the following speeds (in km/h):  10.02

| | | | | | | | | | |
|---|---|---|---|---|---|---|---|---|---|
| 78 | 95 | 64 | 77 | 81 | 84 | 77 | 89 | 90 | 78 |
| 79 | 80 | 82 | 84 | 80 | 79 | 95 | 86 | 84 | 70 |
| 78 | 65 | 82 | 91 | 89 | 60 | 85 | 81 | 78 | 68 |
| 90 | 84 | 69 | 70 | 80 | 91 | 85 | 84 | 80 | 76 |
| 68 | 65 | 85 | 76 | 79 | 83 | 82 | 91 | 84 | 80 |

**a** Sort the data in a frequency table using classes of 60–<70, 70–<80, and so on, and include a column of class centres.

**b** Calculate an estimate for the mean speed.

**c** Find the median class of speeds.

**d** What is the modal class?

**5** The dot plot represents the sum of 2 dice rolled 20 times. 10.02

Find the mean, median and mode of this data.

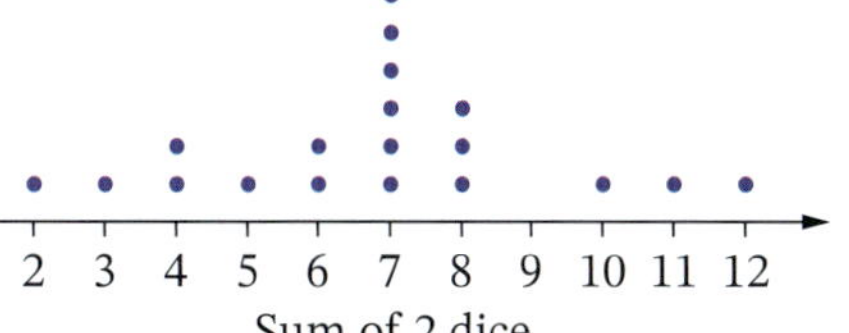


**6** The house prices realised at auction one Saturday in Vincentia were: 10.02

| | | | |
|---|---|---|---|
| \$1 242 000 | \$1 185 000 | \$ 952 000 | \$2 080 000 |
| \$1 305 000 | \$1 015 000 | \$1 280 000 | \$1 340 000 |

**a** Calculate the mean price.

**b** Calculate the median price.

**c** Is the mean or the median the better measure to use as the average price of the houses? Why?

**7** Which measure of centre is most appropriate for describing each average below? Give a reason for each answer.  10.02

**a** The average men's shoe size

**b** The average height of Year 11 students

**c** The average starting salary of an Australian worker

Foundation Mastery Complex

10.02 **8** A grouped data frequency table is shown.

What is the mean? Select **A, B, C** or **D.**

**A** 24.1 **B** 25.3

**C** 26.1 **D** 28.1

| Class interval | Frequency |
|---|---|
| 11–15 | 4 |
| 16–20 | 7 |
| 21–25 | 12 |
| 26–30 | 24 |
| 31–35 | 15 |

10.03 **9** **a** What is the meaning of 'interquartile range'?

**b** A random sample of 15 packets of corn chips had the following masses in grams. Find the range and interquartile range of these masses.

52 51 50 49 50 50 48 51

51 50 49 53 50 49 51

10.03 **10** The stem-and-leaf plot on the right represents the number of points per match scored by the Tigers in a football season. For this data, find:

**a** the range

**b** the interquartile range.

| Stem | Leaf |
|---|---|
| 0 | 6 6 |
| 1 | 2 3 4 4 4 8 8 9 |
| 2 | 0 0 0 5 6 |
| 3 | 0 0 2 4 4 6 7 |
| 4 | 0 |
| 5 | |
| 6 | 2 |

10.04 **11** For quality testing, a manufacturer takes a random sample of 10 screws, each designed to have a length of 2 cm. The actual lengths of the screws, in centimetres, are:

2.00 1.99 1.98 2.01 2.01 1.97 2.03 1.98 2.01 2.00

**a** Find the mean screw length.

**b** Find the standard deviation, correct to 2 decimal places.

10.04, 10.06 **12** Students were surveyed about the number of pairs of shoes they owned, and the results are shown in the table.

| Pairs of shoes | Frequency |
|---|---|
| 5 | 8 |
| 6 | 11 |
| 7 | 10 |
| 8 | 6 |
| 9 | 5 |

**a** Calculate, correct to one decimal place, the mean and the standard deviation of this data.

**b** Copy the table, adding a cumulative frequency column. Then draw a cumulative frequency histogram and polygon.

**c** Use your polygon to calculate:

**i** the median

**ii** the interquartile range.

10.05 **13** In a supermarket, 8 employees earn the following wages per week.

$1026 $874 $950 $950 $980 $1140 $1216 $1710

Is the wage of $1710 an outlier for this set of data? Justify your answer with a calculation.

10.05 **14** Consider this set of data:

4 7 8 8 12 15 19 20

**a** What is the effect on the mean and median if an outlier of 40 is added to this data set?

**b** Is the mean or median a better measure of centre when there is an outlier in the data set?

□ Foundation ○ Mastery ○ Complex

**15** The cumulative frequency graph shows the results of an assignment marked out of 10. 10.06

**a** How many students completed the assignment?

**b** Use the graph to estimate:

**i** the median

**ii** the interquartile range.

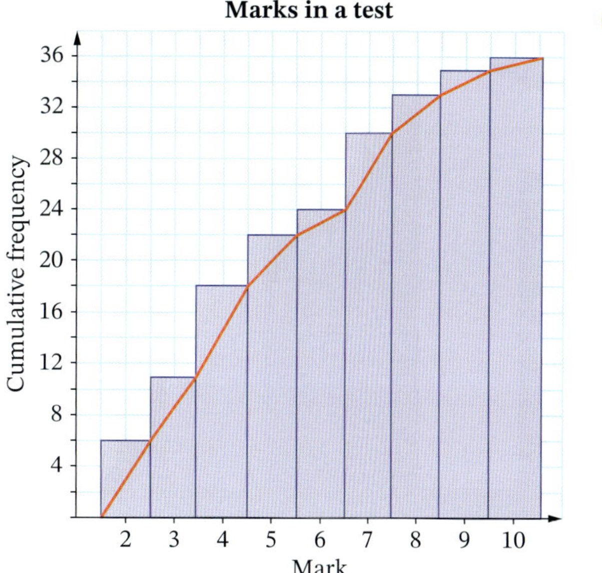


**16** This box plot represents the number of goals scored per game by a hockey team over a season. 10.07

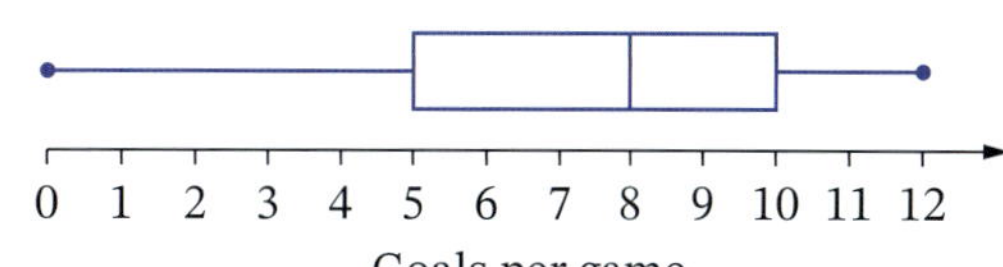


**a** What was the lowest score?

**b** Find the interquartile range.

**c** In what fraction of games were more than 8 goals scored?

**d** In what percentage of games were fewer than 5 goals scored?

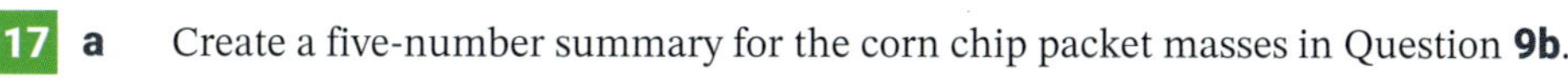

**17** **a** Create a five-number summary for the corn chip packet masses in Question **9b**.

**b** Represent the mass data on a box plot.

**18** This parallel box plot shows the distribution of marks for exams in English and History. 10.09

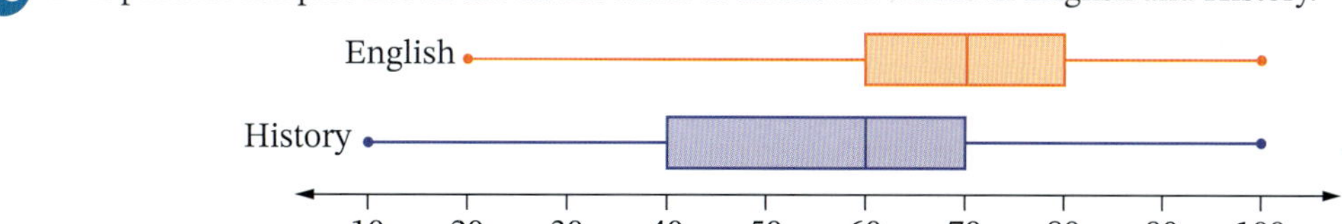


**a** Which subject has the smaller spread of marks? Give reasons.

**b** The number of students who scored 70 or less is the same for both subjects.
If 144 students did the English exam, how many students did the History exam?

**19** The results for the multiple-choice section in 2 tests taken by a Year 11 Mathematics class are shown below. 10.08, 10.09

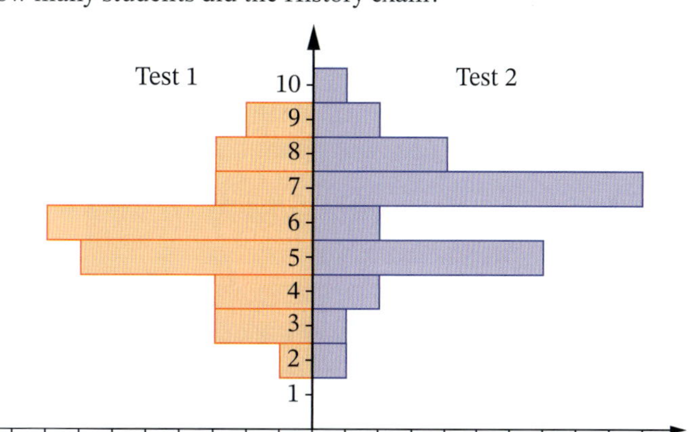


**a** Find the mean, median and mode for each test.

**b** Describe the shape of the data set for each test.

**c** For each test, find:

**i** the range

**ii** the interquartile range

**iii** the standard deviation.

**d** Are there any significant differences in the results of the 2 tests? Justify your answer by referring to the measures of centre and spread of the tests.

Foundation Mastery Complex

# 11.

# WORLD LOCATIONS AND TIMES

The Samoan Islands lie in the south Pacific Ocean and are made up of Samoa and American Samoa, 64 km apart. New Zealand governed Samoa until it became an independent country in 1964, and it was called Western Samoa until 1997. American Samoa is a small territory of the USA. Despite being next to each other, when it is Monday in Samoa, it is still Sunday in American Samoa. How can this be so?

## Chapter outline

Samoa

## In this chapter you will:

- locate positions on Earth using latitude and longitude
- convert between 12-hour and 24-hour time
- calculate differences between times of day
- calculate time differences between locations based on their difference in longitude, using the relation 15° longitude difference = 1-hour time difference
- calculate times in different parts of the world and within Australia using international time zones, taking daylight saving into account
- solve problems involving travelling and communicating across the world using international time zones.

## Videos (6):

**11.01** Latitude and longitude 1 • Latitude and longitude 2

**11.02** Time differences

**11.04** International time zones 1 • Time zones with daylight saving • International time zones 2

## Worksheets (12):

**SkillCheck** Assignment 11

**11.01** Map of the world • Positions on the globe • Latitude and longitude • Australian coordinates

**11.02** 12-hour and 24-hour time • 24-hour time on an analog clock • TV times • Time differences

**11.04** Table of time zones • Map of the world • Applications of time zones

## Puzzle (1):

**Chapter summary** World crossword

Nelson MindTap

To access resources above, visit **cengage.com.au/nelsonmindtap**

## Terminology

| | | |
|---|---|---|
| 12-hour time | 24-hour time | Australian Eastern Standard Time (AEST) |
| Coordinated Universal Time (UTC) | coordinates | daylight saving time |
| Equator | great circle | hemisphere |
| International Date Line (IDL) | latitude | longitude |
| meridian of longitude | parallel of latitude | prime meridian |
| small circle | time zone | |

Worksheet Assignment 11

## SkillCheck Answers on p. 489

1 a Write the coordinates of points $A$ and $B$.

b What is the length of the interval $AB$?

c Write the coordinates of point $C$ if it is 3 units up and 4 units left from $B$.

2 a Plot the points $D(3, 5)$ and $E(3, 2)$ on a number plane.

b What is the length of the interval $DE$?

3 How many hours between:

a 4 am and 1 pm?

b 9 am and 2 pm?

c 12:30 am and 5:30 pm?

d 2:30 pm and 10:30 pm?

4 How many minutes between:

a 3:00 pm and 3:17 pm?

b 6:00 am and 6:44 am?

c 10:35 am and 11:00 am?

d 7:17 pm and 8:00 pm?

5 What is the time:

a 7 hours after 7:00 pm?

b 8 hours before 2:00 pm?

c 11 h 18 min after 1:15 pm?

d 7 h 36 min before 2:20 am?

# Latitude and longitude

11.01

When describing the location of a point on a number plane or map, we use a **coordinate** system involving ordered pairs $(x, y)$.

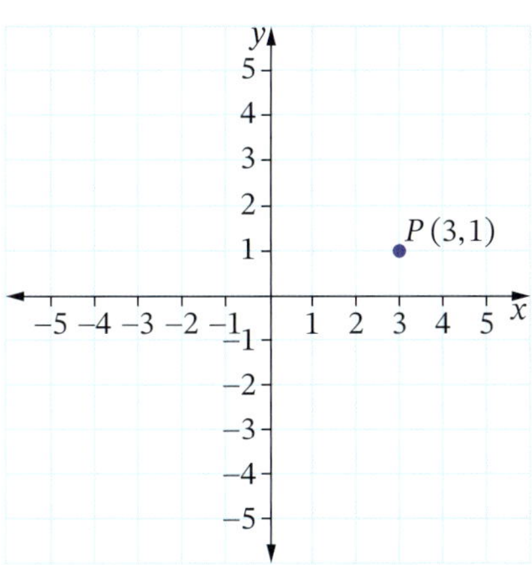


*P* has coordinates (3,1).

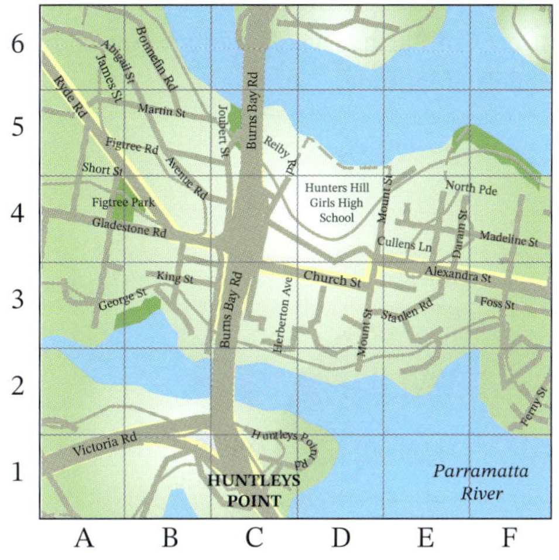


Hunters Hill Girls High School has coordinates D4.

Positions on Earth's surface are described by a coordinate system involving **latitude** and **longitude**. However, because Earth is a sphere, we use a special grid of lines that run across and down a sphere. The diagrams on the next page show this grid on a world globe and a flat world map.

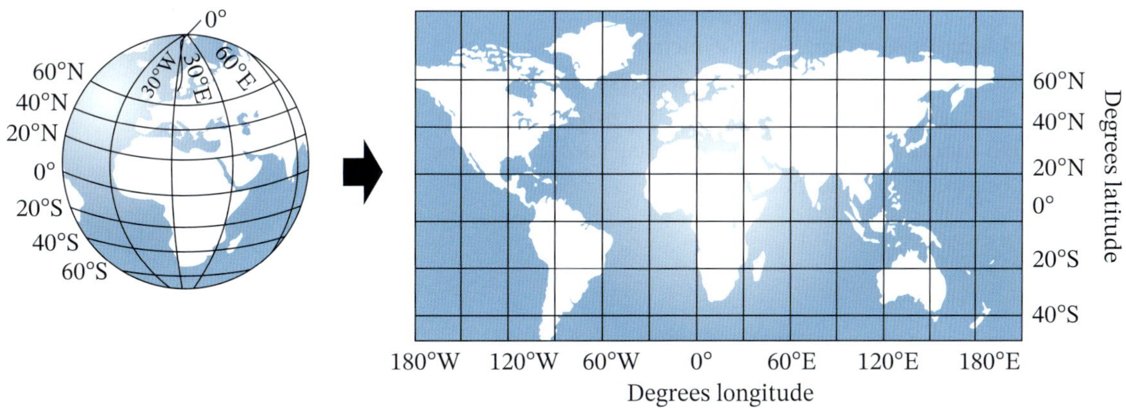


## Great and small circles

If you cut a slice through a sphere, its shape is a circle. A slice through the **centre** of a sphere is called a **great circle**, and its radius is the same as that of the sphere. Any other slice is called a **small circle**, because its radius is smaller than that of a great circle.

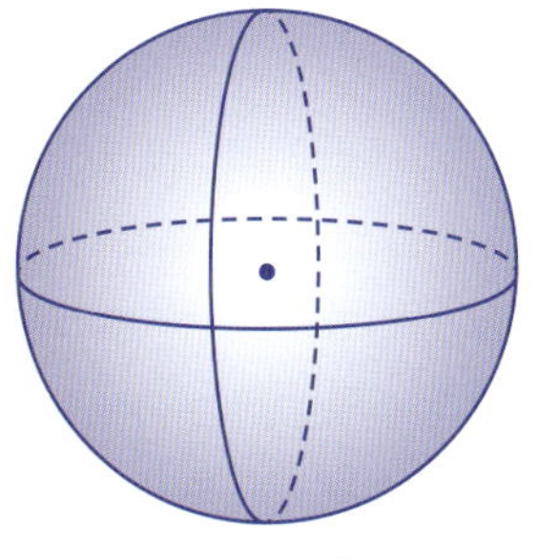

Great circles

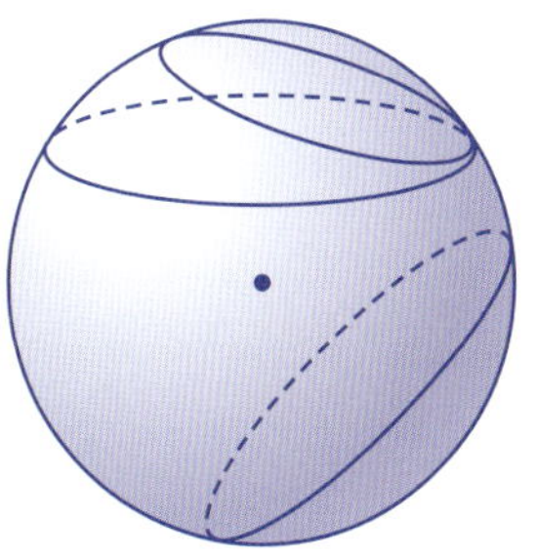

Small circles

**Worksheets**
Map of the world
Positions on the globe
Latitude and longitude
Australian coordinates

## Parallels of latitude

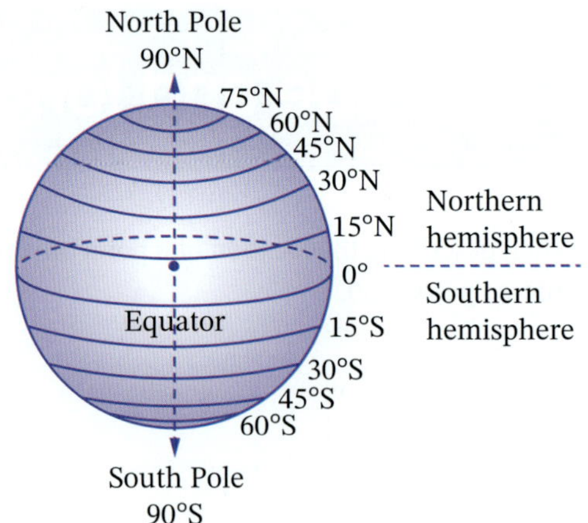


**Parallels of latitude** are imaginary parallel circles that run around Earth. The largest parallel of latitude is the **Equator**, 0°, which is a great circle. The other parallels of latitude are all small circles. The North Pole has a latitude 90° north (90°N). The South Pole has a latitude of 90°S.

The **angle of latitude** is the angle the parallel makes with the equator at the centre, *O*, of Earth. The diagrams below show the 50°N and 35°S parallels of latitude.

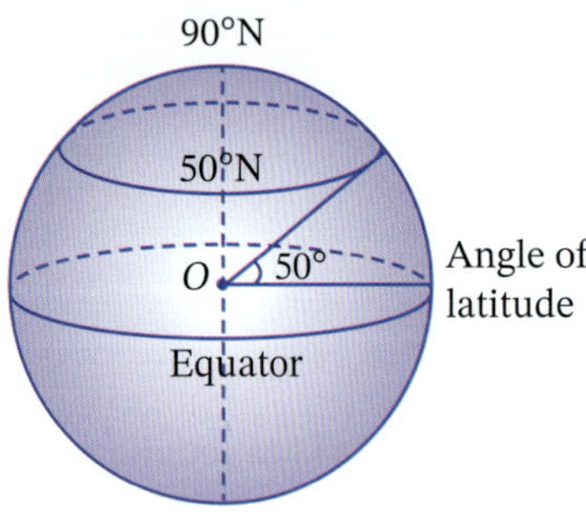


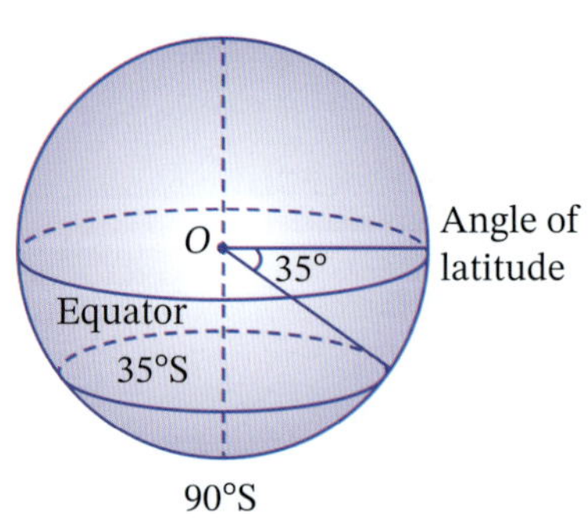


## Meridians of longitude

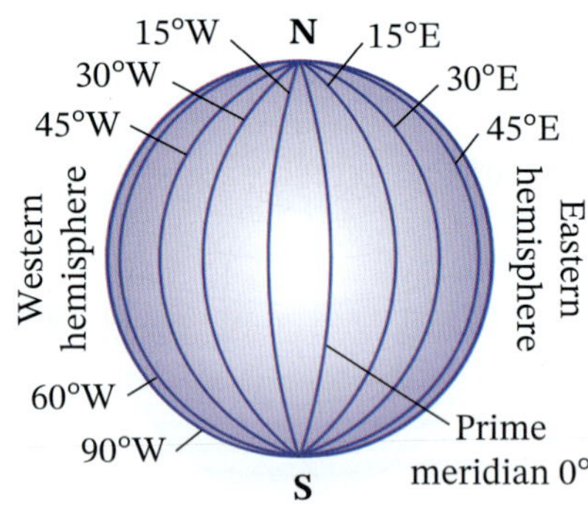


**Meridians of longitude** are imaginary semicircles that run down Earth. They are 'half' great circles that meet at the North and South poles. The main meridian of longitude is the **prime meridian** (or **Greenwich meridian**), 0°. The prime meridian runs through the Greenwich Observatory in London, UK.

Greenwich is pronounced 'gren-itch'.

The **angle of longitude** is the angle the meridian makes with the prime meridian at the centre, *O*, of Earth. The diagram shows the 35°E meridian of longitude.

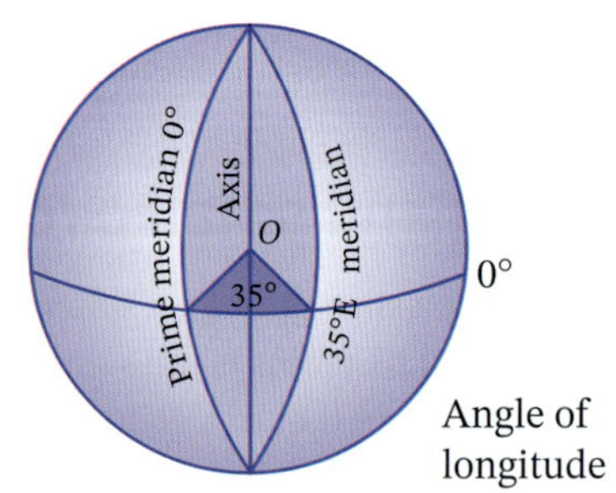


Meridians of longitude range from 180°E to 180°W. 180°E and 180°W are actually the same meridian, on the opposite side of Earth to the prime meridian. It runs through the Pacific Ocean, east of Fiji and west of Hawaii.

## Position coordinates

Locations on Earth are described using latitude (°N or °S) and longitude (°E or °W). For example, Canberra has coordinates (35°S, 149°E), meaning it is 35° south of the equator and 149° east of the prime meridian.

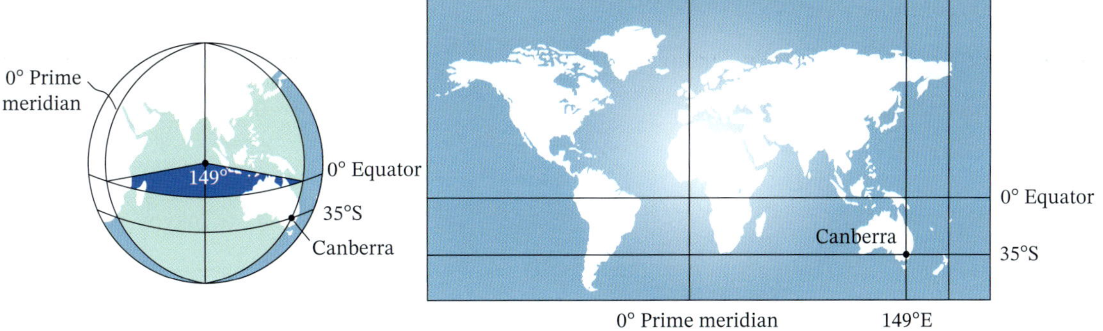


These diagrams show some mnemonics for understanding latitude and longitude.

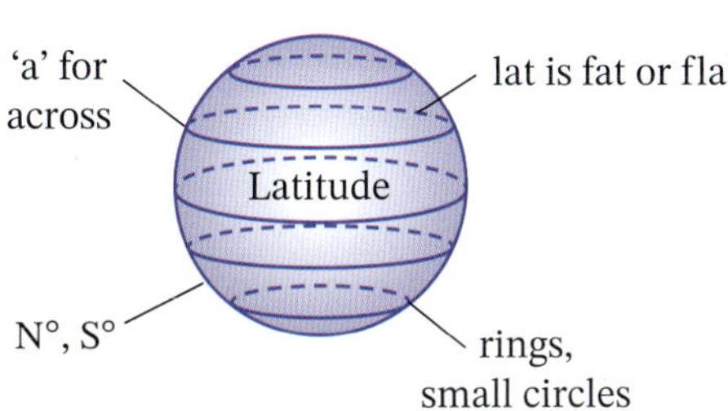


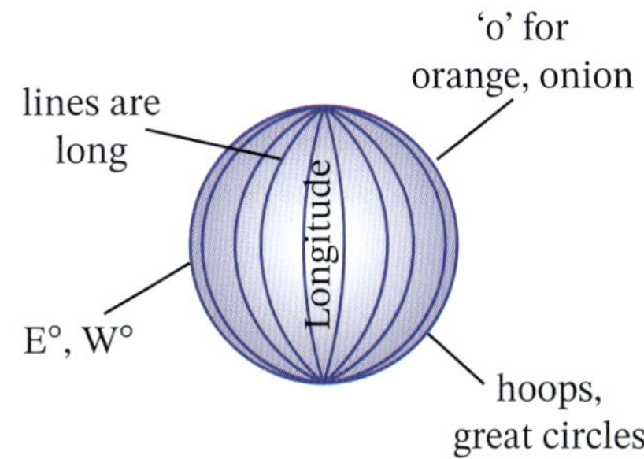


### Example 1

Match each pair of coordinates to a point on Earth.

**a** (50°S, 55°E) **b** (30°N, 55°E)

**c** (75°N, 0°E) **d** (0°, 0°)

**e** (0°, 55°E) **f** (75°N, 55°E)

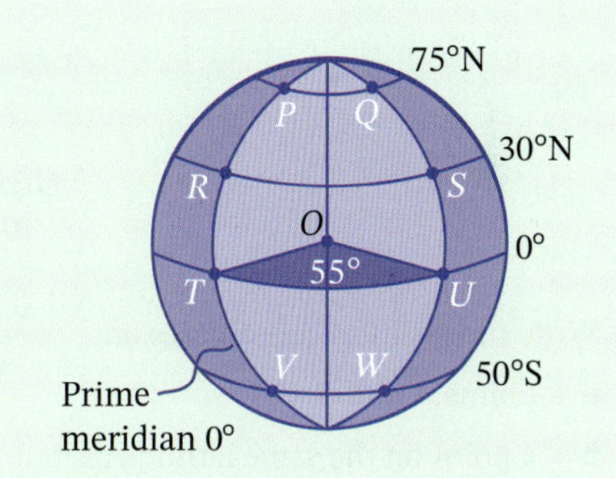

**Videos**
Latitude and longitude 1
Latitude and longitude 2

### Solution

Since $\angle TOU = 55°$, $QSUW$ is the 55°E meridian of longitude.

**a** (50°S, 55°E) is $W$ **b** (30°N, 55°E) is $S$

**c** (75°N, 0°E) is $P$ **d** (0°, 0°) is $T$

**e** (0°, 55°E) is $U$ **f** (75°N, 55°E) is $Q$

## Example 2

Sydney's coordinates are (34°S, 151°E), while Tokyo's are (35°N, 139°E).

**a** Find their differencc in latitude.

**b** Find their difference in longitude.

**c** Which city is further west?

### Solution

Draw a rough sketch to position the cities.

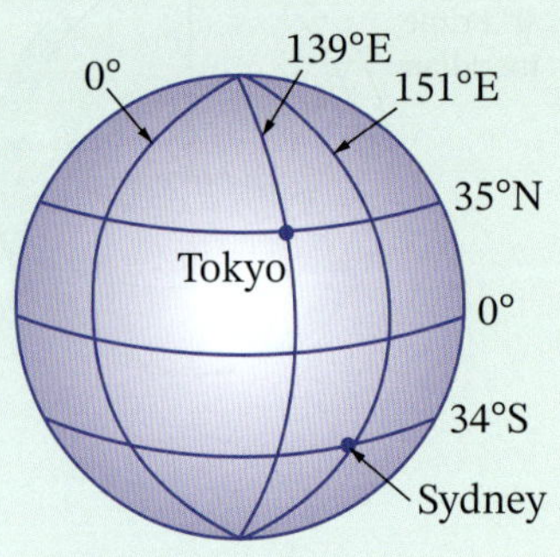


**a** Difference in latitude $= 35° + 34° = 69°$ — difference between 35°N and 34°S

**b** Difference in longitude $= 151° - 139° = 12°$ — difference between 151°E and 139°E

**c** Sydney is further east in longitude, so Tokyo is further west.

## Example 3

A diagram of Earth's surface is shown.

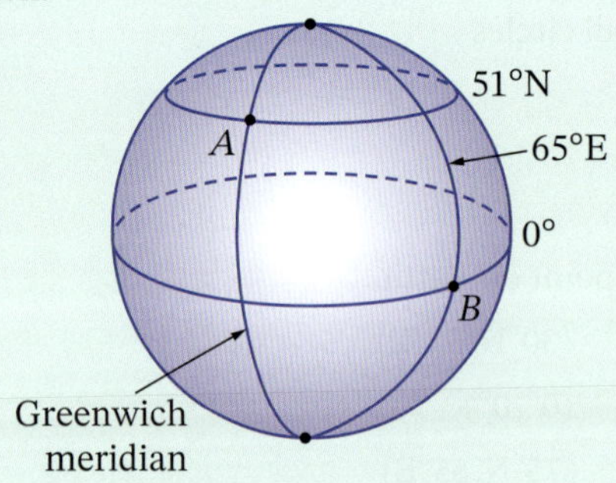


Write the position coordinates of:

**a** points $A$ and $B$.

**b** a point on the same latitude as, but 12° east of, point $A$.

**c** a point on the same longitude as, but 28° south of, point $B$.

### Solution

**a** Point $A$ is on the Greenwich meridian with longitude 0°. Its latitude is 51°N of the equator.

The coordinates of $A$ are (51°N, 0°).

Point $B$ has longitude 65°E of the Greenwich meridian. It lies on the equator, 0°.

The coordinates of $B$ are (0°, 65°E).

**b** 12° east of point $A$ (51° N, 0) is (51°N, 12°E).

**c** 28° south of point $B$ (0°, 65°E) is (28° S, 65°E).

**EXERCISE 11.01** Answers on p. 489

## Latitude and longitude

*Note*: A world map or globe may be helpful for this exercise.

EXAMPLE 1

**1** Match each pair of coordinates to a point on Earth.

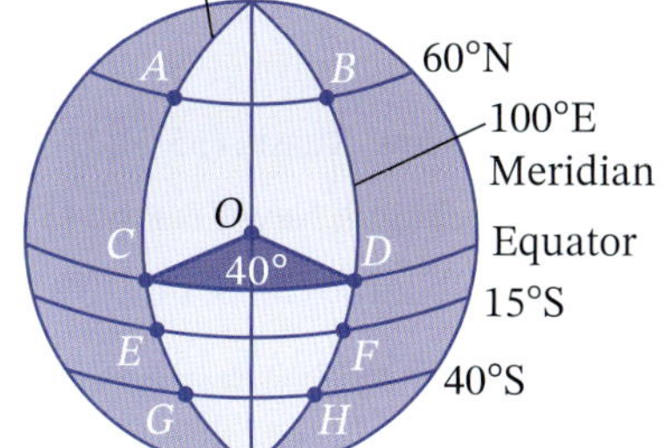


- **a** (0°, 100°E)
- **b** (40°S, 60°E)
- **c** (60°N, 60°E)
- **d** (0°, 60°E)
- **e** (15°S, 100°E)
- **f** (60°N, 100°E)
- **g** (15°S, 60°E)
- **h** (40°S, 100°E)

**2** Match each city to a letter on the map below.

- **a** Moscow (55°N, 37°E)
- **b** Edinburgh (56°N, 3°W)
- **c** Athens (38°N, 23°E)
- **d** San Francisco (38°N, 122°W)
- **e** Perth (32°S, 116°E)
- **f** Beijing (40°N, 116°E)
- **g** Alexandria (31°N, 30°E)
- **h** Tokyo (35°N, 139°E)
- **i** St Petersburg (60°N, 30°E)
- **j** Johannesburg (26°S, 28°E)
- **k** Lima (12°S, 77°W)
- **l** Hong Kong (22°N, 114°E)

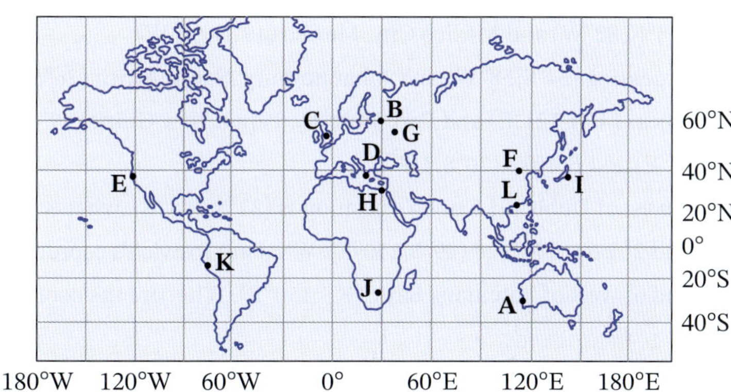


**3** Write the coordinates of each of these points on Earth.

- **a** *V*
- **b** *T*
- **c** *U*
- **d** *Z*
- **e** *R*
- **f** *W*
- **g** *X*
- **h** *S*

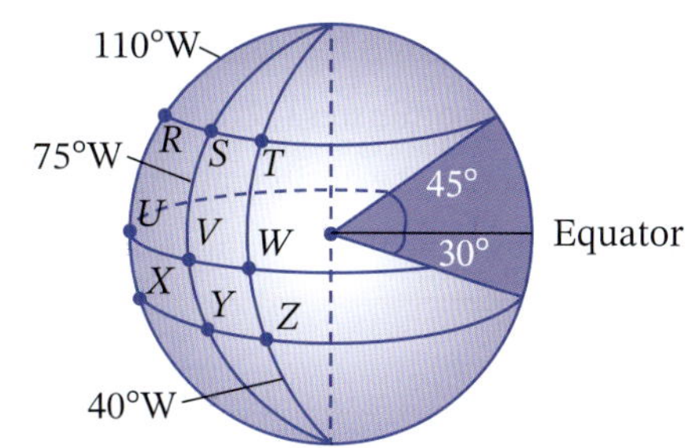


**4** What can be found at the 90°S parallel of latitude?

☐ Foundation ◯ Mastery ◯ Complex

5. Are these small circles or great circles?
   - **a** the 30°N parallel
   - **b** the prime meridian
   - **c** the 145°W meridian
6. Calculate the difference in latitude between each pair of cities, and state whether the second city listed is north or south of the first.
   - **a** Shanghai, China (31°N, 121°E) and New York, USA (40°N, 64°W)
   - **b** Nairobi, Kenya (1°S, 37°E) and Bangkok, Thailand (13°N, 100°E)
   - **c** Moscow, Russia (55°N, 37°E) and London, UK (51°N, 0°)
   - **d** Auckland, New Zealand (37°S, 174°E) and Canberra (35°S, 149°E)
   - **e** Melbourne (37°S, 145°E) and Cairo, Egypt (30°N, 31°E)
   - **f** Newcastle (33°S, 151°E) and West Wyalong (34°S, 147°E)
7. **a** Deniliquin is 1° due south of Hay. If Hay's coordinates are (34°S, 145°E), what are Deniliquin's coordinates? Select **A, B, C** or **D**.

   **A** (34°S, 144°E) **B** (35°S, 145°E) **C** (33°S, 145°E) **D** (34°S, 146°E)

   **b** Lord Howe Island is 6° due east of Forster. If Forster's coordinates are (32°S, 152°E), what are Lord Howe Island's coordinates? Select **A, B, C** or **D**.

   **A** (32°S, 146°E) **B** (38°S, 152°E) **C** (32°S, 158°E) **D** (26°S, 158°E)

8. Calculate the difference in longitude between each pair of cities, and state whether the second city listed is east or west of the first.
   - **a** Budapest, Hungary (47°N, 19°E) and Miami, USA (25°N, 80°W)
   - **b** Athens, Greece (38°N, 23°E) and Paris, France (49°N, 2°E)
   - **c** Havana, Cuba (23°N, 82°W) and Mexico City, Mexico (19°N, 99°W)
   - **d** Buenos Aires, Argentina (34°S, 58°W) and Johannesburg, South Africa (26°S, 28°E)
   - **e** Manila, Philippines (14°N, 121°E) and Port Moresby, Papua New Guinea (9°S, 147°E)
   - **f** Finley (35°S, 145°E) and Bourke (30°S, 146°E)
9. Broken Hill is 7° due west of Dubbo. If Dubbo's position is (32°S, 148°E), what is Broken Hill's position?

EXAMPLE 3

10. Ipswich is 2° north and 2° east of Moree (29°S, 150°E). What are Ipswich's coordinates?
11. Ballarat is 2° south and 6° west of Batemans Bay (35°S, 150°E). What are the coordinates of Ballarat? Select **A, B, C** or **D**.

    **A** (33°S, 144°E) **B** (37°S, 156°E) **C** (33°S, 156°E) **D** (37°S, 144°E)
12. **a** In Australia, do Sydney and Tamworth lie on the same line of latitude or longitude?

    **b** Do Athens, Greece, and Sicily, Italy, lie on the same line of latitude or longitude?
13. Where on Earth is it possible to stand on every meridian of longitude?
14. Perth has coordinates (32°S, 115°E) and Bali has coordinates (8°S, 115°E).
    - **a** Describe the positional relationship on Earth's surface between these 2 cities.
    - **b** Determine the coordinates of a point halfway between these cities.

## Investigation

### International coordinates

Use a website or app to locate the positions of different places around the world. Search 'GPS coordinates' or 'latitude/longitude finder' to find software that will give the coordinates of any city you type in; for example, **Google Earth** lists the coordinates at the bottom of the screen.

**1** Use a world map or globe to estimate the coordinates of each city.

**a** Washington DC, USA **b** Singapore **c** Adelaide, Australia

**d** Dublin, Ireland **e** Lisbon, Portugal **f** Moscow, Russia

**2** Find the actual coordinates of the cities from Question **1**, and then list them in order from:

**a** west to east **b** north to south.

**3** Which city is located at:

**a** (19°N, 155°W)? **b** (14°N, 121°E)? **c** (34°S, 58°W)?

**d** (42°N, 12°E)? **e** (12°S, 131°E)? **f** (42°N, 87°W)?

**4** How does a flat map distort the positions and sizes of places with high latitudes?

**5** Investigate where these regions are in the world and how they got their names.

**a** the Middle East **b** the Orient **c** Western world

**d** the Antipodes **e** Ecuador **f** the West Indies

## Did you know?

### The Mercator projection

Representing Earth's spherical surface on a flat rectangular map is a complex task. Gerardus Mercator (1512–1594) was a Flemish geographer and cartographer who created the Mercator projection in 1569, a world map where the land masses of the world globe are projected onto a flat surface. This breakthrough innovation revolutionised marine navigation and continues to be used in nautical charts today.

**Why was the Mercator map projection Why was the Mercator map projection such an important development and so widely used today? Do an internet search on its important features and uses.**

**What other useful map projections also exist? Research 2 other map projections and their uses, advantages and disadvantages.**

# 11.02 Time

**Worksheets**
12-hour and 24-hour time
24-hour time on an analog clock
TV times
Time differences

## Time notation

- **12-hour time:** using am and pm and the hours 1 to 12
- **24-hour time:** using 4 digits from 0000 to 2359 with no am/pm

With **24-hour time,** the hours are numbered from 0 (for midnight) to 12 (for midday) to 23 (for 11 pm).

| 24-hour time | 12-hour time |
|---|---|
| 0000 | 12 am (midnight) |
| 0100 | 1 am |
| 0200 | 2 am |
| 0300 | 3 am |
| 0400 | 4 am |
| 0500 | 5 am |
| 0600 | 6 am |
| 0700 | 7 am |
| 0800 | 8 am |
| 0900 | 9 am |
| 1000 | 10 am |
| 1100 | 11 am |

| 24-hour time | 12-hour time |
|---|---|
| 1200 | 12 pm (midday) |
| 1300 | 1 pm |
| 1400 | 2 pm |
| 1500 | 3 pm |
| 1600 | 4 pm |
| 1700 | 5 pm |
| 1800 | 6 pm |
| 1900 | 7 pm |
| 2000 | 8 pm |
| 2100 | 9 pm |
| 2200 | 10 pm |
| 2300 | 11 pm |

## Example 4

Convert each 24-hour time to 12-hour time.

**a** 1020 **b** 2335 **c** 0048

### Solution

We know that 1200 is midday, when am becomes pm.

**a** 10 is less than 12, so it is am time.

1020 is 10:20 am

**b** 23 is greater than 12, so it is pm time.

$23 - 12 = 11$

2335 is 11:35 pm

**c** 00 is the midnight hour

0048 is 12:48 am

## Example 5

Convert each 12-hour time to 24-hour time.

**a** 7:18 am **b** 2:50 pm **c** 12:33 am **d** 12:04 pm

### Solution

**a** For am times from 1 am onwards, write the time with 4 digits, starting with 0 if needed.

7:18 am = 0718

**b** For pm times from 1 pm onwards, add 12 to the hour.

$2 + 12 = 14$

2:50 pm = 1450

**c** 12 midnight = 0000

12:33 am = 0033

**d** 12 midday = 1200

12:04 pm = 1204

## Time differences

Video
Time differences

## Example 6

What is the difference in time between 8:35 am and 3:10 pm?

### Solution

Draw a timeline showing the whole hours in between, and calculate the time differences in steps.

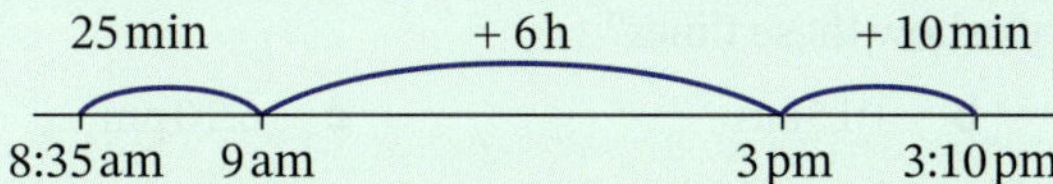


Time difference = 6 h + 25 min + 10 min

= 6 h 35 min

**Alternative method**

Convert to 24-hour time first, then use the calculator's °'" or DMS key to subtract the times:

Time difference = 3:10 pm − 8:35 am

= 1510 − 0835

= 6 h 35 min

15 °'" 10 °'" − 8 °'" 35 °'" =

## Example 7

**a** What is the time 7 hours 40 minutes after 11:52 pm?

**b** What is the time 9 hours and 30 minutes before 7:38 am?

### Solution

**a** Draw a timeline starting at 11:52 pm, then add the hours and minutes in steps.

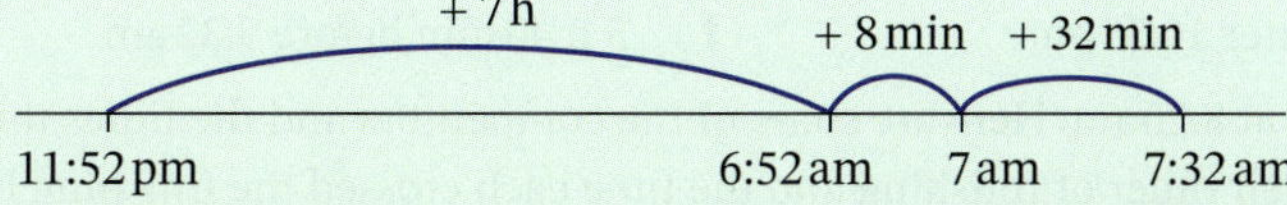


11:52 pm + 7 h = 6:52 am

Add the minutes to the next whole hour:

6:52 am + 8 min = 7 am

Add the extra minutes: 7 am + 32 min = 7:32 am

8 min + 32 min = 40 min

So 7 h 40 min after 11:52 pm is 7:32 am.

**b** Draw a timeline ending at 7:38 am, and subtract the hours and minutes in steps.

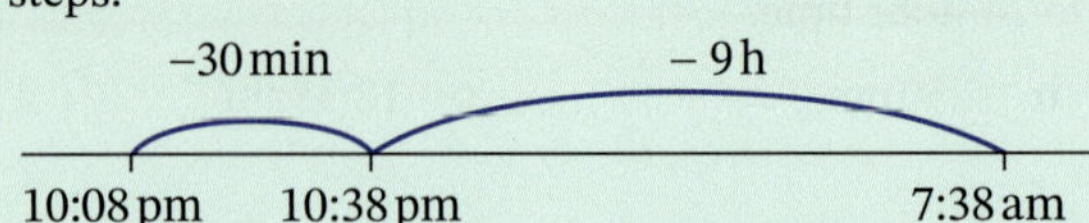


7:38 am − 9 h = 10:38 pm

Subtract 30 minutes: 10:38 pm − 30 min = 10:08 pm

So 9 h 30 min before 7:38 am is 10:08 pm.

## EXERCISE 11.02 Answers on p. 489

## Time

EXAMPLE 4

**1** Convert each 24-hour time to 12-hour time.

**a** 0845 **b** 1750 **c** 0019 **d** 1105 **e** 1628

**f** 0915 **g** 0240 **h** 2000 **i** 0625 **j** 1853

EXAMPLE 5

**2** Convert each 12-hour time to 24-hour time.

**a** 7:35 pm **b** 11:42 am **c** 11:59 pm **d** 12:17 am

**e** 12:30 pm **f** 3:40 am **g** 7:10 am **h** 9:54 pm

**i** 10:18 pm **j** 1:59 am

**3** Write the time shown on this 24-hour clock in 12-hour time.

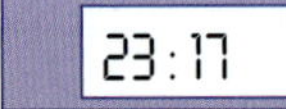


**4** How would a 24-hour clock show these times?

**a** 7:30 am **b** 1:15 am **c** 5:10 pm

**5** Write the time shown on this 12-hour clock in 24-hour time.

**6** How would a 12-hour clock show these times?

**a** 2205 **b** 0545 **c** 1421

EXAMPLE 6

**7** What is the difference in time between 11:42 am and 2:13 pm? Select **A**, **B**, **C** or **D**.

**A** 2 h 31 min **B** 3 h 55 min **C** 7 h 29 min **D** 11 h 55 min

**8** Calculate the time difference between each pair of times.

**a** 6:15 pm and 8:10 pm **b** 1116 and 1206

**c** 4:10 am and 8:55 am **d** 11:25 pm and 3:20 am

**e** 0725 and 1310 **f** 2120 and 0815

**9** A movie starts at 3:17 pm and ends at 5:09 pm. How long is the movie?

EXAMPLE 7

**10** Find the time:

**a** 6 hours after 2 pm **b** 3 hours before 10 am

**c** 20 minutes before 7:15 pm **d** 2 h 32 min after 10:45 am

**e** 3 h 29 min after 10:35 pm **f** 5 h 40 min before 9:32 am

**11** A car rally began at 8:20 am. Here are some of the competitors and the times they took. List the competitors in their order of finishing and the time each crossed the finishing line.

Tom 5:24 (5 h 24 min) Manal 5:23 Eddie 5:44

Robert 6:01 Sarah 5:59 Gianni 5:42

12 Minnie's flight lands in Singapore at 0950 for a stopover. Her next flight leaves for Florida at 1920. How long does Minnie have in Singapore airport?

13 At St Margaret's College, the school day begins at 8:25 am and ends at 3:10 pm. How long is the school day?

14 Klaas wants to go for a run for 1 h 20 min but be back in time to watch Friday Night Football on TV starting at 7:35 pm. What is the latest time he can leave for his run?

15 This is part of a bus timetable showing trips between Apollo Bay and Lorne along the Great Ocean Road in Victoria.

| | | | | |
|---|---|---|---|---|
| Apollo Bay arr | 06:05 | 09:30 | | 14.30 |
| Apollo Bay dep | 06:07 | 09:32 | | 14:32 |
| Skenes Creek | 06:10 | 09:35 | | 14:35 |
| Kennett River | 06:30 | 09:55 | | 14:55 |
| Wye River | 06:35 | 10:00 | | 15:00 |
| Lorne Hotel | 06:55 | 10:20 | 13:32 | 15:20 |
| Lorne arr | | 10:25 | | 15:25 |
| Lorne dep | 07:10 | 10:35 | 13:47 | 15:35 |

**a** How long does the trip take from Wye River to Lorne Hotel?

**b** You live in Apollo Bay and have an appointment in Lorne at 11 am. At what time do you need to catch the bus in Apollo Bay?

**c** Sia boards the bus at Skenes Creek and her trip takes 50 minutes. Where does she stop?

**d** At what time does the 2:32 pm bus from Apollo Bay get to Wye River?

**e** Where is this bus at 2:55 pm?

16 Alexis' netball team flies from Sydney to Brisbane for a tournament. Alexis visited the Nelson Air website and found the following daily flight schedule.

| Flight no. | Sydney | Brisbane |
|---|---|---|
| NA503 | 0905 | 1030 |
| NA511 | 0935 | 1100 |
| NA038 | 1005 | 1130 |
| NA114 | 1040 | 1210 |
| NA514 | 1105 | 1230 |
| NA051 | 1135 | 1300 |

**a** When does flight NA038 leave Sydney and how long is the journey?

**b** What and when is the latest flight before 11 am?

**c** The team needs to be at Sydney airport 60 minutes before a flight takes off. When should they be at the airport to catch the NA514 flight?

**d** The team needs to be at the hotel in Brisbane by 12:30 pm and it takes 35 minutes to drive from Brisbane airport to the hotel. What is the latest flight the team can catch from Sydney?

**e** Which flight takes longer to reach Brisbane than the others? Give 2 reasons why it might take longer.

☐ Foundation ○ Mastery ○ Complex

# 11.03 Longitude and time differences

When the Sun shines directly on a meridian of longitude, it is 12 midday at all places along that meridian (meridian means 'midday' in Latin). Directly on the opposite side of the world, it is 12 midnight. For example, when it is 12 midday in Western Australia on the 120°E meridian, it is 12 midnight in Chile on the 60°W meridian.

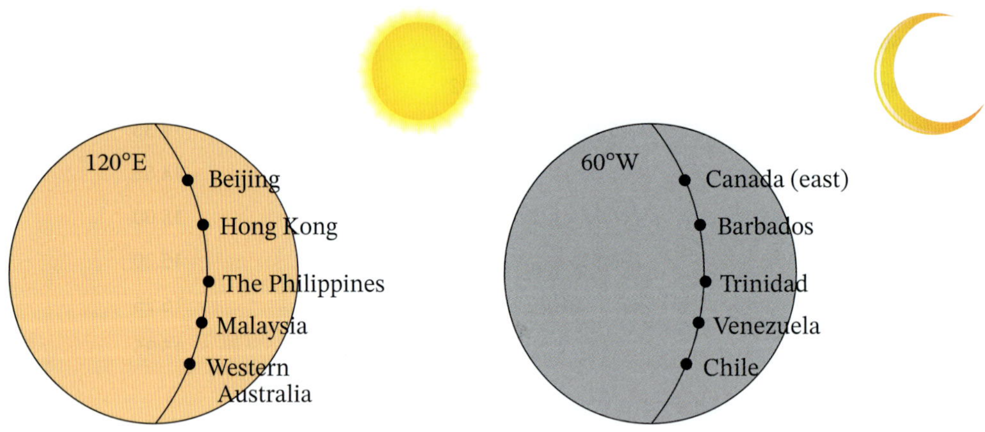


Midday on the 120°E meridian

Midnight on the 60°W meridian

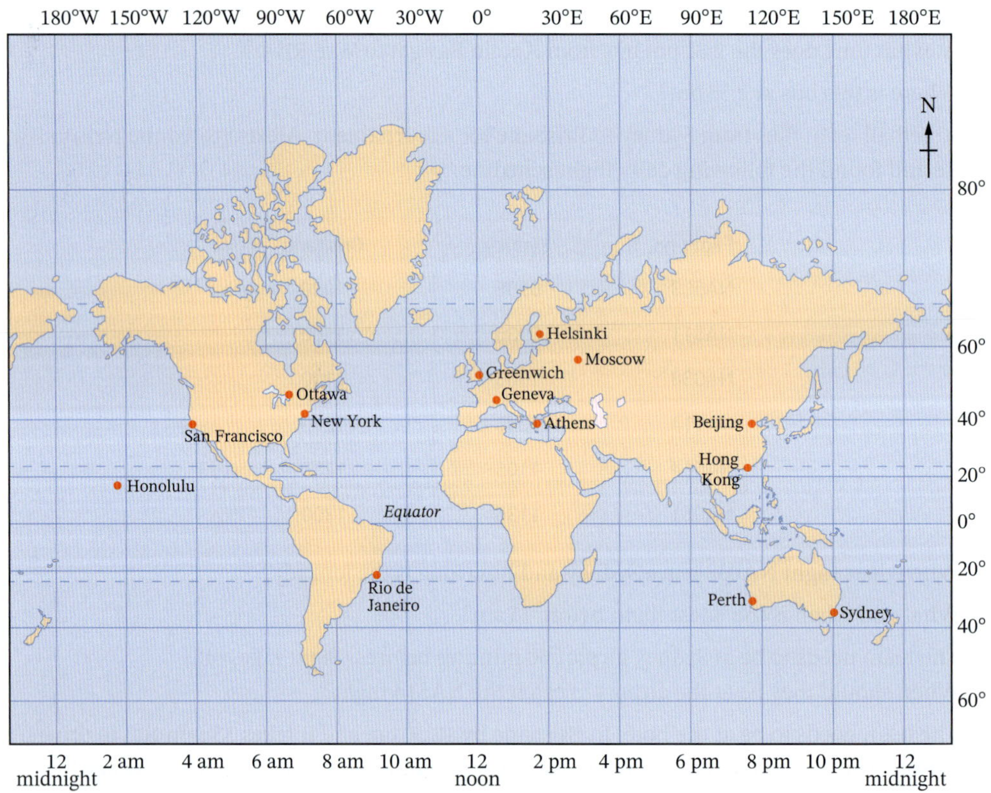


The world spans 360 degrees of longitude from 180°W to 180°E, which you would expect as there are 360° in a circle. Also, the world has 24 hourly time zones, which you would expect as there are 24 hours in one day.

So, 360° of longitude = 24 hours.

Dividing both sides by 24 will give the size of one time zone:

$360° \div 24 = 1$ hour

15° difference in longitude = 1-hour time difference

Notice on the map that 30° difference in longitude = 2 hours' time difference, so this makes sense.

## Longitude and time differences

A 15° change in longitude corresponds to a 1-hour time difference.

## Example 8

The longitude in Perth is 116°E.

**a** Explain why the time in Hong Kong, 22°N, 116°E is the same as the time in Perth.

**b** Bangkok is at 13°N, 101°E. When it is 8 pm in Perth, what time is it in Bangkok?

### Solution

**a** Hong Kong and Perth are both on the 116°E meridian of longitude.
They have the same time because they have the same longitude.

**b** First work out the longitude difference, then determine the time difference.

Difference in longitude = 116° – 101°

= 15°

15° difference in longitude represents a 1-hour time difference.

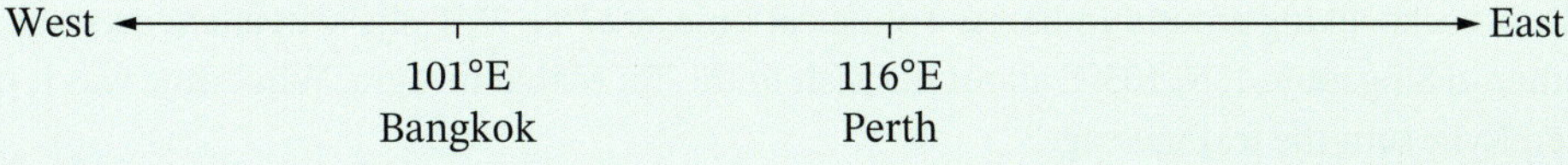


Bangkok (101° E) is west of Perth (116° E), so the time in Perth is 1 hour later than the time in Bangkok.

When it is 8 pm in Perth, it is 7 pm in Bangkok.

## Example 9

Zelko lives in Warsaw, Poland, at 52°N, 20°E. At 11 am local time, he called his sister in Adelaide at 35°S, 140°E. What time was it in Adelaide when Zelko phoned?

### Solution

Difference in longitudes = 140° – 20°

= 120°

Time difference = 120 ÷ 15

= 8 hours

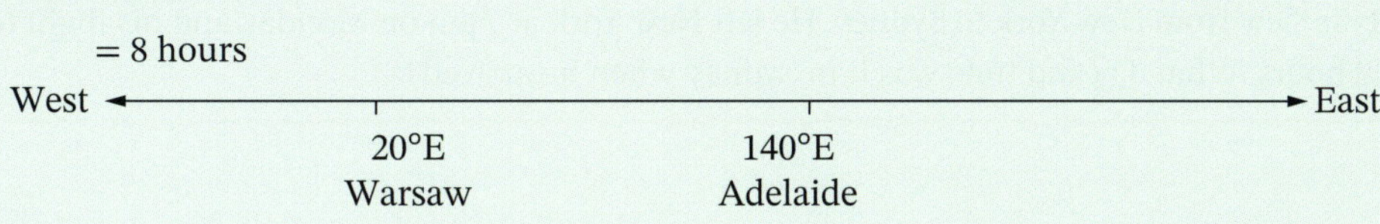


Adelaide (140°E) is east of Warsaw (20°E). The time in Adelaide is later than the time in Warsaw.

Time in Adelaide = 11 am + 8 hours

= 7 pm

When it is 11 am in Warsaw, it is 7 pm in Adelaide.

**EXERCISE 11.03** Answers on p. 489

## Longitude and time differences

1 Explain why Istanbul, 41°N, 30°E and Cairo 30°N, 30°E have the same time.

2 The difference in longitude between 2 locations is 60°. Use a calculation to show that there is a 4-hour time difference between the locations.

3 What is the time difference between 2 cities that have a 150° difference in longitude?

4 It is 10 am in London, 52°N, 0°. What time is it in each of these locations?

- **a** Ephesus, Turkey, 38°N, 30°E
- **b** Shanghai, China, 31°N, 120°E
- **c** Cairo, Egypt, 30°N, 30°E
- **d** New Orleans, USA, 30°N, 90°W
- **e** Townsville, Queensland, 19°S, 150°E
- **f** Buenos Aires, Argentina 34°S, 60°W
- **g** Texas, USA, 29°N, 90°W

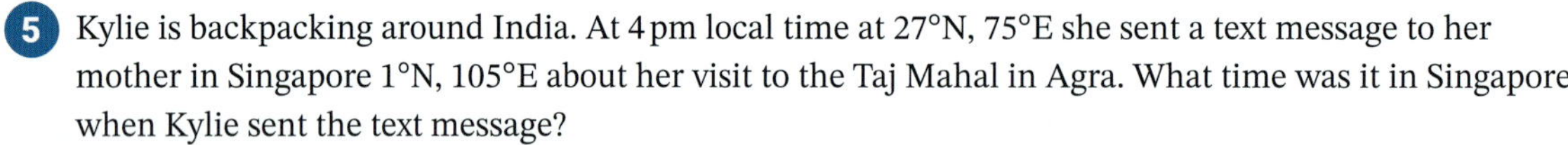

5 Kylie is backpacking around India. At 4 pm local time at 27°N, 75°E she sent a text message to her mother in Singapore 1°N, 105°E about her visit to the Taj Mahal in Agra. What time was it in Singapore when Kylie sent the text message?

6 Phillipa is working in her company's Melbourne office at 38°S, 145°E. At 3 pm she called Bob in the Hawaii office at 20°N, 155°W about a legal problem. Suggest a reason why no one answered the phone.

7 Perth is located at 32°S, 116°E and Johannesburg in South Africa is at 26°S, 26°E.

- **a** When it is 8 pm on Friday in Perth, what day and time is it in Johannesburg?
- **b** Christine caught a flight that left Perth at 8 pm on Friday and flew non-stop to Johannesburg. The flight took 11 hours. What was the day and the time when she landed in Johannesburg?

8 The first cricket test between the West Indies and South Africa started on Thursday at 10 am in Kingston, Jamaica 18°N, 72°W. Albert lives in Cape Town 33°S, 18°E and he is a keen cricket fan. What time should he switch his TV on to watch the start of the test match?

9 Sydney is at 34°S, 151°E and New York is at 41°N, 74°W.

- **a** When it is 7 pm on Monday in New York, what day and time is it in Sydney?
- **b** Ryan flew from New York to Sydney. He left New York at 7 pm on Monday and his flight took 20 hours. What day and time was it in Sydney when he arrived?

Foundation | Mastery | Complex

# International time zones 11.04

Local times around the world are measured relative to the time along the prime meridian 0°, called **Coordinated Universal Time (UTC)**, previously called **Greenwich Mean Time (GMT)** because it is measured from the Greenwich Observatory in London, UK. Places east of the prime meridian are *ahead* of UTC, while places west are *behind* UTC.

The abbreviation for Coordinated Universal Time is UTC instead of 'CUT' so it can be used internationally to cover all languages.

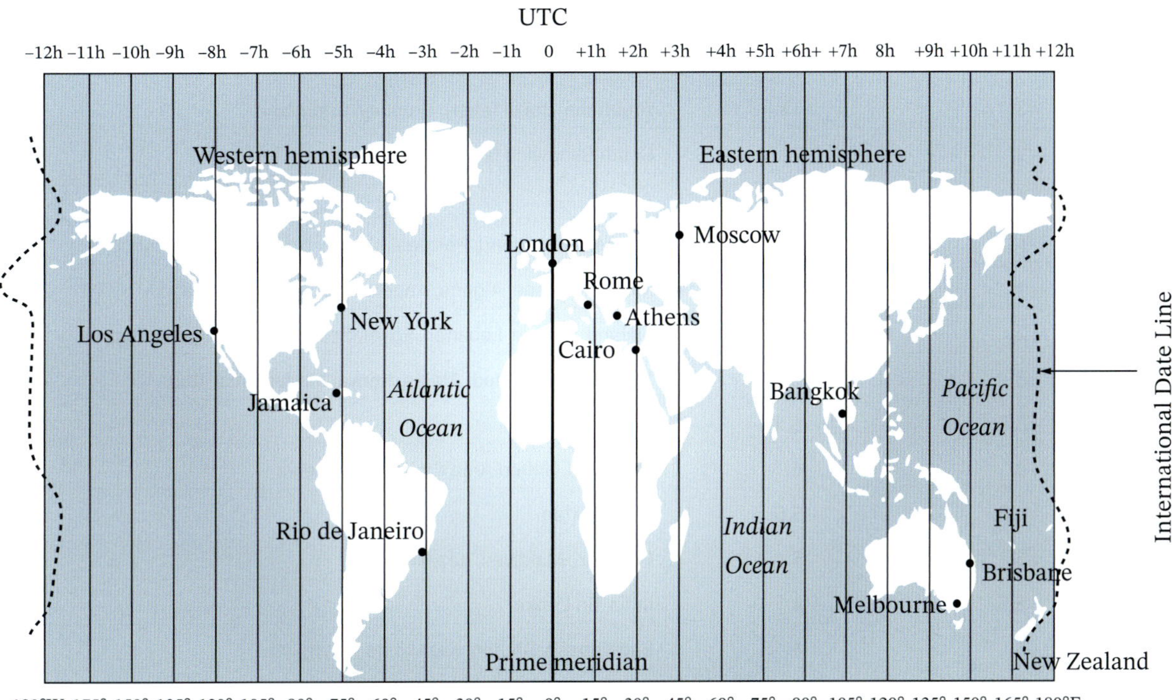


Since 1884, the world has been divided into standard **time zones**, with roughly 1 hour difference for every 15° of longitude. Some countries make variations for geographical and/or administrative reasons. For example, **Australian Central Standard Time (ACST)** is 9.5 hours (not 9 hours) ahead of UTC and China has only one time zone (UTC +8) even though it stretches across five. On the map above, places along the 105°E meridian are 7 hours ahead of UTC, while places along the 150°W meridian are 10 hours behind UTC.

This table lists the international time zones relative to UTC and **Australian Eastern Standard Time (AEST)**, the time zone for eastern Australia, covering Queensland, NSW, ACT, Victoria and Tasmania.

| Hours from UTC | Hours from AEST | Places in this zone |
|---|---|---|
| −11 | −21 | American Samoa, Niue Island, Midway Island |
| −10 | −20 | Hawaii, Cook Islands, Tahiti |
| −9 | −19 | Alaska |
| −8 | −18 | USA/Canada Pacific time (west coast), Los Angeles, Vancouver |
| −7 | −17 | USA/Canada Mountain time, Mexico (west) |
| −6 | −16 | USA/Canada Central time (mid-west), Mexico (east), El Salvador, Nicaragua, Guatemala, Honduras, Belize |
| −5 | −15 | USA/Canada Eastern time (east coast), New York, Cuba, Peru |
| −4 | −14 | Canada Atlantic time, Chile, Barbados, Brazil (west), Bolivia |
| −3 | −13 | Argentina, Brazil (east), Uruguay, Greenland |
| −2 | −12 | South Sandwich Islands |
| −1 | −11 | Azores |
| 0 | −10 | UK, Ireland, Iceland, Portugal, Ghana, Liberia, Mali, Morocco |
| +1 | −9 | Most of Europe, Algeria, Angola, Chad, Bosnia, Croatia |
| +2 | −8 | Finland, Greece, Lebanon, Egypt, South Africa, Bulgaria, Cyprus |
| +3 | −7 | Russia (west), Saudi Arabia, Kenya, Madagascar, Iraq |
| +3.5 | −6.5 | Iran |
| +4 | −6 | Mauritius, United Arab Emirates, Armenia |
| +4.5 | −5.5 | Afghanistan |
| +5 | −5 | Pakistan, Maldives, Kazakhstan |
| +5.5 | −4.5 | India, Sri Lanka |
| +6 | −4 | Bangladesh, Bhutan, Kazakhstan, Kyrgystan |
| +6.5 | −3.5 | Myanmar (Burma), Cocos Islands |
| +7 | −3 | Vietnam, Cambodia, Thailand, Indonesia (west), Laos |
| +8 | −2 | Australian Western Standard Time (WA), Malaysia, China, the Philippines, Indonesia (central), Mongolia, Taiwan |
| +9 | −1 | Japan, North Korea, South Korea, Indonesia (east), East Timor |
| +9.5 | −0.5 | Australian Central Standard Time (NT, SA, Broken Hill) |
| +10 | 0 | Australian Eastern Standard Time, Papua New Guinea, Guam |
| +10.5 | +0.5 | Lord Howe Island |
| +11 | +1 | Vanuatu, New Caledonia, Solomon Islands, Micronesia |
| +11.5 | +1.5 | Norfolk Island |
| +12 | +2 | New Zealand, Fiji, Nauru |
| +13 | +3 | Samoa, Tonga |

## The International Date Line

The **International Date Line (IDL)**, on the opposite side of the world to the prime meridian, runs from the North Pole to the South Pole through the Pacific Ocean between Fiji and Hawaii. It is bent to ensure that it does not run through any countries, as shown by the black line on this map.

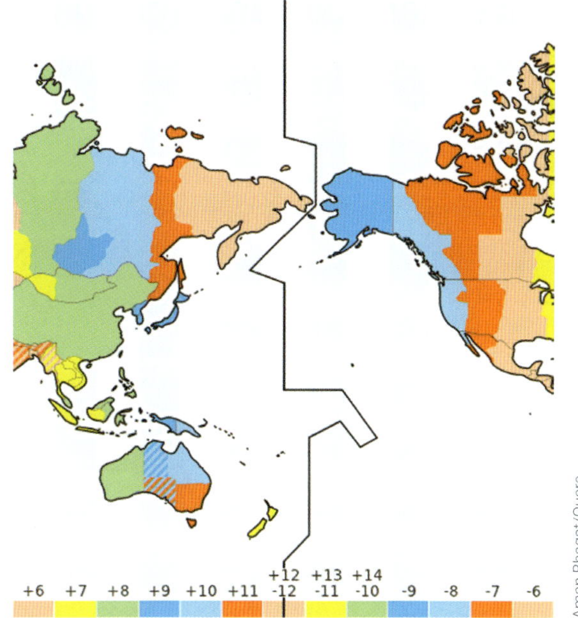

Aman Bhagat/Quora

The IDL also runs between the Samoan islands mentioned at the start of this chapter. Suppose it is 1 am Monday in London, or 1 am UTC. Samoa's time zone is 13 hours *east* of UTC, or UTC +13, so its local time is 2 pm Monday. American Samoa's time zone is 11 hours *west* of UTC, or UTC −11, so its local time is 2 pm *Sunday*. So even though the 2 countries are only 64 km apart, their local times are 24 hours or a day apart.

If you cross the IDL travelling east (such as going from Australia to the USA), today becomes yesterday (for example, Monday becomes Sunday) and you 'gain' a day. The reverse occurs when you cross the IDL travelling west (the opposite direction): today becomes tomorrow (for example, Sunday becomes Monday) and you 'lose' a day.

### Example 10

**Video**
International time zones 2

Habib is flying from New York, USA to Istanbul, Turkey to visit his grandfather. New York is 5 hours behind UTC while Turkey is 2 hours ahead of UTC.

**a** Habib wants to call his grandfather on the phone before he leaves New York. At what time should he call so that it is 6:00 pm in Turkey?

**b** Habib boards the plane at 1:45 pm New York time and the trip lasts 8.5 hours. What is the local time in Istanbul when he arrives?

### Solution

**a** Time difference = 5 + 2 = 7 hours

New York is 7 hours behind Istanbul.

New York — Istanbul

−5 — 0 (UTC) — +2

New York time = 6:00 pm − 7 hours

= 11:00 am

**b** Arrival time = 1:45 pm + 8.5 hours

= 10:15 pm — New York time

= 10:15 pm + 7 hours — Istanbul time

= 5:15 am (next day)

## Australian standard time zones

There are 3 main time zones in Australia (ignoring daylight saving and external island territories), shown on the following map.

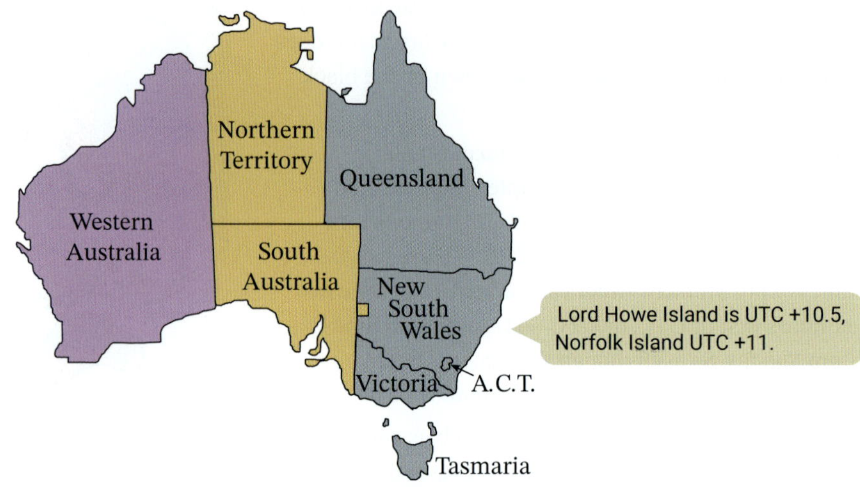


| Australian Western Standard time (AWST) UTC +8 | Australian Central Standard time (ACST) UTC +9.5 | Australian Eastern Standard time (AEST) UTC +10 |
|---|---|---|

These time zones can also be shown on a timeline.

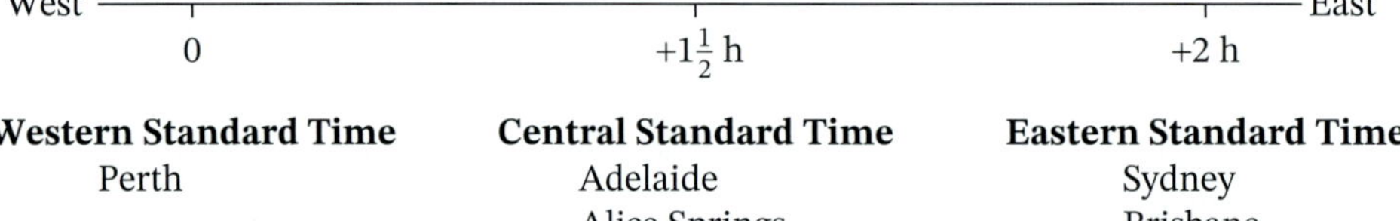


| **Western Standard Time** | **Central Standard Time** | **Eastern Standard Time** |
|---|---|---|
| Perth | Adelaide | Sydney |
| | Alice Springs | Brisbane |
| | Darwin | Canberra |
| | Broken Hill (NSW) | Hobart |

Small communities near the WA/SA border use ACST or UTC +8:45, Australian Central Western Standard time (Eucla, WA, and Border Village, SA).

### Example 11

**a** Pete is in Perth and it is 11:30 am. He wants to ring his mother in Sydney. What time is it in Sydney?

**b** Nic has just finished work at 5 pm in Sydney. What time is it in Adelaide?

### Solution

**a**

| 0 | $+1\frac{1}{2}$h | +2h |
|---|---|---|
| Perth | Adelaide | Sydney |

Moving from west (Perth) to east (Sydney), *add* 2 hours.

Time in Sydney = 11:30 am + 2 hours

= 1:30 pm

It is 1:30 pm in Sydney.

**b** Moving from east (Sydney) to central (Adelaide), *subtract* $\frac{1}{2}$ hour or 30 minutes.

Time in Adelaide = 5 pm − 30 min

= 4:30 pm

It is 4:30 pm in Adelaide.

## Daylight saving time

In summer, many countries adopt **daylight saving time**, or **summer time**, and turn their clocks forward 1 hour to take advantage of the increased hours of sunlight. By 'losing' that hour at the start of daylight saving, people are actually working and sleeping 1 hour earlier and gaining maximum use of the daylight.

Daylight saving has been operating in Australia since 1971, although Queensland, Western Australia and the Northern Territory do not participate in the scheme. Each year, it runs from the first Sunday in October (mid-spring) to the last Sunday in April (mid-autumn).

### Daylight saving time

For daylight saving time, add 1 hour.

This map shows the local times around Australia during daylight saving when it is 12:00 midday in NSW.

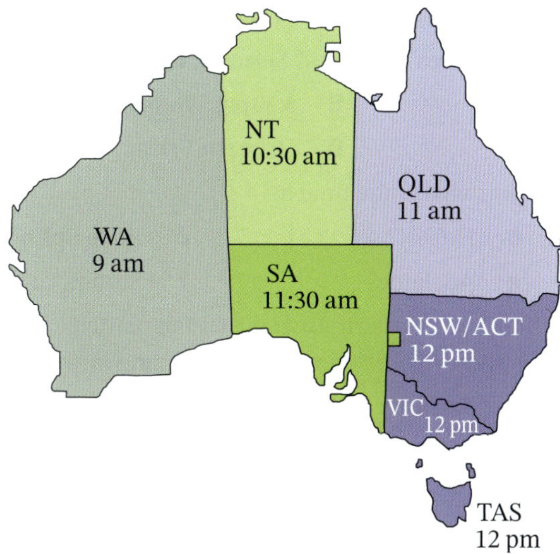


### Example 12

Simone, in Melbourne, wants to chat online with her cousin, Zac, in Vancouver, Canada. Canadian Pacific Standard Time is 8 hours behind UTC while Melbourne is 10 hours ahead of UTC. Daylight saving is operating in Canada. When should Simone log on to reach Zac at 4 pm Friday in Vancouver?

#### Solution

Vancouver time is UTC −8 but adding 1 hour for daylight saving gives UTC −7.

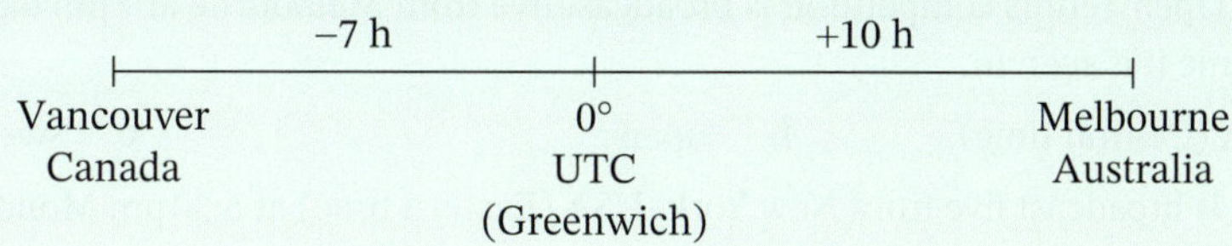


Time difference = 7 + 10 = 17 hours

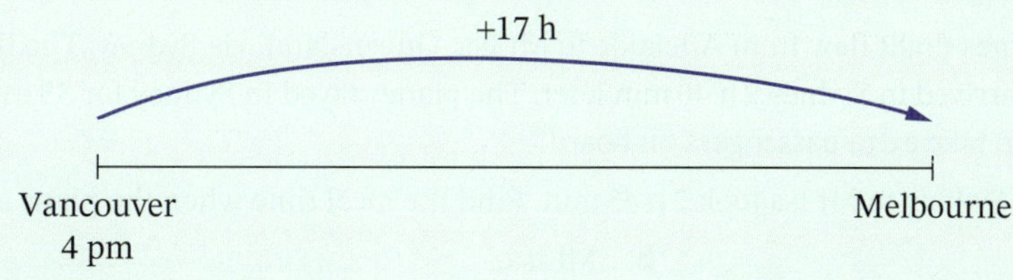


Melbourne is 17 hours ahead of Vancouver.

Log-on time = 4:00 pm Friday + 17 hours

= 9:00 am Saturday

**Worksheet**
Table of time zones

## EXERCISE 11.04 Answers on p. 489

## International time zones

Refer to the table on page 442 or the worksheet 'Table of time zones' for this exercise.

**1** When it is 4:00 pm Saturday in Sydney, what is the time in:

**a** India? **b** Russia (west)? **c** Greece? **d** Tonga?

EXAMPLE 10

**2** When it is 10:00 am Tuesday in London, UK, what is the time in:

**a** Greenland? **b** Bosnia-Herzegovina?

**c** Myanmar? **d** New Caledonia?

**3** What is the time difference (in hours) between:

**a** Malaysia and New Zealand? **b** Brazil (east) and Chile?

**c** Iceland and the Philippines? **d** Algeria and Salt Lake City, USA (Mountain time)?

**4** It takes $8\frac{1}{2}$ hours for a plane to fly from Sydney to Hawaii. If Heath leaves Sydney at 1:30 p.m. Friday, what is the local time when he arrives in Hawaii? Select **A, B, C** or **D**.

**A** 2:00 pm Thursday **B** 5:30 pm Thursday

**C** 2:00 am Friday **D** 10:00 pm Friday

EXAMPLE 11

**5** If the time in Sydney is 6:00 pm, what is the time in:

**a** Melbourne? **b** Hobart? **c** Alice Springs?

**d** Perth? **e** Adelaide? **f** Surfers Paradise?

**6** Queensland, Northern Territory and Western Australia do not participate in daylight saving. If the daylight saving time in Sydney is 10:30 am, what is the time in:

**a** Melbourne? **b** Hobart? **c** Alice Springs?

**d** Perth? **e** Adelaide? **f** Surfers Paradise?

**7** **a** What is the time in Santiago, Chile, when it is 12:00 noon in Perth?

**b** What does this say about the positions of Perth and Santiago on the world globe?

**8** A plane takes 12 hours to fly from Adelaide (UTC +9.5) to Cape Town, South Africa (UTC +2). If Shaunna leaves Adelaide at 10:00 am, what is the local time when she arrives in Cape Town?

EXAMPLE 12

**9** Robert in Tasmania wants to call his brother in Scotland, UK, when the time there is 4:30 pm daylight saving time. At what time should Robert call? Select **A, B, C** or **D**.

**A** 1:30 am **B** 2:30 am **C** 3:30 am **D** 5:30 am

**10** Indira flew from New Delhi, India, to Sydney, leaving at 6:30 pm and arriving the next day at 12:00 pm local time. How long did her plane trip take?

**11** The Australian Open Tennis competition is broadcast live from Melbourne at 7 pm daylight saving time. Calculate the time it is seen in:

**a** Dallas, USA (Central time) **b** Japan **c** Sweden (Europe)

**12** The Superbowl is broadcast live from New York, USA (Eastern time) at 6:30 pm Monday. At what time and day is it seen in:

**a** Sydney (daylight saving time)? **b** Berlin, Germany (Europe)? **c** Dublin, Ireland?

**13** A Wallaby Airlines flight flew from Adelaide to Mt Isa, Queensland, via Sydney. The flight left Adelaide at 2:30 pm and arrived in Sydney 2 h 40 min later. The plane stayed in Sydney for 35 minutes for refuelling and to take extra passengers on board.

The flight from Sydney to Mt Isa took 2 h 45 min. Find the local time when the plane arrived in:

**a** Sydney **b** Mt Isa.

**14** Laire boards a flight in Coffs Harbour at 9:50 am Tuesday morning. The flight to Sydney takes 40 minutes. Laire then waits in Sydney for 1 hour and boards a flight to Perth, which takes 4 hours 25 minutes. What time is it in Perth when the plane lands?

Complex

**15** A plane takes 2 hours and 45 minutes to fly from Wellington, New Zealand (UTC +12) to Sydney (UTC +10). If Paula leaves Wellington at 4:20 pm (daylight saving time) in September, what is the local standard time when she arrives in Sydney (standard time)?

**16** A plane leaves Sydney at 1:00 pm Thursday and flies 20 hours to Paris, France. At what time does the plane arrive in Paris if daylight saving is operating in France but not in Sydney?

**17** A plane leaves Hawaii on Monday at 12 noon and flies 6 hours west to Tokyo, Japan.

**a** When the plane crosses the International Date Line, what day does it become?

**b** What is the local time in Tokyo when the plane arrives?

**18** Liz flew west from Buenos Aires, Argentina (UTC −3) to Melbourne (UTC +10). To her surprise, she discovered that when she arrived in Melbourne it was the same time, 7:30 pm, as when she left Buenos Aires.

**a** How long did her plane journey take?

**b** Was it still the same day? If not, what day was it?

## Investigation

### Researching time zones

1 Why is UTC also known as GMT, Z time, Zulu time or Zebra time?

2 How is time measured at the North and South poles, where all the meridians of longitude meet?

3 How is time measured in Australia's external territories, such as Norfolk Island, Cocos Islands, Christmas Island (Indian Ocean) and in Antarctica?

4 Which country stretches across 11 time zones?

5 Australia, USA and Canada are roughly the same width, but Australia has 3 time zones while USA and Canada have 4 and 5 time zones respectively. Why do you think this is so? (*Hint*: Look at the countries on a world globe or map.)

## Did you know?

### Samoa lost an entire day in 2011

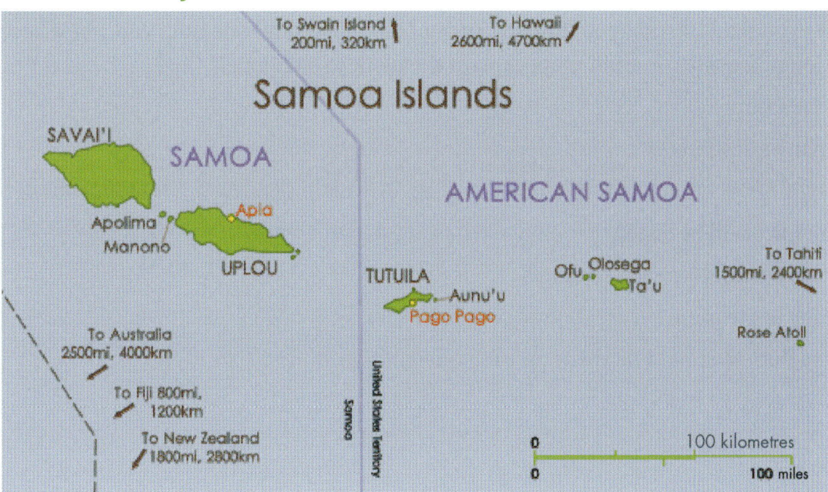


The Sun rises in Australia before most of the rest of the world, and for much of the day, the USA is a day behind Australia. The only places that are ahead of us in time are some Pacific islands, such as New Zealand, Fiji, Vanuatu, Samoa and Tonga.

Samoa has undergone much change in the 21st century. In 2009, it switched from driving on the right side of the road to the left side. At the end of 2011, it changed time zones from being UTC −11 like American Samoa to UTC +13. This meant that the date jumped from 29 December to 31 December, skipping the day Friday, 30 December 2011.

**Find out the main reasons Samoa changed time zones (and therefore day zones) and driving orientation.**

## Investigation

### Plan an overseas holiday

Use a world map or globe to plan a trip to visit up to 5 different places in the world.
Design an itinerary that lists the places you'll be visiting. You will need to take into account:

- distances between places
- mode of transport and travelling speeds: by air, land or sea
- arrival and departure times
- international time zones
- climate/seasonal conditions (these also affect the popularity of destinations).

## Sample HSC problem Answers on p. 490

(4 marks) Crystal is travelling by plane from Seattle, USA (48°N, 122°W) to Sydney (34°S, 151°E) for a holiday.

**a** What is the difference in latitude between Seattle and Sydney? 1 mark

**b** Crystal leaves Seattle at 9:45 am on Tuesday and arrives in Sydney at 4:10 am Wednesday Seattle time. How long was the journey in hours and minutes? 1 mark

**c** Calculate the local time and day in Sydney when Crystal arrives if Seattle is UTC −7 and Sydney is UTC +10. 2 marks

## Study Tip

### Before an exam

- Make a study plan early – don't leave it to the last minute.
- Read and memorise your topic summaries.
- Work on your weak areas – learn from your mistakes.
- Don't spend too much time studying work you know already.
- Use revision exercises and past exams/assignments for practice.
- Vary the way you study so that you don't become bored: have someone quiz you, voice-record your summary, design a poster or mind map, explain the work to someone.
- Anticipate the exam:
  - How many questions?
  - What type of questions: multiple-choice, short answer, long answer, problem-solving?
  - Which topics will be tested?
  - How many marks for each section?
  - How long is the exam?
  - How much time should be spent on each question/section?

# CHAPTER SUMMARY

This chapter, *World locations and times*, focused on position and time measurement of locations around the world. Make sure that all theory, rules, terminology and worked examples are included in your summary.

Before you move on, consider what you learned in this chapter and revisit any sections that may have been unclear.

| What you learned in this chapter... | Section | |
|---|---|---|
| Locate positions on Earth using latitude and longitude | 11.01 | Latitude and longitude |
| Convert between 12-hour and 24-hour time | 11.02 | Time |
| Calculate differences between times of day | 11.02 | Time |
| Calculate time differences between locations based on their difference in longitude, using the relation<br>15° longitude difference = 1-hour time difference | 11.03 | Longitude and time differences |
| Calculate times in different parts of the world and within Australia using international time zones, taking daylight saving into account | 11.04 | International time zones |
| Solve problems involving travelling and communicating across the world using international time zones | 11.04 | International time zones |

To help master this topic, make a summary mind map. Use the chapter outline and the mind map below as a guide. Add your own words, symbols, diagrams, boxes and reminders. The summary should give you a 'whole picture' view of the topic and allow you to identify any weak areas to revisit in your revision.

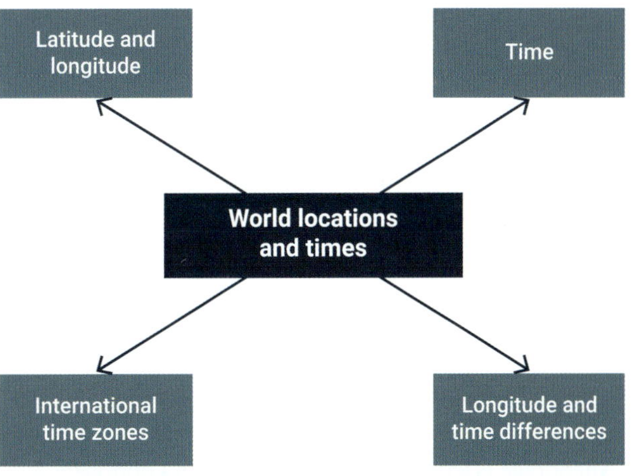

# 11 Test yourself Answers on p. 490

11.01 **1** What are the position coordinates of $A$, $B$, $C$ and $D$ on Earth?

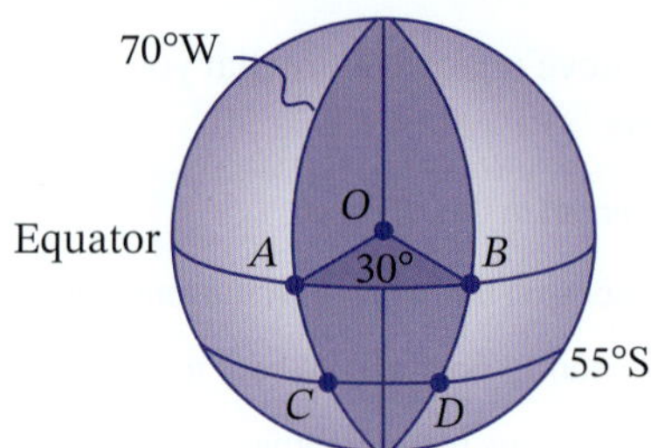


11.01 **2** Albury, NSW has coordinates (36°S, 147°E).

- **a** What is the longitude of Albury?
- **b** How many degrees is Albury north of the South Pole?
- **c** Broome has coordinates (18°S, 122°E). What is the difference in longitude between Albury and Broome?

11.01 **3** Amsterdam, The Netherlands is 15° north and 122° west of Seoul, South Korea (37°N, 127°E). What are the coordinates of Amsterdam?

11.02 **4** Convert each 24-hour time to 12-hour time.

**a** 1436 **b** 0505 **c** 2218 **d** 0027

11.02 **5** Convert each 12-hour time to 24-hour time.

**a** 5:23 am **b** 4:32 pm **c** 9:05 pm **d** 12:55 pm

11.02 **6** Calculate the time of day:

- **a** 2 h 34 min after 6:21 am
- **b** 5 h 8 min before midday
- **c** 7 h 12 min after 0315
- **d** 4 h 27 min before 1532.

11.03 **7** Hanoi, Vietnam, lies on the 105°E meridian while Sydney lies on the 150°E meridian. When the time in Hanoi is 5:30 pm, what is the time in:

**a** Sydney? **b** London?

11.04 **8** The time in Anchorage, Alaska (USA), is 9 hours behind UTC, while the time in Lusaka, Zambia, is 2 hours ahead of UTC.

- **a** When the time in Anchorage is 8:00 am, what is the time in London?
- **b** When the time in Lusaka is 6:30 pm, what is the time in Anchorage?

11.04 **9** What is the time in Brisbane (AEST) when it is:

**a** 4 pm in Perth? **b** 6 am in Adelaide? **c** 2:30 am in Hobart?

11.04 **10** A plane leaves Darwin on Thursday at 6:00 pm and travels east for 8 hours to reach Niue Island.

- **a** When the plane crosses the International Date Line, what day does it become?
- **b** If Darwin is UTC +9.5 and Niue Island is UTC −11, what is the time on Niue Island when the plane arrives there?

11.04 **11** Kerrie, who lives in Perth, wants to call her sister in Los Angeles, USA, when the time there is 12 noon Sunday. At what time should she call if Perth is UTC +8 and Los Angeles is usually UTC −8, but the USA is on daylight saving time?

11.04 **12** Zoe travelled from Sydney to London by plane. She left Sydney at 1:40 pm (UTC +10) on Wednesday and spent 8 hours 15 minutes on the flight from Sydney to Singapore. She waited at Singapore airport for 4 hours 20 minutes, then flew to London for 13 hours 10 minutes. What was the time and day in London (UTC 0) when Zoe arrived?

Foundation | Mastery | Complex

# PRACTICE EXAM 4

Answers on p. 490

**Recommended time: One hour**

**Section 1**
**10 multiple-choice questions: 1 mark each**
**Select the correct answer A, B, C or D.**

**1** What is 12:15 am in 24-hour time? 11.02

**A** 0015 **B** 1215 **C** 1415 **D** 2415

**2** Which statement about the data set {0, 1, 2, 3, 4, 4, 7} is true? 

**A** The mean and mode are equal.

**B** The mean and median are equal.

**C** The median and mode are equal.

**D** The mean, median and mode are not equal.

**3** Port Moresby, Papua New Guinea, has coordinates (9°S, 147°E). If Nyngan, NSW is due south of Port Moresby, which one of the following could be the coordinates of Nyngan? 

**A** (4°S, 147°E) **B** (31°S, 147°E) **C** (9°S, 140°E) **D** (9°S, 151°E)

**4** Which statement is true about these 2 data sets? 

**A** The mean of $Y$ is 6.

**B** $X$ is uniform.

**C** $Y$ is positively skewed.

**D** $X$ and $Y$ are both unimodal.

Set $X$

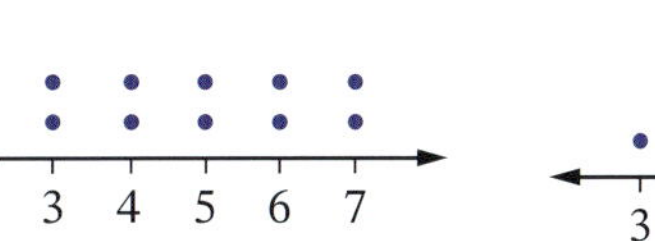


Set $Y$

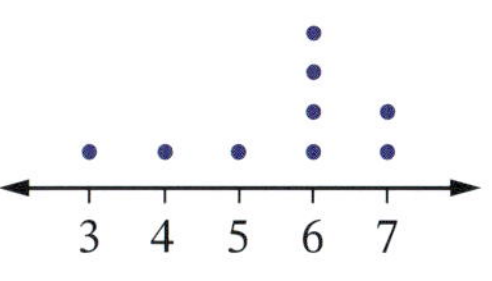


**5** The data sets A and B are displayed in these histograms. 10.03

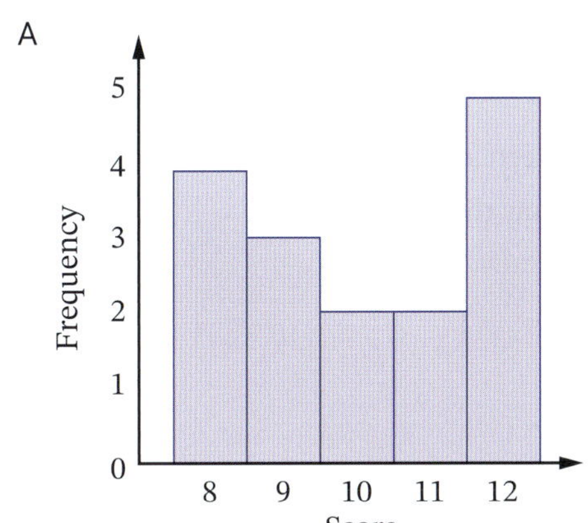


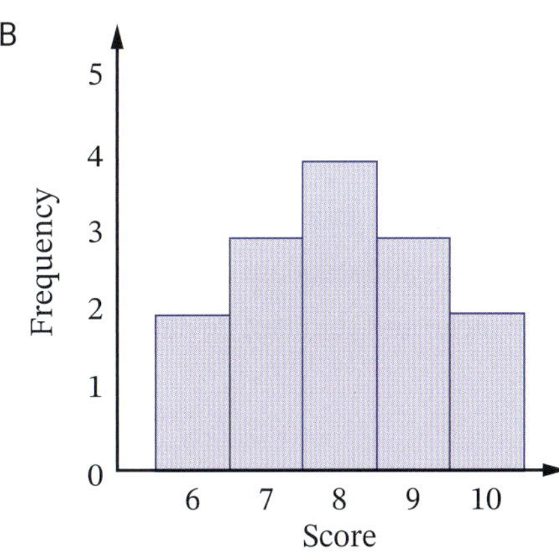


Which one of these statements is true about the mode and range of sets A and B?

**A** A has a higher mode and B has a higher range.

**B** A has a higher mode and both ranges are the same.

**C** Both modes are the same and B has a higher range.

**D** Both modes are the same and both ranges are the same.

**6** Oishi needs to board a plane at Sydney airport that departs at 6:35 am. Passengers must check-in and board the plane 45 minutes before departure. It takes 50 minutes for Oishi to drive to the airport from home. The shuttle bus from the car park to the airport takes up to 20 minutes.
What is the latest time that Oishi can leave home to board the plane on time?

**A** 4:40 am **B** 4:50 am **C** 5:00 am **D** 5:20 am

**7** Ramy leaves Melbourne (UTC +10) at 12 noon Monday and flies for 13 hours to Santiago, Chile (UTC −3). What is the local time and day when Ramy arrives in Santiago?

**A** 12 noon Sunday **B** 11 am Monday
**C** 12 noon Monday **D** 1 pm Tuesday

10.08

**8** Which dot plot illustrates a distribution that is positively skewed?

**A**

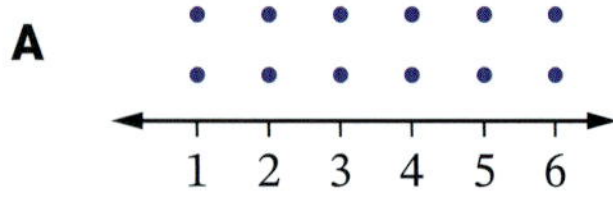


**B**

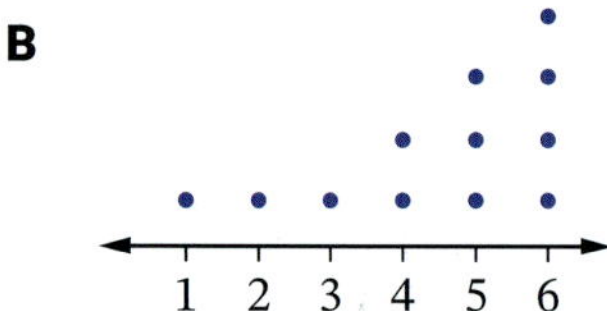


**C**

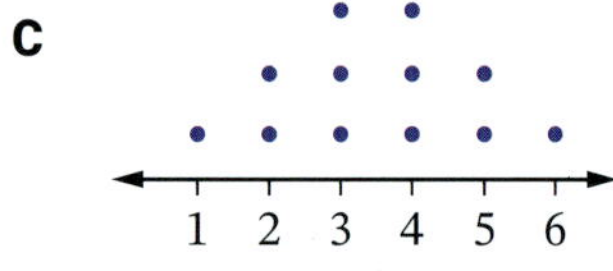


**D**

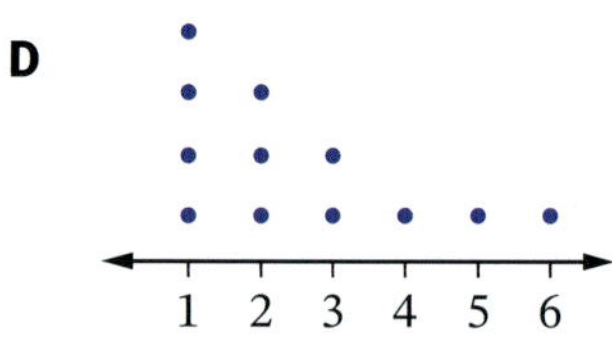

**9** Australian Eastern Standard Time AEST is 10 hours ahead of UTC. Calculate the local time in the United Kingdom when it is 6 pm in NSW and daylight saving is operating in the United Kingdom.

**A** 10 am **B** 3 am **C** 4 am **D** 9 am

10.04

**10** The 6 employees at the Bug and Beef Cafe earn the following weekly wages:

$850 $920 $1010 $630 $820 $960

The mean and standard deviation of this data set were calculated. If a wage of $1220 is added to this data set, how will its mean and standard deviation be affected?

**A** The mean will increase but the standard deviation will decrease.

**B** The mean will increase but the standard deviation will stay the same.

**C** The mean and standard deviation will both increase.

**D** The mean and standard deviation will both decrease.

## Section 2
## 3 questions: 10 marks each

### Question 11 10 marks

11.04

**a** This timetable shows flights from Sydney and Melbourne to Christchurch in New Zealand. 10.05

| | Flight | Depart | Arrive |
|---|---|---|---|
| Sydney – Christchurch | NC482 | 0855 | 1355 |
| | NC490 | 1710 | 2210 |
| | NC511 | 1945 | 0045 |
| Melbourne – Christchurch | NC479 | 0845 | 1405 |
| | NC488 | 0905 | 1425 |
| | NC491 | 1755 | 2315 |

**i** When does the first flight leave Sydney and when does it arrive in Christchurch? Write your answers in 12-hour time. 1 mark

**ii** The arrival times are shown in New Zealand time, so you need to subtract 2 hours for Australian Eastern Standard Time. Calculate how long it takes to fly between Melbourne and Christchurch. Answer in hours and minutes. 2 marks

**iii** Miriam lives in Sydney and needs to be in Christchurch by 10 am on Tuesday for a conference. What is the latest flight she can take? 1 mark

**b** A sample of households was surveyed on the number of motor vehicles owned. 10.01

| Number of vehicles | Frequency | Cumulative frequency |
|---|---|---|
| 0 | 1 | |
| 1 | 14 | |
| 2 | 26 | |
| 3 | 7 | |
| 4 | 2 | |

**i** Copy and complete the cumulative frequency column of the table. 1 mark

**ii** Find the mean for the data. 1 mark

**iii** Find the range for the data. 1 mark

**iv** Construct a cumulative frequency histogram and polygon for the data. 2 marks

**v** Use the cumulative frequency polygon to find the interquartile range. 1 mark

### Question 12 10 marks

11.04

**a** A tennis match is being played at 2 pm on Tuesday in Los Angeles, USA (119°W). Aimee is watching the match live on TV in Sydney (151°E).

What is the day and time in Sydney? 2 marks

**b** The number of road accidents each week at a busy intersection was recorded: 10.01, 10.02

3 5 0 2 6 4 2 0 1 5 4 2

**i** Calculate the mean and standard deviation, correct to 2 decimal places. 2 marks

**ii** Calculate the interquartile range. 1 mark

**c** The number of accidents per month for a period of 10 months on a building site were: 10.04

9 4 3 0 1 7 5 6 5 3

For this data, find:

**i** the mode 1 mark

**ii** the interquartile range 1 mark

**iii** the standard deviation, correct to 2 decimal places. 1 mark

**d** The weights (in kilograms) of a group of rugby players are displayed in this stem-and-leaf plot.

| Stem | Leaf |
|---|---|
| 5 | 9 |
| 6 | 0 4 |
| 7 | 3 7 9 |
| 8 | 0 2 3 6 7 |
| 9 | 1 2 |

**i** Comment on the shape of the data, mentioning skewness and clusters. 1 mark

**ii** Which is the better measure of centre for this data, the mean or median? State a reason for your answer. 1 mark

## Question 13 10 marks

**a** This box plot shows the ages of members at a lawn bowls club.

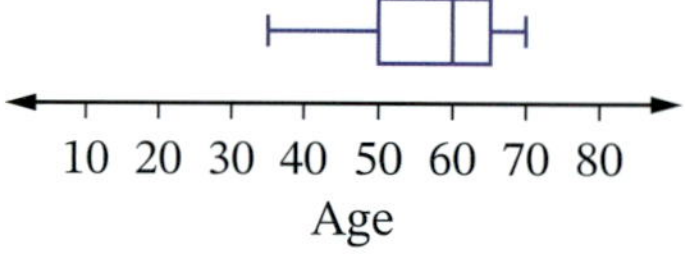


**i** What is the age of the oldest member? 1 mark

**ii** Calculate the interquartile range of ages. 1 mark

**iii** If there are 124 members in the club, how many of them are aged between 50 and 60? 1 mark

10.09 **b** The ages of members of a yoga class are shown in this back-to-back stem-and-leaf plot.

| Women | | Men |
|---|---|---|
| 4 | 2 | 2 6 |
| 2 | 3 | 2 2 5 6 8 |
| 8 3 | 4 | 2 4 5 6 |
| 9 4 0 0 | 5 | 3 |
| 8 5 4 3 3 | 6 | 3 3 4 |

**i** Compare the shapes of age distributions of the women and men, commenting on any skewness, clusters and modality. 2 marks

**ii** Which group was younger on average: women or men? Justify your answer. 1 mark

11.04 **c** A plane flies east from Manila, the Philippines (14°N, 121°E) to Mexico City (19°N, 99°W). It flies for 19 hours, leaving Manila at 6:00 am. Tuesday.

**i** Which city is further south: Manila or Mexico City? 1 mark

**ii** After the plane crosses the International Date Line, what day will it be? 1 mark

**iii** If the Philippines is 8 hours ahead of UTC and Mexico City is 8 hours behind UTC, what is the time and day in Mexico City when the plane arrives? 2 marks

**END OF EXAMINATION**

# ANSWERS

## Chapter 1

### SkillCheck

**1** **a** sector graph **b** stem-and-leaf plot
**c** line graph **d** divided bar graph

**2** **Column graph**

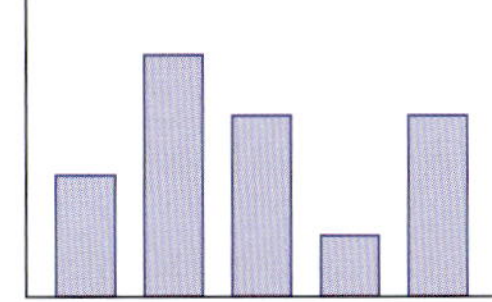

**Dot plot**

0 1 2 3 4 5

**3** **a** 20 **b** 5 **c** 500

**4** **a** $\frac{3}{8}$ **b** $37\frac{1}{2}\%$

**5** **a** $\frac{7}{20}$ **b** 35%

**6** 24.5

### Exercise 1.01

**1** **a** 38% **b** 5% **c** 11 812 **d** 34 000
**e** Close to 1 in 3 as 32% were drinking at a risky level.
**f** 669 600
**g** decrease by 2%

**2** **a** $\frac{1}{4}$ **b** labour **c** $2.63
**d** $6500 **e** 54° **f** $1060

**3** **a** $\frac{27}{50}$ **b** $\frac{3}{50}$ **c** natural
**d** 4464

**4** **a** male **b** 55–64 **c** 8.5%
**d** 150 040 **e** 65–74
**f** It increases to the 55–64 age bracket and then it decreases.

**5** **a** listening to broadcast radio
**b** approx. $\frac{1}{8}$
**c** 200 hours (using 30 days for a month)
**d** 4 h 50 min

**6** **a** yes **b** underweight **c** 18
**d** 26 **e** 16 to 25

**7** **a** 26% **b** 42%
**c** **i** 1110 **ii** 870 **iii** 555

**8** **a** 18–24 **b** 17.5% **c** 35–44
**d** decreases, awareness of health considerations
**e** E-cigarette use is much lower than smoking.

**9** **a** in-store, online via PC
**b** **i** in-store **ii** in-store **iii** online via PC
**c** **i** 17% **ii** 8% **iii** 11%

**10** **a** 5 mm **b** number of days of rain
**c** 19.5 mm; 2.5 **d** January; 29 mm
**e** August and October, 2.8 days
**f** June (lowest rainfall)
**g** 6

**11** A

**12** **a** $63b
**b** **i** 25.4% **ii** 15.2% **iii** 6.2%
**c** $14.5b **d** 17.5%

**13** **a** share prices over 6 days
**b** 10c **c** $4.55
**d** Tuesday **e** It fell, then rose.
**f** Thursday; lowest price **g** NCM Bank

**14** **a** 17 **b** east; 25 on Thursday
**c** Monday; first working day
**d** Wednesday; 6
**e** Saturday; no industry

**15** **a** 84 h 6 min **b** 9 times
**c–d** Teacher to check.

**16** **a** 45%
**b** **i** 28.8 L **ii** 67.2 L
**c** 26.28 kL

### Exercise 1.02

**1** **a** The vertical axis starts at $1.50 so increases in share price appear steeper than they actually are.
**b**

**Spin Industries share price**

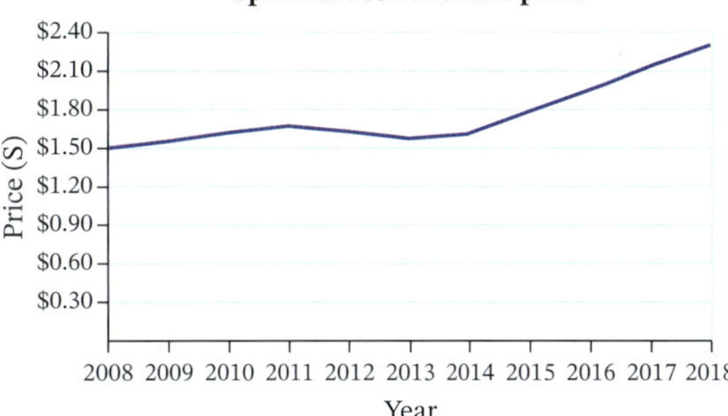


**2** **a** Because the 2026 picture is double the length and double the width, it is actually 4 times the size of the 2024 picture instead of 2 times the size.
**b** **Growth in wheat crops**

2024

2026

OR

**Growth in wheat crops**

2024

2026

**3** **a** a company's sales
**b** The new director has made sales go down.
**c** sales figures and months

**4** **a** The $800 note is double the length and double the width of the $400 note and this makes it four times the size of the $400 note instead of two times the size.
**b**

**5** **a** By starting the scale on the vertical axis at 40.
**b** 7 times
**c**

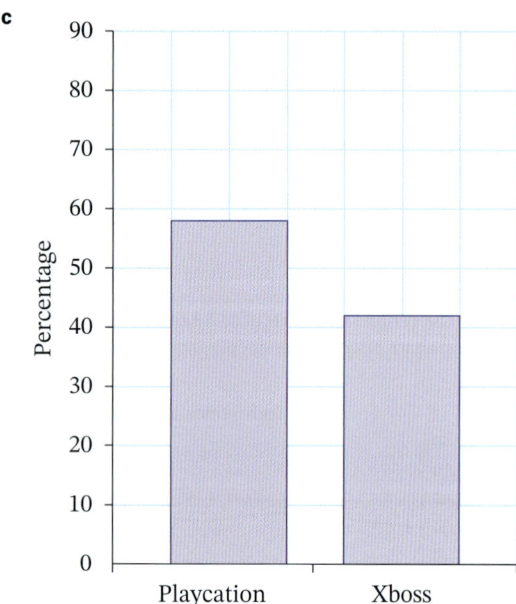


**6** **a** The picture for '$100 of oil today' is actually about $\frac{1}{4}$ the size of the picture for '10 years ago'.
**b**

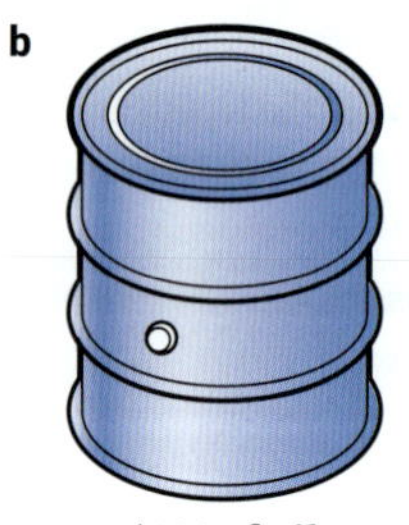
$100 of oil 10 years ago

$100 of oil today

**7** **a** 7
**b** The hours scale does not begin at 0.
**c** Big difference between school or sleeping and the other activities.
**d**

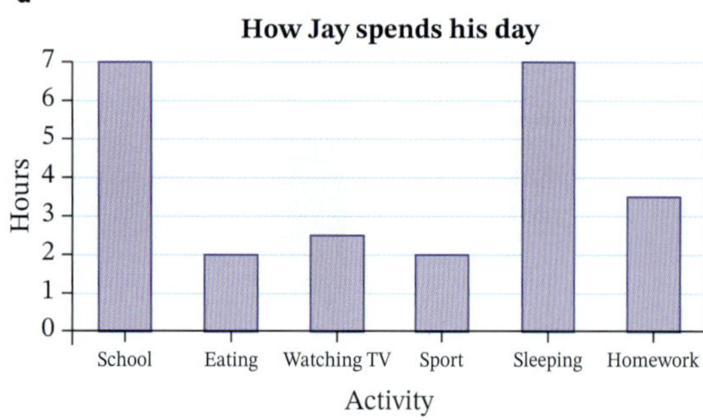


**8** **a** The 3D graph makes the front 8% slice larger than the 8% slice at the back.
**b**

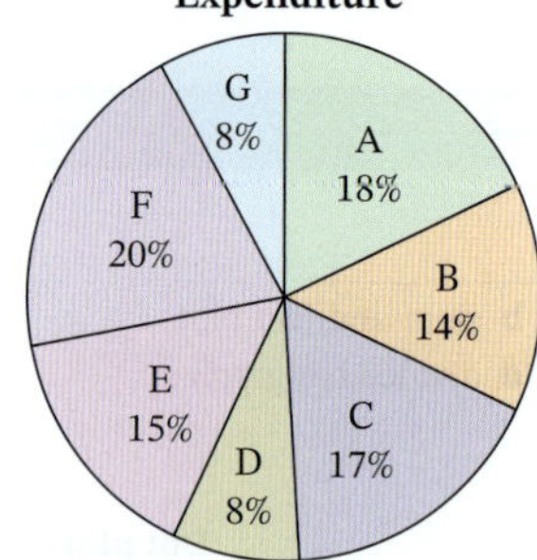


**9** **a** The different-sized icons for the French fries and Pizza indicate that more fries are sold than pizzas.
**b** Icons all need to be the same size and a key should be shown to indicate how many sales an icon represents.

## Exercise 1.03

**1** **a** C **b** N **c** C **d** N
**e** N **f** C **g** C **h** C
**i** N **j** N **k** N **l** C

**2** Categorical data uses words/symbols, for example, type of pet. Numerical data uses quantities/numbers, for example, height.

**3** **a** N **b** O **c** O **d** N
**e** N **f** O **g** O **h** N

**4** **a** C **b** C **c** C **d** C
**e** D **f** C **g** D **h** C

**5** Answers will vary.
**a** HSC subjects
**b** size of hot chips at a fast food store
**c** number of children in a family
**d** heights of boys in Year 11

**6** **a** numerical **b** categorical
**c** continuous **d** nominal

**7** B

**8** **a** Shirt size can be small, medium, large, etc. or it could be by collar size like 30 cm, 32 cm, etc.
**b** Results can just be a grade, A, B, C ..., or it could be a numerical mark.
**c** Fan speed can be low, medium, high or it could be speeds of 1, 2, 3, etc.

**9** Location is categorical, temperature is numerical.

## Exercise 1.04

**1** **a** C **b** S **c** S **d** C
**e** S **f** S **g** C **h** S

**2** 2026, 2031

**3** Teacher to check reasons.
**a** systematic **b** random **c** self-selected
**d** stratified **e** random **f** random
**g** random **h** random **i** random
**j** self-selected

**4** 138 **5** 188 **6** B **7** B

**8** **a** Not all teens read magazines.
**b** Only those diners interested/volunteering to complete the survey are sampled.
**c** Need to select more students and not only those who arrive early.
**d** Need to sample workers as well, not only management, who may have a biased view.
**e** People at a premiere are not the usual moviegoers.
**f** It is restricted to one radio station.
**g** Relies on passengers to reply to the survey, a self-selected sample. May be biased towards passengers who had a bad experience or a very good experience.

**9** **b**, **f** and **g**

**10** **a** systematic sample

**b** shoppers in one store only; only females or males

## Exercise 1.05

**1** **a**

| Score | Frequency |
|---|---|
| 0 | 3 |
| 1 | 6 |
| 2 | 9 |
| 3 | 12 |
| 4 | 7 |
| 5 | 2 |
| 6 | 1 |
| | 40 |

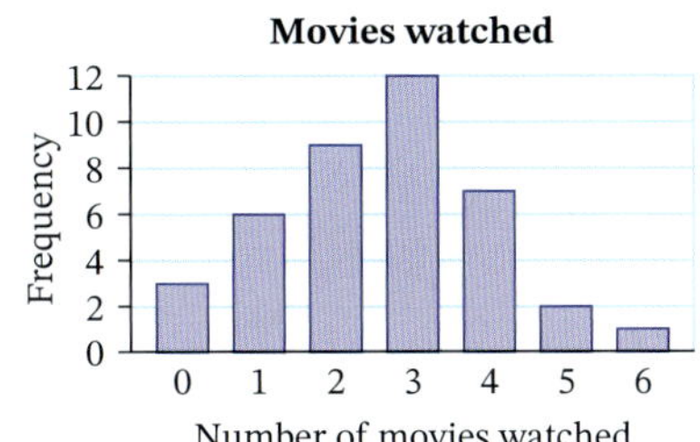


**b** data is discrete (not continuous)

**c** 3 **d** 6; 1 student **e** 7.5%

**2** **a** shows parts of the whole amount

**b** How Georgie spent $800 000

New car; Pay home loan; Home renovations; Travel; Savings

**c** saved it **d** $\frac{9}{50}$ **e** 12%

**3** **a** 14 970 **b** sector; divided bar; column graph

**c**

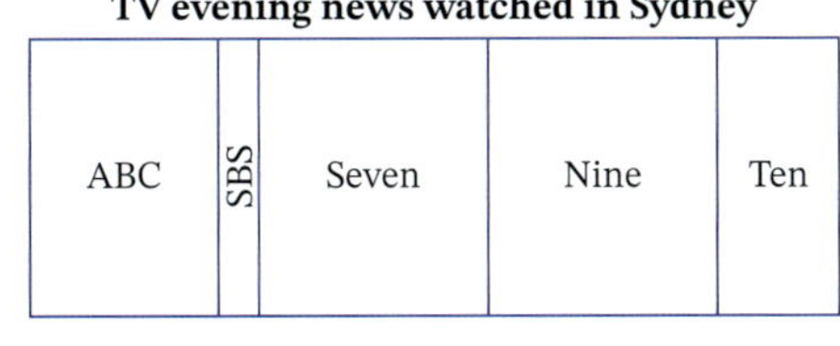


**d** 23% **e** yes

**f** No, since the survey only included Sydney households.

**4** **a** 72 **b** column graph; discrete data

**c** 9.7% **d** $\frac{11}{18}$

**5** D

**6** **a** 46.7%

**b**

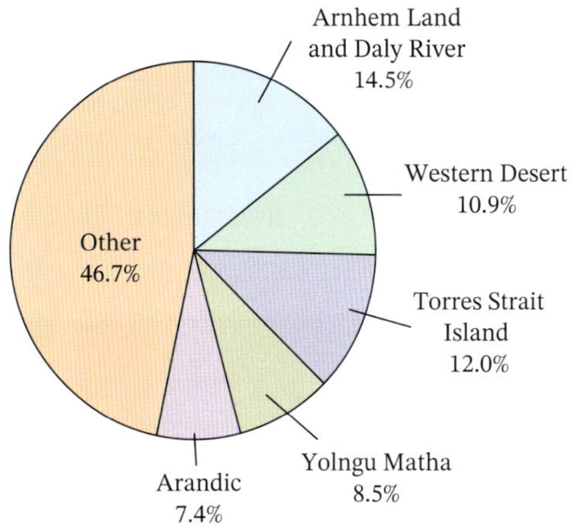


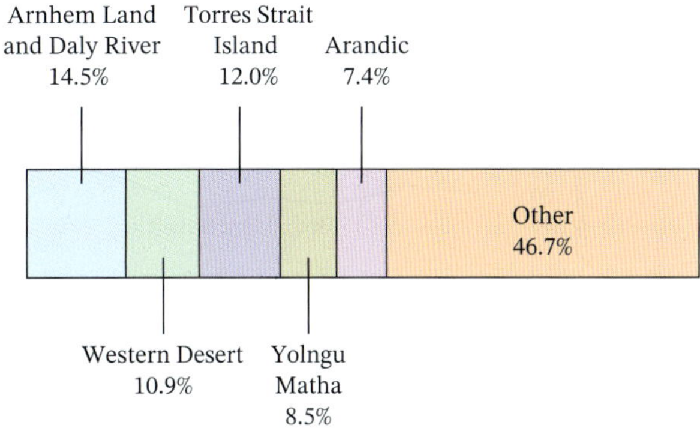


**7** **a** The sum of the percentages is more than 100 since people usually go on more than one social media platform.

**b** Teacher to check – column graph or horizontal bar graph.

**8** **a** Sector graph and divided bar graph since the sum of the percentages is 100. Can also use column graphs.

**b**

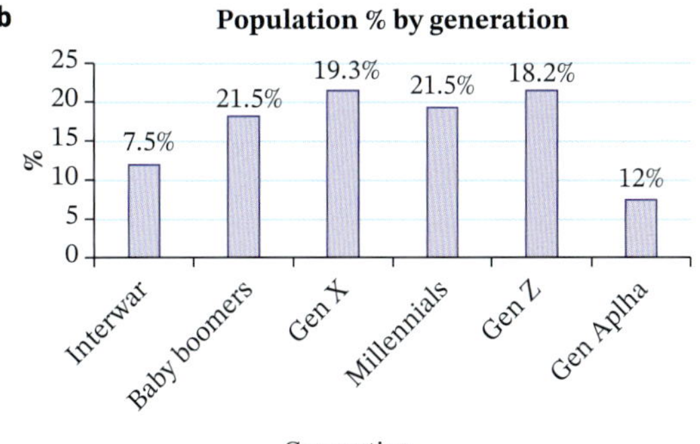


**c** **i** 5 538 586 **ii** 1 932 065

**9** **a** 46.47%

**b**

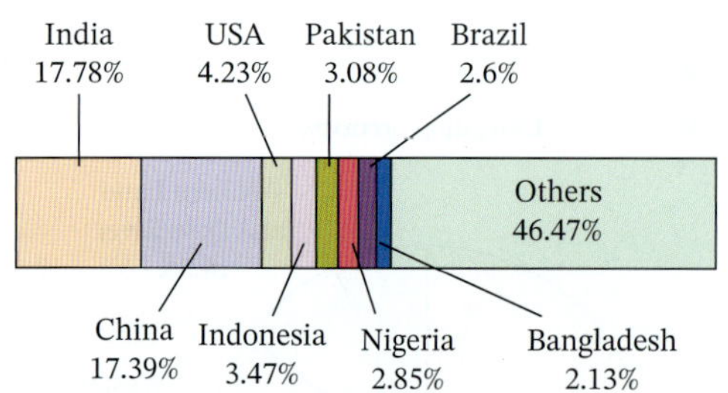


**c** **i** 1 419 109 865 **ii** 283 169 133

**iii** 8 160 493 763

**10** **a**

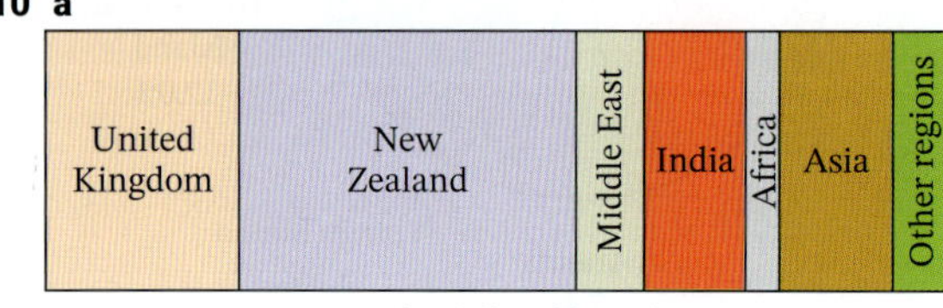


**b** T **c** $\frac{4}{31}$ **d** 8% **e** India

**11** **a**

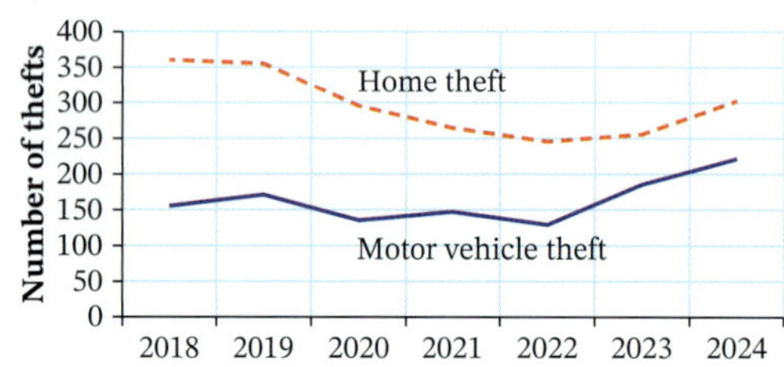


**b** 2018 **c** 2022 **d** home theft

**e** Home thefts decreased until 2022 and then began to increase.

**f** Increased, since the increase began in 2022.

## Exercise 1.06

**1** **a**

| Photo posts/day | Frequency |
|---|---|
| 0 | 2 |
| 1 | 10 |
| 2 | 5 |
| 3 | 8 |
| 4 | 7 |
| 5 | 8 |
| Total | 40 |

**b**

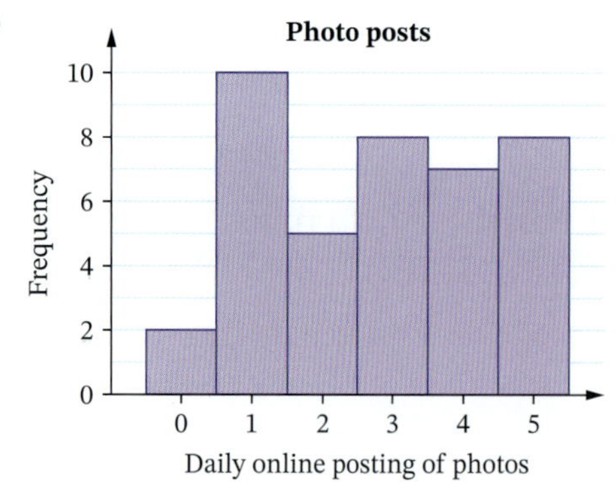


**2** **a**

| Mass (g) | Number of eggs |
|---|---|
| 56 | 2 |
| 57 | 7 |
| 58 | 11 |
| 59 | 6 |
| 60 | 7 |
| 61 | 5 |
| 62 | 4 |
| 63 | 0 |
| 64 | 6 |
| Total | 48 |

**b**

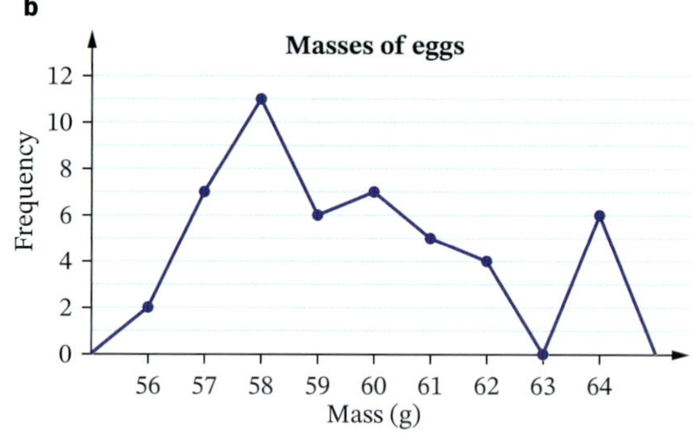


**c** Yes, since 26 out of 48 (about 54%) eggs are lighter than 60 g.

**3** **a** 70

**b**

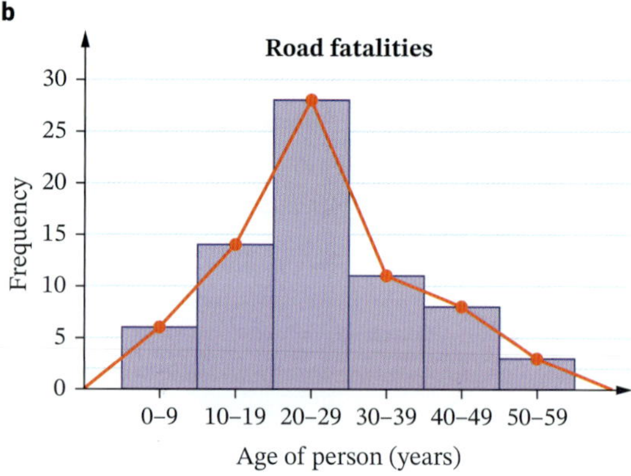


**c** Most road deaths occur in the 10–39 years age group.

**4** **a** discrete

**b**

| Number of words in sentence | Class centre | Tally | Number of sentences |
|---|---|---|---|
| 1–5 | 3 | \|\| | 2 |
| 6–10 | 8 | 卌 卌 | 10 |
| 11–15 | 13 | 卌 卌 | 10 |
| 16–20 | 18 | 卌 | 5 |
| 21–25 | 23 | 卌 \|\|\| | 8 |
| 26–30 | 28 | \| | 1 |
| | | Total | 36 |

**c**

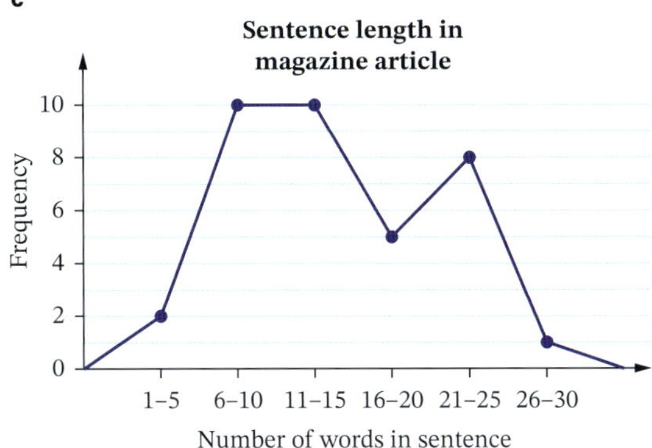


**d** 6–10 and 11–15 are the two modal classes.

**5 a**

| Class | Class centre | Tally | Frequency |
|---|---|---|---|
| 15–19 | 17 | 𝍸 𝍸 𝍸 \| | 16 |
| 20–24 | 22 | 𝍸 𝍸 𝍸 𝍸 | 20 |
| 25–29 | 27 | \|\|\|\| | 4 |
| 30–34 | 32 | 𝍸 | 5 |
| 35–39 | 37 | \| | 1 |
| 40–44 | 42 | \|\| | 2 |
| 45–49 | 47 | \|\| | 2 |
| | | Total | 50 |

**b** 50

**c**

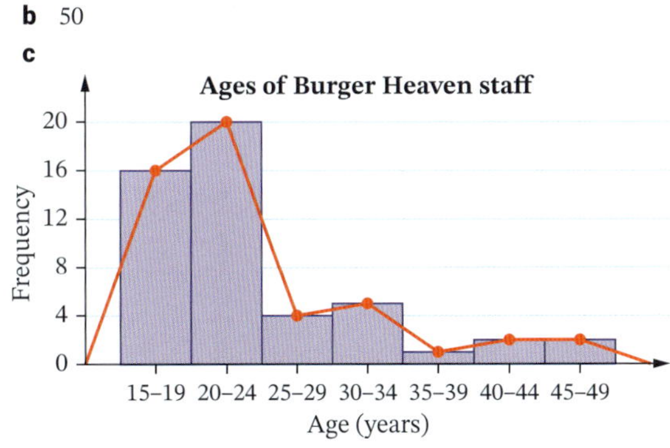


**d** The data is skewed towards younger workers; most of the workers are aged under 25 years.

**6 a** none **b** 43 **c** 16.3%

**d** No, since the last column only tells those who made more than 30 calls.

**7 a** continuous

**b**

| Class | Class centre | Frequency |
|---|---|---|
| 0–<2 | 1 | 5 |
| 2–<4 | 3 | 10 |
| 4–<6 | 5 | 10 |
| 6–<8 | 7 | 6 |
| 8–<10 | 9 | 4 |
| 10–<12 | 11 | 3 |
| 12–<14 | 13 | 2 |
| | Total | 40 |

**c**

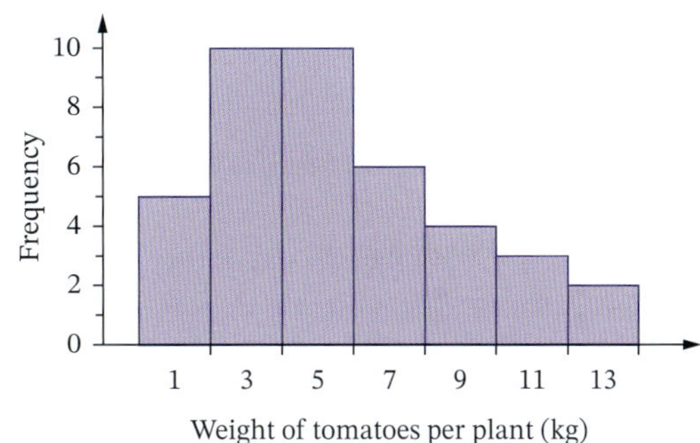


**d** Modal classes are 2–<4 and 4–<6.

**e** The "no dig" method seems effective because the majority of plants produced more than 3 kg of tomatoes, with some even producing 13 kg.

**8 a**

| Height (cm) | Class centre | Tally | Number of applicants |
|---|---|---|---|
| 160–<165 | 162.2 | \|\|\|\| | 4 |
| 165–<170 | 167.5 | 𝍸 | 5 |
| 170–<175 | 172.5 | 𝍸 | 5 |
| 175–<180 | 177.5 | 𝍸 | 5 |
| 180–<185 | 182.5 | \|\| | 2 |
| 185–<190 | 187.5 | 𝍸 \| | 6 |
| | | Total | 27 |

**b**

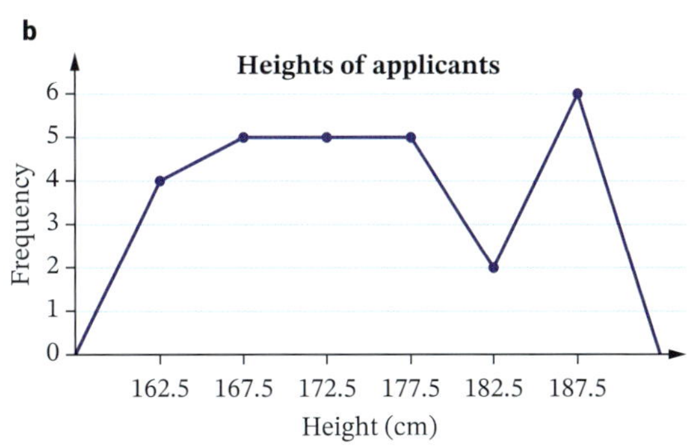


**c** Modal class interval is 185–<190

**d** 160–<165 **e** 180–<185

## Exercise 1.07

**1 a**

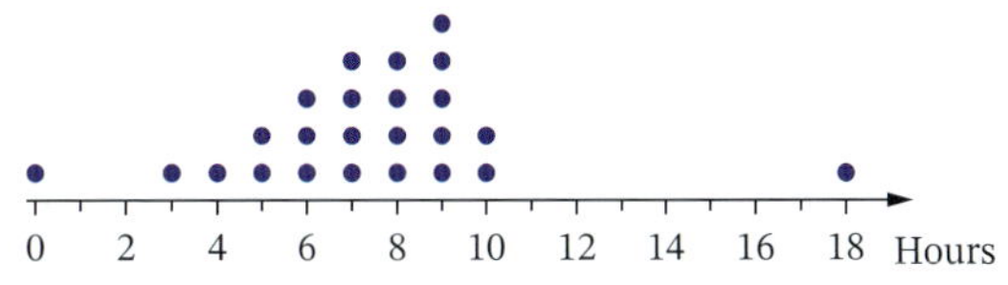


**b** 24 **c** 0 and 18

**d** cluster 6–9; gaps at 1–2, 11–17 hours

**2 a** 10 **b** 7

**c** 12; a student with big feet **d** $\frac{2}{5}$

**3 a** Goals scored per game

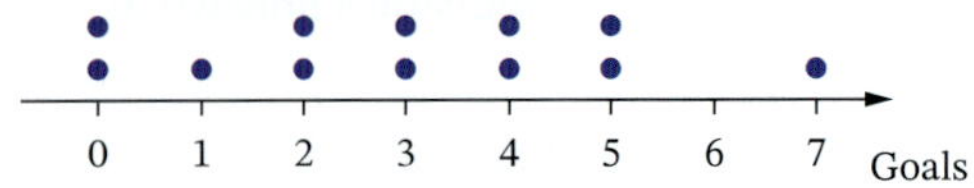


**b** 0

**c** Goals per game ranged from 0 to 7, so the team is not consistent.

**4 a** People per car

1 2 3 4 5 6

**b** 20 **c** 6

**d** no real clusters or outliers

**5 a** 30 **b** 50% **c** around 6–10

**d** mostly good results but 3 quite low scores

**6 a**

| Stem | Leaf |
|---|---|
| 4 | 3 5 9 |
| 5 | 0 2 7 8 |
| 6 | 1 2 4 5 7 8 |
| 7 | 0 2 3 9 |
| 8 | 2 4 9 |

**b** 20 **c** 89 **d** 25%

**7 a**

| Stem | Leaf |
|---|---|
| 2 | 8 |
| 3 | |
| 4 | 2 3 7 |
| 5 | 1 3 7 9 |
| 6 | 1 3 4 6 8 8 |
| 7 | 2 8 9 9 |
| 8 | 1 2 3 4 |
| 9 | 3 4 |
| 10 | 0 3 4 |
| 11 | 0 2 7 |

**b** 87% **c** 28; rainy day and few customers

**8 a** 36 **b** 91 seconds **c** about 28%

**9 a**

| Stem | Leaf |
|---|---|
| 5 | 2 5 6 8 |
| 6 | 0 0 3 3 4 4 4 5 8 8 9 9 |
| 7 | 1 2 2 4 4 6 6 8 9 |
| 8 | 0 1 1 3 3 4 4 5 9 |
| 9 | 0 0 3 4 7 |
| 10 | |
| 11 | 0 |

**b** 10 **c** 110, heart is beating quickly

**d** 4

**10 a** $975 000 **b** 50 **c** $385 000

**11 a**

| Stem | Leaf |
|---|---|
| 11 | 6 8 |
| 12 | 0 1 4 |
| 13 | 6 9 |
| 14 | 3 5 6 6 6 |
| 15 | 4 6 6 7 |
| 16 | 3 5 8 |
| 17 | 2 4 7 |
| 18 | 1 7 |

**b** 24 **c i** 11.6 s **ii** 18.7 s

**d** 50.0% **e** $\frac{5}{24}$

## Sample HSC problem

**a** A

**b** Select some fans from each team's home ground.

**c** 83°

**d** players' salaries

**e** Doesn't show exact values; it is difficult to compare similar values.

## Test yourself 1

**1 a** 2700 **b** liver **c** 6 **d** false

**e** These cancers do not represent parts of a whole.

**2 a** 13 918 647 females

**b** ≈ 154 000 000 km$^2$

**c i** $105 449 million **ii** $76 502.2 million

**iii** $28 946.3 million

**3 a** The vertical scale should start at zero.

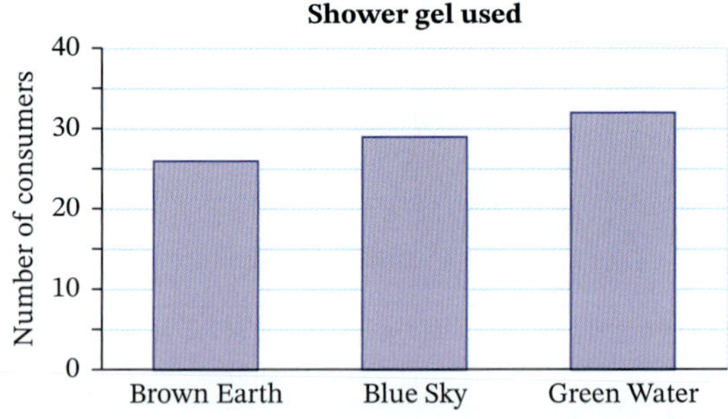


**b** sector graph, divided bar graph

**4** B; In graph A, both dimensions have been tripled so the number of houses represented by the second diagram is 90.

**5 a** N **b** C **c** C

**d** N **e** N **f** C

**6 b** N **c** O **f** N

**7 a** D **d** C **e** D

**8** systematic, for example, take every 10th bottle and test

**9 a** The sample may not necessarily take students from each Year level into account.

**b** 9

**10** Teacher to check.

**11 a** 93

**b** number of countries visited

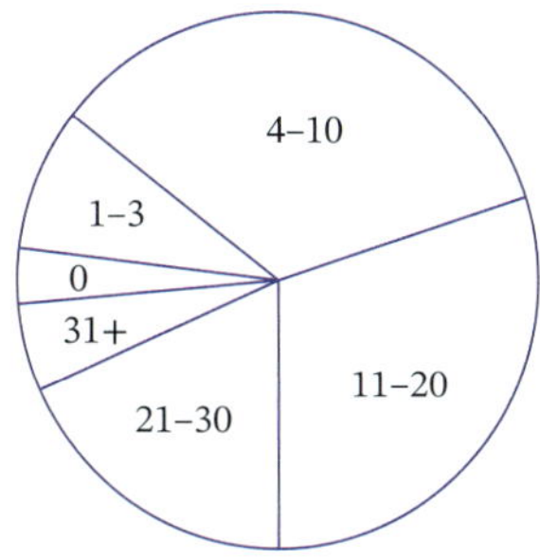


**c** 12° **d** 34.4% **e** T

**f** The data is not continuous; the class intervals are not equal in size.

**12 a** continuous

**b**

| Mass (kg) | Frequency |
|---|---|
| 50–<60 | 4 |
| 60–<70 | 12 |
| 70–<80 | 7 |
| 80–<90 | 8 |
| 90–<100 | 3 |
| 100–<110 | 6 |
| Total | 40 |

**c**

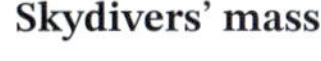


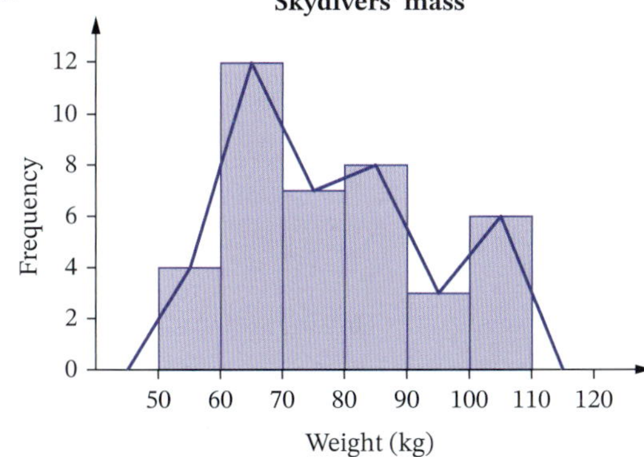


**d** 60–<70 **e** $\frac{1}{5}$

**13** Age of children

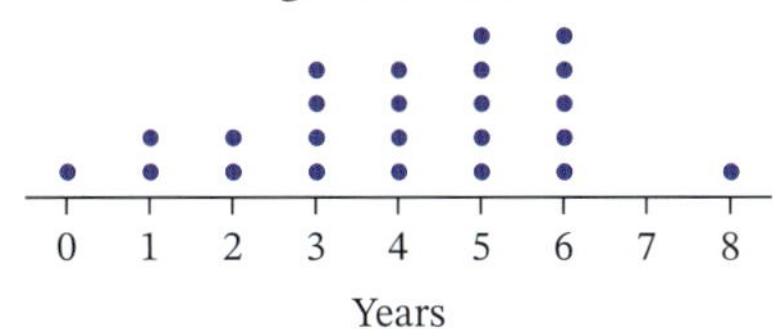


**b** 24 **c** $\frac{1}{4}$

**d** Data are clustered around 3–6 years.

**14 a** 3 years **b** 29 **c** 12 **d** 41%

# Chapter 2

## SkillCheck

**1 a** −27 **b** 6 **c** −8
**d** 25 **e** −4 **f** 2

**2 a** 15 **b** 3 **c** 16
**d** 6 **e** 243 **f** 18

**3 a** 0 **b** −3 **c** 10
**d** −128 **e** 9 **f** 9

**4 a** $x-3$ **b** $20-a$ **c** $-3$

**5 a** $6x-3$ **b** $28a+35$ **c** $-12p+4$

**6 a** $b=-5$ **b** $w=17$ **c** $r=9$

**7** $A=37.5$

## Exercise 2.01

**1 a** $3x^2+5x$ **b** $7ab+a+b$ **c** $2m^2+5$
**d** $-4r+2ar$ **e** $y^2+2y-5$ **f** $a^2+2$
**g** $k^2-4k-1$ **h** $-3de+2d$ **i** $-6p$
**j** $4u-15$ **k** $2t^2+3at$ **l** $18x$

**2 a** $16a^2km$ **b** $-24cd^2$ **c** $\frac{5}{2}t^3w^2$
**d** $25x^6$ **e** $6n^4$ **f** $14k^5p^2$
**g** $36mn$ **h** $\frac{3}{2}e^3$ **i** $4x^2y^4$
**j** $-20a^2b^2$ **k** $-n^6$ **l** $9k^2$

**3 a** 2 **b** $7x$ **c** $-5$
**d** $2tu$ **e** $\frac{3}{x^2}$ or $3x^{-2}$ **f** $\frac{-3r}{q}$ or $3rq^{-1}$
**g** $\frac{3g}{5}$ or $\frac{3}{5}g$ **h** $\frac{1}{a}$ or $a^{-1}$ **i** $\frac{5m^2}{3n}$ or $\frac{5}{3}m^2n^{-1}$
**j** $\frac{-d}{5}$ **k** $\frac{1}{4}$ **l** $\frac{x^2}{4w}$ or $\frac{1}{4}x^2w^{-1}$

**4** A

**5 a** $\frac{45y^2}{4}$ **b** $\frac{8ah}{35}$ **c** 2
**d** $\frac{5}{6}$ **e** $2n^2$ **f** $\frac{5n}{4}$
**g** $\frac{14a}{5h}$ **h** $\frac{16}{15}$ **i** $d$

**6 a** $-k^2-2k+8$ **b** $\frac{1}{2}ay^4$ or $\frac{ay^4}{2}$
**c** $\frac{1}{4}$ **d** $-8fg$
**e** $-10c^2+13c$ **f** $-4z^2+5z+8$
**g** $6a^3$ **h** $\frac{3c}{e}$ or $3ce^{-1}$
**i** $-10m^2$ **j** $\frac{2n^2}{x}$ or $2n^2x^{-1}$
**k** $4x^2+2xy+3x$ **l** $-3a^2$

## Exercise 2.02

**1 a** $3a+6$ **b** $15-10b$ **c** $-4a-2$
**d** $-6b+12$ **e** $3x^2-6x$ **f** $3p^2-3ap$
**g** $-8k-16$ **h** $6t-8t^2$ **i** $-d^2+5d$
**j** $7k-5k^2$ **k** $-9b^2+9b$ **l** $14xy+8y^2$

**2** D **3** B

**4 a** $-24n+6n^2$ **b** $5rx^2+10rx$
**c** $-2a^2+4$ **d** $5a^2b+15b^2-35b$
**e** $-x^2+4x-10$ **f** $3h^2-21eh-12eh^2$
**g** $2y^2+3y-y^3$ **h** $d^3e-2de+de^3$
**i** $9av^2-3v^2+6av$

**5 a** $3x+14$ **b** $d-22$ **c** $2r+80$
**d** $7f+9$ **e** $-9x-32$ **f** $3x^2+27x$
**g** $2b^2+23b$ **h** $3w^2-29w$ **i** $9k+12p$
**j** $4a$ **k** $vx+3x$ **l** $-7t-w$
**m** $4e^2+3e$ **n** $2a-18$ **o** $p^2-q^2$

**6** C

**7 a** 2 and −2; opposites **b** −6 and 6
**c** yes; $-(b-a)=-b+a$
$=a-b$

## Exercise 2.03

**1** **a** 91 **b** $-\frac{8}{9}$ **c** 6 **d** 0
**e** 36 **f** 4 **g** 60 **h** 3
**i** 42 **j** −12 **k** 7 **l** 169
**2** C **3** B **4** 602.63 cm$^3$
**5** **a** 100°C **b** 55°C **c** 28°C
**6** **a** $14.44 **b** $64.62 **c** $3465.60
**7** **a** $v$ **b** 16 m/s
**8** **a** 17 **b** 81 **c** 301
**9** 1.6 m **10** 12 770.1 cm$^3$ **11** 3.05 s
**12** $8775.56 **13** 13.5 m
**14** **a** $B$ **b** 20.8
**c** Increase her weight to around 65 kg.
**15** **a** 25°C **b** 38°C **c** 15°C **d** 37°C
**16** A
**17** **a** 144 mg **b** 129 mg **c** 141 mg
**18** A
**19** **a** 12.5 mg **b** 29.2 mg **c** 45.8 mg
**20** **a** 111 mg **b** 141 mg
**21** 59 km **22** $605.10 **23** 21 m
**24** 207.35 cm$^2$ **25** B

## Exercise 2.04

**1** **a** $d = 6$ **b** $p = 2\frac{1}{2}$ **c** $u = 3\frac{1}{2}$
**d** $a = -3$ **e** $b = -\frac{1}{3}$ **f** $a = 4\frac{1}{2}$
**g** $m = -5$ **h** $h = 12$ **i** $r = 13$
**j** $x = -20$ **k** $y = -5$ **l** $n = -1$
**m** $y = 5$ **n** $c = 7\frac{1}{2}$ **o** $z = 60$
**2** D
**3** **a** $k = 11$ **b** $e = -2$ **c** $f = 6\frac{1}{2}$
**d** $x = 17$ **e** $w = 70$ **f** $d = 3$
**g** $n = \frac{11}{4}$ or $2\frac{3}{4}$ **h** $w = \frac{-17}{5}$ or $-3\frac{2}{5}$
**i** $p = 4$ **j** $z = \frac{-26}{5}$ or $-5\frac{1}{5}$
**k** $b = -3$ **l** $q = 2$
**4** A **5** A

## Exercise 2.05

**1** **a** **i** 13 **ii** 31 **b** **i** 7 **ii** 18
**2** 12 cm
**3** **a** 300 m **b** 60 s
**4** 2 m **5** 21 **6** 8 cm **7** $4387.15
**8** **a** 2 min **b** 3.6 min
**9** **a** 3.11 miles **b** 0.93 miles
**10** **a** **i** 12 **ii** 26 **b** **i** 11 **ii** 19
**11** $4600 **12** A **13** 19.2 km
**14** **a** 18 **b** 27
**15** 3 h
**16** **a** **i** 16th **ii** 4th
**b** **i** 7 s **ii** 12 s
**17** D **18** 3 m/s$^2$ **19** $1625
**20** **a** $5(N + 3) = 33 - N$ **b** $N = 3$
**21** **a** $2(x + 7) = 29 - 4x$ **b** $x = 2.5$
**22** D
**23** **a** $4(15 - k) = 2k + 3$ **b** $k = 9.5$
**24** **a** $5w = 8w + 84$ **b** $w = 28$ **c** area $= 2352$ cm$^2$

## Exercise 2.06

**1** **a** $x = \frac{y-4}{2}$ **b** $x = \frac{T+7}{3}$ **c** $x = 3d - 1$
**d** $x = 18 - p$ **e** $x = \frac{k-r}{4}$ **f** $x = 2(C - n)$
**g** $x = \frac{s}{10b}$ **h** $x = 6V + 5$ **i** $x = \frac{12-z}{a}$
**2** **a** $d = St$ **b** $t = \frac{d}{S}$
**3** D **4** B
**5** **a** 1800° **b** $n = \frac{A}{180} + 2$ or $n = \frac{A+360}{180}$
**c** 9
**6** D
**7** **a** $h = \frac{2A}{a+b}$ **b** $b = \frac{2A}{h} - a$ or $b = \frac{2A-ah}{h}$
**8** **a** $r = \pm\sqrt{\frac{A}{\pi}}$ **b** $m = \frac{y-c}{x}$ **c** $s = \frac{v^2-u^2}{2a}$
**d** $x = sz + m$ **e** $c = \pm\sqrt{\frac{E}{m}}$ **f** $v = \pm\sqrt{\frac{2K}{m}}$
**g** $P = \frac{A}{(1+r)^n}$ **h** $y = \frac{x^2-3}{6}$ **i** $h = \frac{2A}{b}$
**j** $l = \frac{T^2 g}{4\pi^2}$ **k** $R = \frac{V+E}{I}$ **l** $Q = C(e - ir)$
**m** $a = \frac{2(s-ut)}{t^2}$ **n** $d = Kw - m$ **o** $t = \pm\sqrt{\frac{h-15}{4 \cdot 9}}$
**9** $h = \sqrt{\frac{m}{B}}$, 1.84 m or 184 cm **10** 20 cm
**11** $F = \frac{9C}{5} + 32$ or $F = \frac{9C+160}{5}$
**a** 104 **b** 212
**12** $P = \frac{10}{\pi hM}$, 1.2
**13** **a** $r = \sqrt{\frac{V}{\pi h}}$ **b** 2.922 m
**14** $m = \sqrt{\frac{y-d}{3}}$
**15** **a** $t = \pm\sqrt{\frac{100-h}{4 \cdot 9}}$
**b** **i** 2.0 s **ii** 4.3 s
**16** **a** $d = \sqrt{\frac{h-15}{0.008}}$
**b** **i** 50 m **ii** 19 m **iii** 75 m

## Sample HSC problem

**a** $248 **b** $h = \frac{C-110}{92} + 1$
**c** $5\frac{1}{2}$ hours

## Test yourself 2

**1** **a** $t^2 + 9ut$ **b** $15k^3$ **c** $-6d^2$
**d** $32p^5$ **e** $7 - 6h$ **f** $8r^2$
**g** $-27d^6$ **h** $\frac{3}{5}$ **i** $\frac{-vw}{7}$
**j** $8x^2 + 8x$ **k** $\frac{3c}{b}$ or $3cb^{-1}$ **l** $\frac{-1}{a^2}$ or $-a^{-2}$
**m** 2 **n** $\frac{2vy}{3}$ **o** $\frac{x}{12}$
**p** $\frac{5}{3a}$

**2** **a** $10x - 20$ **b** $-3a - 21$ **c** $48t - 4y$
**d** $-9r^2 - 18w$ **e** $8m^2n - 8mn^2$ **f** $-8d^2 + 2d^3$

**3** **a** $14x - 1$ **b** $2n^2 - n - 1$ **c** $24 - 10d$
**d** $-4p$ **e** $11u + 8$ **f** $8h^2 + 26h$

**4** **a** 128 **b** −28 **c** −400 **d** 10

**5** 204.20 cm² **6** $700

**7** **a** $p = 5$ **b** $a = -1$ **c** $b = -9$
**d** $r = \frac{1}{2}$ **e** $n = \frac{55}{4}$ or $13\frac{3}{4}$ **f** $r = -18$
**g** $n = -2$ **h** $t = -12\frac{1}{2}$ **i** $g = -1$

**8** 110.5 m **9** 80

**10** **a** $\frac{d-5}{4} = 2d + 7$ **b** $\frac{-33}{7}$ or $-4\frac{5}{7}$

**11** $y = 2(P - 110)$ or $y = 2P - 220$, 28 years

**12** **a** $y = \frac{a-h}{g}$ **b** $y = \frac{3m-1}{5}$
**c** $y = \pm\sqrt{\frac{4w-3}{k}}$ **d** $y = \pm\sqrt{\frac{2(V-12)}{h}}$

**13** **a** $t = \sqrt{\frac{50-h}{4.9}}$ **b** 2.7 seconds

## Chapter 3

### SkillCheck

**1** **a** 2 **b** 1.74 **c** 2.94
**d** 0.04 **e** 0.07 **f** 0

**2** **a**

| $x$ | 0 | 1 | 2 | 3 | 4 | 5 |
|---|---|---|---|---|---|---|
| $y$ | 4 | 3.7 | 3.4 | 3.1 | 2.8 | 2.5 |

**b** −0.3

**3**

| | i | ii | iii | iv | v |
|---|---|---|---|---|---|
| **a** | 5.5 | 5 | 4, 5 | 9 | 4 |
| **b** | 7.6 | 9 | 12 | 14 | 10.5 |

**4** **a** 28.4 **b** 31 326.7 **c** 37.6

**5** **a** 25 000 m **b** 2.463 km **c** 3.75 h
**d** 5 h 12 min **e** 7 h 20 min **f** $5.\dot{5}$ m/s
**g** $30.\dot{5}$ m/s **h** 12.6 km/h

**6** 20 160 km **7** 65.3 km/h

**8** **a** $5.01 **b** $2525.10 **c** $37.33

### Exercise 3.01

**1** B **2** B

**3** **a** Sonia **b** 6.58 L

**4** C

**5** **a** 2 **b** 3 **c** 73 kg **d** 0.25
**e**

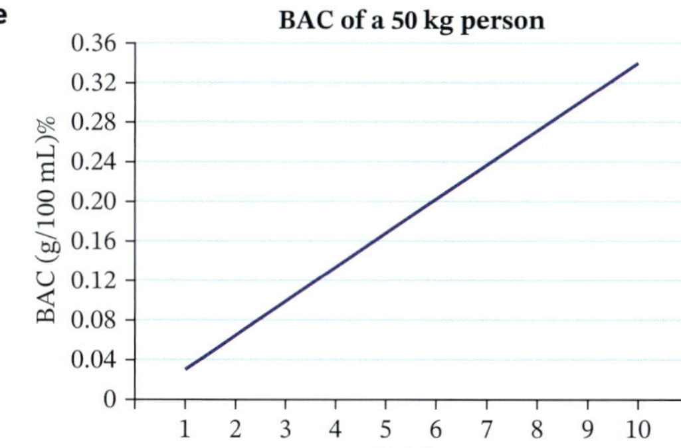


**6** **a** heavy drinker
**b** **i** 0.02 **ii** 0.017
**c** 0.07 **d** about 0.028 **e** about 0.01
**f** A heavy person gets rid of the alcohol in their blood system more quickly.

**7** **a** 0.21 **b** 10 h 30 min **c** 10:30 am the next day

**8** **a** 0.18 **b** 12 h **c** 2 pm

**9** 0.012 per hour

**10** **a**

| Time (hours) | BAC |
|---|---|
| 0.0 | 0.08 |
| 0.5 | 0.072 |
| 1.0 | 0.064 |
| 1.5 | 0.056 |
| 2.0 | 0.048 |
| 2.5 | 0.04 |
| 3.0 | 0.032 |
| 3.5 | 0.024 |
| 4.0 | 0.016 |
| 4.5 | 0.008 |
| 5.0 | 0 |

**b**

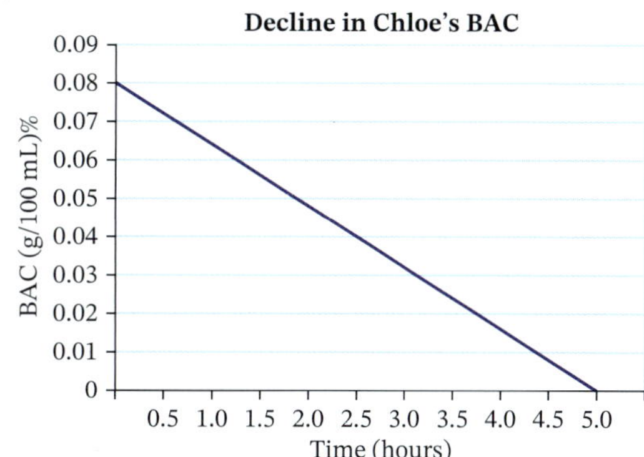


**c** **i** about 1:45 am **ii** 5 am
**d** 0.016, rate at which BAC is decreasing per hour

### Exercise 3.02

**1** **a** $\frac{28}{55}$ **b** 62% **c** 49%

**2** **a** **i** 1129 **ii** 1194
**b** 24.5%
**c** ACT. It is the smallest in area and population.
**d** South Australia
**e** 5.8%
**f**

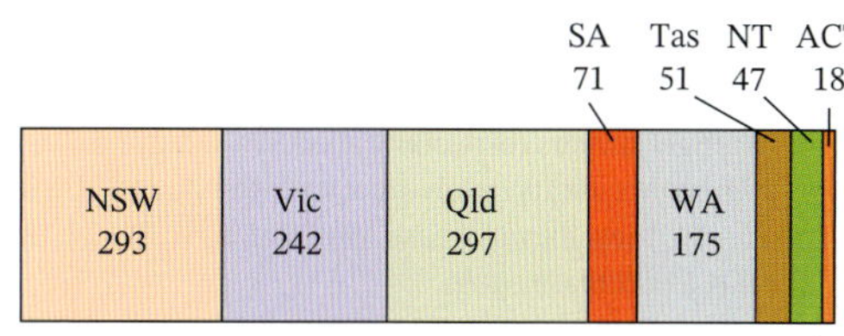


**3** **a** 55.4% × 18 764 ≈ 10 395
**b** 5592 **c** 4597

**4** **a** **i** 180° **ii** 68° **iii** 7°
**b** **i** 84 **ii** 62 **c** 227

**5** **a** 457 **b** 43.5% **c** 21.6%
**d** 65–74; fewer drivers in this category; more careful; less speeding; don't drive as much; travelling shorter distances; retired from work

**e**

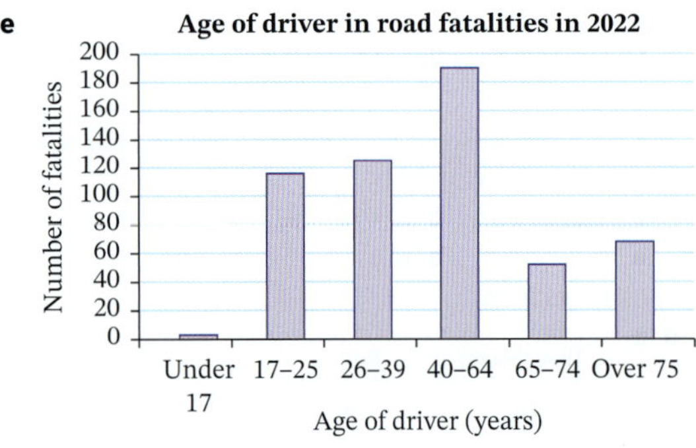


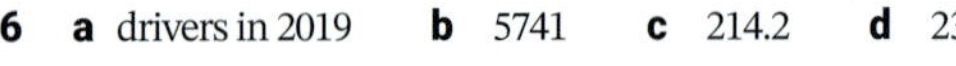

**6** **a** drivers in 2019 **b** 5741 **c** 214.2 **d** 23

**e**

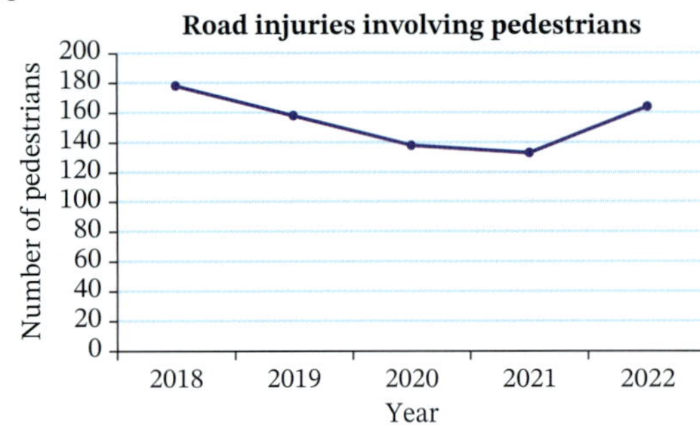


**7** **a** mean: 664.7, range: 123

**b** Vehicles are more likely to crash into stationary objects such as trees and poles than other vehicles because there are more objects

**c** It is gradually decreasing, possibly due to better roads, driver safety awareness and car safety technology.

**8** **a** 910 **b** 280 **c** 1190

**d** Male deaths increasing and more than double the number of female deaths. Female deaths fairly constant, no pattern.

**9** **a** 2 pm–3:59 pm; busy time when people start leaving work or picking up children from school.

**b** 2 am–3:59 am; quietest time of day when fewest vehicles on roads (people still sleeping).

**c** false

**d** **i** The day starts off with low casualties for 6 hours, at its minimum from 2 to 3:59 am After 6 am, the number of casualties begins to rise, reaching a peak from 8 to 9:59 am when people are going to work and school. It levels off between 10 am and 12 midday, then starts to increase as the afternoon peak time begins, at its maximum from 2 to 5:59 pm Then in the final 6 hours of the day until midnight, the number of casualties decreases.

**ii** Yes, because working hours and school hours have not changed, unless more people work from home. The numbers would be different on a weekend though.

**10** **a** 588 **b** 212 **c** 77.5% **d** 22.52%

**e** Male drivers are more than 3.5 times the number of female drivers; males drive more aggressively and faster than females.

**f** Younger drivers (especially aged 17–20) are more likely to have accidents involving speed, the figures dip for ages 26–29, then they increase again for ages 30–39, then start to decrease again.

## Exercise 3.03

**1** C **2** B

**3** **a** 78 km/h **b** 13 h 8 min **c** 247 km

**4** **a** 8 h 54 min **b** 2 h 48 min **c** 7 h 43 min

**5** **a** $3\frac{1}{2}$ h **b** 14 km/h

**6** **a** 34.2 km/h **b** 333 450 m

**7** **a** 1.1 m **b** 1.7 m **c** 2.2 m **d** 2.8 m

**8** **a** 3 s **b** 1472 m

**9** **a** 76.5 m/s **b** 12.6 min

**10** **a** 17.5 min **b** 248 km/h

**11** 41 min **12** 104.2 m **13** 62.1 km/h

## Exercise 3.04

**1** A **2** 197.7 m **3** 5.8 m

**4** **a** $A = 13.9$, $B = 12.6$, $C = 28.3$, $D = 106.5$

**b** **i** 299% **ii** 217%

**c** Yes. It takes 106.5 m to stop.

**5** **a** 7.9 m **b** 11.1 m

**c** Both would stop in time, Hanna in 30.1 m and Freya in 19 m.

**6** **a** 44.4 m **b** 17.7 m **c** Sanjay (52.3 m)

**7** **a** $k = 0.0097$ **b** 106.9 m **c** 88 km/h

**d** **i** 97.6 m **ii** 153.6 m

**8** **a** $k = 0.00678$ **b** 61.1 m **c** 153.5 m

**9** **a** 33.7 m **b** 102 m **c** 345 m **d** 238 m

**10** **a** 1.88 s **b** 99 m

**11** **a** 181 **b** 145 **c** 153 **d** 126

**12** **a** 0.007 46 **b** 90.3 m **c** 151 m

**13** **a** 5500 m **b** 32 500 m **c** 46 500 m

**14** **a** 160 m

**b** **i** 108 m **ii** 133 m **iii** 190 m

**c**

| Road condition | Reaction time | Stopping distance |
|---|---|---|
| dry | 1 | 78 |
| dry | 2 | 105 |
| dry | 4 | 160 |
| wet | 1 | 108 |
| wet | 2 | 133 |
| wet | 4 | 190 |

**d** You can see exact values in a table but a graph gives a good visual overall picture of the situation.

## Sample HSC problem

**a** 0.15 **b** 4 **c** 10

**d** 16 h

## Test yourself 3

**1** **a** 0.073 **b** 4 h 52 min **c** 0.024

**2** **a** **i** 0 **ii** 0.07 **iii** 0.17

**b** 9 h 27 min

**3** **a** Heavier people have more blood and water in their bodies to dilute alcohol.

**b** Females tend to be lighter than males.

**4** **a** 414 **b** 44 **c** 75.8% **d** 50.0%

**e** Fewer passengers killed, but roughly the same number of males and females. Many car accidents happen when the only occupant is the driver (no passengers) so drivers dominate the statistics, or passengers sit in a safer position in the car.

**f** Generally decreasing despite increasing populations. Reasons may include: better road conditions, safer cars, more police controls, better road safety education (around speed, drink- and drug-driving, wearing seat belts, phone use).

**5** **a** 11.1% **b** true

**c** **i** 2017 **ii** 2022

**d** 13.3%

**e**

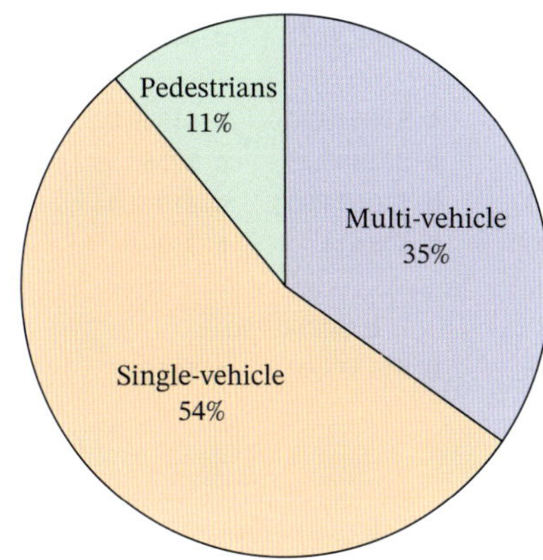


**6** **a** 73 km/h **b** 5 h 55 min **c** 364 km

**7** 935 m

**8** No, he travels 21.7 m before he applies the brakes so his stopping distance is 21.7 m + 24 m = 45.7 m.

**9** **a** same

**b** increases as reaction time increases

**c** $A = 23.7, B = 34.8, C = 57$

**d** No, he needs 57 m to stop.

**e** Drive slower than the speed limit, get someone else to drive you.

**10** **a** about 38 m **b** 10 m

**c** 100 km/h **d** 50 km/h

**e** It takes much longer to stop on a wet road. If the speed is double, the distance is not. For example, at 100 km/h on a wet road it takes 4 times as long to stop as it does at 50 km/h.

## Practice exam 1

**1** B **2** B **3** D **4** B **5** A

**6** C **7** A **8** C **9** D **10** B

**11** **a** $x = 3\frac{3}{5}$

**b** **i** 23 m **ii** 0.0029 **iii** 23.1 m

**c** **i** It could be 16 minutes, 17 minutes or 18 minutes.

**ii** 19 patients

**iii** 21.1%

**iv** 2 minutes

**12** **a** **i** 0.065

**ii** BAC would have been lower

**b** $2p + 3$ **c** 47 days

**d** **i** continuous

**ii**

| Mass (kg) | Frequency |
|---|---|
| 40–<45 | 3 |
| 45–<50 | 4 |
| 50–<55 | 4 |
| 55–<60 | 2 |
| 60–<65 | 2 |
| 65–<70 | 4 |
| 70–<75 | 1 |

**iii**

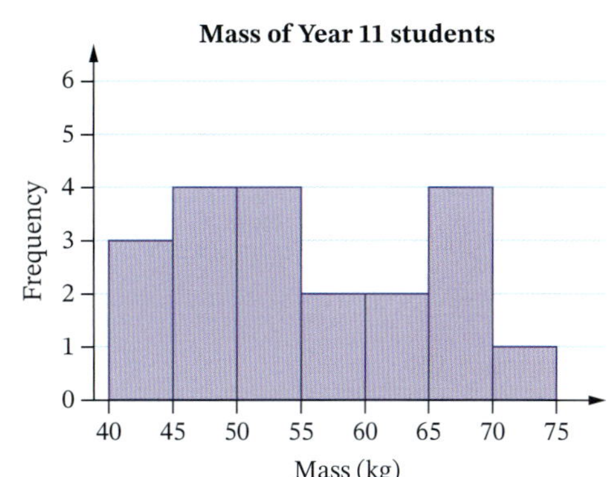


**iv** 35%

**13** **a** **i** $2(y + 5) = 44 - 5y$ **ii** $y = \frac{34}{7}$

**b** **i** $r = \sqrt{\frac{V}{\pi h}}$ **ii** 5.05 cm

**c** **i** 340 **ii** 18.2%

**iii** Drivers are older and may not be as aware of traffic, have slower reflexes.

**iv** 73 fatalities, which is more than the other age groups of 9 years. Drivers in the 17–25 age group are less experienced, more prone to peer pressure and possibly more reckless.

# Chapter 4

## SkillCheck

**1** **a** \$233.75 **b** \$4.65 **c** \$2772

**2** **a** \$14 473.12 **b** \$2942.31 **c** \$1284.44

**3** 6%

**4** **a** 1035 **b** 89 500

**5** **a** 4 h **b** 5 h 30 min **c** 9 h 45 min

## Exercise 4.01

**1** \$1792.88 **2** \$2209.20

**3** C **4** A **5** C

**6** **a** \$80 999.10 **b** \$1557.68 **c** \$40.99

**7** \$1399.44

**8** **a** \$61 318.40 **b** \$5109.87

**9** \$1358.37 **10** \$1176.06 **11** \$2678.00

**12** \$2101.45

**13** **a** \$8986.54 **b** \$19 470.83 **c** \$4493.27

**14** **a** \$6934.58 **b** \$3200.58 **c** \$320.06

**15** **a** \$61.81 **b** \$5192.31 **c** \$11 250

**16** \$986.90 **17** \$1282.56

**18** **a** \$353.21 **b** \$1766.05 **c** \$7652.88

**19** B **20** C

**21** **a** \$32.40 **b** \$1652.40 **c** \$85 924.80

**22** \$1789.44

**23** **a** \$3.70 **b** \$222

**24** \$28.80 **25** 5 hours

## Exercise 4.02

**1** A

**2** **a** 168 **b** \$56.70

**3** D

**4** **a** \$30.10 **b** \$6750

**5** \$9620

**6** **a** \$16 060 **b** \$24 100 **c** \$36 625

**7** \$203.98 **8** \$2714.04

**9** **a** \$341 880 **b** \$14 207 955.46

**c** \$2 117 874.43

**10** **a** $146.19 **b** 2500
**11** $386.10 **12** $3600 **13** $148 637.45
**14** $1215 **15** $3580 **16** $1384
**17** $1077.12 **18** $1678 **19** 8.8%
**20** **a** $644.10 **b** 47
**21** 60 **22** $4016.75
**23** **a** $1100 **b** $75 900
**24** $530.85

## Exercise 4.03

**1** $1308.54 **2** $1632.78
**3** **a** $1426.88 **b** $998.81
**4** $1895.40 **5** $663.06 **6** $334.40
**7** $668.36 **8** $1428
**9** **a** $827.40 **b** $5555.40
**10** B **11** C **12** 38 hours
**13** **a** $2439 **b** $1707.30

## Exercise 4.04

**1** **a** $1144.40 **b** $1453.50 **c** $0 (<67 years)
**2** Because a couple live in the same house and the cost of living is shared.
**3** **a** **i** $778.00 **ii** $20 228 **iii** $1685.67
**b** $9259.90
**c** More. This is fair because they have no partner to share costs.
**4** C
**5** **a** $712.30 **b** $833.20 **c** $1007.50
**d** **i** $21 663.20 **ii** 416.60
**6** B
**7** **a** $663.30 **b** $712.30
**c** $712.30 **d** $410.30

## Exercise 4.05

**1** C
**2** **a** $831.95 **b** 25.4%
**3** **a** $1262.96 **b** $353.63 **c** $792.46
**4** **a** $301.18 **b** 29.09% **c** 3.29%
**5** **a** $340.51 **b** $99.06
**c** $461.51 **d** $776.69
**6** **a** $1070.55 **b** $107.06 **c** $723.27
**7** **a** $2086.58 **b** $1606.58 **c** 14.7%
**8** **a** $1145.82 **b** $286.46 **c** $841.96
**9** **a** $9022.92; $3504.88 **b** 48%
**c** 38.8%

**10**

| | Pay | Deductions | |
|---|---|---|---|
| Normal | $1241.80 | Tax (25%) | $381.41 |
| Time-and-a-half | $212.88 | Superannuation | $60.65 |
| Double time | $70.96 | Union | $18.15 |
| Gross wages | $1525.64 | Health fund | $24.80 |
| Total deductions | $485.01 | | |
| Net wage | $1040.63 | | |

**11**

| | |
|---|---|
| Gross pay | $3347.15 |
| Tax (31%) | $1037.62 |
| Other deductions | $152.21 |
| Net pay | $2157.33 |
| Total deductions | $1189.83 |

## Exercise 4.06

**1** **a** $2070.40 **b** $36 739.21 **c** $0
**2** C
**3** **a** $89 451 **b** $17 623.30 **c** $1789.02
**4** **a** $138 129 **b** $32 445.73 **c** $2762.58
**5** **a** $99 623 **b** $20 674.90 **c** $1992.46
**6** **a** $154 800 **b** $38 614 **c** $3096
**7** $35 572.96
**8** **a** $94 128 **b** $20 908.96
**9** **a** $11 355 **b** $43 337.90 **c** $66 925
**10** 44 cents

## Exercise 4.07

**1** **a** $7696 **b** $45 065.80
**2** **a** $14 352 **b** $1074.54
**3** **a** $1330 **b** $298
**4** **a** $6532.42 **b** $1326 **c** $4976.92
**d** 23.8% **e** $15 912
**5** **a** $2018.04 **b** $292
**c** $1593.34 **d** $7592
**6** **a** $194 026 **b** $53 449.70
**c** $3880.52 **d** refund of $5045.78
**7** **a** $91 525.55 **b** $5572.78 **c** refund of $104.67
**8** **a** $54 520 **b** $7144
**c** $1090.40 **d** refund of $709.60
**9** A **10** A
**11** **a** debt of $8953.10 **b** debt of $11 423.58
**c** refund of $6170
**12** **a** $6590 **b** $1343 **c** $5058.50
**d** $77 486 **e** $15 583.52 **f** tax refund of $532.48

## Sample HSC problem

**a** $12 324 **b** $67 350.40 **c** $66 363
**d** income tax $10 696.90; Medicare levy $1327.26
**e** tax refund of $299.84

## Test yourself 4

**1** **a** $43 219.80 **b** $50 648 **c** $69 432
**2** **a** $2665.60 **b** $3788.58 **c** $21 980.33
**3** $751.50 **4** $28.63
**5** **a** $208.60 **b** 54
**6** 2 cents **7** $11 657.52 **8** $1294.85
**9** $1163.48 **10** D **11** 827.90 each
**12** **a** $21 372 **b** $822 **c** $3644.69 **d** $2765.29
**13** **a** $129 533 **b** $29 647.90
**c** $2590.66 **d** refund of $4793.24
**14** **a** $298 **b** $7344.80 **c** tax refund of $403.20
**15** $2672.15

# Chapter 5

## SkillCheck

**1** **a** 20 830 **b** 97.02 **c** 65 900
**d** 0.725 **e** 0.0104 **f** 0.735
**2** **a** 10 000 000 **b** 10 000 **c** 0.01
**3** **a** 16 cm$^2$ **b** 90 m$^2$
**c** 78.54 cm$^2$ **d** 60 cm$^2$
**4** **a** 180 cm$^2$ **b** 41
**5** **a** $1.28 **b** $2.49
**6** **a** 2.1 m **b** 2100 mm **c** 0.0021 km
**7** **a** 0.0081 kL **b** 8100 mL

## Exercise 5.01

**1** **a** 63 mm **b** 436 cm **c** 7.2 m
**d** 0.285 kg **e** 6900 μm **f** 58 L
**g** 5320 g **h** 3.4 t **i** 4.72 kL
**j** 0.006 kg **k** 27 000 s **l** 940 000 cm

**2** A

**3** **a** 1690 mm **b** 1.69 m

**4** 57 500 g **5** 1.5 km

**6** **a** 97 min **b** 5820 s **c** 1.6 h

**7** 59 000 L

**8** **a** kg **b** cm or m **c** m **d** km
**e** L **f** mg **g** cm **h** kL
**i** t **j** mm

**9** **a** 0.102 kg **b** 0.052 m

**10** 4.9 Mt

**11** **a** 26.25 km **b** 1 h 38 min

**12** 8.56 kL

**13** **a** B **b** D **c** C **d** C
**e** A **f** A

**14** about 1420

**15** 86 400

## Exercise 5.02

**1** B

**2** **a** 3800 **b** 2100 **c** 0.0061
**d** 250 000 **e** 15 000 000 **f** 0.000 47

**3** **a** 130 **b** 4980 **c** 0.0106
**d** 1 360 000 **e** 25 400 000 **f** 0.000 680

**4** **a** 3000 **b** 3 **c** 10 000
**d** 0.005 **e** 20 **f** 0.7

**5** **a** Y **b** N **c** Y
**d** N **e** Y **f** Y

**6** **b, c, d, e, f**

**7** **a** 0.67 **b** 0.0055 **c** 1.3
**d** 0.087 **e** 790 000 **f** 7.3
**g** 24 **h** 2.4 **i** 34

**8** 8 660 000

**9** **a** 352 000 kg **b** 4190 m **c** 67.1
**d** 14.8 mL **e** 150 000 000 km

## Exercise 5.03

**1** **a** $4.213 \times 10^7$ **b** $1.81 \times 10^{-2}$ **c** $3.4 \times 10^3$
**d** $2.0 \times 10^4$ **e** $3.5 \times 10^{-3}$ **f** $2.0 \times 10^{-4}$
**g** $3.3 \times 10^{-1}$ **h** $4.0 \times 10^{-3}$ **i** $2.3 \times 10^2$
**j** $7.23 \times 10^{-5}$ **k** $6.1 \times 10^8$ **l** $8.0 \times 10^{-8}$

**2** **a** $5.3 \times 10^7$ **b** $1.5 \times 10^5$ **c** $2.5 \times 10^3$
**d** $4.6 \times 10^{-4}$ **e** $2.7 \times 10^{-3}$ **f** $1.0 \times 10^{-1}$
**g** $3.3 \times 10^{-5}$ **h** $4.4 \times 10^{-1}$ **i** $6.5 \times 10^0$

**3** B

**4** **a** 740 000 **b** 0.312 **c** 1850
**d** 0.000 66 **e** 0.002 54 **f** 475 100 000
**g** 0.098 **h** 300 **i** 0.054 97
**j** 12 160 **k** 0.802 **l** 6309

**5** **a** $1.37 \times 10^{10}$ **b** $1.0 \times 10^{-6}$

**6** **a** 0.000 000 03 **b** 9 461 000 000
**c** 0.000 002 **d** 152 600 000

**7** $8.19 \times 10^9$

**8** **a** $2.144 \times 10^7$ **b** $3.2 \times 10^5$
**c** $3.5 \times 10^4$ **d** $2.304 \times 10^{-5}$
**e** $5.314\,41 \times 10^{14}$ **f** $3.76 \times 10^1$
**g** $1.26 \times 10^4$ **h** $2.3 \times 10^5$

**9** **a** $3.0 \times 10^4$ **b** $-3.6 \times 10^{-3}$
**c** $4.1 \times 10^{12}$ **d** $3.3 \times 10^{-8}$
**e** $5.9 \times 10^4$ **f** $3.4 \times 10^{-2}$
**g** $9.2 \times 10^2$ **h** $1.2 \times 10^{-4}$

**10** $3.4 \times 10^9$ nm **11** $7.8 \times 10^{13}$ mL

**12** $3.75 \times 10^5$ **13** $4 \times 10^{-12}$ Tm

**14** **a** 0.0048 **b** 43 680 000 000
**c** 0.0094 **d** 187 690 000
**e** 13 640 000 **f** 19 063 000
**g** 0.000 052 **h** 350 000

**15** Answers will vary. Teacher to check.

**16** $2.5 \times 10^6$ μm **17** $4.5 \times 10^9$ μg/L

**18** $4.6 \times 10^{-3}$ TB

## Exercise 5.04

**1** **a** 5.39 m **b** 7.30 m **c** 12.21 cm
**d** 18.73 cm **e** 5.66 cm **f** 7.62 m

**2** **a** 9.2 cm **b** 11.0 cm **c** 30.0 cm
**d** 98.6 m **e** 57.8 cm **f** 33.0 m

**3** 420 cm **4** 461 m **5** C

**6** 120 nautical miles **7** 1080 m

**8** **a** 335 cm **b** 602 cm

**9** 19.4 m **10** 214 mm **11** 396 cm

**12** 110.99 m

**13** **a** 0.9 m **b** 198.0 m

**14** 1.3 m **15** 2.54 m **16** 116.62 m

**17** 58.31 km **18** 21.77 cm

**19** **a**

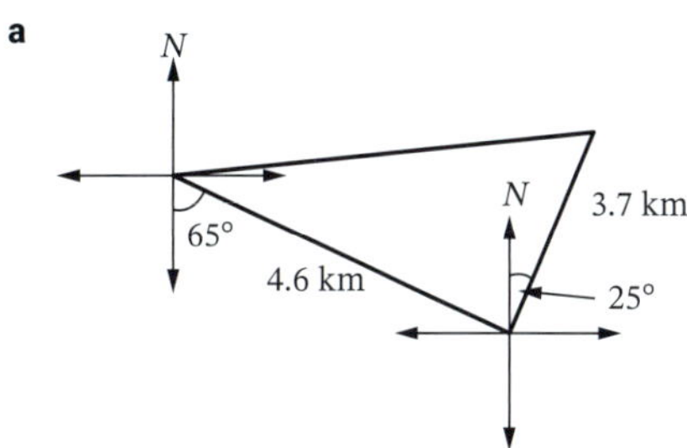


**b** 5.9 km

**20** 0.67

## Exercise 5.05

**1** **a** 30 m **b** 68 cm **c** 66 cm
**d** 54 cm **e** 1720 cm or 17.2 m
**f** 79.67 m **g** 30.6 cm **h** 172 mm
**i** 100 cm **j** 230 mm = 23 cm
**k** 40 cm **l** 58 cm

**2** **a** 53.4 m **b** 10.3 m **c** 21.4 cm
**d** 22.3 cm **e** 100.7 mm **f** 46.6 cm
**g** 51.4 cm **h** 121.1 mm **i** 34.3 m
**j** 94.2 cm **k** 43.2 m **l** 57.1 cm

**3** **a i** 16.8 m **ii** 40.8 m
**b i** 15.1 m **ii** 34.3 m
**c i** 202.6 mm **ii** 374.6 mm

**4** 25.1 m

**5** **a** 45° **b** 41.8 cm

## Exercise 5.06

**1** C **2** A

**3** **a** 50 000 **b** 25 **c** 7.2
**d** 0.68 **e** 3 090 000 **f** 360
**g** 4 730 000 **h** 5.4

**4** $8.016 \times 10^7$

**5** **a** 4 209 000 mm², 4.209 m²
**b** $2.3 \times 1.83 = 4.209$ m²

**6** A **7** C

**8** **a** $64\,m^2$ **b** $4322.5\,m^2$ **c** $24\,m^2$ **d** $40\,m^2$ **e** $375\,m^2$ **f** $4\,523\,893.4\,km^2$ **g** $300\,m^2$ **h** $6\,m^2$ **i** $88.4\,m^2$ **j** $197.9\,m^2$ **k** $189\,m^2$

**9** **a** $93\,m^2$ **b** $36\,m^2$ **c** $7747\,mm^2$ **d** $127\,m^2$ **e** $30\,m^2$ **f** $89\,m^2$

**10** **a** $7\,m^2$ **b** $8482\,cm^2$ **c** $150\,796\,mm^2$

**11** **a** $52\,m^2$ **b** $165\,m^2$ **c** $105\,m^2$ **d** $33\,m^2$ **e** $168\,m^2$ **f** $66\,m^2$

**12** $42.3\,m^2$

**13** **a** $15\,000\,m^2$ **b** 588.5 m **c** \$72 974

**14** **a** 0.4096 **b** \$55.30

**15** **a** **i** $60.7\,cm^2$ **ii** $256\,cm^2$

**b** After inserting the four outer parts into the semicircular gaps, it is equivalent to the area of the dotted square.

**16** $15.52\,m^2$

## Exercise 5.07

**1** **a** $430\,m^2$ **b** $960\,m^2$ **c** $6200\,m^2$ **d** $3800\,m^2$ **e** $2400\,m^2$

**2** $1871\,m^2$

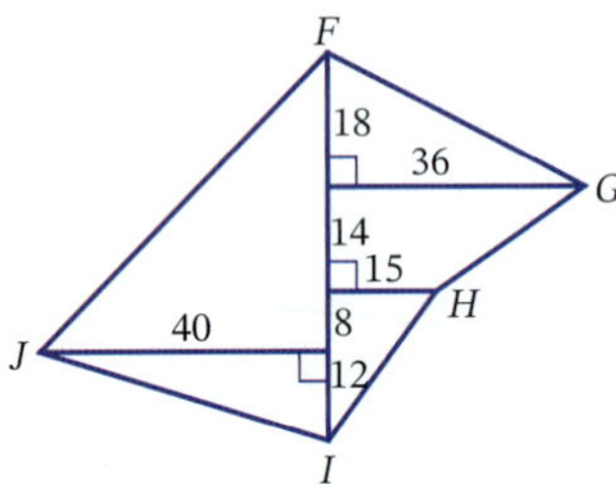


**3** **a** $1161\,m^2$ **b** 148 m

**4** B

**5** **a** $297\,m^2$ **b** $660\,m^2$

**6** **a** $61\,m^2$ **b** \$270

**7** $128\,m^2$

**8** $106.5\,m^2$

**9** **a** $15.9\,m^2$ **b** 19 080 L

**10** **a** $22\,m^2$ **b** 35 200 L

**11** **a** $10\,047\,m^2$ **b** $1266\,m^3$

**12** **a** $21\,375\,m^2$ **b** $17\,000\,m^2$

## Exercise 5.08

**1** **a** $7\,000\,000\,cm^3$ **b** $50\,000\,mm^3$ **c** $0.089\,m^3$ **d** $468\,000\,cm^3$ **e** $2.4\,cm^3$ **f** $5.6\,m^3$ **g** $9100\,cm^3$ **h** $12\,000\,000\,cm^3$ **i** $2.45 \times 10^9\,mm^3$

**2** **a** 680 mL **b** 8.5 L **c** 22 000 L **d** $8\,m^3$ **e** $3.5 \times 10^6$ mL **f** $690\,000\,cm^3$ **g** 55 000 L **h** 4300 kL **i** $9.5\,m^3$ **j** 85 L **k** $4300\,cm^3$ **l** 1000 ML

**3** A

**4** **a** $0.864\,m^3$ **b** 864 L

**5** B

**6** **a** $250\,cm^3$ **b** $1.701\,m^3$ **c** $3036\,cm^3$ **d** $20.16\,m^3$

**7** 249 L

**8** **a** $16.1\,m^3$ **b** 16 100 L

**9** **a** $4\,m^3$ **b** \$264

**10** **a** $2.031 \times 10^{12}$ L **b** $2.031 \times 10^9\,m^3$

**11** C

**12** **a** $236\,m^2$ **b** $347\,m^2$ **c** $205\,m^2$ **d** $317\,m^2$ **e** $736\,m^2$

**13** **a** $1.7\,m^2$ **b** 180 L

**14** **a** $6\,m^2$ **b** $0.8\,m^3$

**15** **a** $1.8\,m^3$ **b** $10.5\,m^2$

**16** **a** $84\,cm^3$ **b** 298

## Exercise 5.09

**1** **a** $1800\,m^3$ **b** $1.0\,m^3$ **c** $2700\,m^3$ **d** $0.27\,m^3$ **e** $43\,m^3$ **f** $3.0\,m^3$

**2** **a** $840\,m^2$ **b** $6.0\,m^2$ **c** $1100\,cm^2$ **d** $3.5\,m^2$ **e** $72\,m^2$ **f** $11\,m^2$

**3** **a** $327\,cm^3$ **b** 2 L

**4** **a** $7100\,cm^3$ **b** $0.85\,m^3$ **c** $31\,m^3$ **d** $9000\,cm^3$

**5** $5451\,cm^2$

**6** **a** $3.1\,m^3$ **b** $13.4\,m^2$

**7** **a** $39\,m^3$ **b** 25 133 L

**8** C

## Exercise 5.10

**1** **a** $360\,cm^3$ **b** $208\,m^3$ **c** $495\,mm^3$ **d** $1560\,cm^3$ **e** $1707\,m^3$ **f** $48\,m^3$

**2** **a** $2261.95\,cm^3$ **b** $87.93\,cm^3$ **c** $1.17\,cm^3$ **d** $61\,261.06\,mm^3$

**3** $96\,mm^3$

**4** **a** $1440\,cm^3$ **b** 1.44 L

**5** $2450\,cm^3$

**6** $2\,420\,000\,m^3$

**7** $576\,m^3$

## Exercise 5.11

**1** **a** $310\,000\,km^3$ **b** $224\,000\,m^3$ **c** $9.05 \times 10^8\,m^3$ **d** $994\,cm^3$ **e** $2.44\,m^3$ **f** $11\,500\,m^3$

**2** **a** **i** $1.011 \times 10^8\,km^2$ **ii** $1.068 \times 10^{12}\,km^3$ **b** $5.592 \times 10^9\,t/km^3$

**3** **a** 262 L **b** $1.57\,m^2$

**4** **a** $113\,cm^3$ **b** $103\,cm^3$ **c** \$16.96

**5** A

**6** **a** $92\,000\,cm^3$ **b** 9852

**7** **a** $2.9\,m^2$ **b** $1.4\,m^2$ **c** $0.46\,m^3$

*Note:* The light is a whole sphere but split in half.

**8** $3769.7\,cm^2$

## Exercise 5.12

**1** **a** $994\,020\,mm^3$ **b** 66.7% **c** $1\,265\,625\,mm^3$

**d** It is the most efficient packaging as it has the least amount of air.

**2** **a** $1\,716\,000\,mm^3$ **b** $1593\,cm^3$ **c** 7% **d** $86\,200\,mm^2$

**3** **a** $8400\,cm^3$ **b** $3200\,cm^2$ **c** $1200\,cm^3$ **d** $410\,cm^2$

**4** **a** 42 103 L; 4800 L **b** 2nd pool by 5897 L **c** $73\,m^2$

**5** **a** $0.31\,m^3$ **b** $33\,414\,cm^2$

**6** $460.8\,m^2$

**7** **a** $3141.6\,cm^2$ **b** $13\,090.0\,cm^3$ **c** 13.1 L

## Sample HSC problem

20 slices

## Test yourself 5

1 a 28 500 b 6400 c 0.34
2 43 000
3 a 39 b 1000 c 0.0087
d 6 600 000
4 $8.35 \times 10^9$
5 a 106.3 m b 7000 $m^2$
6 0.000 004 6
7 a 40 m b 316 cm c 55.00 cm
8 a 84 $cm^2$ b 56 000 m
9 a 44 $m^2$ b 1040 $cm^2$ c 230 $cm^2$
10 a 120.1 cm b 646.3 $cm^2$
11 a 605 $m^2$ b 0.0605 ha
12 a 20 700 $mm^3$ b 1.65 $m^3$
13 a i 750 $cm^3$ ii 480 $cm^2$
b i 0.22 $m^3$ or 220 000 $cm^3$
ii 3.2 $m^2$ or 32 000 $cm^2$
14 a 894 mL b 6500 L
15 a 110 $m^2$ b 187 $m^2$
c 187 $m^2$ is more accurate, because by using 2 applications we use more trapeziums to approximate the area more accurately.
16 a 8.9 $m^2$ b 1.4 $m^3$
17 57 470
18 a 2 574 000 $m^3$ b 218.6m
19 a 1715.31 $cm^3$ b 1.07 L
20 a 204 $m^3$ b 26 L
21 a 4700 $cm^3$ b 52 cm

# Chapter 6

## SkillCheck

1 90 days 2 $5.07 3 $87.86
4 a 0.025 b 0.84 c 1.06 d 0.1875
5 a 2.5 b 0.125 c 0.45 d 0.15
6 a $83.40 b $1145 c $115.50
d $381.20 e $36.48 f $11.30
7 $1393.20
8 a $29.36 b $2375 c $88.70

## Exercise 6.01

1 $56.25 2 $80 3 $7700
4 30 190 321 5 $84 346.35 6 $1607.33
7 $134.52
8 $2523.75 for cash (accept $2523.74)
9 a i $670 ii $549.40
b $494.46
10 a $102.35 b $94.16
c $5.16 increase, 5.8%
11 16.8%
12 a $440 b 36.7%
13 a $889.88 loss b 11.7% loss
14 8.3%
15 a loss of $18 b 4.0%
16 profit of $750

## Exercise 6.02

1 D 2 €1845 3 $154.78
4 A 5 $32.00 6 £105.75
7 46.10 rand 8 21%
9 a €7.97 b 24%
10 a $38 550 b $42 405
11 Shop A, cheaper by £1.40
12 319.80 kronor 13 $814
14 a $74 750 b $65 750 c $1369.79/month
15 $A = \$13.20$, $B = \$8.75$
16 $52.20

## Exercise 6.03

1 A
2 a $720 b $120
3 $677 4 12 weeks
5 a $95.88 b $616
6 a Option 1: $1130, Option 2: $1142
b Option 1 is cheaper by $12.
c Option 1: $94.17, Option 2: $63.44
Jasmine should choose Option 2, as the monthly repayment is $63.44, which is lower than the $94.17 monthly repayment for Option 1.
7 Virat will pay $14.60 more under Plan B.
8 Option 2; it's $36 cheaper.
9 Teacher to check. 10 The plan offered by BUZZ.

## Exercise 6.04

1 B 2 $30 3 A 4 C 5 A
6 a $27 interest b $477 c $59.63
7 a Plan B b Plan B
8 a $4550 b $40 950 c $19 451.25
d $60 401.25 e $1006.69 f $64 951.25
9 a $43 008.32 b $6140.46 c 2.7% p.a.
d Sample answers: Popularity of car, excess stock is cheaper, price reductions on old stock as new models are released.
10 a $43 778 b $1112.13
11 a i $2491.20 ii $2580.48
b i $1593 ii $1622.70
c Local Loans – Positive: 12 months to save up balance
Negative: interest charged per day on amount owing
Feelgood Finance – Positive: 100 days to save up price of computer
Negative: more interest paid as no deposit
12 a $1950 b $1694.80 c $2078.80
13 B
14 a $2000 b $4000 c $540
d $6540 e $252.22
15 a $3200 b $28 800 c 15% and 16% p.a.
d 4-year option: $46 080, 8-year option: $65 664
e 4-year option: $960, 8-year option: $684

## Exercise 6.05

1 a $A = \$118.58$ b $B = \$322.85$
2 $29.07
3 a 21 days
b quarterly
c water and wastewater sewerage service, $180.78
d 18 kL
e no, in cooler months fewer showers, water garden less
f similar usage so can compare savings made for seasonal conditions
g 205 L h $2.276/kL i $18 \times 2.276$ j $138.84
4 a $54.62 b 15.4 kL c $230.10
5 a 255.6 kL b $682.45 c $170.61
d 3.864 kL e $10.32 f $17.89
g $35.51
6 C 7 $81.69 8 B
9 a 7.5 kL b 14 days c $17.07
10 a Baskarans $114.71, Zhongs $145.44, Zhongs pay more
b Baskarans 140 L, Zhongs 118 L, Baskarans use more

## Exercise 6.06

1 **a** $A = 1084.4$, $B = \$156.37$, $C = \$48.86$
**b** $408.50

2 $253.27

3 $561.61

4 **a** quarterly **b** 17.23 kWh
**c** June/July — winter, more electricity used for heating.
**d** 1568 kWh **e** 91 days
**f** 14 November, 58712 kWh **g** increased, $82.92
**h** Yes, because the comparison graphs show home usage for 2- and 4-people households
**i** Peak electricity used when most people use it, 2 pm–7.59 pm; off-peak at lower usage times, 10 pm–6.59 am; "shoulder" at in-between usage times, 7 am–1.59 pm, 8 pm-9.59 pm and weekends 7 am–9.59 pm. Supply charge is a daily charge for accessing electricity.

5 **a** $8.62 **b** 91 days **c** 5074 MJ
**d** 10.98 MJ **e** 26 July and 24 October
**f** $A = \$84.17$, $B = \$61.55$, $C = \$42.93$, $D = \$61.27$
**g** October to April has a lower usage at about 35 MJ/day, which are the warmer months of the year. Increases to about 60 MJ/day in colder months, May–October, where heating is used, cooking more hot food etc.

6 **a** 7.98 kWh/day for peak usage, 4.86 kWh/day for off-peak usage
**b** June to August — winter, when electricity is used for heating.
**c** Daily supply charge of $0.9964. For 90 days, $90 \times \$0.9964 = \$89.676 \approx \$89.68$.
**d** During off-peak shoulder times, people are still home in the day and later at night using appliances, charging devices etc. for longer periods of time, 7 am–2 pm and 8 pm–10 pm; Other: off-peak Mon–Fri 10 pm–7 am: shorter amount of time, more people sleeping usually.
**e** 2 pm–8 pm Mon–Fri
**f** peak
**g** Peak: slightly decreased by 1.8 kWh/day; off-peak usage: increased very slightly by 0.3 kWh/day.

7 $499.62

## Exercise 6.07

1 **a** expense **b** income **c** expense
**d** expense **e** income **f** expense
**g** expense **h** expense **i** income
**j** expense **k** income **l** expense
**m** income **n** income **o** expense

2 fixed: a, f, g, l; discretionary: c, d, h, j, o

3 D **4** B

5 **a** $257.70 **b** discretionary, $158
**c** Sample answers: spend less on clothes, look for grocery specials

6 $3000

7 **a** C **b** A **c** B **d** D

8 **a** $A = \$\,237.50$, $B = \$12\,748.76$ **b** $269.70

9 **a** $96.90 **b** yes **c** no

10 **a** Teacher to check. **b** $140.75

11 **a** $1084.90 **b** $730.77 **c** $354.13

## Sample HSC problem

$26.80

## Test yourself 6

1 $102.56 **2** A **3** 17% profit **4** B

5 **a** $28.75 **b** $40.74

6 **a** $1080 owed **b** 1240

7 **a** $680 **b** $68

8 **a** $970 **b** $80.83

9 C

10 **a** $950 **b** $152 **c** $1102 **d** $45.92

11 $960.47 **12** $272.12

13 **a** Noh: $114.71, Waddock: $145.44; Waddocks pay more
**b** Noh: 140 L/person, $28.68
Waddock: 118.3 L/person, $24.24
The Noh family uses more water per day.

14 $822.21 **15** $803.86

16 **a** $1120 **b** Teacher to check.
**c** sample answer: cut down on coffee and take-away meals

17 **a** $10 250 **b** 21 months

18 $375 per month

## Practice exam 2

1 B **2** C **3** C **4** B

5 D **6** A **7** C **8** A

9 D **10** B

11 **a i** 17.9 kL **ii** $279.17
**b** 1130 m$^2$
**c i** $1025 **ii** 1 **iii** $27.38 **iv** 27

12 **a i** $78 975
**ii** $14 480.50 **iii** $1579.50
**b** 4500 L
**c i** $9.285 \times 10^{11}$ km$^3$ **ii** $4.603 \times 10^{10}$ ha

13 **a** 45 kL
**b i** 424 cm$^2$ **ii** 83 cm
**c** $647.50 **d** $37 872

# Chapter 7

## SkillCheck

1 **a** $1776 **b** $9100 **c** $14 352
**d** $229.50 **e** $42

2 **a** $64 **b** $3328 **c** $277.33

3 **a** $770.40 **b** $1608.00 **c** $1397.00
**d** $102.56 **e** $4454.45 **f** $1374.65

4 **a** 144 **b** 1008
**c** 4380 **d** 52 560

5 $197.10 **6** $56 745

## Exercise 7.01

1 C **2** D **3** $1535.36 **4** A

5 **a** $158.40 **b** $234.50
**c** $385.02 **d** $370.63

6 $217.98 **7** $510.35

8 **a** $470.20 **b** $622.20 **c** $1360

9 **a** $573 **b** $603.90 **c** 524.70
**d** $468.50 **e** $573 **f** $624.25

10 **a** $933.40 **b** $9699 **c** $734.80

## Exercise 7.02

**1** D **2** C

**3** **a** \$900 **b** \$2200 **c** \$2800

**4** **a** \$550 **b** \$1987.50 **c** \$4165

**5** **a** \$320 **b** \$3450 **c** \$7300

**6** **a** \$260 **b** \$440 **c** \$2730

**7** D **8** B

**9** \$70 599 **10** \$32 532.20

**11** **a** \$31 496.58 **b** \$6993.50 **c** \$65 169.22
**d** \$27 516.75 **e** \$19 953.47 **f** \$58 096.55

**12** \$47 732.80

## Exercise 7.03

**1** \$10 538.84

**2** **a** 43 L **b** \$4559.88

**3** **a** \$4680 **b** \$10 570
**c** \$407/fortnight

**4** \$29 097.28

**5** \$12 265.24 **6** \$189.58

**7** \$153.88

## Exercise 7.04

**1** C **2** D

**3** **a** 0.115 L/km **b** 113 L
**c** 478 km **d** 1477 km

**4** **a** 0.55 L **b** 129.09 L

**5** **a** 0.08 L **b** 12 L

**6** **a** Toyota Corolla **b** 1098 km
**c** 120 L **d** \$296.80

**7** **a** \$139.37 **b** 719 km

**8** \$112.50 **9** 9.09 L/100 km

**10** Brodie's (10.4 L/100 km)

**11** 7.2, 383.9, 109, 5, 886.6, 303.9, 9.7

## Exercise 7.05

**1** \$222.87

**2** Car 1: \$105 756
Car 2: \$105 510
Car 2 is slightly cheaper.

**3** \$7.80

**4** **a** \$24 525 **b** \$201.55

**5** A = \$35 096.16, B = \$34 474.44
B is the better option.

**6** The warranty is for 5 years or 100 000 km, whichever comes first.
So if a car with this warranty covers a distance of 100 000 km in 2 years, the warranty has then lasted only 2 years.

**7** Teacher to check.

## Sample HSC problem

**a** \$561
**b** **i** \$19 300 **ii** \$25 920 **iii** \$124.62

## Test yourself 7

**1** **a** \$286.90 **b** \$320.76

**2** **a** \$8.30
**b** Nothing. He was only insured for damage to other people's property. He needs comprehensive insurance to insure his own property.

**3** **a** \$642 **b** \$1323 **c** \$2035 **d** \$3565

**4** \$27 708.96

**5** **a** \$8110.40 **b** \$155.97

**6** **a** 10.74 L/100 km
**b** 5.1, 3.3, 49.7, 221.3, 720.5
**c** Don't speed; accelerate slowly; walk or cycle short distances

**7** **a** 941 km **b** 165 L

**8** \$7.28

# Chapter 8

## SkillCheck

**1** **a**

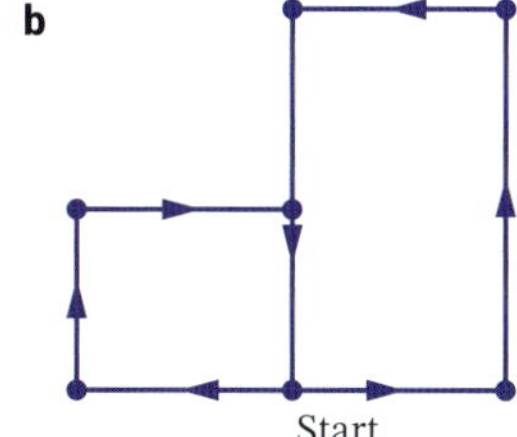

**b**

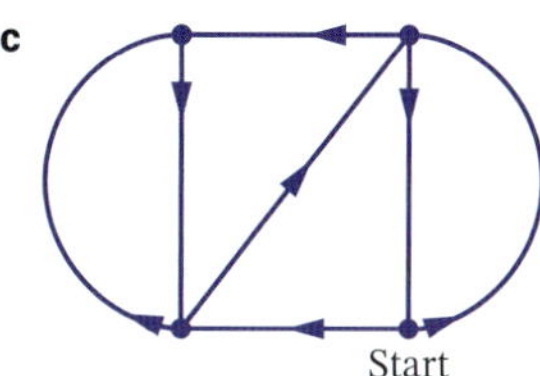


**c**

Start

**2** **a** *AEBDCA* **b** *AB*

**3** 78 km

**4** **a** **i** Ritu, Kristy, Belinda **ii** Madison, Belinda
**b** Madison
**c** Through Madison or Belinda

## Exercise 8.01

**1** **a** **i** 5 **ii** 6
**b** **i** 8 **ii** 11
**c** **i** 5 **ii** 7
**d** **i** 7 **ii** 10
**e** **i** 4 **ii** 5
**f** **i** 4 **ii** 6

**2** B **3** B **4** D **5** D

**6** **a** **i** 111 **ii** 42, *ABGEF*
**iii** 101, *ABGECBDEF*
**b** **i** 216 **ii** 34, *ABCF*
**iii** 160, *ABCDEADF*

**7** D

**8** *DHPYTX*, 71 km

**9** *GFEDFADC* or *GFADFEDC*, 2060 m

**10** **a** **i** $M-2, T-4, X-2, W-3, P-3$
**ii** 3 **iii** $14 = 2 \times 7$
**b** **i** $B-2, C-2, D-2, F-2$ **ii** 4
**iii** $8 = 2 \times 4$
**c** **i** $A-2, B-3, W-2, K-4, R-1, Y-3, X-3, G-2$
**ii** 4 **iii** $20 = 2 \times 10$
**d** **i** $A-1, C-4, P-2, Y-1$
**ii** 2 **iii** $8 = 2 \times 4$

**11** B **12** B **13** C

**14 a**

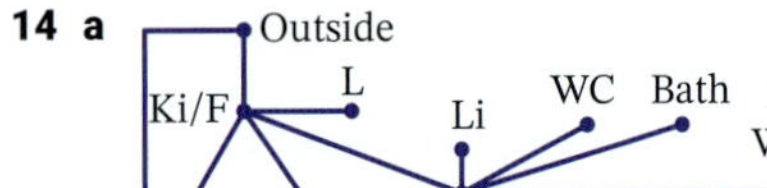

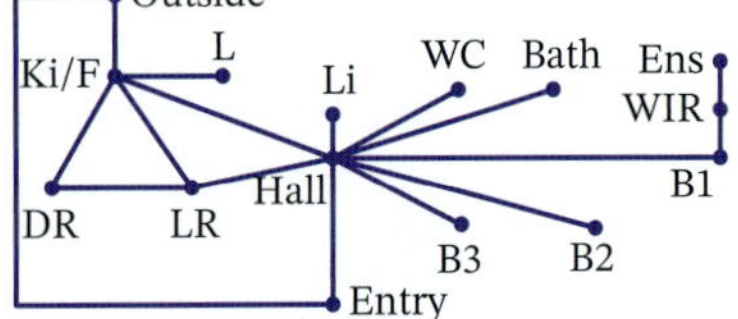


**b**

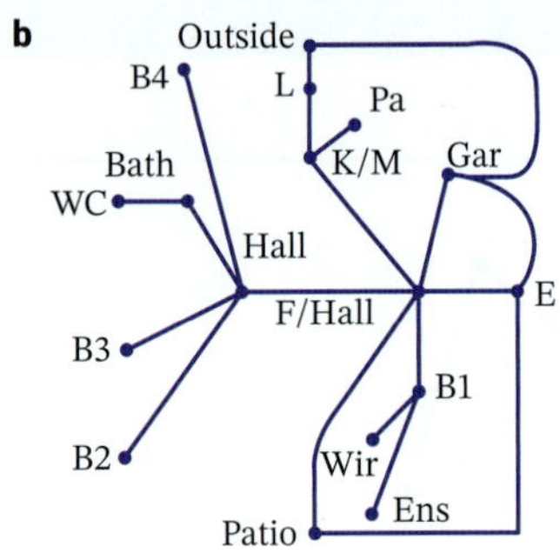


**15**

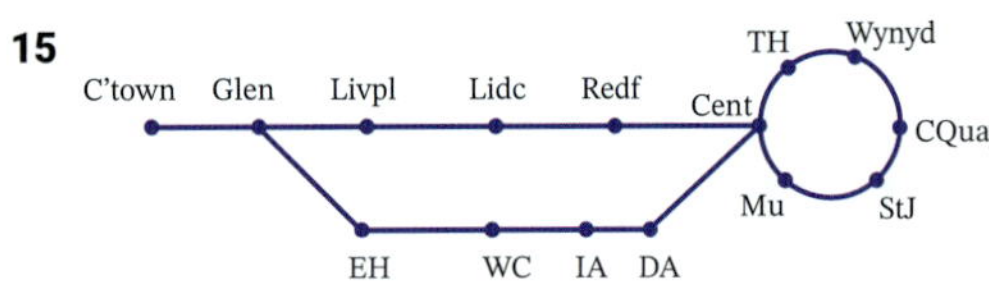


**16 a**

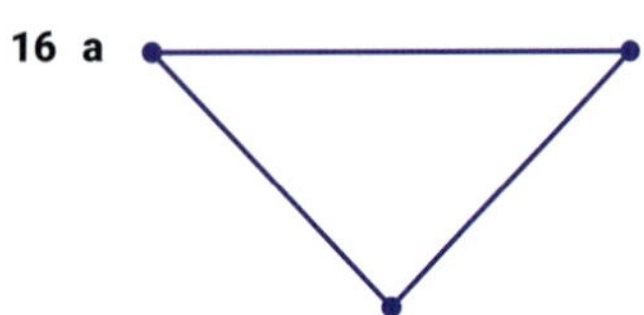

**i** 2 **ii** 3

**b**

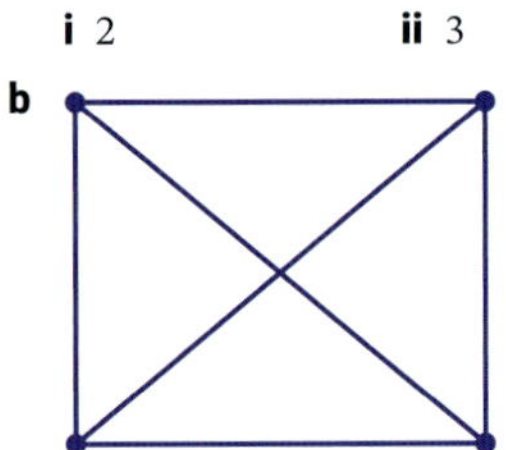

**i** 3 **ii** 6

**c**

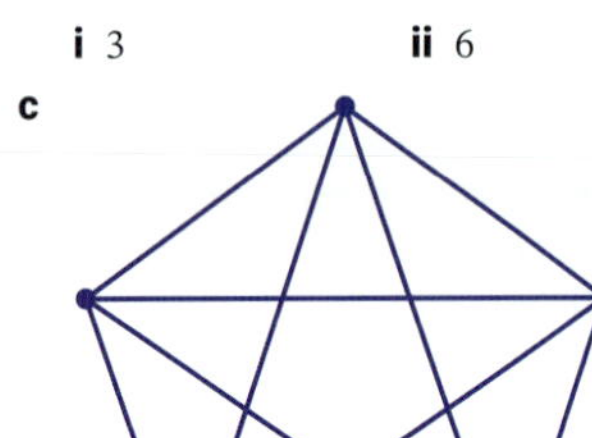

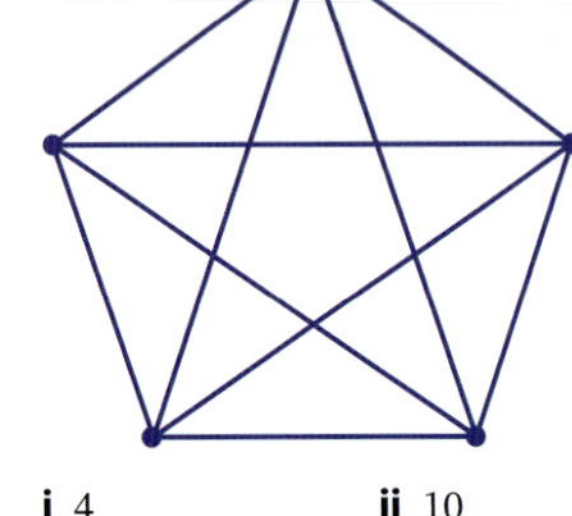

**i** 4 **ii** 10

**d**

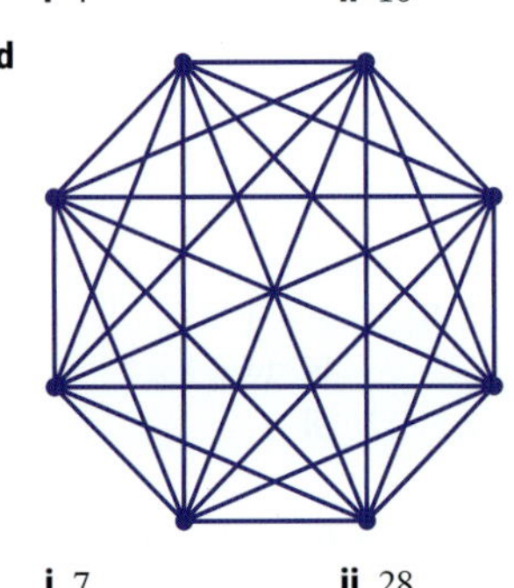

**i** 7 **ii** 28

## Exercise 8.02

**1** **a** There is no edge between the vertices.

**b** $a = 1, b = 0$

**2**

| | A | B | C | D | E |
|---|---|---|---|---|---|
| **A** | 0 | 1 | 0 | 0 | 1 |
| **B** | 1 | 0 | 1 | 1 | 1 |
| **C** | 0 | 1 | 0 | 1 | 0 |
| **D** | 0 | 1 | 1 | 0 | 1 |
| **E** | 1 | 1 | 0 | 1 | 0 |

**3** **a**

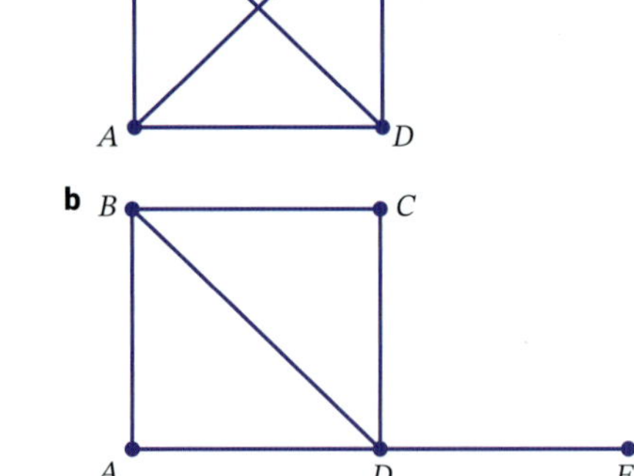


**b**

**4** **a**

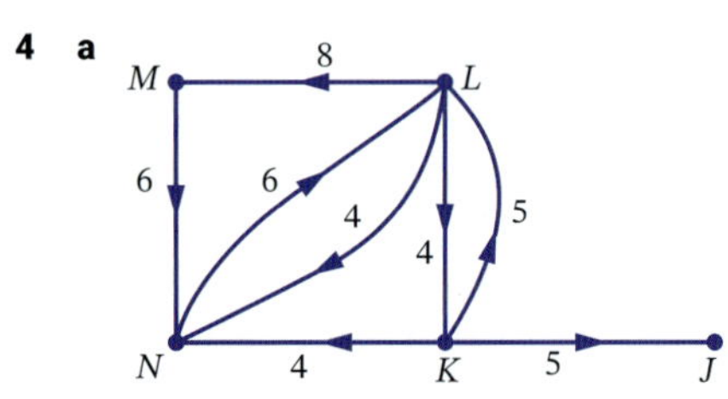


**b**

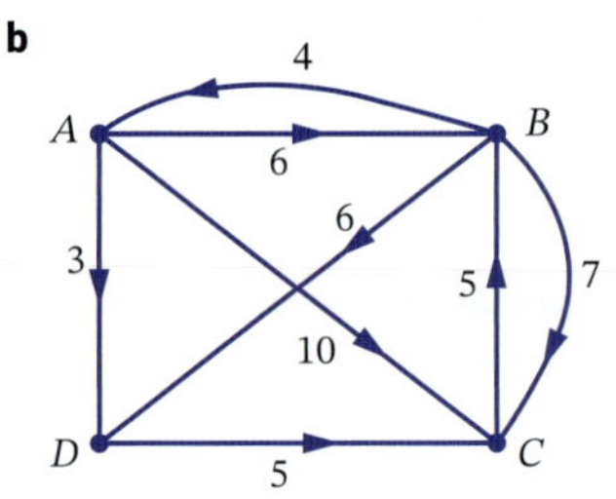


**5**

| | | To | | | |
|---|---|---|---|---|---|
| | | **M** | **P** | **T** | **R** |
| From | **M** | 0 | 3 | 0 | 2 |
| | **P** | 1 | 0 | 1 | 0 |
| | **T** | 1 | 2 | 0 | 2 |
| | **R** | 3 | 0 | 2 | 0 |

**6**

| | A | B | G | H | P | W |
|---|---|---|---|---|---|---|
| **A** | 0 | 15 | 0 | 0 | 18 | 20 |
| **B** | 15 | 0 | 16 | 6 | 0 | 0 |
| **G** | 0 | 16 | 0 | 8 | 14 | 12 |
| **H** | 0 | 6 | 8 | 0 | 0 | 0 |
| **P** | 18 | 0 | 14 | 0 | 0 | 13 |
| **W** | 20 | 0 | 12 | 0 | 13 | 0 |

**7**

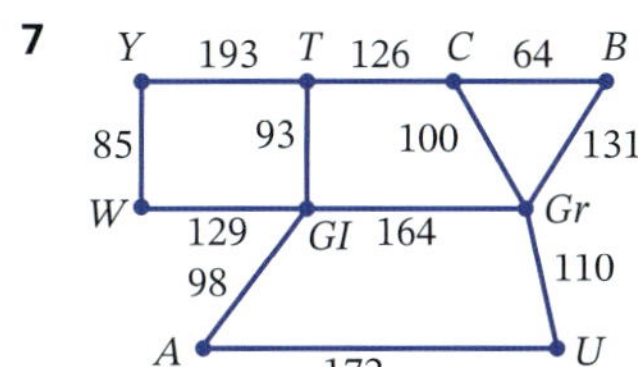


**8 a**

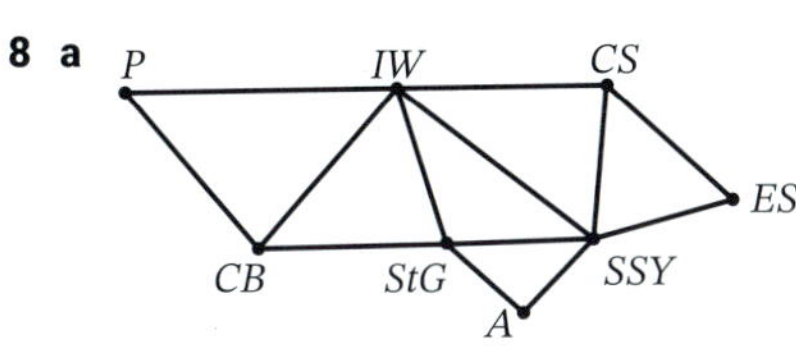


**b** *IW* and *SSY** **c** Inner West and South Sydney

**9 a** $V - 3, W - 2, P - 4, R - 3, T - 2, U - 4$

**b**

**10**

| | B | W | F | S |
|---|---|---|---|---|
| **B** | 0 | 1 | 0 | 1 |
| **W** | 1 | 0 | 2 | 2 |
| **F** | 0 | 2 | 0 | 2 |
| **S** | 1 | 2 | 2 | 0 |

**11** Teacher to check – one example is:

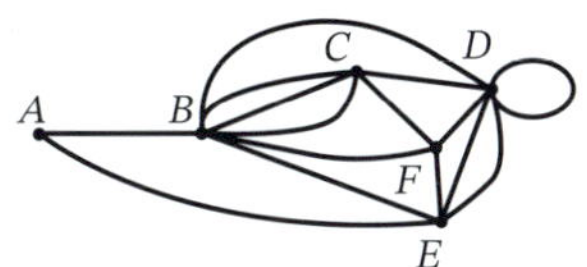


**12 a** *HEAGB* or *HEACB*

**b**

| | | To | | | | | | |
|---|---|---|---|---|---|---|---|---|
| | | **A** | **B** | **C** | **D** | **E** | **G** | **H** |
| From | **A** | – | – | 1 | – | 1 | 1 | – |
| | **B** | 1 | – | 1 | – | – | 1 | – |
| | **C** | – | 1 | – | – | – | – | – |
| | **D** | – | – | – | – | – | 1 | – |
| | **E** | 1 | – | – | – | – | – | 1 |
| | **G** | 1 | 1 | – | – | 1 | – | – |
| | **H** | – | – | – | – | 1 | – | – |

## Exercise 8.03

**1** **a**, **d** are trees as any 2 vertices are connected by exactly one path.

**b**, **e** are not trees as they are not connected networks.

Network **c** is not a tree because it contains a cycle.

**2** **a** **i** 9 **ii** 8

**b** **i** 12 **ii** 11

**c** **i** 8 **ii** 7

**3** B

**4** **a**, **d**, **e** and **f** are spanning trees.

**5** 3 possible spanning trees are:

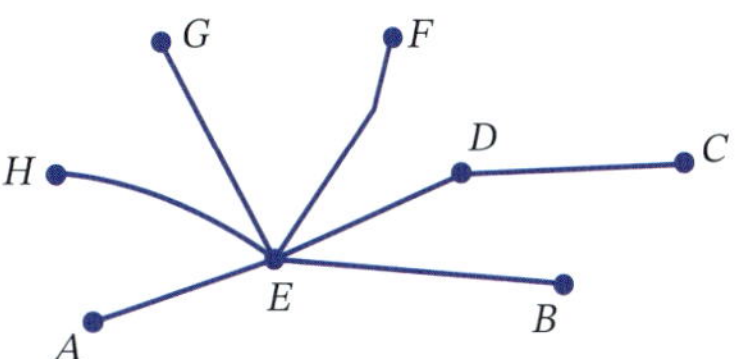


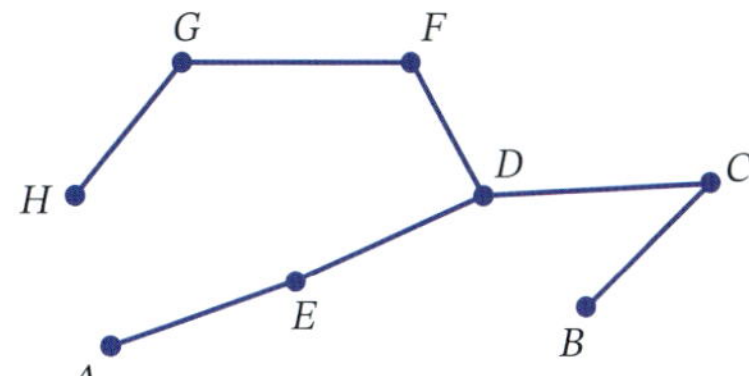


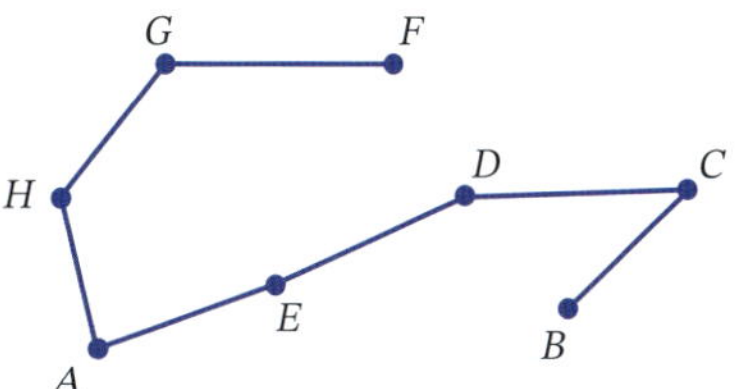


**6** **a** **i** \$7960 **ii** \$8090 **iii** \$7530 **iv** \$7840

**b** Option **iii** is the best.

A better option is the spanning tree shown with a weight of \$7230.

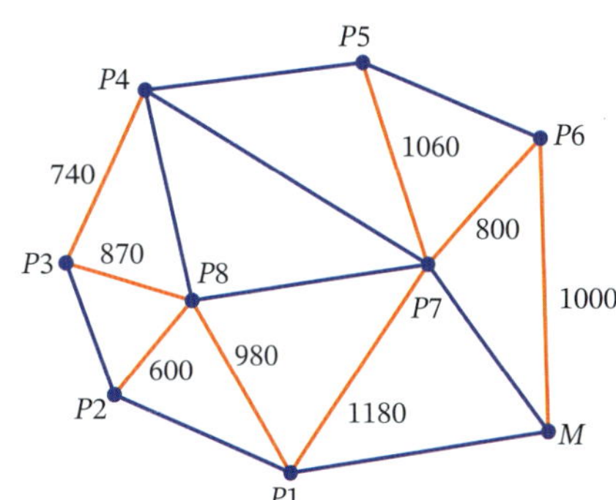


**7**

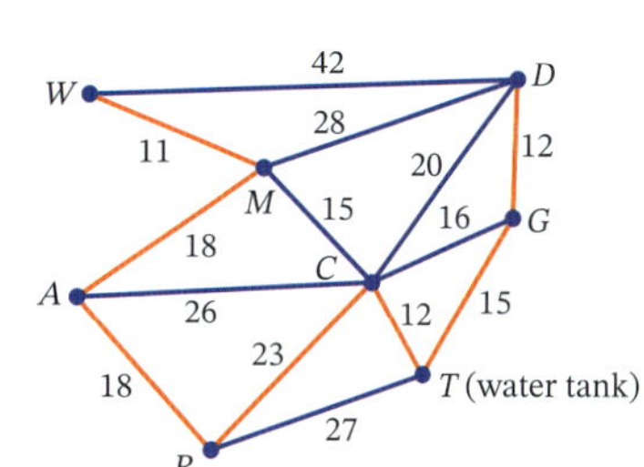


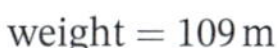
weight = 109 m

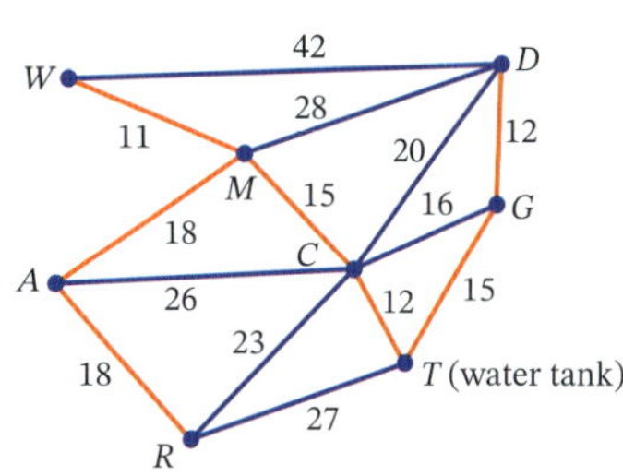


weight = 101 m

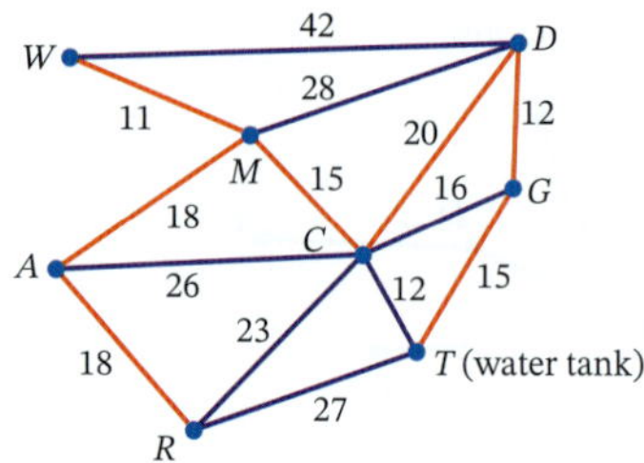


weight = 109 m

Minimum length of water pipe required is 101 metres.

**8** **a**

| Edge | Weight |
|---|---|
| *KM* | 5 |
| *BE* | 6 |
| *EM* | 7 |
| *AK* | 8 |
| *AM* | 9 |
| *AE* | 10 |
| *AB* | 11 |
| *BK* | 12 |

**b** Edges in order: *KM*, *BE*, *EM*, *AK*

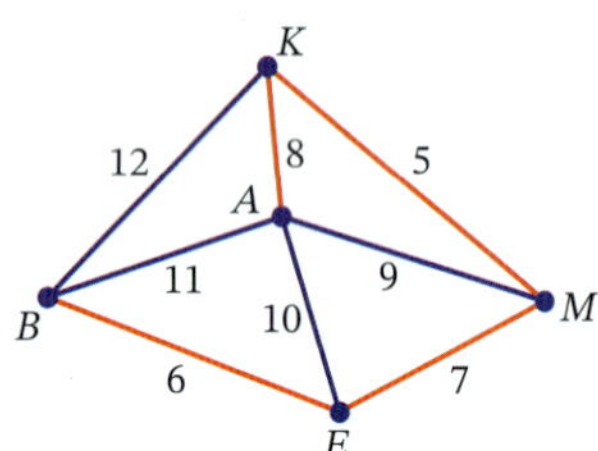


**c** weight = 26

**9** **a** **i** Edges in order: *GM*, *QT*, *AP*, *PW*, *PM*, *AT*; weight = 34

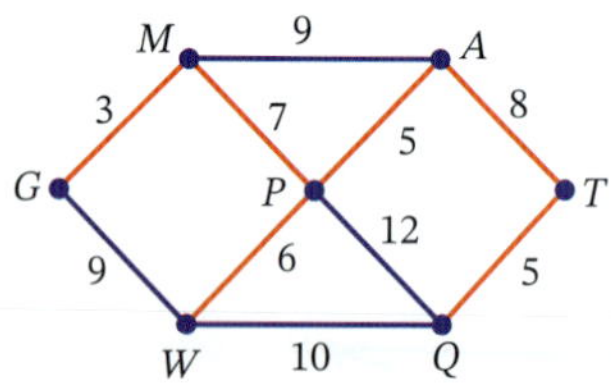


**ii** Spanning tree has 7 vertices, 6 edges.

**b** **i** Edges in order: *CD*, *AE*, *CF*, *BC*, *AB*; weight = 42

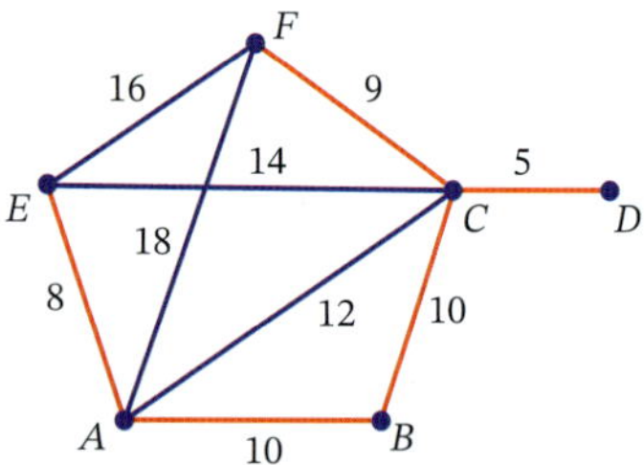


**ii** 6 vertices, 5 edges

**c** **i** Edges in order: *AF*, *CF*, *BC*, *DE*, *CD*; weight = 32

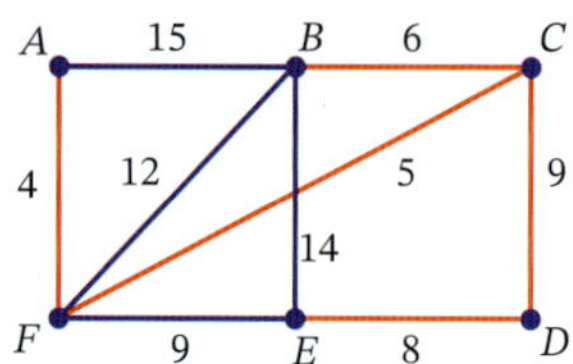


Note: another minimum spanning tree is *AF*, *CF*, *BC*, *DE*, *EF*

**ii** 6 vertices, 5 edges

**d** **i** Edges in order: *KR*, *AD*, *EY*, *PR*, *PY*, *DE*, *HP*, *AW*; weight = 33 (edge *AE* not used as it would have made a cycle)

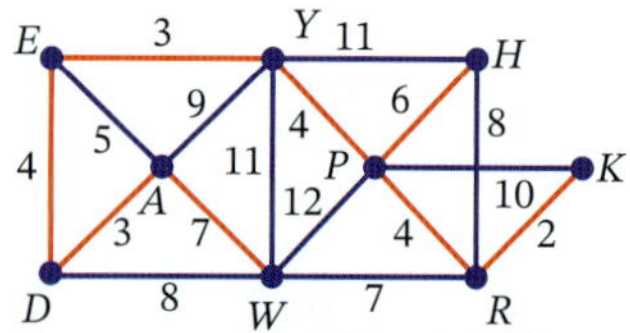


**ii** 9 vertices, 8 edges

**10** **a** *L SCR*, *L IT*, *A S1*, *S2 SCR*, *A S2*

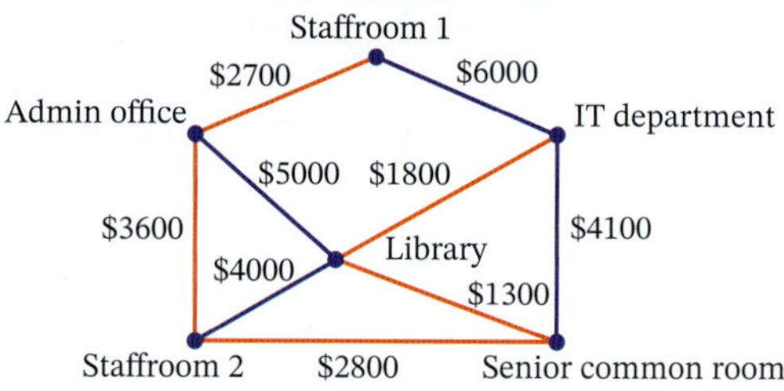


**b** minimum cost = $12 200

**11** A

**12** **a** Starting at *D*, edges in order are *DE*, *DC*, *CB*, *BG*, *GA*.

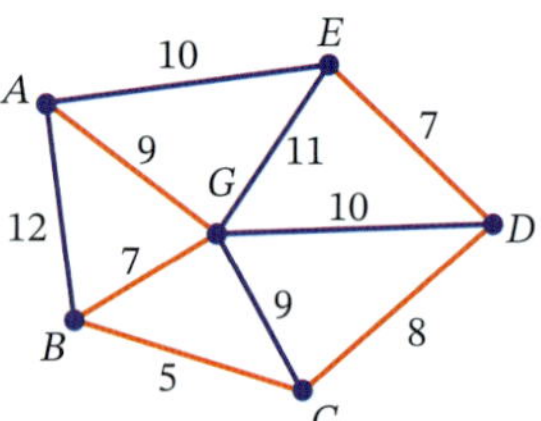


**b** 36

**c** no

**13** **a** **i** Starting at *M*, edges in order are *MS*, *SP*, *PT*, *MR*.

**ii** 45

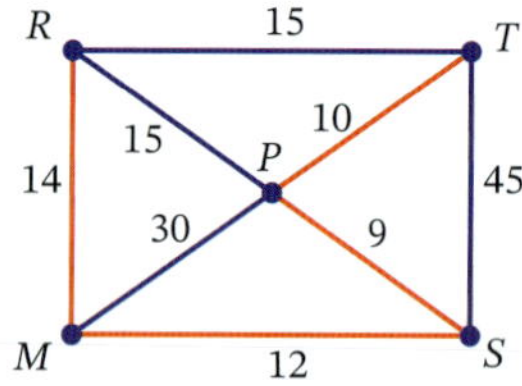


**b** **i** Starting at *A*, edges in order are *AF*, *FE*, *EC*, *CB*, *ED*.

**ii** 43

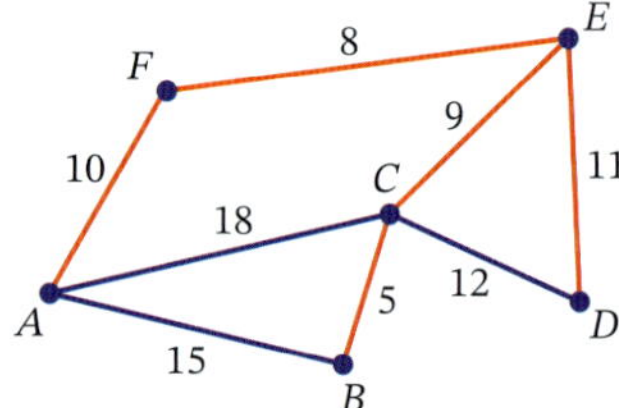


**c** **i** Starting at *G*, edges in order are *GK*, *KE*, *ED*, *DH*, *HF*, *DC*.

**ii** 71

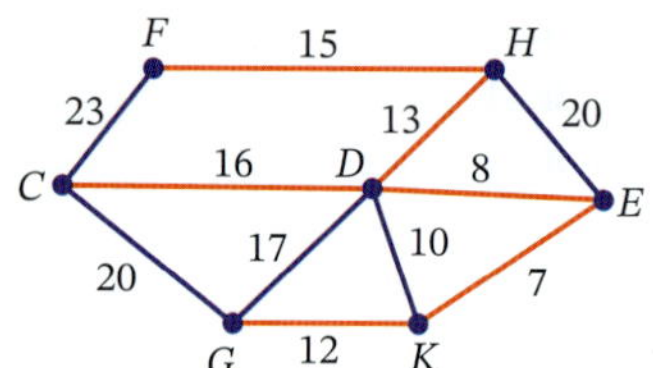

**d i** Starting at *H*, edges in order are *HG*, *GL*, *LM*, *ME*, *HK*.

**ii** 65

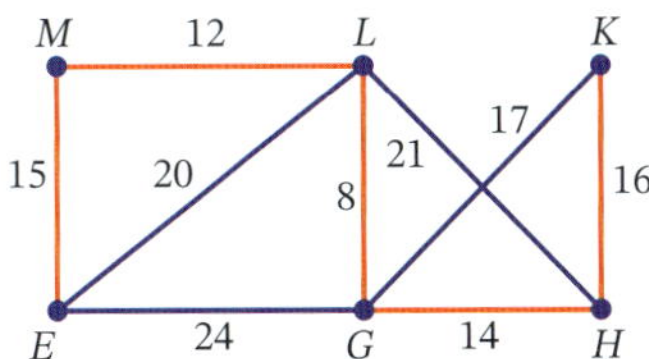


**14 a** Edges in order: *AM*, *MC*, *CS*, *SB*

**b** 2745 km

**15 a i** Edges in order: *CF*, *AG*, *BG*, *HK*, *DE*, *CG*, *EK*, *AH*

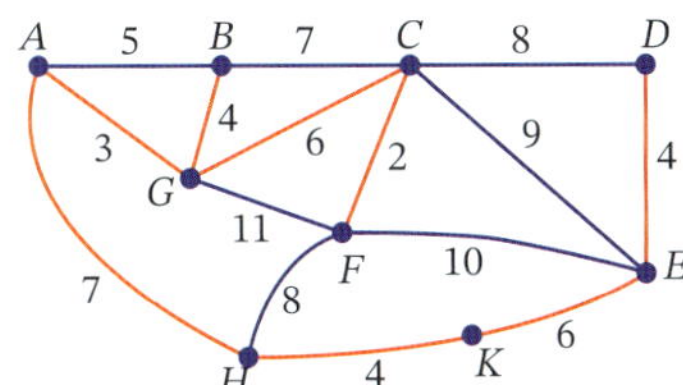


**ii** Edges in order starting from *E* are *ED*, *EK*, *KH*, *HA*, *AG*, *GB*, *GC*, *CF*.

**b** yes

**c** $7 + 3 + 4 + 6 + 2 + 4 + 6 + 4 = 36$ km

**d** Shortest route from *A* to *D* is *AG*, *GC*, *CD* and the length is 17 km.

**16 a**

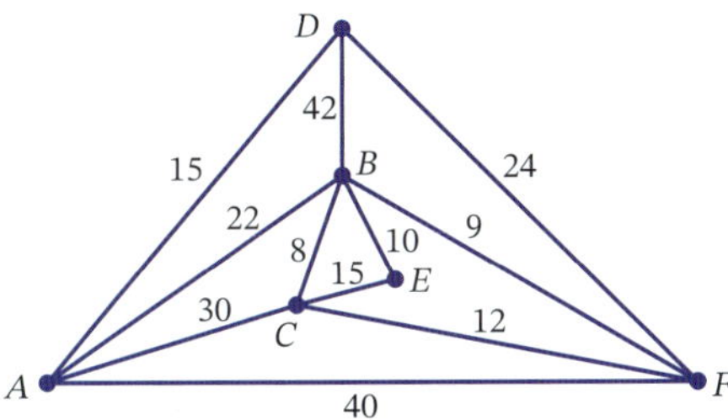


**b i** Edges in order: *BC*, *BF*, *BE*, *AD*, *AB*

**ii** Edges in order starting at *A* are *AD*, *AB*, *BC*, *BF*, *BE*.

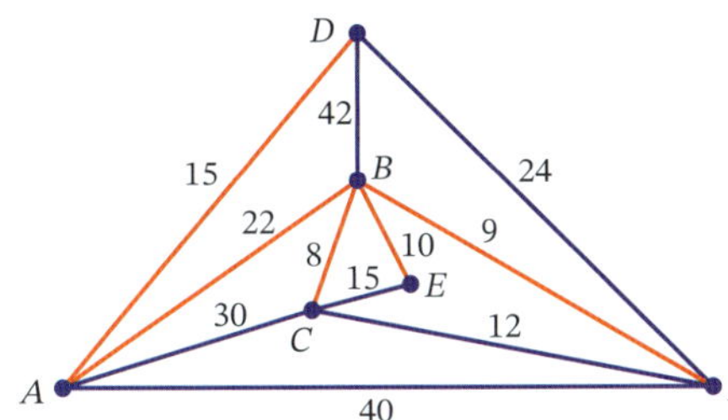


**c** 64 km

**17** A

**18** $m = 1, 2, 3$ or 4

**19 a**

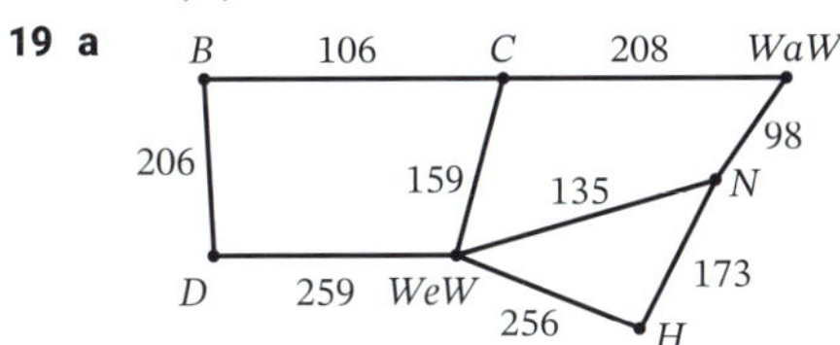


**b**

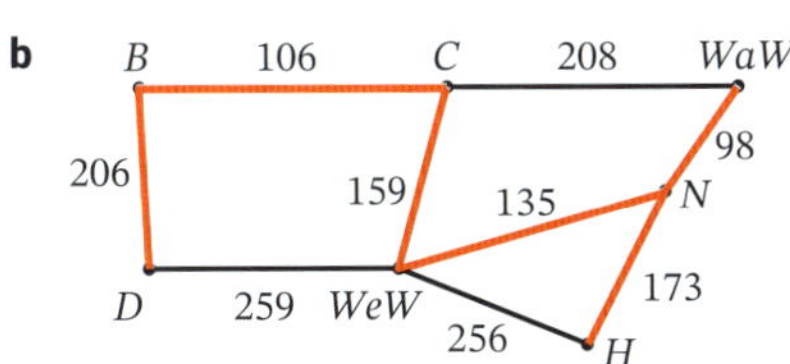


**c** 877 km

**20 a**

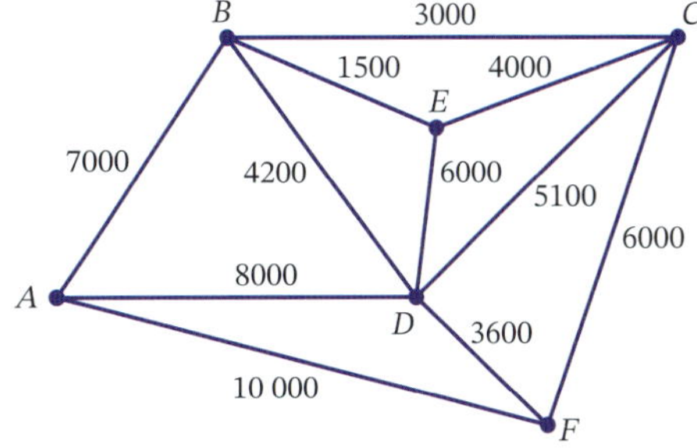


**b** Edges in order: *BE*, *BC*, *DF*, *BD*, *BA*

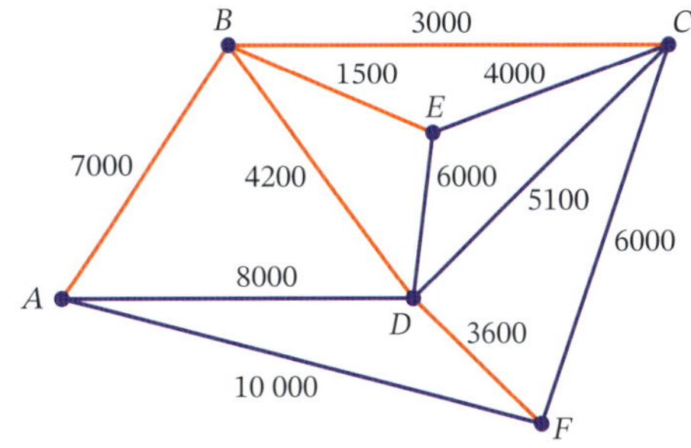


**c** lowest cost = $19 300

## Exercise 8.04

**1** B

**2 a i** Mt Pritchard–Cabramatta–Fairfield, 7.9 km

**ii** Mt Pritchard–Canley Heights–Fairfield, 17 min

**b** no

**3 a**

| Vertex | *B* | *C* | *D* | *E* | *F* | *G* | *H* |
|---|---|---|---|---|---|---|---|
| Path | *AB* | *ABDC* | *ABD* | *ABFGE* | *ABF* | *ABFG* | *AH* |
| Weight | 5 | 14 | 10 | 15 | 9 | 10 | 7 |

**b**

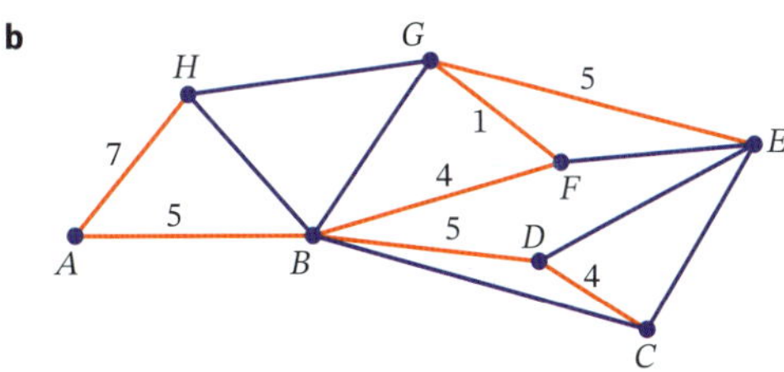


weight $= 7 + 5 + 4 + 1 + 5 + 5 + 4 = 31$

**c**

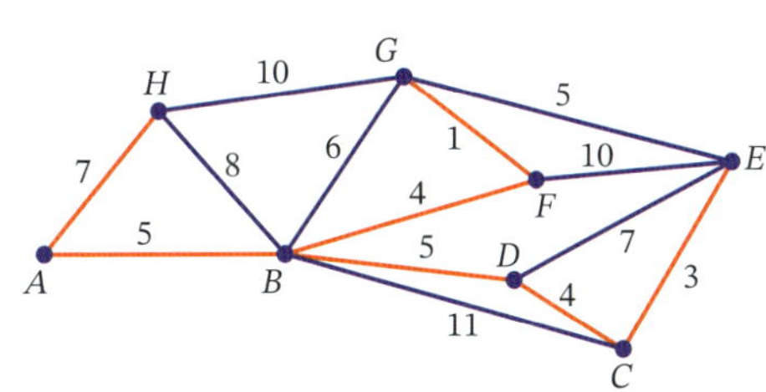


weight of minimum spanning tree = 29

**d** The trees are different and the weight of minimum spanning tree is less than the weight of shortest path spanning tree.

The shortest path from *A* to *E* is not in this minimum spanning tree.

**4 a i**

| Vertex | $MP$ | $S$ | $CH$ | $C$ | $F$ |
|---|---|---|---|---|---|
| Path | $L$–$MP$ | $L$–$CH$–$S$ | $L$–$CH$ | $L$–$C$ | $L$–$C$–$F$ |
| Distance | 4.1 | 10 | 5.3 | 5.1 | 8.7 |

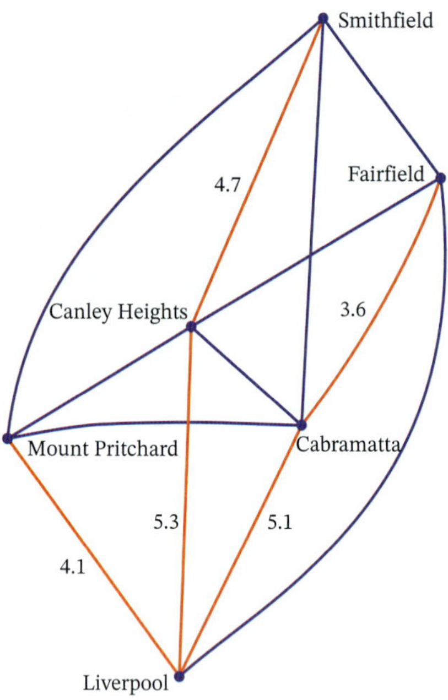


weight = 22.8 km

**ii**

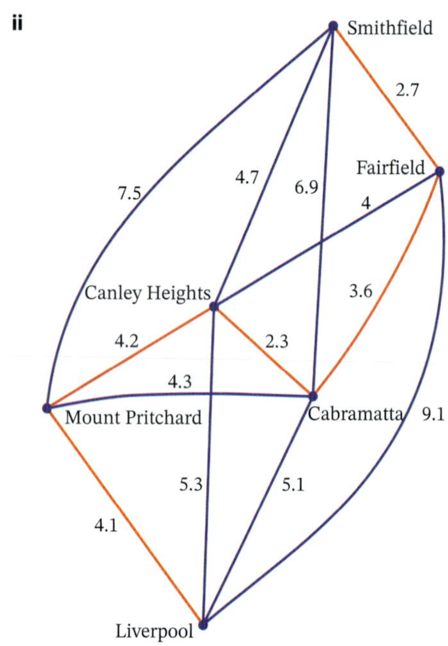


weight of minimum spanning tree = 16.9 km

**b** no

**c** Liverpool, Canley Heights , Cabramatta, Fairfield

**5 a i**

| Vertex | $MP$ | $S$ | $CH$ | $C$ | $F$ |
|---|---|---|---|---|---|
| Path | $L$–$MP$ | $L$–$CH$–$S$ | $L$–$CH$ | $L$–$C$ | $L$–$F$ |
| Distance | 8 | 18 | 11 | 12 | 17 |

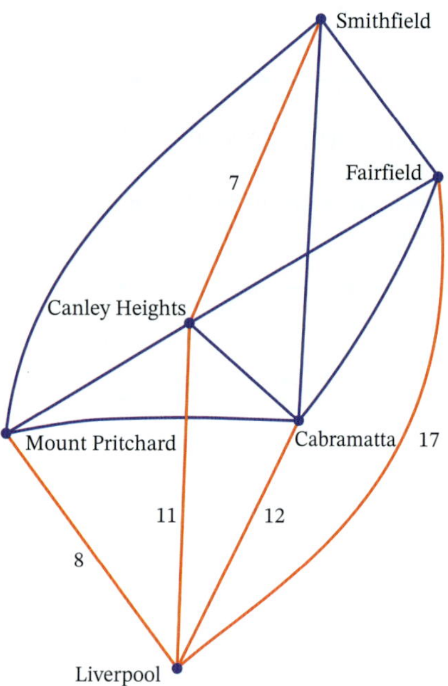


weight = 55 min

**ii**

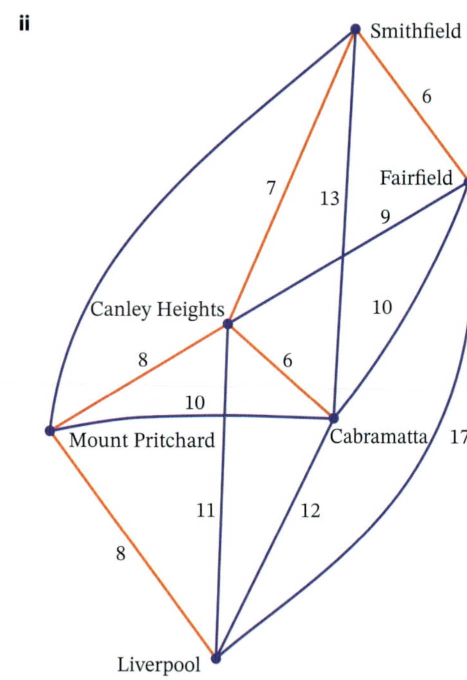


weight of minimum spanning tree = 35 min

**b** no

**c** Liverpool, Canley Heights, Cabramatta, Fairfield

**6**

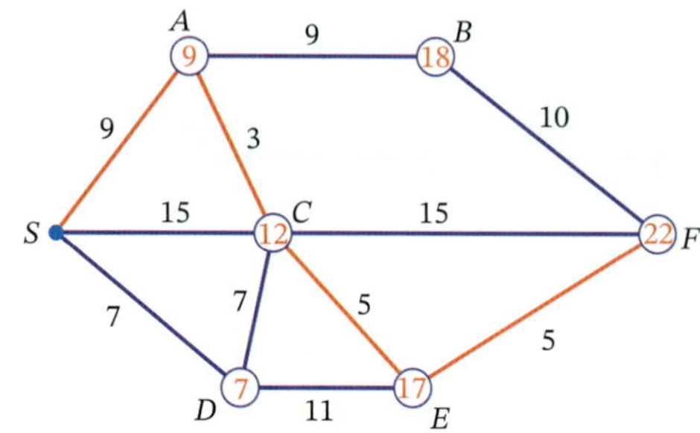


Shortest path from $S$ to $F$ is $SACEF$ and has weight 22.

**7 a** Shortest path is *SPNF*, weight = 11
**b** Shortest path is *SADCF*, weight = 14
**c** Shortest path is *SGDEF*, weight = 11
**d** Shortest path is *SKGEF* or *SAEF*, weight = 46

**8** B

**9** The shortest path is *EDBGC*, length = 3.6 km

**10 a**

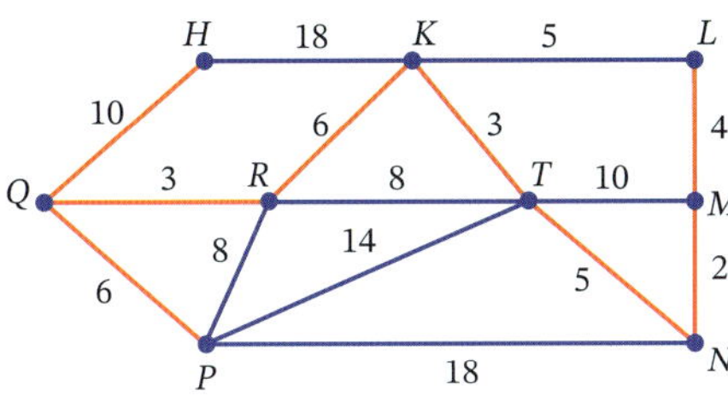


weight = 39

**b** No; the other spanning tree is

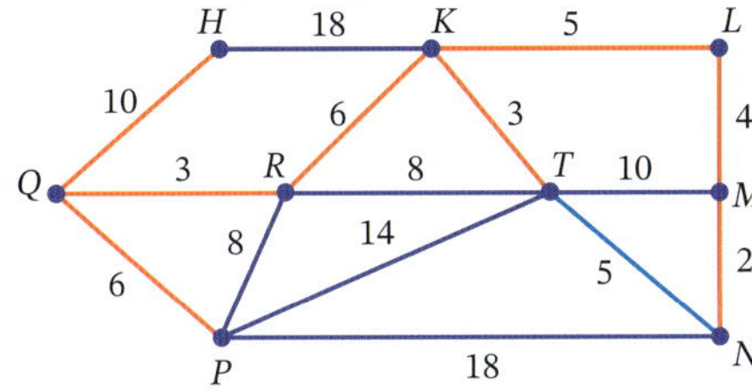


**c i** Shortest path from *Q* to *N* is *QRTN*, weight = 16
**ii** no

**d**

| Vertex | *H* | *K* | *L* | *R* | *T* | *M* | *P* | *N* |
|---|---|---|---|---|---|---|---|---|
| Path | *QH* | *QRK* | *QRKL* | *QR* | *QRT* | *QRKLM* | *QP* | *QRTN* |
| Distance | 10 | 9 | 14 | 3 | 11 | 18 | 6 | 16 |

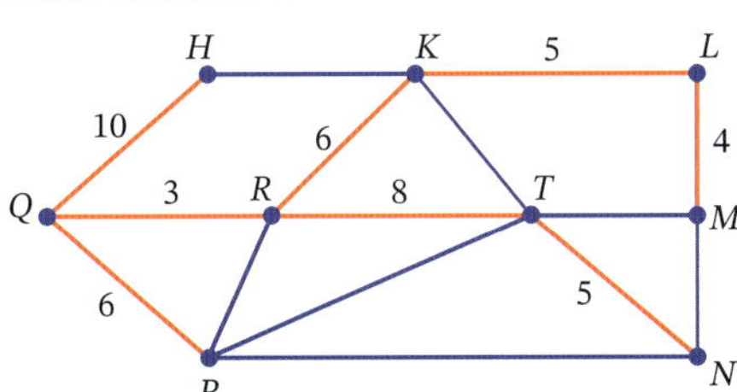


**11 a**

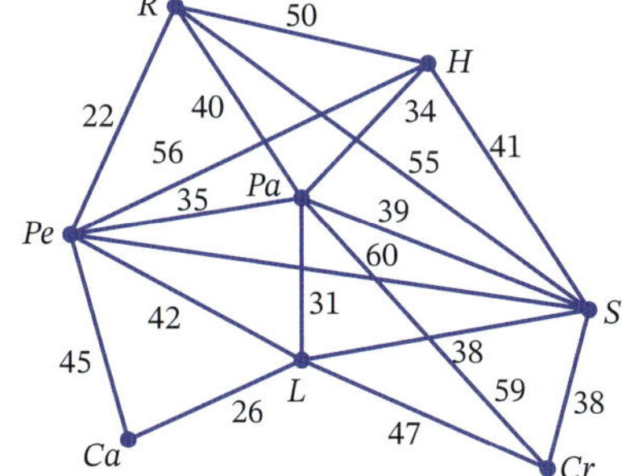


**b** *Cr, S, R*, 93 min

**c**

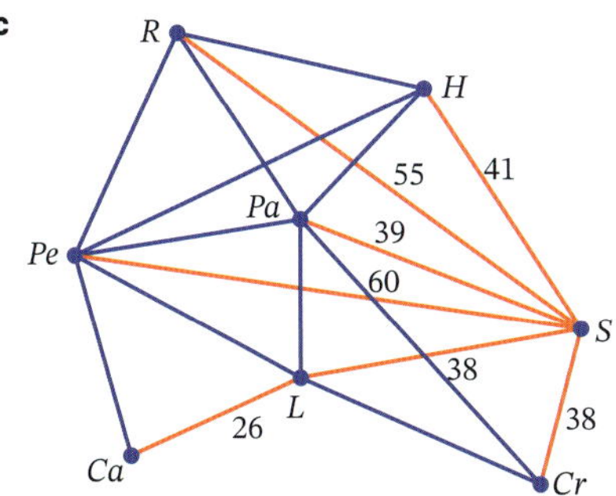


weight = 297 min

**d i** Order of edges is: *PeR, LCa, PaL, PaH, PaPe, SL, SCr*

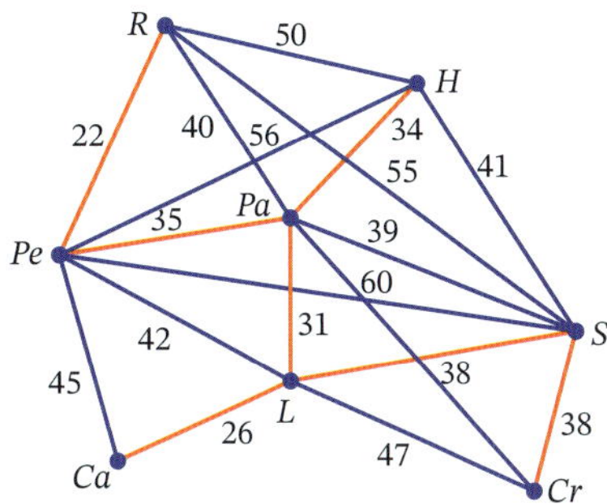


**ii** weight = 224 min

**e** No. The minimum spanning tree does not take into account the motorway from Sydney to Richmond as it is not one of the shorter distances.

## Sample HSC problem

**a**

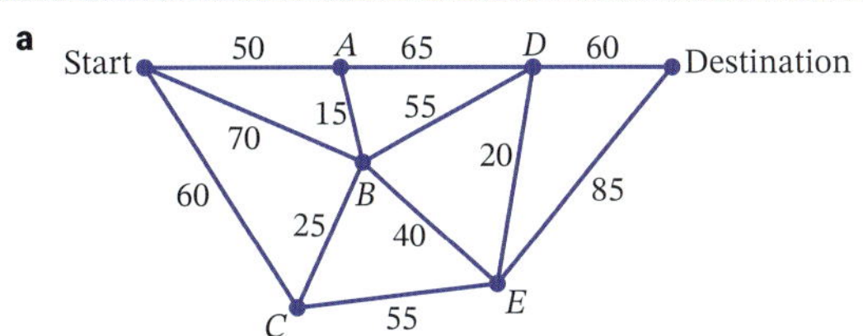


**b**

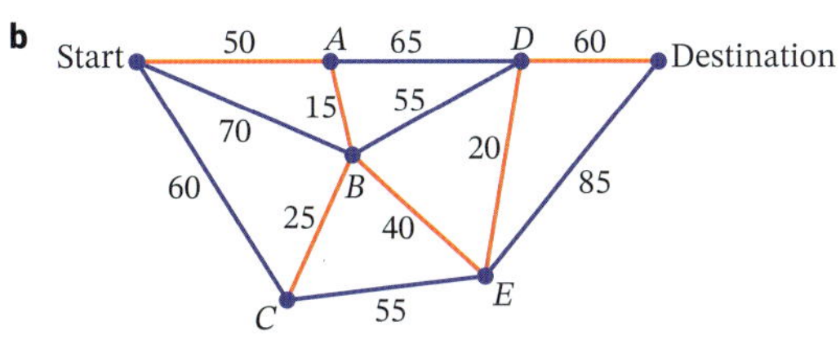


**c** Shortest path is Start, *A, D*, Destination.
**d** no

## Test yourself 8

**1 a i** $A - 2, B - 3, C - 3, D - 2, E - 4$
**ii** number of edges = 7, sum of vertices = 14 = 2 × 7
**b i** $A - 4, B - 3, C - 4, D - 3, E - 4$
**ii** number of edges = 9, sum of vertices = 18 = 2 × 9

**2 a** weight = 58 **b** *LNRW*, weight = 12
**c** *LPNLMNTRW* or *LMNLPNTRW*, weight = 45

**3**

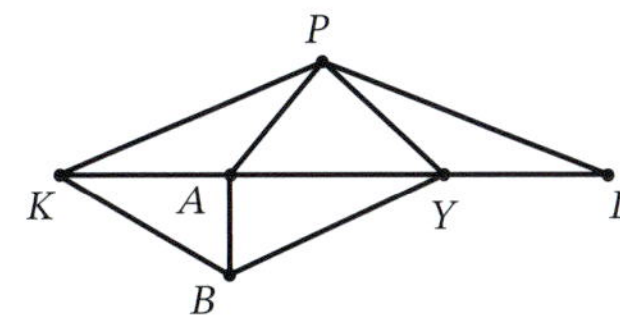


**4**

| | D | E | G | L | P | Y |
|---|---|---|---|---|---|---|
| D | – | 1 | 1 | – | – | 1 |
| E | 1 | – | 1 | – | 1 | 1 |
| G | 1 | 1 | – | 1 | 1 | – |
| L | – | – | 1 | – | 1 | – |
| P | – | 1 | 1 | 1 | – | 1 |
| Y | 1 | 1 | – | – | 1 | – |

**5**

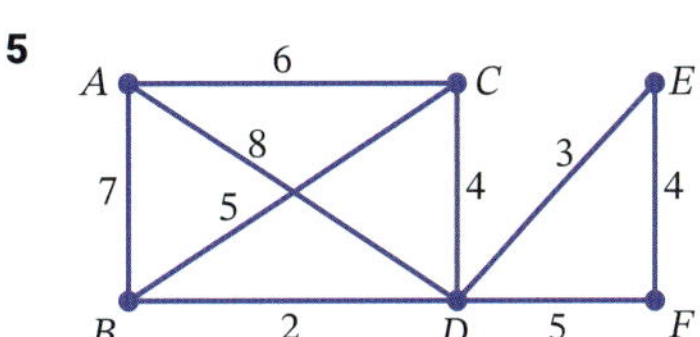

**6** 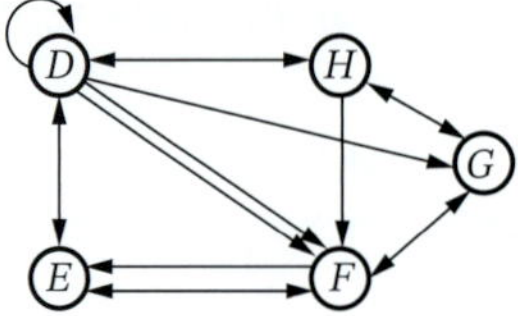


**7** 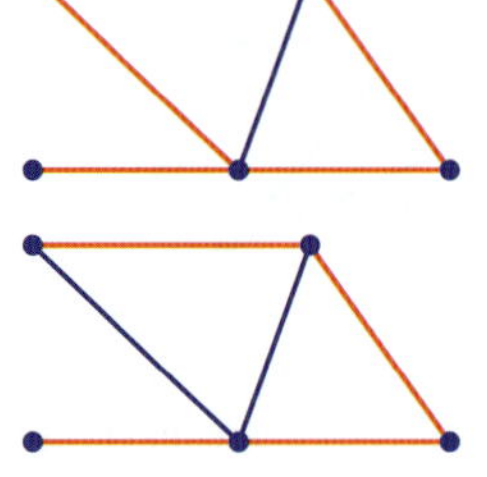

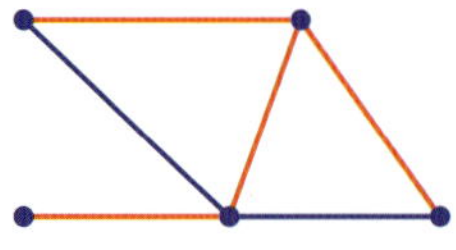

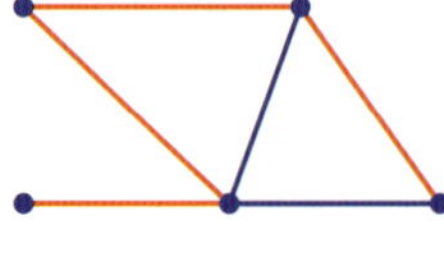

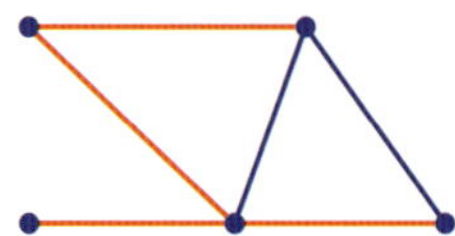

**8** Order of edges: *GH*, *HK*, *FE*, *KD*, *KE*, *BH*, *HC*; weight = 38

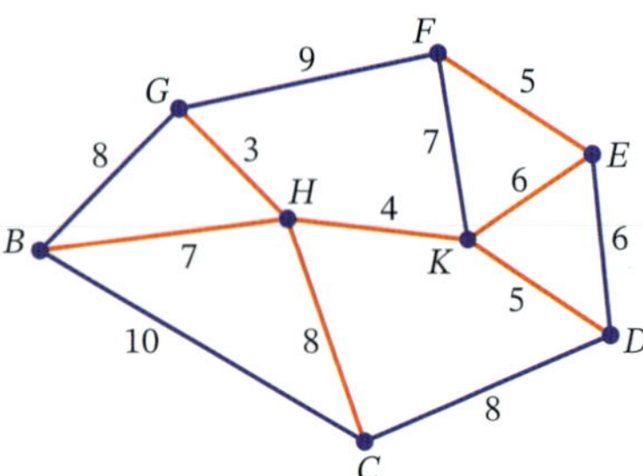


(There is more than one spanning tree.)

**9** Order of edges: *CH*, *HG*, *HK*, *KD*, *DE*, *FE*, *HB*; weight = 38

(The order of edges depends on starting vertex.)

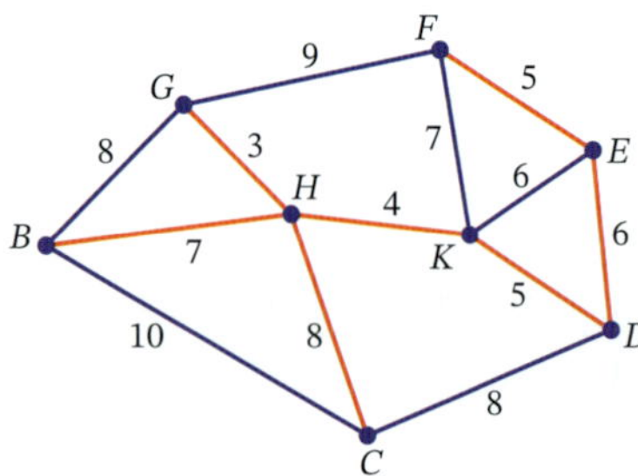


**10 a** 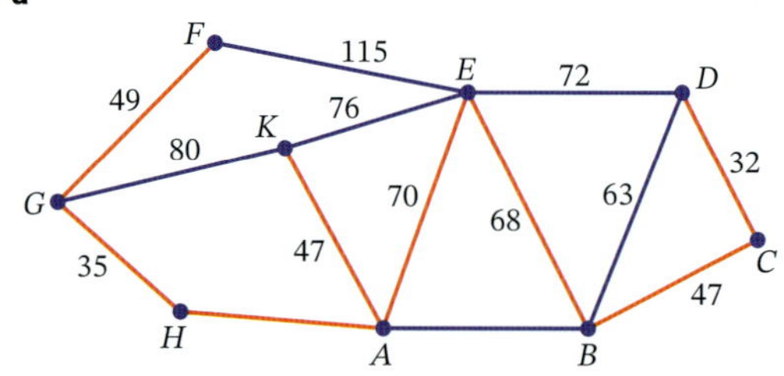


Order of edges: *EB*, *BC*, *CD*, *EA*, *AK*, *AH*, *HG*, *GF*

**b** no

**11** $m = 4, p = 2$

**12 a** 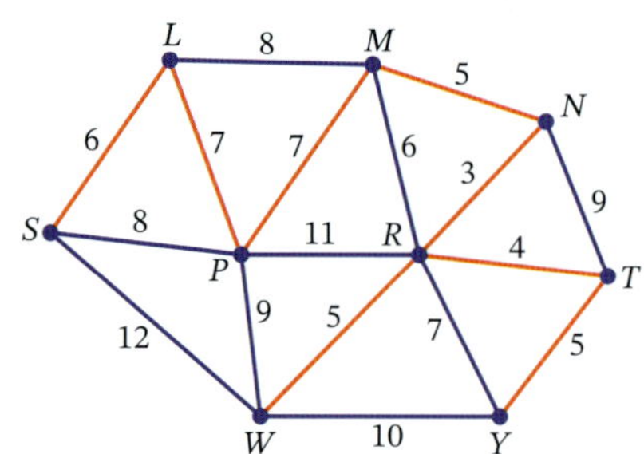


minimum spanning tree length = 4200 m

**b** 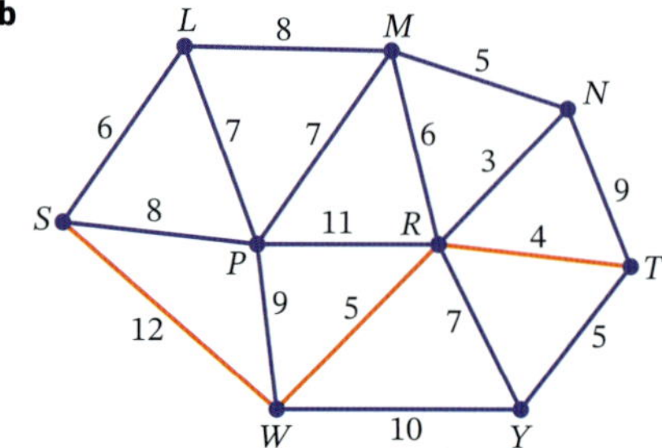


Shortest path from *S* to *T* is *SWRT*, which is not contained in the minimum spanning tree.

**13** The shortest distance is along the path *SADEF* and has weight 23.

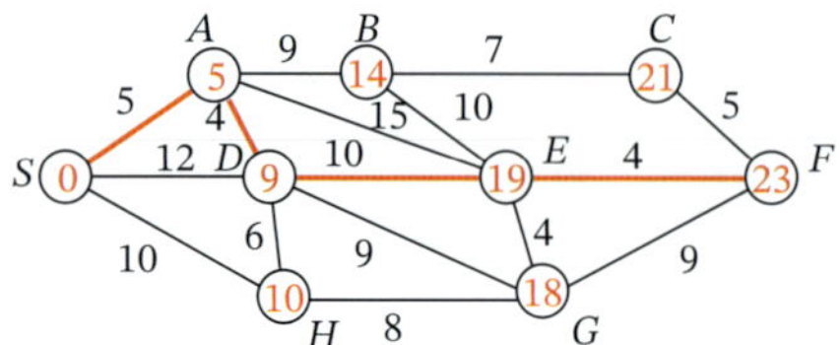


**14 a** 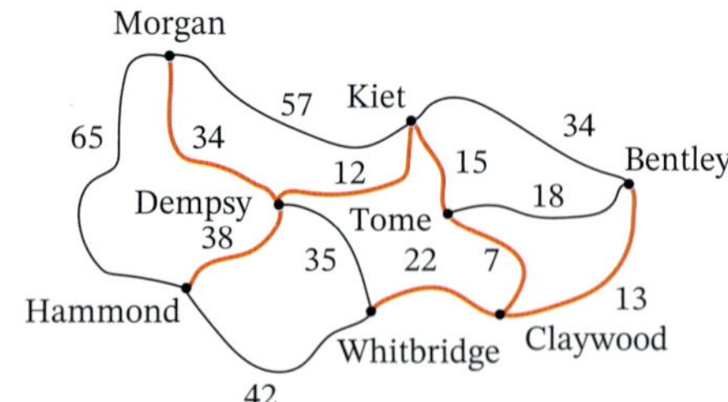


length = 141 km

**b** *MDKTB*, length = 79 km

**c** no

# Chapter 9

## SkillCheck

1

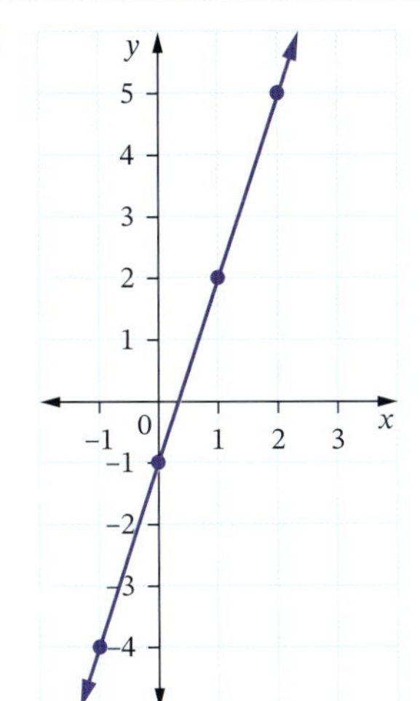

2 **a** $y$: 5, 7, 9, 11 **b** $y$: −3, −4, −5, −6

3 $c = -1$

4 **a** 2.8 **b** 4

5 **a i** 1225 **ii** 685

**b i** 2.5 **ii** 5.7

## Exercise 9.01

1 **a**

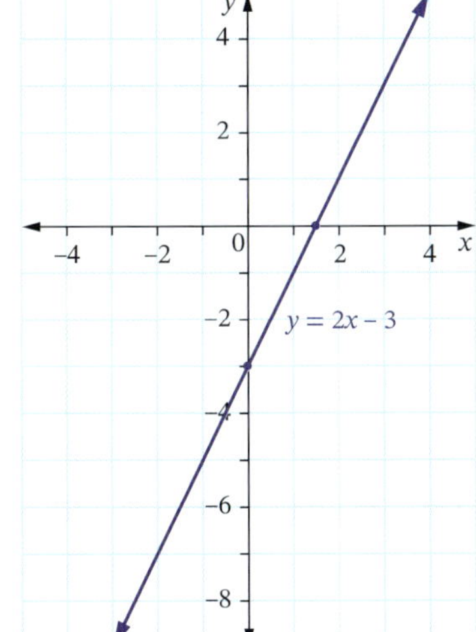


**b**

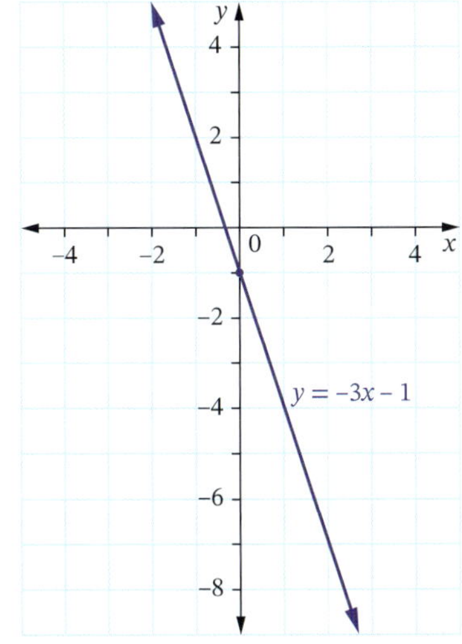


2 **a** $y = x + 2$

| x | −2 | −1 | 0 | 1 | 2 |
|---|---|---|---|---|---|
| y | 0 | 1 | 2 | 3 | 4 |

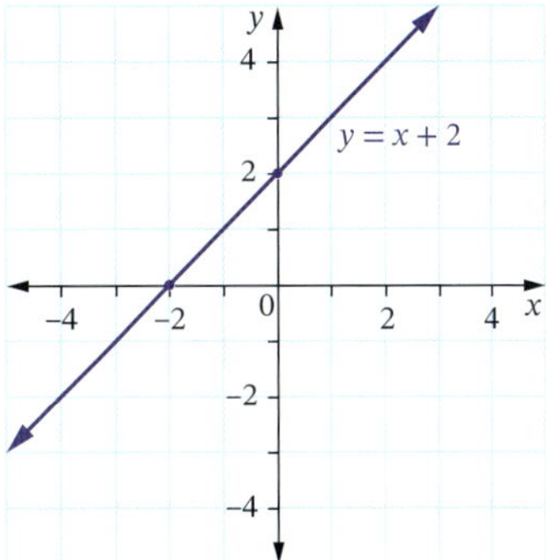


gradient 1, $y$-intercept 2

**b** $y = 2x - 2$

| x | −2 | −1 | 0 | 1 | 2 |
|---|---|---|---|---|---|
| y | −6 | −4 | −2 | 0 | 2 |

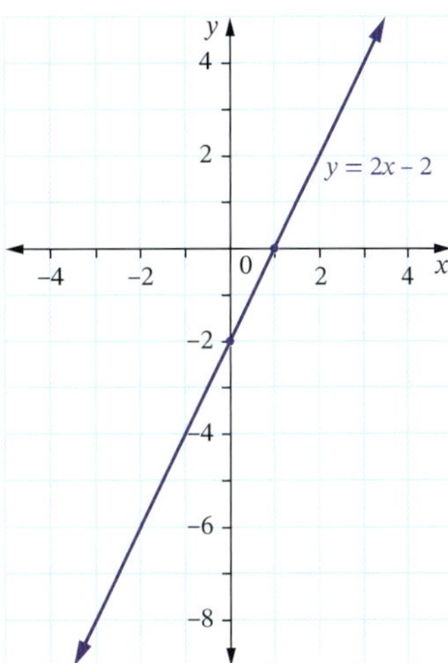


gradient 2, $y$-intercept −2

**c** $y = 3x + 1$

| x | −2 | −1 | 0 | 1 | 2 |
|---|---|---|---|---|---|
| y | −5 | −2 | 1 | 4 | 7 |

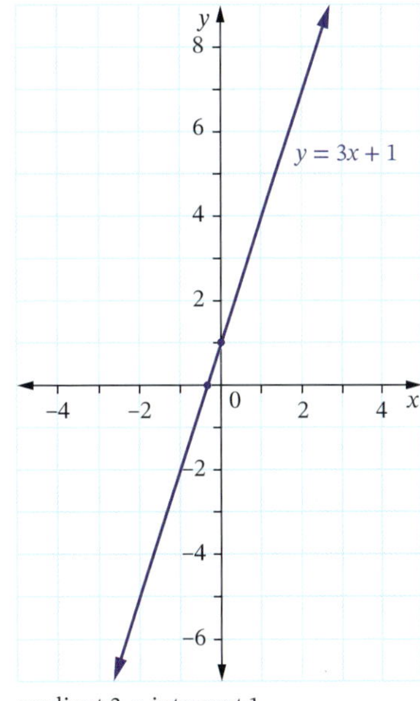


gradient 3, $y$-intercept 1

**d** $y = \frac{x}{2} - 1$

| $x$ | −4 | −2 | 0 | 2 | 4 |
|---|---|---|---|---|---|
| $y$ | −3 | −2 | −1 | 0 | 1 |

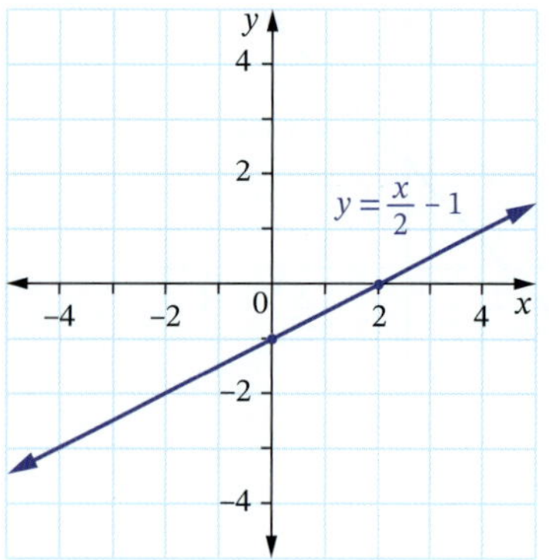


gradient $\frac{1}{2}$, $y$-intercept $-1$

**e** $N = 1 - M$

| $M$ | −2 | −1 | 0 | 1 | 2 |
|---|---|---|---|---|---|
| $N$ | 3 | 2 | 1 | 0 | −1 |

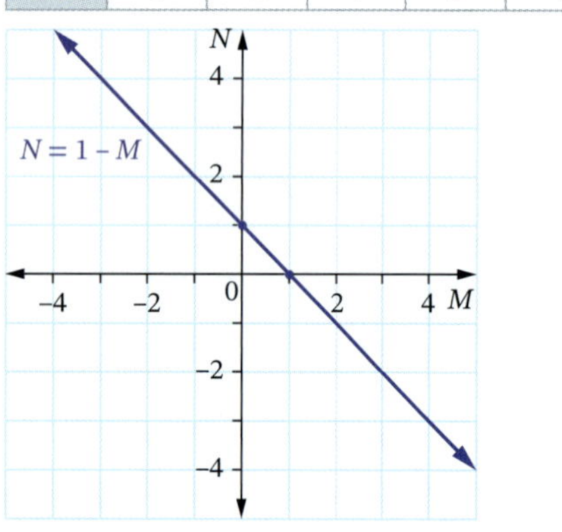


gradient $-1$, $y$-intercept 1

**f** $P = \frac{5t - 1}{2}$

| $t$ | −3 | −1 | 0 | 1 | 3 |
|---|---|---|---|---|---|
| $P$ | −8 | −3 | $-\frac{1}{2}$ | 2 | 7 |

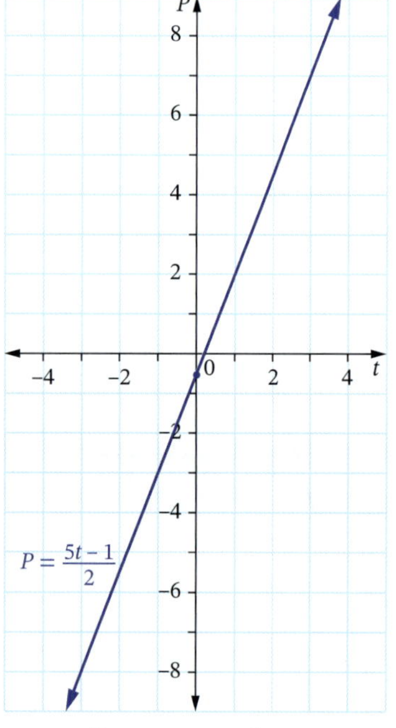


gradient $\frac{5}{2}$, $y$-intercept $-\frac{1}{2}$

**3** **a** $y = 3x + 7$ **b** $y = -2x + 1$

**c** $y = x - 1$ **d** $y = \frac{1}{3}x + \frac{1}{2}$

**e** $y = -\frac{5}{4}x$ **f** $y = 5$

**4** **a**

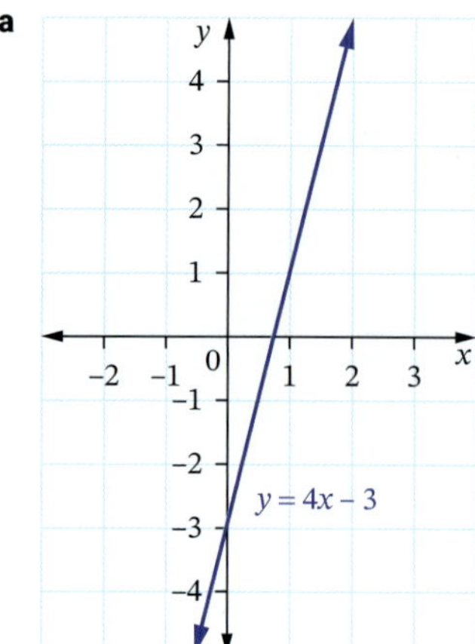


**b**

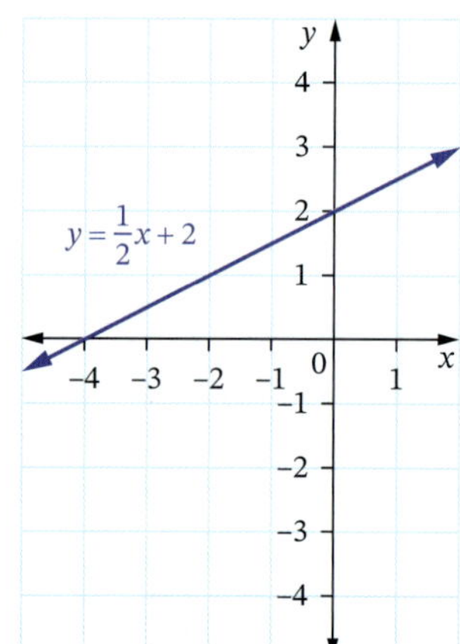


**c**

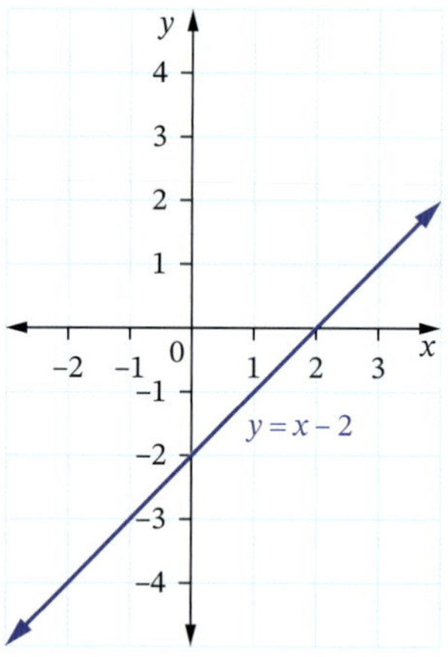

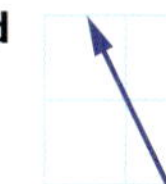

d

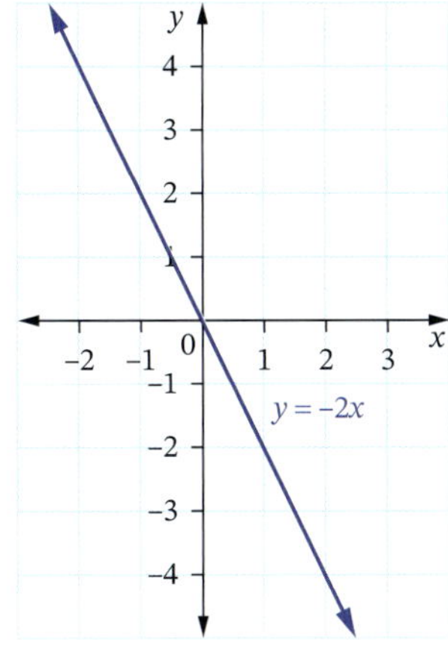

e

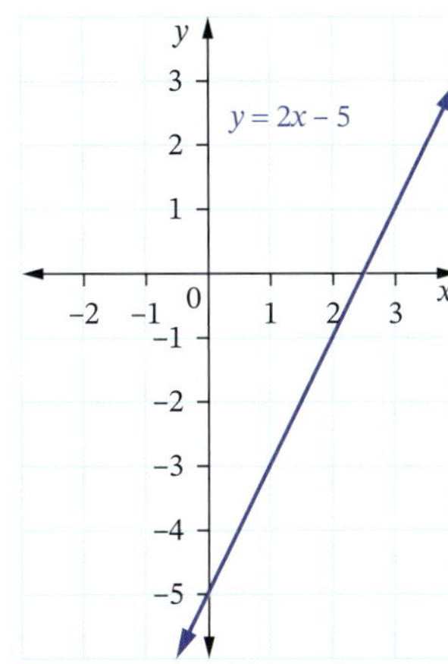


f

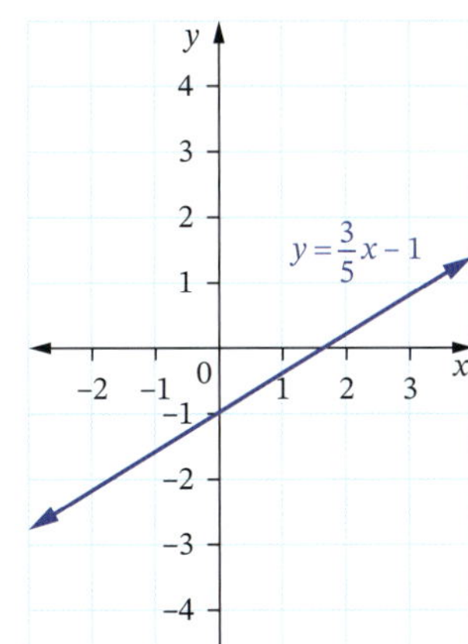

g

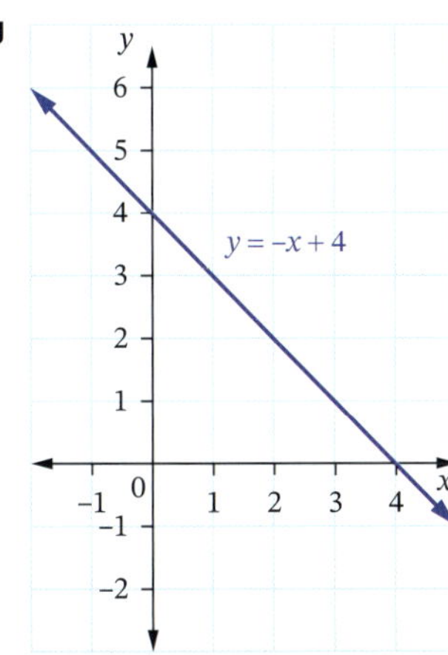


h

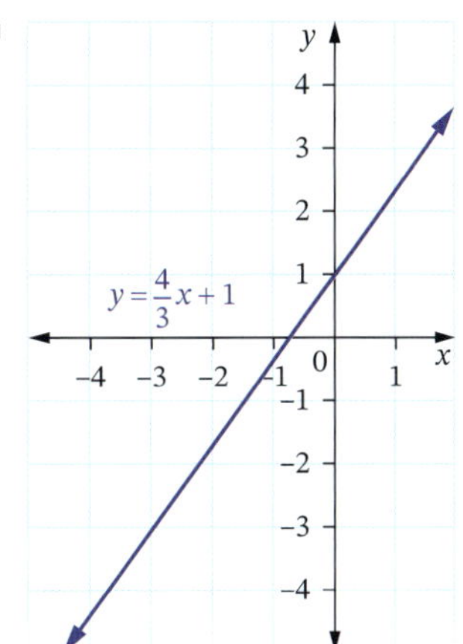


i

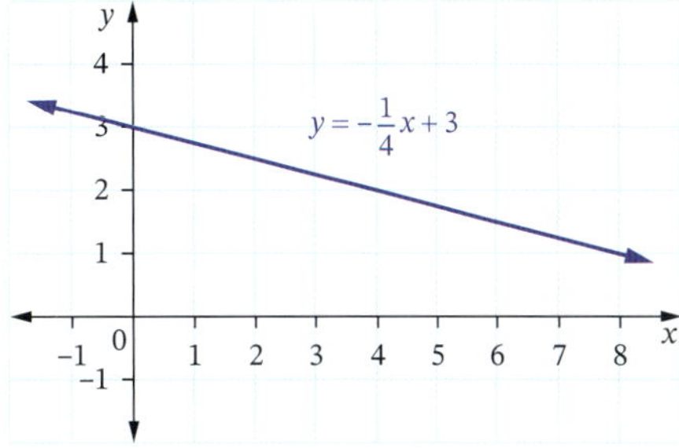


**5** D

**6** **a** $y = 2x + 3$ **b** $y = 3x - 1$

**c** $y = -x + 1$ **d** $y = -x$

**e** $y = \frac{3}{4}x - 2$ **f** $y = x - 3$

**7** A **8** B **9** D

## Exercise 9.02

**1** **a** 2 **b** $-1$ **c** 2

**d** $-\frac{1}{2}$ **e** $\frac{7}{5}$ **f** $-5$

**g** $\frac{3}{4}$ **h** 4 **i** $\frac{2}{9}$

**2** **a** 3 **b** 5 **c** $\frac{1}{4}$

**d** $-2$ **e** $\frac{3}{5}$ **f** $-\frac{1}{2}$

**3** **a** $y = -2x + 2$ **b** $y = \frac{1}{2}x - 1$ **c** $y = 4x$

## Exercise 9.03

**1** **a** weight

**b**

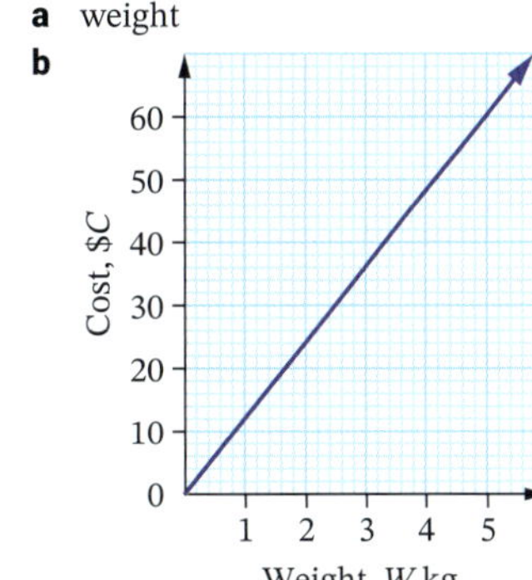


**c** $18.00 **d** 2.5 kg

**2** C

**3** **a** **i** 6.5 **ii** 4.5

**b** $m = \frac{1}{2}, c = 5; y = \frac{1}{2}x + 5$

**c** $x = 10$, extrapolated

**4** **a** $w = 80n + 20$

| $n$ | 0 | 1 | 2 | 3 | 4 | 5 | 6 |
|---|---|---|---|---|---|---|---|
| $\$w$ | 20 | 100 | 180 | 260 | 340 | 420 | 500 |

**b**

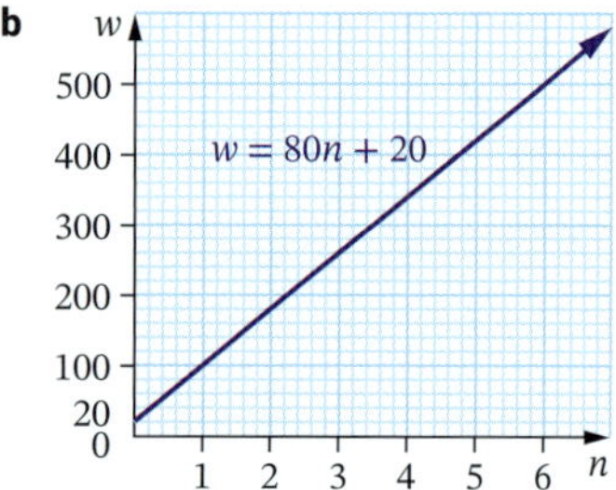


**c** **i** 20

**ii** There is a fixed cost of \$20 no matter how many windows are installed.

**d** **i** \$820, extrapolation

**ii** 7 windows, extrapolation

**5** **a** dependent **b** $S = 4.5n$

**c** 4.5 runs/over

**d** 0, the number of runs scored after 0 overs

**e** **i** 95 runs **ii** 225 runs

**f** **i** 12th over **ii** 40th over

**g** Weaker batters bat later, with lower run rate.

**6** **a**

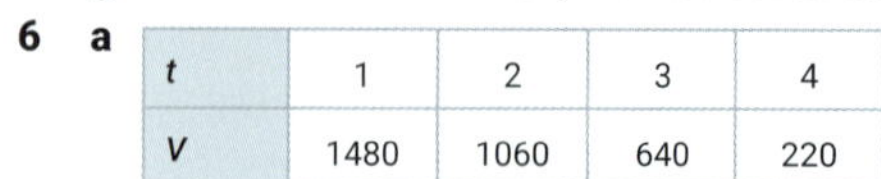

| $t$ | 1 | 2 | 3 | 4 |
|---|---|---|---|---|
| $V$ | 1480 | 1060 | 640 | 220 |

**b**

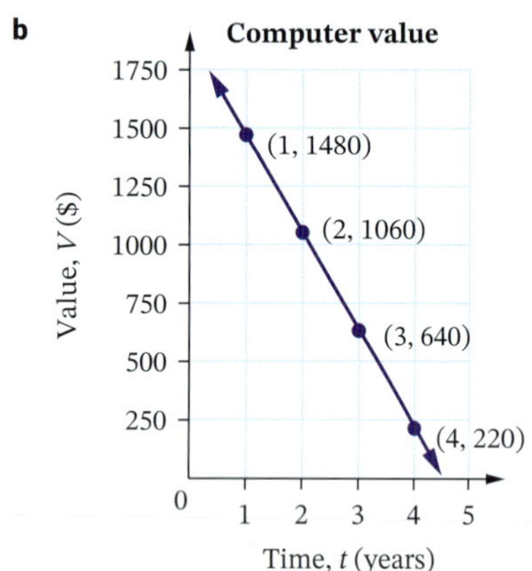


**c** rate of depreciation in dollars per year

**d** \$850 **e** \$1900

**f** $V$ becomes negative.

**g** 4.5 years

**7** **a** $P = 50 + 3n$ **b** $n$

**c** 50, the base pay **d** \$134

**e** 16

**8** **a**

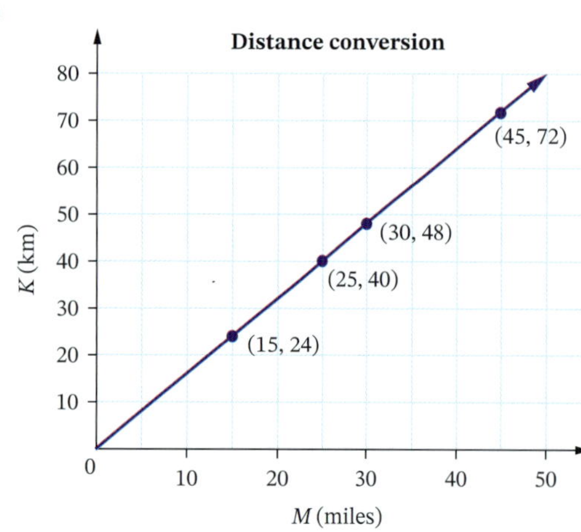


**b** $K = 1.6M$ **c** 0, 0 miles = 0 km

**d** 1.6, the number of kilometres in 1 mile

**e** **i** 160 km **ii** 62.5 miles

**f** **i** 19 km **ii** 13 miles

**9** **a** $C$

**b** gradient 0.7, vertical intercept 4.2

**c** $C = 0.7d + 4.2$ **d** \$3.50

**e** **i** \$18.20 **ii** \$4.20

**f** 48 km

**10** **a** independent **b** $n = 8T - 24$

**c**

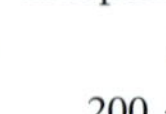

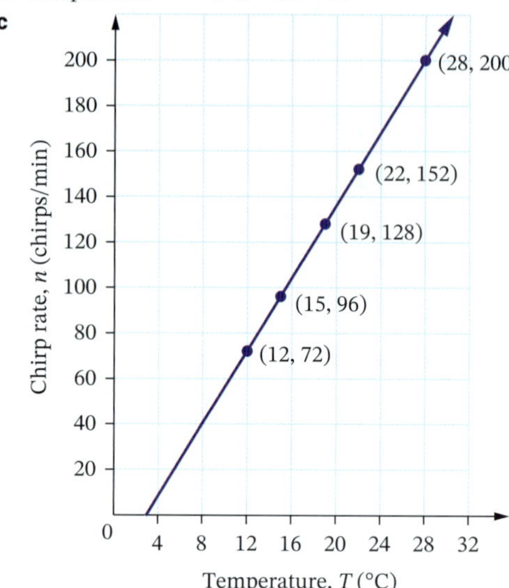


**d** increases by 16 chirps/min

**e** 184 chirps/min, interpolation

**f** 21°C

**g** −24; the number of chirps per minute cannot be below zero.

**11** D

**12** **a** 30; it costs a fixed amount of \$30 to hire the electrician

**b** \$240

**c** 1.5 hours

**d** One limitation is that the electrician has specific job hours and may not be able to work long hours. Or they may charge a lower rate per hour if working beyond 4 hours, as they have saved money and time on travelling to just one address.

**13** **a** **i** 64–79.5 kg **ii** 49–62 kg

**iii** 71–88 kg

**b** **i** 164 –193 cm **ii** 149–170 kg

**iii** 181–207 cm

**c** 0.65, rate of change of weight per height in kg/cm

## Exercise 9.04

**1** $k = 3.5, y = 3.5, 35, 38.5$

**2** **a** $d = 2.75r$ **b** 90.75 m **c** 36

**3** **a** $p = 5.88d$ **b** 235.2 kPa **c** $141\frac{1}{3}$ m

**4** **a** $F = 2.4d$ **b** 7800 kg

**5** **a** $y = 1.6x$ **b** 96 km/h **c** 75 miles/h

**6** A

**7** **a** 64

**b** increase in water volume per minute

**c** 512 L **d** 15 min

**8** **a** 120 min or 2 hours

**b** The marathon runner cannot continue to run forever and keep travelling more distance; he needs to rest.

**9** **a** 22.8 kg **b** 84.1 kg

**10** 8 h 53 min

**11** **a** 64.8 seconds **b** 222 megabytes

**12** B

**13 a** $S = 1.3h$ **b** 317.2 cm **c** 92 cm

**14 a** 117.6 m/s **b** 18 s

**c** The object cannot keep falling forever and will stop when it hits the ground.

**15 a i** \$151 **ii** \$108 **iii** \$131

**iv** \$89 **v** \$43 **vi** \$72

**b** 1.72; the number of Australian dollars per euro, what €1 equals in A\$

**16 a i** 41 kg **ii** 61 kg **iii** 73 kg

**b i** 88 lb **ii** 141 lb **iii** 198 lb

**c** 0.45 kg/lb **d** $K = 0.45P$

**e i** 60.8 kg **ii** 142.2 lb

## Sample HSC problem

**a**

| Time, $t$ (h) | 0 | 1 | 2 | 3 | 4 | 5 | 6 |
|---|---|---|---|---|---|---|---|
| Volume, $V$ (mL) | 750 | 690 | 630 | 570 | 510 | 450 | 390 |

**b**

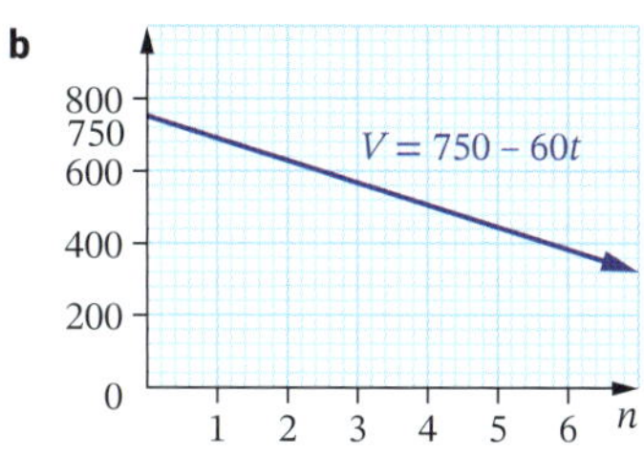


**c** $V = 750 - 60t$

**d** 498 mL

**e** 10 h, extrapolation because it is beyond the values in the table, after 6 hours.

**f** Water from the bucket cannot evaporate indefinitely. Eventually $V = 0$ and this is when $t = 12.5$ h. This linear model also assumes the bucket stays in the same level of direct sunlight, which would not be the case.

## Test yourself 9

**1 a**

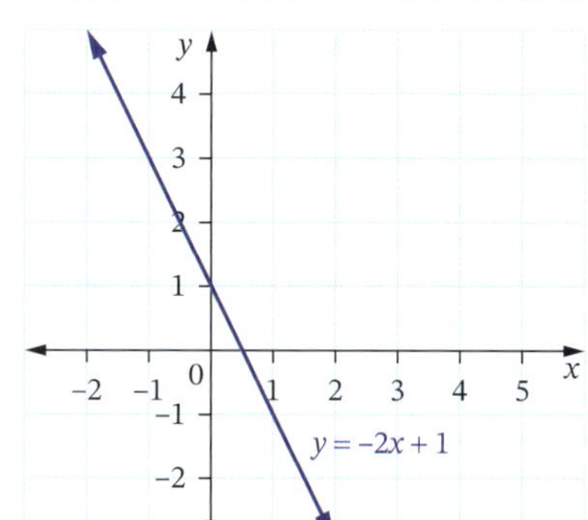


**b**

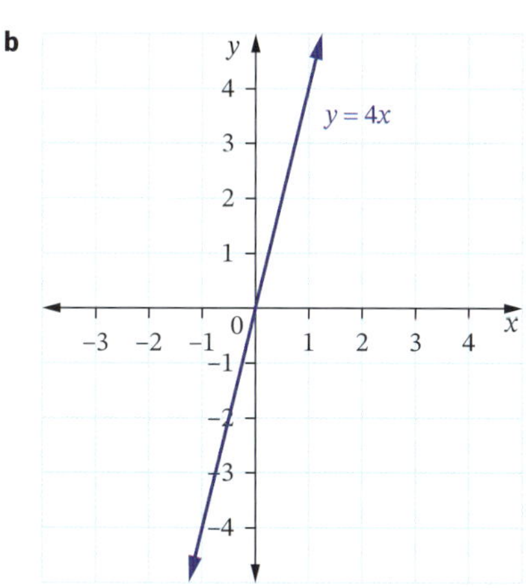


**2 a** $y = -\frac{1}{2}x + 1$ **b** $y = 3x - 3$

**3 a** $2\frac{1}{2}$ **b** $-3$

**4 a** $\frac{1}{2}$ **b** $-4$

**5 a** $S$

**b**

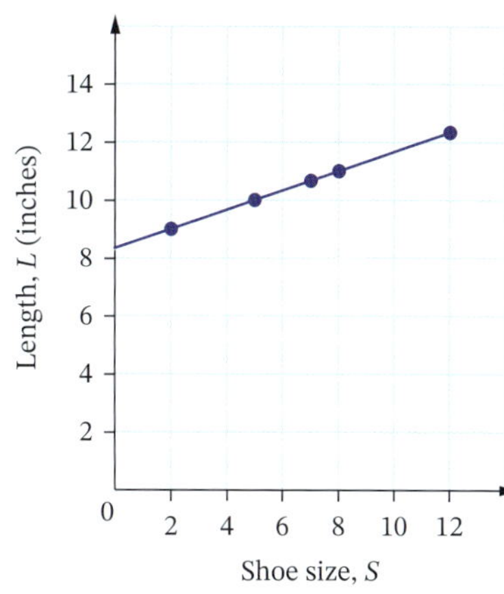


**c** $\frac{1}{3}$, the increase in length per shoe size, in inches/size

**d** $8\frac{1}{3}$ inches, the vertical intercept

**e** $L = \frac{1}{3}S + 8\frac{1}{3}$

**f** $10\frac{5}{6}$ **g** 14

**6 a** $P = 4n - 24$

**b** 4, increase in pocket money in dollars per year of age

**c** \$40 **d** 11

**e i** Values of $P$ will be 0 or less.

**ii** After 18 years, children become adults and either receive no pocket money or an amount based on a different formula.

**7** 18 km

**8 a i** \$38 **ii** \$98

**b i** 1350 baht **ii** 1700 baht

**c** 0.048, the number of Australian dollars per Thai baht, what 1 baht equals in A\$

## Practice exam 3

**1** C **2** D **3** B **4** C

**5** D **6** B **7** B **8** B

**9** C **10** D

**11 a**

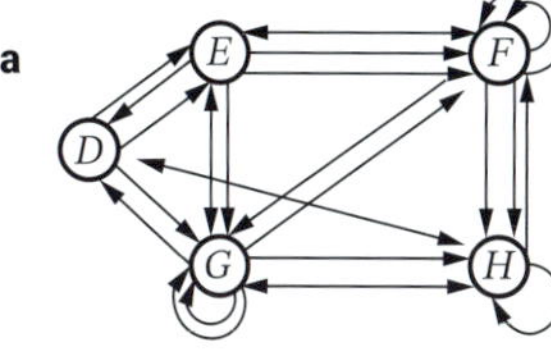


**b i** $m = -\frac{1}{2}$, $y$-intercept, $c = 4$

**ii**

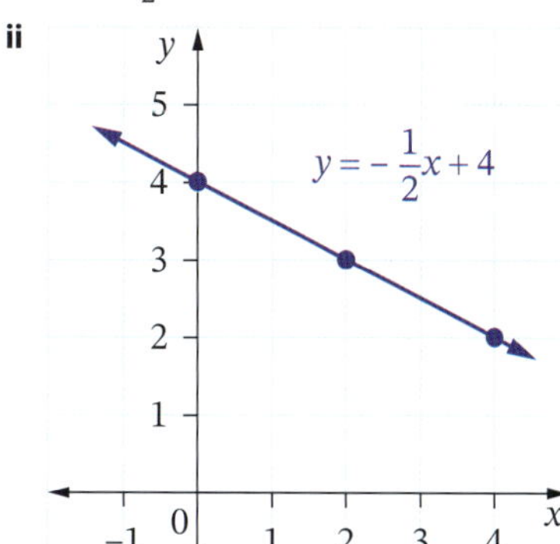

**c** **i** 600, 900, 1200, 1350

**ii**

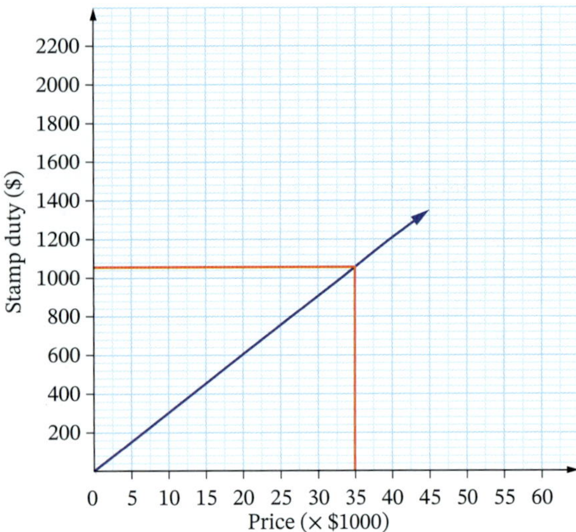


**iii** $1050

**12** **a** CTP insurance protects owners/drivers from legal liability for personal injury or death to any other person. Third Party Property insurance is optional and covers damage to other vehicles and property where you are at fault. It does not cover damage to your vehicle.

**b** **i** $V = -5.5t + 84$

**ii** decrease in water volume in kilolitres per hour

**iii** amount of water in the pool before you start draining it

**iv** 15.3 h

**c** **i** O and $C$ **ii** $OFC$

**iii**

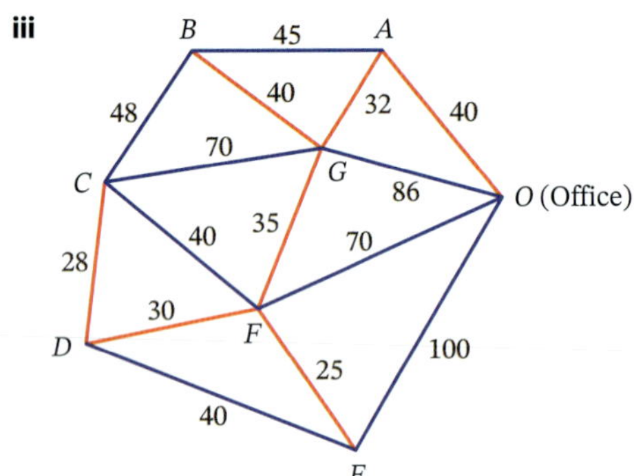


**iv** $32 200

**13** **a** **i** $m = \frac{50}{3}$ **ii** $D = \frac{50f}{3}$ **iii** 54 L

**b** $56 635

**c** **i** 39.48 L **ii** $112.44

# Chapter 10

## SkillCheck

**1** **a** 29 **b** 58 **c** 12

**d** 8 **e** 24

**2** column graph

**3** **a** 10 **b** 7

**c** 12, wears large shoes, has big feet

**d** 0 **e** 30%

**4** **a** 4 **b** 1 **c** 4

**d** 32 **e** 42

**5**

| Mass (kg) | Class centre | Frequency |
|---|---|---|
| 50–<60 | 55 | 4 |
| 60–<70 | 65 | 12 |
| 70–<80 | 75 | 7 |
| 80–<90 | 85 | 8 |
| 90–<100 | 95 | 3 |
| 100–<110 | 105 | 6 |

## Exercise 10.01

**1** **a** C **b** F **c** D

**d** H **e** E **f** I

**g** A **h** G **i** B

**2** **a** Commonwealth Scientific and Industrial Research Organisation

**b** Roads and Maritime Services

**c** United Nations Children's Fund

**d** National Roads and Motorists' Association

**e** Bureau of Meteorology

**f** Australian Securities Exchange

**g** Australian Security Intelligence Organisation

**h** World Health Organization

**i** National Rugby League

**3, 4** Teacher to check.

**5** **a** Stage 5 **b** Stage 1 **c** Stage 4 **d** Stage 2

**6** Teacher to check.

**7** **a** bank account details, amount of transaction, time and date

**b** personal details, email address, interests and preferences

**c** your location, time and date of message

**d** amount of spending, your location

**e** your location, time and date of travelling, account balance

**f** time and date, your location

**8** theft of identity; money stolen from accounts

**9** **a** People could give the age they are turning this year instead of their present age.

**b** ii, iv, v

**c** i, viii

**d** There could be many different types of answers (open-ended).

**e** Some people prefer not to talk about religion (privacy issue).

**f** People are paid differently.

**g** D

**10** **a** biased; invites a 'yes' answer; question not specific enough

**b** too many options; question should have a choice of 4 things

**c** too general; be specific, e.g., 'How many times do you vape per day?'

**d** unclear: including ones you don't use; better to say, 'How many TV's in your home are used?'

**e** too wordy; simplify to say, 'Circle the mode of transport you use to go to work'.

**f** too general; only yes or no answer is possible

**g** too general; express question in the same form as Question **viii** of the census form in Question **9**

**h** too general; express question in the same form as Question **iv** of the census form in Question **9**

**11, 12** Teacher to check.

## Exercise 10.02

**1** **a** **i** 5 **ii** 5 **iii** 1, 5
**b** **i** 34.4 **ii** 34.5 **iii** 31, 32, 35
**c** **i** 36 **ii** 38 **iii** no mode
**d** **i** 9 **ii** 8 **iii** 5, 7, 8

**2** **a** 24 **b** 24 **c** 20 **d** 14, 20

**3** **a** D **b** C **c** A

**4** **a** 50.0 **b** Yes, the mean is 50.0.
**c** 50 **d** 50

**5** **a** 20 **b** 2.6 **c** 2
**d** 2 **e** 8
**f** **i** decreases **ii** no effect **iii** no effect

**6** **a**

| Number of calls, $x$ | Frequency $f$ | $fx$ |
|---|---|---|
| 2 | 1 | 2 |
| 3 | 2 | 6 |
| 4 | 3 | 12 |
| 5 | 5 | 25 |
| 6 | 2 | 12 |
| 7 | 2 | 14 |
| Total | 15 | 71 |

**b** 15 **c** 5 **d** 5 **e** 4.7

**7** **a** 38, 43, 48, 53, 58, 63, $\bar{x} = 44.24$ **b** 149

**8** **a** 40–49 **b** 40–49

**9** **a** −2, 2 **b** 2 **c** 2.6

**10** **a** 1000–<1100 **b** 1000–<1100

**11** **a** mean **b** median **c** median **d** mode
**e** mean **f** mode **g** mean **h** mode

**12** **a** mean **b** mode
**c** median **d** mode or mean
**e** mean **f** mode

**13** **a** 7 **b** \$91 971 **c** \$78 500
**d** Mean; the high salary of the general manager pushes it up.
**e** median

**14** **a** **i** 38.125 **ii** 38 **iii** 32
**b** **i** decrease **ii** decrease **iii** no effect

**15** **a** 39 **b** white
**c** Data is categorical, not numerical.

**16** **a** **i** \$523 **ii** \$481 **iii** no mode
**b** Mean distorted by outlier of \$1027; there is no mode.
**c** mean \$473, median \$458, yes

**17** **a** **i** 7.9 **ii** 7.5 **iii** 7
**b** **i** decrease **ii** no effect
**c** mode

**18** **a** **i** 27°C **ii** 27°C **iii** 27°C
**b** mean

## Exercise 10.03

**1** **a** 9.1, 10.1, 11.0 **b** 49.5, 50, 51
**c** 6.20, 6.50, 6.90 **d** 8, 18, 23

**2** **a** 113.5 **b** 101.5 **c** 125.5

**3** D

**4** **a** 6 **b** 5 **c** 728 **d** 12

**5** **a** 2 **b** 2.5 **c** 258 **d** 3

**6** A

**7** **a** 63 **b** IQR = 75 − 48 = 27

**8** **a** 46 **b** 12
**c** The person who took 15 s to complete the test may know all the work very well or may have guessed the answers.

**9** B **10** 80

## Exercise 10.04

**1** **a** 2 **b** 1.3

**2** **a** 2.25 **b** 3.03 **c** 5.28, −0.78
**d** 6 **e** 50%

**3** **a** 75.00 **b** $\sigma = 9.43$

**4** **a** \$836 **b** \$277

**5** **a** 2.88 **b** 1.8 **c** 27 **d** 54%

**6** **a** 6.8 **b** 1.86 **c** 15 **d** 75%

**7** **a** 950, 1050, 1150, 1250, 1350, 1450
**b** **i** \$1164.89 **ii** \$119.36

**8** **a** Males: $\bar{x} = 175.5$, $\sigma = 7.53$
Females: $\bar{x} = 164.6$, $\sigma = 6.93$
**b** Males are significantly taller than females (difference in means is 10.9 cm).
The spread of heights of females is slightly less than the spread of heights of males.

**9** **a** Test 1: $\bar{x} = 65.95$, $\sigma = 14.36$
Test 2: $\bar{x} = 71.55$, $\sigma = 14.45$
**b** The mean for Test 2 is 5.6 marks higher than the mean for Test 1, which is significant, but the standard deviations are nearly equal, indicating the spread for both tests is the same.
**c** Because the mean for Test 2 is significantly higher, the students performed better in Test 2.

**10** **a** Men: $\bar{x} = 340.55$ $\sigma = 436.03$
Women $\bar{x} = 399.65$ $\sigma = 559.58$
**b** Men: $\bar{x} = 224.22$ $\sigma = 269.11$
Women $\bar{x} = 234.06$ $\sigma = 248.21$
**c** For this group, women make longer calls than men (by 60 s) and the spread of the time for their calls is much more than the time for calls made by men. However, if the outliers are excluded, there is no significant difference, as the mean time of calls for women is only 10 s longer than that for men. Also, the spread of times for calls is now more for men (their standard deviation is 20 more than that for women) but again it is not significant.

## Exercise 10.05

**1** **a** IQR = 1.5
**b** **i** −0.25 **ii** 5.75
**c** Yes, since 8 is greater than 5.75 ($= Q_3 + 1.5 \times \text{IQR}$).

**2** **a** No, since outliers must be less than −1.25 or greater than 16.75.
**b** Yes, 9 is an outlier since it is less than 11 ($= Q_1 - 1.5 \times \text{IQR}$).
**c** No, since outliers must be less than −11 and greater than 81.
**d** 9 is the outlier since it is greater than 7.5.

**3** **a** \$831.43 **b** \$820
**c** IQR = 910 − 720 = 190
**d** \$1200, $Q_3 + 1.5 \times \text{IQR} = 910 + 1.5 \times 190 = 1195$, so \$1200 is an outlier.
**e** Mean, median and interquartile range would be lower (\$770, \$785, \$170 respectively).
**f** \$914.57, \$902, \$209 respectively; yes

**4** **a** 20 **b** 8 since $8 > 5 + 1.5 \times 1.5$
**c** **i** mean 4.35, range 6, standard deviation 1.24
**ii** mean 4.16, range 3, standard deviation 0.93
**d** They all increase.

**5** D

**6** **a** The Wombats: $\bar{x} = 16.8$ points; median = 18 points; range = 18 points
The Possums: $\bar{x} = 16$ points; median = 16 points; range = 3 points
The Koalas: $\bar{x} = 17.2$ points; median = 14 points; range = 28 points

- **b** The Possums, as mean and median are the same.
- **c** $\bar{x} = 18.8$ points; median = 18 points; range = 10 points
- **d** Still The Possums, with The Wombats a close second.

**7** **a** Sam: 3 copiers; Terri: 3 copiers
- **b** They are equally good salespeople.
- **c** Sam: 5.5 copiers; Terri: 17 copiers
- **d** Sam: 7.2 copiers; Terri: 16.6 copiers
- **e** Median, because Sam has an outlier of 25.
- **f** Terri, because her median sales are much higher than Sam's.

**8** **a** mean = 2.92; median = 3, mode = 3, range = 9, IQR = 3, standard deviation = 2.40
- **b** 9 since $9 > 4 + 1.5 \times 3 = 8.5$ ($Q_3 + 1.5 \times \text{IQR}$)
- **c** mean = 2.36, median = 3, mode = 3, range = 5, IQR = 3, standard deviation = 1.61
- **d** Increases the mean, range and standard deviation, no effect on the mode, median and IQR.

**9** **a** mean = \$60 570; median = \$64 200; mode = \$66 500
- **b** mode, the highest measure of centre
- **c** median, as it is not affected by the outlier wage of \$173 800.

## Exercise 10.06

**1** **a** 1, 8, 17, 23, 23, 24, median = 3
- **b**

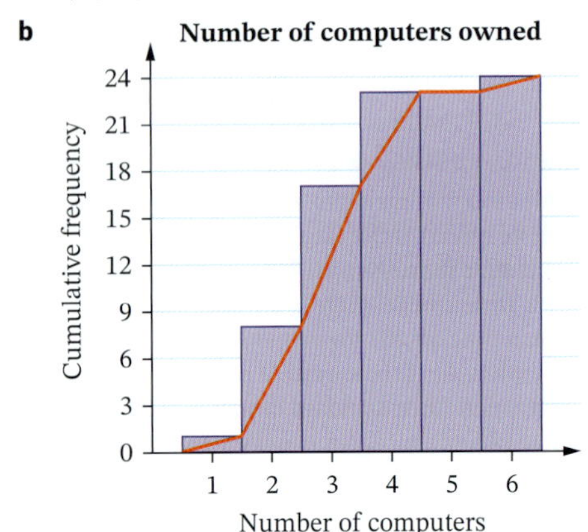


- **c** **i** 3 **ii** 2

**2** **a** 25 **b** 45
- **c** 56 − 27 = 29

**3** **a** 6, 40, 96, 124, 129, 130
- **b**

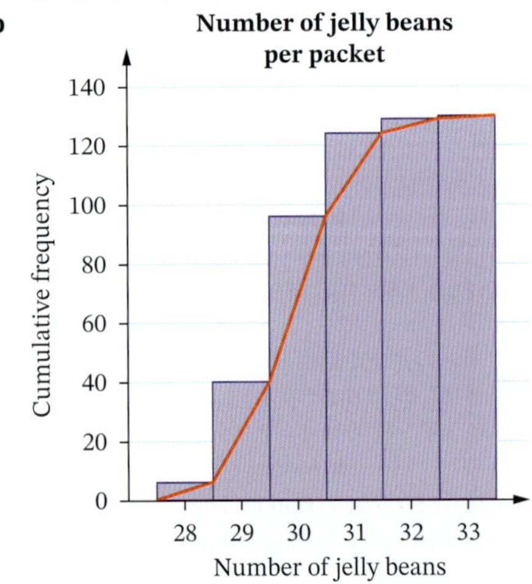


**i** 30 **ii** 31 − 29 = 2

**4** **a** Class centres: 137.5, 144.5, 151.5, 158.5, 165.5, 172.5,179.5

Cumulative frequencies: 2, 5, 9, 22, 37, 48, 50
- **b** 162–<169 **c** 162–<169
- **d** Students' heights

**i** 164 **ii** 171 − 158 = 13

## Exercise 10.07

**1** **a** 3, 5, 6.5, 9, 11
- **b**

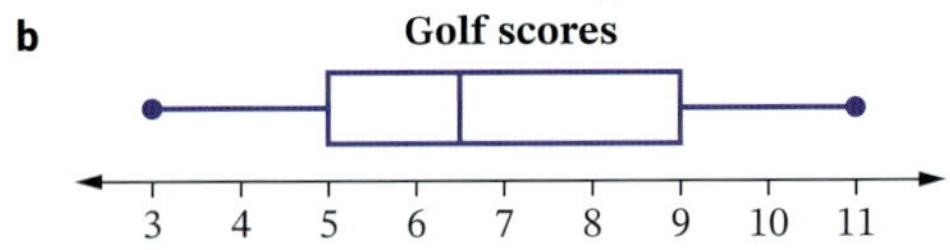


**2**

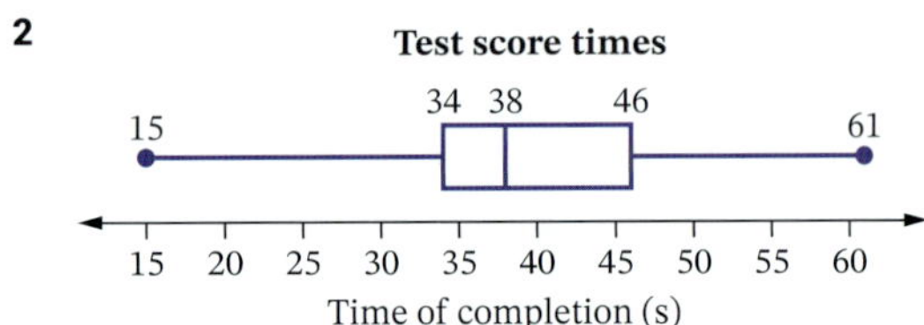


**3** Ages of cinema patrons

28.5 36 44.5

14 69

15 20 25 30 35 40 45 50 55 60 65 70

**4** **a** 20 **b** 12 **c** 3 **d** 15 **e** 30

**5** **a**

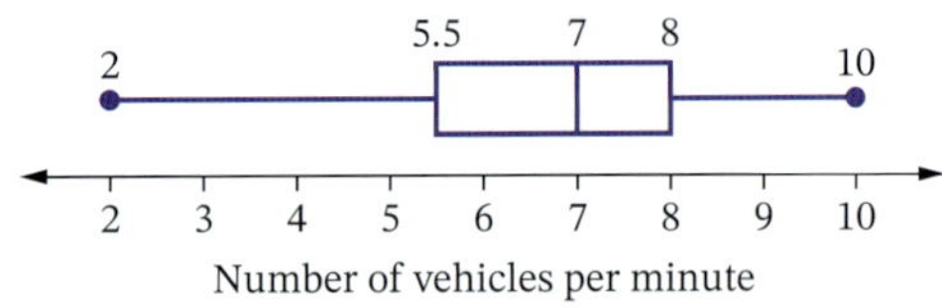


- **b** Different answers possible: The dot plot shows individual data values and any clusters and outliers, while the box-and-whisker plot identifies the quartiles, including the median.

**6** **a** \$75 000 **b** \$41 000 **c** \$41 000
- **d** \$58 000 **e** \$28 000, \$51 000
- **f** 25%

**7** **a** Geography: 45, 55, 60, 75, 90

Modern History: 40, 60, 70, 75, 80
- **b** Geography: **i** 45 **ii** 20

Modern History: **i** 45 **ii** 15
- **c** Geography 60, Modern History 70
- **d** Modern History as range for both subjects is the same but IQR for History is lower.
- **e** **i** 10 **ii** 20
- **f** Modern History as 30 students scored above 60 compared with 20 students in Geography.

**8** **a** Males: 58, 60, 68, 75, 106

Females: 55, 68, 74, 82, 120

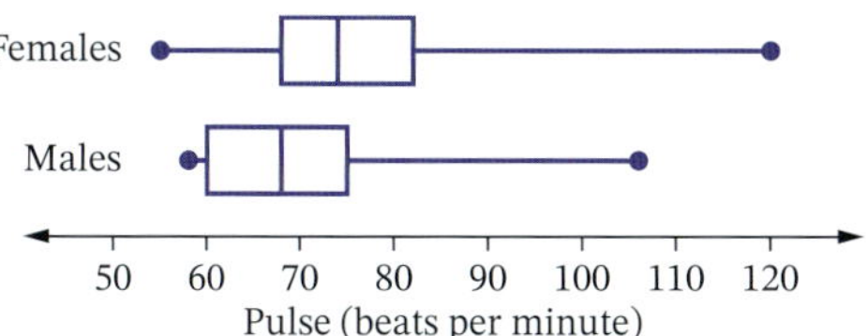


**b** Males: range = 48, IQR = 15

Females: range = 65, IQR = 14

**c** Range of females is higher due to the outlier of 120. The IQRs are similar, so no significant difference in the spread of either group.

**d** Males: 25% of males had pulses above 75 while 50% of female pulses were above 74.

**9** **a** Dominant: 0.29, 0.32, 0.38, 0.44, 0.52

Non-dominant: 0.30, 0.35, 0.40, 0.45, 0.9

Non-dominant

Dominant

0.20 0.30 0.40 0.50 0.60 0.70 0.80 0.90

Reaction time (seconds)

**b** Dominant: range = 0.23, IQR = 0.12

Non-dominant: range = 0.60, IQR = 0.10

**c** If outlier of 0.9 is omitted, the spreads of the groups are similar, with the dominant hand having slightly better (shorter) reaction times.

**10** D

## Exercise 10.08

**1** C

**2** **a** **i** negatively skewed

**ii** unimodal, peak at 6

**iii** cluster from 6–8

**b** **i** positively skewed or symmetrical

**ii** multimodal

**iii** no clusters

**c** **i** positively skewed

**ii** unimodal, peak at 16

**iii** clustering in the 10s and 30s

**d** **i** negatively skewed

**ii** multimodal, peaks at 12, 14, 16 and 18

**iii** no clusters

**e** **i** symmetrical

**ii** multimodal, modes at 55, 65 and 85

**iii** clustering in the 60s

**f** **i** not symmetrical, not skewed

**ii** bimodal, peaks at 15–20 and 30–35

**iii** clustering at 10–20, 30–40

**3** **a** 36

**b** negatively skewed

**c** 70s and 80s **d** 88

**4** **a**

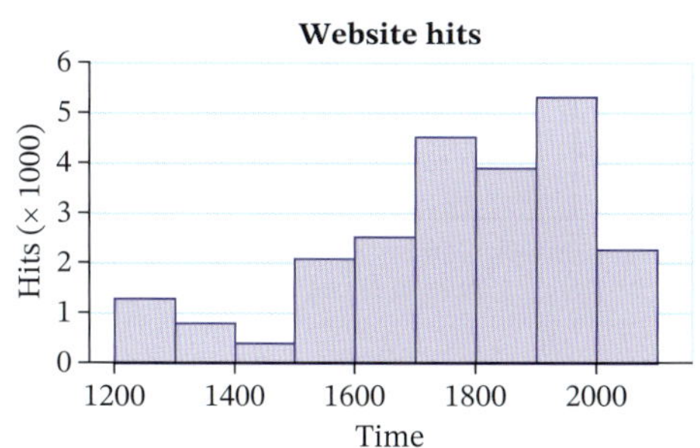


**b** negatively skewed, cluster from 1700 to 2000, peak 1900–2000

**c** Most visits to internet sites occur after work hours.

**5** B

**6** **a**

| Stem | Leaf |
|---|---|
| 1 | 5 5 5 6 6 7 8 9 9 |
| 2 | 0 1 2 2 |
| 3 | 5 6 |
| 4 | 3 9 |
| 5 | 0 5 9 |

**b** positively skewed; cluster 15–19 years; peak mode of 15

**7** **a** positively skewed

**b** 3

**c** 2.9; it is slightly less than mode.

**8** **a** 12W: 50 12X: 40

**b** 12W: range = 38, IQR = 12

12X: range = 56, IQR = 28

**c** Both sets of data are positively skewed.

**d** 12W had the better results as the median was 50 compared to 40 for 12X. More than 75% of students from 12W scored more than 50% of students from 12X.

## Exercise 10.09

**1** **a** A: 90, 58.5, 56.5 B: 91, 18, 53

**b** Test B as it has a smaller interquartile range

**2** **a**

| Netball | | Hockey |
|---|---|---|
| 8 | 3 | 2 |
| 2 2 | 6 | 4 4 |
| 9 7 7 5 2 2 | 7 | 3 4 5 9 |
| 0 | 8 | 1 1 2 6 |
| | 9 | 2 |

**3** A

**b** Netball: negative skew, cluster in 70s

Hockey: negative skew, cluster in 70s and 80s

**c** 71.4, 75.25 **d** 73.5, 77

**e** Median since data is skewed

**4** B

**5** **a** **i** 12 cm **ii** 26 cm

**b** Estimates: range = 34 cm, IQR = 17.5 cm

Actual: range = 11 cm, IQR = 3.5 cm

**c** Definitely agree; the box plot for estimates has a much greater spread as well as a much lower median, even though some people did overestimate.

**6** **a** Boys: 31, 50, 55.5, 65, 66

Girls: 31, 38, 43, 47, 69

**b** Spread of data values, five-number summaries (highest and lowest values, quartiles, median)

**7** **a** **i** Darwin **ii** Brisbane

**iii** Darwin **iv** Canberra

**b** Sydney (lower median price)

**8** **a** 5

**b** Those with the lowest pulse did not exercise hard enough and pulse stayed the same; or someone who had a higher pulse due to rushing to class could have reduced their pulse rate by not working hard in the class.

**c** Two people exercised very hard and got their pulse rates up very high.

**d** 22 beats

**e** 6 beats/min

**9** **a** Five-number summaries are:

| Before | 9 | 15 | 21 | 35 | 45 |
|---|---|---|---|---|---|
| 6 week later | 6 | 14.5 | 18.5 | 26 | 32 |

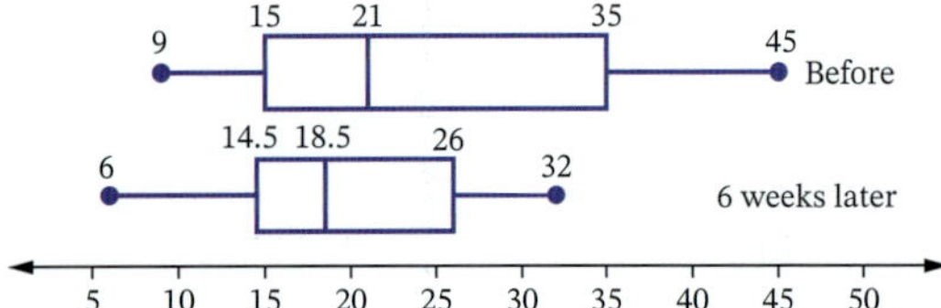


**b** 20, 11.5

**c** Yes; interquartile range and median decreased.

**10** **a**

| Sydney | | Melbourne |
|---|---|---|
| | 0 | 7 8 9 |
| 3 2 2 2 2 2 2 2 1 0 0 0 | 1 | 1 2 2 4 4 4 5 5 6 |

**b**

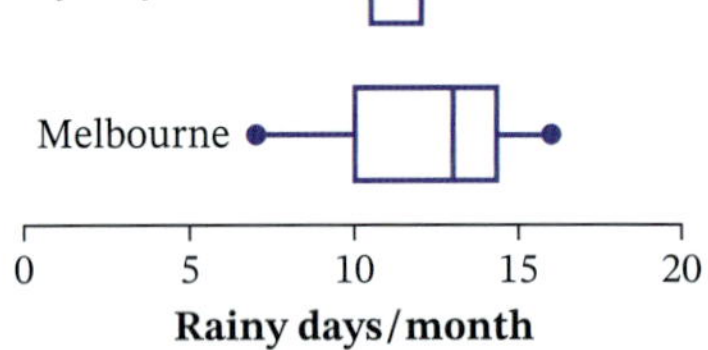


**c** Stem-and-leaf plot: Sydney values cluster around 12 days.
Box plots: Sydney has a much smaller spread than Melbourne; median and upper quartile for Sydney are equal.

**d** Disagree with 'more'; the median number of rainy days per month is only 1 day greater for Melbourne (13) than for Sydney (12).

**11** **a** Australians: mean = 61.9, median = 59, modes = 47, 56, 57
New Zealanders: mean = 65, median = 66.5, modes = 64, 71

**b** Australians: range = 47, IQR = 24, $\sigma = 12.4$
New Zealanders: range = 50, IQR = 15.5, $\sigma = 12.5$

**c** Australians: positively skewed
New Zealanders: negatively skewed

**d** New Zealanders were older since the mean and median are both greater than the mean and median for Australians (by 3.1, 7.5 respectively). Although New Zealanders had the larger range, the IQR was less and the standard deviation was similar, so the spreads of ages were approximately the same.

**12** **a**

| Estimates | | Test results |
|---|---|---|
| 3 | 3 | 2 |
| | 4 | 6 9 |
| 8 | 5 | 2 |
| 5 5 | 6 | 4 |
| 8 7 7 5 3 0 | 7 | 2 3 3 3 8 9 9 9 |
| 9 9 8 7 7 6 5 3 2 2 2 0 | 8 | 0 2 5 6 6 7 7 |
| 3 2 | 9 | 1 1 1 5 |

**b** The estimates are negatively skewed, has one mode (82) and are clustered in the 80s. The test results are also negatively skewed, have 3 modes (73, 79 and 91) and are clustered in the 70s and 80s.

**c** Estimates: **i** 78.2 **ii** 82 **iii** 82
Test results: **i** 75.4 **ii** 79 **iii** 73, 79, 91

**d** Estimates: **i** 60 **ii** 13 **iii** 12.9
Test results: **i** 63 **ii** 14 **iii** 15.8

**e** The students did overestimate their results but only by 3 when comparing means and medians. Also, the spread of the 2 sets of data are similar with the ranges (60 and 63) and IQRs (13 and 14) being very close.

## Sample HSC problem

**a** 39.25 **b** 31.5

**c** Mean; involves every data value and there are no outliers.

**d** $51 - 18.5 = 32.5$

**e**

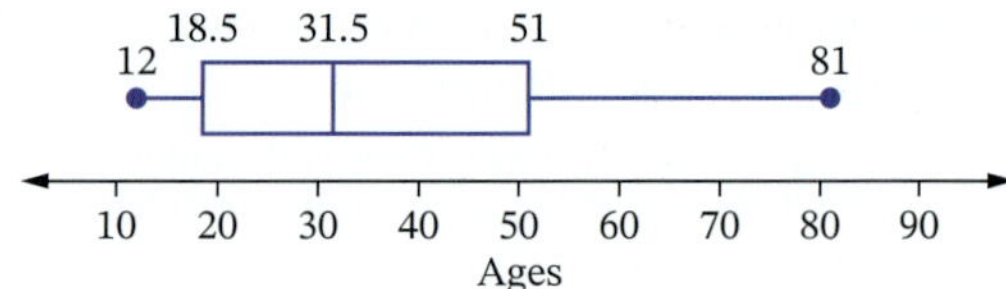


## Test yourself 10

**1** Teacher to check.

**2** **a** not precise; should give a range of answer options

**b** too broad/general, open-ended

**c** biased towards answering yes

**3** **a** 173.2 **b** 171.5 **c** 168, 170

**4** **a**

| Speed (km/h) | Class centre | Frequency |
|---|---|---|
| 60–<70 | 65 | 7 |
| 70–<80 | 75 | 13 |
| 80–<90 | 85 | 23 |
| 90–<100 | 95 | 7 |
| | | 50 |

**b** 81 km/h **c** 80–90 **d** 80–90

**5** mean = 6.8, median = 7, mode = 7

**6** **a** mean $1 299 875

**b** median $1 261 000

**c** median; the outlier of 2 080 000 does not affect its value as it does the mean.

**7** **a** mode : shows most common and not affected by outliers

**b** mean: includes all students

**c** median: not affected by outliers

**8** C

**9** **a** range of middle 50% of data

**b** range = 53 − 48 = 5, IQR = 51 − 49 = 2

**10** **a** 56 **b** 19

**11** **a** 1.998 cm **b** 0.02 cm

**12** **a** mean = 6.7; standard deviation = 1.3

**b** Cumulative frequency: 8, 19, 29, 35, 40

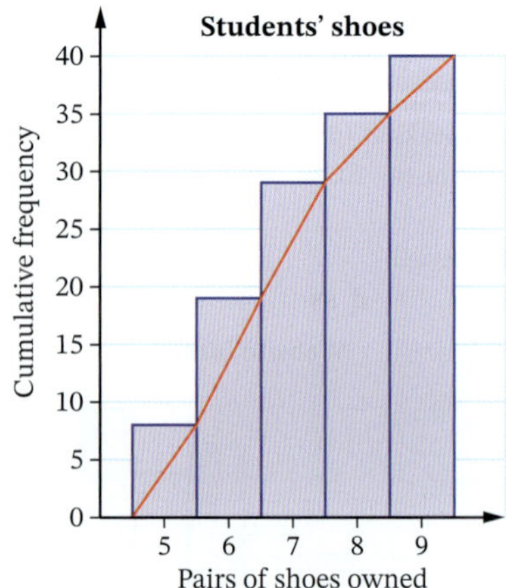


**c** **i** 7 **ii** $8 - 6 = 2$

**13** $Q_3 + 1.5 \times \text{IQR} = 1178 + 1.5 \times 228 = 1520$
So, 1710 is an outlier.

**14 a** Increases the mean from 11.625 to 14.778 and also increases the median from 10 to 12.

**b** The median; less affected by the outlier.

**15 a** 36

**b i** 4.5 **ii** $7 - 3 = 4$

**16 a** 0 **b** 5 **c** $\frac{1}{2}$ **d** 25%

**17 a** 48, 49, 50, 51, 53

**b** Masses of chip packets

48 49 50 51 52 53

Grams

**18 a** English; smaller range and IQR

**b** 96

**19 a** Test 1: mean = 5.6, median = 6, mode = 6

Test 2: mean = 6.3, median = 7, mode = 7

**b** Test 1: Results are symmetrical.

Test 2: Results are negatively skewed.

**c** Test 1: **i** 7 **ii** 2 **iii** 1.74

Test 2: **i** 8 **ii** 2 **iii** 1.80

**d** Results of Test 2 are just better than Test 1 as mean, mode and median of Test 2 are higher than for Test 1. The spreads for the two tests are similar as there is only a difference of 1 between ranges, the IQRs are equal and the standard deviations are approximately equal.

# Chapter 11

## SkillCheck

**1 a** $A(-3, 1)$, $B(2, 1)$

**b** 5 units

**c** $C(-2, 4)$

**2 a**

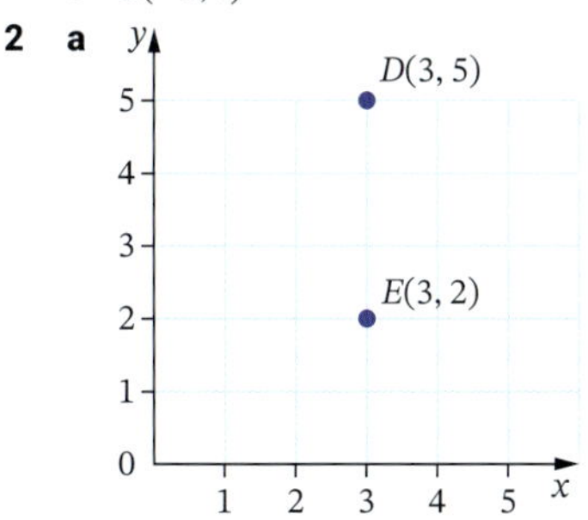


**b** 3 units

**3 a** 9 **b** 5 **c** 17 **d** 8

**4 a** 17 **b** 44 **c** 25 **d** 43

**5 a** 2:00 am **b** 6:00 am

**c** 12:33 am **d** 6:44 pm

## Exercise 11.01

**1 a** D **b** G **c** A **d** C

**e** F **f** B **g** E **h** H

**2 a** G **b** C **c** D **d** E

**e** A **f** F **g** H **h** I

**i** B **j** J **k** K **l** L

**3 a** (0°, 75°W) **b** (45°N, 40°W)

**c** (0°, 110°W) **d** (30°S, 40°W)

**e** (45°N, 110°W) **f** (0°, 40°W)

**g** (30°S, 110°W) **h** (45°N, 75°W)

**4** The South Pole

**5 a** small **b** great **c** great

**6 a** 9°, north **b** 14°, north **c** 4°, south

**d** 2°, north **e** 67°, north **f** 1°, south

**7 a** B **b** C

**8 a** 99°, west **b** 21°, west **c** 17°, west

**d** 86°, east **e** 26°, east **f** 1°, east

**9** (32°S, 141°E) **10** (27°S, 152°E)

**11** D

**12 a** longitude **b** latitude

**13** North and South poles

**14 a** Perth and Bali are on the same meridian of longitude 115°E.

**b** (20°S, 115°E)

## Exercise 11.02

**1 a** 8:45 am **b** 5:50 pm **c** 12:19 am

**d** 11:05 am **e** 4:28 pm **f** 9:15 am

**g** 2:40 am **h** 8 pm

**i** 6:25 am **j** 6:53 pm

**2 a** 1935 **b** 1142 **c** 2359

**d** 0017 **e** 1230 **f** 0340

**g** 0710 **h** 2154

**i** 2218 **j** 0159

**3** 11:17 pm

**4 a** 07:30 **b** 01:15 **c** 17:10

**5** 1621

**6 a** PM AM 10:05 **b** PM AM .05:45

**c** PM AM 02:21

**7** A

**8 a** 1 h 55 min **b** 50 min **c** 4 h 45 min

**d** 3 h 55 min **e** 5 h 45 min **f** 10 h 55 min

**9** 1 h 52 min

**10 a** 8 pm **b** 7 am **c** 6:55 pm

**d** 1:17 pm **e** 2:04 am **f** 3:52 am

**11** Manal 1:43 pm; Tom 1:44 pm; Gianni 2:02 pm; Eddie 2:04 pm; Sarah 2:19 pm; Robert 2:21 pm

**12** 9 h 30 min **13** 6 h 45 min **14** 6:15 pm

**15 a** 20 min **b** 9:32 am **c** Lorne

**d** 3:00 pm **e** Kennett River

**16 a** 1005, 1 h 25 min **b** NA114 at 1040

**c** 1005 **d** NA038 at 1005

**e** NA114 takes 5 minutes longer. Smaller plane, different flight path

## Exercise 11.03

**1** They have the same longitude, 30°E.

**2** $60 \div 15 = 4$ **3** 10 hours

**4 a** 12 noon **b** 6 pm **c** 12 noon **d** 4 am

**e** 8 pm **f** 6 am **g** 4 am

**5** 6 pm

**6** It's 7 pm in Hawaii. It's too late for anyone to be in the office.

**7 a** 2 pm Friday **b** 1 am Saturday

**8** 4 pm

**9 a** 10 am Tuesday **b** 6 am Wednesday

## Exercise 11.04

**1 a** 11:30 am **b** 9:00 am

**c** 8:00 am **d** 7:00 pm

**2 a** 7:00 am **b** 11:00 am

**c** 4:30 pm **d** 9:00 pm

**3 a** 4 h **b** 1 h

**c** 8 h **d** 8 h

**4** C

**5 a** 6:00 pm **b** 6:00 pm **c** 5:30 pm

**d** 4:00 pm **e** 5:30 pm **f** 6:00 pm

**6 a** 10:30 am **b** 10:30 am **c** 9:00 am

**d** 7:30 am **e** 10:00 am **f** 9:30 am

**7 a** 12 midnight **b** directly opposite sides

**8** 2:30 pm **9** A **10** 13 hours

**11** **a** 2 am **b** 5 pm **c** 9 am

**12** **a** 10:30 am Tuesday

**b** 12:30 am Tuesday

**c** 11:30 pm Monday

**13** **a** 5:40 pm **b** 9:00 pm

**14** 1:55 pm

**15** 4:05 pm **16** 1:00 am Friday

**17** **a** Tuesday **b** 1:00 pm Tuesday

**18** **a** 11 h **b** No; the next day

## Sample HSC problem

**a** 82° **b** 18 h 25 min

**c** 9:10 pm Wednesday

## Test yourself 11

**1** $A$ (0°, 70°W), $B$ (0°, 40°W), $C$ (55°S, 70°W), $D$ (55°S, 40°W)

**2** **a** 147°E **b** 54° **c** 25°

**3** (52°N, 5°E)

**4** **a** 2:36 pm **b** 5:05 am

**c** 10:18 pm **d** 12:27 am

**5** **a** 0523 **b** 1632

**c** 2105 **d** 1255

**6** **a** 8:55 am **b** 6:52 am

**c** 1027 **d** 1105

**7** **a** 8:30 pm **b** 10:30 am

**8** **a** 5:00 pm **b** 7:30 am

**9** **a** 6:00 pm **b** 6:30 am **c** 2:30 am

**10** **a** Wednesday **b** 5:30 am Thursday

**11** 3:00 am Monday

**12** 5:25 am Thursday

## Practice exam 4

**1** A **2** B **3** B **4** B **5** B

**6** A **7** C **8** D **9** D **10** C

**11** **a** **i** 8:55 am, 1:55 pm **ii** 3 h 20 min

**iii** NC511 on Monday night

**b** **i** 1, 15, 41, 48, 50 **ii** 1.9 **iii** 4

**iv**

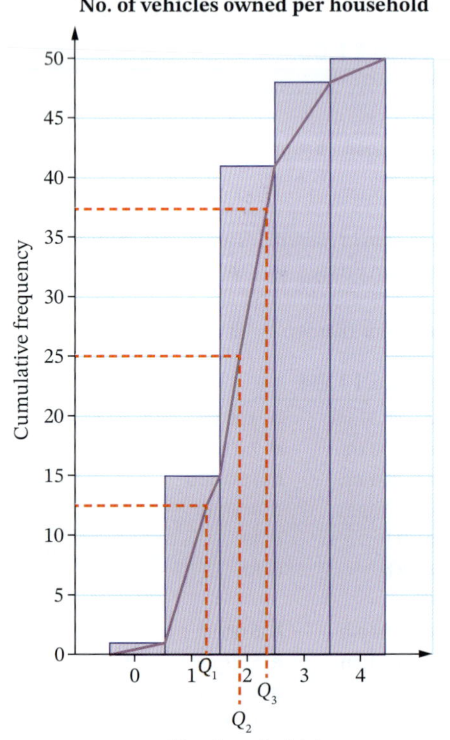


**v** $2 - 1 = 1$

**12** **a** 8 am Wednesday

**b** **i** mean 2.83, standard deviation 1.91

**ii** 3

**c** **i** 3 and 5 **ii** $6 - 3 = 3$ **iii** 2.57

**d** **i** negatively skewed with a cluster in the 80s

**ii** median because it is skewed

**13** **a** **i** 70 **ii** 15 **iii** 31

**b** **i** Women: negatively skewed, clustered 50s–60s unimodal; Men: cluster 30s–40s, multimodal

**ii** The men were younger. More ages under 50.

**c** **i** Manila **ii** Monday

**iii** 9 am Tuesday

# GLOSSARY AND INDEX

**12-hour time** Time of day written in the usual way using am or pm and the hours 1 to 12; for example, 9:27 pm. (p. 434) See also **24-hour time**.

**24-hour time** Time of day written using 4 digits (instead of am or pm) and the hours 0 to 23. For example, 1745 is the 24-hour time for 5:45 pm. (p. 434) See also **12-hour time.**

## A

**allowable tax deduction** A part of a person's yearly income that is not taxed, such as work-related expenses or donations to charities. All deductions are subtracted from yearly income to determine **taxable income**. (p. 140)

**annual leave loading** Extra payment to a worker based on a percentage (usually 17.5%) of 4 weeks annual leave. (p. 130)

**arc** (of a network) See **edge**.

**arc** Part of the circumference of a circle. (p. 173)

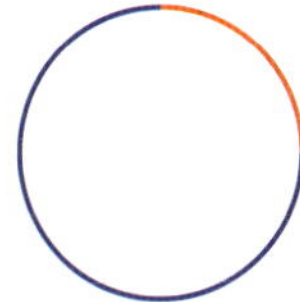

**Australian Central Standard Time (ACST)** Standard time zone (UTC +9.5) for central Australia: the Northern Territory and South Australia. (p. 441)

**Australian Eastern Standard Time (AEST)** Standard time zone (UTC +10) for eastern Australia: NSW, Queensland, Victoria, ACT and Tasmania. (p. 442)

**Australian Western Standard Time (AWST)** Standard time zone (UTC +8) for Western Australia. (p. 444)

## B

**bar chart** See **column graph.**

**base** (in index notation) A number being raised to a power. For example, in $2^5$, the base is 2. (p. 57)

**base** (of a prism) One of the parallel end faces of a prism. (p. 192)

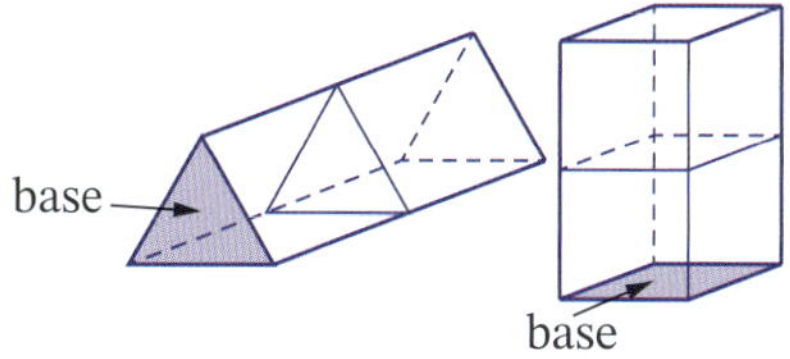


**bias** In statistics, an unwanted influence that stops a sample from being representative of a population. (p. 25)

**biased sample** A sample that is not truly representative of the population and affects the quality or reliability of a statistical study. (p. 25)

**blood alcohol content (BAC)** The concentration of alcohol in a person's blood, measured in g/100 mL. (p. 86)

**bonus** Extra pay for achieving a high quality or volume of work, such as meeting an important quota, goal or deadline. (p. 130)

**box plot** A diagram that displays the quartiles of a set of data as a box and the extremes as whiskers. (p. 401)

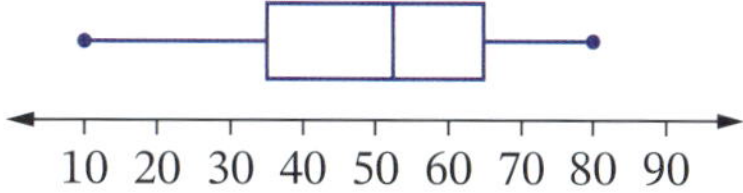


**braking distance** The distance travelled between when a driver applies the brakes and when the vehicle stops completely. (p. 102) See also **reaction distance** and **stopping distance**.

**budget** A plan for managing money. (p. 240)

## C

**capacity** Maximum volume of liquid that can be held by a container, usually measured in millilitres (mL), litres (L) or kilolitres (kL). (p. 191)

**categorical data** Information or data represented as a category rather than as a number (for example, makes of cars, colour of eyes). Differs from **numerical data**. (p. 21)

**census** Collection of information about every member of a population. (p. 24)

**circumference** The perimeter of a circle. $C = \pi d$ or $C = 2\pi r$, where $C$ is the circumference, $\pi$ is pi (3.14159...), $d$ is the **diameter** and $r$ is the **radius**. (p. 172)

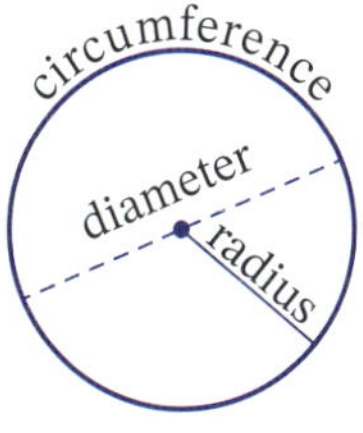


**class centre** The centre of a class interval. For example, the class centre of the class interval 10–19 is 14.5. (pp. 36, 375)

**class interval** In statistics, when there are many data values, they may be grouped into class intervals. For example, ages of people may be grouped into class intervals of 1–10, 11–20, 21–30, and so on. (pp. 36, 375)

**cluster** A group of data values that are bunched or close together. (p. 43)

**clustered column graph** A bar chart that compares the data of 2 or more categories. (p. 12)

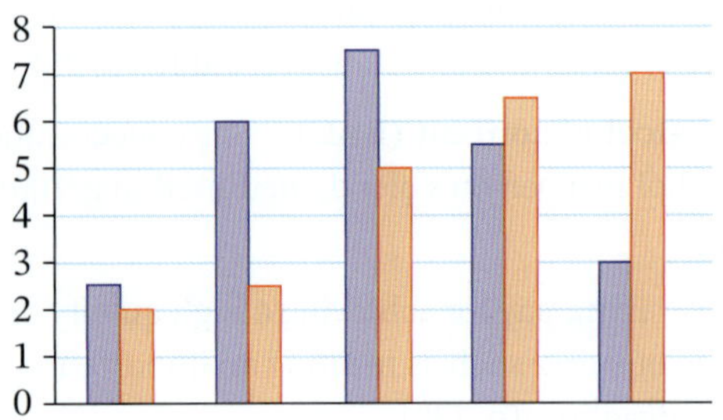


**column graph (or bar chart)** A graph consisting of vertical bars of equal width. (p. 6)

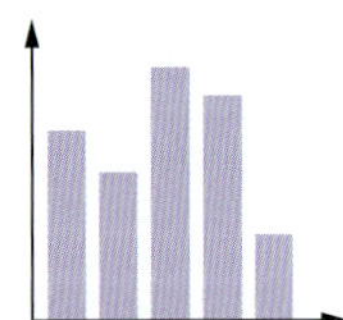

**commission** The earnings of a sales person or agent; usually a percentage of the value of items sold. (p. 127)

**comprehensive insurance** Motor insurance that covers all damage to vehicles and property, including your own, in an accident in which you are at fault. (p. 257) See also **CTP insurance** and **Third Party Property insurance**.

**Compulsory Third Party (CTP) insurance** Motor insurance that covers personal injury or death to another person ('third party') in an accident in which you are at fault. Also called '**green slip**' because the insurance certificate is green. (p. 257)
See also **comprehensive insurance** and **Third Party Property insurance**.

**constant** A value that does not change. (p. 345) See also **variable**.

**constant of variation** (or **constant of proportionality**) The constant in a variation equation. For example, if $y$ varies as $x$, the equation is $y = kx$ and the constant of variation is $k$. (p. 345)

**continuous data** Numerical (quantitative) data that can be measured on a smooth scale of values (without gaps), such as the heights of people. (p. 21)

**Coordinated Universal Time (UTC)** See **UTC**.

**coordinates** A pair of points (e.g. in $x$ and $y$ planes) that describe the location of a point on a number plane, map or involving **latitude** and **longitude**. (pp. 332, 427)

**cumulative frequency** A running total of frequencies. (p. 374)

**cumulative frequency histogram** A histogram in which the height of each column represents the cumulative frequency at each data value. (p. 397)

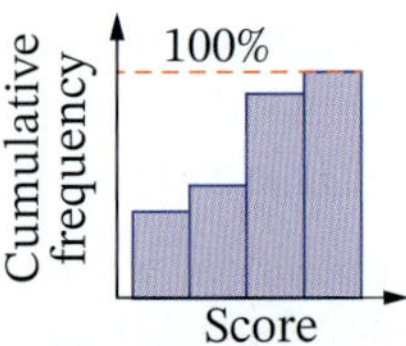


**cumulative frequency polygon** A line graph formed by joining the ends of the tops of the columns of the cumulative frequency histogram. Also called an **ogive**. (p. 397)

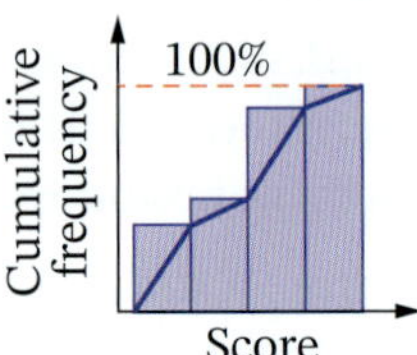


## D

**data** Observations or facts which, when collected, organised and evaluated, become information. (p. 6)

**data set** Collection or group of data values. (p. 36)

**daylight saving time** Scheme where clocks are turned forward an hour to take advantage of increased hours of daylight during summer. (p. 445)

**deduction (tax)** See **allowable tax deduction**.

**deductions (from pay)** Amounts taken out of a person's gross pay (for example, union fees, superannuation, and income tax). (p. 136)

**dependent variable** The variable in a function that depends on another variable for its value. For example, if $y = 3x + 5$, $y$ is the dependent variable because its value depends on the value of $x$. Differs from an **independent variable**. (p. 336)

**degree** (of a vertex) The number of edges connected to a vertex. (p. 283)

**diameter** The length of the interval passing through the centre of a circle and joining two points on the **circumference** of the circle. The diameter is double the **radius**. (p. 172)

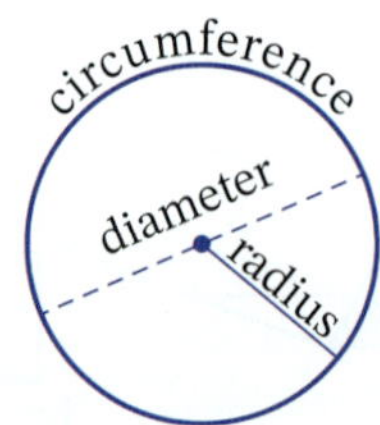


**direct linear variation** (or **direct proportion**) The relationship between two variables (say $x$ and $y$) is represented by an equation of the form $y = kx$, where $k$ is the constant of variation. (p. 345)

**directed network** A network with arrows on the edges to indicate that movement can only be in the direction of the arrows. (p. 282)

**discrete data** Quantitative (numerical) data that can be counted and whose values are separate and distinct, such as the numbers of pets owned, or the numbers of people in families. (p. 21)

**discretionary spending** Money spent on items other than necessities. (p. 240)

**distribution** (or **frequency distribution**) The way the values of a data set are arranged, especially when graphed. (p. 406)
See also **shape of a distribution**.

**divided bar graph** A graph representing the parts of a whole by a rectangle divided into proportionately sized sections. Displays the same kind of data as a **sector graph**. (p. 6)

**dosage** A prescribed amount of medicine to be taken. (p. 63)

**dot plot** A graph that uses dots to show frequencies of data values (for example, the temperatures, in °C, of 10 hospital patients). (p. 43)

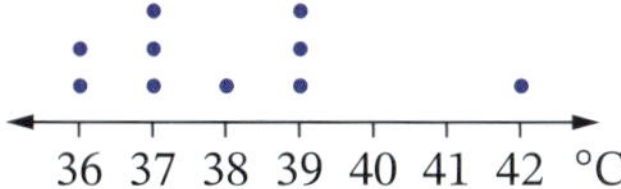


**double time** Wages paid at twice the normal rate (for example, for working on a Sunday or a public holiday). (p. 123)

## E

**edge** A line joining 2 vertices in a network. (p. 282)

**equation** A mathematical statement that 2 quantities are equal. An equation contains an equals sign (for example, $4x - 5 = 11$). (p. 68)

**Equator** The 0° parallel of latitude, the great circle running around the middle of Earth. The latitude of any point on Earth's surface is measured north or south of the Equator. (p. 428)

**excess** Amount the person taking out the insurance agrees to pay upfront in the event of an accident before the company pays the rest. (p. 258)

**expense** The cost of spending. (p. 240)

**extrapolate** To estimate the value of a quantity that is outside the given range of values. (p. 337)

## F

**fatality** An accident resulting in death, or a person who dies in an accident. (p. 94)

**field diagram** A diagram produced from a land survey. (p. 183)

**five-number summary** Lowest value, lower quartile, median, upper quartile and highest value of a data set. These 5 numbers are used to draw a **box plot**. (p. 401)

**fixed spending** Money spent on necessities such as food, clothes, fuel and household bills. (p. 240)

**formula** A rule written as an algebraic equation, using variables. For example, the formula for the area of a triangle is $A = \frac{1}{2}bh$. (p. 62)

**fortnightly** Every 2 weeks. (p. 124)

**frequency** The number of times a value or class interval of values occurs in a data set. (p. 30)

**frequency distribution** See **distribution**.

**frequency histogram** A histogram in which the height of each column represents the frequency of a value or class interval of values. (p. 36)

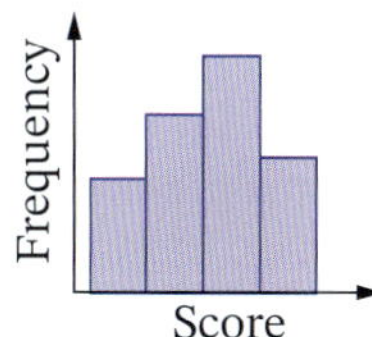


**frequency polygon** A line graph formed by joining the midpoints of the tops of the columns of a frequency histogram. (p. 36)

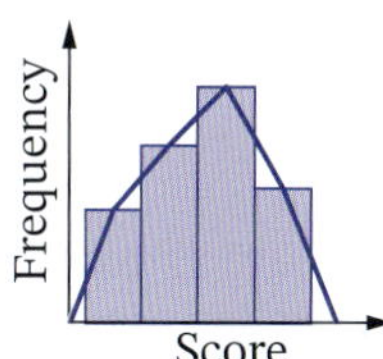


**fuel consumption** The rate at which fuel is used by a vehicle, usually measured in L/100 km. (p. 267)

**function** The relationship between 2 variables (for example, $y = 3x - 5$). (p. 325)

## G

**giga-** (symbol G) A prefix meaning 'one billion'. (p. 159)

**gigawatt** (GW) 1000 MW or 1 000 000 000 W. (p. 234)

**government allowance** Money paid by the government to support individuals for specific purposes (for example, to support the aged, unemployed, disabled, students and parents). (p. 131)

**goods and services tax (GST)** The tax a consumer pays on any purchased item or service (for example buying a car, or hiring a painter). (p. 220)

**gradient** (symbol $m$) Slope of a line. (p. 325)

$$\text{gradient} = \frac{\text{rise}}{\text{run}} = \frac{\text{change in } y}{\text{change in } x}.$$

**great circle** A slice through the centre of a sphere; its radius is the same as that of the sphere. (p. 427)

**green slip** See **Compulsory Third Party (CTP) insurance**.

**Greenwich Mean Time (GMT)** See **UTC**.

**Greenwich meridian** See **prime meridian**.

**gross income** Earnings before deductions have been made. (p. 140)

**gross pay** (or **gross wage**) A person's pay before tax is deducted. Different from **net pay**. (p. 136)

## H

**hectare** A land measure equal to the area of a square 100 m by 100 m. 1 ha = 10 000 m$^2$. (p. 176)

**hemisphere** Half a sphere, or half of Earth; for example, the southern hemisphere. (p. 204)

**histogram** See **frequency histogram**.

## I

**income** Money that is earned or gained (usually regularly). (p. 127)

**income tax** A tax on a person's income, paid to the government. (p. 140)

**independent variable** A variable in a function whose values do not depend on any other variable. For example, if $A = \pi r^2$, $r$ is the independent variable. (p. 336)

**infographic** A display that combines images, graphs and text to illustrate data, statistics and information. (p. 6)

**insurance** A scheme in which you pay an insurance company regularly in return for being paid a larger amount if an accident occurs. (p. 257)

**interest** Money earned on an investment, or money paid to a financial institution for borrowing. (p. 140)

**International Date Line (IDL)** The imaginary line that runs through the Pacific Ocean and is approximately the 180° meridian of longitude. A day is either gained or lost when this line is crossed. (p. 443)

**interpolate** To estimate the value of a quantity that is within the given range of values. (p. 337)

**interquartile range (IQR)** The difference between the upper quartile and lower quartile of a data set ($Q_3 - Q_1$). It is a measure of the spread of the data. (p. 383)

## K

**kilo-** (symbol k) A prefix meaning 'one thousand'. For example, a kilolitre is 1000 litres. (p. 159)

**kilojoule** (kJ) A unit of energy equal to 1000 joules. (p. 159)

**kilowatt** (kW) A unit of power equal to 1000 watts. (p. 234)

**kilowatt-hour** (kWh) A unit of electrical energy equivalent to that used by 1 kilowatt of power in 1 hour. (p. 234)

**Kruskal's algorithm** A method of finding a minimum spanning tree for a weighted network. (p. 298) See also **Prim's algorithm**.

## L

**land survey** See **offset survey**.

**latitude** The angular distance north or south of the Equator of a point on Earth's surface; the size of the angle made between the point and the Equator at Earth's centre. (p. 427)

**LHS** Left-hand side (of an equation). (p. 68)

**like terms** Terms that have exactly the same variable part, even though the numeral part may differ. (p. 57)

**line graph** A graph made up of a line or several line intervals, often showing how a quantity is changing over time. (p. 6)

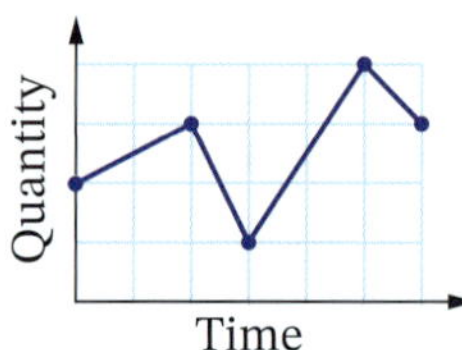


**linear** A word used to describe something to do with a line. (p. 325)

**linear function** A function of the form $y = mx + c$, whose graph is a straight line. (p. 325)

**linear modelling** Using a linear function to approximate a real-life situation. (p. 336)

**longitude** The angular distance east or west of the prime meridian of a point on Earth's surface; the size of the angle made between the point and the prime meridian at Earth's centre. (p. 427)

**loop** An edge that connects a vertex to itself. (p. 282)

**loss** The amount lost when selling an item at a lower price, when the selling price is less than the cost price. The opposite of **profit**. (p. 217)

## M

**mean** The average of a set of data values.

$$\text{mean (or } \bar{x}) = \frac{\text{sum of values}}{\text{number of values}} = \frac{\Sigma x}{n} = \frac{\Sigma fx}{\Sigma f}. \text{ (p. 371)}$$

**measure of centre** (or **measure of central tendency**) A statistical value, such as the mean, median or mode, which describes the centre or average of a set of data. (p. 371)

**measure of spread** A statistical value, such as the range, interquartile range or standard deviation, which describes the spread of a set of data. (p. 383)

**median** The middle value of a data set when the data are arranged in ascending order. If there are 2 middle values, the median is the average of them. (p. 371)

**median class** The class interval that contains the median. (p. 376)

**Medicare levy** A tax to cover the costs of the public health system, calculated as a percentage (usually 2%) of a person's income. (p. 141)

**mega-** (symbol M) A prefix meaning 'one million'. For example, a megatonne is 1 000 000 tonnes. (p. 159)

**megajoule** (MJ) 1000 kJ or 1 000 000 J. (p. 235)

**megalitre** (ML) 1000 kL or 1 000 000 L. (p. 156)

**megatonne** (Mt) 1 000 000 tonnes. (p. 156)

**megawatt** (MW) 1000 kW or 1 000 000 W. (p. 234)

**meridian of longitude** Great semicircle running down Earth's surface from the North Pole to the South Pole, measured east or west of the prime meridian (0° longitude). (p. 428)

**micro-** (symbol $\mu$) A prefix meaning 'one millionth'. For example, a micrometre is one-millionth of a metre. (p. 159)

**microgram** (μg) One-millionth of a gram. 1 g = 1 000 000 μg. (p. 159)

**milligram** (mg) One-thousandth of a gram. 1 g = 1000 mg. (p. 156)

**minimum spanning tree** A spanning tree in a network where the sum of the weights is smallest. (p. 298) See also **spanning tree**.

**modal class** The class interval with the highest frequency. (p. 376)

**mode** The most common or frequent value(s) or category(ies) in a set of data. (p. 371)

**modelling** Using mathematics to describe a real-life pattern or relationship. (p. 336)

## N

**net pay** (or **net wage**) A person's pay after tax has been deducted. Differs from **gross pay**. (p. 136)

**network** A collection of objects connected to each other in some way, represented by a diagram involving vertices and edges. (p. 282)

**nominal data** Categorical data that cannot be ordered; for example, colour of eyes. (p. 21)

**numerical data** (or **quantitative data**) Data that involve numbers, such as heights, or the number of contacts on a phone. (p. 21)

## O

**offset survey** A method of measuring lengths in an irregularly-shaped field, using offsets to each corner of the field from a traverse line. (p. 183)

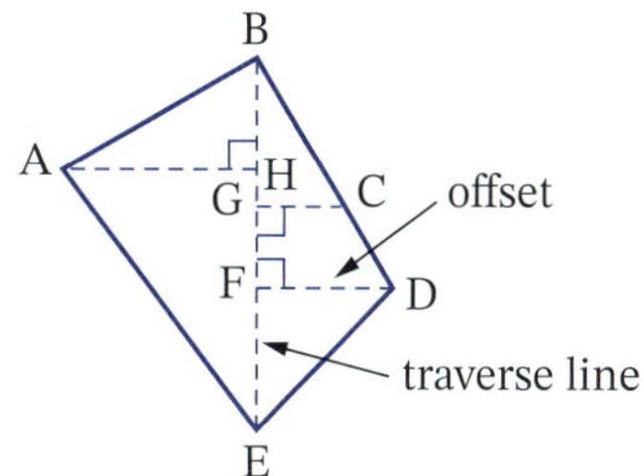


**ogive** See **cumulative frequency polygon**.

**on-road costs** After purchasing a motor vehicle, the additional costs required before it can be used on the road, such as registration, stamp duty and CTP insurance. (p. 261)

**ordinal data** Categorical data that can be ordered; for example, level of swimming class. (p. 21)

**outlier** An extreme (high or low) value in a data set that is much different than the other values, either less than $Q_1 - 1.5 \times$ IQR or greater than $Q_3 + 1.5 \times$ IQR. (pp. 43, 393)

**overtime** Time worked beyond usual working hours, usually paid at a higher rate. (p. 123)

## P

**parallel of latitude** Small circle running around Earth's surface, parallel to and measured north or south of the Equator (0° latitude). (p. 428)

**Pay As You Go (PAYG) tax** Income tax deducted from pay in instalments each payday by the employer or at set intervals (such as quarterly) by a self-employed business owner. (p. 143)

**peak** A high point of a distribution; the highest peak shows the mode. (p. 407)

**penalty rate** A rate of pay for working overtime, such as time-and-a-half or double time. (p. 123)

**per annum (p.a.)** Per year. (p. 62)

**pie chart** See **sector graph**.

**piecework** Earnings based on the number of items processed, made or delivered, paid at a rate per item (rather than per hour). (p. 127)

**population** All of the items under investigation. (p. 24) See also **sample**.

**power (index)** The number of times a base is multiplied by itself. In $2^5$, the power is 5. (p. 57)

**premium** The regular cost of an insurance policy. (p. 257)

**Prim's algorithm** A method of finding a minimum spanning tree for a weighted network. (p. 298) See also **Kruskal's algorithm**.

**prime meridian** (or **Greenwich meridian**) The 0° meridian of longitude that passes through Greenwich in London. The longitude of any point on Earth's surface is measured east or west of this meridian. (p. 428)

**prism** A solid with flat faces and a uniform cross-section. (p. 192)

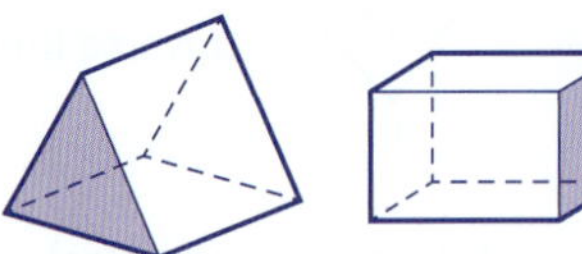

**profit** The amount made when selling an item, when the selling price is more than the cost price. The opposite of **loss**. (p. 217)

**proportional to** In the same ratio as. If $y$ is proportional to $x$, we say $y = kx$ and $x$ and $y$ have a direct linear variation. (p. 345) See also **direct linear variation**.

**pyramid** A solid with a polygon as a base (for example, a rectangle, or a triangle) and triangular side faces that meet at a point (called the apex). (p. 199)

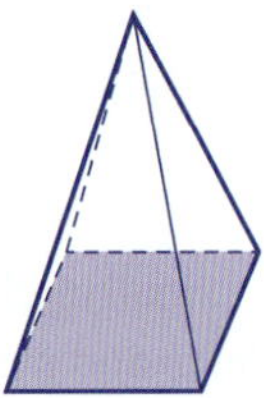

**Pythagoras' theorem** A theorem that states that in a right-angled triangle, the square of the hypotenuse is equal to the sum of the squares of the 2 shorter sides. (p. 166)

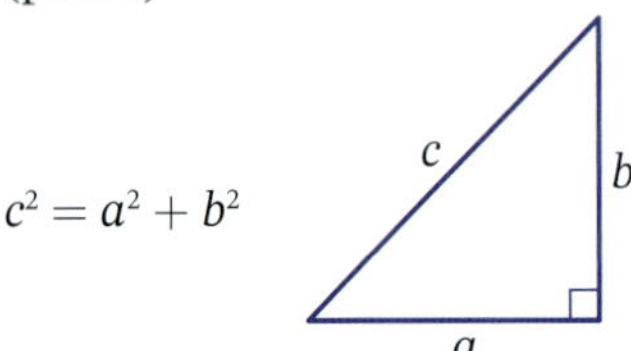


## Q

**quadrant** Quarter of a circle. (p. 173)

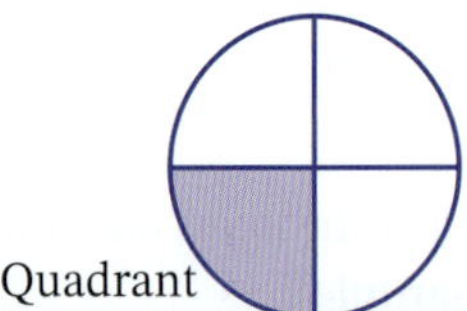


**quantitative data** See **numerical data**.

**quarterly** Every 3 months. (p. 231)

**quartiles (upper, middle and lower)** The upper quartile is the 3rd quartile ($Q_3$), which cuts off the top 25% of values in a data set; , the middle quartile ($Q_2$) is the median, and separates the 2 middle quarters; the lower quartile is the 1st quartile ($Q_1$), which cuts off the bottom 25% of values. (p. 383)

## R

**radius** (plural: **radii**) The length of the interval joining the centre of a circle to the circumference. The radius is half of the **diameter**. (p. 172)

**random sample** A sample for which every member of a population has an equal chance of selection. (p. 24)

**range** The difference between the highest value and the lowest value in a set of data. (p. 383)

**reaction distance** The distance travelled during a driver's **reaction time**. (p. 102) See also **braking distance** and **stopping distance**.

**reaction time** The time between when a driver senses the need to stop and when the brakes are applied. (p. 102)

**registration** See **vehicle registration**.

**repayment** (or **instalment**) Amount paid regularly to pay off a loan. (p. 139)

**retainer** A fixed amount paid to a salesperson before commission is added. (p. 127)

**RHS** Right-hand side (of an equation). (p. 68)

**rise** The vertical change in position between two points on a line; the number of units going up. Used with the **run** to calculate **gradient**. (p. 325)

**royalties** Income earned by recording artists and authors, based on the number of copies of their work that are sold. (p. 127)

**run** The horizontal change in position between two points on a line; the number of units going to the right. Used with the **rise** to calculate **gradient**. (p. 325)

## S

**salary** Fixed earnings quoted as a yearly amount, but paid weekly, fortnightly or monthly. (p. 124) See also **wage**.

**sample** A group of items selected from a population. (p. 24)

**sample size** The number of items in a **sample**. (p. 24)

**scientific notation** A way of writing very large or very small numbers. For example, 98 000 000 = $9.8 \times 10^7$. (p. 162)

**sector** Part of a circle bounded by an arc and 2 radii. (p. 173)

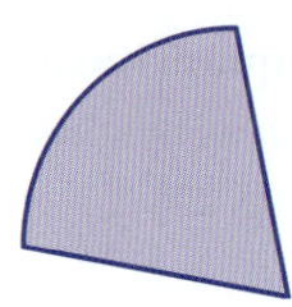

**sector graph** (or **pie chart**) A graph representing the parts of a whole population using a circle divided into proportionately sized sectors. Displays the same kind of data as a **divided bar graph**. (p. 6)

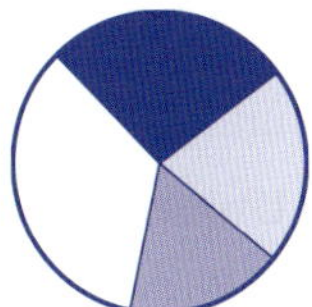

**self-selected sample** A sample in which people volunteer to be part of the sample, such as an SMS poll or website survey, so it is not really random. (p. 25)

**sewerage** The system of removing waste water from a home or building. (p. 230)

**shape of a distribution** The way the data in a frequency distribution is spread, can be symmetrical, positively skewed or negatively skewed. (p. 406)

**shortest path** The path in a weighted network where the sum of the weights is minimised. (p. 309)

**significant figures** Meaningful digits in a numeral that tell 'how many'. For example, 98 000 000 has 2 significant figures: 9 and 8. (p. 159)

**simple interest** (or **flat rate interest**) Interest earned or charged only on the original amount of money (principal) invested or borrowed. (p. 225)

**simplify** In algebra, the process of making an algebraic expression less complex or complicated, which mainly involves collecting like terms. (p. 57)

**skewed** The shape of a statistical distribution when most of the data values are either low (positively skewed) or high (negatively skewed). The tail indicates the direction of the skew. (p. 406)

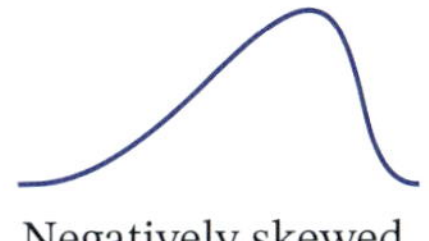
Negatively skewed

Positively skewed

**small circle** A slice through a sphere that does not pass through the centre. It has a smaller radius than a **great circle**. (p. 427)

**sole trader** A person who owns their own small business and works for themself. (p. 128)

**spanning tree** A tree in a network that includes all the vertices of the network. (p. 296) See also **tree** and **minimum spanning tree**.

**speed** A rate that compares distance travelled with time taken. Speed is often measured in kilometres per hour (km/h) or metres per second (m/s).

$$\text{average speed} = \frac{\text{distance travelled}}{\text{time taken}} \text{ (p. 86)}$$

**stamp duty** A tax paid to the Office of State Revenue when buying a vehicle, calculated on the market value of the vehicle. (p. 261)

**standard deviation** (symbol $\sigma$) A statistical measure of the spread of a set of numerical data. (p. 383)

**standard drink** A drink that contain 10 grams of alcohol, such as a small glass of beer or wine. (p. 86)

**standard form notation** See **scientific notation**.

**statistics** A collection, tabulation or display of data for the purpose of analysis. (p. 24)

**stem-and-leaf plot** A 'number graph' that lists all the data values, in groups. This stem-and-leaf plot shows 12 test scores, from 42 to 82. (p. 43)

| Stem | Leaf |
|---|---|
| 4 | 2 5 |
| 5 | 0 2 8 |
| 6 | 6 7 |
| 7 | 3 5 7 7 |
| 8 | 2 |

**stopping distance** The distance travelled between when a driver senses the need to stop and when the vehicle stops completely, equal to the sum of the **reaction distance** and **braking distance**. (p. 102)

**stratified sample** A sample consisting of a percentage of items from each strata or layer of a population. For example, a stratified sample from a population of 35% children and 65% adults should contain 35% children and 65% adults. (p. 24)

**subject** (of a formula) The variable on its own on the left side of the = sign of a formula; what the formula describes. For example, $A$ (for area) in the formula $A = \frac{1}{2}bh$. (p. 62)

**substitute** To replace a variable with a number. (p. 71)

**summary statistic** A calculated value that represents or summarises a set of data (for example, the mean or the standard deviation). (p. 371)

**summer time** See **daylight saving time**.

**superannuation** A retirement fund for employees. Every payday, an employee and his/her employer invest a part of the employee's pay into a fund in order to provide income for the employee during retirement. (p. 141)

**survey (land)** To measure lengths and angles on a field and perform calculations on these measurements. (p. 183)

**survey (statistical)** To gather information for statistical purposes. (p. 8)

**symmetrical** A distribution is symmetrical if the data are balanced or evenly spread about the centre of the distribution, with the mean, median and mode being equal. (p. 406)

**systematic sample** A sample chosen by using a set pattern (for example, choosing every 10th number in a phone book). (p. 24)

## T

**tail** The long slope part of a frequency distribution curve. (p. 406)

**tax debt** The amount by which the amount of **PAYG tax** already paid is below the amount of tax due. This is owed by the taxpayer to the Australian Office of Taxation. (p. 143)

**tax deduction** See **allowable tax deduction**.

**tax refund** The amount by which the **PAYG tax** already paid is above the amount of tax due; given back to the taxpayer. (p. 143)

**tax return** A form completed at the end of a financial year, to account for income earned, allowable deductions and tax already paid. Used to calculate a **tax refund** or **tax debt**. (p. 143)

**taxable income** The part of a person's income that is taxed, equal to yearly income minus allowable deductions. (p. 140)

**Third Party Property insurance** Motor insurance that covers damage to another person's ('third party') vehicle or property in an accident in which you are at fault. It does not cover damage to your vehicle. (p. 257) See also **comprehensive insurance** and **Compulsory Third Party (CTP) insurance**.

**time-and-a-half** Wages paid at 1.5 times the normal rate (for example, for working on a Saturday). (p. 123)

**time zone** A zone of the world in which the time is the same for all places. (p. 441)

**tonne** (t) A unit of mass equal to 1000 kg. (p. 156)

**trapezoidal rule** Formula for finding the approximate area of an irregular-shaped block using the area of a trapezium.

$$A \approx \frac{h}{2}\left(d_f + d_l\right) \qquad \text{(p. 184)}$$

**traverse survey** See **offset survey**.

**tree** A network in which any 2 vertices are connected by exactly one path, which means it is connected with no cycles. (p. 296) See also **spanning tree** and **minimum spanning tree**.

## U

**uniform (distribution)** A distribution where the frequencies of scores are similar, so its shape is flat or rectangular. (p. 407)

**unitary method** A way of calculating an amount by first finding one unit then multiplying. (p. 218)

**UTC (Coordinated Universal Time)** or **Greenwich Mean Time (GMT)** Local time at the prime meridian (0° longitude) time zone, from which other times in the world are measured. (p. 441)

## V

**value-added tax (VAT)** A tax on goods or services used in some countries; similar to the GST in Australia. (p. 220)

**variable** A pronumeral that can take a range of values (that is, it is not a **constant**). (p. 57)

**vehicle registration** The yearly process of paying for and obtaining permission for using a vehicle on public roads. (p. 261)

**vertex** (**of a network**, plural: **vertices**) A point (or node) in a network where edges meet. A vertex can be odd or even, depending on the number of edges connected to it. (p. 282)

**vertical intercept** (or *y*-**intercept**) The value at which a graph cuts the vertical axis. For example, the vertical intercept of this graph is 3 or (0, 3). (p. 337)

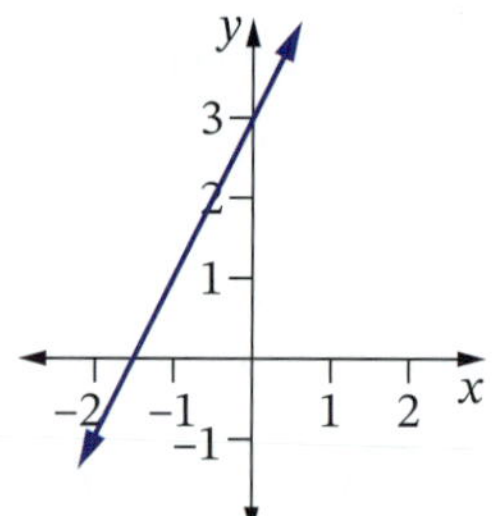


**volume** The amount of space occupied by a solid, measured in cubic units. (p. 190)

## W

**wage** The amount earned by an employee for a set number of working hours, usually paid weekly. (p. 123) See also **overtime** and **salary**.

**watt** A unit of power equal to 1 joule of energy per second. (p. 234)

**weighted network** A network with edges labelled with numbers called **weights**. The numbers could represent distances, times, costs or other quantities. (p. 282)

## Y

*y*-**intercept** See **vertical intercept**.